Encyclopaedia of Mathematical Sciences

Volume 91

Editor-in-Chief: R.V. Gamkrelidze

Springer
Berlin
Heidelberg
New York
Barcelona
Hong Kong
London
Milan
Paris
Singapore
Tokyo

M. M. Postnikov

Geometry VI

Riemannian Geometry

Springer

Title of the Russian edition:
Rimanova Geometriya
Publisher Faktorial, Moscow 1998

Mathematics Subject Classification (2000):
53-XX

ISSN 0938-0396
ISBN 978-3-642-07434-9

Springer-Verlag Berlin Heidelberg New York
a member of BertelsmannSpringer Science+Business Media GmbH

Printed in Germany
Printed on acid-free paper

List of Editor, Author and Translator

Editor-in-Chief

R. V. Gamkrelidze, Russian Academy of Sciences, Steklov Mathematical Institute, ul. Gubkina 8, 117966 Moscow; Institute for Scientific Information (VINITI), ul. Usievicha 20a, 125219 Moscow, Russia; e-mail: gam@ipsun.ras.ru

Author

M. M. Postnikov, MIRAN, ul. Gubkina 8, 117966 Moscow, Russia; e-mail: mmpostn@mi.ras.ru

Translator

S. A. Vakhrameev, Department of Mathematics, VINITI, ul. Usievicha 20a, 125219 Moscow, Russia

Preface

The original Russian edition of this book is the fifth in my series "Lectures on Geometry." Therefore, to make the presentation relatively independent and self-contained in the English translation, I have added supplementary chapters in a special addendum (Chaps. 30–36), in which the necessary facts from manifold theory and vector bundle theory are briefly summarized without proofs as a rule.

In the original edition, the book is divided not into chapters but into lectures. This is explained by its origin as classroom lectures that I gave. The principal distinction between chapters and lectures is that the material of each chapter should be complete to a certain extent and the length of chapters can differ, while, in contrast, all lectures should be approximately the same in length and the topic of any lecture can change suddenly in the middle. For the series "Encyclopedia of Mathematical Sciences," the origin of a book has no significance, and the name "chapter" is more usual. Therefore, the name of subdivisions was changed in the translation, although no structural surgery was performed. I have also added a brief bibliography, which was absent in the original edition.

The first ten chapters are devoted to the geometry of affine connection spaces. In the first chapter, I present the main properties of geodesics in these spaces. Chapter 2 is devoted to the formalism of covariant derivatives, torsion tensor, and curvature tensor. The major part of Chap. 3 is devoted to the geometry of submanifolds of affine connection spaces (Gauss–Weingarten formulas, etc.).

In Chap. 4, Cartan structural equations in polar coordinates are deduced. The second half of this chapter is devoted to locally symmetric affine connection spaces. Globally symmetric spaces are considered in Chap. 5 (and the beginning of Chap. 6). In particular, their coincidence with symmetric space in the Loos sense is proved. In the major part of Chap. 6, the general theory is illustrated by examining Lie groups. In Chap. 7, the language of categories and functors is explained (this material is set in a smaller font), and also the main theorems on the relation between Lie groups and Lie algebras are presented in essence without proofs. In Chaps. 8 and 9, these theorems are generalized to the case of symmetric spaces; in Chap. 10, they are generalized to the case of finite-dimensional Lie algebras of vector fields.

Chapters 13 and 14 are mainly devoted to the theory of elementary surfaces. The main focus is on their isothermal coordinates and minimal surfaces. In Chap. 15, the main properties of the curvature tensor are established. The main topic of Chap. 16 is the Gauss–Bonnet theorem. In Chap. 17, its generalizations to Riemannian spaces of large dimension are presented without proof. In the same chapter, the Ricci tensor of a Riemannian space is considered, and Einstein spaces are introduced.

Chapter 18 is devoted to conformal transformations of a metric. The main focus is on the case where $n = 2$. In the first half of Chap. 19, isometries and Killing fields are considered; the rest of this chapter is devoted to the specialization of constructions in Chap. 3 to the case of submanifolds of a Riemannian space. In Chap. 20, certain specific classes of submanifolds (locally symmetric and compact ones) are considered, and consideration of the theory of hypersurfaces is started; all of Chap. 21 is devoted to this topic.

Chapters 22 and 23 are devoted to spaces of constant curvature, Chap. 24 is devoted to four-dimensional Riemannian spaces, and Chaps. 25 and 26 are devoted to invariant metrics on Lie groups. Chapter 27 is devoted to the Jacobi theory of the second variation, and the last two chapters are devoted to its applications (in particular, the Mayers theorem and the Cartan–Hadamard theorem are proved). Chapter 29 concludes with the proof of the Bochner theorem on the finiteness of the isometry group of a compact Riemannian space with a negative-definite Ricci tensor (the general topological theorem on the compactness of the isometry group of an arbitrary compact metric space, which is needed for this proof, is also proved).

M. M. Postnikov Moscow, January 2000

Contents

Chapter 1
Affine Connections

§1. Connection on a Manifold

Let $\mathcal{X}$ be an arbitrary smooth manifold of dimension $n > 0$, and let $\tau_{\mathcal{X}} = (\mathsf{T}\mathcal{X}, \pi, \mathcal{X})$ be its tangent bundle. As we know (see Chap. 34), each chart $(U, h) = (U, x^1, \ldots, x^n)$ of the manifold $\mathcal{X}$ defines a chart $(\mathsf{T}U, \mathsf{T}h)$ of the manifold $\mathsf{T}\mathcal{X}$ for which $\mathsf{T}U = \pi^{-1}U$. The coordinates of the vector $A \in \mathsf{T}U$ in this chart are the coordinates $x^1, \ldots, x^n$ of the point $p = \pi A$ in the chart (U, h) and the coordinates of this vector in the basis

$$\left(\frac{\partial}{\partial x^1}\right)_p, \quad \ldots, \quad \left(\frac{\partial}{\partial x^n}\right)_p$$

of the linear space $\mathsf{T}_p\mathcal{X}$. The latter coordinates are denoted by $\dot{x}^1, \ldots, \dot{x}^n$, and the point $(\mathsf{T}h)(A) \in \mathbb{R}^{2n} = \mathbb{R}^n \times \mathbb{R}^n$ is accordingly denoted by

$$(x^1, \ldots, x^n, \dot{x}^1, \ldots, \dot{x}^n) = (\boldsymbol{x}, \dot{\boldsymbol{x}}),$$

where $\boldsymbol{x} = (x^1, \ldots, x^n)$ and $\dot{\boldsymbol{x}} = (\dot{x}^1, \ldots, \dot{x}^n)$. We stress that, in general, the vector $\boldsymbol{x}$ (ranging the open set $U \subset \mathbb{R}^n$) is not related to the vector $\dot{\boldsymbol{x}}$ (ranging the whole space $\mathbb{R}^n$).

The diffeomorphism $\mathsf{T}h$ defines a trivialization of the bundle $\tau_{\mathcal{X}}$ over U, which, as a basis of the $\mathsf{F}U$-module $\Gamma(\tau_{\mathcal{X}}|_U) = \mathfrak{a}U$ of all vector fields[1] on U, is just the coordinate basis

$$\frac{\partial}{\partial x^1}, \quad \ldots, \quad \frac{\partial}{\partial x^n} \tag{1}$$

of this module corresponding to the local coordinates $x^1, \ldots, x^n$. Bases of the module $\mathfrak{a}U$ of form (1) are also called *holonomic bases* (or *holonomic trivializations*). As a rule, we use only these bases in what follows.

If $(U, h) = (U, x^1, \ldots, x^n)$ and $(U', h') = (U', x^{1'}, \ldots, x^{n'})$ are two charts in $\mathcal{X}$ and $x^{i'} = x^{i'}(\boldsymbol{x})$, $1 \le i' \le n$, are the corresponding transition functions, then the transition functions of the charts $(\mathsf{T}U, \mathsf{T}h)$ and $(\mathsf{T}U', \mathsf{T}h')$ have the following forms on $\mathsf{T}U \cap \mathsf{T}U' = \mathsf{T}(U \cap U')$:

$$x^{i'} = x^{i'}(\boldsymbol{x}),$$

$$\dot{x}^{i'} = \frac{\partial x^{i'}}{\partial x^i}\dot{x}^i,$$

[1] We recall that for any $\mathcal{X}$, the symbol $\mathsf{F}\mathcal{X}$ denotes the algebra of all smooth functions on $\mathcal{X}$ and the symbol $\mathfrak{a}\mathcal{X}$ denotes the $\mathsf{F}\mathcal{X}$-module (Lie algebra) of all smooth vector fields on $\mathcal{X}$.

where $1 \leq i' \leq n$ (up to the notation, these are formulas (5) and (6) in Chap. 34).

In particular, this means that the transition matrix $\|\varphi_i^{i'}\|$ from one trivialization of the bundle $\tau_{\mathcal{X}}$ to another is the Jacobi matrix

$$\frac{\partial h'}{\partial h} = \left\| \frac{\partial x^{i'}}{\partial x^i} \right\|.$$

In the addendum, we introduce the concept of a *connection* on an arbitrary vector bundle and study it in detail. In the particular case of a vector bundle $\tau_{\mathcal{X}}$ (its sections are vector fields on $\mathcal{X}$), each connection ∇ on $\tau_{\mathcal{X}}$ (in one of many equivalent interpretations) is a mapping $X \mapsto \nabla_X$ that sets the operator of *covariant differentiation with respect to* X,

$$\nabla_X : \mathfrak{a}\mathcal{X} \to \mathfrak{a}\mathcal{X},$$

in correspondence to each vector field $X \in \mathfrak{a}\mathcal{X}$. This operator has the following properties (see Proposition 35.9):

1. The operator ∇_X is linear over $\mathbb{R}$.
2. For any function $f \in \mathsf{F}\mathcal{X}$ and any vector field $Y \in \mathfrak{a}\mathcal{X}$, the *Leibnitz formula*
$$\nabla_X(fY) = Xf \cdot Y + f\nabla_X Y$$
holds.
3. The operation ∇_X over $\mathsf{F}\mathcal{X}$ depends linearly on X, i.e.,
$$\nabla_{X+Y} = \nabla_X + \nabla_Y, \qquad \nabla_{fX}Y = f\nabla_X Y$$
for any fields $X, Y \in \mathfrak{a}\mathcal{X}$ and any function $f \in \mathsf{F}\mathcal{X}$.

Definition 1.1. A connection ∇ on the bundle $\tau_{\mathcal{X}}$ is called a *connection on the manifold* $\mathcal{X}$.

Traditionally, connections on manifolds are also called *affine connections* (although the current preference is to call these connections *linear connections* and reserve the term *affine connection* for the associated connection with the affine structure group). A manifold $\mathcal{X}$ with a given connection on it is accordingly called an *affine connection space.*

We note that *there is at least one affine connection on any paracompact Hausdorff manifold* $\mathcal{X}$ (According to Corollary 35.1). In what follows, we assume that all affine connection spaces are Hausdorff and paracompact.

In each trivialization of the vector bundle $\tau_{\mathcal{X}}$, an affine connection is given by n^3 functions Γ^i_{kj}, $i, j, k = 1, \ldots, n$, called the *coefficients* of this connection (see Chap. 35). In holonomic trivialization (1), they are expressed by

$$\Gamma^i_{kj} = \left(\nabla_k \frac{\partial}{\partial x^j}\right)^i, \quad i, j, k = 1, \ldots, n, \tag{2}$$

where $\nabla_k = \nabla_{\partial/\partial x^k}$ is the operator of covariant differentiation in the coordinate x^k, $k = 1, \ldots, n$. In two distinct trivializations, these coefficients are connected by the formula (see formula (11) in Chap. 35)

$$\Gamma^{i'}_{k'j'} = \frac{\partial x^{i'}}{\partial x^i}\frac{\partial x^j}{\partial x^{j'}}\frac{\partial x^k}{\partial x^{k'}}\Gamma^i_{kj} + \frac{\partial x^{i'}}{\partial x^i}\frac{\partial^2 x^i}{\partial x^{j'}\partial x^{k'}}; \tag{3}$$

in the matrix form, this becomes

$$\omega' = \frac{\partial h'}{\partial h}\,\omega\,\frac{\partial h}{\partial h'} + \frac{\partial h'}{\partial h}\,d\,\frac{\partial h}{\partial h'}, \tag{3'}$$

where, as always, $\omega = \|\omega^i_j\|$ is the matrix of *connection forms* $\omega^i_j = \Gamma^i_{kj}\,dx^k$ (see formula (13) in Chap. 35). Of course, transformation formulas (3) only hold for coefficients Γ^i_{kj} of form (2), i.e., evaluated in holonomic trivializations.

The components $(\nabla_Y X)^i$ of the covariant derivatives are expressed via the coefficients Γ^i_{kj} by

$$(\nabla_Y X)^i = \left(\frac{\partial X^i}{\partial x^k} + \Gamma^i_{kj}X^j\right)Y^k, \quad i = 1, \ldots, n. \tag{4}$$

§2. Covariant Differentiation and Parallel Translation Along a Curve

For each curve $\gamma\colon I \to \mathcal{X}$, its arbitrary lift to $\mathbf{T}\mathcal{X}$ is just a vector field

$$X\colon t \mapsto X(t) \in \mathbf{T}_{\gamma(t)}\mathcal{X}, \quad t \in I,$$

on the curve γ, and to each such field, the operation ∇/dt of *covariant differentiation along the curve* (see § 35.7) assigns the field $\nabla X/dt$ with the components

$$\left(\frac{\nabla X}{dt}\right)^i(t) = \frac{dX^i(t)}{dt} + \Gamma^i_{kj}(x(t))X^j(t)\dot{x}^k(t),$$

where $x^i = x^i(t)$ are equations of the curve γ and $X^i(t)$ are components of the field X in a given local coordinate system.

A field X on a curve γ is said to be *covariantly constant* (to consist of *parallel vectors*) if $\nabla X/dt = 0$ (i.e., if it is horizontal when it is considered as a lift of the curve γ). For any point $t_0 \in I$ and any vector $A \in \mathbf{T}_{\gamma(t_0)}\mathcal{X}$ on the curve γ, there exists a unique covariantly constant field X for which $X(t_0) = A$. The vectors $X(t)$ composing this field tree said to be *parallel to the vector A along the curve* γ, and the mapping

$$\Pi_\gamma: \mathbf{T}_{\gamma(t_0)}\mathcal{X} \to \mathbf{T}_{\gamma(t)}\mathcal{X}$$

defined by

$$\Pi_\gamma: A \mapsto X(t),$$

is called the *parallel translation along* γ.

Exercise 1.1. Show that for any $t \in I$,

$$\frac{\nabla X}{dt}(t) = \lim_{h\to 0} \frac{\Pi^t_{t+h}X(t+h) - X(t)}{h},$$

where Π^t_{t+h} denotes the parallel translation operator $\Pi_\gamma: \mathbf{T}_{\gamma(t+h)}\mathcal{X} \to \mathbf{T}_{\gamma(t)}\mathcal{X}$.

§3. Geodesics

An example of a vector field on γ is given by the field $\dot\gamma$ consisting of tangent vectors $\dot\gamma(t)$. This field is called the *tangent vector field* on γ (or the *natural lift* of the curve γ to $\mathbf{T}\mathcal{X}$).

Definition 1.2. A curve $\gamma: I \to \mathcal{X}$ in an affine connection space $\mathcal{X}$ is called a *geodesic* if its tangent field consists of parallel vectors (is covariantly constant), i.e., if its natural lift is horizontal.

In a visual form, the essential property of a geodesic γ is that in the process of motion along this curve, its tangent vectors $\dot\gamma(t)$ are transported parallelly, i.e., the curve is not curved. In this sense, geodesics are a generalization of straight lines in affine geometry.

Geodesics are characterized analytically by the equation

$$\frac{\nabla\dot\gamma}{dt}(t) = 0,$$

i.e., in local coordinates, they are characterized by the equations

$$\ddot x^i(t) + \Gamma^i_{kj}(x(t))\dot x^j(t)\dot x^k(t) = 0, \quad i = 1, \dots, n. \tag{5}$$

These are second-order differential equations that are resolved with respect to higher-order derivatives. Therefore, by standard theorems in the theory of ordinary differential equations, for any point $p \in \mathcal{X}$ and for any vector $A \in \mathbf{T}_p\mathcal{X}$, there exists a maximal (not continued to any larger interval of the real line $\mathbb{R}$) geodesic γ for which $\gamma(0) = p$ and $\dot{\gamma}(0) = A$. By the usual arguments, *this geodesic is unique.* (We recall that the manifold $\mathcal{X}$ is assumed to be Hausdorff.) We let $\gamma_{p,A}$ denote it, but, as a rule, we write $\gamma_p(t, A)$ instead of $\gamma_{p,A}(t)$. The interval of the real line $\mathbb{R}$ on which the geodesic $\gamma_{p,A}$ is defined is denoted by $I_{p,A}$ or $I_p(A)$.

The functions $x^i(t)$ that give the geodesic $\gamma_{p,A}$ in local coordinates certainly depend on the point p and the vector A, i.e., they are in fact functions of $2n+1$ arguments: the number t, n coordinates of the point p, and n coordinates of the vector A. According to the well-known *theorem on the smooth dependence of solutions of differential equations on initial data*, these functions are smooth functions of all $2n+1$ arguments. In this sense, the *geodesic $\gamma_{p,A}$ depends smoothly on the point p and the vector A.* We note that similar assertions are not true for the interval $I_{p,A}$ (i.e., for abscissas of its endpoints).

Example 1.1. Let the manifold $\mathcal{X}$ be the (u, v)-plane $\mathbb{R}^2$ from which the two semicircles $u^2 + v^2 = 1$, $v \geq 0$, and $u^2 + v^2 = 4$, $v \leq 0$, are removed. Geodesics $\gamma_{p,A}$ passing through the point $p(0,0)$ are rectilinear intervals with

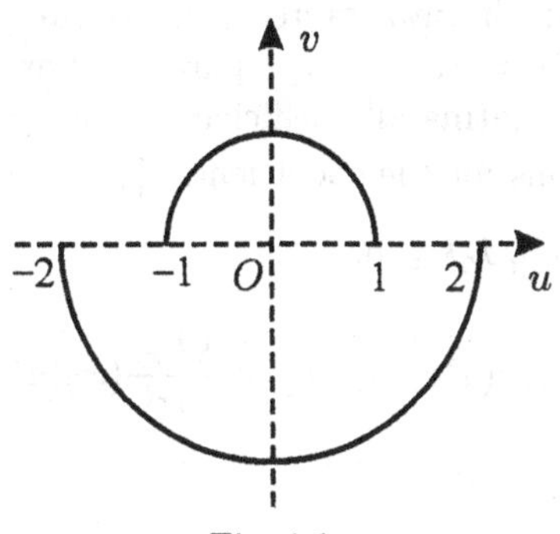

Fig. 1.1.

the directing vector A whose endpoints lie on these semicircles. Therefore, if φ is the inclination angle to the abscissa of the vector A (which is assumed to be a unit vector for definiteness), then

$$I_{p,A} = \begin{cases} (-2, 1) & \text{if } 0 < \varphi < \pi, \\ (-1, 2) & \text{if } \pi < \varphi < 2\pi, \\ (-1, 1) & \text{if } \varphi = 0 \text{ or } \varphi = \pi. \end{cases}$$

Therefore, for $A = (\pm 1, 0)$, the function $A \mapsto I_{p,A}$ has a discontinuity.

If $I_{p,A} = \mathbb{R}$ for any point $p \in \mathcal{X}$ and any vector $A \in \mathbf{T}_p\mathcal{X}$, then the manifold $\mathcal{X}$ with the connection ∇ (and also the connection ∇ itself) is said to be *geodesically complete*. The manifold $\mathcal{X}$ in Example 1.1 is not geodesically complete.

If the functions $x^i(t)$ satisfy Eqs. (5), then the functions $y^i(t) = x^i(\lambda t)$ also satisfy Eqs. (5) for any $\lambda \in \mathbb{R}$ because of the quadratic dependence of the left-hand sides of these equations on the first derivatives. Because $\dot{y}^i(0) = \lambda\dot{x}^i(0)$, this implies

$$\gamma_p(\lambda t, A) = \gamma_p(t, \lambda A). \tag{6}$$

This means that *under the parameter change* $t \mapsto \lambda t$, *the geodesic* $\gamma_{p,A}$ *transforms into the geodesic* $\gamma_{p,\lambda A}$. Of course, in this case,

$$t \in I_p(\lambda A) \iff \lambda t \in I_p(A).$$

The Taylor series for each of the functions $x^i(t)$, which assign the geodesic $\gamma_{p,A}$ in local coordinates, has the form

$$x^i(t) = x_0^i + a^i t + c_2^i t^2 + \cdots + c_m^i t^m + \cdots, \tag{7}$$

where x_0^i, $1 \le i \le n$, are coordinates of the point p and $a^i = \dot{x}^i(0)$, $1 \le i \le n$, are coordinates of the vector A. (For simplicity, we assume here that the manifold $\mathcal{X}$ is analytic (of class C^ω). For manifolds of class C^r for $r = \infty$ or for a finite (but sufficiently large) r, instead of the Taylor series, we should consider its segments, which unwarrantedly complicates the statements.)

Substituting series (7) in Eqs. (5) (preparatorily expanding the functions Γ_{jk}^i in $x^1, \ldots, x^n$) and equating all coefficients by all powers of t to zero, we obtain the set of equations for the coefficients c_m^i

$$\begin{aligned} &2c_2^i + \Gamma_{kj}^i(x_0)a^j a^k = 0, \\ &6c_3^i + 2\Gamma_{kj}^i(x_0)(a^j c_2^k + a^k c_2^j) + \frac{\partial \Gamma_{kj}^i}{\partial x^l}(x_0)a^j a^k a^l = 0, \\ &\ldots, \end{aligned} \tag{8}$$

from which these coefficients can be found step-by-step (in analysis, this procedure has the technical name the *indeterminate-coefficient method*).

Equations (8) obviously imply that *all the coefficients* c_m^i *are polynomials in* $a^1, \ldots, a^n$. Moreover, a slightly more careful analysis of these equations shows that *each polynomial* c_m^i, $i = 1, \ldots, n$, *is homogeneous and its degree equals* m. This latter fact can be obtained more simply by observing that relation (6) is equivalent to the identities

$$c_m^i(\lambda a) = \lambda^m c_m^i(a), \quad 1 \le i \le n, \quad 2 \le m < \infty,$$

which should hold for any sufficiently small $|\lambda|$.

§4. Exponential Mapping and Normal Neighborhoods

Let p_0 be an arbitrary point of an affine connection space $\mathcal{X}$, and let $(U_0', h) = (U_0', x^1, \ldots, x^n)$ be a chart centered at the point p_0 for which the set $h(U_0') \in \mathbb{R}^n$ is an open ball of the space $\mathbb{R}^n$ centered at the point 0. For each point $p \in U_0'$, we assume that the linear space $\mathbf{T}_p\mathcal{X}$ is equipped with the Euclidean structure with respect to which the basis

$$\left(\frac{\partial}{\partial x^1}\right)_p, \quad \ldots, \quad \left(\frac{\partial}{\partial x^n}\right)_p \tag{9}$$

of this linear space is orthonormalized. This structure depends on the choice of the chart (U_0', h) and has no intrinsic geometric meaning. We only need it for analytic estimates.

Lemma 1.1. *There exist a number $\varepsilon > 0$ and a neighborhood $U_0 \subset U_0'$ of the point p_0 such that for any point $p \in U_0$ and any vector $A \in \mathbf{T}_p\mathcal{X}$ with $|A| < \varepsilon$, the geodesic $\gamma_{p,A}$ is defined for $|t| < 2$ (the interval $(-2,2)$ is contained in the interval $I_{p,A}$).*

Proof. Because the geodesic $\gamma_{p,A}$ smoothly (and therefore continuously) depends on p and A, there exist a neighborhood $U_0 \subset U_0'$ of the point p_0 and numbers $\varepsilon_1 > 0$ and $\varepsilon_2 > 0$ such that for each point $p \in U_0$ and for any vector $A \in \mathbf{T}_p\mathcal{X}$ with $|A| < \varepsilon_1$, the geodesic $\gamma_{p,A}$ is defined for $|t| < 2\varepsilon_2$ (on the interval $(-2\varepsilon_2, 2\varepsilon_2)$). Let $0 < \varepsilon < \varepsilon_1\varepsilon_2$. Then for $|A| < \varepsilon$, we have the inequality $|\varepsilon_2^{-1}A| < \varepsilon_1$, and the geodesic $t \mapsto \gamma_p(t, \varepsilon_2^{-1}A)$ is therefore defined for $|t| < 2\varepsilon_2$, i.e., for $|\varepsilon_2^{-1}t| < 2$. Because $t \in I_p(A) \iff \varepsilon_2 t \in I_p(\varepsilon_2^{-1}A)$, Lemma 1.1 is proved. □

Reducing, if necessary, the neighborhood U_0, we can assume without loss of generality that the set $h(U_0)$ is an open ball of the space $\mathbb{R}^n$ centered at the point 0.

Definition 1.3. A vector $A \in \mathbf{T}_p\mathcal{X}$ is said to be *exponentiable* if the geodesic $\gamma_{p,A}$ is defined for $t = 1$ (i.e., if $1 \in I_p(A)$). For any exponentiable vector $A \in \mathbf{T}_p\mathcal{X}$, the point

$$\exp_p A = \gamma_p(1, A)$$

is called the *exponential* of the vector A. The mapping

$$\exp_p \colon A \mapsto \exp_p A$$

is called the *exponential mapping.*

We note that the zero $0 = 0_p$ of the linear space $\mathbf{T}_p\mathcal{X}$ is exponentiable and

$$\exp_p 0 = p.$$

The set of all exponentiable vectors of the linear space $\mathbf{T}_p\mathcal{X}$ (the domain of the mapping $\exp_p$) is denoted by O_p. It contains the vector 0, and, as is directly implied by (6), it is *star-shaped* with respect to this point, i.e., if $A \in O_p$, then $\lambda A \in O_p$ for any λ, $0 \leq \lambda \leq 1$. The relation $O_p = \mathbf{T}_p\mathcal{X}$ holds for all $p \in \mathcal{X}$ iff the manifold $\mathcal{X}$ is geodesically complete.

According to Lemma 1.1, for any point $p \in U_0$, the set O_p contains an open ball $|A| < \varepsilon$ of radius ε centered at the point 0. In particular, this implies that the *set* $\operatorname{Int} O_p$ *contains the point* 0 (and is therefore its neighborhood). Because the point p_0 is arbitrary, *the same is true for any point* $p \in \mathcal{X}$.

Formula (7) for $t = 1$ implies that the coordinates $(\exp_p A)^i$, $1 \leq i \leq n$, of the point $\exp_p A$ (if, of course, these coordinates are defined, i.e., if $\exp_p A \in U_0$) are expressed by

$$(\exp_p A)^i = x^i + a^i + c_2^i(\boldsymbol{x}, \boldsymbol{a}) + \cdots + c_m^i(\boldsymbol{x}, \boldsymbol{a}) + \cdots, \tag{10}$$

where x^i, $1 \leq i \leq n$, are coordinates of the point p in the chart (U_0, h), $a^i = \dot{x}^i$, $1 \leq i \leq n$, are coordinates of the vector A in basis (9), and $c_m^i(\boldsymbol{x}, \boldsymbol{a})$, $1 \leq i \leq n$, $2 \leq m < \infty$, are smooth functions of $\boldsymbol{x} = (x^1, \ldots, x^n)$ and $\boldsymbol{a} = (a^1, \ldots, a^n)$, which are homogeneous polynomials of degree m in $a^1, \ldots, a^n$. Therefore, for any $i, j = 1, \ldots, n$,

$$\frac{\partial(\exp_p A)^i}{\partial a^j} = \delta_j^i + \frac{\partial c_2^i(\boldsymbol{x}, \boldsymbol{a})}{\partial a^j} + \cdots + \frac{\partial c_m^i(\boldsymbol{x}, \boldsymbol{a})}{\partial a^j} + \cdots,$$

where $\partial c_m^i(\boldsymbol{x}, \boldsymbol{a})/\partial a^j$ are homogeneous polynomials of positive degree $m - 1$ in $a^1, \ldots, a^m$ (and therefore vanish for $a^1 = 0, \ldots, a^n = 0$). This proves that *at the point* $0 \in \mathbf{T}_p\mathcal{X}$, *the Jacobi matrix of functions* (10) *is the identity matrix* $E = \|\delta_j^i\|$.

Therefore, the mapping $\exp_p$ is étale at the point 0 (see Definition 31.5), and the point 0 has a fundamental system of neighborhoods in $\mathbf{T}_p\mathcal{X}$ (that are contained in O_p), on each of which the mapping $\exp_p$ is its diffeomorphism onto a certain neighborhood of the point p.

Definition 1.4. A neighborhood $U^{(0)}$ of the vector 0 in $\mathbf{T}_p\mathcal{X}$ and a neighborhood U of a point p in $\mathcal{X}$ are called *normal neighborhoods* (of the vector 0 and the point p respectively) if the neighborhood $U^{(0)}$ is star-shaped and the mapping $\exp_p$ is its diffeomorphism onto U.

According to what was said above, *normal neighborhoods form a fundamental system (base) of neighborhoods* (of the vector $0 \in \mathbf{T}_p\mathcal{X}$ and the point $p \in \mathcal{X}$ respectively).

Exercise 1.2. Prove that *any open star-shaped subset of a linear n-dimensional space is diffeomorphic to the ball* $\mathbb{B}^n$.

In particular, we see that *each normal neighborhood is diffeomorphic to the ball* $\mathbb{B}^n$.

The coordinates $x^1, \dots, x^n$ in a normal neighborhood U are called *normal coordinates* if the diffeomorphism $\exp_p^{-1}$ transforms them into linear coordinates on $\mathbf{T}_p\mathcal{X}$ (more precisely, on $U^{(0)}$). A characteristic property of normal coordinates is that geodesics passing through a point p have the equations

$$x^i = a^i t, \quad i = 1, \dots, n, \tag{11}$$

in these coordinates.

Exercise 1.3. Prove that the coordinates $x^1, \dots, x^n$ centered at p are normal iff

$$\Gamma^i_{kj} x^j x^k = 0, \quad i = 1, \dots, n, \tag{12}$$

identically in $x^1, \dots, x^n$. [*Hint*: Solutions of Eq. (5) have form (11) iff condition (12) holds.]

§5. Whitehead Theorem

The results obtained can be improved essentially. Let (U_0, h_0) be a chart as in Lemma 1.1. In the manifold $\mathbf{T}\mathcal{X}$, we consider a neighborhood of zero 0_{p_0} in the linear space $\mathbf{T}_{p_0}\mathcal{X}$ consisting of all those vectors $A \in \mathbf{T}_p\mathcal{X}$, $p \in U_0$, for which $|A| < \varepsilon$ and the mapping f of this neighborhood into the manifold $\mathcal{X} \times \mathcal{X}$ defined by

$$f(A) = (\exp_p A, p).$$

In the coordinates $x_1^1 = x^1 \circ \mathrm{pr}_1, \dots, x_1^n = x^1 \circ \mathrm{pr}_1$, $x_2^1 = x^1 \circ \mathrm{pr}_2, \dots, x_2^n = x^n \circ \mathrm{pr}_2$, defined in the neighborhood $U_0 \times U_0$ of the point (p_0, p_0) in the manifold $\mathcal{X} \times \mathcal{X}$, where $\mathrm{pr}_1\colon (p,q) \mapsto p$ and $\mathrm{pr}_2\colon (p,q) \mapsto q$ are the natural projections, and in the coordinates $x^1, \dots, x^n$, $a^1 = \dot{x}^1, \dots, a^n = \dot{x}^n$, defined in the neighborhood $\mathbf{T}U_0$ of the point 0_{p_0} of the manifold $\mathbf{T}\mathcal{X}$, the mapping f is written as

$$\begin{aligned} x_1^i &= x^i + a^i + c_2^i(\boldsymbol{x}, \boldsymbol{a}) + \dots + c_m^i(\boldsymbol{x}, \boldsymbol{a}) + \cdots, \\ x_2^i &= x^i, \end{aligned}$$

where $i = 1, \dots, n$.

Because all the derivatives $\partial c_m^i / \partial x^j$ are homogeneous polynomials of degree m in $a^1, \dots, a^n$ (and therefore vanish for $\boldsymbol{a} = 0$), the Jacobi matrix of the mapping f at the point $0_{p_0} \in \mathbf{T}\mathcal{X}$ becomes

$$\begin{Vmatrix} E & E \\ E & 0 \end{Vmatrix}$$

and is therefore nonsingular. Therefore, on a certain neighborhood $W^{(0)}$ of the vector 0_{p_0} in $\mathbf{T}\mathcal{X}$, the mapping f is a diffeomorphism onto a neighborhood $W = f(W^{(0)})$ of the point (p_0, p_0). In this case, we can assume without loss of generality that the neighborhood $W^{(0)}$ consists of all vectors $A \in \mathbf{T}_p\mathcal{X}$

for which $\pi A \in U'$ and $|A| < \delta$, where U' is a neighborhood of the point p_0 (which is contained in the neighborhood U_0) and δ is a positive number (not exceeding the number ε in Lemma 1.1). In other words, we can assume that

$$W^{(0)} = \bigcup_{p \in U'} U^{(0)}_{\delta,p},$$

where $U^{(0)}_{\delta,p}$ is the ball of radius δ centered at the point 0_p of the space $\mathbf{T}_p\mathcal{X}$.

Because the mapping f is a diffeomorphism on $W^{(0)}$, it follows directly that for any point $p \in U'$, the mapping $\exp_p$ is a diffeomorphism of the neighborhood $U^{(0)}_{\delta,p}$ onto a certain neighborhood $U_{\delta,p}$ of the point p, i.e., the neighborhoods $U^{(0)}_{\delta,p}$ and $U_{\delta,p}$ are normal (the ball $U^{(0)}_{\delta,p}$ is certainly star-shaped). We note that the neighborhoods $U^{(0)}_{\delta,p}$ and $U_{\delta,p}$ *depend* on the auxiliary Euclidean metric on the linear spaces $\mathbf{T}_p\mathcal{X}$, $p \in U_0$ (i.e., on the chart (U_0, h)). Moreover, $U_{\delta,p} \subset U_0$. On the other hand, by the definition of the topology

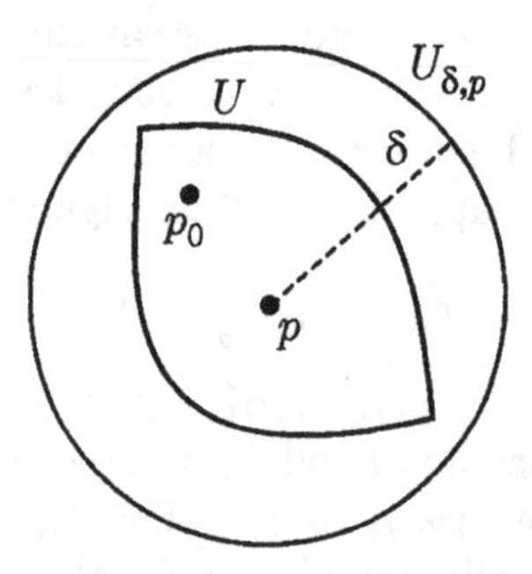

Fig. 1.2.

of the manifold $\mathcal{X} \times \mathcal{X}$, the point p_0 has a neighborhood $U \subset U'$ in $\mathcal{X}$ such that $U \times U \subset W$. Moreover, because $\mathrm{pr}_2 \circ f = \pi$, we have $U \subset U_{\delta,p}$ for any point $p \in U'$ and, in particular, for any point $p \in U$. This proves the following statement.

Proposition 1.1. *For any point p_0 of an affine connection space $\mathcal{X}$, there exist a number $\delta > 0$ and a neighborhood U of the point p_0 such that for each point $p \in U$, the neighborhood U is contained in a normal δ-neighborhood $U_{\delta,p}$ of the point p.*

Without loss of generality, we can assume here that the set $h(U) \subset h(U_0)$ is a ball in $\mathbb{R}^n$ (concentric to the ball $h(U_0)$). However, the neighborhood U is not star-shaped with respect to the point $p \in U$ in general and is therefore not a normal neighborhood of this point. We consider this question in more detail.

Because $U \subset U_{\delta,p}$ and the neighborhood $U_{\delta,p}$ is normal, the set $U^{(0)} = \exp_p^{-1} U \subset U^{(0)}_{\delta,p}$ is defined, and the neighborhood U is normal iff this set is star-shaped. Let $q \in U$. Because $U \subset U_{\delta,p}$ and the neighborhood $U_{\delta,p}$ is normal, there exists a vector $A \in \mathbf{T}_p\mathcal{X}$ such that $\exp_p A = q$ and the segment

$$\gamma_{p,q}: t \mapsto \exp_p tA, \quad 0 \le t \le 1, \tag{13}$$

of the geodesic $\gamma_{p,A}$ connects the point p with the point q in $U_{\delta,p}$. It is clear that the set $U^{(0)}$ is star-shaped (the neighborhood U is normal) iff for any point $q \in U$, segment (13) is contained in U (i.e., if $\gamma_{p,q}(t) \in U$ for any t, $0 \le t \le 1$).

Proposition 1.2 (Whitehead theorem). *Each point p_0 of an affine connection space $\mathcal{X}$ has a neighborhood U such that for any points $p, q \in U$, the segment $\gamma_{p,q}$ is contained in U.*

Proof. On a coordinate neighborhood U_0, we consider the matrix

$$\left\| \delta_{kj} - \sum_{i=1}^{n} \Gamma^i_{kj} x^i \right\|, \quad j, k = 1, \dots, n.$$

Reducing (if necessary) the neighborhood U_0, we can assume without loss of generality that *this matrix is positive definite at each point of the neighborhood* U_0.

We show that under this condition, the neighborhood U of the point p_0, which is presupposed by Proposition 1.1 (and such that its image $h(U)$ is a ball in $\mathbb{R}^n$) has the required property. Let

$$x^i = x^i(t), \quad i = 1, \dots, n, \quad 0 \le t \le 1,$$

be parametric equations of the segment of the geodesic $\gamma_{p,q}$, $p, q \in U$, in the chart (U_0, h). (By construction, the segment $\gamma_{p,q}$ obviously is contained in U_0.) Then for the function $f(t) = \sum_{i=1}^n (x^i(t))^2$, $0 \le t \le 1$, which is equal to the squared distance in $\mathbb{R}^n$ from the point 0 to the point $(h \circ \gamma_{p,q})(t))$, we have

$$\begin{aligned} \ddot{f}(t) &= 2\sum_{i=1}^{n} [(\dot{x}^i(t))^2 + \ddot{x}^i(t)x^i(t)] \\ &= 2\sum_{i=1}^{n} [(\dot{x}^i(t))^2 - \Gamma^i_{kj}(x(t))\dot{x}^j(t)\dot{x}^k(t)x^i(t)] \\ &= 2\left[\delta_{jk} - \sum_{i=1}^{n} \Gamma^i_{kj} x^i\right]_{\gamma(t)} \dot{x}^j(t)\dot{x}^k(t) \ge 0; \end{aligned}$$

as is shown in analysis, this implies

$$f(t) \le (1-t)f(0) + tf(1)$$

for any t, $0 \leq t \leq 1$, and therefore

$$f(t) \leq \max[f(0), f(1)], \quad 0 \leq t \leq 1.$$

For completeness, we present the proof of this fact. According to the Lagrange formula, there exist numbers ξ_0 and ξ_1 such that $0 \leq \xi_0 \leq t \leq \xi_1 \leq 1$ and

$$f(0) = f(t) - t\dot{f}(\xi_0), \quad f(1) = f(t) + (1-t)\dot{f}(\xi_1).$$

Multiplying the first relation by $1-t$ and the second by t and adding the results, we obtain the identity

$$(1-t)f(0) + tf(1) = f(t) + (1-t)t[\dot{f}(\xi_1) - \dot{f}(\xi_0)].$$

It remains to note that if $\ddot{f} \geq 0$ on $[0,1]$, then $\dot{f}(\xi_1) - \dot{f}(\xi_0) \geq 0$.

Because $p, q \in U$, the numbers $f(0)$ and $f(1)$ are less than the radius r of the ball $h(U)$. Therefore, $f(t) < r$ for any t, $0 \leq t \leq 1$, and the point $\gamma_{p,q}(t)$ hence belongs to the neighborhood U. □

Corollary 1.1. *Each point p_0 of an affine connection space has a neighborhood U that is a normal neighborhood of any of its points.*

Proof. As was observed above, the neighborhood U in Proposition 1.2 has this property. □

Because the relation $\gamma_p(t, A) = \gamma_p(1, tA)$ holds for any $t \in I_p(A)$ in accordance with (6), we have $t \in I_p(A)$ iff $tA \in O_p$ (the point $\exp_p tA$ is defined), and in this case,

$$\gamma_p(t, A) = \exp_p tA. \tag{14}$$

In particular, we see that the path $t \mapsto \exp_p tA$, $0 \leq t \leq 1$, is the restriction of the geodesic $\gamma_{p,A}$ to the closed interval $[0,1]$. This path is called a *geodesic segment* connecting the point p with the point $q = \exp_p A$.

We stress that *a geodesic segment is a path* (a curve defined on the closed interval $I = [0,1]$) by definition. In particular, this segment is defined when the point q belongs to a certain normal neighborhood U of the point p. In this case, we call it a *normal geodesic segment* from p to q and let $\gamma_{p,q}$ denote it. It is contained in the neighborhood U and is a unique geodesic segment in U that connects the point p with the point q. Because this is true for any neighborhood U (containing the point q), we find that *$\gamma_{p,q}$ is a unique normal geodesic segment connecting the point p with the point q.*

Of course, the points p and q can also be connected by other geodesic segments, but none of these segments can be contained in a normal neighborhood of the point p. Moreover, according to Proposition 1.1, *for each point $p_0 \in \mathcal{X}$, there exists a neighborhood U such that for any points $p, q \in U$, there exists a normal geodesic segment $\gamma_{p,q}$.* We note that the segment $\gamma_{p,q}$, in general, *is not contained in the neighborhood U.*

§6. Normal Convex Neighborhoods

Definition 1.5. A neighborhood U of a point p_0 is called a *normal convex neighborhood* if for any points $p, q \in U$, the segment $\gamma_{p,q}$ is defined and is contained in U.

We note that *a normal convex neighborhood is a normal neighborhood of each of its points* and is diffeomorphic to the ball $\mathbb{B}^n$ (see Exercise 1.1). The following proposition improves Proposition 1.1 and is known as the *strong form of the Whitehead theorem.*

Proposition 1.3. *Each point p_0 of an affine connection space $\mathcal{X}$ has a normal convex neighborhood.*

The proof of this proposition is essentially the same as the proof of Proposition 1.2.

§7. Existence of Leray Coverings

An open covering $\{U_k\}$ of a manifold $\mathcal{X}$ is called a *Leray covering* if for any $k_0, \ldots, k_m$, the intersection $U_{k_0} \cap \ldots \cap U_{k_m}$ is either empty or diffeomorphic to the ball $\mathbb{B}^n$.

Corollary 1.2. *On each Hausdorff paracompact manifold $\mathcal{X}$, there exists a Leray covering. A finite Leray covering exists on a compact manifold.*

Proof. We introduce an arbitrary connection on $\mathcal{X}$. Then, according to Proposition 1.2, normal convex neighborhoods also form an open covering of the manifold $\mathcal{X}$. Because each normal convex neighborhood is diffeomorphic to the ball $\mathbb{B}^n$(see above) and the intersection of any finite family of such neighborhoods is obviously a normal convex neighborhood (if it is not empty), this covering and also any finite subcovering, which exists if the manifold $\mathcal{X}$ is compact, is a Leray covering. $\square$

Chapter 2
Covariant Differentiation. Curvature

§1. Covariant Differentiation

We now pass directly to covariant differentiations that correspond to affine connections. As already stated in Chap. 1, covariant differentiations ∇_X with respect to an arbitrary connection ∇ on a manifold $\mathcal{X}$ are operators

$$\nabla_X : \mathfrak{a}\mathcal{X} \to \mathfrak{a}\mathcal{X} \tag{1}$$

defined on the linear space $\mathfrak{a}\mathcal{X}$ and having properties 1, 2, and 3 indicated in Chap. 1. Moreover, because for $\xi = \tau_{\mathcal{X}}$, the vector bundle $\mathbf{T}^s_r\xi$ is just the tensor bundle $\tau^s_r\mathcal{X}$ over the manifold $\mathcal{X}$, in accordance with the general results presented in Chap. 34, operators (1) are naturally extended to operators

$$\nabla_X : \mathbf{T}^s_r\mathcal{X} \to \mathbf{T}^s_r\mathcal{X} \tag{2}$$

on the linear spaces $\mathbf{T}^s_r\mathcal{X} = \Gamma(\tau^s_r\mathcal{X})$ of tensor fields. In general, the operators ∇_X assign a derivation of the algebra of tensor fields on $\mathcal{X}$ that commutes with contractions.

We recall (see Chap. 33) that a *derivation of the algebra of tensor fields* is a family of linear operators

$$D : \mathbf{T}^s_r\mathcal{X} \to \mathbf{T}^s_r\mathcal{X}$$

defined for all r and s such that

$$D(S \otimes T) = DS \otimes T + S \otimes DT \tag{3}$$

for any tensor fields S and T. For $r = s = 0$, each such derivation is just a derivation of the algebra of smooth functions $\mathbf{F}\mathcal{X} = \mathbf{T}^0_0\mathcal{X}$ in the sense of Definition 33.1, i.e., a certain vector field $X \in \mathfrak{a}\mathcal{X}$ (see Theorem 33.1). For $S = f$ and $T = Y$, where $f \in \mathbf{F}\mathcal{X}$ and $Y \in \mathfrak{a}\mathcal{X}$, relation (3) becomes

$$D(fY) = Xf \cdot Y + fDY, \tag{4}$$

where D is a linear operator $\mathfrak{a}\mathcal{X} \to \mathfrak{a}\mathcal{X}$. (See property 2 of the operators ∇_X in Chap. 1.)

Exercise 2.1. Prove that each linear (over the field $\mathbb{R}$) operator

$$D : \mathfrak{a}\mathcal{X} \to \mathfrak{a}\mathcal{X}$$

satisfying relation (4) (for a certain $X \in \mathfrak{a}\mathcal{X}$) is uniquely extended to a derivation D (coinciding with X on $\mathbf{F}\mathcal{X}$) of the algebra of tensor fields that commutes with contractions.

On linear differential forms α, the derivation D is defined by

$$(D\alpha)(Y) = X(\alpha(Y)) - \alpha(DY), \quad Y \in \mathfrak{a}\mathcal{X}.$$

In the particular case where $X = 0$, condition (4) means that D is a linear operator over the algebra $\mathbf{F}\mathcal{X}$. Therefore, *any $\mathbf{F}\mathcal{X}$-linear operator $D: \mathfrak{a}\mathcal{X} \to \mathfrak{a}\mathcal{X}$ is uniquely extended to a derivation of the algebra of tensor fields on the manifold $\mathcal{X}$ (denoted by the same symbol D) that commutes with contractions. The derivation equals zero on $\mathbf{F}\mathcal{X}$ and acts on linear differential forms by the formula*

$$(D\alpha)(Y) = -\alpha(DY), \quad Y \in \mathfrak{a}\mathcal{X}.$$

Exercise 2.2. State and prove a general theorem whose particular cases are Proposition 35.12 and the assertion of Exercise 2.1.

In each chart $(U, x^1, \dots, x^n)$, operator (2) is a linear combination $\nabla_X = X^k \nabla_k$ of the operators $\nabla_k = \nabla_{\partial/\partial x^k}$ of partial covariant derivatives,

$$\nabla_k : \mathbf{T}^s_r\mathcal{X} \to \mathbf{T}^s_r\mathcal{X},$$

and, moreover, for each tensor field S in $\mathbf{T}^s_r\mathcal{X}$ (and even in $\mathbf{T}^s_r U$), the components $(\nabla_k S)^{i_1\cdots i_s}_{j_1\cdots j_r}$ of the tensor field $\nabla_k S$ (we stress that it is defined only on U) are expressed through the components $S^{i_1\cdots i_s}_{j_1\cdots j_r}$ of the field S by

$$(\nabla_k S)^{i_1\cdots i_s}_{j_1\cdots j_r} = \frac{\partial S^{i_1\cdots i_s}_{j_1\cdots j_r}}{\partial x^k} + \sum_{a=1}^{s} \Gamma^{i_a}_{kp} S^{i_1\cdots p\cdots i_s}_{j_1\cdots\cdots\cdots j_r} - \sum_{b=1}^{r} \Gamma^{q}_{kj_b} S^{i_1\cdots\cdots\cdots i_s}_{j_1\cdots q\cdots j_r} \tag{5}$$

(see (39) in Chap. 35); we stress that summation with respect to p and q is assumed. In particular,

$$(\nabla_k Y)^i = \frac{\partial Y^i}{\partial x^k} + \Gamma^i_{kj} Y^j \tag{6}$$

for any vector field $Y \in \mathfrak{a}\mathcal{X}$ (see Eq. (4) in Chap. 1).

A tensor field S is said to be *covariantly constant* if $\nabla_X S = 0$ for each vector field X on $\mathcal{X}$ or, equivalently, if $\nabla_k S = 0$ for any $k = 1, \dots, n$ in each chart.

Example 2.1. The formula

$$\nabla_k \delta^i_j = \frac{\partial \delta^i_j}{\partial x^k} + \Gamma^i_{kp}\delta^p_j - \Gamma^q_{kj}\delta^i_q = 0$$

shows that the *Kronecker tensor δ^i_j is covariantly constant.*

It is clear that all covariantly constant tensor fields of a given type form a linear subspace of the linear space of all tensor fields.

§2. The Case of Tensors of Type $(r, 1)$

The case of tensor fields of type $(r, 1)$ is of special interest. As is known, such fields are in a natural bijective correspondence with $\mathbf{F}\mathcal{X}$-multilinear mappings of the form

$$\underbrace{\mathfrak{a}\mathcal{X} \times \cdots \times \mathfrak{a}\mathcal{X}}_{r} \to \mathfrak{a}\mathcal{X} \tag{7}$$

(mapping (7) that corresponds to the tensor field S sets the convolution $S(X_1, \ldots, X_r)$ of the field S with the fields $X_1, \ldots, X_r$ in correspondence with the vector fields $X_1, \ldots, X_r$, where S is the vector field with the components $S^i_{j_1 \cdots j_r} X_1^{j_1}, \ldots, X_r^{j_r}$). As a rule, tensor fields of type $(r, 1)$ and the corresponding mappings (7) are identified.

Using this identification, we can set the tensor field ∇S of type $(r+1, 1)$ in correspondence with each tensor field S of type $(r, 1)$ by setting

$$(\nabla S)(X_1, \ldots, X_r, X) = (\nabla_X S)(X_1, \ldots, X_r) \tag{8}$$

for any fields $X_1, \ldots, X_r, X \in \mathfrak{a}\mathcal{X}$. The tensor field ∇S is called the *covariant differential* of the tensor field S.

Exercise 2.3. Generalize this construction to tensor fields of an arbitrary type (r, s).

Exercise 2.4. The concept of covariant differential in the general case of an arbitrary bundle is introduced in Chap. 35. Because $\tau^*_{\mathcal{X}} \otimes \tau^s_r\mathcal{X} = \tau^s_{r+1}\mathcal{X}$, this covariant differential for $\xi = \tau^s_r\mathcal{X}$ is a mapping $\mathbf{T}^s_r\mathcal{X} \to \mathbf{T}^s_{r+1}\mathcal{X}$. Show that for $s = 1$, this is exactly mapping (8) (and its generalization from Exercise 2.3 for any s).

The components $(\nabla S)^i_{j_1 \cdots j_r j}$ of the tensor field ∇S in an arbitrary chart are the components

$$(\nabla_j S)^i_{j_1 \cdots j_r} = \frac{\partial S^i_{j_1 \cdots j_r}}{\partial x^j} + \Gamma^i_{jp} S^p_{j_1 \cdots j_r} - \sum_{b=1}^{r} \Gamma^q_{j j_b} S^i_{j_1 \cdots q \cdots j_r}$$

of the covariant partial derivatives $\nabla_j S$ of the field S (see Eq. (5)). Therefore, for any fields $X_1, \ldots, X_r, X$, we have

$$\begin{aligned}
[(\nabla S)(X_1, \ldots, X_r, X)]^i &= (\nabla S)^i_{j_1 \cdots j_r j} X_1^{j_1} \cdots X_r^{j_r} X^j \\
&= \frac{\partial S^i_{j_1 \cdots j_r}}{\partial x^j} X_1^{j_1} \cdots X_r^{j_r} X^j + \Gamma^i_{jp} S^p_{j_1 \cdots j_r} X_1^{j_1} \cdots X_r^{j_r} X^j \\
&\quad - \sum_{b=1}^{r} \Gamma^q_{j j_b} S^i_{j_1 \cdots q \cdots j_r} X_1^{j_1} \cdots X_r^{j_r} X^j
\end{aligned}$$

$$= \frac{\partial S^i_{j_1\cdots j_r} X^{j_1}_1 \cdots X^{j_r}_r}{\partial x^j} X^j - \sum_{b=1}^r S^i_{j_1\cdots j_b\cdots j_r} X^{j_1}_1 \cdots \frac{\partial X^{j_b}_b}{\partial x^j} \cdots X^{j_r}_r X^j$$

$$+ \Gamma^i_{jp} S^p_{j_1\cdots j_r} X^{j_1}_1 \cdots X^{j_r}_r X^j - \sum_{b=1}^r \Gamma^q_{jj_b} S^i_{j_1\cdots q\cdots j_r} X^{j_1}_1 \cdots X^{j_r}_r X^j$$

$$= \left(\frac{\partial S(X_1, \dots, X_r)^i}{\partial x^j} + \Gamma^i_{jp} S(X_1, \dots, X_r)^p \right) X^j$$

$$- \sum_{b=1}^r S^i_{j_1\cdots j_b\cdots j_r} X^{j_1}_1 \cdots \left(\frac{\partial X^{j_b}_b}{\partial x^j} + \Gamma^{j_b}_{jq} X^q_b \right) \cdots X^{j_r}_r X^j$$

$$= [\nabla_X S(X_1, \dots, X_r)]^i - \sum_{b=1}^r S(X_1, \dots, \nabla_X X_b, \dots, X_r)^i,$$

and hence

$$(\nabla S)(X_1, \dots, X_r, X) = \nabla_X S(X_1, \dots, X_r) - \sum_{b=1}^r S(X_1, \dots, \nabla_X X_b, \dots, X_r). \tag{9}$$

For example, for $r = 1, 2, 3$, we have

$$(\nabla S)(X, Y) = \nabla_Y S(X) - S(\nabla_Y X),$$
$$(\nabla S)(X, Y, Z) = \nabla_Z S(X, Y) - S(\nabla_Z X, Y) - S(X, \nabla_Z Y),$$
$$(\nabla S)(X, Y, Z, T) = \nabla_T S(X, Y, Z) - S(\nabla_T X, Y, Z) - S(X, \nabla_T Y, Z) - S(X, Y, \nabla_T Z)$$

or, in another notation,

$$\begin{aligned} (\nabla_Y S)(X) &= \nabla_Y S(X) - S(\nabla_Y X), \\ (\nabla_Z S)(X, Y) &= \nabla_Z S(X, Y) - S(\nabla_Z X, Y) - S(X, \nabla_Z Y), \\ (\nabla_T S)(X, Y, Z) &= \nabla_T S(X, Y, Z) - S(\nabla_T X, Y, Z) \\ &\quad - S(X, \nabla_T Y, Z) - S(X, Y, \nabla_T Z). \end{aligned} \tag{10}$$

In the first of these formulas, the symbols $\nabla_Y S$, S, and ∇_Y denote operators on $\mathfrak{a}\mathcal{X}$. The first summand in the right-hand side is the result of applying the operator S and then the operator ∇_Y to the field X, and the second one is the result of applying the operator ∇_Y and then the operator S. Therefore, using the sign $\circ$ (and replacing Y with X) for the composition of operators in order to avoid confusion, we can write this formula in the form of the relation between operators

$$\nabla_X S = \nabla_X \circ S - S \circ \nabla_X.$$

By definition, this means that *the operator* $\nabla_X S$ *is the commutator* $[\nabla_X, S]$ *of the operators* ∇_X *and* S:

$$\nabla_X S = [\nabla_X, S]. \tag{11}$$

Other formulas in (10) admit a similar interpretation. For example, each tensor field S of type (3,1) can be identified with the mapping $\mathfrak{a}\mathcal{X} \times \mathfrak{a}\mathcal{X} \to \operatorname{Hom}(\mathfrak{a}\mathcal{X}, \mathfrak{a}\mathcal{X})$ that sets the linear operator

$$S(X, Y): \mathfrak{a}\mathcal{X} \to \mathfrak{a}\mathcal{X}, \quad Z \mapsto S(X, Y, Z),$$

in correspondence with the two vector fields $X, Y \in \mathfrak{a}\mathcal{X}$. By this identification, the third formula in (10) (after the corresponding renaming of the arguments) becomes

$$(\nabla_X S)(Y, Z) = [\nabla_X, S(Y, Z)] - S(\nabla_X Y, Z) - S(Y, \nabla_X Z), \tag{12}$$

where X, Y, and Z are arbitrary vector fields on the manifold $\mathcal{X}$.

We note that a tensor field S of type $(r, 1)$ *is covariantly constant iff* $\nabla S = 0$. Therefore, in particular (see Eq. (11)), a tensor field S of type $(1, 1)$ *is covariantly constant iff* it commutes with all operators ∇_X when it is considered as an $\mathsf{F}\mathcal{X}$-linear operator $\mathfrak{a}\mathcal{X} \to \mathfrak{a}\mathcal{X}$.

§3. Torsion Tensor and Symmetric Connections

Because the operator ∇_X for a connection X on a manifold $\mathcal{X}$ is defined on the same linear space $\mathfrak{a}\mathcal{X}$ to which X belongs, we can interchange X and Y in the expression $\nabla_X Y$ for any fields $X, Y \in \mathfrak{a}\mathcal{X}$ (in the general case, this operation has no sense). This allows introducing the vector field

$$T(X, Y) = \nabla_X Y - \nabla_Y X - [X, Y].$$

The mapping

$$T: \mathfrak{a}\mathcal{X} \times \mathfrak{a}\mathcal{X} \to \mathfrak{a}\mathcal{X}, \quad (X, Y) \mapsto T(X, Y), \tag{13}$$

is obviously skew-symmetric, i.e.,

$$T(X, Y) = -T(Y, X),$$

and $\mathbb{R}$-linear in each of the arguments ($\mathbb{R}$-bilinear). Moreover, because the relation

$$[fX, Y] = f[X, Y] - Yf \cdot X$$

holds for any function $f \in \mathsf{F}\mathcal{X}$ (see § 33.3), we have

$$\begin{aligned} T(fX, Y) &= \nabla_{fX} Y - \nabla_Y (fX) - [fX, Y] = \\ &= f\nabla_X Y - f\nabla_Y X - f[X, Y] = fT(X, Y), \end{aligned}$$

and mapping (13) is therefore $\mathsf{F}\mathcal{X}$-bilinear. It therefore corresponds to a certain tensor field on $\mathcal{X}$ of type (2,1) that is skew-symmetric with respect to the subscripts.

Definition 2.1. A tensor field T is called the *torsion tensor* of an affine connection ∇. In the case where this tensor vanishes, i.e., when

$$\nabla_X Y - \nabla_Y X = [X, Y] \tag{14}$$

for any fields $X, Y \in \mathfrak{a}\mathcal{X}$, the connection ∇ is called *symmetric.*

In each coordinate neighborhood, the components T^i_{jk} of the tensor T are expressed as

$$T^i_{jk} = T\left(\frac{\partial}{\partial x^j}, \frac{\partial}{\partial x^k}\right)^i = \left(\nabla_j \frac{\partial}{\partial x^k}\right)^i - \left(\nabla_k \frac{\partial}{\partial x^j}\right)^i,$$

i.e., by

$$T^i_{jk} = \Gamma^i_{jk} - \Gamma^i_{kj} = \Gamma^i_{[jk]}, \quad i, j, k = 1, \ldots, n, \tag{15}$$

(see Eq. (2) in Chap. 1). In particular, we see that a connection ∇ *is symmetric iff* its connection coefficients Γ^i_{kj} in any chart are symmetric with respect to the subscripts,

$$\Gamma^i_{kj} = \Gamma^i_{jk}$$

for any $i, j, k = 1, \ldots, n$.

Proposition 2.1. *If a connection ∇ on a manifold $\mathcal{X}$ is symmetric, then for any point $p_0 \in \mathcal{X}$, there exist local coordinates $x^1, \ldots, x^n$ centered at p_0 in which the connection coefficients Γ^i_{kj} at the point p_0 vanish:*

$$(\Gamma^i_{kj})_{p_0} = 0. \tag{16}$$

These coordinates are are normal at the point p_0.

Proof. It follows from the assertion in Exercise 1.3 that if the coordinates $x^1, \ldots, x^n$ are normal at the point p_0, then

$$\Gamma^i_{kj} x^j x^k = 0$$

identically in $x^1, \ldots, x^n$. In particular, this implies that for $x^i = a^i t$,

$$\Gamma^i_{kj}(t\boldsymbol{a}) a^j a^k = 0, \quad \boldsymbol{a} = (a^1, \ldots, a^n),$$

for any $a^1, \ldots, a^n$ and t. For $t = 0$, we obtain

$$(\Gamma^i_{kj})_{p_0} a^j a^k = 0,$$

which, by the symmetry property of the coefficients $(\Gamma^i_{kj})_{p_0}$ in k and j and by the arbitrariness of the numbers $a^1, \ldots, a^n$, is possible only if these coefficients are zero. □

§4. Geometric Meaning of the Symmetry of a Connection

What is the geometric meaning of the torsion tensor and the symmetry of a connection? Let $p_0 \in \mathcal{X}$ and $A, B \in \mathbf{T}_{p_0}\mathcal{X}$. We then have a vector field X on the manifold $\mathcal{X}$ such that $X_{p_0} = A$ (see the addendum). Moreover, if $(U, h) = (U, x^1, \ldots, x^n)$ is an arbitrary chart containing the point p_0, then the field X can be chosen such that its components X^i in the chart (U, h) are constant (and are therefore equal to the coordinates a^i of the vector A in the basis $(\partial/\partial x^1)_{p_0}, \ldots, (\partial/\partial x^n)_{p_0}$).

If the field X is chosen as above, then the integral curve $u: t \mapsto u(t)$ of the field X passing through the point p_0 for $t = 0$ is given by the following linear functions in the chart (U, h):

$$x^i(t) = x_0^i + a^i t, \quad i = 1, \ldots, n,$$

where x_0^i, $1 \le i \le n$, are the coordinates of the point p_0. Therefore, for the coordinates $b^i(t)$ of the vector $B(t)$, which is obtained from the vector B by a parallel translation along the curve u at the point $u(t)$, we have

$$b^i(t) = b^i - \int_0^t \Gamma_{kj}^i(\boldsymbol{x}(t)) b^j(t) a^k dt, \quad \boldsymbol{x}(t) = (x^1(t), \ldots, x^n(t)),$$

where b^i, $1 \le i \le n$, are the coordinates of the vector B. This directly implies that

$$b^i(t) = b^i - (\Gamma_{kj}^i)_0 b^j a^k t + O(t^2),$$

where $(\Gamma_{kj}^i)_0$ are the values of the connection coefficients Γ_{kj}^i at the point p_0.

Now let t be fixed, and let Y be a vector field on $\mathcal{X}$ whose components in the chart (U, h) are constant and equal to $b^i(t)$. Then the integral curve $s \mapsto v(s)$ of the field Y passing through the point $u(t)$ for $s = 0$ is given by the functions

$$s \mapsto x^i(t) + b^i(t)s, \quad i = 1, \ldots, n,$$

which are linear in s. Letting p_t denote the point $v(s)$ for $s = t$, we therefore obtain the relation

$$x_t^i = x_0^i + (a^i + b^i)t - (\Gamma_{kj}^i)_0 b^j a^k t^2 + O(t^3)$$

for the coordinates x_t^i of this point. We can visualize the point p_t as the result of a translation of the point p_0 through the distance t first in direction of the vector A and then in direction of the vector B.

Similar formulas (with the interchange of the coordinates a^i and b^i) also hold for the point q_t obtained by a translation of the point p_0 first in direction of the vector B and then in direction of the vector A. Therefore, the difference of the coordinates of the points q_t and p_t equals

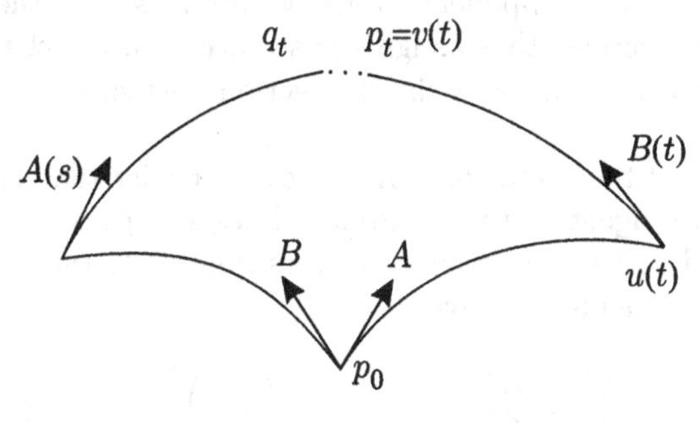

Fig. 2.1.

$$\begin{aligned}[(\Gamma^i_{kj})_0 b^j a^k - (\Gamma^i_{kj})_0 a^j b^k]t^2 + O(t^3) &= [((\Gamma^i_{kj})_0 - (\Gamma^i_{jk})_0)a^k b^j]t^2 + O(t^3) \\ &= T(A,B)^i t^2 + O(t^3),\end{aligned}$$

where $T(A,B)^i = (\Gamma^i_{[kj]})_0 a^k b^j$ are the components of the field $T(X,Y)$ at the point p_0 (depending on the vectors A and B only).

In the infinitesimal language, this means that an attempt to construct an infinitely small parallelogram in a manifold $\mathcal{X}$ spanned by two vectors A and B leads to a pentagon whose closing side is an infinitely small value of the second order with accuracy up to infinitely small values of the third order, and it is equal to $T(A,B)$.

This yields a visual geometric interpretation of the torsion tensor and, in particular, shows that *an affine connection is symmetric iff each infinitely small parallelogram is closed with accuracy up to infinitely small values of the third order.*

§5. Commutativity of Second Covariant Derivatives

An interpretation of the symmetry of a connection that is slightly different from the formal standpoint but is essentially the same can be obtained in the language of covariant derivatives. Let W be an open set in the (s,t)-plane $\mathbb{R}^2$, and let $\varphi: W \to \mathcal{X}$ be a smooth mapping (an *elementary surface* in $\mathcal{X}$). Further, let $X: (s,t) \mapsto X(s,t)$ be a smooth mapping $W \to \mathbf{T}\mathcal{X}$ such that

$$X(s,t) \in \mathbf{T}_{\varphi(s,t)}\mathcal{X}$$

for any point $(s,t) \in W$ (a *vector field on the surface* φ). Then we have two vector fields $\nabla X/\partial t$ and $\nabla X/\partial s$ on φ whose components in each chart $(U,h) = (U, x^1, \dots, x^n)$ are given by

$$\left(\frac{\nabla X}{\partial t}\right)^i = \frac{\partial X^i}{\partial t} + \Gamma^i_{kj} X^j \frac{\partial x^k}{\partial t}, \qquad \left(\frac{\nabla X}{\partial s}\right)^i = \frac{\partial X^i}{\partial s} + \Gamma^i_{kj} X^j \frac{\partial x^k}{\partial s},$$

where $X^i = X^i(s,t)$ are components of the vector $X(s,t)$ in the chart (U,h), $x^k = x^k(s,t)$ are functions that assign the surface φ in the chart (U,h), and $\Gamma^i_{kj} = \Gamma^i_{kj}(s,t)$ are the values of the connection coefficients ∇ at the point $\varphi(s,t)$.

As an example of the vector field X, we can consider the field $\partial\varphi/\partial t$ consisting of vectors tangent to the coordinate lines $s = \text{const}$ of the surface φ and a similar field $\partial\varphi/\partial s$ consisting of vectors tangent to the coordinate lines $t = \text{const}$. For these fields, we have

$$\left(\frac{\partial\varphi}{\partial t}\right)^i = \frac{\partial x^i}{\partial t} \quad \text{and} \quad \left(\frac{\partial\varphi}{\partial s}\right)^i = \frac{\partial x^i}{\partial s},$$

and the relations

$$\left(\frac{\nabla}{\partial s}\partial\right)^i = \frac{\partial^2 x^i}{\partial s\,\partial t} + \Gamma^i_{kj}\frac{\partial x^j}{\partial t}\frac{\partial x^k}{\partial s},$$

$$\left(\frac{\nabla}{\partial t}\frac{\partial\varphi}{\partial s}\right)^i = \frac{-\partial^2 x^i}{t\,\partial s} + \Gamma^i_{kj}\frac{\partial x^j}{\partial s}\frac{\partial x^k}{\partial t}$$

therefore hold. Therefore, a connection ∇ *is symmetric iff*

$$\frac{\nabla}{\partial s}\frac{\partial\varphi}{\partial t} = \frac{\nabla}{\partial t}\frac{\partial\varphi}{\partial s} \tag{17}$$

(identically with respect to s and t) for any surface $\varphi\colon W \to \mathcal{X}$.

§6. Curvature Tensor of an Affine Connection

As we know (see Chap. 36), for any connection in a vector bundle over a manifold $\mathcal{X}$ and for any affine connection on $\mathcal{X}$ in particular, the *curvature tensor* R of this connection is defined; its components $R^i_{j,kl}$ in each chart are expressed by the formula

$$R^i_{j,kl} = \frac{\partial\Gamma^i_{lj}}{\partial x^k} - \frac{\partial\Gamma^i_{kj}}{\partial x^l} + \Gamma^i_{kp}\Gamma^p_{lj} - \Gamma^i_{lp}\Gamma^p_{kj}.$$

For an affine connection, the transformation law of these components under a change of coordinates has the form

$$R^{i'}_{j',k'l'} = \frac{\partial x^{i'}}{\partial x^i}\frac{\partial x^j}{\partial x^{j'}}\frac{\partial x^k}{\partial x^{k'}}\frac{\partial x^l}{\partial x^{l'}}R^i_{j,kl}$$

(compare with (40) in Chap. 36), and R is therefore a *tensor of type* $(3,1)$ on the manifold $\mathcal{X}$. Considered as a mapping

$$R\colon \mathfrak{a}\mathcal{X} \times \mathfrak{a}\mathcal{X} \times \mathfrak{a}\mathcal{X} \to \mathfrak{a}\mathcal{X},$$

this tensor sets the vector field $R(X,Y)Z$, where

$$R(X,Y) = \nabla_X\nabla_Y - \nabla_Y\nabla_X - \nabla_{[X,Y]} \tag{18}$$

in correspondence with the vector fields $X, Y, Z \in \mathfrak{a}\mathcal{X}$ (see (41) in Chap. 36). In each chart, the field $R(X,Y)Z$ has the components $R^i_{j,kl}X^kY^lZ^j$, where X^k, Y^l, and Z^j are components of the vector fields X, Y, and Z.

Exercise 2.5. Show that for an arbitrary elementary surface φ, we have

$$\frac{\nabla}{\partial s}\frac{\nabla}{\partial t} - \frac{\nabla}{\partial t}\frac{\nabla}{\partial s} = R_{\varphi(s,t)}\left(\frac{\partial\varphi}{\partial s}, \frac{\partial\varphi}{\partial t}\right),$$

i.e., in more detail,

$$\frac{\nabla}{\partial s}\frac{\nabla}{\partial t}X(s,t) - \frac{\nabla}{\partial t}\frac{\nabla}{\partial s}X(s,t) = R_{\varphi(s,t)}\left(\frac{\partial\varphi}{\partial s}, \frac{\partial\varphi}{\partial t}\right)X(s,t)$$

for any vector field X on φ.

According to the assertion in Exercise 2.1 (i.e., its particular case concerning $\mathsf{F}\mathcal{X}$-linear operators D) the curvature operator

$$R(X,Y)\colon \mathfrak{a}\mathcal{X} \to \mathfrak{a}\mathcal{X}$$

for any vector fields $X, Y \in \mathfrak{a}\mathcal{X}$ is uniquely extended up to a certain derivation (denoted by the same symbol) of the algebra of tensor fields on the manifold $\mathcal{X}$; because the commutator of two derivations is also a derivation (see Chap. 33), the right-hand side of (18) is the restriction of the derivation $\nabla_X\nabla_Y - \nabla_Y\nabla_X - \nabla_{[X,Y]}$ of the algebra of tensor fields on the manifold $\mathcal{X}$ to $\mathfrak{a}\mathcal{X}$. As the derivation $R(X,Y)$, the latter derivation commutes with convolutions and vanishes on $\mathsf{F}\mathcal{X}$. Therefore, *both sides of* (18) *coincide as derivations of tensor fields on* $\mathcal{X}$ *as well.* In particular, this implies that *for any tensor* S *of type* $(r,1)$ *and for any vector fields* $X, Y, X_1, \ldots, X_r \in \mathfrak{a}\mathcal{X}$, *the formula*

$$\begin{aligned}(R(X,Y)S)(X_1,\ldots X_r) = R(X,Y)S(X_1,\ldots,X_r)\\ - \sum_{b=1}^{r} S(X_1,\ldots,R(X,Y)X_b,\ldots,X_r)\end{aligned} \tag{19}$$

holds. (It suffices to apply formula (9), with the left-hand side rewritten in the form $(\nabla_X S)(X_1,\ldots,X_r)$, to the tensor $\nabla_Y S$, alternate the result in X and Y, subtract formula (9) for the field $[X,Y]$, and replace $\nabla_X\nabla_Y - \nabla_Y\nabla_X - \nabla_{[X,Y]}$ with $R(X,Y)$ everywhere.)

For $r = 1$, formula (19) becomes

$$R(X,Y)S = [R(X,Y), S] \tag{20}$$

(compare with (11)), where S is an arbitrary $\mathsf{F}\mathcal{X}$-linear operator $\mathfrak{a}\mathcal{X} \to \mathfrak{a}\mathcal{X}$ (a tensor of type $(1,1)$). For $r = 3$ (see (12)), it becomes

$$\begin{aligned}(R(X,Y)S)(U,V) = [R(X,Y), S(U,V)]\\ - S(R(X,Y)U,V) - S(U,R(X,Y)V),\end{aligned} \tag{21}$$

where S is an arbitrary tensor of type $(3,1)$ (and $S(U,V)$, where $U, V \in \mathfrak{a}\mathcal{X}$, is therefore a certain $\mathsf{F}\mathcal{X}$-linear operator $\mathfrak{a}\mathcal{X} \to \mathfrak{a}\mathcal{X}$).

§7. Space with Absolute Parallelism

Definition 2.2. A connected affine connection space $\mathcal{X}$ is called a *space with absolute parallelism* if for any two points $p, q \in \mathcal{X}$ and for any path $u: I \to \mathcal{X}$ connecting these points, the parallel translation

$$\Pi_u: \mathbf{T}_p\mathcal{X} \to \mathbf{T}_q\mathcal{X}, \quad p = u(0), \quad q = u(1),$$

(see Chap. 1) does not depend on the choice of this path (i.e., if the connection on $\mathcal{X}$ is a connection with absolute parallelism.)

In a space with absolute parallelism, the translation is denoted by Π_q^p, and two vectors $A \in \mathbf{T}_p\mathcal{X}$ and $B \in \mathbf{T}_q\mathcal{X}$ that are related by $B = \Pi_q^p A$ are said to be *parallel.* A vector field $X \in \mathfrak{a}\mathcal{X}$ on a space with absolute parallelism $\mathcal{X}$ is called a *field of parallel vectors* if for any points $p, q \in \mathcal{X}$, the vectors X_p and X_q are parallel. All such fields obviously form an n-dimensional linear subspace of the space $\mathfrak{a}\mathcal{X}$.

According to the general results, *if an affine connection space is a space with absolute parallelism, then its curvature tensor R vanishes* (connections with the curvature tensor identically equal to zero are said to be flat; however, as applied to affine connections, the latter term is used only under the additional assumption that the connection is symmetric). This necessary condition is sufficient if the manifold $\mathcal{X}$ is simply connected.

On the other hand, it is easy to see (prove this!) that *for an affine connection space $\mathcal{X}$*, the curvature tensor vanishes identically iff the space $\mathfrak{c}\mathcal{X}$ of covariantly constant vector fields is n-dimensional (where $n = \dim \mathcal{X}$ as usual). In particular, $\dim \mathfrak{c}\mathcal{X} = n$ for each space $\mathcal{X}$ with absolute parallelism.

In general, the dimension of the space $\mathfrak{c}\mathcal{X}$ is equal to the dimension of the space of all vectors that are invariant under the restricted holonomy group transformations. (By the general results presented in Chap. 36, the latter group is trivial iff $R = 0$.)

§8. Bianci Identities

The covariant derivative $\nabla_Z R$ of a tensor R is also a tensor of type (3,1), and it therefore sets the operator

$$(\nabla_Z R)(X, Y): \mathfrak{a}\mathcal{X} \to \mathfrak{a}\mathcal{X}$$

in correspondence with any vector fields $X, Y \in \mathfrak{a}\mathcal{X}$. For this operator, Eq. (12) holds (where S should certainly be replaced with R).

Proposition 2.2. *If a connection on a manifold $\mathcal{X}$ is symmetric, then for any fields $X, Y, Z \in \mathfrak{a}\mathcal{X}$, we have*

$$R(X, Y)Z + R(Y, Z)X + R(Z, X)Y = 0, \tag{22}$$

$$(\nabla_X R)(Y, Z) + (\nabla_Y R)(Z, X) + (\nabla_Z R)(X, Y) = 0. \tag{23}$$

Proof. For any function S of the arguments X, Y, and Z, we say that the sum

$$S(X,Y,Z) + S(Y,Z,X) + S(Z,X,Y)$$

is obtained *by cycling* $S(X,Y,Z)$. In this terminology, formula (22) asserts that *cycling the field* $R(X,Y)Z$ *yields zero.* Because the cycling operation is obviously linear, we see that to prove this formula, it suffices to represent $R(X,Y)Z$ as a sum of expressions for which cycling each yields zero in advance.

But for the field

$$R(X,Y)Z = \nabla_X \nabla_Y Z - \nabla_Y \nabla_X Z - \nabla_{[X,Y]} Z$$

in the case where the connection ∇ is symmetric, we have the relation

$$R(X,Y)Z = \nabla_X(\nabla_Z Y + [Y,Z]) - \nabla_Y \nabla_X Z - (\nabla_Z [X,Y] + [[X,Y],Z])$$

according to (14), i.e., the relation

$$\begin{aligned} R(X,Y)Z = &(\nabla_X \nabla_Z Y - \nabla_Y \nabla_X Z) \\ &+ (\nabla_X [Y,Z] - \nabla_Z [X,Y]) - [[X,Y],Z]. \end{aligned} \tag{24}$$

On the other hand, an automatic computation shows that *cycling any function of the form*

$$S(X,Y,Z) - S(Y,Z,X)$$

yields zero. Indeed,

$$\begin{aligned} &S(X,Y,Z) - S(Y,Z,X) + S(Y,Z,X) \\ &-S(Z,X,Y) + S(Z,X,Y) - S(X,Y,Z) = 0 \end{aligned}$$

(we note that the same is also true for any function of the form $S(X,Y,Z) - S(X',Y',Z')$, where X',Y',Z' is an arbitrary even permutation of the arguments X,Y,Z). In particular, cycling each set of parentheses in the right-hand side of formula (24) therefore yields zero. Moreover, according to the Jacobi identity (see (19) in Chap. 33), cycling the third summand $[[X,Y],Z]$ also yields zero. This proves (22).

Similarly, according to (12) and (15),

$$\begin{aligned} (\nabla_X R)(Y,Z) &= [\nabla_X, R(Y,Z)] - R(\nabla_X Y, Z) - R(Y, \nabla_X Z) \\ &= [\nabla_X, R(Y,Z)] - R(\nabla_Y X, Z) \\ &\quad - R([X,Y],Z) - R(Y, \nabla_X Z) \\ &= ([\nabla_X, R(Y,Z)] - R([X,Y],Z)) \\ &\quad + (R(Z, \nabla_Y X) - R(Y, \nabla_X Z)), \end{aligned}$$

where the second set of parentheses is of the form $S(Z,Y,X) - S(Y,X,Z)$, and cycling it therefore yields zero (the triple Y, X, Z is obtained from the triple Z, Y, X by an even permutation). On the other hand,

$$\begin{aligned}[\nabla_X, R(Y,Z)] - R([X,Y],Z) = [\nabla_X, [\nabla_Y, \nabla_Z]] - [\nabla_X, \nabla_{[Y,Z]}] \\ - [\nabla_{[X,Y]}, \nabla_Z] + \nabla_{[[X,Y],Z]},\end{aligned}$$

where cycling the first and last summands yields zero by the Jacobi identity and cycling the sum of the middle summands yields zero because this sum can be written in the form

$$[\nabla_{[Y,Z]}, \nabla_X] - [\nabla_{[X,Y]}, \nabla_Z].$$

Exercise 2.6. Show that the following relations hold for an arbitrary affine connection on a manifold:

$$\begin{aligned}\mathfrak{S}\{R(X,Y)Z\} &= \mathfrak{S}\{T(T(X,Y),Z)\} + \mathfrak{S}\{\nabla_X T(Y,Z)\}, \\ \mathfrak{S}\{(\nabla_X R)(Y,Z)\} &= \mathfrak{S}\{R(X,T(Y,Z))\},\end{aligned}$$

where $\mathfrak{S}$ is the cycling operator.

Written in terms of components, relations (22) and (23) become

$$R^i_{j,kl} + R^i_{k,lj} + R^i_{l,jk} = 0, \tag{22'}$$

$$\nabla_s R^i_{j,kl} + \nabla_k R^i_{j,ls} + \nabla_l R^i_{j,sk} = 0, \tag{23'}$$

Using the notation introduced in the addendum, we can write these formulas in the shorter form

$$R^i_{(j,kl)} = 0, \tag{22''}$$

$$\nabla_{(s} R^i_{|j|,kl)} = 0. \tag{23''}$$

It is interesting to compare (23″) with formula (50) in Chap. 36 (written for the case $\xi = \tau_{\mathcal{X}}$, i.e., for the case of a connection on a manifold). Substantively, these formulas are distinct because the symbol ∇_s in (23″) denotes the operator of partial covariant differentiation of tensors of type (3,1) on the manifold $\mathcal{X}$ (when computing it, we alternately use the contraction of connection coefficients Γ^i_{kj} with all subscripts and superscripts k, l, i, and j), and in formula (?) in the supplement, this symbol denotes the operator of partial covariant differentiation of ξ-tensor fields of type (1, 1) (we use the contraction by only the superscript i and subscript j). Nevertheless, it turns out that *these formulas imply one another* (which, in particular, gives a new proof of (23″) because the results of applying both operators ∇_s to the tensor R have components that differ by the expression

$$\Gamma^p_{sk} R^i_{j,pl} + \Gamma^p_{sl} R^i_{j,kp} = \Gamma^p_{sk} R^i_{j,pl} - \Gamma^p_{sl} R^i_{j,pk},$$

which yields zero when cycled with respect s, k, and l.

Formulas (22″) and (23″) (and also (22) and (23), which are equivalent to them) are called the *first* and *second Bianci identities* for symmetric connections. The first Bianci identity is also called the *Ricci identity*, and the second one is also called the *Bianchi–Padov identity.*

§9. Trace of the Curvature Tensor

Because for any vector fields $X, Y \in \mathfrak{a}\mathcal{X}$, the curvature operator $R(X,Y)$ of an affine connection space $\mathcal{X}$ is a linear operator $\mathfrak{a}\mathcal{X} \to \mathfrak{a}\mathcal{X}$ and the $\mathsf{F}\mathcal{X}$-module $\mathfrak{a}\mathcal{X}$ is a free module of a finite rank over each coordinate neighborhood, we can introduce the trace $\operatorname{Tr} R(X,Y)$ of the operator $R(X,Y)$ into consideration. This trace is $\mathsf{F}\mathcal{X}$-linear in X, Y, i.e., the correspondence

$$\operatorname{Tr} R\colon (X,Y) \mapsto \operatorname{Tr} R(X,Y)$$

is a skew-symmetric tensor of type $(2,0)$ on the space $\mathcal{X}$ or, in another terminology, a differential form of the second degree on $\mathcal{X}$.

In each chart $(U, x^1, \dots, x^n)$, this tensor has the components

$$R^i_{i,kl} = \frac{\partial \Gamma^i_{li}}{\partial x^k} - \frac{\partial \Gamma^i_{ki}}{\partial x^l}$$

(because $\Gamma^i_{kp}\Gamma^p_{li} = \Gamma^i_{lp}\Gamma^p_{ki}$). Therefore, considered as a differential form, the tensor $\operatorname{Tr} R$ is the exterior differential $d\gamma$ of the linear differential form

$$\gamma = \Gamma^i_{ki} dx^k. \tag{25}$$

We stress that in contrast to $\operatorname{Tr} R$, the form γ *depends* on the choice of local coordinates $x^1, \dots, x^n$; changing these coordinates, we add a form df, where f is a certain function, to γ.

Exercise 2.7. Prove the latter assertion. [*Hint*: The function f equals the logarithm of the determinant of the transition matrix.]

§10. Ricci Tensor

Another approach to the construction of a tensor of type (2,0) from the curvature tensor, which seems not so natural at first glance, consists in the contraction of i not with j but with one of the subscripts k or l (with accuracy up to the sign, the choice of subscript is not essential; we choose k).

Definition 2.3. A tensor of type $(2,0)$ with the components

$$R_{ij} = R^k_{i,kj}$$

is called the *Ricci tensor* of an affine connection space $\mathcal{X}$. In the modern literature, it is usually denoted by $\operatorname{Ric} \mathcal{X}$ (however, the classical notation R_{ij} is preserved for its components).

To describe the Ricci tensor in invariant, coordinate-free terms, we note that for any vector fields $X, Y \in \mathfrak{a}\mathcal{X}$, the correspondence $Z \mapsto R(Z,Y)X$, $Z \in \mathfrak{a}\mathcal{X}$, defines a certain linear operator on the $\mathsf{F}\mathcal{X}$-module $\mathfrak{a}\mathcal{X}$. Because the $\mathsf{F}\mathcal{X}$-module $\mathfrak{a}\mathcal{X}$ is a free module of a finite rank on each coordinate neighborhood, the trace $\operatorname{Tr}[Z \mapsto R(Z,Y)X]$, which is a function from $\mathsf{F}\mathcal{X}$, is defined, and a direct comparison of the definitions shows that this function is just the value $\operatorname{Ric}(X,Y)$ of the tensor Ric at the fields X, Y (interpreted as the mapping $\mathfrak{a}\mathcal{X} \times \mathfrak{a}\mathcal{X} \to \mathsf{F}\mathcal{X}$):

$$\operatorname{Ric}(X,Y) = \operatorname{Tr}[Z \mapsto R(Z,Y)X]. \tag{26}$$

The components R_{ij} of the Ricci tensor are expressed through the connection coefficients by the formula

$$R_{ij} = \frac{\partial \Gamma^k_{ji}}{\partial x^k} - \frac{\partial \Gamma^k_{ki}}{\partial x^j} + \Gamma^k_{kp}\Gamma^p_{ji} - \Gamma^k_{jp}\Gamma^p_{ki}, \tag{27}$$

and in each chart, we have

$$\operatorname{Ric}(X,Y) = R_{ij}X^iY^j, \quad X, Y \in \mathfrak{a}\mathcal{X}, \tag{28}$$

where X^i and Y^j are the respective components of the fields X and Y in a given chart.

Exercise 2.8. Show that *if a connection* ∇ *is symmetric, then*

$$\operatorname{Ric}(X,Y) - \operatorname{Ric}(Y,X) = \operatorname{Tr} R(X,Y)$$

for any vector fields X *and* Y. [*Hint*: For components, this formula asserts that $R^k_{i,kj} - R^k_{j,ki} = R^k_{k,ij}$.]

In particular, this implies that *the Ricci tensor of a symmetric connection is symmetric iff the tensor* $\operatorname{Tr} R$ *is identically zero,* i.e.,

$$d\gamma = 0$$

(the differential form γ is closed).

Chapter 3
Affine Mappings. Submanifolds

§1. Affine Mappings

Let $\mathcal{X}$ and $\mathcal{Y}$ be affine connection spaces with the connections $\nabla^{\mathcal{X}}$ and $\nabla^{\mathcal{Y}}$ (to simplify formulas, we often write ∇ instead of $\nabla^{\mathcal{X}}$ and $\widehat{\nabla}$ instead of $\nabla^{\mathcal{Y}}$). On each coordinate neighborhood U of the manifold $\mathcal{X}$ (coordinate neighborhood V of the manifold $\mathcal{Y}$), the connection $\nabla^{\mathcal{X}}$ (connection $\nabla^{\mathcal{Y}}$) is given by the matrix $\omega = \omega^{\mathcal{X}}$ (matrix $\widehat{\omega} = \omega^{\mathcal{Y}}$) of connection forms. The connection $\nabla^{\mathcal{X}}$ sets the horizontal subspace $H_A^{\mathcal{X}}$ of the tangent space $\mathbf{T}_A(\mathbf{T}\mathcal{X})$ in correspondence with each tangent vector A (point of the total space $\mathbf{T}\mathcal{X}$ of the tangent bundle $\tau_{\mathcal{X}}$). Similarly, the connection $\nabla^{\mathcal{Y}}$ sets the horizontal subspace $H_B^{\mathcal{Y}} \subset \mathbf{T}_B(\mathbf{T}\mathcal{Y})$ in correspondence with each point $B \in \mathbf{T}\mathcal{Y}$.

Let $f\colon \mathcal{X} \to \mathcal{Y}$ be an arbitrary smooth mapping. Two charts $(U,h) = (U, x^1, \dots, x^n)$ and $(V,k) = (V, y^1, \dots, y^m)$ of manifolds $\mathcal{X}$ and $\mathcal{Y}$ are said to be *f-related* if $fU \subset V$. In such charts, the mapping f (or, more precisely, the mapping $U \to V$ induced by it) is given by functions of the form

$$y^a = f^a(x^1, \dots, x^n), \quad a = 1, \dots, m.$$

The Jacobi matrix

$$J_f = \left\| \frac{\partial f^a}{\partial x^i} \right\|, \quad 1 \le i \le n, \quad 1 \le a \le m,$$

of these functions is called the *Jacobi matrix of the mapping f* in the charts (U,h) and (V,k).

We recall that the vector fields $X \in \mathfrak{a}\mathcal{X}$ and $\widehat{X} \in \mathfrak{a}\mathcal{Y}$ are said to be *f-related* if

$$(df)_p X_p = \widehat{X}_{f(p)} \tag{1}$$

for any point $p \in \mathcal{X}$, i.e., if for any pair of f-related charts (U,h) and (V,k), we have

$$\frac{\partial f^a}{\partial x^i} X^i = \widehat{X}^a \circ f, \quad 1 \le i \le n, \quad 1 \le a \le m,$$

where X^i and $\widehat{X}^a$ are components if the fields X and $\widehat{X}$ in the charts (U,h) and (V,k).

On the other hand, it is clear that the formula

$$(\mathbf{T}f)A = (df)_p A, \quad A \in \mathbf{T}\mathcal{X},$$

where $p \in \mathcal{X}$ is a point such that $A \in \mathbf{T}_p\mathcal{X}$, correctly defines the smooth mapping

$$\mathbf{T}f\colon \mathbf{T}\mathcal{X} \to \mathbf{T}\mathcal{Y}$$

of manifolds of tangent vectors. Because this mapping is smooth, its differential at an arbitrary point $A \in \mathbf{T}\mathcal{X}$

$$(d\mathbf{T}f)_A : \mathbf{T}_A(\mathbf{T}\mathcal{X}) \to \mathbf{T}_B(\mathbf{T}\mathcal{Y})$$

is defined, where $B = (\mathbf{T}f)A$.

Proposition 3.1. *The following properties of a smooth mapping $f: \mathcal{X} \to \mathcal{Y}$ are equivalent:*

1. *If two fields $X, Y \in \mathfrak{a}\mathcal{X}$ are f-related to the fields $\widehat{X}, \widehat{Y} \in \mathfrak{a}\mathcal{Y}$, then the field $\nabla_X Y$ is f-related to the field $\widehat{\nabla}_{\widehat{X}}\widehat{Y}$.*
2. *For any f-related charts (U, h) and (V, k),*

$$J_f\omega = f^*\widehat{\omega}J_f + dJ_f \quad \text{on } U. \tag{2}$$

3. *For any curve $\gamma: I \to \mathcal{X}$ and any vector field $X: t \mapsto X(t)$ on γ,*

$$\frac{\widehat{\nabla}}{dt}[(df)_{\gamma(t)}X(t)] = (df)_{\gamma(t)}\frac{\nabla}{dt}X(t), \quad t \in I. \tag{3}$$

4. *For any curve $\gamma: I \to \mathcal{X}$, the diagram*

$$\begin{array}{ccc} \mathbf{T}_{p_0}\mathcal{X} & \xrightarrow{(df)_{p_0}} & \mathbf{T}_{q_0}\mathcal{Y} \\ \Pi_\gamma \downarrow & & \downarrow \widehat{\Pi}_{f\circ\gamma} \\ \mathbf{T}_{p_1}\mathcal{X} & \xrightarrow{(df)_{p_1}} & \mathbf{T}_{q_1}\mathcal{Y} \end{array} \tag{4}$$

   is commutative, where p_0 and p_1 are the respective initial point and endpoint of the curve γ, $q_0 = f(p_0)$, $q_1 = f(p_1)$, and Π_γ and $\widehat{\Pi}_{f\circ\gamma}$ are parallel translations along the curves γ and $f \circ \gamma$.
5. *At each point $A \in \mathbf{T}\mathcal{X}$,*

$$(d\mathbf{T}f)_A H_A^{\mathcal{X}} \subset H_B^{\mathcal{Y}}, \quad B = (\mathbf{T}f)A.$$

Proof. If the fields $X, Y \in \mathfrak{a}\mathcal{X}$ and $\widehat{X}, \widehat{Y} \in \mathfrak{a}\mathcal{Y}$ are f-related, then for any f-related charts (U, h) and (V, k),

$$X^i\frac{\partial f^a}{\partial x^i} = \widehat{X}^a \circ f, \qquad Y^i\frac{\partial f^a}{\partial x^i} = \widehat{Y}^a \circ f \quad \text{on } U.$$

On the other hand, if Γ^i_{kj} and $\widehat{\Gamma}^a_{cb}$ are coefficients of the connections ∇ and $\widehat{\nabla}$ in these charts, then

$$(\nabla_X Y)^i = \left(\frac{\partial Y^i}{\partial x^k} + \Gamma^i_{kj}Y^j\right)X^k, \qquad (\widehat{\nabla}_{\widehat{X}}\widehat{Y})^a = \left(\frac{\partial \widehat{Y}^a}{\partial y^c} + \widehat{\Gamma}^a_{cb}\widehat{Y}^b\right)\widehat{X}^c.$$

Therefore,

$$
\begin{aligned}
(\widehat{\nabla}_{\widehat{X}}\widehat{Y})^a \circ f &= \left(\left(\frac{\partial \widehat{Y}^a}{\partial y^c} \circ f\right) + (\widehat{\Gamma}^a_{cb} \circ f)(\widehat{Y}^b \circ f)\right)(\widehat{X}^c \circ f) \\
&= \left(\left(\frac{\partial \widehat{Y}^a}{\partial y^c} \circ f\right)\frac{\partial f^c}{\partial x^k} + (\widehat{\Gamma}^a_{cb} \circ f)\frac{\partial f^b}{\partial x^j} Y^j \frac{\partial f^c}{\partial x^k}\right) X^k \\
&= \left(\frac{\partial(\widehat{Y}^a \circ f)}{\partial x^k} + (\widehat{\Gamma}^a_{cb} \circ f)\frac{\partial f^b}{\partial x^j}\frac{\partial f^c}{\partial x^k} Y^j\right) X^k \\
&= \left(\frac{\partial}{\partial x^k}\left(Y^i \frac{\partial f^a}{\partial x^i}\right) + (\widehat{\Gamma}^a_{cb} \circ f)\frac{\partial f^b}{\partial x^j}\frac{\partial f^c}{\partial x^k} Y^j\right) X^k \\
&= \left(\frac{\partial Y^i}{\partial x^k}\frac{\partial f^a}{\partial x^i} + Y^i \frac{\partial^2 f^a}{\partial x^i \partial x^k} + (\widehat{\Gamma}^a_{cb} \circ f)\frac{\partial f^b}{\partial x^j}\frac{\partial f^c}{\partial x^k} Y^j\right) X^k \\
&= (\nabla_X Y)^i \frac{\partial f^a}{\partial x^i} \\
&\quad + \left(-\frac{\partial f^a}{\partial x^i}\Gamma^i_{kj} + \frac{\partial^2 f^a}{\partial x^j \partial x^k} + (\widehat{\Gamma}^a_{cb} \circ f)\frac{\partial f^b}{\partial x^j}\frac{\partial f^c}{\partial x^k}\right) Y^j X^k.
\end{aligned}
$$

Because the relation $(\nabla_X Y)^i(\partial f^a/\partial x^i) = (\widehat{\nabla}_{\widehat{X}}\widehat{Y})^a \circ f$ means that the fields $\nabla_X Y$ and $\widehat{\nabla}_{\widehat{X}}\widehat{Y}$ are f-related and the relation

$$
\frac{\partial f^a}{\partial x^i}\Gamma^i_{kj} = \frac{\partial^2 f^a}{\partial x^j \partial x^k} + (\widehat{\Gamma}^a_{cb} \circ f)\frac{\partial f^b}{\partial x^j}\frac{\partial f^c}{\partial x^k}
$$

transforms into (2) after multiplication by dx^k, this proves the equivalence of properties 1 and 2.

Exercise 3.1. Prove the equivalence of properties 2, 3, and 4.

Exercise 3.2. Prove the equivalence of properties 2 and 5. [*Hint*: In the charts $(\mathbf{T}U, x^1, \dots, x^n, \dot{x}^1, \dots, \dot{x}^n)$ and $(\mathbf{T}V, y^1, \dots, y^m, \dot{y}^1, \dots, \dot{y}^m)$ (see Chap. 1), the Jacobi matrix of the mapping $\mathbf{T}f$ becomes

$$
\begin{Vmatrix} J_f & 0 \\ K_f & J_f \end{Vmatrix}, \quad \text{where } K_f = \begin{Vmatrix} \frac{\partial^2 f^a}{\partial x^i \partial x^k}\dot{x}^k \end{Vmatrix}.
$$

On the other hand, the subspace $H^{\mathcal{X}}_A$ is generated by the vectors $(\partial/\partial x^k)_A - \Gamma^i_{kj}\dot{x}^j(\partial/\partial \dot{x}^i)_A$, and the subspace $H^{\mathcal{Y}}$ is generated by the vectors $(\partial/\partial y^c)_B - \Gamma^a_{cb}\dot{y}^b(\partial/\partial \dot{y}^a)_B$, where $1 \le k \le n$ and $1 \le c \le n$.]

This proves Proposition 3.1. □

Definition 3.1. A smooth mapping $f\colon \mathcal{X} \to \mathcal{Y}$ having properties 1–5 is called an *affine mapping*.

Exercise 3.3. Each interval I of the real axis $\mathbb{R}$ is an affine connection space with respect to the trivial connection $\nabla_{\partial/\partial t} = \partial/\partial t$. Therefore, it is reasonable to speak about curves $\gamma\colon \mathbb{R} \to \mathcal{X}$ that are affine mappings. Show that they are exactly the geodesics of the space $\mathcal{X}$.

It is clear that the affinity property of a mapping is a local one, i.e., *a mapping $f: \mathcal{X} \to \mathcal{Y}$ is affine if it is affine on a certain neighborhood of any point $p \in \mathcal{X}$.* Moreover, property 4 of affine mappings directly implies that *each affine mapping transforms geodesics into geodesics and can therefore be written in terms of linear functions in normal coordinates.* This is a base for proving the following proposition.

Proposition 3.2 (uniqueness of an affine mapping). *Let affine mappings $f, g: \mathcal{X} \to \mathcal{Y}$ coincide at a point $p_0 \in \mathcal{X}$, and let their differential coincide at this point:*

$$f(p_0) = g(p_0), \qquad (df)_{p_0} = (dg)_{p_0}.$$

Then $f = g$ on a component of the manifold $\mathcal{X}$ that contains the point p_0.

Proof. Let

$$C = \{p \in \mathcal{X}\colon\ f(p) = g(p),\ (df)_p = (dg)_p\}$$

be the set of all points $p \in \mathcal{X}$ at which the mappings f and g and their differentials coincide. Because the mappings f and g are continuous (and the manifold $\mathcal{Y}$ is Hausdorff), the set C is closed. On the other hand, because each affine mapping in normal coordinates is written in terms of linear functions, the normal neighborhood of any point $p \in C$ is contained in C. Therefore, the set C is open.Being an open-closed set that contains the point p_0, the set C contains the component $\mathcal{X}_0$ of this point. Therefore, $f = g$ on $\mathcal{X}_0$. □

§2. Affinities

It is clear that a composition of affine mappings is an affine mapping and that all affine connection spaces and all their affine mappings therefore form a category (see Chap. 7 below). As is easily seen, isomorphisms in this category are affine mappings that are diffeomorphisms. We call them *affine isomorphisms* or *affine diffeomorphisms* (and isomorphisms $\mathcal{X} \to \mathcal{X}$ of an affine connection space $\mathcal{X}$ onto itself are called *affine automorphisms*). Affine diffeomorphisms are also called *affinities.*

In the case where f is a diffeomorphism, we can rewrite condition (2) in the form

$$\omega = J_f^{-1}(f^*\widehat{\omega})J_f + J_f^{-1}dJ_f. \tag{2'}$$

In the particular case where the diffeomorphism f acts with respect to the equality of coordinates (i.e., it transforms each point $p \in U$ into a point $q \in V$ having the same coordinates in the chart (V, k) as the point p has in the chart (U, h)), this condition becomes $\omega = f^*\widehat{\omega}$, which means that $\widehat{\omega}$ transforms into ω under the substitution $y^i = x^i$, $1 \le i \le n$ (the forms ω and $\widehat{\omega}$ differ only by the notation of variables).

Each diffeomorphism $f: \mathcal{X} \to \mathcal{Y}$ defines a bijective mapping $f_*: \mathfrak{a}\mathcal{X} \to \mathfrak{a}\mathcal{Y}$ by

$$(f_*X)_q = (df)_p X_p, \quad p = f^{-1}(q);$$

this mapping is such that the vector fields X and f_*X are f-related. (This is exactly the mapping $(f^{-1})^*$ in the addendum.) Therefore (see property 1 in Proposition 3.1), a *diffeomorphism* $f: \mathcal{X} \to \mathcal{Y}$ *is an affinity iff for any field* $X \in \mathfrak{a}\mathcal{X}$, *the diagram*

$$\begin{array}{ccc} \mathfrak{a}\mathcal{X} & \xrightarrow{f_*} & \mathfrak{a}\mathcal{Y} \\ \Big\downarrow{\scriptstyle \nabla_X} & & \Big\downarrow{\scriptstyle \widehat{\nabla}_{f_*X}} \\ \mathfrak{a}\mathcal{X} & \xrightarrow{f_*} & \mathfrak{a}\mathcal{Y} \end{array} \tag{5}$$

is commutative.

This directly implies that *each affinity preserves the torsion and curvature tensors* or, more exactly, for any affinity $f: \mathcal{X} \to \mathcal{Y}$, we have

$$T^{\mathcal{X}} = f^* T^{\mathcal{Y}} \qquad \text{and} \qquad R^{\mathcal{X}} = f^* R^{\mathcal{Y}}, \tag{6}$$

where $T^{\mathcal{X}}$ and $R^{\mathcal{X}}$ are the torsion and curvature tensors of the space $\mathcal{X}$ and $T^{\mathcal{Y}}$ and $R^{\mathcal{Y}}$ are those of the space $\mathcal{Y}$. (For tensor fields, the mapping f^* is defined in the addendum.)

Therefore, relations (6) are necessary for a diffeomorphism $f: \mathcal{X} \to \mathcal{Y}$ to be an affinity, but they are not sufficient in general. (Some textbooks and monographs contain a mistake at exactly this point.) We discuss this question in the next chapter and now pass to coverings and submanifolds of affine connection spaces.

§3. Affine Coverings

Let $\mathcal{X}$ be an arbitrary smooth manifold (without an affine connection), and let $f: \mathcal{X} \to \mathcal{Y}$ be its étale mapping onto an affine connection space $\mathcal{Y}$.

Exercise 3.4. Show that *on $\mathcal{X}$, we have a unique connection with respect to which the mapping f is affine.* [*Hint*: If $(U, x^1, \dots, x^n)$ and $(V, y^1, \dots, y^n)$ are charts of the manifolds $\mathcal{X}$ and $\mathcal{Y}$ such that $V = fU$ and f acts with respect to the equality of coordinates (i.e., $x^i = y^i \circ f$ for all $i = 1, \dots, n$), then the connection forms of $\nabla^{\mathcal{X}}$ on U are the forms $f^*\widehat{\omega}^i_j$, where $\widehat{\omega}^i_j$ are connection forms of $\nabla^{\mathcal{Y}}$ on $\mathcal{Y}$ (the expressions of the forms $f^*\widehat{\omega}^i_j$ and $\widehat{\omega}^i_j$ in coordinates differ only by the notation of variables).]

In particular, we see that *for any smooth covering mapping $\pi: \widetilde{\mathcal{X}} \to \mathcal{X}$ of the affine connection space $\mathcal{X}$, there exists a unique connection on $\widetilde{\mathcal{X}}$ with respect to which the mapping π is affine.* A smooth covering $(\widetilde{\mathcal{X}}, \pi, \mathcal{X})$ whose projection π is an affine mapping is called an *affine covering*.

If the manifold $\widetilde{\mathcal{X}}$ is an affine connection space in a smooth covering $(\widetilde{\mathcal{X}}, \pi, \mathcal{X})$, then, in general, there is no connection on the manifold $\mathcal{X}$ with respect to which this covering is affine. We consider this question in more detail. Let Aut $\widetilde{\mathcal{X}}$ be an automorphism (gliding) group of a covering $(\widetilde{\mathcal{X}}, \pi, \mathcal{X})$. We recall (see the addendum) that this group acts on $\widetilde{\mathcal{X}}$ discretely and smoothly.

Exercise 3.5. Show that for any affine covering $(\widetilde{\mathcal{X}}, \pi, \mathcal{X})$, the group $\operatorname{Aut}\widetilde{\mathcal{X}}$ consists of affine automorphisms of the space $\widetilde{\mathcal{X}}$.

Therefore, to introduce a connection on the manifold $\mathcal{X}$ with respect to which the covering $(\widetilde{\mathcal{X}}, \pi, \mathcal{X})$ is affine, it is necessary that the group $\operatorname{Aut}\widetilde{\mathcal{X}}$ consist of affine automorphisms. When is this necessary condition sufficient?

We recall (see the addendum) that for any group Γ that acts on a smooth manifold $\widetilde{\mathcal{X}}$ discretely and smoothly, the triple $(\widetilde{\mathcal{X}}, \pi, \mathcal{X})$, where $\mathcal{X} = \widetilde{\mathcal{X}}/\Gamma$ is the orbit space and π is the canonical mapping $\widetilde{\mathcal{X}} \to \mathcal{X}$, is a smooth regular covering with $\operatorname{Aut}\widetilde{\mathcal{X}} = \Gamma$ and that any regular covering is isomorphic to a covering of such form.

Exercise 3.6. Show that for any group Γ of affine automorphisms that acts on an affine connection space $\widetilde{\mathcal{X}}$, we can find a unique affine connection on the manifold $\mathcal{X} = \widetilde{\mathcal{X}}/\Gamma$ with respect to which the covering $(\widetilde{\mathcal{X}}, \pi, \mathcal{X})$ is affine.

We see that the following statement therefore holds.

Proposition 3.3. *If the manifold $\widetilde{\mathcal{X}}$ is an affine connection space in a smooth regular covering $(\widetilde{\mathcal{X}}, \pi, \mathcal{X})$, then $\mathcal{X}$ admits an affine connection with respect to which this covering is affine if the group* $\operatorname{Aut}\widetilde{\mathcal{X}}$ *consists of affine automorphisms of the space $\widetilde{\mathcal{X}}$.*

In what follows, we also need the following simple proposition, which is essentially obvious.

Proposition 3.4. *In the affine covering $(\widetilde{\mathcal{X}}, \pi, \mathcal{X})$, the space $\widetilde{\mathcal{X}}$ is geodesically complete iff the space $\mathcal{X}$ is geodesically complete.*

Proof. Let the space $\mathcal{X}$ be geodesically complete, and let

$$\widetilde{p} \in \widetilde{X}, \qquad \widetilde{A} \in \mathbf{T}_{\widetilde{p}}\widetilde{\mathcal{X}}, \qquad p = \pi(\widetilde{p}), \qquad \text{and} \qquad A = (d\pi)_{\widetilde{p}}\widetilde{A}.$$

We consider a geodesic $\gamma: t \mapsto \exp_p tA$. By the condition, this geodesic is defined for all $t \in \mathbb{R}$. According to a general result on coloring mappings (or, more precisely, according to its variant concerning curves of the form $\mathbb{R} \to \mathcal{X}$, which it obviously implies), we can find a curve $\widetilde{\gamma}: \mathbb{R} \to \widetilde{\mathcal{X}}$ on the manifold $\widetilde{\mathcal{X}}$ such that $\pi \circ \widetilde{\gamma} = \gamma$ and $\widetilde{\gamma}(0) = \widetilde{p}$. But then $(d\pi)_{\widetilde{\gamma}(t)}(\dot{\widetilde{\gamma}}(t)) = \dot{\gamma}(t)$ for any $t \in \mathbb{R}$, and, in particular, $(d\pi)_{\widetilde{p}}(\dot{\widetilde{\gamma}}(0)) = \dot{\gamma}(0) = A = (d\pi)_{\widetilde{p}}(\widetilde{A})$. Therefore, $\dot{\widetilde{\gamma}}(0) = \widetilde{A}$. In addition, because the mapping π is affine, the curve $\widetilde{\gamma}$ is locally (and therefore globally) a geodesic. This proves that for any point $\widetilde{p} \in \widetilde{X}$ and for any vector $\widetilde{A} \in \mathbf{T}_{\widetilde{p}}\widetilde{\mathcal{X}}$, the maximal geodesic of the space $\widetilde{\mathcal{X}}$ that passes through the point $\widetilde{p}$ for $t = 0$ and has the tangent vector $\widetilde{A}$ at this point is defined for all t. Therefore, the space $\widetilde{\mathcal{X}}$ is geodesically complete.

The proof of the converse statement is easier. Let the space $\widetilde{\mathcal{X}}$ be geodesically complete, and let $p \in \mathcal{X}$ and $A \in \mathbf{T}_p\mathcal{X}$. Because the mappings π and $(d\pi)_{\widetilde{p}}$ are surjective, there exist a point $\widetilde{p} \in \widetilde{X}$ and a vector $\widetilde{A} \in \mathbf{T}_{\widetilde{p}}\widetilde{\mathcal{X}}$ such that $p = \pi(\widetilde{p})$ and $A = (d\pi)_{\widetilde{p}}\widetilde{A}$; because the mapping π is affine, the curve $\pi \circ \gamma_{\widetilde{p},\widetilde{A}}$ that passes through the point p for $t = 0$, has the tangent vector A at this point, and is defined for all $t \in \mathbb{R}$ is a geodesic. Therefore, the space $\mathcal{X}$ is geodesically complete. □

§4. Restriction of a Connection to a Submanifold

We now pass to embedded submanifolds, i.e., to injective and moneomorphic immersions $\mathcal{X} \to \mathcal{Y}$. However, because any immersion $\mathcal{X} \to \mathcal{Y}$ is locally injective and moneomorphic (in a neighborhood of any point $p_0 \in \mathcal{X}$), all local results can be applied to any immersion $\mathcal{X} \to \mathcal{Y}$, or (conventionally, but not precisely, speaking) to any *immersed submanifolds $\mathcal{X}$ of the manifold $\mathcal{Y}$*. This means that we admit that $\mathcal{X}$ has *self-intersections* (probably multiple ones); therefore, such $\mathcal{X}$ are called *submanifolds with self-intersections.* (We note that the concept of an immersed submanifold here *differs* from the concept of an immersed submanifold in the sense in Chap. 32. Unfortunately, both definitions are equally common. When distinguishing them is necessary, we talk about immersed submanifolds with or without self-intersections.)

We first consider this problem in full generality for connections on bundles. Let $\mathcal{Y}$ be an arbitrary smooth manifold, $\xi = (\mathcal{E}, \pi, \mathcal{Y})$ be a smooth vector bundle over $\mathcal{Y}$, and ∇ be a connection on ξ. Then for any submanifold $\mathcal{X} \subset \mathcal{Y}$, the bundle

$$\xi|_{\mathcal{X}} = (\mathcal{E}_{\mathcal{X}}, \pi_{\mathcal{X}}, \mathcal{X}),$$

where $\mathcal{E}_{\mathcal{X}} = \pi^{-1}\mathcal{X}$ and $\pi_{\mathcal{X}} = \pi|_{\mathcal{E}_{\mathcal{X}}}$, is defined and is called the *restriction of the bundle* ξ to $\mathcal{X}$. We note that because the mapping π is a submersion, the subspace $\mathcal{E}_{\mathcal{X}}$ is an embedded submanifold of the manifold $\mathcal{E}$.

Exercise 3.7. Prove that the following assertions hold:

1. The bundle $\xi|_{\mathcal{X}}$ is a smooth vector bundle. [*Hint*: Over each point $b \in \mathcal{X}$, the fiber $\pi_{\mathcal{X}}^{-1}(b)$ of the bundle $\xi|_{\mathcal{X}}$ is the fiber $\mathcal{F}_b = \pi^{-1}(b)$ of the bundle ξ.]
2. The bundle $\xi|_{\mathcal{X}}$ is isomorphic to the bundle $\iota^*\xi$, where $\iota: \mathcal{X} \to \mathcal{Y}$ is an embedding.

At each point $p \in \mathcal{E}_{\mathcal{X}}$, the tangent space $\mathbf{T}_p\mathcal{E}_{\mathcal{X}}$ is a subspace of the tangent space $\mathbf{T}_p\mathcal{E}$ that contains the vertical subspace $\mathbf{T}_p\mathcal{F}_b$, $b = \pi(p)$. This implies (prove this!) that if H_p are horizontal subspaces of the connection ∇, then the subspaces

$$H'_p = H_p \cap \mathbf{T}_p\mathcal{E}_{\mathcal{X}}, \quad p \in \mathcal{E}_{\mathcal{X}}, \tag{7}$$

compose a connection on $\xi|_{\mathcal{X}}$. This connection is denoted by $\nabla|_{\mathcal{X}}$ and is called the *restriction of the connection* ∇ *to* $\mathcal{X}$.

Exercise 3.8. Show that $\nabla|_{\mathcal{X}} = \iota^*\nabla$, where $\iota: \mathcal{X} \to \mathcal{Y}$ is an embedding.

Let $(U, x^1, \dots, x^n)$ and $(V, y^1, \dots, y^m)$ be charts of the manifolds $\mathcal{X}$ and $\mathcal{Y}$ such that $U = V \cap \mathcal{X}$ and

$$y^1|_V = x^1, \quad \dots, \quad y^n|_V = x^n, \; y^{n+1}|_V = 0, \quad \dots, \quad y^m|_V = 0 \tag{8}$$

(see Chap. 30); therefore, $n = \dim \mathcal{X}$ and $m = \dim \mathcal{Y}$. Moreover, let the bundle ξ be trivializable over the neighborhood V and the forms $\omega^i_j = \Gamma^i_{kj}\, dx^k$ of the connection ∇ therefore be defined. Then the bundle $\xi|_{\mathcal{X}}$ is trivializable over U, and the connection forms of $\nabla|_{\mathcal{X}}$ are the restrictions $\iota^*\omega^i_j = \omega^i_j|_U$ of the forms ω^i_j to U. In particular, the coefficients of the connection $\nabla|_{\mathcal{X}}$ are the restrictions $\Gamma^i_{kj}|_U$ of the coefficients Γ^i_{kj} of the connection ∇ with $k = 1, \dots, n$ (the coefficients Γ^i_{kj} with $k = n+1, \dots, m$ are not related to the connection $\nabla|_{\mathcal{X}}$).

The operation of restriction of sections $s \mapsto s|_U$ defines a homomorphism $\Gamma(\xi, V) \to \Gamma(\xi|_{\mathcal{X}}, U)$ of modules of sections that is compatible with the restriction homomorphism $\mathbf{F}V \to \mathbf{F}U$, i.e., it is such that

$$(fs)|_U = (f|_U)(s|_U)$$

for any section $s \in \Gamma(\xi, V)$ and any function $f \in \mathbf{F}V$. This homomorphism transforms the basis $\{s_i\}$ of the $\mathbf{F}V$-module $\Gamma(\xi, V)$ into the basis $\{s_i|_U\}$ of the $\mathbf{F}U$-module $\Gamma(\xi|_{\mathcal{X}}, U)$ and is therefore an *epimorphism.* (Indeed, if $s \in \Gamma(\xi|_{\mathcal{X}}, U)$ and $s = s^i s_i|_U$, where $s^i = s^i(x^1, \dots, x^n)$ on U, then $s = s'|_U$, where $s' = s^i(y^1, \dots, y^n)s_i$ on V.)

In particular, this holds for the bundle $\tau_{\mathcal{Y}}$, and each vector field X on U is therefore the restriction of a certain vector field X' to V. In this case, because the relations

$$\left.\frac{\partial f}{\partial y^k}\right|_U = \frac{\partial(f|_U)}{\partial x^k}, \quad k = 1, \dots, n,$$

hold for any function f on V, we have

$$(\nabla|_{\mathcal{X}})_X\, s = \nabla_{X'}\, s'|_U \tag{9}$$

for any section $s \in \Gamma(\xi|_{\mathcal{X}}, U)$, as an obvious computation shows (we again return to sections of an arbitrary bundle ξ); here, as above, s' is a section of the bundle ξ over V such that $s'|_U = s$). We can regard (9) as a definition of the connection $\nabla|_{\mathcal{X}}$ (of course, in this case, we must prove its correctness).

In what follows, as a rule, we write s and X instead of s' and X' and $|_{\mathcal{X}}$ instead of $|_U$. In this notation, which is ambiguous but convenient, formula (9) becomes

$$(\nabla|_{\mathcal{X}})_X\, s = (\nabla_X\, s)|_{\mathcal{X}}, \tag{9'}$$

where one should keep in mind that s and X are defined over U in the left-hand side and over V in the right-hand side.

Exercise 3.9. Show that any section s (in particular, any vector field X) over $\mathcal{X}$ is the restriction of a certain section s' (vector field X') over $\mathcal{Y}$. [*Hint*: The section s' can be constructed in a neighborhood of any point $p \in \mathcal{X}$. These local continuations are glued together using a partition of unity.]

§5. Induced Connection on a Normalized Submanifold

Of course, all this can be automatically applied to the case where ξ is the tangent bundle $\tau_{\mathcal{Y}}$ (and ∇ is therefore a connection on $\mathcal{Y}$). In particular, (9′) in this case becomes

$$(\nabla|_{\mathcal{X}})_X Y = (\nabla_X Y)|_{\mathcal{X}}, \tag{9''}$$

where X and Y are vector fields defined on $\mathcal{X}$ (or on U) in the left-hand side and their extensions to $\mathcal{Y}$ (to V) in the right-hand side. But in this case, the restriction $\tau_{\mathcal{Y}}|_{\mathcal{X}}$ is not the tangent bundle $\tau_{\mathcal{X}}$ (and the connection $\nabla|_{\mathcal{X}}$ is therefore not a connection on $\mathcal{X}$, and Y and $(\nabla|_{\mathcal{X}})_X Y$ are not vector fields on $\mathcal{X}$). We can only assert that for any point $p \in \mathcal{X}$, the fiber $\mathbf{T}_p\mathcal{X}$ of the bundle $\tau_{\mathcal{X}}$ is a subspace of the fiber $\mathbf{T}_p\mathcal{Y}$ of the bundle $\tau_{\mathcal{Y}}|_{\mathcal{X}}$, i.e., that the bundle $\tau_{\mathcal{X}}$ is a *subbundle* of the bundle $\tau_{\mathcal{Y}}|_{\mathcal{X}}$.

Exercise 3.10. Show that the following properties hold:

1. For any subbundle η of a vector bundle ξ on an arbitrary (paracompact and Hausdorff) manifold $\mathcal{X}$, there exists a subbundle ζ such that
$$\xi = \eta \oplus \zeta \tag{10}$$
(i.e., $\mathcal{F}_p^\xi = \mathcal{F}_p^\eta \oplus \mathcal{F}_p^\zeta$ for any point $p \in \mathcal{X}$). [*Hint*: Introducing a metric in ξ, set $\mathcal{F}_p^\zeta = (\mathcal{F}_p^\eta)^\perp$.]
2. All such subbundles ζ are isomorphic. [*Hint*: Define the *quotient bundle* ξ/η, and show that it is isomorphic to the bundle ζ.]

In particular, this implies that in the bundle $\tau_{\mathcal{Y}}|_{\mathcal{X}}$, we have a unique (up to an isomorphism) subbundle ν for which

$$\tau_{\mathcal{Y}}|_{\mathcal{X}} = \tau_{\mathcal{X}} \oplus \nu. \tag{11}$$

Definition 3.2. Each subbundle ν satisfying relation (11) is called a *normal* (or, more precisely, an *affine normal*) bundle over the submanifold $\mathcal{X}$ in the manifold $\mathcal{Y}$. Its fiber $\mathcal{F}_p^\nu$ over the point $p \in \mathcal{X}$ is called the *normal space* of the submanifold $\mathcal{X}$ at the point p and is denoted by $\mathbf{N}_p\mathcal{X}$.

Therefore,

$$\mathbf{T}_p\mathcal{Y} = \mathbf{T}_p\mathcal{X} + \mathbf{N}_p\mathcal{X}$$

for any point $p \in \mathcal{X}$. The rank $m-n$ of the bundle ν is called the *codimension* of the submanifold $\mathcal{X}$. A submanifold in $\mathcal{X}$ for which a certain normal bundle ν is chosen and fixed is said to be *normalized*. Sections of the bundle ν are called *fields of normal vectors* on $\mathcal{X}$. (Such fields are also called *normal vector fields on* $\mathcal{X}$ although, strictly speaking, they are not vector fields on $\mathcal{X}$.)

Each decomposition (10) defines an obvious morphism $\xi \mapsto \eta$ whose kernel (in a clear sense) is the bundle ζ. (We note that not every morphism of vector bundles admits a kernel.) For decomposition (11), this morphism is called a *normalizing morphism*. It is uniquely defined by the normal bundle ν, and, vice versa, it uniquely defines this bundle.

By construction, the normalizing morphism P is a mapping of the smooth manifold $\mathcal{E}_{\mathcal{X}} = \mathcal{E}(\tau_{\mathcal{Y}}|_{\mathcal{X}})$ onto the manifold $\mathcal{E}(\tau_{\mathcal{X}}) = \mathbf{T}\mathcal{X}$. It is easy to see that *this mapping is smooth* (prove this!). Therefore, at any point $A \in \mathcal{E}_{\mathcal{X}}$ (which is a tangent vector to the manifold $\mathcal{Y}$ at the point $p = \pi(A)$ of the submanifold $\mathcal{X}$), its differential $(dP)_A$ is defined; it is a linear mapping of the linear space $\mathbf{T}_A\mathcal{E}_{\mathcal{X}}$ onto the linear space $\mathbf{T}_B(\mathbf{T}\mathcal{X})$, where $B = P(A)$ is a vector tangent to the submanifold $\mathcal{X}$ at the point p. But in the linear space $\mathbf{T}_A\mathcal{E}_{\mathcal{X}}$, we have isolated the linear subspace H_A consisting of horizontal vectors with respect to the connection ∇. Therefore, in the space $\mathbf{T}_B(\mathbf{T}\mathcal{X})$, we have the subspace $(dP)_A H_A$.

Exercise 3.11. Show that the following properties hold:

1. The subspace $(dP)_A H_A$ depends only on the point $B \in \mathbf{T}\mathcal{X}$.
2. The field of subspaces
$$B \mapsto (dP)_A H_A \tag{12}$$
on $\mathbf{T}\mathcal{X}$ is a connection on $\mathcal{X}$.

Connection (12) is denoted by $\nabla^{\mathcal{X}}$, and we say that this connection is *induced on* $\mathcal{X}$ by the connection $\nabla = \nabla^{\mathcal{Y}}$ on $\mathcal{Y}$. We stress that the induced connection $\nabla^{\mathcal{X}}$ is only defined for normalized submanifolds. Under a change of normalization (normal bundle), generally speaking, it changes.

§6. Gauss Formula and the Second Fundamental Form of a Normalized Submanifold

For any vector field X on $\mathcal{X}$ and for any section s of the bundle $\tau_{\mathcal{Y}}|_{\mathcal{X}}$, the value $[(\nabla|_{\mathcal{X}})_X\, s]_p$ of the covariant derivative $(\nabla|_{\mathcal{X}})_X\, s$ at the point $p \in \mathcal{X}$ is a vector of the space $\mathbf{T}_p\mathcal{Y}$. Projecting it to the subspace $\mathbf{T}_p\mathcal{X}$ along the the subspace $\mathbf{N}_p\mathcal{X}$ (i.e., applying the linear mapping P_p), we obtain a vector $P_p[(\nabla|_{\mathcal{X}})_X\, s]_p$ in $\mathbf{T}_p\mathcal{X}$. In particular, this construction is applicable in the case where the section s assumes its values in $\mathbf{T}\mathcal{X} \subset \mathcal{E}_{\mathcal{X}}$, i.e., when this section is a vector field Y on $\mathcal{X}$. On the other hand, in this case, the vector $(\nabla^{\mathcal{X}}_X Y)_p$, which is the value of the covariant derivative $\nabla^{\mathcal{X}}_X Y$ at the point p, is defined on the space $\mathbf{T}_p\mathcal{X}$.

Exercise 3.12. Show that
$$(\nabla^{\mathcal{X}}_X Y)_p = P_p[(\nabla|_{\mathcal{X}})_X Y]_p \tag{13}$$
for any point $p \in \mathcal{X}$ and any vector fields X and Y on $\mathcal{X}$.

Formula (13) can be accepted as a definition of the connection $\nabla^{\mathcal{X}}$. We can rewrite it in the form

$$(\nabla|_{\mathcal{X}})_X Y = \nabla^{\mathcal{X}}_X Y + h(X,Y), \tag{14}$$

where $h(X,Y)$ is a certain normal vector field on $\mathcal{X}$. In this form, it is called the *Gauss formula.*

It is easy to see that the *field* $h(X,Y)$ *linearly* (*over* $\mathsf{F}\mathcal{X}$) *depends on the fields* X *and* Y, i.e., the correspondence $X,Y \mapsto h(X,Y)$ is an $\mathsf{F}\mathcal{X}$-linear mapping

$$\mathfrak{a}\mathcal{X} \otimes \mathfrak{a}\mathcal{X} \to \Gamma\nu.$$

Indeed, the $\mathbb{R}$-linearity in Y and the $\mathsf{F}\mathcal{X}$-linearity in X is implied by the corresponding properties of the covariant derivatives. Moreover, for any function f on $\mathcal{X}$,

$$(\nabla|_{\mathcal{X}})_X(fY) = Xf \cdot Y + f \cdot (\nabla|_{\mathcal{X}})_X Y$$

and

$$\nabla^{\mathcal{X}}_X(fY) = Xf \cdot Y + f \cdot \nabla^{\mathcal{X}}_X Y.$$

Therefore,

$$\begin{aligned} h(X,fY) &= (\nabla|_{\mathcal{X}})_X(fY) - \nabla^{\mathcal{X}}_X(fY) \\ &= f[(\nabla|_{\mathcal{X}})_X Y - \nabla^{\mathcal{X}}_X Y] = fh(X,Y). \end{aligned}$$

A variant of (14) also holds for vector fields on curves.

Exercise 3.13. For any vector field $X: t \mapsto X_t$ on a curve $\gamma: t \mapsto \gamma(t)$ of a manifold $\mathcal{X}$, its covariant derivatives $\nabla^{\mathcal{X}} X/dt$ and $\nabla^{\mathcal{Y}} X/dt$ along γ with respect to the connections $\nabla^{\mathcal{X}}$ and $\nabla^{\mathcal{Y}}$ are defined (on $\mathcal{X}$ and $\mathcal{Y}$ respectively). Show that

$$\frac{\nabla^{\mathcal{Y}} X}{dt} = \frac{\nabla^{\mathcal{X}} X}{dt} + h(\dot\gamma, X). \tag{14'}$$

In particular,

$$\frac{\nabla^{\mathcal{Y}} \dot\gamma}{dt} = \frac{\nabla^{\mathcal{X}} \dot\gamma}{dt} + h(\dot\gamma, \dot\gamma). \tag{14''}$$

for each curve γ on $\mathcal{X}$.

For any point $p \in \mathcal{X}$ and any vectors $A, B \in \mathbf{T}_p\mathcal{X}$, we set

$$h_p(A,B) = h(X,Y)(p),$$

where X and Y are arbitrary vector fields on $\mathcal{X}$ for which $X_p = A$ and $Y_p = B$.

Exercise 3.14. Show that this construction correctly defines a smooth field $p \mapsto h_p$ on $\mathcal{X}$, i.e., the vector $h_p(A,B)$ does not depend on the choice of the vector fields X and Y.

Traditionally, this field is called the *second fundamental form* (or the *second fundamental tensor*) of the normalized submanifold $\mathcal{X}$.

If the connection $\nabla = \nabla^{\mathcal{Y}}$ is symmetric, i.e.,

$$\nabla_X Y - \nabla_Y X = [X, Y]$$

for any fields X and Y on $\mathcal{Y}$, then

$$(\nabla_X Y)|_{\mathcal{X}} - (\nabla_Y X)|_{\mathcal{X}} = [X, Y]|_{\mathcal{X}},$$

and therefore (see (9''))

$$(\nabla|_{\mathcal{X}})_X Y - (\nabla|_{\mathcal{X}})_Y X = [X, Y]$$

for any fields X and Y on $\mathcal{X}$. Applying (14) and equating the tangent components, we immediately obtain

$$\nabla_X^{\mathcal{X}} Y - \nabla_Y^{\mathcal{X}} X = [X, Y],$$

i.e., the connection $\nabla^{\mathcal{X}}$ is symmetric. Equating the normal components, we obtain

$$h(X, Y) = h(Y, X).$$

Therefore, *if the connection $\nabla^{\mathcal{Y}}$ is symmetric, then the connection $\nabla^{\mathcal{X}}$ is symmetric, and the second fundamental form $h(X, Y)$ is symmetric in X and Y.*

§7. Totally Geodesic and Auto-Parallel Submanifolds

It is easy to see that *each geodesic γ of the manifold $\mathcal{Y}$ (with respect to the connection $\nabla^{\mathcal{Y}}$) that lies in $\mathcal{X}$ is a geodesic of the submanifold $\mathcal{X}$ (with respect to the connection $\nabla^{\mathcal{X}}$).* (Indeed, if the left-hand side of (14'') vanishes, then both summands in the right-hand side also vanish because they are orthogonal to one another.) The converse is not true in general: a geodesic of the submanifold $\mathcal{X}$ is not necessarily a geodesic in the whole manifold $\mathcal{Y}$ (it is iff $h(\dot{\gamma}, \dot{\gamma}) = 0$.)

Definition 3.3. A normalized submanifold $\mathcal{X}$ is said to be *totally geodesic* if each geodesic γ (with respect to the connection $\nabla^{\mathcal{X}}$) is also a geodesic of the whole manifold $\mathcal{Y}$ (with respect to the connection $\nabla^{\mathcal{Y}}$).

Proposition 3.5. *A normalized submanifold $\mathcal{X}$ is a totally geodesic submanifold of a space $\mathcal{Y}$ with a symmetric affine connection iff its second fundamental form h is identically zero,*

$$h(X, Y) = 0 \tag{15}$$

for any fields $X, Y \in \mathfrak{a}\mathcal{X}$.

Proof. If condition (15) holds, then it follows from (14″) that

$$\frac{\nabla^{\mathcal{Y}}}{dt}\dot{\gamma}(t) = \frac{\nabla^{\mathcal{X}}}{dt}\dot{\gamma}(t)$$

for any curve $\gamma: t \mapsto \gamma(t)$ on $\mathcal{X}$ (where the curve γ in the left-hand side is considered as a curve in $\mathcal{Y}$). Therefore, if the curve γ is a geodesic in $\mathcal{X}$, then it is also a geodesic in $\mathcal{Y}$.

Conversely, let the submanifold $\mathcal{X}$ be totally geodesic. Then for any point $p \in \mathcal{X}$ and for any vector $A \in \mathbf{T}_p\mathcal{X}$, the geodesic $\gamma_{p,A}$ in $\mathcal{X}$ (i.e., a geodesic γ with $\gamma(0) = p$ and $\dot{\gamma}(0) = A$) is also the geodesic $\gamma_{p,A}$ in $\mathcal{Y}$. Because the geodesic γ is regular, we can find a vector field X on $\mathcal{X}$ such that $X_{\gamma(t)} = \dot{\gamma}(t)$ for all sufficiently small $|t|$. Therefore, at the point p, the relation

$$(\nabla^{\mathcal{X}}_X X)_p = \frac{\nabla^{\mathcal{X}}\dot{\gamma}}{dt}(0) = 0,$$

holds on $\mathcal{X}$ and

$$(\nabla^{\mathcal{Y}}_X X)_p = \frac{\nabla^{\mathcal{Y}}\dot{\gamma}}{dt}(0) = 0$$

holds on $\mathcal{Y}$ (where X is certainly understood as its extension X' to $\mathcal{Y}$). Therefore, $h_p(A, A) = h(X, X)(p) = 0$. Because this is true for any vector $A \in \mathbf{T}_p\mathcal{X}$ and for any point $p \in \mathcal{X}$, we have $h(X, X) = 0$ for any field $X \in \mathfrak{a}\mathcal{X}$; this is possible for the symmetric form h only when $h = 0$. □

Definition 3.4. A submanifold $\mathcal{X}$ of an affine connection space $\mathcal{Y}$ is said to be *auto-parallel* if for any two points $p_0, p_1 \in \mathcal{X}$ and for any curve γ in $\mathcal{X}$ connecting these points, the parallel translation along γ,

$$\Pi_\gamma: \mathbf{T}_{p_0}\mathcal{Y} \to \mathbf{T}_{p_0}\mathcal{Y}$$

maps the subspace $\mathbf{T}_{p_0}\mathcal{X} \subset \mathbf{T}_{p_0}\mathcal{Y}$ onto the subspace $\mathbf{T}_{p_1}\mathcal{X} \subset \mathbf{T}_{p_1}\mathcal{Y}$ (i.e., a vector tangent to $\mathcal{X}$ remains a tangent vector).

Exercise 3.15. Prove that a submanifold $\mathcal{X}$ is auto-parallel iff for any two vector fields $X, Y \in \mathfrak{a}\mathcal{X}$ and for any point $p \in \mathcal{X}$, the vector $[(\nabla|_{\mathcal{X}})_X Y]_p$ belongs to $\mathbf{T}_p\mathcal{X}$.

This implies that for a auto-parallel submanifold $\mathcal{X}$, the formula

$$(\nabla^{\mathcal{X}}_X Y)_p = [(\nabla|_{\mathcal{X}})_X Y]_p, \quad p \in \mathcal{X}, \quad X, Y \in \mathfrak{a}\mathcal{X},$$

defines a certain connection $\nabla^{\mathcal{X}}$ on $\mathcal{X}$. This connection coincides with connection (13), which corresponds to an arbitrary normalization of the manifold $\mathcal{X}$ (and therefore does not depend on its choice).

Exercise 3.16. Show that any auto-parallel submanifold is totally geodesic and, conversely, if a connection $\nabla^{\mathcal{Y}}$ is symmetric, then any totally geodesic submanifold is auto-parallel.

In particular, we see that *the connection induced by a symmetric connection on a totally geodesic manifold does not depend on the choice of normalization.*

§8. Normal Connection and the Weingarten Formula

Now let s be an arbitrary normal vector field on $\mathcal{X}$ (a section of the bundle ν). Because s is automatically a section of the bundle $\tau_{\mathcal{Y}}|_{\mathcal{X}}$, the section $(\nabla|_{\mathcal{X}})_X\, s$ of the latter bundle is defined for any vector field X on $\mathcal{X}$. Decomposing it into the normal and tangent components, we obtain

$$(\nabla|_{\mathcal{X}})_X\, s = -A_s X + D_X\, s, \tag{16}$$

where $A_s X$ is the tangent vector field and $D_X s$ is the normal vector field on $\mathcal{X}$. Formula (16) is called the *Weingarten formula*.

According to (16), each tangent vector field $X \in \mathfrak{a}\mathcal{X}$ defines an embedding $D_X\colon s \mapsto D_X s$ of the module Γ_ν in itself, and each normal vector field $s \in \Gamma_\nu$ defines a mapping $X \mapsto A_s X$ of the module $\mathfrak{a}\mathcal{X} = \Gamma(\tau_{\mathcal{X}})$ into itself. The fields $A_s X$ and $D_X s$ obviously depend $\mathbb{R}$-linearly on s and $\mathbf{F}\mathcal{X}$-linearly on X. Moreover, because

$$\begin{aligned}-A_{fs}X + D_X(fs) &= (\nabla|_{\mathcal{X}})_X(fs)\\ &= Xf\cdot s + f\cdot(\nabla|_{\mathcal{X}})_X s = -fA_sX + [Xf\cdot s + fD_X s]\end{aligned}$$

for any function $f \in \mathbf{F}\mathcal{X}$,

$$A_{fs}X = fA_sX \qquad \text{and} \qquad D_X(fs) = (Xf)s + fD_X s.$$

Therefore, the field $A_s X$, in fact, depends $\mathbf{F}\mathcal{X}$-linearly on s, and the operators D_X have properties 1, 2, and 3 in Chap. 1 and are the operators of covariant differentiation with respect to a certain connection D on ν.

The connection D is called the *normal connection* on $\mathcal{X}$ induced by the connection ∇. (We note that just as normal vector fields on $\mathcal{X}$ are not vector fields on $\mathcal{X}$, the normal connection is similarly not a connection on $\mathcal{X}$.)

§9. Van der Waerden–Bortolotti Connection

We can unite the connections $\nabla^{\mathcal{X}}$ and D into one connection. Any two connections ∇_1 and ∇_2 on vector bundles ξ_1 and ξ_2 (over the same manifold $\mathcal{X}$) naturally define a connection $\nabla_1 \oplus \nabla_2$ on the Whitney sum $\xi_1 \oplus \xi_2$ of these bundles. In particular, on the vector bundle $\tau_{\mathcal{Y}}|_{\mathcal{X}} = \tau_{\mathcal{X}} \oplus \nu$ (see (11)), the connection

$$\overline{\nabla} = \nabla^{\mathcal{X}} \oplus D \tag{17}$$

is defined. Connection (17) is called the *Van der Waerden–Bortolotti connection*.

Sections of bundles of the form

$$\tau_{\mathcal{X}}^* \otimes \cdots \otimes \tau_{\mathcal{X}}^* \otimes \nu^* \otimes \cdots \otimes \nu^* \otimes \tau_{\mathcal{X}} \otimes \cdots \otimes \tau_{\mathcal{X}} \otimes \nu \otimes \cdots \otimes \nu \tag{18}$$

are called *mixed* (τ, ν)*-tensor fields* on $\mathcal{X}$. Because $\tau\mathcal{Y}|_{\mathcal{X}}$-tensor fields of a given type (r, s)) are just sections of the bundle

$$\begin{aligned}\mathbf{T}^s_r\tau\mathcal{Y}|_{\mathcal{X}} &= \mathbf{T}^s_r(\tau_{\mathcal{X}} \oplus \nu)\\ &= \underbrace{(\tau_{\mathcal{X}} \oplus \nu)^* \otimes \cdots \otimes (\tau_{\mathcal{X}} \oplus \nu)^*}_{r \text{ times}} \otimes \underbrace{(\tau_{\mathcal{X}} \oplus \nu) \otimes \cdots \otimes (\tau_{\mathcal{X}} \oplus \nu)}_{s \text{ times}},\end{aligned}$$

which is a direct sum of bundles of form (18), *any* $\tau\mathcal{Y}|_{\mathcal{X}}$*-tensor field uniquely decomposes into a sum of mixed* (τ, ν)*-tensor fields*, and the covariant differentiations corresponding to connection (17) are therefore uniquely defined by their actions on mixed fields.

An example of a mixed field is given (because of obvious identifications) by the second fundamental form or, in general, by an arbitrary $\mathbf{F}\mathcal{X}$-linear mapping

$$h\colon \mathfrak{a}\mathcal{X} \otimes \mathfrak{a}\mathcal{X} \to \Gamma\nu. \tag{19}$$

Therefore, for any such field h and any vector field $X \in \mathfrak{a}\mathcal{X}$, the field $\overline{\nabla}_X h$ is defined (and is also a mapping of form (19)).

Exercise 3.17. Prove that

$$(\overline{\nabla}_X h)(Y, Z) = D_X h(Y, Z) - h(\nabla_X Y, Z) - h(Y, \nabla_X Z) \tag{20}$$

for any vector fields X, Y, and Z on $\mathcal{X}$ (where $\nabla = \nabla^{\mathcal{X}}$).

Exercise 3.18. Prove that a similar formula holds for mixed (τ, ν)-tensor fields and write it in coordinate form.

Connection (17) allows introducing a number of interesting classes of submanifolds. For example, we can consider normalized submanifolds with a covariantly constant (with respect to connection (17)) second fundamental form, i.e. (see (20)), such that

$$D_X h(Y, Z) = h(\nabla_X Y, Z) + h(Y, \nabla_X Z)$$

for any vector fields X, Y, and Z on $\mathcal{X}$, or normalized submanifolds with a covariantly constant (in the same sense) curvature tensor of connection, and so on.

Chapter 4
Structural Equations. Local Symmetries

§1. Torsion and Curvature Forms

As we know (see Chap. 36), instead of the curvature tensor, it is convenient to consider the *curvature forms*

$$\Omega^i_j = \sum_{k<l} R^i_{j,kl} dx^k \wedge dx^l = \frac{1}{2} R^i_{j,kl} dx^k \wedge dx^l.$$

The passage to the forms Ω^i_j is adequate because of the skew-symmetry of the components $R^i_{j,kl}$ in the subscripts k and l.

We recall that the curvature forms Ω^i_j of a connection ∇ on a vector bundle ξ are differential forms over a trivializing neighborhood U and are expressed through the forms ω^i_j of the connection ∇ in accordance with the Cartan structural equation

$$\Omega = d\omega + \omega \wedge \omega, \quad \Omega = \|\Omega^i_j\|, \quad \omega = \|\omega^i_j\|. \tag{1}$$

In turn, the forms ω^i_j are defined (see § 36.8) by

$$\nabla s_j = \omega^i_j \otimes s_i, \quad i, j = 1, \dots, n,$$

where $s_1, \dots, s_n$ is a basis of the module of sections $\Gamma\xi$ of the bundle ξ over U (the trivialization of the bundle ξ over U).

In the case of an affine connection, the basis $s_1, \dots, s_n$, is usually assumed to be the holonomic basis

$$\frac{\partial}{\partial x^1}, \quad \dots, \quad \frac{\partial}{\partial x^n} \tag{2}$$

of the module of vector fields $\mathfrak{a}\mathcal{X} = \Gamma\tau_{\mathcal{X}}$ over U that corresponds to a given chart $(U, x^1, \dots, x^n)$. However, it is very often convenient not to restrict ourselves to the holonomic property and to consider an arbitrary basis

$$X_1, \quad \dots, \quad X_n \tag{3}$$

of the module $\mathfrak{a}\mathcal{X}$ over U instead of basis (2). It is preferable to take the specific properties of the bundle $\tau_{\mathcal{X}}$ into account in this case by expressing the differential forms ω^i_j and Ω^i_j in the basis

$$\theta^1, \quad \dots, \quad \theta^n \tag{4}$$

of the module $\Omega^1\mathcal{X}$ over U that is dual to the basis (3). (By definition, $\theta^i(X_j) = \delta^i_j$; for basis (2), $\theta^i = dx^i$.) Therefore, using basis (3), we set

$$\omega_j^i = \Gamma_{kj}^i \theta^k \quad \text{on } U \tag{5}$$

and

$$\Omega_j^i = \sum_{k<l} R_{j,kl}^i \theta^k \wedge \theta^l. \tag{6}$$

The functions Γ_{kj}^i (functions $R_{j,kl}^i$) in this case are called the coefficients of the affine connection ∇ (components of the curvature tensor R) *in basis* (3).

The functions Γ_{kj}^i are coefficients of the connection ∇, and the functions $R_{j,kl}^i$ are components of the tensor R only in the case where basis (3) is holonomic (is of form (2)).

Exercise 4.1. Prove the formulas

$$\begin{aligned} &\Gamma_{kj}^i = \theta^i(\nabla_{X_k} X_j), \qquad \omega_j^i(X) = \theta^i(\nabla_X X_j), \\ &R_{j,kl}^i = \theta^i(R(X_k, X_l)X_j), \end{aligned} \tag{7}$$

and

$$\begin{aligned} R &= R_{j,kl}^i X_i \otimes \theta^j \otimes \theta^k \otimes \theta^l = \\ &= \sum_{k<l} R_{j,kl}^i X_i \otimes \theta^j \otimes (\theta^k \wedge \theta^l) = (X_i \otimes \theta^j) \otimes \Omega_j^i. \end{aligned} \tag{8}$$

Similarly, the functions

$$T_{jk}^i = \theta^i(T(X_j, X_k))$$

are called the components of the torsion tensor T in basis (3). In this case,

$$T = T_{jk}^i X_i \otimes \theta^j \otimes \theta^k = \sum_{j<k} T_{jk}^i X_i \otimes (\theta^j \wedge \theta^k) = X_i \otimes \Theta^i, \tag{9}$$

where

$$\Theta^i = \sum_{j<k} T_{jk}^i \theta^j \wedge \theta^k = \frac{1}{2} T_{jk}^i \theta^j \wedge \theta^k \tag{10}$$

are *torsion forms* of the connection ∇ in basis (3).

The expression of the components T_{jk}^i through the connection coefficients Γ_{jk}^i in the case of an arbitrary basis becomes

$$T_{jk}^i = \Gamma_{[jk]}^i - c_{jk}^i, \tag{11}$$

where c_{jk}^i are coefficients of the expressions of the fields $[X_j, X_k]$ in terms of fields (3),

$$[X_j, X_k] = c_{jk}^i X_i.$$

Indeed, by definition,

$$T(X_j, X_k) = \nabla_{X_j} X_k - \nabla_{X_k} X_j - [X_j, X_k]$$

and hence

$$\theta^i(T(X_j, X_k)) = \theta^i(\nabla_{X_j} X_k) - \theta^i(\nabla_{X_k} X_j) - \theta^i([X_j, X_k]).$$

Multiplying (11) by $\theta^j \wedge \theta^k$, we obtain the curvature form Θ^i in the left-hand side. The first term in the right-hand side equals

$$\sum_{j<k} \Gamma^i_{[jk]} \theta^j \wedge \theta^k = \frac{1}{2} \Gamma^i_{[jk]} \theta^j \wedge \theta^k$$

$$= \frac{1}{2} \Gamma^i_{jk} \theta^j \wedge \theta^k - \frac{1}{2} \Gamma^i_{kj} \theta^j \wedge \theta^k$$

$$= \Gamma^i_{jk} \theta^j \wedge \theta^k = \omega^i_k \wedge \theta^k.$$

It is easy to see that the second term

$$-\sum_{j<k} c^i_{jk} \theta^j \wedge \theta^k$$

equals the differential $d\theta^i$ of the form θ^i. (Indeed, if

$$d\theta^i = \sum_{j<k} a^i_{jk} \theta^j \wedge \theta^k,$$

then $a^i_{jk} = (d\theta^i)(X_j, X_k)$. On the other hand, as we know,

$$(d\theta^i)(X_j, X_k) = X_j \theta^i(X_k) - X_k \theta^i(X_j) - \theta^i[X_j, X_k].$$

Because $\theta^i(X_k) = \delta^i_k = \text{const}$, we have $X_j \theta^i(X_k) = 0$, and similarly $-X_k \theta^i(X_j) = 0$. Moreover,

$$\theta^i[X_j, X_k] = \theta^i(c^p_{jk} X_p) = c^i_{jk}.$$

Therefore, $a^i_{jk} = -c^i_{jk}$.)

Consequently,

$$\Theta^i = d\theta^i + \omega^i_j \wedge \theta^j, \quad i = 1, \ldots, n, \tag{12}$$

or, in the matrix form,

$$\Theta = d\theta + \omega \wedge \theta, \tag{12'}$$

where Θ and θ are matrix columns,

$$\Theta = (\Theta^1, \ldots, \Theta^n)^\top, \qquad \theta = (\theta^1, \ldots, \theta^n)^\top.$$

Equation (12′) is called the *Cartan structural equation for the torsion forms.* In holonomic basis (2), this equation is reduced to the relations $\Theta^i = \omega^i_j \wedge dx^j$, which are equivalent to formulas (15) in Chap. 2.

§2. Cartan Structural Equations in Polar Coordinates

In the case where U is a normal neighborhood of a point $p_0 \in \mathcal{X}$, each point $p \in U$ can be connected with the point p_0 by a geodesic segment of the form $t \mapsto \exp_{p_0} tA$, where $A \in \mathbf{T}_{p_0}\mathcal{X}$ and $0 \le t \le 1$. This segment is denoted by $\gamma_{p_0 p}$.

A basis $X_1, \dots, X_n$ of the module $\mathfrak{a}\mathcal{X}$ over U is said to be *adapted* if for any point $p \in U$, the basis $(X_1)_p, \dots, (X_n)_p$ of the linear space $\mathbf{T}_p\mathcal{X}$ is obtained from the basis $(X_1)_{p_0}, \dots, (X_n)_{p_0}$ of the linear space $\mathbf{T}_{p_0}\mathcal{X}$ by a parallel translation along the segment $\gamma_{p_0 p}$.

Exercise 4.2. Prove that for any basis $A_1, \dots, A_n$ of the linear space $\mathbf{T}_{p_0}\mathcal{X}$, there exists a unique adapted basis $X_1, \dots, X_n$ of the module $\mathfrak{a}\mathcal{X}$ over U for which

$$(X_1)_{p_0} = A_1, \quad \dots, \quad (X_n)_{p_0} = A_n. \tag{13}$$

[*Hint*: The construction of the vector fields $X_1, \dots, X_n$ is obvious. One must only prove that they are smooth.]

In this case, we say that the basis $X_1, \dots, X_n$ *is adapted to the basis* $A_1, \dots, A_n$.

Let $\theta^1, \dots, \theta^n$ be a basis over U of the module $\Omega^1\mathcal{X}$ that is dual to an adapted basis $X_1, \dots, X_n$ of the module $\mathfrak{a}\mathcal{X}$, and let

$$\Theta^i = \sum_{j<k} T^i_{jk}\theta^j \wedge \theta^k, \qquad \Omega^i_j = \sum_{k<l} R^i_{j,kl}\theta^k \wedge \theta^l$$

be the corresponding torsion and curvature forms as above. Here, the technical trick of introducing overdetermined *polar coordinates* $t, x^1, \dots, x^n$ is appropriate. By definition, a point $p \in U$ has the coordinates $t, x^1, \dots, x^n$ if $p = \exp_{p_0}(tx^i A_i)$. We therefore have the following:

1. If $t, x^1, \dots, x^n$ are coordinates of a point p, then for any $\lambda \ne 0$, the numbers $t\lambda^{-1}, \lambda x^1, \dots, \lambda x^n$ are also coordinates of the point p; for $p \ne p_0$, this is the only arbitrariness in the choice of the coordinates $t, x^1, \dots, x^n$.
2. If $x^1, \dots, x^n$ are normal coordinates of a point p that correspond to a basis $A_1, \dots, A_n$, then the numbers $1, x^1, \dots, x^n$ are the polar coordinates of this point.
3. The point p_0 has the coordinates $t, 0, \dots, 0$, where t is arbitrary, and $0, x^1, \dots, x^n$, where $x^1, \dots, x^n$ are arbitrary.

If $U^{(0)}$ is a subset of the space $\mathbb{R}^{n+1} = \mathbb{R} \times \mathbb{R}^n$ that consists of the points $(t, \boldsymbol{x}) = (t, x^1, \dots, x^n)$ for which the point $\exp_{p_0}(tx^i A_i)$ is defined and belongs to U, then the formula

$$h(t, \boldsymbol{x}) = \exp_{p_0}(tx^i A_i) \tag{14}$$

defines a smooth mapping $h: U^{(0)} \to U$, and for any form α on U, the form $h^*\alpha$ is therefore defined on $U^{(0)}$. We say that the form $h^*\alpha$ is the *expression of the form* α *in the coordinates* $t, x^1, \dots, x^n$; it is denoted by the same symbol α. (We note that the restriction of the form $h^*\alpha$ for $t = 1$ equals α.)

In particular, let

$$\theta^i = g^i dt + \beta^i, \quad i = 1, \dots, n,$$

be the expression of the form θ^i in the coordinates $t, x^1, \dots, x^n$, where β^i is a form independent of dt and g^i is a certain function on $U^{(0)}$. The coefficients of the form β^i depend on t in general; for $t = 1$, this form is just the form θ^i.

We see that the function g^i is the value of the form θ^i (or, more precisely, of the form $h^*\theta^i$) at the vector field $\partial/\partial t$, and in order to compute it at the point $(t, \boldsymbol{x})$, we must consider the geodesic segment $\gamma = \gamma_{p_0 p}$ connecting the point p_0 with the point $p = h(t, \boldsymbol{x})$ in U. Because the tangent vector $\dot{\gamma}(t)$ to the geodesic γ is parallel to the vector $\dot{\gamma}(0) = x^i A_i$ along γ and the vectors $(X_i)_{\gamma(t)}$ are obtained from the vectors A_i via a parallel translation along γ by construction, we have $\dot{\gamma}(t) = x^i (X_i)_{\gamma(t)}$ and therefore

$$\theta^i(\dot{\gamma}(t)) = x^i, \quad i = 1, \dots, n.$$

On the other hand, because $\gamma(t) = h(t, \boldsymbol{x})$, we have $\gamma^*\theta^i = g^i dt$ and therefore

$$g^i(t, \boldsymbol{x}) = (\gamma^*\theta^i)\left(\frac{\partial}{\partial t}\right)_{(t,\boldsymbol{x})} = \theta^i(\dot{\gamma}(t)).$$

This proves that *the expression of the form θ^i in the coordinates $t, x^1, \dots, x^n$ becomes*

$$\theta^i = x^i dt + \beta^i, \quad i = 1, \dots, n,$$

where β^i is a form independent of dt (and such that $\beta^i\big|_{t=1} = \theta^i$).

Similarly, the value of the coefficient by dt at the point $(t, \boldsymbol{x})$ in the expression of the form ω^i_j through the coordinates $t, x^1, \dots, x^n$ equals

$$(\gamma^*\omega^i_j)\left(\frac{\partial}{\partial t}\right)_{(t,\boldsymbol{x})} = \omega^i_j(\dot{\gamma}(t)) = \theta^i_{\gamma(t)}\left(\frac{\nabla(X_j)_{\gamma(t)}}{dt}\right)$$

(see the second formula in (7)); therefore, because the field $t \mapsto (X_j)_{\gamma(t)}$ is covariantly constant on the curve γ, this value equals zero. Therefore, *the expression of the form ω^i_j through $t^1, x^1, \dots, x^n$ does not contain dt.*

We recall that for any form α that does not contain dt, the symbol $\partial\alpha/\partial t$ denotes the form obtained from the form α by differentiating all its coefficients with respect to t. This form is connected with the exterior differential $d\alpha$ by the relation

$$d\alpha = \partial\alpha + dt \wedge \frac{\partial\alpha}{\partial t},$$

where $\partial\alpha$ is the differential of the form α computed under the assumption that t is constant (therefore, similar to the form α, it does not contain dt).

In particular,

$$d\beta^i = \partial\beta^i + dt \wedge \frac{\partial\beta^i}{\partial t},$$

and therefore

$$d\theta^i = dx^i \wedge dt + dt \wedge \frac{\partial \beta^i}{\partial t} + \partial \beta^i.$$

By Cartan structural equation (12′), this implies that the expression of the torsion form Θ^i through $t, x^1, \dots, x^n$ has the form

$$\Theta^i = \left(dx^i - \frac{\partial \beta^i}{\partial t} + \omega^i_j x^j \right) \wedge dt + \dots,$$

where the dots denote terms not containing dt. On the other hand, because

$$\begin{aligned}\Theta^i &= \frac{1}{2} T^i_{jk}(x^j dt + \beta^j) \wedge (x^k dt + \beta^k) = \\ &= \frac{1}{2}(T^i_{jk}\beta^j x^k - T^i_{jk} x^j \beta^k) \wedge dt + \dots = -T^i_{jk} x^j \beta^k \wedge dt + \dots,\end{aligned}$$

where T^i_{jk} denote the functions $h^* T^i_{jk} = T^i_{jk} \circ h$ of $t, x^1, \dots, x^n$, we therefore have

$$dx^i - \frac{\partial \beta^i}{\partial t} + \omega^i_j x^j = -T^i_{jk} x^j \beta^k,$$

i.e.,

$$\frac{\partial \beta^i}{\partial t} = dx^i + x^j \omega^i_j + T^i_{jk} x^j \beta^k, \quad i = 1, \dots, n. \tag{15}$$

Similarly, in the coordinates $t, x^1, \dots, x^n$, Cartan structural equation (1) for curvature forms is

$$\sum_{k<l} R^i_{j,kl}(x^k dt + \beta^k) \wedge (x^l dt + \beta^l) = \left(\partial \omega^i_j + dt \wedge \frac{\partial \omega^i_j}{\partial t} \right) + \omega^i_k \wedge \omega^k_j,$$

i.e.,

$$dt \wedge R^i_{j,kl} x^k \beta^l + \dots = dt \wedge \frac{\partial \omega^i_j}{\partial t} + \dots,$$

where, as above, the dots denote terms that do not contain dt. Therefore,

$$\frac{\partial \omega^i_j}{\partial t} = R^i_{j,kl} x^k \beta^l, \quad i, j = 1, \dots, n. \tag{16}$$

Equations (15) and (16) together compose a closed system of ordinary differential equations in the coefficients of the forms β^i and ω^i_j, which are considered as functions of t (for fixed $x^1, \dots, x^n$). The initial conditions for these equations are

$$\beta^i|_{t=0} = 0 \qquad \text{and} \qquad \omega^i_j|_{t=0} = 0. \tag{17}$$

(Indeed, because $h(0, \boldsymbol{x}) = p_0$ for any $\boldsymbol{x}$, we have

$$(dh)_{(0,\boldsymbol{x})} \left(\frac{\partial}{\partial x^k} \right)_{(0,\boldsymbol{x})} = 0, \quad k = 1, \dots, n,$$

and each of the forms like $h^*\alpha$ therefore vanishes at the vectors $(\partial/\partial x^k)_{(0,x)}$. This yields the first relation in (17) for $\alpha = \theta^i$ and the second one for $\alpha = \omega^i_j$.)

Equations (15) and (16) are called the *Cartan structural equations in polar coordinates.* Using them, it is easy to answer the question posed in Chap. 3 concerning the conditions that ensure the affinity of a given diffeomorphism $f\colon \mathcal{X} \to \mathcal{Y}$. In this case, it is convenient to first restrict ourselves to a local case and slightly change the statement of the question.

§3. Existence of Affine Local Mappings

Let $p_0 \in \mathcal{X}$, $q_0 \in \mathcal{Y}$, and

$$\varphi\colon \mathbf{T}_{p_0}\mathcal{X} \to \mathbf{T}_{q_0}\mathcal{Y} \tag{18}$$

be an arbitrary isomorphism of the linear space $\mathbf{T}_{p_0}\mathcal{X}$ onto the linear space $\mathbf{T}_{q_0}\mathcal{Y}$. Because normal coordinates compose a fundamental system of neighborhoods, we have normal neighborhoods of zero $U^{(0)}$ and $V^{(0)}$ in the linear spaces $\mathbf{T}_{p_0}\mathcal{X}$ and $\mathbf{T}_{q_0}\mathcal{Y}$ such that $V^{(0)} = \varphi U^{(0)}$.

Let $U = \exp_{p_0} U^{(0)}$ and $V = \exp_{q_0} V^{(0)}$ be the corresponding normal neighborhoods of the points p_0 and q_0 in the manifolds $\mathcal{X}$ and $\mathcal{Y}$. Choosing a basis $A_1, \ldots, A_n$ of the linear space $\mathbf{T}_{p_0}\mathcal{X}$, we let $x^1, \ldots, x^n$ and $y^1, \ldots, y^n$ denote the normal coordinates in the neighborhoods U and V that correspond to the bases $A_1, \ldots, A_n$ and $B_1 = \varphi A_1, \ldots, B_n = \varphi A_n$ of the linear spaces $\mathbf{T}_{p_0}\mathcal{X}$ and $\mathbf{T}_{q_0}\mathcal{Y}$. Then the mapping

$$f\colon U \to V \tag{19}$$

(given by the formulas $y^1 = x^1, \ldots, y^n = x^n$) is a diffeomorphism with respect to equality of coordinates of the neighborhood U onto the neighborhood V that satisfies the relation $(df)_{p_0} = \varphi$.

Further, let $X_1, \ldots, X_n$ be a basis of the module $\mathfrak{a}\mathcal{X}$ over U that is adapted to the basis $A_1, \ldots, A_n$, and let $Y_1, \ldots, Y_n$ be a basis of the module $\mathfrak{a}\mathcal{Y}$ over V that is adapted to the basis $B_1, \ldots, B_n$. Finally, let T^i_{jk} and $R^i_{j,kl}$ be the components of the tensors $T^{\mathcal{X}}$ and $R^{\mathcal{X}}$ in the basis $X_1, \ldots, X_n$, and let $\widehat{T}^i_{jk}$ and $\widehat{R}^i_{j,kl}$ be components of the tensors $T^{\mathcal{Y}}$ and $R^{\mathcal{Y}}$ in the basis $Y_1, \ldots, Y_n$ in which the substitution $y^1 = x^1, \ldots, y^n = x^n$ is carried out (i.e., the compositions of these components with the diffeomorphism $f\colon U \to V$). We stress that all the functions $\widehat{T}^i_{jk}$ and $\widehat{R}^i_{j,kl}$ are therefore functions on U similar to T^i_{jk} and $R^i_{j,kl}$.

Proposition 4.1. *Mapping* (19) *is an affine mapping iff the relations*

$$\widehat{T}^i_{jk} = T^i_{jk} \qquad \text{and} \qquad \widehat{R}^i_{j,kl} = R^i_{j,kl} \tag{20}$$

hold on U.

(We note that relations (20) express something different from (6) in Chap. 3 (for $\mathcal{X} = U$ and $\mathcal{Y} = V$).)

Proof. That mapping (19) is affine means that the affine connection spaces $\mathcal{X}$ and $\mathcal{Y}$ only differ in the notation of their coordinates. Therefore, if mapping (19) is affine, then it transforms the basis $X_1, \dots, X_n$ into the basis $Y_1, \dots, Y_n$ and relations (20) are hence exactly equivalent to (6) in Chap. 3. Therefore, if mapping (19) is affine, then relations (20) hold.

Conversely, if (20) hold, then structural equations (15) and (16) for the connections $\nabla^{\mathcal{X}}$ and $\nabla^{\mathcal{Y}}$ differ only in the notation of their coordinates (and pass into one another under the substitution $y^1 = x^1, \dots, y^n = x^n$). Therefore, their solutions β^i and ω^i_j satisfying initial conditions (17) also differ only in the notation of their coordinates. Applied to the forms ω^i_j for $t = 1$, this means that we have $\omega^{\mathcal{X}} = f^*\omega^{\mathcal{Y}}$ for connection forms of $\nabla^{\mathcal{X}}$ and $\nabla^{\mathcal{Y}}$. Therefore, mapping (19) is affine. □

Of course, conditions (20) should be made more manageable. This can be done by setting additional conditions on the connections.

§4. Locally Symmetric Affine Connection Spaces

Definition 4.1. An affine connection space $\mathcal{X}$ is said to be *locally symmetric* if

1. the connection on $\mathcal{X}$ is symmetric (the torsion tensor T vanishes) and
2. the curvature tensor R is covariantly constant,

$$\nabla R = 0 \qquad (\nabla_s R^i_{j,kl} = 0 \text{ componentwise}).$$

Exercise 4.3. Show that for $\nabla R = 0$ the parallel translation Π_γ along an arbitrary path $\gamma: I \to \mathcal{X}$ preserves the curvature tensor, i.e., for any vectors $A, B \in \mathbf{T}_{p_0}\mathcal{X}$, $p_0 = \gamma(0)$, the diagram

$$\begin{array}{ccc} \mathbf{T}_{p_0}\mathcal{X} & \xrightarrow{R_{p_0}(A,B)} & \mathbf{T}_{p_0}\mathcal{X} \\ \Pi_\gamma \downarrow & & \downarrow \Pi_\gamma \\ \mathbf{T}_{p_1}\mathcal{X} & \xrightarrow{R_{p_1}(\Pi_\gamma A, \Pi_\gamma B)} & \mathbf{T}_{p_1}\mathcal{X} \end{array},$$

where $p_1 = \gamma(1)$, is commutative.

Conversely, if this diagram is commutative for any geodesic γ, then $\nabla R = 0$.

We say that affine connection spaces $\mathcal{X}$ and $\mathcal{Y}$ have *the same curvature at the points $p_0 \in \mathcal{X}$ and $q_0 \in \mathcal{Y}$* if there exists an isomorphism

$$\varphi: \mathbf{T}_{p_0}\mathcal{X} \to \mathbf{T}_{q_0}\mathcal{Y} \tag{21}$$

transforming the tensor $R^{\mathcal{X}}$ at the point p_0 into the tensor $R^{\mathcal{Y}}$ at the point q_0, i.e., if for any vectors $A, B \in \mathbf{T}_{p_0}\mathcal{X}$, the diagram

$$\begin{array}{ccc} \mathbf{T}_{p_0}\mathcal{X} & \xrightarrow{R^{\mathcal{X}}_{p_0}(A,B)} & \mathbf{T}_{p_0}\mathcal{X} \\ \downarrow \varphi & & \downarrow \varphi \\ \mathbf{T}_{q_0}\mathcal{Y} & \xrightarrow{R^{\mathcal{Y}}_{q_0}(\varphi A,\varphi B)} & \mathbf{T}_{q_0}\mathcal{Y} \end{array}$$

is commutative.

The assertion of Exercise 4.3 directly implies that *each locally symmetric connection space $\mathcal{X}$ has the same curvature at all its points.*

We say that locally symmetric connection spaces $\mathcal{X}$ and $\mathcal{Y}$ *have the same curvature* if for certain points $p_0 \in \mathcal{X}$ and $q_0 \in \mathcal{Y}$ (and therefore for all), they have the same curvature at the points p_0 and q_0.

Proposition 4.2. *If locally symmetric affine connection spaces $\mathcal{X}$ and $\mathcal{Y}$ have the same curvature, then any points $p_0 \in \mathcal{X}$ and $q_0 \in \mathcal{Y}$ have neighborhoods U and V that are affinely isomorphic. Moreover, for any isomorphism (21) (i.e., isomorphism transforming $R^{\mathcal{X}}_{p_0}$ into $R^{\mathcal{Y}}_{q_0}$), there exists an affinity*

$$f: U \to V \tag{22}$$

such that $f(p_0) = q_0$ and $(df)_{p_0} = \varphi$.

Proof. As mapping (22), we accept mapping (19) constructed according to isomorphism (21). By Proposition 4.1, it therefore suffices to verify the condition $\widehat{R}^i_{j,kl} = R^i_{j,kl}$ (see (20)); the condition $\widehat{T}^i_{jk} = T^i_{jk}$ holds automatically by the symmetry of the connections $\nabla^{\mathcal{X}}$ and $\nabla^{\mathcal{Y}}$.

But the condition of covariant constancy of a tensor field is obviously equivalent to its components in an arbitrary adapted basis being constants. Therefore, for locally symmetric spaces $\mathcal{X}$ and $\mathcal{Y}$, the functions $R^i_{j,kl}$ and $\widehat{R}^i_{j,kl}$ are constant on U. Therefore, the relation $\widehat{R}^i_{j,kl} = R^i_{j,kl}$ holds on the whole neighborhood U if it holds at the point p_0.

Because the fulfillment of this relation at the point p_0 is exactly equivalent to the spaces $\mathcal{X}$ and $\mathcal{Y}$ having the same curvature at the points p_0 and q_0, Proposition 4.2 is proved. □

Remark 4.1. Proposition 4.2 remains valid (together with its proof) if the condition $T = 0$ for the connections $\nabla^{\mathcal{X}}$ and $\nabla^{\mathcal{Y}}$ is replaced by the weaker condition $\nabla T = 0$.

In an invariant form, mapping (22) is given by

$$f(\exp_{p_0} A) = \exp_{q_0} \varphi A, \quad A \in U^{(0)}. \tag{23}$$

If the manifold $\mathcal{Y}$ is geodesically complete, then this formula has sense for the points $\exp_{p_0} A$ of an *arbitrary* normal neighborhood U of the point p_0 (but the mapping $f: U \to \mathcal{Y}$ thus obtained is not injective in general).

§5. Local Geodesic Symmetries

We apply Proposition 4.2 to the case where $\mathcal{X} = \mathcal{Y}$, $p_0 = q_0$, $U = V$, and φ is the central symmetry $A \mapsto -A$, i.e., to the case where the mapping $f: U \to U$ is given by

$$\exp_{p_0} A \mapsto \exp_{p_0}(-A). \tag{24}$$

Mapping (24) is called a *local geodesic symmetry* at the point p_0; symmetry (24) is denoted by s_{p_0}.

Proposition 4.3. *An affine connection space $\mathcal{X}$ is locally symmetric iff local geodesic symmetry* (24) *is an affine mapping for any point $p_0 \in \mathcal{X}$.*

Proof. Because the valence (number of subscripts and superscripts) of the curvature tensor is even, the symmetry s_{p_0} transforms it into itself. Therefore, if the space $\mathcal{X}$ is locally symmetric, then, as was shown in proving Proposition 4.2, local symmetry (24) is an affine mapping.

Conversely, for all points p_0 of an affine connection space $\mathcal{X}$, let the local symmetry s_{p_0} be an affine mapping (of a certain normal neighborhood U of the point p_0 onto itself). Because affine mappings transform the torsion tensor into the torsion tensor and the curvature tensor into the curvature tensor (see (6) in Chap. 3), we have the relations $s_{p_0}^* T = T$ and $s_{p_0}^* R = R$ on U. By definition, this means that

$$(ds_{p_0})_p T_p(A, B) = T_q((ds_{p_0})_p A, (ds_{p_0})_p B)$$

and

$$(ds_{p_0})_p \circ R_p(A, B) = R_q((ds_{p_0})_p A, (ds_{p_0})_p B) \circ (ds_{p_0})_p$$

for any point $p \in U$ and any vectors $A, B \in \mathbf{T}_p\mathcal{X}$, where $q = s_{p_0} p$.

For $p_0 = p$, the first relation implies

$$-T_{p_0}(A, B) = T_{p_0}(-A, -B) = T_{p_0}(A, B),$$

and therefore $T = 0$ (at the point p_0 and hence on the whole $\mathcal{X}$ because this point is arbitrary).

On the other hand, by definition, the relation $s_{p_0} p = q$ means that there exists a geodesic γ given on the closed interval $[-1, 1]$ such that $p = \gamma(-1)$, $p_0 = \gamma(0)$, and $q = \gamma(1)$. Moreover, because the symmetry s_{p_0} is an affine mapping, the diagram

$$\begin{array}{ccc} \mathbf{T}_p\mathcal{X} & \xrightarrow{(ds_{p_0})_p} & \mathbf{T}_q\mathcal{X} \\ \Pi_{\gamma_0} \downarrow & & \downarrow \Pi_{s_{p_0} \circ \gamma_0} \\ \mathbf{T}_{p_0}\mathcal{X} & \xrightarrow{(ds_{p_0})_{p_0}} & \mathbf{T}_{p_0}\mathcal{X} \end{array},$$

where γ_0 is the segment $\gamma|_{[-1,0]}$ of the geodesic γ, is commutative (see property 4 in Proposition 3.1). But it is clear that $s_{p_0} \circ \gamma_0 = \gamma_1^{-1}$, where γ_1 is the complementing segment $\gamma|_{[0,1]}$ of the geodesic γ. In addition, by definition,

$$(ds_{p_0})_{p_0} A = -A$$

for any vector $A \in \mathbf{T}_{p_0}\mathcal{X}$. Because $\Pi_\gamma = \Pi_{\gamma_1} \circ \Pi_{\gamma_0}$, this proves that

$$(ds_{p_0})_p = -\Pi_\gamma \quad \text{on } \mathbf{T}_p\mathcal{X}.$$

Therefore, for any vectors $A, B \in \mathbf{T}_p\mathcal{X}$,

$$\Pi_\gamma \circ R_p(A,B) = R_q(\Pi_\gamma A, \Pi_\gamma B) \circ \Pi_\gamma,$$

and therefore (see Exercise 4.3) $\nabla R = 0$ (on U and hence on the whole $\mathcal{X}$). □

Proposition 4.3 explains why affine connection spaces with $T = 0$ and $\nabla R = 0$ are said to be locally symmetric.

§6. Semisymmetric Spaces

The condition $\nabla R = 0$ means that $\nabla_X R = 0$ for any field $X \in \mathfrak{a}\mathcal{X}$. Therefore, according to (18) in Chap. 2,

$$R(X,Y)R = 0 \tag{25}$$

for any fields $X, Y \in \mathfrak{a}\mathcal{X}$. (Here, $R(X,Y)$ is considered a derivation of the algebra of tensor fields on $\mathcal{X}$; see Chap. 2.) Therefore, *the curvature tensor of any locally symmetric space satisfies condition* (25). Affine connection spaces that have this property (and such that $T = 0$) are said to be *semisymmetric.*

According to (21) in Chap. 2 (applied to the tensor $S = R$), condition (25) is equivalent to the identity

$$\begin{aligned} R(X,Y) \circ R(U,V) - R(U,V) \circ R(X,Y) \\ = R(R(X,Y)U,V) + R(U,R(X,Y)V), \end{aligned} \tag{26}$$

which should hold for any vector fields $X, Y, U, V \in \mathfrak{a}\mathcal{X}$. The advantage of conditions (25) and (26) as compared with the condition $\nabla R = 0$ is that they are algebraic.

Chapter 5
Symmetric Spaces

§1. Globally Symmetric Spaces

Proposition 3.2 implies that for any point $p \in \mathcal{X}$ of a locally symmetric connection space $\mathcal{X}$, there exists at most one affine mapping $\mathcal{X} \to \mathcal{X}$ that coincides with a locally geodesic symmetry s_p on a certain normal neighborhood of the point p. This mapping (when it exists) is called a *geodesic symmetry at the point* p and is denoted by s_p, as before.

Definition 5.1. An affine connection space $\mathcal{X}$ is called a *globally symmetric space* if it is connected and for any point $p \in \mathcal{X}$, there exists a (unique, as was proved) geodesic symmetry $s_p \colon \mathcal{X} \to \mathcal{X}$.

Of course, each globally symmetric space is locally symmetric (and its curvature tensor is therefore covariantly constant). The converse statement holds in the following formulation.

Theorem 5.1. *A geodesically complete, connected, and simply connected locally symmetric space $\mathcal{X}$ is globally symmetric.*

Exercise 5.1. Prove that any globally symmetric space is geodesically complete. [*Hint*: Each geodesic defined on a finite interval can be extended using an appropriate symmetry.]

Therefore, in Theorem 5.1, only the condition of simple connectedness is unnecessary.

§2. Germs of Smooth Mappings

To prove Theorem 5.1, it is convenient to introduce a general mathematical concept. Let $\mathcal{X}$ and $\mathcal{Y}$ be two manifolds and p be a point of the manifold $\mathcal{X}$. We consider the set of all possible smooth mappings $f \colon U \to \mathcal{Y}$, where U is an arbitrary neighborhood of the point p in the manifold $\mathcal{X}$. Two elements $f_1 \colon U_1 \to \mathcal{Y}$ and $f_2 \colon U_2 \to \mathcal{Y}$ of this set are *equivalent* if there exists a neighborhood U of the point p such that $U \subset U_1 \cap U_2$ and $f_1|_U = f_2|_U$. The corresponding equivalence classes $[f]_p$ are called the *germs at the point $p \in \mathcal{X}$ of smooth mappings of $\mathcal{X}$ into $\mathcal{Y}$*. The set of these germs is denoted by $G_p(\mathcal{X}, \mathcal{Y})$, and the disjoint union of all the sets $G_p(\mathcal{X}, \mathcal{Y})$, $p \in \mathcal{X}$, is denoted by $G(\mathcal{X}, \mathcal{Y})$:

$$G(\mathcal{X}, \mathcal{Y}) = \coprod_{p \in \mathcal{X}} G_p(\mathcal{X}, \mathcal{Y}).$$

We define (obviously, correctly) two surjective mappings

$$\alpha: G(\mathcal{X}, \mathcal{Y}) \to \mathcal{X} \quad \text{and} \quad \beta: G(\mathcal{X}, \mathcal{Y}) \to \mathcal{Y}$$

by

$$\alpha[f]_p = p \quad \text{and} \quad \beta[f]_p = f(p).$$

Each mapping $f: U \to \mathcal{Y}$ of an open set $U \subset \mathcal{X}$ into $\mathcal{Y}$ defines a section (on U)

$$\sigma_f: p \mapsto [f]_p, \quad p \in U,$$

of the mapping α. It is easy to see that the set $\{\sigma_f U\}$ of all subsets of the set $G(\mathcal{X}, \mathcal{Y})$ having the form $\sigma_f U$ together with the empty set is closed with respect to intersections. Indeed, if for two mappings $f: U \to \mathcal{Y}$ and $g: V \to \mathcal{Y}$, the intersection $\sigma_f U \cap \sigma_g V$ is not empty, then there exist points $p \in U \cap V$ for which $[f]_p = [g]_p$. The set W of all such points is open, and

$$\sigma_f U \cap \sigma_g V = \sigma_h W,$$

where $h = f|_W$ (or, equivalently, $h = g|_W$). Therefore, the set $\{\sigma_f U\}$ is a base of a certain topology on $G(\mathcal{X}, \mathcal{Y})$.

In what follows, we always assume that $G(\mathcal{X}, \mathcal{Y})$ is a topological space endowed with this topology. It is clear that *the mappings α and β are continuous.* (We note that $f = \beta \circ \sigma_f$ on U for any $f: U \to \mathcal{Y}$.)

Exercise 5.2. Show that following properties hold:

1. The mapping α is a local homeomorphism. (Each point $[f] \in G(\mathcal{X}, \mathcal{Y})$ has a neighborhood that α homeomorphically maps onto a neighborhood of the point $p = \alpha[f]$.)
2. Each fiber $G(\mathcal{X}, \mathcal{Y})_p = \alpha^{-1}(p)$ of the mapping α is a discrete subspace of the space $G(\mathcal{X}, \mathcal{Y})$.
3. The topological space $G(\mathcal{X}, \mathcal{Y})$ is not Hausdorff (although the manifolds $\mathcal{X}$ and $\mathcal{Y}$ are Hausdorff).

Property 3 shows that in spite of properties 1 and 2, the mapping α is not a covering. However, it can be a covering on certain subspaces of the space $G(\mathcal{X}, \mathcal{Y})$.

§3. Extensions of Affine Mappings

Let $\mathcal{X}$ and $\mathcal{Y}$ be affine connection spaces of the same dimension n, and let $A(\mathcal{X}, \mathcal{Y})$ be the subspace of the space $G(\mathcal{X}, \mathcal{Y})$ consisting of germs of affine mappings $U \to \mathcal{Y}$.

Lemma 5.1. *If $f: U \to \mathcal{Y}$ and $g: V \to \mathcal{Y}$ are two affine mappings such that*

$$[f]_{p_0} = [g]_{p_0}$$

for a certain point p_0, then $f = g$ on a component of the intersection $U \cap V$ that contains the point p_0.

Proof. This follows immediately from Proposition 3.2. □

Lemma 5.2. *Let the spaces $\mathcal{X}$ and $\mathcal{Y}$ be connected and locally symmetric, and let them have the same curvature. In addition, let the space $\mathcal{Y}$ be geodesically complete. Then the space $A(\mathcal{X}, \mathcal{Y})$ is not empty, and on each of its components, the mapping*

$$\alpha: A(\mathcal{X}, \mathcal{Y}) \to \mathcal{X}, \qquad [f]_p \mapsto p,$$

is a covering.

Proof. It suffices to show that for each point $p_0 \in \mathcal{X}$, there exists a neighborhood that is exactly covered by the mapping α. It turns out that *a neighborhood U that is a normal neighborhood of each of its points has this property.* Indeed, we consider all possible sets of the form $\sigma_f U$, where f is an arbitrary affine mapping $U \to \mathcal{Y}$. All these sets are open, contained in $\alpha^{-1}U$, and by Lemma 5.1, any two of them either coincide or are disjoint. (We note that the neighborhood U, being star-shaped, is connected.) Moreover, each of these sets is homeomorphically mapped by the mapping α onto U (prove this!). Therefore, we only need to show that these sets exhaust the whole set $\alpha^{-1}U$ in $A(\mathcal{X}, \mathcal{Y})$, i.e., that for any germ $[g]_p \in A(\mathcal{X}, \mathcal{Y})$, where $p \in U$, there exists an affine mapping $f: U \to \mathcal{Y}$ such that $[g]_p \in \sigma_f U$ (and therefore such that $[g]_p = [f]_p$).

Let $q = g(p)$, and let $\varphi: \mathbf{T}_p\mathcal{X} \to \mathbf{T}_q\mathcal{Y}$ be the differential of the mapping g at the point p. Because the mapping g (defined on a certain neighborhood V of the point p, which we can assume to be connected) is affine, the mapping φ transforms the tensor $R_p^{\mathcal{X}}$ into the tensor $R_q^{\mathcal{Y}}$. Therefore, according to the assertion in Exercise 4.4, there exists an affine mapping $f: U \to \mathcal{Y}$ such that $f(p) = q$ and $(df)_p = \varphi$. To complete the proof, it suffices to note that according to Lemma 5.1, this mapping coincides with the mapping g on the neighborhood V. □

Proposition 5.1. *Let $\mathcal{X}$ and $\mathcal{Y}$ be connected locally symmetric spaces, and let U be an open connected subset of the space $\mathcal{X}$. If the space $\mathcal{X}$ is simply connected and the space $\mathcal{Y}$ is geodesically complete, then for any affine mapping $f: U \to \mathcal{Y}$, there exists an affine mapping $F: \mathcal{X} \to \mathcal{Y}$ such that*

$$f = F|_U.$$

Proof. The set $\sigma_f U \subset A(\mathcal{X}, \mathcal{Y})$, being a continuous image of the connected space U, is connected. Therefore, it is contained in a uniquely defined component $A_f(\mathcal{X}, \mathcal{Y})$ of the space $A(\mathcal{X}, \mathcal{Y})$. The mapping

$$\alpha: A_f(\mathcal{X}, \mathcal{Y}) \to \mathcal{X},$$

being a covering according to Lemma 5.2, is a homeomorphism because the space $\mathcal{X}$ is simply connected. Therefore, the formula

$$F = \beta \circ \alpha^{-1}$$

correctly defines a continuous mapping $F: \mathcal{X} \to \mathcal{Y}$ that makes the diagram

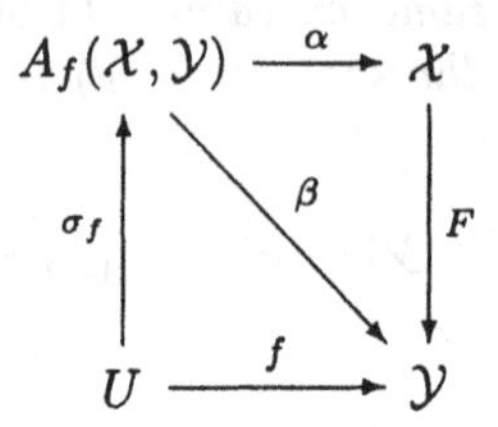

commutative.

Because $\beta \circ \sigma_f = f$ and $\alpha \circ \sigma_f = i$, where i is an embedding $U \to \mathcal{X}$, we have

$$F|_U = F \circ i = \beta \circ \alpha^{-1} \circ \alpha \circ \sigma_f = f.$$

Moreover, because the mapping α is bijective, we see that for any point $p \in \mathcal{X}$, there exists an affine mapping $h: V \to \mathcal{Y}$ (unique with accuracy up to an equivalence), where V is a certain connected neighborhood of the point p, such that $[h]_p \in A_f(\mathcal{X}, \mathcal{Y})$. Because the neighborhood V is connected, the germ $[h]_q = \sigma_h(q)$ belongs to $A_f(\mathcal{X}, \mathcal{Y})$ for any point $q \in V$, and therefore

$$h(q) = \beta[h]_q = (F \circ \alpha)[h]_q = F(q).$$

This proves that near every point $p \in \mathcal{X}$, the mapping F coincides with a certain affine mapping. Therefore, the mapping F is also affine. □

We note that according Lemma 5.1, *such a mapping F is unique.*

Exercise 5.3. Let $\mathcal{X}$, $\mathcal{Y}$, and U be as in Proposition 5.1. Moreover, let $\mathcal{S}$ be an arbitrary manifold, and let $f: \mathcal{S} \times U \to \mathcal{Y}$ be a smooth mapping such that for any $s \in \mathcal{S}$, the mapping $f^s: p \mapsto f(s,p)$, $p \in U$, is affine. According to Proposition 5.1, there exists a (unique!) mapping $F: \mathcal{S} \times \mathcal{X} \to \mathcal{Y}$ such that for any $s \in \mathcal{S}$, the mapping $F^s: p \mapsto F(s,p)$, $p \in \mathcal{X}$, is affine and coincides with the mapping f^s on U. Prove that the mapping F is smooth.

We can now prove Theorem 5.1.

Proof (of Theorem 5.1). It suffices to apply Proposition 5.1 to the case where $\mathcal{Y} = \mathcal{X}$ and f is a local geodesic symmetry. □

§4. Uniqueness Theorem

Theorem 5.2. *Let $\mathcal{X}$ be a simply connected globally symmetric space. Then for any geodesically complete locally symmetric space $\mathcal{Y}$ of the same curvature, for any points $p_0 \in \mathcal{X}$ and $q_0 \in \mathcal{Y}$, and for any isomorphism*

$$\varphi: \mathbf{T}_{p_0}\mathcal{X} \to \mathbf{T}_{q_0}\mathcal{Y}$$

of tangent spaces that transforms the tensor $R^{\mathcal{X}}$ at the point p_0 into the tensor $R^{\mathcal{Y}}$ at the point q_0, there exists a unique affine mapping

$$f: \mathcal{X} \to \mathcal{Y}$$

for which $f(p_0) = q_0$ and $(df)_{p_0} = \varphi$.

Proof. It suffices to compare Proposition 5.1 with Proposition 4.4. □

Corollary 5.1 (uniqueness theorem). *Simply connected globally symmetric spaces with the same curvature are affinely isomorphic.*

§5. Reduction of Locally Symmetric Spaces to Globally Symmetric Spaces

Because the local symmetry property is local, *if one of the spaces $\widetilde{\mathcal{X}}$ or $\mathcal{X}$ is locally symmetric in an affine covering $\widetilde{\mathcal{X}}, \pi, \mathcal{X})$, then the other is also locally symmetric.* Similarly, because a curve γ in $\widetilde{\mathcal{X}}$ is a geodesic iff its projection $\pi \circ \gamma$ is a geodesic, the same is true for the (nonlocal!) geodesic completeness property, i.e., *if one of the spaces $\widetilde{\mathcal{X}}$ or $\mathcal{X}$ is geodesically complete in an affine covering $(\widetilde{\mathcal{X}}, \pi, \mathcal{X})$, then the other is also geodesically complete.*

Combining these assertions and noting that for any connected manifold $\mathcal{X}$, there exists a covering $(\widetilde{\mathcal{X}}, \pi, \mathcal{X})$ with the simply connected manifold $\widetilde{\mathcal{X}}$, we immediately obtain from Theorem 5.1 that *for each connected and geodesically complete locally symmetric space, there exists a globally symmetric covering space (which can even be chosen in the class of simply connected spaces).*

Exercise 5.4. Prove that if a space $\mathcal{Y}$ is connected, then the mapping f in Theorem 5.2 is a covering.

Because for any locally (and, in particular, globally) symmetric space $\widetilde{\mathcal{X}}$ and for any discrete group Γ of its affine automorphisms, the orbit manifold $\mathcal{X} = \widetilde{\mathcal{X}}/\Gamma$ is naturally defined as a locally symmetric space (which is geodesically complete iff the space $\widetilde{\mathcal{X}}$ is geodesically complete), we obtain the following proposition.

Proposition 5.2. *Connected and geodesically complete symmetric spaces are exactly the affine connection spaces isomorphic to spaces of the form $\mathcal{X}/\Gamma$, where $\mathcal{X}$ is a simply connected symmetric space and Γ is a discrete group of its affine automorphisms.*

Therefore, the description of all geodesically complete locally (and, in particular, globally) symmetric spaces is reduced to the description of all simply connected geodesically complete globally symmetric spaces and to the classification of all discrete groups of their affine automorphisms.

§6. Properties of Symmetries in Globally Symmetric Spaces

The following proposition is the basis of the theory of globally symmetric spaces.

Proposition 5.3. *In each globally symmetric space $\mathcal{X}$, the symmetries*

$$s_p: \mathcal{X} \to \mathcal{X}, \quad p \in \mathcal{X},$$

have the following properties:

1. *The mapping*

$$\mathcal{X} \times \mathcal{X} \to \mathcal{X}, \qquad (p, q) \mapsto s_p q, \tag{1}$$

is smooth.

2. *Each symmetry s_p is an involutive transformation, i.e., $s_p \circ s_p = \mathrm{id}$.*
3. *The point p is an isolated fixed point of the symmetry s_p, i.e., $s_p p = p$ and there exists a neighborhood U of the point p such that $s_p q \neq q$ for any point $q \in U \setminus \{p\}$.*
4. *For any two points $p, q \in \mathcal{X}$, we have*

$$s_q \circ s_p \circ s_q = s_{p'}, \quad \textit{where } p' = s_q p.$$

Proof. Let U be an open set of the space $\mathcal{X}$ that is a normal neighborhood of each of its points. Then the mapping

$$U \times U \to \mathcal{X}, \qquad (p, q) \mapsto s_p q,$$

is obviously smooth. Therefore, according to the assertion in Exercise 5.3, mapping (1) is smooth on $U \times \mathcal{X}$. Because the sets U cover $\mathcal{X}$, this proves property 1.

By Proposition 3.2, property 2 follows from the involutiveness of the central symmetry $A \mapsto -A$, $A \in \mathbf{T}_p\mathcal{X}$.

Property 3 follows because in the neighborhood U, the symmetry s_p is the local geodesic symmetry

$$\exp_{p_0} A \mapsto \exp_{p_0}(-A).$$

To prove property 4, we consider the composition mapping $f = s_q \circ s_p \circ s_q$ and the point $p' = s_q p$. Because $s_q p' = p$, we have $f(p') = (s_q \circ s_p)(p) = s_q p = p' = s_{p'} p'$. Similarly, $(df)_{p'} A = -A = (ds_{p'})_{p'} A$ for any vector $A \in \mathbf{T}_{p'}\mathcal{X}$. Therefore, according to Proposition 3.2, we have $f = s_{p'}$.

§7. Symmetric Spaces

Proposition 5.3 provides the motive for the following definition.

Definition 5.2. A connected Hausdorff smooth manifold $\mathcal{X}$ is called a *symmetric space* if a diffeomorphism

$$s_p: \mathcal{X} \to \mathcal{X}$$

corresponds to each point $p \in \mathcal{X}$, and these diffeomorphisms have properties 1–4 in Proposition 5.3.

We note that no differential-geometric concepts are used in this definition. It was suggested by Loos (who, by the way, did not require the connectedness condition).

A smooth mapping $f: \mathcal{X} \to \mathcal{Y}$ of symmetric spaces is called a *morphism* if for any point $p \in \mathcal{X}$, the diagram

$$\begin{array}{ccc} \mathcal{X} & \xrightarrow{f} & \mathcal{Y} \\ {\scriptstyle s_p}\downarrow & & \downarrow{\scriptstyle s_q} \\ \mathcal{X} & \xrightarrow{f} & \mathcal{Y} \end{array}$$

is commutative (where $q = f(p)$. It is clear that all symmetric spaces and all their morphisms form a category (the identity mapping and the composition of two morphisms are morphisms; see Chap. 7).

Remark 5.1. Mapping (1) is just a multiplication on $\mathcal{X}$. Therefore, symmetric spaces can be defined as connected Hausdorff smooth manifolds $\mathcal{X}$ on which a smooth multiplication

$$\mathcal{X} \times \mathcal{X} \to \mathcal{X}, \qquad (p, q) \mapsto p \cdot q, \tag{1'}$$

is given, and this multiplication is such that

1. $p \cdot p = p$, and there exists a neighborhood U of the point p such that $p \cdot q \neq q$ for any point $q \in U \setminus \{p\}$;
2. $p \cdot (p \cdot q) = q$;
3. $q \cdot (p \cdot (q \cdot r)) = (q \cdot p) \cdot r$.

From this viewpoint, morphisms of symmetric spaces are merely their homomorphisms in the abstract algebra sense (smooth mappings $f: \mathcal{X} \to \mathcal{Y}$ for which $f(p \cdot q) = f(p) \cdot f(q)$, $p, q \in \mathcal{X}$). In particular, this proves that *a morphism of symmetric spaces is an isomorphism iff it is a diffeomorphism.*

§8. Examples of Symmetric Spaces

Exercise 5.5 (Example 1). Verify that *each connected Lie group $\mathcal{G}$ is a symmetric space with the symmetries*

$$s_p q = pq^{-1}p, \quad p, q \in \mathcal{G}. \tag{2}$$

[*Hint*: There exists a neighborhood U of the identity e of the group $\mathcal{G}$ such that $q^2 = e$, $q \in U$, only if $q = e$.]

For $\mathcal{G} = \mathbb{R}^n$, symmetries (2) are the usual central symmetries $\boldsymbol{x} \mapsto 2\boldsymbol{a} - \boldsymbol{x}$.

Exercise 5.6 (Example 2). In an $(n+1)$-dimensional linear space $\mathcal{V}$, consider a nondegenerate (in general, not positive definite) inner product $\boldsymbol{x}, \boldsymbol{y} \mapsto \boldsymbol{xy}$ (in another terminology, the pseudo-Euclidean structure on $\mathcal{V}$). Let $R \neq 0$, and let $\mathcal{X}$ be an arbitrary component of the sphere $\{\boldsymbol{x}: \boldsymbol{x}^2 = R\}$ in the space $\mathcal{V}$ (in the standard Euclidean structure on $\mathcal{V}$, this is either an ellipsoid or half a hyperboloid). Show that *$\mathcal{X}$ is a symmetric space with the symmetries*

$$s_{\boldsymbol{x}}\boldsymbol{y} = \frac{2\boldsymbol{xy}}{R}\boldsymbol{x} - \boldsymbol{y}, \quad \boldsymbol{x}, \boldsymbol{y} \in \mathcal{X}. \tag{3}$$

(In the Euclidean space, these are symmetries with respect to straight lines.)

Exercise 5.7 (Example 3). Let $\mathbb{K} = \mathbb{R}$, $\mathbb{C}$, or $\mathbb{H}$ as usual, and let $G_{\mathbb{K}}(m, n)$ be the manifold of all m-dimensional subspaces of the space $\mathbb{K}^n$. Show that *$G_{\mathbb{K}}(m, n)$ is a symmetric space whose symmetries are induced by symmetries $\sigma_{\mathcal{P}}$ of the space $\mathbb{K}^n$ with respect to subspaces $\mathcal{P} \in G_{\mathbb{K}}(m, n)$* (the symmetry $\sigma_{\mathcal{P}}$ leaves all vectors in $\mathcal{P}$ fixed and replaces each vector orthogonal to $\mathcal{P}$ by its opposite).

Exercise 5.8 (Example 4). Let $\mathcal{G}$ be a connected Lie group, $\sigma: \mathcal{G} \to \mathcal{G}$ be its involutive automorphism, $\mathrm{Fix}(\sigma)$ be the subgroup of the group $\mathcal{G}$ consisting of all elements of the form $a \in \mathcal{G}$ for which $\sigma a = a$, and let $\mathcal{G}_\sigma$ be the subset of the group $\mathcal{G}$ consisting of all elements of the form $a\sigma(a)^{-1}$, $a \in \mathcal{G}$. Because the subgroup $\mathrm{Fix}(\sigma)$ is obviously closed and is therefore a Lie subgroup, the quotient space $\mathcal{G}/\mathrm{Fix}(\sigma)$ has a natural smoothness. Because, as is easily seen, the mapping $a \mapsto a\sigma(a)^{-1}$, $a \in \mathcal{G}$, induces a bijective mapping $\mathcal{G}/\mathrm{Fix}(\sigma) \to \mathcal{G}_\sigma$, this smoothness is naturally transported to $\mathcal{G}_\sigma$. Verify that *the smooth manifold $\mathcal{G}_\sigma$ is a symmetric space with the symmetries*

$$s_p q = c\sigma(c)^{-1}, \tag{4}$$

where $p = a\sigma(a)^{-1}$, $q = b\sigma(b)^{-1}$, and $c = a\sigma(a^{-1}b)$. (Symmetries (4) are just the restrictions of symmetries (2) to $\mathcal{G}_\sigma$.)

Also, show that *$\mathcal{G}_\sigma$ is a closed submanifold of the group $\mathcal{G}$.*

Exercise 5.9 (Example 5). Under the conditions in Exercise 5.8 (Example 4), consider an arbitrary subgroup $\mathcal{H}$ of the group $\mathcal{G}$ satisfying the relations

$$\mathrm{Fix}(\sigma)_e \subset \mathcal{H} \subset \mathrm{Fix}(\sigma). \tag{5}$$

Show that the following assertions hold:

1. *The subgroup $\mathcal{H}$ is closed* (it is therefore a Lie subgroup, and the quotient space $\mathcal{G}/\mathcal{H}$ is a smooth manifold).

2. *The quotient space* $\mathcal{G}/\mathcal{H}$ *is a symmetric space with the symmetries*

$$s_p q = a\sigma(a^{-1}b)\mathcal{H}, \quad p = a\mathcal{H}, \quad q = b\mathcal{H}. \tag{6}$$

This space is denoted by $(\mathcal{G}/\mathcal{H})_\sigma$. For $\mathcal{H} = \mathrm{Fix}(\sigma)$, it is isomorphic to the space $\mathcal{G}_\sigma$ in Exercise 5.8 (Example 4).

We see in Chap. 9 that spaces of the form $(\mathfrak{G}/\mathcal{H})_\sigma$ in effect exhaust all symmetric spaces.

§9. Coincidence of Classes of Symmetric and Globally Symmetric Spaces

According to Proposition 5.3, *each globally symmetric space is a symmetric space.* It turns out that the converse is also true.

Proposition 5.4. *Each symmetric space* $\mathcal{X}$ *has a symmetric connection* ∇ *with respect to which it is a globally symmetric space with the symmetries* s_p. *This connection is unique.*

Therefore, Definition 2, seemingly very general, in fact gives nothing new. The connection ∇ in Proposition 5.4 is called the *canonical connection* on the symmetric space $\mathcal{X}$. To prove Proposition 5.4, we need the following lemma.

Lemma 5.3. *For any point* p_0 *of a symmetric space* $\mathcal{X}$, *the relation*

$$(ds_{p_0})_{p_0} = -\mathrm{id}$$

holds in the linear space $\mathbf{T}_{p_0}\mathcal{X}$.

Proof. Because the mapping s_{p_0} is involutive and $s_{p_0}p_0 = p_0$, the linear operator $(ds_{p_0})_{p_0}$ is also involutive, and the linear space $\mathbf{T}_{p_0}\mathcal{X}$ is therefore a direct sum of two subspaces where the operator $(ds_{p_0})_{p_0}$ is identical on one of them and is equal to $-\mathrm{id}$ on the other. Therefore, to prove Lemma 5.3, it suffices to prove that if the relation $(ds_{p_0})_{p_0}A = A$ holds for the vector $A \in \mathbf{T}_{p_0}\mathcal{X}$, then $A = 0$.

Introducing an arbitrary connection ∇' on $\mathcal{X}$ (however, it suffices to introduce the connection ∇' on a neighborhood of the point p_0), we consider the connection

$$\nabla = \frac{\nabla' + s_{p_0}^* \nabla'}{2}.$$

For this connection, $s_{p_0}^*\nabla = \nabla$, i.e., the mapping s_{p_0} is affine with respect to ∇. Therefore, the mapping s_{p_0} transforms each geodesic $t \mapsto \exp_{p_0} tA$ of the connection ∇ passing through the point p_0 for $t = 0$ into the geodesic $t \mapsto \exp_{p_0} tB$, where $B = (ds_{p_0})_{p_0}A$. In particular, for $B = A$, the mapping s_{p_0} leaves all the points $\exp_{p_0} tA$ fixed, which, for $A \neq 0$, contradicts the fact that the fixed point p_0 is isolated. Therefore, $A = 0$. □

Remark 5.2. The argument used in the proof of this assertion also proves that if on a symmetric space $\mathcal{X}$, we introduce a connection ∇ with respect to which all the symmetries s_p, $p \in \mathcal{X}$, are affine mappings, then these symmetries are local geodesic symmetries, and, therefore, the space $\mathcal{X}$ is a globally symmetric space.

Taking all these into account, we can now pass directly to the proof of Proposition 5.4.

Proof (of Proposition 5.4). We consider an arbitrary chart $(U, x^1, \ldots, x^n)$ of the manifold $\mathcal{X}$. Lemma 5.3 implies that for any point $p_0 \in U$, the symmetry s_{p_0} in the chart $(U, x^1, \ldots, x^n)$ is given by functions of the form

$$y^i = 2x_0^i - x^i + (b^i_{jk})_0(x^j - x_0^j)(x^k - x_0^k) + o(|x - x_0|^2), \qquad (7)$$

$i = 1, \ldots, n$, where $x_0^1, \ldots, x_0^n$ are coordinates of the point p_0 and $(b^i_{jk})_0$ are values of smooth functions b^i_{jk}, $1 \le i, j, k \le n$, defined on the neighborhood U at the point p_0. In a conditional, but graphic, notation,

$$b^i_{jk} = \frac{1}{2}\frac{\partial^2 s^i_p}{\partial x^j \partial x^k}$$

on U.

By the involutiveness, x^i is expressed through y^j in exactly the same way as y^i is expressed through x^j:

$$x^i = 2x_0^i - y^i + (b^i_{jk})_0(y^j - x_0^j)(y^k - x_0^k) + \cdots$$

(here and in what follows, we replace the o term by dots).

Let the connection ∇ exist, Γ^i_{kj} be coefficients of the connection ∇ in the chart $(U, x^1, \ldots, x^n)$, and $(\Gamma^i_{kj})_0$ be their values at the point p_0. Because the mapping s_{p_0} is affine with respect to the connection (see property 2 in Proposition 3.1), we have the following identity on U:

$$\Gamma^i_{kj} \circ s_{p_0} = \frac{\partial y^i}{\partial x^r}\frac{\partial x^t}{\partial y^k}\frac{\partial x^s}{\partial y^j}\Gamma^r_{ts} + \frac{\partial y^i}{\partial x^r}\frac{\partial^2 x^r}{\partial y^k \partial y^j}.$$

In particular,

$$\begin{aligned}(\Gamma^i_{kj})_0 &= \left(\frac{\partial y^i}{\partial x^r}\right)_0\left(\frac{\partial x^t}{\partial y^k}\right)_0\left(\frac{\partial x^s}{\partial y^j}\right)_0(\Gamma^r_{ts})_0 + \left(\frac{\partial y^i}{\partial x^r}\right)_0\left(\frac{\partial^2 x^r}{\partial y^k \partial y^j}\right)_0 \\ &= (-\delta^i_r)(-\delta^t_k)(-\delta^s_j)(\Gamma^r_{ts})_0 + (-\delta^i_r)(2b^r_{kj})_0 = -(\Gamma^i_{kj})_0 - 2(b^i_{kj})_0,\end{aligned}$$

and therefore

$$(\Gamma^i_{kj})_0 = -(b^i_{kj})_0.$$

By the arbitrariness of the point $p_0 \in U$, this proves that

$$\Gamma^i_{kj} = -b^i_{kj} \quad \text{on } U. \qquad (8)$$

Therefore, the coefficients of the connection ∇ are expressed through the coefficients of series (7) that assigns the symmetries s_p. Therefore, the connection ∇ is unique.

To prove the existence of the connection ∇, we define the functions Γ^i_{kj} in each chart $(U, x^1, \dots, x^n)$ by formula (8). Proposition 5.4 is proved if we show that the following assertions hold.

1. *These functions are coefficients of a certain connection*, i.e. (see (3) in Chap. 1), for any other chart $(U', x^{1'}, \dots, x^{n'})$, we have

$$\Gamma^{i'}_{k'j'} = \frac{\partial x^{i'}}{\partial x^i}\frac{\partial x^k}{\partial x^{k'}}\frac{\partial x^j}{\partial x^{j'}}\Gamma^i_{kj} + \frac{\partial x^{i'}}{\partial x^i}\frac{\partial^2 x^i}{\partial x^{k'}\partial x^{j'}} \tag{9}$$

   on the intersection $U \cap U'$, where $\Gamma^{i'}_{k'j'}$ are functions (8) constructed for the chart $(U', x^{1'}, \dots, x^{n'})$.
2. *The symmetries s_{p_0} are affine mappings with respect to this connection.*

On $U \cap U'$, let

$$x^{i'} = x^{i'}_0 + c^{i'}_i(x^i - x^i_0) + c^{i'}_{jk}(x^j - x^j_0)(x^k - x^k_0) + \cdots$$

and

$$x^i = x^i_0 + c^i_{i'}(x^{i'} - x^{i'}_0) + c^i_{j'k'}(x^{j'} - x^{j'}_0)(x^{k'} - x^{k'}_0) + \dots,$$

where

$$c^{i'}_i = \left(\frac{\partial x^{i'}}{\partial x^i}\right)_0, \qquad c^{i'}_{jk} = \frac{1}{2}\left(\frac{\partial^2 x^{i'}}{\partial x^j \partial x^k}\right)_0,$$

$$c^i_{i'} = \left(\frac{\partial x^i}{\partial x^{i'}}\right)_0, \qquad c^i_{j'k'} = \frac{1}{2}\left(\frac{\partial^2 x^i}{\partial x^{j'} \partial x^{k'}}\right)_0.$$

Then, as we know, the matrices $\|c^{i'}_i\|$ and $\|c^i_{i'}\|$ are mutually inverse. Moreover, as an obvious computation shows, we have

$$c^{i'}_{jk} = -c^{i'}_i c^i_{j'k'} c^{j'}_j c^{k'}_k.$$

In the chart $(U', x^{1'}, \dots, x^{n'})$, the mapping s_{p_0} is given by

$$y^{i'} = 2x^{i'}_0 - x^{i'} + (b^{i'}_{j'k'})_0(x^{j'} - x^{j'}_0)(x^{k'} - x^{k'}_0) + \dots,$$

where $b^{i'}_{j'k'}$ are the functions b^i_{jk} constructed for the chart $(U', x^{1'}, \dots, x^{n'})$. On the other hand,

$$y^{i'} = x^{i'}_0 + c^{i'}_i(y^i - x^i_0) + c^{i'}_{jk}(y^j - x^j_0)(y^k - x^k_0) + \dots,$$

and therefore

$$\begin{aligned}
y^{i'} &= x_0^{i'} + c_i^{i'}(-(x^i - x_0^i) + (b_{jk}^i)_0(x^j - x_0^j)(x^k - x_0^k) + \cdots) \\
&\quad + c_{jk}^{i'}(x^j - x_0^j)(x^k - x_0^k) + \cdots \\
&= x_0^{i'} + c_i^{i'}[-c_{j'}^i(x^{j'} - x_0^{j'}) - c_{j'k'}^i(x^{j'} - x_0^{j'})(x^{k'} - x_0^{k'}) - \cdots \\
&\quad + (b_{jk}^i)_0 c_{j'}^j c_{k'}^k (x^{j'} - x_0^{j'})(x^{k'} - x_0^{k'}) + \cdots] \\
&\quad + c_{jk}^{i'} c_{j'}^j c_{k'}^k (x^{j'} - x_0^{j'})(x^{k'} - x_0^{k'}) + \cdots .
\end{aligned}$$

Therefore,

$$(b_{j'k'}^{i'})_0 = -c_i^{i'} c_{j'k'}^i + c_i^{i'} (b_{jk}^i)_0 c_{j'}^j c_{k'}^k + c_{jk}^{i'} c_{j'}^j c_{k'}^k .$$

Because

$$c_{jk}^{i'} c_{j'}^j c_{k'}^k = -c_i^{i'} c_{p'q'}^i c_j^{p'} c_k^{q'} c_{j'}^j c_{k'}^k = -c_i^{i'} c_{j'k'}^i ,$$

this proves that

$$(b_{j'k'}^{i'})_0 = c_i^{i'} c_{j'}^j c_{k'}^k (b_{jk}^i)_0 - 2c_i^{i'} c_{j'k'}^i .$$

To complete the proof of assertion 1, it remains to note that this is relation (9) written at the point p_0 with accuracy up to the notation.

Now let $\mathcal{X}$ and $\mathcal{Y}$ be isomorphic symmetric spaces, and let $\varphi\colon \mathcal{X} \to \mathcal{Y}$ be an arbitrary isomorphism. It is easy to see that *with respect to the canonical connections on $\mathcal{X}$ and $\mathcal{Y}$* (constructed above), *the isomorphism φ is affine.*

Indeed, let $(U, x^1, \ldots, x^n)$ and $(V, y^1, \ldots, y^n)$ be charts of the manifolds $\mathcal{X}$ and $\mathcal{Y}$ such that $\varphi U = V$, and let the isomorphism φ be given on U by

$$y^1 = x^1, \quad \ldots, \quad y^n = x^n \tag{10}$$

(acts with respect to the equality of coordinates). Then, for any point $p_0 \in U$, the symmetries s_{p_0} and s_{q_0}, where $q_0 = \varphi(p_0)$, are expressed in the charts $(U, x^1, \ldots, x^n)$ and $(V, y^1, \ldots, y^n)$ by the same formulas (which only differ in the notation of coordinates). Therefore, for both symmetries, the functions b_{jk}^i also coincide and the canonical connections in the charts U and V therefore have the same coefficients (more precisely, the coefficients of one connection pass into the coefficients of the other one under substitutions (10)). Therefore, the mapping φ is affine. For $\mathcal{X} = \mathcal{Y}$ and $\varphi = s_{p_0}$, this yields assertion 2. Proposition 5.4 is thus completely proved. □

Chapter 6
Connections on Lie Groups

§1. Invariant Construction of the Canonical Connection

We can also describe the canonical connection on a symmetric space $\mathcal{X}$ (or, more precisely, the corresponding covariant derivatives) in an invariant way not using local coordinates. Each smooth function F on $\mathcal{X} \times \mathcal{X}$ and each point $q \in \mathcal{X}$ define the smooth function

$$F_q^{\mathrm{I}}: p \mapsto F(p, q), \quad p \in \mathcal{X},$$

on $\mathcal{X}$ (*the trace of the function F on the qth level*). We can use this to set the vector field X^{I} on $\mathcal{X} \times \mathcal{X}$ in correspondence with each vector field X on $\mathcal{X}$, defining its action on an arbitrary function $F \in \mathbf{F}(\mathcal{X} \times \mathcal{X})$ by

$$(X^{\mathrm{I}} F)(p, q) = (X F_q^{\mathrm{I}})(p), \quad (p, q) \in \mathcal{X} \times \mathcal{X}.$$

(Here, we interpret the vector fields as derivations of the algebra of functions). Similarly, we define the vector field X^{II} by

$$(X^{\mathrm{II}} F)(p, q) = (X F_p^{\mathrm{II}})(q), \quad (p, q) \in \mathcal{X} \times \mathcal{X},$$

where F_p^{II} is the function $q \mapsto F(p, q)$ on $\mathcal{X}$.

Because the composition $f \circ \mu$ for each function f on $\mathcal{X}$, where μ is the multiplication on $\mathcal{X}$ (see Remark 5.1) is a function on $\mathcal{X} \times \mathcal{X}$, this construction allows constructing the following function on $\mathcal{X} \times \mathcal{X}$ for any two vector fields $X, Y \in \mathfrak{a}\mathcal{X}$ and for any function $f \in \mathbf{F}\mathcal{X}$:

$$X^{\mathrm{I}} Y^{\mathrm{II}}(f \circ \mu) = Y^{\mathrm{II}} X^{\mathrm{I}}(f \circ \mu).$$

The composition $X^{\mathrm{I}} Y^{\mathrm{II}}(f \circ \mu) \circ \Delta$ of this function with the diagonal mapping $\Delta: \mathcal{X} \to \mathcal{X} \times \mathcal{X}, p \mapsto (p, p)$, is therefore a function on $\mathcal{X}$. We define the operator $X \cdot Y: \mathbf{F}\mathcal{X} \to \mathbf{F}\mathcal{X}$ by setting

$$(X \cdot Y) f = X^{\mathrm{I}} Y^{\mathrm{II}}(f \circ \mu) \circ \Delta \tag{1}$$

for any function $f \in \mathbf{F}\mathcal{X}$. We find the expression for this operator in coordinates.

Let $(U, x^1, \ldots, x^n)$ be an arbitrary chart of the manifold $\mathcal{X}$. As we know, the functions

$$x^1 \circ \mathrm{pr}_1, \quad \ldots, \quad x^n \circ \mathrm{pr}_1, \quad x^1 \circ \mathrm{pr}_2, \quad \ldots, \quad x^n \circ \mathrm{pr}_2$$

are coordinates on the product $U \times U \subset \mathcal{X} \times \mathcal{X}$. The coordinates $x^1 \circ \mathrm{pr}_1, \ldots, x^n \circ \mathrm{pr}_1$ are denoted by $x^1, \ldots, x^n$, and the coordinates $x^1 \circ \mathrm{pr}_2, \ldots, x^n \circ$

pr_2 by $y^1, \dots, y^n$. Then for a vector field $X = X^i(\partial/\partial x^i)$ on U, the vector fields X^{I} and X^{II} on $U \times U$ are given by

$$X^{\mathrm{I}} = X^i(\boldsymbol{x})\frac{\partial}{\partial x^i} \qquad \text{and} \qquad X^{\mathrm{II}} = X^i(\boldsymbol{y})\frac{\partial}{\partial y^i},$$

and the multiplication μ is written in coordinates as the vector-valued function $\mu(\boldsymbol{x}, \boldsymbol{y})$ with the components

$$\mu^i(\boldsymbol{x}, \boldsymbol{y}) = 2x^i - y^i + b^i_{jk}(\boldsymbol{x})(y^j - x^j)(y^k - x^k) + \cdots$$

(this is exactly formula (7) in Chap. 5 with x_0^i replaced by x^i and x^i by y^i). Therefore,

$$\begin{aligned} X^{\mathrm{I}}Y^{\mathrm{II}}(f \circ \mu) &= X^i(\boldsymbol{x})\frac{\partial}{\partial x^i}\left(Y^j(\boldsymbol{y})\frac{\partial \mu^k}{\partial y^j}\frac{\partial f}{\partial x^k}\right) \\ &= X^i(\boldsymbol{x})Y^j(\boldsymbol{y})\left(\frac{\partial^2 \mu^k}{\partial x^i \partial y^j}\frac{\partial f}{\partial x^k} + \frac{\partial \mu^l}{\partial x^i}\frac{\partial \mu^k}{\partial y^j}\frac{\partial^2 f}{\partial x^l \partial x^k}\right), \end{aligned}$$

where

$$\begin{aligned} \frac{\partial \mu^l}{\partial x^i} &= 2\delta^l_i - 2b^l_{ij}(y^j - x^j)\dots, \\ \frac{\partial \mu^k}{\partial y^j} &= -\delta^k_j + 2b^k_{ij}(y^i - x^i)\dots, \\ \frac{\partial^2 \mu^k}{\partial x^i \partial y^j} &= -2b^k_{ij} + \dots, \end{aligned}$$

and therefore

$$\begin{aligned} X^{\mathrm{I}}Y^{\mathrm{II}}(f \circ \mu) \circ \Delta &= X^iY^j\left(-2b^k_{ij}\frac{\partial f}{\partial x^k} - 2\delta^l_i\delta^k_j\frac{\partial^2 f}{\partial x^l \partial x^k}\right) \\ &= -2X^iY^j\left(b^k_{ij}\frac{\partial f}{\partial x^k} + \frac{\partial^2 f}{\partial x^i \partial x^j}\right), \end{aligned}$$

i.e.,

$$X \cdot Y = -2X^iY^j\left(\frac{\partial^2}{\partial x^i \partial x^j} + b^k_{ij}\frac{\partial}{\partial x^k}\right) \quad \text{on } U. \tag{2}$$

We see that $X \cdot Y$ is a differential operator of the *second order*.

To obtain a differential operator of the first order (a vector field), we note that

$$XY = X^iY^j\frac{\partial^2}{\partial x^i \partial x^j} + X^i\frac{\partial Y^j}{\partial x^i}\frac{\partial}{\partial x^j}.$$

Therefore,

$$XY + \frac{1}{2}X \cdot Y = \left(X^i\frac{\partial Y^k}{\partial x^i} - b^k_{ij}X^iY^j\right)\frac{\partial}{\partial x^k}$$

and (see (8) in Chap. 5 and (4) in Chap. 1)

$$\nabla_X Y = XY + \frac{1}{2} X \cdot Y \tag{3}$$

(on U and hence on the entire $\mathcal{X}$ because of the arbitrariness of U). Taking (1) into account, we can accept (3) as the definition of the canonical connection on $\mathcal{X}$.

Exercise 6.1. Prove that the canonical curvature operator acts by

$$R(X,Y)Z = \frac{1}{4}(X \cdot (Y \cdot Z) - Y \cdot (X \cdot Z)), \quad X, Y, Z \in \mathfrak{a}\mathcal{X}, \tag{4}$$

where the right-hand side is defined in an obvious way. [*Hint*: First prove that $X(Y \cdot Z) = XY \cdot Z + Y \cdot XZ$.]

§2. Morphisms of Symmetric Spaces as Affine Mappings

We now compare morphisms and affine mappings of symmetric spaces.

Proposition 6.1. *A mapping $f: \mathcal{X} \to \mathcal{Y}$ of symmetric spaces is a morphism iff it is an affine mapping (with respect to the canonical connections on $\mathcal{X}$ and $\mathcal{Y}$).*

Proof. If the mapping f is affine, then for any point $p \in \mathcal{X}$, the mappings $f \circ s_p$ and $s_{f(p)} \circ f$ are also affine (because each symmetry is an affine mapping with respect to the canonical connection). On the other hand, because $(ds_p)_p = -\operatorname{id}$ and $(ds_{f(p)})_{f(p)} = -\operatorname{id}$, we have

$$d(f \circ s_p)_p = -(df)_p = d(s_{f(p)} \circ f)_p,$$

and these mappings therefore coincide according to Proposition 3.2:

$$f \circ s_p = s_{f(p)} \circ f,$$

i.e., f is a morphism.

Now let $f: \mathcal{X} \to \mathcal{Y}$ be an arbitrary morphism of symmetric spaces.

Exercise 6.2. Prove that if two fields $\widehat{X}, \widehat{Y} \in \mathfrak{a}\mathcal{Y}$ are f-related to the fields $X, Y \in \mathfrak{a}\mathcal{X}$, then the operator $\widehat{X} \cdot \widehat{Y}$ is f-related to the operator $X \cdot Y$ (i.e.,

$$(\widehat{X} \cdot \widehat{Y})\varphi \circ f = (X \cdot Y)(\varphi \circ f)$$

for any smooth function φ on $\mathcal{Y}$).

Because a similar property obviously also holds for the operator XY, this proves (by (3)) that the mapping f satisfies condition 1 of Proposition 3.1 and is therefore affine. □

§3. Left-Invariant Connections on a Lie Group

Interesting examples of connections arise in Lie group theory. Let $\mathcal{G}$ be an arbitrary Lie group.

Definition 6.1. A connection ∇ on the Lie group $\mathcal{G}$ is said to be *left invariant* if for any two left-invariant vector fields X and Y (elements of the Lie algebra $\mathfrak{g} = \mathfrak{l}\mathcal{G}$), the field $\nabla_X Y$ is also left invariant (belongs to the Lie algebra $\mathfrak{g}$).

Exercise 6.3. Show that *a connection ∇ on a Lie group $\mathcal{G}$ is left invariant iff each left translation*

$$L_a: \mathcal{G} \to \mathcal{G}, \quad p \mapsto ap, \quad a, p \in \mathcal{G},$$

is affine. [*Hint*: See property 1 of affine mappings in Proposition 3.1.]

For each left-invariant connection ∇, the formula

$$\alpha(X, Y) = \nabla_X Y, \quad X, Y \in \mathfrak{g},$$

defines a certain mapping

$$\alpha: \mathfrak{g} \times \mathfrak{g} \to \mathfrak{g}. \tag{5}$$

This mapping is obviously $\mathbb{R}$-bilinear, i.e., is a multiplication on the linear space $\mathfrak{g}$.

Exercise 6.4. Show that *each basis*

$$X_1, \quad \dots, \quad X_n \tag{6}$$

of the algebra $\mathfrak{g}$ (over the field $\mathbb{R}$) is also a basis of the module $\mathfrak{a}\mathcal{G}$ (over the algebra $\mathsf{F}\mathcal{G}$). [*Hint*: For each point $a \in \mathcal{G}$, the vectors $(X_1)_a, \dots, (X_n)_a$ compose a basis of the linear space $\mathsf{T}_a\mathcal{G}$.]

Therefore, each left-invariant connection on $\mathcal{G}$ is uniquely defined by the fields $A_{ij} = \alpha(X_i, X_j)$: if $X = f^i X_i$ and $Y = g^j X_j$, where $f^i, g^j \in \mathsf{F}\mathcal{G}$, then

$$\nabla_X Y = f^i\left[(X_i g^j) X_j + g^j A_{ij}\right]. \tag{7}$$

In this case, as an automatic verification shows, formula (7) for any fields $A_{ij} \in \mathfrak{g}$, $i, j = 1, \dots, n$, defines a left-invariant covariant differentiation on $\mathcal{G}$ for which $\alpha(X_i, X_j) = A_{ij}$.

Therefore, *the correspondence*

$$\text{connection } \nabla \quad \Longleftrightarrow \quad \text{multiplication } \alpha$$

is a bijective correspondence between left-invariant connections on a Lie group $\mathcal{G}$ and multiplications on the Lie algebra $\mathfrak{g} = \mathfrak{l}\mathcal{G}$. It is clear that each multiplication (5) is uniquely represented in the form

$$\alpha = \alpha' + \alpha'',$$

where α' is a commutative (symmetric) multiplication and α'' is an anticommutative (skew-symmetric) multiplication, because it suffices to set

$$\alpha'(X,Y) = \frac{\alpha(X,Y) + \alpha(Y,X)}{2}, \qquad \alpha''(X,Y) = \frac{\alpha(X,Y) - \alpha(Y,X)}{2}$$

for any fields $X, Y \in \mathfrak{g}$.

Exercise 6.5. Show that *a multiplication α is anticommutative (i.e., $\alpha' = 0$) iff $\alpha(X,X) = 0$ for any field $X \in \mathfrak{g}$.* [*Hint*: Consider the field $\alpha(X+Y, X+Y)$.]

The assertion of Exercise 6.4 implies that each $\mathsf{F}\mathcal{G}$-linear mapping

$$T\colon \mathfrak{a}\mathcal{G} \times \mathfrak{a}\mathcal{G} \to \mathfrak{a}\mathcal{G}$$

(a tensor of type (2,1) on $\mathcal{G}$) is uniquely defined by its values $T(X,Y)$ on fields $X, Y \in \mathfrak{g}$. In particular, this is true for the torsion tensor of an arbitrary left-invariant connection ∇. But because we have

$$\begin{aligned} T(X,Y) &= \nabla_X Y - \nabla_Y X - [X,Y] \\ &= \alpha(X,Y) - \alpha(Y,X) - [X,Y] = 2\alpha''(X,Y) - [X,Y], \quad X, Y \in \mathfrak{g}, \end{aligned}$$

for this tensor, *a left-invariant connection ∇ on a Lie group $\mathcal{G}$ is symmetric iff*

$$\alpha''(X,Y) = \frac{1}{2}[X,Y]$$

for any fields $X, Y \in \mathfrak{g}$.

§4. Cartan Connections

For a given connection on $\mathcal{G}$ and for each vector $A \in \mathbf{T}_e\mathcal{G}$, we have two curves passing through the point e of the group $\mathcal{G}$ that have the tangent vector A at this point; they are the one-parameter subgroup β_A and the geodesic γ_A.

Definition 6.2. A left-invariant connection ∇ on a Lie group $\mathcal{G}$ is called a *Cartan connection* if for any vector $A \in \mathbf{T}_e\mathcal{G}$, the curves β_A and γ_A coincide,

$$\beta_A = \gamma_A, \quad A \in \mathbf{T}_e\mathcal{G}.$$

Exercise 6.6. Prove that *for any point $a \in \mathcal{G}$ and any vector $A \in \mathbf{T}_a\mathcal{G}$, the geodesic $\gamma_{a,A}$ of a Cartan connection passing through the point a for $t = 0$ and having the tangent vector A at this point is the curve*

$$R_a \circ \beta_{A_0}\colon t \mapsto \beta_{A_0}(t)a,$$

where $A_0 = (dR_{a^{-1}})_a A$.

In particular, this implies that *each Lie group is geodesically complete with respect to an arbitrary Cartan connection.*

As we know, each one-parameter subgroup $\beta = \beta_A$ is an integral curve of a left-invariant vector field $X \in \mathfrak{g}$ for which $X_e = A$, i.e., the restriction of the field X to β is the field of tangent vectors $t \mapsto \dot{\beta}(t)$. Therefore,

$$\frac{\nabla}{dt}\dot{\beta}(t) = (\nabla_X X)_{\beta(t)} = \alpha(X, X)_{\beta(t)} \quad \text{for any } t \in \mathbb{R};$$

this implies (because a left-invariant field identically vanishes iff it vanishes at least at one point) that *the subgroup β_A is a geodesic of a left-invariant connection ∇ iff $\alpha(X, X) = 0$.* Therefore (see Exercise 6.5), *a left-invariant connection ∇ is a Cartan connection iff the multiplication α corresponding to this connection is anticommutative.*

It is easiest to assign such a multiplication by

$$\alpha(X, Y) = \lambda[X, Y], \quad X, Y \in \mathfrak{g},$$

where λ is a certain number. The torsion tensor of the corresponding connection assumes the value

$$T(X, Y) = (2\lambda - 1)[X, Y], \quad X, Y \in \mathfrak{g},$$

at fields $X, Y \in \mathfrak{g}$, and for the curvature tensor, we have

$$R(X, Y)Z = (\lambda^2 - \lambda)[[X, Y], Z], \quad X, Y, Z \in \mathfrak{g}.$$

For $\lambda = 1$ and $\lambda = 0$, we obtain from this that *on any Lie group $\mathcal{G}$, there exist two natural Cartan connections whose curvature tensors identically vanish.* For $X, Y \in \mathfrak{g}$, we have

$$\nabla_X Y = [X, Y] \tag{8}$$

for the first connection and

$$\nabla_X Y = 0 \tag{9}$$

for the second; for the torsion tensors of these connections, we have

$$T(X, Y) = \varepsilon[X, Y],$$

where $\varepsilon = 1$ for connection (8) and $\varepsilon = -1$ for connection (9).

Moreover, we see that *on any Lie group $\mathcal{G}$, there exists a unique symmetric Cartan connection.* For this connection, we have

$$\nabla_X Y = \frac{1}{2}[X, Y] \quad \text{for } X, Y \in \mathfrak{g} \tag{10}$$

and

$$R(X, Y)Z = -\frac{1}{4}[[X, Y], Z] \quad \text{for } X, Y, Z \in \mathfrak{g}. \tag{11}$$

Exercise 6.7. Prove that *for connections* (8) *and* (9), *the torsion tensor* T *is covariantly constant,*

$$\nabla T = 0.$$

[*Hint*: Use general formula (9) in Chap. 2 (for $r = 2$).]

Connection (10) is the half-sum of connections (8) and (9) and is therefore sometimes called the *mean connection.*

That one-parameter subgroups are geodesics of a Cartan connection explains the phenomenon noted in Remark 1.1.

Exercise 6.8. Being a symmetric space (see Exercise 5.5), the group $\mathcal{G}$ has a canonical connection. Show that *mean connection* (10) *is this connection.*

In particular, this implies that *the curvature tensor of connection* (10) *is covariantly constant,*

$$\nabla R = 0.$$

Exercise 6.9. Deduce this property directly from (11).

§5. Left Cartan Connection

It is clear that the condition $\nabla_X Y = 0$ holds for all fields $X \in \mathfrak{a}\mathcal{G}$ iff it holds for all left-invariant fields $X \in \mathfrak{l}\mathcal{G}$. Therefore, when dealing with connection (9), we see that the field $Y = f^i X_i$ (where $X_1, \ldots, X_n$ is a basis and $f^1, \ldots, f^n \in \mathsf{F}\mathcal{G}$) is covariantly constant (the condition $\nabla_X Y = 0$ holds for all fields $X \in \mathfrak{a}\mathcal{X}$) iff $Xf^i = 0$ for any $i = 1, \ldots, n$ and any field $X \in \mathfrak{l}\mathcal{G}$. But if $Xf^i = 0$ for any field $X \in \mathfrak{l}\mathcal{G}$, then, obviously, $Xf^i = 0$ for any field $X \in \mathfrak{a}\mathcal{G}$ (see Exercise 6.4), and therefore $f^i = \text{const}$, i.e., $Y \in \mathfrak{l}\mathcal{G}$. This proves that *the fields* $Y \in \mathfrak{a}\mathcal{G}$ *that are covariantly constant with respect to connection* (9) *are exactly the left-invariant fields* $Y \in \mathfrak{l}\mathcal{G}$.

In particular, this implies that for any path $u: I \to \mathcal{G}$ connecting a point p with a point q, the parallel translation Π_u corresponding to connection (9) is such that $\Pi_u X_p = X_q$ for any field $X \in \mathfrak{l}\mathcal{G}$. But it is clear that, first, this property uniquely defines the linear mapping $\Pi_u: \mathsf{T}_p\mathcal{G} \to \mathsf{T}_q\mathcal{G}$ and, second, the linear mapping $(dL_a)_p: \mathsf{T}_p\mathcal{G} \to \mathsf{T}_q\mathcal{G}$, where $a = qp^{-1}$, also has this property. Therefore, $\Pi_u = (dL_a)_p$, and hence Π_u does not depend on the choice of the path u. This means that *with respect to connection* (9), *the Lie group* $\mathcal{G}$ *is a space with absolute parallelism* (even if the group $\mathcal{G}$ is not simply connected!). Parallel translations with respect to this connection are differentials of left translations L_a. Connection (9) is therefore usually called the *left* Cartan connection on the Lie group $\mathcal{G}$.

§6. Right-Invariant Vector Fields

A vector field Y on a Lie group $\mathcal{G}$ is said to be *right invariant* if for any two elements $a, b \in \mathcal{G}$, we have

$$(dR_{a^{-1}b})Y_a = Y_b, \tag{12}$$

where $R_{a^{-1}b}: \mathcal{G} \to \mathcal{G}$ is the right translation $x \mapsto xa^{-1}b$ by the element $a^{-1}b$.

The inversion diffeomorphism $\nu: x \mapsto x^{-1}$ transforms right translations to left ones and therefore transforms right-invariant vector fields into left-invariant ones. Consequently, the properties of right-invariant fields are completely similar to those of left-invariant ones. In particular, we have the following assertions:

1. *Trajectories of right-invariant vector fields passing through the point* e *are exactly one-parameter subgroups of the group* $\mathcal{G}$ *(trajectories of left-invariant vector fields passing through the point* e*).*
2. *Right-invariant vector fields form a subalgebra* $\mathfrak{r}\mathcal{G}$ *of the Lie algebra* $\mathfrak{a}\mathcal{G}$ *of all smooth vector fields on* $\mathcal{G}$*.*
3. *The mapping* $Y \mapsto Y_e$ *is an isomorphism of the linear space* $\mathfrak{r}\mathcal{G}$ *onto the tangent space* $\mathbf{T}_e\mathcal{G}$ *(with the inverse isomorphism* $B \mapsto Y$, $B \in \mathbf{T}_e\mathcal{G}$, *where* $Y_a = (dR_a)_e B$ *for any point* $a \in \mathcal{G}$*).*
4. *Each basis*
$$Y_1, \quad \ldots, \quad Y_n \tag{13}$$
of the linear space $\mathfrak{r}\mathcal{G}$ *(over the field* $\mathbb{R}$*) is a basis (over the algebra* $\mathbf{F}\mathcal{G}$*) of the module* $\mathfrak{a}\mathcal{G}$*.*

Exercise 6.10. Prove assertions 1–4 directly, without reference to the theory of left-invariant fields.

Exercise 6.11. Prove that *the algebra* $\mathfrak{r}\mathcal{G}$ *is isomorphic to the algebra* $\mathfrak{l}\mathcal{G}$. [*Hint*: The isomorphism is given by the mapping $\nu^*: \mathfrak{a}\mathcal{G} \to \mathfrak{a}\mathcal{G}$, where $\nu: \mathcal{G} \to \mathcal{G}$ is the inversion diffeomorphism $x \mapsto x^{-1}$.]

For any vectors $A, B \in \mathbf{T}_e\mathcal{G}$, we set

$$[A, B]_l = [X, Y]_e,$$

where X and Y are left-invariant vector fields on $\mathcal{G}$ such that $X_e = A$ and $Y_e = B$. (The operation $[\,,\,]_l$ is just the operation $[\,,\,]$ of the Lie algebra $\mathfrak{l}\mathcal{G}$ that is transported to the linear space $\mathbf{T}_e\mathcal{G}$ via the isomorphism $\mathfrak{l}\mathcal{G} \to \mathbf{T}_e\mathcal{G}$.) Similarly in $\mathbf{T}_e\mathcal{G}$, the Lie operation is transported from the Lie algebra $\mathfrak{r}\mathcal{G}$:

$$[A, B]_r = [X, Y]_e,$$

where X and Y are right-invariant vector fields on $\mathcal{G}$ such that $X_e = A$ and $Y_e = B$.

Exercise 6.12. Show that

$$[A,B]_r = -[A,B]_l.$$

Proposition 6.2. *A vector field Y on a Lie group $\mathcal{G}$ is right-invariant iff*

$$[X,Y] = 0 \tag{14}$$

for any left-invariant field $X \in \mathfrak{l}\mathcal{G}$.

Proof. We recall (see the addendum) that for any vector field $Y \in \mathfrak{a}\mathcal{G}$ and for any smooth function $f \in \mathsf{F}\mathcal{G}$, the function Yf can be defined by

$$(Yf)(p) = \lim_{t\to 0} \frac{f(\varphi_p(t)) - f(p)}{t}, \quad p \in \mathcal{G},$$

where $\varphi_p: t \mapsto \varphi_p(t)$ is the trajectory of the field Y passing through the point p for $t = 0$. In the case where the field Y is right invariant, the trajectory φ_p is given by

$$\varphi_p(t) = \beta(t)p = L_{\beta(t)}p,$$

where $\beta: t \mapsto \beta(t)$ is the trajectory passing through the point e for $t = 0$ (i.e., a certain one-parameter subgroup). Therefore, in this case,

$$Yf = \lim_{t\to 0} \frac{f \circ L_{\beta(t)} - f}{t},$$

where the limit is understood as the pointwise limit of functions on the group $\mathcal{G}$. In particular, applying this formula to the function Xf, where $X \in \mathfrak{a}\mathcal{G}$, we obtain

$$YXf = \lim_{t\to 0} \frac{(Xf) \circ L_{\beta(t)} - Xf}{t}.$$

On the other hand, because the operator X obviously commutes with the pointwise limit, we have

$$XYf = \lim_{t\to 0} \frac{X(f \circ L_{\beta(t)}) - Xf}{t}.$$

But, if the field X is left invariant, then $(Xf) \circ L_{\beta(t)} = X(f \circ L_{\beta(t)})$, and therefore $XYf = YXf$ in this case, i.e., $[X,Y]f = 0$. This proves that any right-invariant field satisfies condition (14).

Conversely, let Y be an arbitrary field on the group $\mathcal{G}$ that satisfies condition (14). Choosing basis (9) in the linear space $\mathfrak{r}\mathcal{G}$, we represent the field Y in the form f^iY_i, where $f^i \in \mathsf{F}\mathcal{G}$. For any field $X \in \mathfrak{l}\mathcal{G}$, we then have

$$[X,Y] = (Xf^i)Y_i + f^i[X,Y_i] = (Xf^i)Y_i$$

(because $[X,Y_i] = 0$ by what was just proved). Because $[X,Y] = 0$, this is only possible for $Xf^i = 0$, i.e., for $f^i = \text{const}$. Therefore, $Y \in \mathfrak{r}\mathcal{G}$. □

§7. Right Cartan Connection

It is now easy to see that *a vector field Y on a Lie group $\mathcal{G}$ is covariantly constant with respect to connection* (8) *iff it is right invariant.* Indeed, if $X \in \mathfrak{l}\mathcal{G}$, then the covariant derivative $\nabla_X Y$ with respect to connection (8) of an arbitrary field $Y = f^i X_i$, where $X_1, \ldots, X_n$ is basis (6) and $f^1, \ldots, f^n \in \mathbf{F}\mathcal{G}$, is expressed as

$$\nabla_X Y = (Xf^i)X_i + f^i[X, X_i] = [X, f^i X_i] = [X, Y],$$

and therefore $\nabla_X Y = 0$ iff $[X, Y] = 0$. On the other hand, by the assertion in Exercise 6.4, the relation $\nabla_X Y = 0$ holds for any field $X \in \mathfrak{a}\mathcal{G}$ iff it holds for $X \in \mathfrak{l}\mathcal{G}$.

Therefore (see a similar argument for connection (9) above), *with respect to connection* (8) *the Lie group $\mathcal{G}$ is the space with absolute parallelism in which the differentials of the right translations R_a serve as parallel translations.* Connection (8) is therefore called the *right* Cartan connection on the Lie group.

Exercise 6.13. Prove the following:

1. *A mapping $f: \mathcal{G} \to \mathcal{G}$ is affine with respect to one of the Cartan connections* (8), (9), *or* (10) *iff it is affine with respect to each of them.*
2. *The left and right translations L_a, $R_a: \mathcal{G} \to \mathcal{G}$ are affine with respect to each of the connections* (8), (9), *and* (10).

Remark 6.1. As was already noted, for connection (9), the relation $\nabla_X Y = 0$ holds for any field $X \in \mathfrak{a}\mathcal{G}$ iff $Y \in \mathfrak{l}\mathcal{G}$. Similarly for connection (8), the relation $\nabla_X Y = [X, Y]$ holds for any field $Y \in \mathfrak{a}\mathcal{G}$ iff $X \in \mathfrak{l}\mathcal{G}$.

Chapter 7
Lie Functor

§1. Categories

The main goal of this chapter is to present the procedure for reconstructing a Lie group from its Lie algebra. Moreover, incidentally, we here present certain general mathematical concepts that were already mentioned repeatedly in passing.

Experience in constructing mathematical theories showed long ago that along with objects of a theory, usually sets equipped with one or another structure, the same and sometimes more important role is played by mappings that preserve this structure: in algebra, they are homomorphisms; in topology, they are continuous mappings; and in manifold theory, they are smooth mappings. This fact allows explicating the intuitive concept of the field of operation of a mathematical theory abstractly.

Let $\mathbf{C}$ be a class that is a disjoint union of two classes $\operatorname{Ob}\mathbf{C}$ and $\operatorname{Ar}\mathbf{C}$. Elements of the class $\operatorname{Ob}\mathbf{C}$ are called *objects*, and elements of the class $\operatorname{Ar}\mathbf{C}$ are called *arrows* or *morphisms*.

We assume that two objects A and B correspond to each morphism $f \in \operatorname{Ar}\mathbf{C}$, which is written as $f: A \to B$. All morphisms of the form $f: A \to B$ with a given A and B form a set, which is denoted by $\mathbf{C}(A, B)$ or $\operatorname{Hom}_{\mathbf{C}}(A, B)$ (the notation $\operatorname{Mor}_{\mathbf{C}}(A, B)$ is also used). We say that morphisms from $\mathbf{C}(A, B)$ are *morphisms from A into B*.

Further, we assume that for any three objects $A, B, C \in \operatorname{Ob}\mathbf{C}$, we have a mapping

$$\mathbf{C}(A, B) \times \mathbf{C}(B, C) \to \mathbf{C}(A, C)$$

that sets a morphism from A into C in correspondence to every two morphisms $f: A \to B$ and $g: B \to C$, which is denoted by $g \circ f$ (note the order) and called the *composition* of the morphisms f and g. The operation $\circ$ should be associative, i.e., for any objects A, B, C, and D and for any morphisms $f: A \to B$, $g: B \to C$, and $h: C \to D$, the relation

$$h \circ (g \circ f) = (h \circ g) \circ f$$

holds.

Finally, we assume that for any object $A = \operatorname{Ob}\mathbf{C}$, the set $\mathbf{C}(A, A)$ (also denoted by $\operatorname{End}_{\mathbf{C}} A$) contains a certain element id_A that has the property that for any objects B and C and for any morphisms $f: A \to B$ and $g: C \to A$,

$$f \circ \mathrm{id}_A = f, \qquad \text{and} \qquad \mathrm{id}_A \circ g = g$$

hold. The morphism id_A is called the *identity* of A and is often denoted by 1_A. Instead of id_A (or 1_A), we usually simply write id (or 1). A class $\mathbf{C}$ with the described structure is called a *category*.

Examples of Categories

1. The category LIN($\mathbb{K}$) of finite-dimensional linear spaces over a field $\mathbb{K}$ and their linear mappings.
2. The category TOP of topological spaces and their continuous mappings.
3. The category DIFF of smooth manifolds and their smooth mappings.

4. The category GROUPS of all groups and their homomorphisms.
5. The category LIE of all Lie groups and their homomorphisms (smooth mappings $\mathcal{G} \to \mathcal{H}$ that preserve the multiplication).
6. The category lie = lie($\mathbb{R}$) of all finite-dimensional Lie algebras of the field $\mathbb{R}$ and their homomorphisms.

... And so on.

In examples 1–6, the objects are sets equipped with one or another structure, and the arrows are mappings of sets. In general, this is not required for a category. In the case where the objects are sets and the arrows are their mappings, to verify that we deal with a category, it suffices to verify that the identities and compositions of morphisms are morphisms.

§2. Functors

Among various mathematical constructions (setting objects of another or the same category in correspondence to objects of one category), we isolate the constructions that seem to be "natural." Intuitively, their characteristic is the absence of elements of arbitrariness in them. However, formalizing this point of view is not easy. About fifty years ago, McLane and Eilenberg observed that natural constructions over objects can always be extended to morphisms and suggested considering this property as a basis for formally explicating the concept of natural construction.

Let $\mathbf{C}$ and $\mathbf{D}$ be two categories. A mapping

$$\Phi\colon \operatorname{Ob}\mathbf{C} \to \operatorname{Ob}\mathbf{D} \tag{1}$$

is called a *natural construction* if there exists a mapping

$$\Phi\colon \operatorname{Ar}\mathbf{C} \to \operatorname{Ar}\mathbf{D} \tag{1'}$$

that satisfies one of the two sets of conditions:

1. If $f\colon A \to B$, then $\Phi(f)\colon \Phi(A) \to \Phi(B)$
2. If $f = \mathrm{id}_A$, then $\Phi(f) = \mathrm{id}_{\Phi(A)}$.
3. If $f = h \circ g$, then $\Phi(f) = \Phi(h) \circ \Phi(g)$,

1′. If $f\colon A \to B$, then $\Phi(f)\colon \Phi(B) \to \Phi(A)$.
2′. If $f = \mathrm{id}_A$, then $\Phi(f) = \mathrm{id}_{\Phi(A)}$.
3′. If $f = h \circ g$, then $\Phi(f) = \Phi(g) \circ \Phi(h)$.

We say that mapping (1′) satisfying conditions 1, 2, and 3 or 1′, 2′, and 3′ has the *functorial property*, and the mapping (1) and (1′) taken together are called a *functor* (if conditions 1, 2, and 3 hold) or a *cofunctor* (if conditions 1′, 2′, and 3′ hold) from the category $\mathbf{C}$ to the category $\mathbf{D}$. In this case, mapping (1) is called the *object part*, and mapping (1′) is called the *arrow part* of the functor (cofunctor) Φ. Therefore, a construction is natural if it is the object part of a certain functor or cofunctor. We note that the arrow part of a functor (cofunctor) uniquely defines its object part. Functors are also called *covariant functors* and cofunctors *contravariant functors*.

§3. Lie Functor

We consider the naturalness of the construction

$$\text{Lie group } \mathcal{G} \quad \Longrightarrow \quad \text{Lie algebra } \mathfrak{l}\mathcal{G}. \tag{2}$$

Let $f\colon \mathcal{G} \to \mathcal{H}$ be an arbitrary homomorphism of Lie groups. We recall (see Chap. 3) that two fields $X \in \mathfrak{a}\mathcal{G}$ and $Y \in \mathfrak{a}\mathcal{H}$ are said to be *f-related* if

$$Y_{f(a)} = (df)_a X_a$$

for any point $a \in \mathcal{G}$.

Proposition 7.1. *For any left-invariant field $X \in \mathfrak{l}\mathcal{G}$, there exists a unique left-invariant field $Y \in \mathfrak{l}\mathcal{H}$ that is f-related to the field X.*

Proof. If the field Y exists, then $Y_e = (df)_e X_e$ and $Y_b = (dL_b)_e Y_e$ for any point $b \in \mathcal{H}$, i.e.,

$$Y_b = (dL_b)_e (df)_e X_e, \quad b \in \mathcal{H}. \tag{3}$$

This proves the uniqueness of the field Y.

To prove the existence of the field Y, we define it by (3), i.e., by

$$Y_b = d(L_b \circ f)_e X_e, \quad b \in \mathcal{H}.$$

It is clear that this field is left invariant (belongs to $\mathfrak{l}\mathcal{H}$). Moreover, because $L_{f(a)} \circ f = f \circ L_a$ for any point $a \in \mathcal{G}$ (because $f(ax) = f(a)f(x)$), we have

$$Y_{f(a)} = d(f \circ L_a)_e X_e = (df)_a (dL_a)_e X_e = (df)_a X_a,$$

and the field Y is therefore f-related to the field X. □

Letting $\mathfrak{l}(f)X$ denote the field Y, we thus obtain a certain (obviously linear) mapping

$$\mathfrak{l}(f)\colon \mathfrak{l}\mathcal{G} \to \mathfrak{l}\mathcal{H},$$

which is a homomorphism of Lie algebras. Because the relation $f \mapsto \mathfrak{l}(f)$ has properties 1, 2, and 3 (as is easily seen), this proves that *construction* (2) *is natural.*

Under the identification of the algebras $\mathfrak{l}\mathcal{G}$ and $\mathfrak{l}\mathcal{H}$ with the tangent spaces $\mathbf{T}_e\mathcal{G}$ and $\mathbf{T}_e\mathcal{H}$, the homomorphism $\mathfrak{l}(f)$ is just the differential

$$(df)_e\colon \mathbf{T}_e\mathcal{G} \to \mathbf{T}_e\mathcal{H} \tag{4}$$

of the mapping f at the point e.

The functor $\mathcal{G} \mapsto \mathfrak{l}\mathcal{G}$, $f \mapsto \mathfrak{l}(f)$, which we have just constructed, from the category LIE into the category lie is called the (*left*) *Lie functor.*

Exercise 7.1. Prove that if elements of the Lie algebras $\mathfrak{l}\mathcal{G}$ and $\mathfrak{l}\mathcal{H}$ are interpretated as one-parameter subgroups, then the homomorphism $\mathfrak{l}(f)$ is given by the correspondence

$$\beta \mapsto f \circ \beta, \qquad \beta: \mathbb{R} \to \mathcal{G}. \tag{5}$$

[*Hint*: Formula (5) means that for any homomorphism $f: \mathcal{G} \to \mathcal{H}$ of Lie groups, the diagram

$$\begin{array}{ccc} \mathfrak{g} & \xrightarrow{\mathfrak{l}(f)} & \mathfrak{h} \\ {\scriptstyle \exp}\downarrow & & \downarrow{\scriptstyle \exp} \\ \mathcal{G} & \xrightarrow{f} & \mathcal{H} \end{array} \tag{6}$$

is commutative, where $\mathfrak{g} = \mathfrak{l}\mathcal{G}$ and $\mathfrak{h} = \mathfrak{l}\mathcal{H}$.]

Similarly, we construct the *right Lie functor* $\mathcal{G} \mapsto \mathfrak{r}\mathcal{G}$, $f \mapsto \mathfrak{r}(f)$.

Exercise 7.2. Show that the homomorphism $\mathfrak{r}(f)$ is given by the same formulas (4) and (5), mutatis mutandis. (The passage from $\mathfrak{l}\mathcal{G}$ to $\mathfrak{r}\mathcal{G}$ only changes the direction of motion on one-parameter subgroups; compare with Exercise 6.11.)

§4. Kernel and Image of a Lie Group Homomorphism

Being a closed subgroup of a Lie group $\mathcal{G}$, the kernel $\operatorname{Ker} f$ of an arbitrary homomorphism $f: \mathcal{G} \to \mathcal{H}$ of Lie groups is a Lie subgroup of this group. Its Lie algebra consists of fields $X \in \mathfrak{g}$ such that for all t, we have $f(\exp tX) = e$, i.e., $\exp(t\mathfrak{l}(f)X) = e$ (see diagram (6)). Because the relation $\exp tY = e$ holds for all t iff $Y = 0$, this proves that *the Lie algebra of the kernel of the homomorphism f is the kernel of the homomorphism $\mathfrak{l}(f)$*:

$$\mathfrak{l}(\operatorname{Ker} f) = \operatorname{Ker} \mathfrak{l}(f).$$

Similarly, let $\operatorname{Im} f$ be the image of the homomorphism f (in contrast to the kernel, it is generally not a closed subgroup of the group $\mathcal{H}$), and let $\mathcal{J}$ be a connected Lie subgroup of the group $\mathcal{H}$ that corresponds to the subalgebra $\operatorname{Im} \mathfrak{l}(f) = \mathfrak{l}(f)\mathcal{G}$ of the Lie algebra $\mathfrak{h} = \mathfrak{l}\mathcal{H}$. Because the subgroup $\mathfrak{J}$ is generated by elements of the form $\exp \mathfrak{l}(f)X = f(\exp X)$, $X \in \mathfrak{g}$, and the component of the identity $\mathcal{G}_e$ of the group $\mathcal{G}$ is generated by elements $\exp X$, we see that f maps $\mathcal{G}_e$ onto $\mathcal{J}$. Because

$$\mathcal{G} = \coprod_{a \in A} a\mathcal{G}_e,$$

where A is the set of representatives of cosets of the factor group of $\mathcal{G}$ by the subgroup $\mathcal{G}_e$, we therefore have

$$\operatorname{Im} f = \coprod_{b \in B} b\mathcal{J},$$

where B is the family of representatives of cosets of the quotient of the abstract group $\operatorname{Im} f$ by its subgroup $\mathcal{J}$. We transport the smoothness from $\mathcal{J}$ to $b\mathcal{J}$,

$b \in B$, via the bijective mapping $L_b: x \mapsto bx$, $x \in \mathcal{J}$, and then introduce the topology (and smoothness) on $\operatorname{Im} f$ assuming all smooth manifolds $b\mathcal{J}$ to be connected components of the group $\operatorname{Im} f$. It is clear that on $\operatorname{Im} f$, we thus correctly introduce the structure of a smooth subgroup of the Lie group $\mathcal{H}$ with respect to which the mapping $\mathcal{G} \to \operatorname{Im} f$, $a \mapsto f(a)$, $a \in \mathcal{G}_0$, is smooth. Therefore, *the image* $\operatorname{Im} f$ *of any homomorphism* $f: \mathcal{G} \to \mathcal{H}$ *of Lie groups is a Lie subgroup of the Lie group* $\mathcal{H}$.

In this case, $(\operatorname{Im} f)_e = \mathcal{J}$, and *the Lie algebra of the subgroup* $\operatorname{Im} f$ *is therefore the subalgebra* $\operatorname{Im} \mathfrak{l}(f)$ *of the Lie algebra* $\mathfrak{h} = \mathfrak{l}\mathcal{H}$:

$$\mathfrak{l}(\operatorname{Im} f) = \operatorname{Im} \mathfrak{l}(f).$$

(In general, the subgroup $\operatorname{Im} f$ *is not* a subspace of the topological space $\mathcal{H}$.)

It is known that on the quotient space $\mathcal{G}/\mathcal{H}$ for any closed subgroup $\mathcal{H}$ of the Lie group $\mathcal{G}$, a smooth structure can be naturally introduced with respect to which the canonical mapping

$$\mathcal{G} \to \mathcal{G}/\mathcal{H}, \qquad a \mapsto a\mathcal{H}, \quad a \in \mathcal{G}, \tag{7}$$

is smooth. On the other hand, according to the general theory of groups, if the subgroup $\mathcal{H}$ is invariant, then the set $\mathcal{G}/\mathcal{H}$ is a group with respect to the operation

$$a\mathcal{H} \cdot b\mathcal{H} = ab\mathcal{H}, \quad a, b \in \mathcal{G}, \tag{8}$$

and mapping (7) is a homomorphism.

Exercise 7.3. Show that the following assertions hold:

1. With respect to multiplication (8), the smooth manifold $\mathcal{G}/\mathcal{H}$ is a Lie group (and mapping (7) is therefore a homomorphism of Lie groups).

2. The Lie algebra of the Lie group $\mathcal{G}/\mathcal{H}$ is the quotient algebra $\mathfrak{g}/\mathfrak{h}$:

$$\mathfrak{l}(\mathcal{G}/\mathcal{H}) = \mathfrak{g}/\mathfrak{h}, \quad \mathfrak{g} = \mathfrak{l}\mathcal{G}, \quad \mathfrak{h} = \mathfrak{l}\mathcal{H}.$$

[We recall from the algebra course that for any algebra $\mathcal{A}$ and any (two-sided) ideal $\mathcal{B}$ of it, the quotient space $\mathcal{A}/\mathcal{B}$ is an algebra with respect to the operation $(x + \mathcal{B})(y + \mathcal{B}) = xy + \mathcal{B}$, $x, y \in \mathcal{A}$. This algebra is called the *quotient algebra* of the algebra $\mathcal{A}$ by the ideal $\mathcal{B}$. In the case where $\mathcal{A}$ is a Lie algebra, the algebra $\mathcal{A}/\mathcal{B}$ is also a Lie algebra.]

The Lie group $\mathcal{G}/\mathcal{H}$ is called the *factor group* of the Lie group $\mathcal{G}$ by its invariant closed subgroup $\mathcal{H}$.

Exercise 7.4. Show that the canonical mapping

$$\mathcal{G}/\operatorname{Ker} f \to \operatorname{Im} f, \qquad a \operatorname{Ker} f \mapsto f(a),$$

is an isomorphism of Lie groups.

§5. Campbell–Hausdorff Theorem

As is known, the Lie operation $[\,,\,]$ in the Lie algebra $\mathfrak{l}\mathcal{G} = \mathsf{T}_e\mathcal{G}$ can be used to completely reconstruct the multiplication in a Lie group $\mathcal{G}$. We describe this reconstruction in more detail.

A *Lie monomial of degree n in X, Y* is an expression of the form

$$[\dots[[X_1, X_2], X_3], \dots, X_n],$$

where each X_i, $i = 1, \dots, n$, is either X or Y; a *Lie polynomial in X, Y* (over the field $\mathbb{Q}$) is a finite formal sum

$$\mathfrak{f} = \sum a_i \varphi_i,$$

where φ_i are Lie monomials and a_i are rational numbers. If all Lie monomials φ_i for which $a_i \neq 0$ have the same degree n, then the Lie polynomial $\mathfrak{f}$ is said to be *homogeneous,* and n is called its *degree.*

A *formal Lie series* is an infinite formal sum of the form

$$\mathfrak{f} = \mathfrak{f}_0 + \mathfrak{f}_1 + \dots + \mathfrak{f}_n + \dots, \tag{9}$$

where $\mathfrak{f}_n$ is a homogeneous Lie polynomial of degree n.

If X and Y are elements of a certain algebra $\mathfrak{g}$ (over an arbitrary field $\mathbb{K}$ of zero characteristic), then its *value* $\mathfrak{f}(X, Y)$ is defined for any Lie polynomial $\mathfrak{f}$ in X, Y and is an element of the algebra $\mathfrak{g}$. For series (9), this yields an infinite series

$$\mathfrak{f}_1(X, Y) + \dots + \mathfrak{f}_n(X, Y) + \cdots$$

in $\mathfrak{g}$. If this series converges (for $\mathbb{K} = \mathbb{R}$ or $\mathbb{C}$), then its sum is denoted by $\mathfrak{f}(X, Y)$ and is called the *value* of series (9) at the elements X, Y. Admitting a conventional inaccuracy, we also use the symbol $\mathfrak{f}(X, Y)$ to designate the series $\mathfrak{f}$.

Theorem 7.1. *There exists a formal Lie series*

$$\daleth(X, Y) = \daleth_1(X, Y) + \dots + \daleth_n(X, Y) + \dots, \tag{10}$$

such that for any Lie group $\mathcal{G}$ and for any elements $X, Y \in \mathfrak{l}\mathcal{G}$ in a certain normal neighborhood U_0 of zero of the Lie algebra $\mathfrak{l}\mathcal{G} = \mathfrak{g}$, the value $\daleth(X, Y)$ of series (10) *is defined and*

$$\exp X \cdot \exp Y = \exp \daleth(X, Y). \tag{11}$$

This theorem is known as the *Campbell–Hausdorff theorem.* We do not prove it here.

§6. Dynkin Polynomials

The initial terms of series (10) have the forms

$$\gimel_1(X,Y) = X + Y, \qquad \gimel_2(X,Y) = \frac{1}{2}[X,Y],$$
$$\gimel_3(X,Y) = \frac{1}{12}([[X,Y],Y] - [[X,Y],X]). \tag{12}$$

Using the same method used to find these terms, we can find other terms, but when the degree increases, the computation quickly becomes difficult. However, it is possible to give an explicit formula for $\gimel_n(X,Y)$. This formula is due to Dynkin and has the form

$$\gimel_n(X,Y) = \frac{1}{n}\sum_{k=1}^{n}\frac{(-1)^{k-1}}{k}\sum\frac{[X^{p_1}Y^{q_1}\dots X^{p_k}Y^{q_k}]}{p_1!q_1!\cdots p_k!q_k!}, \tag{13}$$

where the range of the internal summation is all nonnegative exponents $p_1, q_1, \dots, p_k, q_k$ for which

$$p_1 + q_1 > 0, \quad \dots, \quad p_k + q_k > 0,$$
$$(p_1 + q_1) + \dots + (p_k + q_k) = n,$$

and

$$[X^{p_1}Y^{q_1}\dots X^{p_k}Y^{q_k}] =$$
$$= [\dots\underbrace{[X,X],X,\dots,X]}_{p_1}\underbrace{Y],\dots,Y]}_{q_1},\dots,\underbrace{X],\dots,X]}_{p_k},\underbrace{Y],\dots,Y]}_{q_k}.$$

Remark 7.1. Assuming the variables X and Y do not commute, we consider a formal series $\ln(e^X e^Y)$, where

$$e^X = \sum_{n=0}^{\infty}\frac{X^n}{n!}, \qquad e^Y = \sum_{n=0}^{\infty}\frac{Y^n}{n!}, \qquad \ln Z = \sum_{n=1}^{\infty}\frac{(-1)^{n-1}}{n}(Z-1)^n.$$

Let

$$\ln(e^X e^Y) = \sum_{n=1}^{\infty} z_n(X,Y),$$

where $z_n(X,Y)$ are certain homogeneous polynomials of degree n in noncommutative variables X and Y. If $[z_n(X,Y)]$ is the Lie polynomial that is obtained from the polynomial $z_n(X,Y)$ via the substitutions

$$X^{p_1}Y^{q_1}\cdots X^{p_k}Y^{q_k} \implies [X^{p_1}Y^{q_1}\cdots X^{p_k}Y^{q_k}],$$

then

$$\gimel_n(X,Y) = \frac{1}{n}[z_n(X,Y)].$$

This example provides an easy mnemonic for the Dynkin formula (and gives a hint of its proof).

In general, formula (13) contains many noncollected similar terms (including zero terms). For $n = 3$ for example, this formula yields

$$\beth_3(X,Y) = \frac{1}{3}\left(\frac{[[X,Y],Y]}{1!2!} - \frac{1}{2}\left(\frac{[[Y,X],Y]}{0!1!1!1!} + \frac{[[Y,X],X]}{0!1!2!0!}\right.\right.$$
$$+\frac{[[X,Y],Y]}{1!0!0!2!} + \frac{[[X,Y],Y]}{1!1!0!1!} + \left.\frac{[[X,Y],X]}{1!1!1!0!}\right)$$
$$\left.+\frac{1}{3}\left(\frac{[[Y,X],Y]}{0!1!1!0!0!1!} + \frac{[[Y,X],X]}{0!1!1!0!1!0!} + \frac{[[X,Y],Y]}{1!0!0!1!0!1!} + \frac{[[X,Y],X]}{1!0!0!1!1!0!}\right)\right),$$

(we do not write zero terms), and the formula for $\beth_3(X,Y)$ in (12) is obtained from this only after collecting similar terms.

§7. Local Lie Groups

Exercise 7.5. Prove the following properties:

1. In any Lie algebra $\mathfrak{g}$, there exists a neighborhood U of zero such that $\beth(X,Y)$ is defined for all $X, Y \in U$.

2. The relations

$$\beth(X,\beth(Y,Z)) = \beth(\beth(X,Y),Z),$$
$$\beth(X,-X) = \beth(-X,X) = 0,$$
$$\beth(X,0) = \beth(0,X) = X$$

hold (assume that all elements in these formulas are defined). [*Hint*: Use Remark 7.1.]

The assertions in Exercise 7.5 mean that for $\beth(X,Y) \in U$, the formula

$$X \cdot Y = \beth(X,Y) \tag{14}$$

defines a multiplication on U that satisfies all meaningful group identities. (In this case, the element 0 is the unity, and the inverse element is defined by $X^{-1} = -X$.) A set U having such a structure is called a *local Lie group*. Exercise 7.5 yields a method for constructing a local Lie group $U = U(\mathfrak{g})$ for any Lie algebra $\mathfrak{g}$.

The construction of the algebra $\mathfrak{l}\mathcal{G}$ is literally valid for any local Lie group and gives us a Lie algebra that is called the *Lie algebra of this local group*.

Exercise 7.6. Show that the Lie algebra of the local Lie group $U(\mathfrak{g})$ is isomorphic to the initial Lie algebra $\mathfrak{g}$.

Lie himself first constructed a local Lie group for a given Lie algebra $\mathfrak{g}$ (using another method). Cartan later essentially proved that this local group is embeddable in the Lie group, i.e., is isomorphic (in a clear sense) to a local group that is a neighborhood of the identity of a certain Lie group (which therefore has the Lie algebra $\mathfrak{g}$). (Similarly to Theorem 7.1, we leave the Cartan theorem without proof.) The multiplication given by (14) is a real analytic function of the factors (i.e., the coordinates of the product in an arbitrary basis of the algebra $\mathfrak{g}$ are real analytic functions of the coordinates of the factors), and the Lie group $\mathcal{G}$ in the Cartan theorem is therefore a *real analytic group*, i.e., a manifold of class C^ω in which the multiplication $\mathcal{G} \times \mathcal{G} \to \mathcal{G}$ is also a mapping of class C^ω. At the same time, the construction of the Lie algebra $\mathfrak{l}\mathcal{G}$ obviously has sense for an arbitrary Lie group $\mathcal{G}$ of class C^r, $r \geq 2$. This easily implies that *any Lie group of class C^r, $r \geq 2$, is isomorphic (in the category of Lie groups of class C^r) to a real analytic Lie group.* Therefore, in Lie group theory, we can consider only real analytic groups without loss of generality.

§8. Bijectivity of the Lie Functor

Let U be a neighborhood of the identity of a Lie group $\mathcal{G}$. A smooth mapping $f: U \to \mathcal{H}$ of a neighborhood U into a Lie group $\mathcal{H}$ is called a *local homomorphism* if we have

$$f(ab) = f(a)f(b)$$

for any elements $a, b \in U$ with $ab \in U$. It is clear that the construction of the homomorphism $\mathfrak{l}(f)$ is meaningful for any local homomorphism $f: U \to \mathcal{H}$ (and yields the homomorphism $\mathfrak{l}(f): \mathfrak{l}\mathcal{G} \to \mathfrak{l}\mathcal{H}$ of Lie algebras).

Exercise 7.7. Show that the following assertions hold:
1. For $\exp X \in U$, we have

$$f(\exp X) = \exp(\varphi X), \quad X \in \mathfrak{l}\mathcal{G}, \tag{15}$$

where $\varphi = \mathfrak{l}(f)$ (compare with diagram (6)).
2. For any normal neighborhood U of the identity of the group $\mathcal{G}$ and for any Lie algebra homomorphism

$$\varphi: \mathfrak{l}\mathcal{G} \to \mathfrak{l}\mathcal{H},$$

formula (15) correctly defines a local homomorphism

$$f: U \to \mathcal{H}$$

for which $\mathfrak{l}(f) = \varphi$.

This means that the formula $f \mapsto \mathfrak{l}(f)$ *defines a bijective correspondence between local homomorphisms from $\mathcal{G}$ into $\mathcal{H}$* (more precisely, between their kernels at e) and Lie algebra homomorphisms $\mathfrak{l}\mathcal{G} \to \mathfrak{l}\mathcal{H}$. Moreover, formula (15) immediately implies that if $\mathfrak{l}(f) = \mathfrak{l}(g)$ for homomorphisms $f, g: \mathcal{G} \to \mathcal{H}$, then $f = g$ on any normal neighborhood U of the identity of the group $\mathcal{G}$. Because a connected Lie group is generated by each neighborhood of the identity, this relation holds on the whole component of the identity of the group $\mathcal{G}$. Therefore, *if a group $\mathcal{G}$ is connected , then $f = g$ iff $\mathfrak{l}(f) = \mathfrak{l}(g)$.*

Proposition 7.2. *If a group $\mathcal{G}$ is connected and simply connected, then for any Lie algebra homomorphism*

$$\varphi: \mathfrak{l}\mathcal{G} \to \mathfrak{l}\mathcal{H},$$

there exists a Lie group homomorphism

$$f: \mathcal{G} \to \mathcal{H}$$

such that $\mathfrak{l}(f) = \varphi$. This homomorphism is unique.

Therefore, if a group $\mathcal{G}$ is connected and simply connected, then the bijectivity of the correspondence $f \mapsto \mathfrak{l}(f)$ also holds for Lie group homomorphisms $f: \mathcal{G} \to \mathcal{H}$. This means that *on the category* $\mathrm{LIE}^{(0)}$ *of connected and simply connected Lie groups, the Lie functor is a bijection on morphisms.*

We omit the proof of Proposition 7.2.

Exercise 7.8. Prove that the homomorphism f in Proposition 7.2 is an isomorphism iff the homomorphism φ is an isomorphism. [*Hint*: Use the uniqueness of the homomorphism f.]

In particular, we see that the group $\operatorname{Aut} \mathcal{G}$ of automorphisms of an arbitrary connected and simply connected Lie group $\mathcal{G}$ *is canonically isomorphic to the group* $\operatorname{Aut} \mathfrak{g}$ *of automorphisms of its Lie algebra* $\mathfrak{g} = \mathfrak{l}\mathcal{G}$ and is therefore a Lie group.

Moreover, simply connected Lie groups $\mathcal{G}$ and $\mathcal{H}$ are isomorphic iff their Lie algebras $\mathfrak{l}\mathcal{G}$ and $\mathfrak{l}\mathcal{H}$ are isomorphic. Because of the Cartan theorem, *the Lie functor is also bijective on objects in the category* $\mathrm{LIE}^{(0)}$ (but only up to an isomorphism). On objects of the category LIE, the Lie functor is bijective up to a local isomorphism.

Chapter 8
Affine Fields and Related Topics

As Exercise 5.5 shows, Lie groups are a particular case of symmetric spaces. This gives us an idea to generalize the construction of the Lie algebra of a Lie group to symmetric spaces. This can be done, but instead of Lie algebras, we obtain more general algebraic objects, as should be expected.

§1. Affine Fields

First, let $\mathcal{X}$ be an arbitrary affine connection space.

Definition 8.1. A vector field X on the affine connection space $\mathcal{X}$ is said to be *affine* if the maximal flow $\{\varphi_t\}$ it generates consists of affine mappings, i.e., for any $t \in \mathbb{R}$, the mapping φ_t defined on a nonempty open subset $D_t \subset \mathcal{X}$ is an affine mapping $D_t \to \mathcal{X}$.

Proposition 8.1. *A vector field $X \in \mathfrak{a}\mathcal{X}$ is affine iff*

$$[X, \nabla_Y Z] = \nabla_Y [X, Z] + \nabla_{[X,Y]} Z \tag{1}$$

for any fields $Y, Z \in \mathfrak{a}\mathcal{X}$.

Proof. As we know, we have

$$[X, Y] = \lim_{t \to 0} \frac{\varphi_t^* Y - Y}{t}$$

for the commutator $[X, Y]$ of two fields $X, Y \in \mathfrak{a}\mathcal{X}$, where $\{\varphi_t\}$ is the flow generated by the field X. On the other hand, if the mapping φ_t is affine, then we have

$$\varphi_t^* \nabla_Y Z = \nabla_{Y_t} Z_t \tag{2}$$

for any fields $Y, Z \in \mathfrak{a}\mathcal{X}$, where $Y_t = \varphi_t^* Y$ and $Z_t = \varphi_t^* Z$ (see diagram (5) in Chap. 3); therefore,

$$\varphi_t^* \nabla_Y Z - \nabla_Y Z = \nabla_{Y_t} [Z_t - Z] + \nabla_{Y_t} Z - \nabla_Y Z.$$

If the field X is affine, we therefore have

$$\begin{aligned}[X, \nabla_Y Z] &= \lim_{t \to 0} \nabla_{Y_t} \left(\frac{Z_t - Z}{t} \right) + \lim_{t \to 0} \frac{\nabla_{Y_t} Z - \nabla_Y Z}{t} \\ &= \nabla_Y \lim_{t \to 0} \frac{\varphi_t^* Z - Z}{t} + \nabla_{\lim_{t \to 0} \frac{\varphi_t^* Y - Y}{t}} Z = \nabla_Y [X, Z] + \nabla_{[X,Y]} Z.\end{aligned}$$

To prove the converse statement, we set

$$\nabla_{Y(t)} Z = \varphi^*_{-t} \nabla_{Y_t} Z_t.$$

We then have

$$\begin{aligned}
&\frac{\nabla_{Y(t+s)} Z - \nabla_{Y(t)} Z}{s} \\
&= \varphi^*_{-t-s} \frac{\nabla_{\varphi^*_s Y_t}(\varphi^*_s Z_t) - \nabla_{Y_t} Z_t}{s} + \varphi^*_{-t} \frac{\varphi^*_{-s} \nabla_{Y_t} Z_t - \nabla_{Y_t} Z_t}{s} \\
&= \varphi^*_{-t-s} \left[\frac{\nabla_{\varphi^*_s Y_t}(\varphi^*_s Z_t - Z_t)}{s} + \nabla_{\frac{\varphi^*_s Y_t - Y_t}{s}} Z_t \right] \\
&\quad + \varphi^*_{-t} \frac{\varphi^*_{-s} \nabla_{Y_t} Z_t - \nabla_{Y_t} Z_t}{s}.
\end{aligned}$$

for any s and t. Therefore,

$$\lim_{s \to 0} \frac{\nabla_{Y(t+s)} Z - \nabla_{Y(t)} Z}{s} = \varphi^*_{-t}(\nabla_{Y_t}[X, Z_t] + \nabla_{[X,Y_t]} Z_t) - \varphi^*_{-t}[X, \nabla_{Y_t} Z_t].$$

If relation (1) holds (for any fields X, Y, and Z and therefore for the fields X, Y_t, and Z_t in particular), then

$$\lim_{s \to 0} \frac{\nabla_{Y(t+s)} Z - \nabla_{Y(t)} Z}{s} = 0.$$

For components of the field $\nabla_{Y(t)} Z$ considered as functions of t, this means that their derivatives are identically equal to zero. Therefore, these components and hence the fields $\nabla_{Y(t)} Z$ themselves are in fact independent of t. This proves that

$$\nabla_{Y(t)} Z = \nabla_{Y(0)} Z = \nabla_Y Z.$$

Because this is exactly (2) (in another notation), the mappings φ_t are affine. Therefore, the field X is also affine. □

In the operator form, condition (1) becomes

$$\nabla_{(\operatorname{ad} X)Y} = [\operatorname{ad} X, \nabla_Y], \tag{3}$$

which should hold for any field $Y \in \mathfrak{a}\mathcal{X}$.

Corollary 8.1. *The set* $\operatorname{aff} \mathcal{X}$ *of all affine fields* $X \in \mathfrak{a}\mathcal{X}$ *is a Lie subalgebra of* $\mathfrak{a}\mathcal{X}$.

Proof. If two fields X_1 and X_2 satisfy relation (1) (for any fields $Y, Z \in \mathfrak{a}\mathcal{X}$), then any linear combination of them obviously satisfies this relation. Moreover, according to the Jacobi identity, we have

$$[[X_1, X_2], \nabla_Y Z] = [X_1, [X_2, \nabla_Y Z]] - [X_2, [X_1, \nabla_Y Z]];$$

at the same time,

$$\begin{aligned}[X_1,[X_2,\nabla_Y Z]] &= [X_1,\nabla_Y[X_2,Z]] + [X_1,\nabla_{[X_2,Y]}Z] \\ &= \nabla_Y[X_1,[X_2,Z]] + \nabla_{[X_1,Y]}[X_2,Z] \\ &\quad + \nabla_{[X_2,Y]}[X_1,Z] + \nabla_{[X_1,[X_2,Y]]}Z\end{aligned}$$

and, similarly,

$$\begin{aligned}[X_2,[X_1,\nabla_Y Z]] &= [X_2,\nabla_Y[X_1,Z]] + [X_2,\nabla_{[X_1,Y]}Z] \\ &= \nabla_Y[X_2,[X_1,Z]] + \nabla_{[X_2,Y]}[X_1,Z] \\ &\quad + \nabla_{[X_1,Y]}[X_2,Z] + \nabla_{[X_2,[X_1,Y]]}Z.\end{aligned}$$

Therefore,

$$\begin{aligned}[[X_1,X_2],\nabla_Y Z] &= \nabla_Y([X_1,[X_2,Z]] - [X_2,[X_1,Z]]) \\ &\quad + \nabla_{[X_1,[X_2,Y]]-[X_2,[X_1,Y]]}Z \\ &= \nabla_Y[[X_1,X_2],Z] + \nabla_{[[X_1,X_2],Y]}Z.\end{aligned}$$

Therefore, the field $[X_1, X_2]$ also satisfies relation (1). This proves that $\mathfrak{aff}\,\mathcal{X}$ is a subalgebra. □

Exercise 8.1. Show that a vector field X on a Lie group $\mathcal{G}$ is affine with respect to the left, mean, or right Cartan connection ∇ on $\mathcal{G}$ iff for any left-invariant vector field Y, the field $[X,Y]$ is also left invariant.

[*Hint:* Let $Z_1,\dots,Z_n$ be a basis of the Lie algebra $\mathfrak{l}\mathcal{G}$, and let $Z = h^i Z_i$. Show that the fields X, Y, and Z satisfy condition (1) iff the fields X, Y, and Z_i, $1 \le i \le n$, satisfy it. On the other hand, if the field Z is left invariant and the connection ∇ is left, then condition (1) for the fields X, Y, and Z is reduced to the relation $\nabla_Y[X,Z] = 0$, which, as we know, holds for all $Y \in \mathfrak{a}\mathcal{G}$ iff $[X,Z] \in \mathfrak{l}\mathcal{G}$. Similarly, if the connection ∇ is right, then for $Y \in \mathfrak{l}\mathcal{G}$ and $[X,Y] = g^i Z_i$, condition (1) is reduced to the relation $Zg^i \cdot Z_i = 0$, which holds for all Z iff $g^i = \text{const}$.]

We note that *all three Cartan connections therefore have the same affine fields.* Compare this with Exercise 6.12.

§2. Dimension of the Lie Algebra of Affine Fields

Proposition 8.2. *If a manifold $\mathcal{X}$ is connected, then the Lie algebra* $\mathfrak{aff}\,\mathcal{X}$ *is finite dimensional, and its dimension does not exceed $n + n^2$, where $n = \dim\mathcal{X}$.*

Proof. The function $Xf \in \mathsf{F}\mathcal{X}$ is defined for any field $X \in \mathfrak{a}\mathcal{X}$ and any function $f \in \mathsf{F}\mathcal{X}$. Therefore, fixing a point $p_0 \in \mathcal{X}$ for what follows, we can set the number AXf in correspondence to each vector $A \in \mathbf{T}_{p_0}\mathcal{X}$. We thus obtain a certain (obviously linear) functional $AX\colon f \mapsto AXf$ on $\mathcal{X}$, i.e., an element of the dual linear space $(\mathsf{F}\mathcal{X})^*$. The correspondence $A \mapsto AX$ is therefore a linear mapping $\mathbf{T}_{p_0}\mathcal{X} \to (\mathbf{F}\mathcal{X})^*$, i.e., an element of the linear (infinite-dimensional!) space $\mathrm{Hom}(\mathbf{T}_{p_0}\mathcal{X}, (\mathbf{F}\mathcal{X})^*)$. Letting $l_{p_0}X$ denote this mapping, we obtain the mapping (again linear)

$$l_{p_0}\colon \mathfrak{a}\mathcal{X} \to \mathrm{Hom}(\mathbf{T}_{p_0}\mathcal{X}, (\mathbf{F}\mathcal{X})^*), \quad X \mapsto l_{p_0}X.$$

Let $(U, x^1, \dots, x^n)$ be an arbitrary chart of the manifold $\mathcal{X}$ centered at the point p_0, and let $X = X^i\partial/\partial x^n$ on U. Because the vectors

$$\left(\frac{\partial}{\partial x^1}\right)_{p_0}, \quad \dots, \quad \left(\frac{\partial}{\partial x^n}\right)_{p_0} \tag{4}$$

form a basis of the space $\mathbf{T}_{p_0}\mathcal{X}$, the mapping $l_{p_0}X\colon A \mapsto AX$ is uniquely given by its values

$$(l_{p_0}X)\left(\frac{\partial}{\partial x^i}\right)_{p_0}\colon \mathbf{F}\mathcal{X} \to \mathbb{R}, \quad i = 1, \dots, n,$$

on these vectors. By definition, we have

$$\begin{aligned}(l_{p_0}X)\left(\frac{\partial}{\partial x^i}\right)_{p_0} f &= \left(\frac{\partial}{\partial x^i}\right)_{p_0} Xf = \left(\frac{\partial}{\partial x^i}\right)_{p_0}\left(X^j\frac{\partial f}{\partial x^j}\right) \\ &= \left(\frac{\partial X^j}{\partial x^i}\right)_{p_0}\left(\frac{\partial f}{\partial x^j}\right)_{p_0} + X^j(p_0)\left(\frac{\partial^2 f}{\partial x^i \partial x^j}\right)_{p_0},\end{aligned}$$

for any function $f \in \mathbf{F}\mathcal{X}$, i.e.,

$$(l_{p_0}X)\left(\frac{\partial}{\partial x^i}\right)_{p_0} = \left(\frac{\partial X^j}{\partial x^i}\right)_{p_0}\left(\frac{\partial}{\partial x^j}\right)_{p_0} + X^j(p_0)\left(\frac{\partial^2}{x^i \partial x^j}\right)_{p_0}. \tag{5}$$

This proves that the image of the mapping l_{p_0} *belongs to the subspace of the linear space* $\mathrm{Hom}(\mathbf{T}_{p_0}\mathcal{X}, (\mathbf{F}\mathcal{X})^*)$ *consisting of mappings that transform the linear space* $\mathbf{T}_{p_0}\mathcal{X}$ *into the linear span of the functionals*

$$\left(\frac{\partial}{\partial x^i}\right)_{p_0}, \quad \left(\frac{\partial^2}{\partial x^i \partial x^j}\right)_{p_0}, \quad 1 \le i, j \le n \tag{6}$$

Because the dimension of the linear span of functionals (6) does not exceed their number $n + n^2$, to prove Proposition 8.2, it only remains to prove that the mapping l_{p_0} *is injective on the subalgebra* $\mathfrak{aff}\,\mathcal{X}$, i.e., that for $X \in \mathfrak{aff}\,\mathcal{X}$, the relation $l_{p_0}X = 0$ is possible only for $X = 0$.

But according to (5), the relation $l_{p_0}X = 0$ means that

$$X^i(p_0) = 0, \qquad \left(\frac{\partial X^j}{\partial x^i}\right)_{p_0} = 0$$

for all $i, j = 1, \dots, n$. In particular, if $l_{p_0}X = 0$, then $X_{p_0} = 0$, and therefore $\varphi_t(p_0) = p_0$ for any $t \in \mathbb{R}$ (for which the point $\varphi_t(p_0)$ is defined), where, as above, $\{\varphi_t\}$ is the flow generated by the field X. Therefore, on the linear space $\mathbf{T}_{p_0}\mathcal{X}$, the linear operators

$$(d\varphi_t)_{p_0}\colon \mathbf{T}_{p_0}\mathcal{X} \to \mathbf{T}_{p_0}\mathcal{X} \tag{7}$$

are defined and compose a one-parameter subgroup of the group $\operatorname{Aut}\mathbf{T}_{p_0}\mathcal{X}$ of all nonsingular linear operators $\mathbf{T}_{p_0}\mathcal{X} \to \mathbf{T}_{p_0}\mathcal{X}$. This one-parameter subgroup is completely determined by its initial vector

$$\left.\frac{d}{dt}(d\varphi_t)_{p_0}\right|_{t=0}, \tag{8}$$

which is a linear operator from $\operatorname{End}\mathbf{T}_{p_0}\mathcal{X}$. We stress that this operator is defined for any vector field $X \in \mathfrak{a}\mathcal{X}$ for which $X_{p_0} = 0$.

Exercise 8.2. Show that in basis (4) of the space $\mathbf{T}_{p_0}\mathcal{X}$, linear operator (8) has the matrix

$$\left\| \left(\frac{\partial X^i}{\partial x^j}\right)_{p_0} \right\|.$$

For $l_{p_0}X = 0$, this matrix equals zero, and the one-parameter subgroup (8) therefore consists of the identity mappings

$$(d\varphi_t)_{p_0} = \mathrm{id} \quad \text{for any } t \in \mathbb{R}.$$

Now, if $X \in \mathfrak{aff}\,\mathcal{X}$ and all the mappings φ_t are therefore affine, then, by Proposition 3.2, the relations $\varphi_t(p_0) = p_0$ and $(d\varphi_t)_{p_0} = \mathrm{id}$ imply $\varphi_t = \mathrm{id}$ (because the manifold $\mathcal{X}$ is connected) and therefore $X = 0$. Therefore, the mapping l_{p_0} is injective on $\mathfrak{aff}\,\mathcal{X}$, and hence $\dim \mathfrak{aff}\,\mathcal{X} \le n + n^2$. □

§3. Completeness of Affine Fields

The case where an affine connection space $\mathcal{X}$ is geodesically complete is of great importance.

Proposition 8.3. *On a geodesically complete affine connection space $\mathcal{X}$, each affine vector field X is complete.*

We need the following lemma.

Lemma 8.1. *Let $\{\varphi_t\}$ be the maximal flow on a manifold $\mathcal{X}$. If there exists a number $\varepsilon_0 > 0$ such that the point $\varphi_t(p)$ is defined for any point $p \in \mathcal{X}$ and any t with $|t| < \varepsilon_0$, then the flow $\{\varphi_t\}$ is complete (the point $\varphi_t(p)$, $p \in \mathcal{X}$, is defined for all $t \in \mathbb{R}$).*

Proof. Let $0 < \varepsilon < \varepsilon_0$. For any $t \in \mathbb{R}$, there exists a uniquely defined integer m such that $t = m\varepsilon + s$, where $0 \le s < \varepsilon$. We set

$$\psi_t = \underbrace{\varphi_\varepsilon \circ \cdots \circ \varphi_\varepsilon}_{m \text{ times}} \circ \varphi_s.$$

Exercise 8.3. Verify that the mappings $\psi_t: \mathcal{X} \to \mathcal{X}$, $t \in \mathbb{R}$, compose a flow on $\mathcal{X}$.

By definition, the flow $\{\psi_t\}$ is complete. In addition, because $\psi_t = \varphi_t$ for $|t| < \varepsilon$ and the flow $\{\varphi_t\}$ is maximal, we have $\psi_t = \varphi_t$ for all $t \in \mathbb{R}$. Therefore, the flow $\{\varphi_t\}$ is complete. □

Remark 8.1. This lemma directly implies that each vector field $X \in \mathfrak{a}\mathcal{X}$ that vanishes outside a compact set $C \subset \mathcal{X}$ is a complete field. In particular, *each vector field on a compact manifold is complete.*

Proof (of Proposition 8.3). Let $\mathcal{X}_0$ be an arbitrary component of the manifold $\mathcal{X}$, and let $p_0 \in \mathcal{X}_0$. We consider the flow $\{\varphi_t\}$ generated by the field X. By the theorem on the smooth dependence of solutions to differential equations on initial data, there exist $\varepsilon_0 > 0$ and a neighborhood V_0 of the point p_0 such that the point $\varphi_t(q)$ is defined for $q \in V_0$ and $|t| < \varepsilon_0$. Fixing this ε_0, we consider the set $W \subset \mathcal{X}$ of all points $p \in \mathcal{X}$ for which there exists a neighborhood V such that the point $\varphi_t(q)$ is defined for $q \in V$ and $|t| < \varepsilon_0$. This set is open and nonempty (contains the point p_0).

Lemma 8.2. *Each normal neighborhood of any point of the set W belongs to W.*

Assuming that this lemma is already proved, we consider an arbitrary point p of the closure $\overline{W}$ of the set W. By definition, each neighborhood U of the point p intersects the set W. In particular, this is true for the neighborhood U that is a normal neighborhood of each of its points (see Chap. 1). But by Lemma 8.2, such a neighborhood U is entirely contained in W. Therefore, $p \in W$, and the set W is hence closed.

Being an open-closed set that contains the point p_0, the set W contains the whole component $\mathcal{X}_0$ of the point p_0. This means that the point $\varphi_t(q)$ is defined for $|t| < \varepsilon_0$ and for each point $q \in \mathcal{X}_0$. Therefore, according to Lemma 8.1, the flow $\{\varphi_t\}$ on $\mathcal{X}_0$ is complete. Therefore, the flow $\{\varphi_t\}$ is complete on each component of the manifold $\mathcal{X}$. Therefore, it is also complete on the whole $\mathcal{X}$. □

It remains to prove Lemma 8.2.

Proof (of Lemma 8.2). Let U be a normal neighborhood of a point $p \in W$, and let V be its neighborhood such that the point $\varphi_t(q)$ is defined for $q \in V$ and $|t| < \varepsilon_0$. Further, let $p_1 \in U$ and $\gamma_0: I \to \mathcal{X}$ be a geodesic segment connecting the point p with the point p_1 in U, and let $A_0 = \dot{\gamma}_0(0)$. Finally, let $\pi: \mathbf{T}\mathcal{X} \to \mathcal{X}$ be the projection of the tangent bundle on the manifold $\mathcal{X}$. Because the affine connection space $\mathcal{X}$ is geodesically complete, for any vector $A \in \mathbf{T}\mathcal{X}$, a geodesic $\gamma_A: \mathbb{R} \to \mathcal{X}$ is defined for which $\gamma_A(0) = \pi A$ and $\dot{\gamma}_A(0) = A$. (In particular, $\gamma_0 = \gamma_{A_0}|_I$.) We define the mapping

$$h: \mathbf{T}\mathcal{X} \to \mathbf{T}\mathcal{X}$$

by

$$h(A) = \dot{\gamma}_A(1).$$

(The vector $h(A)$ is just the result of the parallel translation of the vector A along the geodesic segment $\gamma_A|_I$.) It is easy to see that this mapping is a diffeomorphism. In particular, we can therefore see that the set $h(\pi^{-1}V)$ is open in $\mathbf{T}\mathcal{X}$ and the set

$$V_1 = (\pi \circ h)(\pi^{-1}V)$$

is therefore open in $\mathcal{X}$. Moreover, because $h(A_0) = \dot{\gamma}_{A_0}(1) = \dot{\gamma}_0(1)$, we have $p_1 \in V_1$, i.e., V_1 is a neighborhood of the point p_1. Therefore, if we show that for $|t| < \varepsilon_0$, the point $\varphi_t(q_1)$ is defined for any point $q_1 \in V_1$, then Lemma 8.2 is proved. We now do this by indicating an explicit formula for the point $\varphi_t(q_1)$.

Because the field X is affine and all the mappings φ_t are therefore affine, the parallel translations commute with the differentials of these mappings (see diagram (4) in Chap. 3). In particular, this means that for any point $q \in V$, any t, $|t| < \varepsilon_0$, and any vector $A \in \mathbf{T}_q\mathcal{X}$, we have

$$(d\varphi_t)_{q_1} h(A) = h((d\varphi_t)_q A), \quad \text{where } q_1 = \gamma_A(1).$$

(We note that when q and A are chosen, it is possible to represent any point of the neighborhood V_1 in the form $q_1 = \gamma_A(1)$.)

But, it is clear that for the mapping

$$d\varphi_t \colon \mathbf{T}\mathcal{X} \to \mathbf{T}\mathcal{X}$$

(acting according to the formula $A \mapsto (d\varphi_t)_q A$, where $q = \pi A$), we have the commutative diagram

$$\begin{array}{ccc} \mathbf{T}\mathcal{X} & \xrightarrow{d\varphi_t} & \mathbf{T}\mathcal{X} \\ {\scriptstyle\pi}\downarrow & & \downarrow{\scriptstyle\pi} \\ \mathcal{X} & \xrightarrow{\varphi_t} & \mathcal{X} \end{array}.$$

Therefore,

$$\varphi_t(q_1) = \pi((d\varphi_t)_{q_1} h(A)) = (\pi \circ h)((d\varphi_t)_q A) \quad \text{for } |t| < \varepsilon_0,$$

which completes the proof. □

§4. Mappings of Left and Right Translation on a Symmetric Space

Now let $\mathcal{X}$ be a symmetric space. Interpreting $\mathcal{X}$ as a space with a multiplication (see Remark 5.1), we set two mappings

$$L_p, R_p: \mathcal{X} \to \mathcal{X}$$

in correspondence to each point $p \in \mathcal{X}$ that act according to the formulas

$$L_p q = pq, \qquad R_p q = qp, \quad q \in \mathcal{X},$$

i.e., by

$$L_p q = s_p q, \qquad R_p q = s_q p$$

(we omit the dot in the notation of the product $pq = s_p q$).

The mapping L_p is just the symmetry s_p and therefore satisfies the relations

$$L_p \circ L_p = \mathrm{id}, \qquad L_{qp} = L_q \circ L_p \circ L_q \tag{9}$$

(see properties 2 and 5 in Proposition 5.3). In addition (see Lemma 5.3),

$$(dL_p)_p = -\,\mathrm{id}\,. \tag{10}$$

Exercise 8.4. Prove that

$$R_{pq} = L_p \circ R_q \circ L_p, \quad p, q \in \mathcal{X}. \tag{11}$$

We recall that for any two manifolds $\mathcal{X}$ and $\mathcal{Y}$ and for any point $(p, q) \in \mathcal{X} \times \mathcal{Y}$, each vector $C \in \mathbf{T}_{(p,q)}(\mathcal{X}, \mathcal{Y})$ is uniquely represented in the form (A, B), where $A \in \mathbf{T}_p\mathcal{X}$, $B \in \mathbf{T}_q\mathcal{Y}$, and

$$A = (d\,\mathrm{pr}_1)_{(p,q)}C, \qquad B = (d\,\mathrm{pr}_2)_{(p,q)}C \tag{12}$$

In this case, for any mapping $\mu: \mathcal{X} \times \mathcal{Y} \to \mathcal{Z}$, we have

$$(d\mu)_{(p,q)}C = (dR_q)_p A + (dL_p)_q B \tag{13}$$

where R_q and L_p are the mappings $r \mapsto \mu(r, q)$ and $r \mapsto \mu(p, r)$, $r \in \mathcal{X}$.

Let $\mathcal{Y} = \mathcal{X}$, and let $\Delta: \mathcal{X} \to \mathcal{X} \times \mathcal{X}$ be the diagonal mapping $p \mapsto (p, p)$. Because $\mathrm{pr}_1 \circ \Delta = \mathrm{pr}_2 \circ \Delta = \mathrm{id}$, formula (12) directly implies

$$(d\Delta)_p A = (A, A)$$

for any vector $A \in \mathbf{T}_p\mathcal{X}$, and therefore

$$d(\mu \circ \Delta)_p A = (dR_p)_p A + (dL_p)_p A. \tag{14}$$

For $\mathcal{Z} = \mathcal{X}$ (when μ is the multiplication $\mathcal{X} \times \mathcal{X} \to \mathcal{X}$) and for $\mu \circ \Delta = \mathrm{id}$ (i.e., under the assumption that the multiplication μ is idempotent), this implies

$$A = (dR_p)_p A + (dL_p)_p A.$$

In particular, this formula holds in a symmetric space $\mathcal{X}$. But then $(dL_p)_p A = -A$ (see (10)), and we therefore have

$$(dR_p)_p A = 2A \tag{15}$$

for any vector $A \in \mathbf{T}_p\mathcal{X}$ in a symmetric space $\mathcal{X}$.

§5. Derivations on Manifolds with Multiplication

Definition 8.2. A vector field $X \in \mathfrak{a}\mathcal{X}$ is called a *derivation on* $\mathcal{X}$ if the flow $\{\varphi_t\}$ it generates consists of automorphisms (local ones in the general case where the field X need not be complete).

This definition is meaningful for any manifold $\mathcal{X}$ with multiplication. In the case where $\mathcal{X}$ is a symmetric space, Proposition 6.1 implies that *derivations on a symmetric space* $\mathcal{X}$ *are exactly affine vector fields on* $\mathcal{X}$ (*with respect to the canonical connection*). Therefore, by Proposition 8.3 (and by the assertion in Exercise 5.1), *each derivation on a symmetric space* $\mathcal{X}$ *is a complete field* (and for symmetric spaces, the field X in Definition 8.2 can therefore be assumed to be complete without loss of generality).

Exercise 8.5. As we know, vector fields on an arbitrary algebra $\mathcal{A}$ are naturally identified with mappings $\mathcal{A} \to \mathcal{A}$. In particular, we can assume that an arbitrary derivation $\mathcal{A} \to \mathcal{A}$ is a vector field $\mathcal{A}$. Show that *the flow* $\{\varphi_t\}$ *generated by this field consists of automorphisms of the algebra* $\mathcal{A}$.

This explains our terminology.

Proposition 8.4. *A vector field* $X \in \mathfrak{a}\mathcal{X}$ *is a derivation iff*

$$X_{pq} = (dL_p)_q X_q + (dR_q)_p X_p \tag{16}$$

for any points $p, q \in \mathcal{X}$.

Proof. Let $\{\varphi_t\}$ be the flow generated by the field X. We compute the vector $\dot{u}(0)$ tangent to the curve $u\colon t \mapsto \varphi_t(p)\varphi_t(q)$ at $t = 0$. By definition, for any function $f \in \mathbf{F}\mathcal{X}$, we have

$$\begin{aligned}
\dot{u}(0)f &= \lim_{t\to 0} \frac{f(\varphi_t(p)\varphi_t(q)) - f(pq)}{t} \\
&= \lim_{t\to 0} \frac{f(\varphi_t(p)\varphi_t(q)) - f(\varphi_t(p)q)}{t} + \lim_{t\to 0} \frac{f(\varphi_t(p)q) - f(pq)}{t} \\
&= \lim_{t\to 0} \frac{(f \circ L_{\varphi_t(p)})(\varphi_t(q)) - (f \circ L_{\varphi_t(p)})(q)}{t} \\
&\quad + \lim_{t\to 0} \frac{(f \circ R_q)(\varphi_t(p)) - (f \circ R_q)(p)}{t} \\
&= X_q(f \circ L_p) + X_p(f \circ R_q) = ((dL_p)_q X_q) f + ((dR_q)_p X_p) f,
\end{aligned}$$

and therefore

$$\dot{u}(0) = (dL_p)_q X_q + (dR_q)_p X_p.$$

Therefore, condition (16) means that the curve u has the same tangent vector X_{pq} for $t = 0$ as the trajectory $v\colon t \mapsto \varphi_t(pq)$ of the field X passing through the point pq for $t = 0$. Therefore, if the flow $\{\varphi_t\}$ consists of automorphisms (and therefore $u = v$), then condition (16) holds.

Conversely, let condition (16) hold. For any $s \in \mathbb{R}$, we consider the curve

$$u_s: t \mapsto u(t+s) = \varphi_{t+s}(p)\varphi_{t+s}(q) = \varphi_t(\varphi_s(p))\varphi_t(\varphi_s.(q)).$$

This is the curve u constructed for the points $\varphi_s(p)$ and $\varphi_s(q)$. Therefore, by what was proved, the vector $\dot{u}_s(0)$ tangent to this curve for $t = 0$ coincides with the tangent vector of the trajectory

$$t \mapsto \varphi_t(\varphi_s(p)\varphi_s(q)) = \varphi_t(u(s))$$

of the field X passing through the point $u(s)$ for $t = 0$, i.e., with the vector $X_{u(s)}$. Because the vector $\dot{u}_s(0)$ is obviously a vector $\dot{u}(s)$ tangent to the curve u for $t = s$, this proves that

$$\dot{u}(s) = X_{u(s)},$$

i.e., the curve u is a trajectory v of the field X passing through the point pq for $t = 0$. Therefore, $u(t) = v(t)$ for all t, and the flow $\{\varphi_t\}$ hence consists of automorphisms. □

We note that the assertion in Proposition 8.4 is meaningful (and holds) for *any* smooth manifold $\mathcal{X}$ on which a multiplication is given (for example, for Lie algebras or Lie groups).

Exercise 8.6. Prove that the following assertions hold:

1. In the case where $\mathcal{X}$ is an algebra $\mathcal{A}$, condition (16) means that a vector field X (interpreted as a mapping $\mathcal{A} \to \mathcal{A}$) is a derivation.

2. In the case where $\mathcal{X}$ is a Lie group $\mathcal{G}$, any field X satisfying condition (16) is an affine field with respect to the Cartan connections in Chap. 6. [*Hint*: Verify that for any left-invariant field Y, the field $[X, Y]$ is also left invariant; see Exercise 8.1.]

§6. Lie Algebra of Derivations

Comparison of the definitions immediately shows that for any field $X \in \mathfrak{a}\mathcal{X}$, the field $X^{\mathrm{I}} \in \mathfrak{a}(\mathcal{X} \times \mathcal{X})$ (see Chap. 6) is given by $(p, q) \mapsto (X_p, 0)$ and the field X^{II} is given by $(p, q) \mapsto (0, X_q)$. Therefore, for $A = X_p$ and $B = X_q$, relation (13) yields

$$(d\mu)_{(p,q)}(X^{\mathrm{I}} + X^{\mathrm{II}})_{(p,q)} = (dR_q)_p X_p + (dL_p)_q X_q.$$

Condition (16) is therefore equivalent to the relation

$$(d\mu)_{(p,q)}(X^{\mathrm{I}} + X^{\mathrm{II}})_{(p,q)} = X_{pq}, \quad pq = \mu(p, q),$$

which means that the fields $X^{\mathrm{I}} + X^{\mathrm{II}}$ and X are μ-related by definition.

Therefore, we see that a vector field X on a manifold $\mathcal{X}$ with multiplication $\mu: \mathcal{X} \times \mathcal{X} \to \mathcal{X}$ *is a derivation iff it is μ-related to the field $X^{\mathrm{I}} + X^{\mathrm{II}}$ on $\mathcal{X} \times \mathcal{X}$.*

But it is easy to verify that we have

$$[X^{\mathrm{I}}, Y^{\mathrm{I}}] = [X, Y]^{\mathrm{I}}, \qquad [X^{\mathrm{I}}, Y^{\mathrm{II}}] = 0, \qquad [X^{\mathrm{II}}, Y^{\mathrm{II}}] = [X, Y]^{\mathrm{II}}$$

for any two fields X and Y on $\mathcal{X}$, and therefore

$$[X^{\mathrm{I}} + X^{\mathrm{II}}, Y^{\mathrm{I}} + Y^{\mathrm{II}}] = [X, Y]^{\mathrm{I}} + [X, Y]^{\mathrm{II}}.$$

Therefore, if two fields X and Y are μ-related to the respective fields $X^{\mathrm{I}} + X^{\mathrm{II}}$ and $Y^{\mathrm{I}} + Y^{\mathrm{II}}$ (are derivations), then the field $[X, Y]$ is μ-related to the field $[X, Y]^{\mathrm{I}} + [X, Y]^{\mathrm{II}}$ (is also a derivation). This means that *the set $\mathfrak{d}\mathcal{X}$ of all derivations on $\mathcal{X}$ is a subalgebra of the Lie algebra $\mathfrak{a}\mathcal{X}$* (and is therefore a Lie algebra). For a symmetric space $\mathcal{X}$, this is certainly only another statement of Corollary 8.1, which is thus once more proved in this case.

§7. Involutive Automorphism of the Derivation Algebra of a Symmetric Space

In what follows, we choose and fix a certain point p_0 in a symmetric space $\mathcal{X}$; we set the field σX defined by

$$(\sigma X)_p = (dL_{p_0})_{p_0 p} X_{p_0 p}, \quad p \in \mathcal{X}, \tag{17}$$

in correspondence to each field $X \in \mathfrak{a}\mathcal{X}$. This field is obviously smooth also in the case where X is a derivation and satisfies the relation

$$\begin{aligned}
&(dR_q)_p(\sigma X)_p + (dL_p)_q(\sigma X)_q \\
&\quad = d(R_q \circ L_{p_0})_{p_0 p} X_{p_0 p} + d(L_p \circ L_{p_0})_{p_0 q} X_{p_0 q} \\
&\quad = d(L_{p_0} \circ R_{p_0 q})_{p_0 p} X_{p_0 p} + d(L_{p_0} \circ L_{p_0 p})_{p_0 q} X_{p_0 q} \\
&\quad = (dL_{p_0})_{(p_0 p)(p_0 q)} [(dR_{p_0 q})_{p_0 p} X_{p_0 p} + (dL_{p_0 p})_{p_0 q} X_{p_0 q}] \\
&\quad = (dL_{p_0})_{(p_0 p)(p_0 q)} X_{(p_0 p)(p_0 q)} = (dL_{p_0})_{p_0 (pq)} X_{p_0 (pq)} = (\sigma X)_{pq},
\end{aligned}$$

i.e., it is also a derivation. This means that the correspondence $\sigma \colon X \to \sigma X$ assigns a linear (and obviously) involutive mapping

$$\sigma \colon \mathfrak{d}\mathcal{X} \to \mathfrak{d}\mathcal{X} \tag{18}$$

of the Lie algebra $\mathfrak{d}\mathcal{X}$ onto itself.

Because the mapping σ is involutive, formula (17) exactly means that the field σX is L_{p_0}-related to the field X. Therefore, for any fields $X, Y \in \mathfrak{a}\mathcal{X}$, we have

$$[\sigma X, \sigma Y] = \sigma[X, Y],$$

which means that mapping (18) *is an automorphism of the Lie algebra $\mathfrak{d}\mathcal{X}$.*

§8. Symmetric Algebras and Lie Ternaries

We consider the general case.

Definition 8.3. A *symmetric Lie algebra* is a pair $(\mathfrak{g}, \sigma)$ consisting of a Lie algebra $\mathfrak{g}$ and an involutive automorphism

$$\sigma\colon \mathfrak{g} \to \mathfrak{g}.$$

The automorphism σ is called the *structural automorphism* of a symmetric Lie algebra. As a rule, it is presupposed everywhere in the notation. For example, instead of $(\mathfrak{g}, \sigma)$, $\mathfrak{g}$ is usually written.

According to the general theorems in linear algebra, each symmetric Lie algebra $\mathfrak{g}$ decomposes into the direct sum

$$\mathfrak{g} = \mathfrak{g}^{(+)} \oplus \mathfrak{g}^{(-)}$$

of the eigenspace $\mathfrak{g}^{(+)}$ of the automorphism σ corresponding to the eigenvalue $+1$ and the eigenspace $\mathfrak{g}^{(-)}$ corresponding to the eigenvalue -1.

The eigenspace $\mathfrak{g}^{(+)}$ is called the *cosocle* of the symmetric Lie algebra $\mathfrak{g}$ and is also denoted by $\mathfrak{h}$ (or $\mathfrak{h}(\mathfrak{g})$), and the eigenspace $\mathfrak{g}^{(-)}$ is called its *socle* and is denoted by $\mathfrak{s}$ (or $\mathfrak{s}(\mathfrak{g})$).

Let $X, Y \in \mathfrak{g}$. Obviously, we have

$$\begin{aligned} &[X,Y] \in \mathfrak{g}^{(+)} \quad \text{if } X, Y \in \mathfrak{g}^{(+)} \text{ or } X, Y \in \mathfrak{g}^{(-)}, \\ &[X,Y] \in \mathfrak{g}^{(-)} \quad \text{if } X \in \mathfrak{g}^{(+)}, Y \in \mathfrak{g}^{(-)} \text{ or } X \in \mathfrak{g}^{(-)}, Y \in \mathfrak{g}^{(+)}, \end{aligned}$$

i.e., in another notation,

$$[\mathfrak{h}, \mathfrak{h}] \subset \mathfrak{h}, \qquad [\mathfrak{s}, \mathfrak{s}] \subset \mathfrak{h}, \qquad [\mathfrak{h}, \mathfrak{s}] \subset \mathfrak{s}. \tag{19}$$

Let a Lie algebra $\mathfrak{g}$ be decomposed into the direct sum of two subspaces satisfying relations (19),

$$\mathfrak{g} = \mathfrak{h} \oplus \mathfrak{s}. \tag{20}$$

We define the linear involutive mapping $\sigma\colon \mathfrak{g} \to \mathfrak{g}$ by setting

$$\sigma X = \begin{cases} X & \text{if } X \in \mathfrak{h}, \\ -X & \text{if } X \in \mathfrak{s}. \end{cases}$$

Exercise 8.7. Show that the pair $(\mathfrak{g}, \sigma)$ is a symmetric Lie algebra with the cosocle $\mathfrak{h}$ and the socle $\mathfrak{s}$.

Therefore, symmetric Lie algebras $\mathfrak{g}$ *are exactly Lie algebras in which decomposition* (20) *satisfying* (19) *is given.* This viewpoint is often more convenient than the initial one. The first relation in (19) means that the cosocle $\mathfrak{h}$ of a symmetric Lie algebra $\mathfrak{g}$ *is a subalgebra.* As for the socle $\mathfrak{s}$, relations (19) directly imply that for any elements $X, Y, Z \in \mathfrak{s}$, the element

$$[X,Y,Z]=[[X,Y],Z] \tag{21}$$

also belongs to $\mathfrak{s}$, i.e., the linear space $\mathfrak{s}$ is closed with respect to the ternary operation

$$X,Y,Z \mapsto [X,Y,Z].$$

It is clear that

$$[X,X,X]=0 \tag{22}$$

$$[X,Y,Z]+[Y,Z,X]+[Z,X,Y]=0 \tag{23}$$

for any elements $X,Y,Z \in \mathfrak{s}$.

Exercise 8.8. Show that

$$[X,Y,[U,V,W]]=[[X,Y,U],V,W]+[U,[X,Y,V],W]+[U,V,[X,Y,W]]. \tag{24}$$

Definition 8.4. A linear space $\mathfrak{s}$ in which a trilinear ternary operation

$$X,Y,Z \mapsto [X,Y,Z] \tag{25}$$

satisfying identities (22), (23), and (24) is given is called a *Lie ternary* (or a *triple Lie system*; the latter name, is more widespread but is unfortunately long and nonexpressive).

Therefore, by this definition, the socle of any symmetric Lie algebra $\mathfrak{g}$ *is a Lie ternary* (with respect to operation (21)). In fact, this yields all Lie ternaries.

Proposition 8.5. *For any Lie ternary $\mathfrak{s}$, there exists a symmetric Lie algebra $\mathfrak{g}$ with the socle $\mathfrak{s}$.*

Proof. Let $\operatorname{ad}(X,Y)$ be the linear operator $Z \mapsto [X,Y,Z]$, $X,Y,Z \in \mathfrak{s}$, and let $\operatorname{ad}\mathfrak{s}$ be the subspace of the space of all linear operators $\mathfrak{s} \to \mathfrak{s}$ generated by all operators of the form $\operatorname{ad}(X,Y)$, $X,Y \in \mathfrak{s}$. Further, let $\mathfrak{g}=\operatorname{ad}\mathfrak{s}\oplus\mathfrak{s}$. In $\mathfrak{g}$, we define the bilinear operation $[\,,\,]$ and the linear operator $\sigma\colon \mathfrak{g}\to\mathfrak{g}$ by

$$[A,B]=\begin{cases} AB-BA & \text{if } A,B\in\operatorname{ad}\mathfrak{s},\\ AX & \text{if } A\in\operatorname{ad}\mathfrak{s},\ B=X,\ X\in\mathfrak{s},\\ -BX, & \text{if } A=X,\ X\in\mathfrak{s},\ B\in\operatorname{ad}\mathfrak{s},\\ \operatorname{ad}(X,Y) & \text{if } A=X,\ B=Y,\ X,Y\in\mathfrak{s},\end{cases}$$

and

$$\sigma A=\begin{cases} A & \text{if } A\in\operatorname{ad}\mathfrak{s},\\ -X & \text{if } A=X,\ X\in\mathfrak{s}.\end{cases}$$

Exercise 8.9. Verify that $\mathfrak{g}$ is a Lie algebra and the mapping σ is its involutive automorphism.

This proves Proposition 8.4. □

Obviously, the dimension of the Lie algebra $\mathfrak{g}$ does not exceed n^2+n, where n is the dimension of the Lie ternary $\mathfrak{s}$.

Remark 8.2. For any Lie algebra $\mathfrak{g}$, formula (21) defines a ternary operation on $\mathfrak{g}$ with respect to which the algebra $\mathfrak{g}$ is a Lie ternary. However, it is more convenient to define the structure of a Lie ternary on $\mathfrak{g}$ by

$$[X,Y,Z] = \frac{1}{4}[[X,Y],Z], \quad X,Y,Z \in \mathfrak{g}. \tag{26}$$

In what follows, we proceed in this way.

§9. Lie Ternary of a Symmetric Space

By construction, for any symmetric space $\mathcal{X}$, the Lie algebra $\mathfrak{d}\mathcal{X}$ of its derivations is a symmetric Lie algebra with structural automorphism (18). Its socle is denoted by $\mathfrak{s}\mathcal{X}$ and is called the *Lie ternary* of the symmetric space $\mathcal{X}$. Vector fields X from $\mathfrak{s}\mathcal{X}$ are called *essential derivations* of the symmetric space $\mathcal{X}$.

We note that the cosocle $\mathfrak{h}(\mathfrak{d}\mathcal{X})$ of the Lie algebra $\mathfrak{d}\mathcal{X}$ consists of derivations X for which

$$X_{p_0} = 0.$$

Exercise 8.10. Prove that for derivations $X,Y,Z \in \mathfrak{s}\mathcal{X}$, we have

$$[X,Y,Z] = -\frac{1}{4}(X \cdot (Y \cdot Z) - Y \cdot (X \cdot Z))$$

and therefore (see Exercise 6.1)

$$R(X,Y)Z = -[X,Y,Z], \quad X,Y,Z \in \mathfrak{s}\mathcal{X}, \tag{27}$$

where R is the curvature tensor of the canonical connection on $\mathcal{X}$.

Remark 8.3. In each affine connection space, the formula

$$[X,Y,Z] = R(X,Y)Z \tag{28}$$

defines a ternary operation that satisfies conditions (22) and (23). As for condition (24), it has the following form for operation (28):

$$\begin{aligned} R(X,Y)R(U,V)W = R(R(X,Y)U,V)W \\ + R(U,R(X,Y)V)W + R(U,V)R(X,Y)W, \end{aligned}$$

which is equivalent to identity (26) in Chap. 4. This means that an affine connection space *is semisymmetrical iff it is a Lie ternary with respect to operation* (28) (and the condition $T = 0$ holds).

Chapter 9
Cartan Theorem

§1. Functor $\mathfrak{s}$

The Lie ternary $\mathfrak{s}\mathcal{X}$ constructed in the previous chapter depends on the choice of the point $p_0 \in \mathcal{X}$, i.e., it is a function of the pair $(\mathcal{X}, p_0)$. Such pairs are called *punctured symmetric spaces.* A *morphism* $f: (\mathcal{X}, p_0) \to (\mathcal{Y}, q_0)$ of punctured spaces is a morphism $f: \mathcal{X} \to \mathcal{Y}$ such that $f(p_0) = q_0$. It is clear that all punctured symmetric spaces and their morphisms form a category.

Proposition 9.1. *If $f: (\mathcal{X}, p_0) \to (\mathcal{Y}, q_0)$ is an arbitrary morphism of punctured symmetric spaces, then for any essential derivation $X \in \mathfrak{s}\mathcal{X}$, there exists a unique essential derivation $Y \in \mathfrak{s}\mathcal{Y}$ that is f-related to X.*

Preparatorily, we prove two lemmas.

Lemma 9.1. *For any essential derivation $X \in \mathfrak{a}\mathcal{X}$ and for any point $p \in \mathcal{X}$, we have*

$$X_p = \frac{1}{2}(dR_{p_0 p})_{p_0} X_{p_0}. \tag{1}$$

Proof. The condition $\sigma X = -X$ characterizing essential derivations means that

$$X_{p_0 p} = -(dL_{p_0})_p X_p \tag{2}$$

for any point $p \in \mathcal{X}$. Therefore,

$$\begin{aligned} X_p &= -(dL_{p_0})_{p_0 p} X_{p_0 p} \\ &= -(dL_{p_0})_{p_0 p}((dR_p)_{p_0} X_{p_0} + (dL_{p_0})_p X_p) \\ &= -d(L_{p_0} \circ R_p)_{p_0} X_{p_0} - X_p. \end{aligned}$$

On the other hand, because $(dL_{p_0})_{p_0} X_{p_0} = -X_{p_0}$ and $L_{p_0} \circ R_p = R_{p_0 p} \circ L_{p_0}$ (see (10) and (11) in Chap. 6), we have

$$d(L_{p_0} \circ R_p)_{p_0} X_{p_0} = -(dR_{p_0 p})_{p_0} X_{p_0}.$$

Therefore, $X_p = (dR_{p_0 p})_{p_0} X_{p_0} - X_p$, which is equivalent to (1). $\square$

Lemma 9.2. *For any vector $A \in \mathbf{T}_{p_0}\mathcal{X}$, the formula*

$$X_p = \frac{1}{2}(dR_{p_0 p})_{p_0} A, \quad p \in \mathcal{X}, \tag{3}$$

defines an essential derivation X on $\mathcal{X}$ for which $X_{p_0} = A$.

Proof. It is clear that the field X given by (3) is smooth. Moreover, formula (15) in Chap. 8 (and the relation $p_0p_0 = p_0$) directly implies $X_{p_0} = A$, and the formula $L_{p_0} \circ R_{p_0p} = R_p \circ L_{p_0}$ (see (11) and (10) in Chap. 8) implies

$$(dL_{p_0})_p X_p = \frac{1}{2} d(R_p \circ L_{p_0})_{p_0} A = -\frac{1}{2}(dR_p)_{p_0} A = -X_{p_0 p},$$

i.e., $\sigma X = -X$. Therefore, we need only prove that the field X is a derivation. For this, we consider the diagram

$$\begin{array}{ccccc} \mathcal{X}\times\mathcal{X}\times\mathcal{X} & \xrightarrow{\Delta\times\mathrm{id}\times\mathrm{id}} & \mathcal{X}\times\mathcal{X}\times\mathcal{X}\times\mathcal{X} & \xrightarrow{\mathrm{id}\times T\times\mathrm{id}} & \mathcal{X}\times\mathcal{X}\times\mathcal{X}\times\mathcal{X} \\ \downarrow{\scriptstyle \mathrm{id}\times\mu} & & & & \downarrow{\scriptstyle \mu\times\mu} \\ \mathcal{X}\times\mathcal{X} & \xrightarrow{\mu} & \mathcal{X} & \xleftarrow{\mu} & \mathcal{X}\times\mathcal{X} \end{array},$$

where T is the coordinate transposition mapping $(p,q) \mapsto (q,p)$ and μ is the multiplication $(p,q) \mapsto pq$.

When we move clockwise from the upper-left corner of this diagram to the center $\mathcal{X}$ in the lower row, the point $(p,q,r) \in \mathcal{X}\times\mathcal{X}\times\mathcal{X}$ passes into the point $(pq)(pr) \in \mathcal{X}$, and when we move counterclockwise, it passes into the point $p(qr) \in \mathcal{X}$. Therefore, for each symmetric space $\mathcal{X}$, this diagram is commutative. Therefore, the corresponding diagram of tangent spaces and differentials is also commutative, i.e., for any vectors $A \in \mathbf{T}_p\mathcal{X}$, $B \in \mathbf{T}_q\mathcal{X}$, and $C \in \mathbf{T}_r\mathcal{X}$, we have

$$\begin{aligned}(dR_{qr})_pA+(dL_p)_{qr}((dR_r)_qB+(dL_q)_rC)&=(dR_{pr})_{pq}((dR_q)_pA+(dL_p)_qB)\\ &\quad+(dL_{pq})_{pr}((dR_r)_pA+(dL_p)_rC)\end{aligned}$$

(see general formula (13) in Chap. 8). For $B = C = 0$, this yields the relation

$$(dR_{qr})_pA = (dR_{pr})_{pq}(dR_q)_pA + (dL_{pq})_{pr}(dR_r)_pA.$$

(For $A = C = 0$ and $A = B = 0$, we obtain consequences of relation (11) and the second relation in (9) in Chap. 8.)

For p, q, and r respectively equal to p_0, p_0p, and p_0q, this implies

$$\frac{1}{2}(dR_{(p_0p)(p_0q)})_{p_0}A = (dR_q)_pX_p + (dL_p)_qX_q$$

(by the relations $p_0(p_0q) = q$ and $p_0(p_0p) = p$). Because $(p_0p)(p_0q) = p_0(pq)$ and therefore

$$\frac{1}{2}(dR_{(p_0p)(p_0q)})_{p_0}A = \frac{1}{2}(dR_{p_0(pq)})_{p_0}A = X_{pq},$$

this proves that X is a derivation. □

Proof (of Proposition 9.1). If the derivation Y exists, then $Y_{q_0} = (df)_{p_0} X_{p_0}$, and according to (1), we therefore have

$$Y_q = \frac{1}{2}(dR_{q_0 q})_{q_0}(df)_{p_0} X_{p_0} \tag{4}$$

for any point $q \in \mathcal{Y}$. This proves the uniqueness of the field Y.

To prove its existence, we define the field Y on $\mathcal{Y}$ by (4). According to Lemma 9.2, this field belongs to $\mathfrak{s}\mathcal{Y}$ and satisfies the relation $Y_{q_0} = (df)_{p_0} X_{p_0}$. On the other hand, because f is a morphism, $f \circ R_{p_0 p} = R_{q_0 q} \circ f$, where $q = f(p)$, and therefore

$$(df)_p \circ (dR_{p_0 p})_{p_0} = (dR_{q_0 q})_{q_0} \circ (df)_{p_0}.$$

Therefore, $Y_{f(p)} = (1/2)(df)_p (dR_{p_0 p})_{p_0} X_{p_0} = (df)_p X_p$, and the field Y is hence p-related to the field X. □

Setting $Y = \mathfrak{s}(f)X$, we therefore obtain a certain mapping

$$\mathfrak{s}(f)\colon \mathfrak{s}\mathcal{X} \to \mathfrak{s}\mathcal{Y},$$

which is obviously a homomorphism of Lie ternaries. Of course, this mapping has the functorial property, i.e., the correspondence $\mathcal{X} \mapsto \mathfrak{s}\mathcal{X}$, $f \mapsto \mathfrak{s}(f)$, is a functor from the category of punctured symmetric spaces into the category of Lie ternaries.

§2. Comparison of the Functor $\mathfrak{s}$ with the Lie Functor $\mathfrak{l}$

In the case where a symmetric space $\mathcal{X}$ is a Lie group $\mathcal{G}$ (see Exercise 5.5) and $p_0 = e$, where e is the identity of the group $\mathcal{G}$ as usual, the operator R_q is expressed by the formula

$$R_q = \mu \circ (\mathrm{id} \times L^{\mu}_{q'}) \circ \Delta,$$

where μ is the multiplication in the group $\mathcal{G}$, $q' = q^{-1}$, and $L^{\mu}_{q'}$ is the left translation by q' with respect to the multiplication μ. Therefore, by general formulas (13) and (14) in Chap. 8, we have

$$(dR_q)_p A = (dR^{\mu}_{q'p})_p A + (dL^{\mu}_{pq'})_p A$$

for any vector $A \in \mathbf{T}_p\mathcal{G}$, where $q' = q^{-1}$ and $R^{\mu}_{q'p}$ is the right translation by $q'p$ with respect to the multiplication μ. In particular, for $p = e$,

$$(dR_q)_e A = (dR^{\mu}_{q'})_e A + (dL^{\mu}_{q'})_e A, \quad A \in \mathbf{T}_e\mathcal{G}.$$

We set $q = e \cdot p = p^{-1}$, and formula (3) in the case considered then becomes

$$X_p = \frac{(dR^{\mu}_p)_e A + (dL^{\mu}_p)_e A}{2},$$

i.e.,

$$X = \frac{X^L + X^R}{2}, \tag{3'}$$

where X^L is a left-invariant vector field and X^R is a right-invariant vector field on $\mathcal{G}$ for which $X_e^L = X_e^R = X_e$. Therefore, formula (3′) yields the general form of the fields $X \in \mathfrak{s}\mathcal{G}$.

Exercise 9.1. Prove that for any Lie group $\mathcal{G}$, the correspondence $X \mapsto X^L$ defines an isomorphism of the Lie ternary $\mathfrak{s}\mathcal{G}$ onto the Lie algebra $\mathfrak{l}\mathcal{G}$, which is considered as a Lie ternary (with the operation given by formula (26) in Chap. 8).

In this sense, the functor $\mathfrak{s}: \mathcal{X} \mapsto \mathfrak{s}\mathcal{X}$ is a generalization of the functor $\mathfrak{l}: \mathcal{G} \mapsto \mathfrak{l}\mathcal{G}$.

§3. Properties of the Functor $\mathfrak{s}$

The properties of the functor $\mathfrak{s}$ are similar to the properties of the functor $\mathfrak{l}$. For example, Lemmas 9.1 and 9.2 directly imply that as for the functor $\mathfrak{l}$, *the correspondence $X \mapsto X_e$ defines an isomorphism of the linear space $\mathfrak{s}\mathcal{X}$ and the tangent space $\mathbf{T}_{p_0}\mathcal{X}$* (the inverse isomorphism $A \mapsto X$ is given by (3)). As a rule, we use this isomorphism to identify $\mathfrak{s}\mathcal{X}$ and $\mathbf{T}_{p_0}\mathcal{X}$.

Exercise 9.2. Prove that for any morphism $f: \mathcal{X} \to \mathcal{Y}$ of symmetric spaces, the homomorphism $\mathfrak{s}(f): \mathfrak{s}\mathcal{X} \to \mathfrak{s}\mathcal{Y}$ of Lie ternaries is just the differential $(df)_{p_0}: \mathbf{T}_{p_0}\mathcal{X} \to \mathbf{T}_{q_0}\mathcal{Y}$ of the morphism f at the point p_0 because of this identification.

The bijectivity properties of the functor $\mathfrak{l}$ (see Chap. 7) are also preserved for the functor $\mathfrak{s}$.

Theorem 9.1. *For any morphism $\varphi: \mathfrak{s}\mathcal{X} \to \mathfrak{s}\mathcal{Y}$ of Lie ternaries, there exists not more than one morphism $f: \mathcal{X} \to \mathcal{Y}$ of punctured symmetric spaces for which*

$$\mathfrak{s}(f) = \varphi.$$

If the space $\mathcal{X}$ is simply connected and the homomorphism φ is an isomorphism, then such a morphism exists (and is a covering).

Proof. According to the assertion in Exercise 9.2, the relation $\mathfrak{s}(f) = \mathfrak{s}(g)$ for morphisms $f, g: \mathcal{X} \to \mathcal{Y}$ is equivalent to the relation $(df)_{p_0} = (dg)_{p_0}$, and according to Proposition 6.1, the morphisms f and g are affine mappings. Therefore, if $\mathfrak{s}(f) = \mathfrak{s}(g)$, then $f = g$ according to Proposition 3.2 (we recall that all symmetric spaces are connected by definition). This proves the first assertion in the theorem.

Now let φ be an isomorphism. Formula (27) in Chap. 8 implies that the isomorphism φ considered as a mapping $\mathrm{T}_{p_0}\mathcal{X} \to \mathrm{T}_{q_0}\mathcal{Y}$ transforms the curvature tensor $R_{p_0}^{\mathcal{X}}$ of the space $\mathcal{X}$ at the point p_0 into the curvature tensor $R_{q_0}^{\mathcal{Y}}$ of the space $\mathcal{Y}$ at the point $q_0 = f(p_0)$. Therefore, according to Theorem 5.2

(and the assertion in Exercise 5.4), there exists an affine covering $f: \mathcal{X} \to \mathcal{Y}$ for which $(df)_e = \varphi$. It remains to note that according to Proposition 6.1, this covering is a morphism of symmetric spaces. □

§4. Computation of the Lie Ternary of the Space $(\mathcal{G}/\mathcal{H})_\sigma$

Theorem 9.2. *For any Lie ternary $\mathfrak{s}$ (which is finite dimensional and is over the field $\mathbb{R}$), there exists a simply connected punctured symmetric space $\mathcal{X}$ with $\mathfrak{s}\mathcal{X} = \mathfrak{s}$. Any other punctured symmetric space $\mathcal{X}'$ with $\mathfrak{s}\mathcal{X}' = \mathfrak{s}$ is covered by the space $\mathcal{X}$.*

In particular, this implies that the space $\mathcal{X}$ *is uniquely defined up to an isomorphism.*

Exercise 9.3. Prove the second assertion in this theorem. [*Hint*: Use the second assertion in Theorem 9.1.]

To prove the first assertion in Theorem 9.2, we must first compute the Lie ternary of a symmetric space of the form $(\mathcal{G}/\mathcal{H})_\sigma$ (see Exercise 5.9).

Let $\mathcal{G}$ be a connected Lie group, let σ be its involutive automorphism, and let $\mathcal{H}$ be a (not necessarily closed) subgroup of the group $\mathcal{G}$ such that

$$(\operatorname{Fix}\sigma)_e \subset \mathcal{H} \subset \operatorname{Fix}\sigma. \tag{5}$$

The corresponding symmetric space $(\mathcal{G}/\mathcal{H})_\sigma$ is just the quotient space $\mathcal{G}/\mathcal{H}$ with the distinguished point $p_0 = e\mathcal{H}$ in which the symmetries act according to the formula

$$s_p q = a\sigma(a^{-1}b)\mathcal{H}, \quad p = a\mathcal{H}, \quad q = b\mathcal{H}.$$

To compute the Lie ternary of this symmetric space, we introduce the Lie algebra $\mathfrak{g} = \mathfrak{r}\mathcal{G}$ of *right-invariant* vector fields on the Lie group $\mathcal{G}$. With respect to the automorphism $\mathfrak{r}(\sigma)$ induced by the automorphism σ of this algebra (which is denoted by σ as previously for simplicity), it is a symmetric Lie algebra.

Exercise 9.4. Show that the cosocle $\mathfrak{h}$ of the symmetric Lie algebra $\mathfrak{g}$ is isomorphic to the Lie algebra of the subgroup $\mathcal{H}$. (It is clear that all subgroups satisfying condition $\mathcal{H}$ have isomorphic Lie algebras (5).)

Let $\mathfrak{s}$ be the socle of a symmetric Lie algebra $\mathfrak{g}$. According to the results obtained above, for each field $X \in \mathfrak{s}$, there exists a unique derivation $X^\sharp \in \mathfrak{s}\mathcal{X}$, $\mathcal{X} = (\mathcal{G}/\mathcal{H})_\sigma$, for which

$$X^\sharp_{p_0} = (dj)_e X_e,$$

where j is the canonical mapping

$$j: \mathcal{G} \to \mathcal{X}, \qquad a \mapsto a\mathcal{H}, \quad a \in \mathcal{G}.$$

This derivation is given by

$$X_p^\sharp = \frac{1}{2}(dR_{pop})_{po}(dj)_e X_e = \frac{1}{2}d(R_{pop}\circ j)_e X_e,$$

where the mapping $R_{pop}\circ j\colon \mathcal{G}\to\mathcal{X}$ acts according to the formula

$$(R_{pop}\circ j)(x) = j(x\sigma(x^{-1}\sigma(a))) = j(x\sigma(x^{-1})a), \quad x\in\mathcal{G}, \tag{6}$$

where a is an element of the group $\mathcal{G}$ such that $p = j(a)$ (and therefore $pop = j(\sigma(a))$).

Formula (6) means that

$$R_{pop}\circ j = j\circ\mu\circ(\mathrm{id}\times T_a)\circ\Delta,$$

where Δ is the diagonal mapping $x\mapsto(x,x)$, μ is the multiplication $(x,y)\mapsto xy$, and T_a is the mapping

$$R_a\circ\sigma\circ\nu\colon x\mapsto\sigma(x^{-1})a, \quad x\in\mathcal{G}$$

(here R_a is the right translation $x\mapsto xa$ on the group $\mathcal{G}$ and ν is the inversion diffeomorphism $x\mapsto x^{-1}$). Therefore,

$$d(R_{pop}\circ j)_e = (dj)_a\circ(d\mu)_{(e,a)}\circ(\mathrm{id}\times(dT_a)_e)\circ(d\Delta)_e,$$

where

$$(dT_a)_e = -(dR_a)_e\circ(d\sigma)_e$$

because $(d\nu)_e = -\mathrm{id}$.

In particular, we see that if a vector $A\in\mathbf{T}_e\mathcal{G}$ has the form X_e, where $X\in\mathfrak{s}$ (and therefore $(d\sigma)_e A = -A$), then

$$(dT_a)_e A = (dR_a)_e A.$$

Because $(d\Delta)_e A = (A, A)$ and

$$(d\mu)_{(e,a)}(A,B) = (dR_a)_e A + (dL_e)_a B = (dR_a)_e A + B$$

for any $A\in\mathbf{T}_e\mathcal{G}$ and $B\in\mathbf{T}_a\mathcal{G}$ (see formula (13) in Chap. 8), this proves that for $A = X_e$, $X\in\mathfrak{s}$, we have

$$d(R_{pop}\circ j)_e A = (dj)_a((dR_a)_e A + (dR_a)_e A) = 2(dj)_a(dR_a)_e A = 2(dj)_a X_a$$

and therefore

$$X_p^\sharp = (dj)_a X_a.$$

Because $p = j(a)$, this means that *the fields X and $X^\sharp$ are j-related* by definition.

Exercise 9.5. Show that $X^\sharp$ is the unique field in $\mathfrak{s}\mathcal{X}$ that is j-related to the field X.

Because the field $[X, Y, Z]$ is j-related to the field $[X^\sharp, Y^\sharp, Z^\sharp]$ for any fields $X, Y, Z \in \mathfrak{s}$, this implies

$$[X, Y, Z]^\sharp = [X^\sharp, Y^\sharp, Z^\sharp],$$

i.e., the mapping $X \mapsto X^\sharp$ is an isomorphism of the Lie ternary $\mathfrak{s}$ onto the Lie ternary $\mathfrak{s}\mathcal{X}$. Therefore, *the Lie ternary* $\mathfrak{s}\mathcal{X}$ *of the symmetric space* $\mathcal{X} = (\mathcal{G}/\mathcal{H})_\sigma$ *is naturally isomorphic to the cosocle* $\mathfrak{s} = \mathfrak{s}(\mathfrak{g})$ *of the symmetric Lie algebra* $\mathfrak{g} = \mathfrak{r}\mathcal{G}$.

Of course, the Lie algebra $\mathfrak{r}\mathcal{G}$ can be replaced with the algebra $\mathfrak{l}\mathcal{G}$ here, which is isomorphic to it and is more usual. [*Question*: What is changed when we deal with the algebra $\mathfrak{l}\mathcal{G}$ from the very beginning?]

Exercise 9.6. Assign the isomorphism $\mathfrak{s}\mathcal{X} \to \mathfrak{s}(\mathfrak{l}\mathcal{G})$ by an explicit formula.

§5. Fundamental Group of the Quotient Space

For what follows, we must consider how the fundamental groups $\pi_1\mathcal{G}$ and $\pi_1(\mathcal{G}/\mathcal{H})$ of the Lie group $\mathcal{G}$ (which is assumed to be connected) and the quotient space $\mathcal{G}/\mathcal{H}$ of the Lie group (which is also connected because the group $\mathcal{G}$ is connected) by its closed subgroup $\mathcal{H}$ (not necessary connected) are related.

Because the component of the identity $\mathcal{H}_e$ of the group $\mathcal{H}$ is its invariant subgroup, the set

$$\pi_0\mathcal{H} = \mathcal{H}/\mathcal{H}_e$$

of all connected components of the group $\mathcal{H}$ is a group with respect to the multiplication

$$a\mathcal{H}_e \cdot b\mathcal{H}_e = ab\mathcal{H}_e, \quad a, b \in \mathcal{H}.$$

As above, let $j: \mathcal{G} \to \mathcal{G}/\mathcal{H}$ be the canonical mapping $a \mapsto a\mathcal{H}$, $a \in \mathcal{G}$, and let

$$j_*: \pi_1\mathcal{G} \to \pi_1(\mathcal{G}/\mathcal{H}) \tag{7}$$

be the homomorphism of fundamental groups induced by this mapping.

Because the mapping j is a bundle in the sense of Hurewicz, any loop $u: I \to \mathcal{G}/\mathcal{H}$ of the space $\mathcal{G}/\mathcal{H}$ at the point $p_0 = e\mathcal{H}$ is covered by a certain path $v: I \to \mathcal{G}$ of the group $\mathcal{G}$ that starts from the point e (and ends at a certain point of the subgroup $\mathcal{H}$). (For (piecewise) smooth loops u, the existence of the covering path v is also implied by the existence of at least one connection for the smooth principal bundle $(\mathcal{G}, j, \mathcal{G}/\mathcal{H})$. This case is sufficient for our purposes.)

If $v_1: I \to \mathcal{G}$ is another path covering the loop u (and also starting from the point e), then the formula

$$w(t) = v(t)^{-1}v_1(t), \quad 0 \le t \le 1,$$

defines a path $w: I \to \mathcal{H}$ starting from the point e and therefore such that $w(t) \in \mathcal{H}_e$. Because $v_1(1) = v(1)w(1)$, this proves that *the element* $v(1)\mathcal{H}_e$ *of the group* $\pi_0\mathcal{H}$ *does not depend on the choice of the covering path* v.

Exercise 9.7. Prove that the formula $\delta[u] = v(1)\mathcal{H}_e$ correctly defines the homomorphism

$$\delta: \pi_1(\mathcal{G}/\mathcal{H}) \to \pi_0\mathcal{H} \tag{8}$$

of the group $\pi_1(\mathcal{G}/\mathcal{H})$ into the group $\pi_0\mathcal{H}$.

Proposition 9.2. *Homomorphism* (8) *is an epimorphism. Its kernel is the image* $\operatorname{Im} j_*$ *of homomorphism* (7).

In the language of exact sequences, Proposition 9.2 asserts that *the exact sequence*

$$\pi_1\mathcal{G} \xrightarrow{j_*} \pi_1(\mathcal{G}/\mathcal{H}) \xrightarrow{\delta} \pi_0\mathcal{H} \to 1 \tag{9}$$

holds.

Proof. For any point $a \in \mathcal{H}$ in $\mathcal{G}$, there exists a path v connecting the point e with the point a because the Lie group $\mathcal{G}$ is connected. Then the path $j \circ v: I \to \mathcal{G}/\mathcal{H}$ is a loop at the point p_0, and the relation

$$\delta[j \circ v] = a\mathcal{H}_e$$

holds in the group $\pi_0\mathcal{H}$. Therefore, mapping (8) is an epimorphism.

If $u = j \circ v$, where v is a loop in $\mathcal{G}$, then $\delta[u] = e\mathcal{H}_e = \mathcal{H}_e$, and therefore $[u] \in \operatorname{Ker}\delta$. This proves that $\operatorname{Im} j_* \subset \operatorname{Ker}\delta$.

Conversely, let $[u] \in \operatorname{Ker}\delta$, i.e., for u, let there exist a covering path v such that $v(1) \in \mathcal{H}_e$. Then we can construct the path $v * v_0$, where v_0 is an arbitrary path in $\mathcal{H}_e$ that connects the point $v(1)$ with the point e. This path covers the path $u * (j \circ v_0) = j \circ (v * v_0)$, where $j \circ v_0$ is a constant path at the point p_0. Because $[u] = [u * (j \circ v_0)]$ in the group $\pi_1(\mathcal{G}/\mathcal{H})$, we therefore have $[u] = [j \circ (v * v_0)] = j_*[v * v_0] \in \operatorname{Im} j_*$. Therefore, $\operatorname{Ker}\delta \subset \operatorname{Im} j_*$, and hence $\operatorname{Ker}\delta = \operatorname{Im} j_*$. □

Corollary 9.1. *If the group* $\mathcal{G}$ *is simply connected, then the group* $\pi_1(\mathcal{G}/\mathcal{H})$ *is isomorphic to the group* $\pi_0\mathcal{H}$.

Corollary 9.2. *If the subgroup* $\mathcal{H}$ *is connected, then the group* $\pi_1(\mathcal{G}/\mathcal{H})$ *is a homomorphic image of the group* $\pi_1\mathcal{G}$.

Corollary 9.3. *If the subgroup* $\mathcal{H}$ *is connected and the group* $\mathcal{G}$ *is simply connected, then the space* $\mathcal{G}/\mathcal{H}$ *is simply connected.*

Exercise 9.8. Prove that the kernel of homomorphism (7) is the image of the group $\pi_1\mathcal{H} = \pi_1\mathcal{H}_e$ under the homomorphism $i_*: \pi_1\mathcal{H} \to \pi_1\mathcal{G}$ induced by an embedding $i: \mathcal{H} \to \mathcal{G}$.

In particular, this implies that if the subgroup $\mathcal{H}$ is connected and simply connected, then the group $\pi_1(\mathcal{G}/\mathcal{H})$ is isomorphic to the group $\pi_1\mathcal{G}$.

Exercise 9.9. Construct a homomorphism

$$\delta\colon \pi_2(\mathcal{G}/\mathcal{H}) \to \pi_1\mathcal{H} \tag{10}$$

and prove that $\operatorname{Im}\delta = \operatorname{Ker} i_*$.

Remark 9.1. It can be proved that $\pi_2\mathcal{G} = 0$ for any simply connected Lie group $\mathcal{G}$ (this is a difficult theorem!). Therefore, by the general assertion on the exactness of the bundle homotopic sequence, homomorphism (10) is a monomorphism. Taken together with the assertions in Exercises 9.8 and 9.9, this means that *exact sequence* (9) *is a segment of the exact sequence*

$$0 \to \pi_2(\mathcal{G}/\mathcal{H}) \xrightarrow{\delta} \pi_1\mathcal{H} \xrightarrow{i_*} \pi_1\mathcal{G} \xrightarrow{j_*} \pi_1(\mathcal{G}/\mathcal{H}) \xrightarrow{\delta} \pi_0\mathcal{H} \to 1.$$

In particular, if the group $\mathcal{G}$ is simply connected, then $\pi_1\mathcal{H} \approx \pi_2(\mathcal{G}/\mathcal{H})$.

§6. Symmetric Space with a Given Lie Ternary

We are now ready to prove Theorem 9.2.

Proof (of Theorem 9.2). We need only prove the first assertion of this theorem (see Exercise 9.3), i.e., that for any Lie ternary $\mathfrak{s}$ (finite-dimensional and considered over the field $\mathbb{R}$), there exists a simply connected punctured symmetric space $\mathcal{X}$ with $\mathfrak{s}\mathcal{X} \approx \mathfrak{s}$. For this, we consider an arbitrary symmetric Lie algebra $\mathfrak{g}$ with $\mathfrak{s}(\mathfrak{g}) \approx \mathfrak{s}$ and a simply connected Lie group $\mathcal{G}$ with $\mathfrak{l}\mathcal{G} \approx \mathfrak{g}$. (The existence of the Lie algebra $\mathfrak{g}$ is ensured by Proposition 8.5, and the existence of the Lie group $\mathcal{G}$ is ensured by the Cartan theorem; see Chap. 7). According to Proposition 7.2, there exists an involutive automorphism of the group $\mathcal{G}$ (which is denoted by the old symbol σ) that induces a structural automorphism σ of the symmetric Lie algebra $\mathfrak{g}$, and the punctured symmetric space $\mathcal{X} = (\mathcal{G}/\operatorname{Fix}(\sigma)_e)_\sigma$ is therefore well defined. Because the group $\mathcal{G}$ is simply connected and the group $\operatorname{Fix}(\sigma)_e$ is connected, this space is simply connected. On the other hand, according to the above calculation, the Lie ternary $\mathfrak{s}\mathcal{X}$ of this space is isomorphic to the socle of the Lie algebra $\mathfrak{g}$, and it is therefore isomorphic to the Lie ternary $\mathfrak{s}$. □

§7. Coverings

For a disconnected subgroup $\mathcal{H}$, the mapping

$$\mathcal{G}/\mathcal{H}_e \to \mathcal{G}/\mathcal{H}, \tag{11}$$

which is induced by cosets, is a locally trivial bundle with the discrete fiber $\mathcal{H}/\mathcal{H}_e$, i.e., it is a covering. According to Corollary 9.1, covering (11) is universal in the case where the group $\mathcal{G}$ is simply connected. In general, if the quotient space $\mathcal{H}_2/\mathcal{H}_1$ is discrete for subgroups $\mathcal{H}_1$ and $\mathcal{H}_2$ with $\mathcal{H}_1 \subset \mathcal{H}_2$, then we have a covering

$$\mathcal{G}/\mathcal{H}_1 \to \mathcal{G}/\mathcal{H}_2$$

with the fiber $\mathcal{H}_2/\mathcal{H}_1$.

Exercise 9.10. Show that the quotient space is discrete iff the components of the identities of the subgroups $\mathcal{H}_1$ and $\mathcal{H}_2$ coincide,

$$(\mathcal{H}_1)_e = (\mathcal{H}_2)_e.$$

This implies that for any involutive automorphism σ of a connected group $\mathcal{G}$, *each symmetric space of the form* $(\mathcal{G}/\mathcal{H})_\sigma$ *is covered by the symmetric space* $(\mathcal{G}/\operatorname{Fix}(\sigma)_e)_\sigma$ (which is usually denoted by $\mathcal{G}^\sigma$) *and covers the symmetric space* $\mathcal{G}/\operatorname{Fix}(\sigma)$ (we recall that it is isomorphic to the symmetric space $\mathcal{G}_\sigma$ from Exercise 5.8).

From the topological standpoint, the construction of the space $\mathcal{G}^\sigma$ is simplest (for example, it is simply connected for the simply connected group $\mathcal{G}$); however, it is often more convenient to use the space $\mathcal{G}_\sigma$ with a larger fundamental group, because it is embedded in the Lie group $\mathcal{G}$ as a totally geodesic submanifold (as is seen in what follows).

§8. Cartan Theorem

Now let $\mathcal{X}$ again be an arbitrary symmetric space. Applying the construction that was used above to prove Theorem 9.2 to the Lie ternary $\mathfrak{s}\mathcal{X}$ and using Theorem 9.1 (more precisely, its second assertion), we immediately establish that for any symmetric space $\mathcal{X}$, there exists an affine universal covering of the form

$$\mathcal{G}^\sigma \to \mathcal{X}.$$

In particular, this proves that *each simply connected space is isomorphic to a space of the form* $\mathcal{G}^\sigma$. It turns out that the condition of simple connectedness can be omitted here.

Theorem 9.3 (Cartan). *Each symmetric space* $\mathcal{X}$ *is isomorphic to a space of the form* $(\mathcal{G}/\mathcal{H})_\sigma$.

Exercise 9.11. Verify that Theorem 9.3 holds for the symmetric spaces from Exercises 5.5–5.7. [*Hint*: For any Lie group $\mathcal{G}$, we have $\mathcal{G} = (\mathcal{G} \times \mathcal{G})/\Delta)_\sigma$, where Δ is the *diagonal subgroup* consisting of elements of the form $(a, a) \in \mathcal{G} \times \mathcal{G}$, $a \in \mathcal{G}$, and $\sigma\colon (a, b) \mapsto (b, a)$, $a, b \in \mathcal{G}$. On the other hand,

$$G_{\mathbb{R}}(m, n) = (\mathrm{SO}\,(n)/\mathcal{H})_\sigma,$$

where $\mathcal{H} = \operatorname{Fix}(\sigma)$, $\sigma = \operatorname{int}_J\colon A \mapsto JAJ^{-1}$, and $J = (-E_m) \oplus E_{n-m}$.]

§9. Identification of Homogeneous Spaces with Quotient Spaces

To prove Theorem 9.3, we need the following general lemma.

Lemma 9.3. *Each smooth manifold $\mathcal{X}$ on which a Lie group $\mathcal{G}$ acts smoothly and transitively is diffeomorphic to the quotient manifold $\mathcal{G}/\mathcal{H}$, where $\mathcal{H}$ is the stabilizer of an arbitrarily chosen point $p_0 \in \mathcal{X}$. The diffeomorphism*

$$\varphi: \mathcal{G}/\mathcal{H} \to \mathcal{X}$$

is given by

$$\varphi(a\mathcal{H}) = ap_0, \quad a \in \mathcal{G}. \tag{12}$$

(By definition, the subgroup $\mathcal{H}$ consists of all elements $a \in \mathcal{G}$ for which $ap_0 = p_0$. This subgroup is obviously closed, and it is therefore a Lie subgroup. Being a quotient space by a closed subgroup, the space $\mathcal{G}/\mathcal{H}$ is a smooth manifold.)

Proof. The bijectivity and smoothness of the mapping φ are obvious. Because a surjective smooth mapping that is an immersion is necessarily étale, we need only prove that the mapping φ is an immersion. Of course, it suffices to do this only at the point $e\mathcal{H}$. But it immediately follows from the definition of charts of the manifold $\mathcal{G}/\mathcal{H}$ that the mapping φ is an immersion at the point $e\mathcal{H}$ iff the differential

$$(d\psi)_e\colon \mathbf{T}_e\mathcal{G} \to \mathbf{T}_{p_0}\mathcal{X}$$

of the mapping

$$\psi\colon \mathcal{G} \to \mathcal{X}, \quad a \mapsto ap_0, \quad a \in \mathcal{G},$$

at the point e has the property that its kernel $\operatorname{Ker}(d\psi)_e$ coincides with the tangent subspace $\mathbf{T}_e\mathcal{H}$ of the subgroup $\mathcal{H}$. Because the inclusion $\mathbf{T}_e\mathcal{H} \subset \operatorname{Ker}(d\psi)_e$ is obviously valid, we therefore need only prove that the relation $(d\psi)_e A = 0$, $A \in \mathbf{T}_e\mathcal{G}$, implies the inclusion $A \in \mathbf{T}_e\mathcal{H}$.

Let $\beta\colon t \mapsto \exp tA$ be a one-parameter subgroup of the group $\mathcal{G}$ with the initial vector A. Because $(d\psi)_e\dot\beta(0) = (d\psi)_e A$ by definition and $(d\psi)_e A = 0$ by the given condition of the lemma, we see that for any function f that is smooth near the point p_0,

$$\left.\frac{df[(\exp tA)p_0]}{dt}\right|_{t=0} = \left.\frac{(df \circ \psi)(\beta(t))}{dt}\right|_{t=0} = [(d\psi)_e\dot\beta(0)]f = 0.$$

In particular, this holds for the function f_s defined by

$$f_s(p) = f((\exp sA)p),$$

where s is an arbitrary real number that is sufficiently close to zero. Because

$$\left.\frac{df[(\exp tA)p_0]}{dt}\right|_{t=s} = \left.\frac{df_s[(\exp tA)p_0]}{dt}\right|_{t=0},$$

the function $t \mapsto f((\exp tA)p_0)$ is constant near the point $t = 0$. By the arbitrariness of the function f, this is possible only if $(\exp tA)p_0 = p_0$, i.e., if $\exp tA \in \mathcal{H}$. Therefore, $A \in \mathbf{T}_e\mathcal{H}$. $\square$

§10. Translations of a Symmetric Space

By a clear association, automorphisms of a symmetric space $\mathcal{X}$ that are compositions of an *even* number of symmetries s_p, $p \in \mathcal{X}$, are called its *translations.* All translations compose a subgroup $\operatorname{Trans}\mathcal{X}$ of the group $\operatorname{Aut}\mathcal{X}$ of automorphisms of the space $\mathcal{X}$, the *translation group.*

The group $\operatorname{Trans}\mathcal{X}$, being a subgroup of the automorphism group, naturally acts on the manifold $\mathcal{X}$. It is easy to see that *this action is transitive,* i.e., for any two points $p, q \in \mathcal{X}$, there exists an element a of the group $\operatorname{Trans}\mathcal{X}$ such that $ap = q$. Indeed, because any point of the space $\mathcal{X}$ has a normal neighborhood (we recall that the space $\mathcal{X}$ is connected), the points p and q can be connected by a piecewise smooth curve composed of normal geodesic segments and the composition of symmetries at the midpoints of these segments obviously transforms p into q. (If the number of these segments is odd, then a symmetry s_p, for example, should be added to them.)

Proposition 9.3. *The group* $\operatorname{Trans}\mathcal{X}$ *has a smoothness with respect to which it becomes a connected Lie group that acts smoothly on the manifold* $\mathcal{X}$.

We prove this proposition in the next chapter.

§11. Proof of the Cartan Theorem

Proof (of Theorem 9.3). Let $\mathcal{G} = \operatorname{Trans}\mathcal{X}$. Choosing a point $p_0 \in \mathcal{X}$, we consider the mapping $\sigma\colon \mathcal{G} \to \mathcal{G}$ defined by

$$\sigma a = s_{p_0} a s_{p_0}, \quad a \in \mathcal{G}.$$

It is clear that σ is an involutive automorphism of the group $\mathcal{G}$. Let $\mathcal{H}$ be the stabilizer of the point p_0 in the group $\mathcal{G}$. Because the relation $\varphi \circ s_p \circ \varphi^{-1} = s_q$, where $q = \varphi(p)$, holds for any automorphism $\varphi\colon \mathcal{X} \to \mathcal{X}$ and any point $p \in \mathcal{X}$, we have $s_{p_0} = a s_{p_0} a^{-1}$ for $a \in \mathcal{H}$ and therefore $\sigma a = a$. Therefore, $\mathcal{H} \subset \operatorname{Fix}(\sigma)$.

Let $A \in \mathbf{T}_e(\operatorname{Fix}(\sigma))$. Then $(d\sigma)_e A = A$ and therefore

$$\sigma(\exp tA) = \exp tA$$

for any $t \in \mathbb{R}$. Therefore,

$$s_{p_0}((\exp tA)p_0) = (s_{p_0} \circ \exp tA)p_0 = (\exp tA \circ s_{p_0})p_0 = (\exp tA)p_0,$$

i.e., the point $(\exp tA)p_0$ is a fixed point of the symmetry s_{p_0}. Therefore, because the fixed point p_0 is isolated, we have

$$(\exp tA)p_0 = p_0$$

(for small and therefore for any $t \in \mathbb{R}$). This means that $\exp tA \in \mathcal{H}$, which implies $\operatorname{Fix}(\sigma)_e \subset \mathcal{H}$ by the arbitrariness of the vector A and the connectedness of the group $\mathrm{Fix}(\sigma)_e$. Therefore, conditions (5) hold for the automorphism σ and the subgroup $\mathcal{H}$, and the symmetric space $(\mathcal{G}/\mathcal{H})_\sigma$ is hence well defined.

On the other hand, because the Lie group $\mathcal{G}$ acts smoothly and transitively on the manifold $\mathcal{X}$, formula (12) correctly defines the diffeomorphism

$$\varphi: (\mathcal{G}/\mathcal{H})_\sigma \to \mathcal{X}$$

according to Lemma 9.3. In this case, because

$$a\sigma(a^{-1}b) = as_{p_0}a^{-1}bs_{p_0} = s_{ap_0}bs_{p_0}, \quad a, b \in \mathcal{G},$$

we have

$$a\sigma(a^{-1}b)p_0 = s_{ap_0}(bp_0).$$

Therefore, the diffeomorphism φ is an automorphism of symmetric spaces. □

Chapter 10
Palais and Kobayashi Theorems

§1. Infinite-Dimensional Manifolds and Lie Groups

In general, the concept of differentiability (smoothness) does not require finite dimensionality and can be defined for mappings of open sets of an arbitrary linear topological space $\mathcal{E}$. This allows defining *smooth manifolds with charts in* $\mathcal{E}$ in an obvious way: it suffices to replace open sets of the space $\mathbb{R}^n$ with those of the space $\mathcal{E}$ everywhere in the usual definition of a smooth manifold (see the addendum). We obtain Hilbert, Banach, locally convex, etc., manifolds depending on the type of the space $\mathcal{E}$. All such manifolds are conventionally called *infinite-dimensional manifolds*, although this term seems not very appropriate. The theory of such manifolds (under one or another condition on the space $\mathcal{E}$) almost literally repeats the finite-dimensional smooth manifold theory in its initial part, but, for example, a smooth vector field on an infinite-dimensional manifold might have no integral curves.

Infinite-dimensional Lie groups are defined as usual, as manifolds $\mathcal{G}$ with a smooth multiplication $\mathcal{G}\times\mathcal{G} \to \mathcal{G}$. The *Lie algebra* $\mathfrak{l}\mathcal{G}$ of each such group is also defined as usual (this algebra as a linear space is isomorphic to the base space $\mathcal{E}$). For *Banach Lie groups* (over a Banach space $\mathcal{E}$), the theory can be well developed in parallel to finite-dimensional Lie group theory. For more general groups, the parallelism is violated, and little is known about them. It is even unknown if there always exist one-parameter subgroups with a given initial vector, i.e., if the exponential mapping $\mathfrak{l}\mathcal{G} \to \mathcal{G}$ is always defined (although a counterexample is also apparently unknown). If the exponential mapping exists, it is not injective in general and does not cover any neighborhood of the identity.

Of course, the absence of the general theory does not prevent the study of one or another class of infinite-dimensional Lie groups (moreover, it even stimulates special attention to them). For example, the theory of *current groups*, whose elements are smooth mappings $\mathcal{X} \to \mathcal{G}$ of a given compact manifold $\mathcal{X}$ (the case $\mathcal{X} = \mathbb{S}^1$ is the most interesting and well developed) into a given finite-dimensional group $\mathcal{G}$, has been intensively developed recently. Another important class of infinite-dimensional non-Banach Lie groups, about which essentially less is known, consists of diffeomorphism groups Diff $\mathcal{X}$ of finite-dimensional smooth manifolds $\mathcal{X}$.

The Lie algebra of the group Diff $\mathcal{X}$ is the algebra $\mathfrak{a}\mathcal{X}$ of all vector fields on $\mathcal{X}$, and the Lie algebras of subgroups of the group Diff $\mathcal{X}$ are therefore subalgebras of the algebra $\mathfrak{a}\mathcal{X}$. As in the case of finite-dimensional Lie groups, it is natural to expect the existence (probably, with certain conditions) of a bijective correspondence between subgroups of the group Diff $\mathcal{X}$ and subalgebras of the algebra $\mathfrak{a}\mathcal{X}$. We consider this question for finite-dimensional subgroups (i.e., for subgroups that are usual Lie groups) and finite-dimensional algebras for which the study can be entirely performed in the framework of finite-dimensional manifolds.

Of course, the restriction to only finite-dimensional manifolds distorts the whole picture and excludes the true perspective from the presentation, but we must follow this line of reasoning. In what follows, we do not explicitly mention infinite-dimensional manifolds, and their theme continues only sotto voce. In particular, we do not introduce a topology and smoothness on the group Diff $\mathcal{X}$.

§2. Vector Fields Induced by a Lie Group Action

If a Lie group $\mathcal{G}$ acts smoothly and effectively on a smooth manifold $\mathcal{X}$, then for any element $a \in \mathcal{G}$, the mapping

$$L_a\colon p \mapsto ap, \quad p \in \mathcal{X},$$

is a diffeomorphism of the manifold $\mathcal{X}$ onto itself, and the correspondence $L\colon a \mapsto L_a$ is a monomorphic mapping of the group $\mathcal{G}$ into the group Diff $\mathcal{X}$ of all diffeomorphisms $\mathcal{X} \to \mathcal{X}$. Therefore, the group $\mathcal{G}$ can be considered as a subgroup of the (abstract) group Diff $\mathcal{X}$. Under this embedding, each one-parameter subgroup $\beta_X\colon t \mapsto \exp tX$, $X \in \mathfrak{l}\mathcal{G}$, of the group $\mathcal{G}$ turns out to be a flow on the manifold $\mathcal{X}$ that is defined for all $t \in \mathbb{R}$. The vector field generating this flow is denoted by $-X^*$. (We emphasize the sign: to avoid introducing it, right-invariant instead of left-invariant fields X should be considered.) As a derivation of the algebra of smooth functions $\mathsf{F}\mathcal{X}$, the field X^* acts according to the formula

$$(X^* f)(p) = \lim_{t\to 0} \frac{f(p) - f((\exp tX)p)}{t}, \quad p \in \mathcal{X}, \quad f \in \mathsf{F}\mathcal{X}.$$

Therefore, for any elements X and Y of the Lie algebra, we have

$$\begin{aligned}(X^*Y^*f)(p) &= \lim_{s\to 0} \frac{(Y^*f)(p) - (Y^*f(\exp sX)p)}{s} \\ &= \lim_{\substack{s\to 0\\ t\to 0}} \frac{f(p) - f((\exp tY)p - f((\exp sX)p) + f((\exp tY)(\exp sX)p)}{st} \\ &= \lim_{t\to 0} \frac{f((\exp tY)(\exp tX)p) - f((\exp tX)p) - f((\exp tY)p) + f(p)}{t^2}\end{aligned}$$

and therefore

$$\begin{aligned}([X^*, Y^*]f)(p) &= (X^*Y^*f)(p) - (Y^*X^*f)(p) \\ &= -\lim_{t\to 0} \frac{f((\exp tX)(\exp tY)p) - f((\exp tY)(\exp tX)p)}{t^2}.\end{aligned}$$

Let $t^1, \ldots, t^m$ be normal coordinates in a neighborhood of the identity of the group $\mathcal{G}$ that correspond to a basis $X_1, \ldots, X_m$ of the Lie algebra $\mathfrak{g}$, and let $x^1, \ldots, x^n$ be local coordinates on $\mathcal{X}$ defined in a neighborhood of the point p and vanishing at p. Further, let $f = f(\boldsymbol{x})$ be the expression of the function f in the coordinates $x^1, \ldots, x^n$, and let $\boldsymbol{x} = \boldsymbol{x}(\boldsymbol{t})$ be the vector-valued function assigning the mapping $a \mapsto ap$ in the coordinates $t^1, \ldots, t^m$ and $x^1, \ldots, x^n$. Because $ep = p$, the components $x^i(\boldsymbol{t})$, $i = 1, \ldots, n$, of the latter function have the form

$$x^i(\boldsymbol{t}) = c^i_a t^a + c^i_{ab} t^a t^b + \ldots,$$

where the dots denote terms of degree ≥ 3 with respect to $t^1, \ldots, t^m$ (and $c^i_{ab} = c^i_{ba}$). Therefore, for any element $X = t^a X_a$ of the Lie algebra $\mathfrak{g}$, we have

$$\begin{aligned} f((\exp X)p) &= f(\boldsymbol{x}(\boldsymbol{t})) \\ &= f(p) + \left(\frac{\partial f}{\partial x^i}\right)_p x^i(\boldsymbol{t}) + \left(\frac{\partial^2 f}{\partial x^i \partial x^j}\right)_p x^i(\boldsymbol{t}) x^j(\boldsymbol{t}) + \cdots \\ &= f(p) + \left(\frac{\partial f}{\partial x^i}\right)_p (c^i_a t^a + c^i_{ab} t^a t^b + \cdots) \\ &\quad + \left(\frac{\partial^2 f}{\partial x^i \partial x^j}\right)_p (c^i_a c^j_b t^a t^b + \ldots) + \ldots, \end{aligned}$$

and therefore

$$(X^* f)(p) = \lim_{t \to 0} \frac{f(p) - f((\exp tX)p)}{t} = -\left(\frac{\partial f}{\partial x^i}\right)_p c^i_a X^a,$$

where we write X^a instead of t^a. Moreover (see (11) and (12) in Chap. 7), we have

$$f((\exp tX)(\exp tY)p) = f(\exp(t(X+Y) + \frac{1}{2} t^2 [X, Y] + \cdots)p),$$

which also implies (in a clear notation)

$$\begin{aligned} f((\exp tX)(\exp tY)p) = f(p) &+ t \left(\frac{\partial f}{\partial x^i}\right)_p c^i_a (X^a + Y^a) \\ &+ t^2 \left[\frac{1}{2}\left(\frac{\partial f}{\partial x^i}\right)_p c^i_a [X, Y]^a \right. \\ &+ \left(\frac{\partial f}{\partial x^i}\right)_p c^i_{ab} (X^a + Y^a)(X^b + Y^b) \\ &+ \left. \left(\frac{\partial^2 f}{\partial x^i \partial x^j}\right)_p c^i_a c^j_b (X^a + Y^a)(X^b + Y^b) \right] + \ldots, \end{aligned}$$

where, as above, the dots stand for terms of degree ≥ 3 with respect to t, and therefore

$$\begin{aligned} f((\exp tX)(\exp tY)p)& \\ - f((\exp tY)(\exp tX)p) &= \frac{t^2}{2}\left(\frac{\partial f}{\partial x^i}\right)_p c^i_a([X,Y]^a - [Y,X]^a) + \cdots \\ &= t^2\left(\frac{\partial f}{\partial x^i}\right)_p c^i_a[X,Y]^a + \cdots \\ &= -t^2([X,Y]^* f)(p) + \dots . \end{aligned}$$

Therefore,

$$\begin{aligned} ([X^*, Y^*]f)(p) &= -\lim_{t\to 0} \frac{f((\exp tX)(\exp tY)p) - f((\exp tY)(\exp tX)p)}{t^2} \\ &= ([X,Y]^* f)(p). \end{aligned}$$

This proves that $[X,Y]^* = [X^*, Y^*]$, i.e., that *the correspondence* $X \mapsto X^*$ *is a homomorphism of the Lie algebra* $\mathfrak{g} = \mathfrak{l}\mathcal{G}$ *of the Lie group* $\mathcal{G}$ *into the Lie algebra* $\mathfrak{a}\mathcal{X}$ *of vector fields on* $\mathcal{X}$.

If $X^* = 0$, then $\exp(-tX)p = p$ for any point $p \in \mathcal{X}$ and any $t \in \mathbb{R}$; because of the effectiveness of the action of the group $\mathcal{G}$ on the manifold $\mathcal{X}$, this is possible only for $X = 0$. Therefore, *the homomorphism* $X \mapsto X^*$ *is a monomorphism.*

This means that identifying X with X^*, we can (and do) consider the Lie algebra $\mathfrak{g}$ a subalgebra of the Lie algebra $\mathfrak{a}\mathcal{X}$. We stress that in contrast to the whole algebra $\mathfrak{a}\mathcal{X}$, this subalgebra is finite dimensional. Moreover, it consists of *complete fields*, i.e., those fields for which the corresponding maximal flow is defined for all $t \in \mathbb{R}$. It turns out that these properties completely characterize subalgebras of the Lie algebra $\mathfrak{a}\mathcal{X}$ that are Lie algebras of Lie groups acting smoothly and effectively on the manifold $\mathcal{X}$. Moreover, this assertion can be slightly strengthened.

§3. Palais Theorem

A subset S of a Lie algebra $\mathfrak{g}$ *generates* $\mathfrak{g}$ if each subalgebra of the algebra $\mathfrak{g}$ containing S coincides with $\mathfrak{g}$. Similarly, a set S *linearly generates* $\mathfrak{g}$ if it contains a basis of the algebra $\mathfrak{g}$. For each complete field $X \in \mathfrak{a}\mathcal{X}$, the flow on $\mathcal{X}$ induced by this field is denoted by $\{\varphi^X_t\}$.

Theorem 10.1. *Let* $\mathfrak{g}$ *be a subalgebra of the Lie algebra* $\mathfrak{a}\mathcal{X}$, *and let* $\mathcal{G}$ *be a subgroup of the group* Diff $\mathcal{X}$ *generated by all diffeomorphisms of the form* φ^X_t, *where* $t \in \mathbb{R}$ *and* X *is an arbitrary complete field from* $\mathfrak{g}$. *If the Lie algebra* $\mathfrak{g}$

1. *is finite dimensional and*
2. *is generated by a set consisting of only complete fields,*

then the group $\mathcal{G}$ admits a smoothness with respect to which it is a connected Lie group that acts smoothly and effectively on the manifold $\mathcal{X}$; moreover, the monomorphism $X \mapsto X^*$ corresponding to this action is an isomorphism of the algebra $\mathfrak{l}\mathcal{G}$ onto the subalgebra $\mathfrak{g}$.

Proof. According to the Cartan theorem, there exists a connected and simply connected Lie group $\widetilde{\mathcal{G}}$ whose Lie algebra $\mathfrak{l}\widetilde{\mathcal{G}}$ is isomorphic to the algebra $\mathfrak{g}$. An element of the Lie algebra $\mathfrak{l}\widetilde{\mathcal{G}}$ (left-invariant vector field on $\widetilde{\mathcal{G}}$) corresponding to the vector field $X \in \mathfrak{g}$ under this isomorphism is denoted by $\widetilde{X}$.

At each point $(a,p) \in \widetilde{\mathcal{G}} \times \mathcal{X}$ of the direct product $\widetilde{\mathcal{G}} \times \mathcal{X}$, the tangent space $\mathsf{T}_{(a,p)}(\widetilde{\mathcal{G}} \times \mathcal{X})$ is naturally decomposed into the direct sum of the tangent spaces $\mathsf{T}_a\widetilde{\mathcal{G}}$ and $\mathsf{T}_p\mathcal{X}$. Therefore, each field $X \in \mathfrak{g}$ defines the vector $(\widetilde{X}_a, X_p)$ in $\mathsf{T}_{(a,p)}(\widetilde{\mathcal{G}} \times \mathcal{X})$ (we recall that $\mathfrak{g} \subset \mathfrak{a}\mathcal{X}$), and all such vectors form a subspace $\mathcal{D}_{(a,p)}$ in $\mathsf{T}_{(a,p)}(\widetilde{\mathcal{G}} \times \mathcal{X})$.

Exercise 10.1. Show that the following statements hold:

1. The subspaces $\mathcal{D}_{(a,p)}$ smoothly depend on the point (a,p), i.e., they compose a distribution $\mathcal{D}$ on $\widetilde{\mathcal{G}} \times \mathcal{X}$.

2. The distribution $\mathcal{D}$ is involutive.

[*Hint*: The vector fields $(a,p) \mapsto (\widetilde{X}_a, X_p)$ generate a $\mathsf{F}(\widetilde{\mathcal{G}} \times \mathcal{X})$-module $\mathfrak{a}\mathcal{D}$.]

Therefore, according to the Frobenius theorem, a unique maximal integral submanifold of the distribution $\mathcal{D}$ passes through any point $(a,p) \in \widetilde{\mathcal{G}} \times \mathcal{X}$. For brevity, these integral manifolds are called *leaves*. We note that by definition *each leaf is a connected submanifold of the manifold* $\widetilde{\mathcal{G}} \times \mathcal{X}$ (in general, it is only immersed). A leaf passing through the point (e,p), where e is the identity of the group $\widetilde{\mathcal{G}}$, is denoted by $\mathcal{L}_p$.

The formula

$$a(b,p) = (ab,p), \quad a, b \in \widetilde{\mathcal{G}}, \quad p \in \mathcal{X},$$

obviously defines a smooth and effective (even free) action of the group $\widetilde{\mathcal{G}}$ on the manifold $\widetilde{\mathcal{G}} \times \mathcal{X}$. For any element $a \in \widetilde{\mathcal{G}}$, the corresponding left translation

$$L_a\colon (b,p) \mapsto (ab,p), \quad (b,p) \in \widetilde{\mathcal{G}} \times \mathcal{X},$$

is such that its differential

$$(dL_a)_{(b,p)}\colon \mathsf{T}_{(b,p)}(\widetilde{\mathcal{G}} \times \mathcal{X}) \to \mathsf{T}_{(ab,p)}(\widetilde{\mathcal{G}} \times \mathcal{X})$$

acts on the vectors $(\widetilde{X}_b, X_p)$ by

$$(dL_a)_{(b,p)}(\widetilde{X}_b, X_p) = ((dL_a)_b\widetilde{X}_b, X_p) = (\widetilde{X}_{ab}, X_p),$$

where L_a is the left translation in the group $\mathcal{G}$. Therefore,

$$(dL_a)_{(b,p)}\mathcal{D}_{(b,p)} = \mathcal{D}_{(ab,p)}$$

for any point $(b,p) \in \widetilde{\mathcal{G}} \times \mathcal{X}$ and any element $a \in \widetilde{\mathcal{G}}$ (by definition, this means that the distribution $\mathcal{D}$ is *invariant* with respect to the action of the group $\widetilde{\mathcal{G}}$

on $\widetilde{\mathcal{G}}\times\mathcal{X}$). This directly implies that *for any leaf* $\mathcal{L}$*, the submanifold* $a\mathcal{L} = L_a\mathcal{L}$ *is also a leaf.*

Therefore, the correspondence $\mathcal{L} \mapsto a\mathcal{L}$ defines the action of the group $\widetilde{\mathcal{G}}$ on the set $\{\mathcal{L}\}$ of all leaves. If $(a, p) \in \mathcal{L}$, then $(e, p) \in a^{-1}\mathcal{L}$ and therefore $a^{-1}\mathcal{L} = \mathcal{L}_p$, i.e., $\mathcal{L} = a\mathcal{L}_p$. This proves that *any leaf* $\mathcal{L}$ *has the form* $a\mathcal{L}_p$*, where* $a \in \widetilde{\mathcal{G}}$ and $p \in \mathcal{X}$. By definition, this means that each orbit of the action of the group $\mathcal{G}$ on the set $\{\mathcal{L}\}$ contains a leaf of the form $\mathcal{L}_p$.

We note that because each leaf is a conservative submanifold (a submanifold $\mathcal{Y}$ of a manifold $\mathcal{X}$ is said to be *conservative* if for any smooth manifold $\mathcal{Z}$, the mapping $\varphi: \mathcal{Z} \to \mathcal{Y}$ iff it is smooth as a mapping into $\mathcal{X}$, i.e., a smooth mapping $\imath \circ \varphi: \mathcal{Z} \to \mathcal{X}$, where $\imath: \mathcal{Y} \to \mathcal{X}$ is an immersion), *the mapping*

$$\mathcal{L} \to a\mathcal{L}, \quad (b, p) \mapsto (ab, p), \quad (b, p) \in \mathcal{L},$$

of the leaf $\mathcal{L}$ *onto the leaf* $a\mathcal{L}$ *induced by the left translation* L_a *is a diffeomorphism.* Let $\pi_p: \mathcal{L}_p \to \widetilde{\mathcal{G}}$ be the restriction of the projection

$$\mathrm{pr}_1: \widetilde{\mathcal{G}} \times \mathcal{X} \to \widetilde{\mathcal{G}}, \quad (a, q) \mapsto a,$$

to $\mathcal{L}_p$. Because the submanifold $\mathcal{L}_p$ is conservative, *the mapping* π_p *is smooth.*

Because $(d\,\mathrm{pr}_1)_{(a,p)}(\widetilde{X}_a, X_q) = \widetilde{X}_a$ for any field $X \in \mathfrak{g}$ and any point $(a, q) \in \widetilde{\mathcal{G}} \times \mathcal{X}$, the differential $(d\pi_p)_{(a,q)}$ of the mapping π_p at the point $(a, q) \in \mathcal{L}_p$ is an isomorphism of the space $\mathcal{D}_{(a,q)} = \mathbf{T}_{(a,q)}\mathcal{L}_p$ onto the space $\mathbf{T}_a\widetilde{\mathcal{G}}$. Therefore, *the mapping* π_p *is étale*, and it is therefore invertible on a neighborhood V_p of the point (e, p) in $\mathcal{L}_p$, i.e., there exists a neighborhood U_p of the identity e of the group $\widetilde{\mathcal{G}}$ and a smooth mapping

$$\varphi_p: U_p \to \mathcal{L}_p \tag{1}$$

such that $\pi_p \circ \varphi_p = \mathrm{id}$ on U_p.

Lemma 10.1. *The neighborhood* U_p *can be chosen one and the same for all points* $p \in \mathcal{X}$.

The neighborhood U_p chosen in such a way is denoted by U. It is clear that we can assume without loss of generality that *the neighborhood* U *is connected.*

We prove Lemma 10.1 below in order not to interrupt the proof of the theorem.

Let $m \geq 1$, and let U^m be the set of all elements of the group $\widetilde{\mathcal{G}}$ of the form $a_1 \cdot \dots \cdot a_m$, where $a_1, \dots, a_m \in U$. We suppose that $U^m \subset \pi_p\mathcal{L}_p$. (Because $\pi_p \circ \varphi_p = \mathrm{id}$ on U, this holds for $m = 1$.) By definition, the inclusion $U^m \subset \pi_p\mathcal{L}_p$ means that for any element $a \in U^m$, there exists a point $q \in \mathcal{X}$ such that $(a, q) \in \mathcal{L}_p$, i.e., $\mathcal{L}_p = a\mathcal{L}_q$. But because $\pi_q \circ \varphi_q = \mathrm{id}$ on U, we have $U \subset \pi_q\mathcal{L}_q$ and therefore

$$aU \subset a(\pi_q\mathcal{L}_q) = \pi_p\mathcal{L}_p.$$

Therefore, because the element $a \in U^m$ is arbitrary, we have $U^{m+1} = U^m U \subset \pi_p \mathcal{L}_p$. By induction, this proves that $U^m \subset \pi_p \mathcal{L}_p$ for any $m \geq 1$ and therefore (because the group $\widetilde{\mathcal{G}}$ is generated by the neighborhood U) $\widetilde{\mathcal{G}} = \pi_p \mathcal{L}_p$. Therefore, *the mapping* $\pi_p\colon \mathcal{L}_p \to \widetilde{\mathcal{G}}$ *is surjective.*

In particular, this implies that for any element $a \in \widetilde{\mathcal{G}}$, we have a point of the form (a^{-1}, q), where $q \in \mathcal{X}$, in the leaf $\mathcal{L}_p$. Moreover, $(e, q) \in a\mathcal{L}_p$ and therefore $\mathcal{L}_q = a\mathcal{L}_p$. Because the leaves $a\mathcal{L}_p$ are (as we know) all leaves of the distribution $\mathcal{D}$, this proves that *any leaf of the distribution* $\mathcal{D}$ *has the form* $\mathcal{L}_p$, $p \in \mathcal{X}$, i.e., the mapping $p \mapsto \mathcal{L}_p$ of the manifold $\mathcal{X}$ onto the set $\{\mathcal{L}_p\}$ is surjective.

We now more carefully consider the open set $V_p = \varphi_p U$ (in $\mathcal{L}_p$ and therefore in $\pi_p^{-1} U = \mathcal{L}_p \cap (U \times \mathcal{X})$) that is diffeomorphically projected on U. Let (a, q) be a point in $\mathcal{L}_p$ such that $\pi_p(a, q) \in U$ but $(a, q) \notin V_p$ (i.e., such that $a \in U$ but $(a, q) \neq \varphi_p(a)$). Because the manifold $\widetilde{\mathcal{G}} \times \mathcal{X}$ is Hausdorff, the manifold $\mathcal{L}_p$ is also Hausdorff (prove this!), and the points (a, q) and $\varphi_p(a)$ therefore admit disjoint neighborhoods $\mathcal{O}_1$ and $\mathcal{O}_2 \subset V_p$ in $\mathcal{L}_p$. Because $\pi_p(a, q) = \pi_p(\varphi_p(a))$, we can assume without loss of generality that $\pi_p \mathcal{O}_1 = \pi_p \mathcal{O}_2$, and therefore $\mathcal{O}_1 \cap V_p = \emptyset$ (because the mapping π_p is injective on V_p). Therefore, each point $(a, q) \in \pi_p^{-1} U$ that does not belong to V_p admits a neighborhood in $\mathcal{L}_p$ that does not intersect V_p. This means that the set V_p is not only open but also closed in $\pi_p^{-1} U$. Because the set V_p (being diffeomorphic to the neighborhood U) is connected, this proves that *the set* V_p *is a connected component of the set* $\pi_p^{-1} U$.

This component is characterized by the fact that $(e, p) \in V_p$. Any other component of the set $\pi_p^{-1} U$ contains a point of the form (e, q) and therefore has the form V_q, where q is a point for which $\mathcal{L}_q = \mathcal{L}_p$. Because V_q is also diffeomorphically projected on U, this proves that *the mapping* π_p *exactly covers the neighborhood* U.

For any point $a \in \widetilde{\mathcal{G}}$, the set aU is its neighborhood, and because any connected component of the set $\pi_p^{-1}(aU)$ has the form aV_q, where $(a, q) \in \mathcal{L}_p$, and, moreover, $\pi_p|_{aV_q} = a \circ \pi_q|_{V_q} \circ a^{-1}$, the mapping π_p regularly covers this neighborhood. Therefore, *the mapping* π_p *is a covering.* (We recall that the leaf $\mathcal{L}_p$, as well as the group $\widetilde{\mathcal{G}}$, is a connected manifold.) Because the group $\widetilde{\mathcal{G}}$ is simply connected by assumption and therefore has only trivial coverings, *the mapping* π_p *is a diffeomorphism.*

In particular, this means that any leaf $\mathcal{L} = \mathcal{L}_p$ contains only one point of the form (e, q), $q \in \mathcal{X}$ (namely, the point (e, p)), i.e., *the mapping* $p \mapsto \mathcal{L}_p$ *of the manifold* $\mathcal{X}$ *onto the set* $\{\mathcal{L}\}$ *of leaves of the distribution* $\mathcal{D}$ *is bijective.*

Therefore, the action of the group $\widetilde{\mathcal{G}}$ on the set $\{\mathcal{L}\}$ is transported to $\mathcal{X}$. (For any element $a \in \widetilde{\mathcal{G}}$ and any point $p \in \mathcal{X}$, the point $ap \in \mathcal{X}$, by definition, is a point q in $\mathcal{X}$ such that $\mathcal{L}_q = a\mathcal{L}_p$.) To study this action, we need to make the choice of the neighborhood U more precise and describe the mappings φ_p explicitly.

Lemma 10.2. *The Lie algebra $\mathfrak{g}$ admits a basis*

$$X_1, \quad \dots, \quad X_m, \tag{2}$$

consisting of smooth vector fields.

We also defer the proof of Lemma 10.2.

This lemma implies that the formulas

$$\beta(t_1X_1 + \cdots + t_mX_m) = \exp t_1\widetilde{X}_1 \cdot \cdots \cdot \exp t_m\widetilde{X}_m,$$
$$\varphi(t_1X_1 + \cdots + t_mX_m) = \varphi_{t_m}^{X_m} \circ \cdots \circ \varphi_{t_1}^{X_1}$$

correctly define certain mappings

$$\beta: \mathfrak{g} \to \widetilde{\mathcal{G}} \quad \text{and} \quad \varphi: \mathfrak{g} \to \mathcal{G}.$$

It is easy to see that the mapping β is étale at the point 0, i.e., this point admits a neighborhood $U^{(0)}$ in the algebra $\mathfrak{g}$ on which the mapping β is a diffeomorphism onto a certain neighborhood U of the identity of the group $\widetilde{\mathcal{G}}$ (we see below that the latter neighborhood is the neighborhood U_p for any point $p \in \mathcal{X}$; therefore (as the notation hints), we can consider it the neighborhood U in Lemma 10.1).

Lemma 10.3. *The formula*

$$\varphi_p(a) = (a, ((\varphi \circ \beta^{-1})a)p), \quad p \in \mathcal{X}, \quad a \in U, \tag{3}$$

defines a smooth mapping

$$\varphi_p: U \to \mathcal{L}_p.$$

We also defer the proof of this lemma.

Because $\varphi_p(e) = (e, p)$ and $\pi_p \circ \varphi_p = \text{id}$, Lemma 10.1 is a direct consequence of Lemma 10.3 (and the mappings φ_p are mappings (1)).

If $q = ((\varphi \circ \beta^{-1})a)^{-1}p$, then $\varphi_q(a) = (a, p)$ and hence $(a, p) \in \mathcal{L}_q$, i.e., $(e, p) \in a^{-1}\mathcal{L}_q$. Therefore, $\mathcal{L}_q = a\mathcal{L}_p$, i.e., $q = ap$. This means that the action

$$\widetilde{\mathcal{G}} \times \mathcal{X} \to \mathcal{X}, \qquad (a, p) \mapsto ap \tag{4}$$

on $U \times \mathcal{X}$ is given by

$$ap = ((\varphi \circ \beta^{-1})a)^{-1}p,$$

which directly implies that *this action is smooth* (on U and therefore on the whole group $\widetilde{\mathcal{G}}$).

Moreover, we see that for and $a = \beta(t_1X_1 + \cdots + t_mX_m) \in U$, the formula

$$L_a = \varphi_{t_1}^{-X_1} \circ \cdots \circ \varphi_{t_m}^{-X_m}$$

holds for the mapping $L_a: p \mapsto ap$. In particular, this shows that $L_a \in \mathcal{G}$. Because the neighborhood U generates a connected group $\widetilde{\mathcal{G}}$, this inclusion also holds for any element $a \in \widetilde{\mathcal{G}}$. Therefore, *the formula*

$$L(a) = L_a, \quad a \in \widetilde{\mathcal{G}},$$

assigns a mapping

$$L\colon \widetilde{\mathcal{G}} \to \mathcal{G}$$

(which is obviously homomorphic).

For $a = \exp t\widetilde{X}_i$, we have $L_a = \varphi_t^{-X_i}$. By definition, this means that the monomorphism $\mathfrak{l}\widetilde{\mathcal{G}} \to \mathfrak{a}\mathcal{X}$ induced by action (4) is given by $\widetilde{X} \mapsto X$ (on the base vectors and therefore everywhere), i.e., it coincides with the initial isomorphism $\mathfrak{l}\widetilde{\mathcal{G}} \to \mathfrak{g}$. Therefore, first, all vector fields $X \in \mathfrak{g}$ are complete, and, second,

$$L(\exp t\widetilde{X}) = \varphi_t^{-X}$$

for each such field. Because all diffeomorphisms φ_t^X, $t \in \mathbb{R}$, $X \in \mathfrak{g}$, generate the group $\mathcal{G}$ by assumption, this implies that *the homomorphism L is an epimorphism* (and therefore generates the isomorphism $\lambda\colon \widetilde{\mathcal{G}}/K \to \mathcal{G}$, where $K = \operatorname{Ker} L$).

If $\exp t\widetilde{X} \in K$ for all $t \in \mathbb{R}$, then $\varphi_t^X = \mathrm{id}$; this is possible only if $X = 0$. Therefore, $\mathfrak{l}K = 0$, and K is hence a discrete invariant subgroup of the group $\widetilde{\mathcal{G}}$. In particular, the subgroup K is closed, and the quotient group $\widetilde{\mathcal{G}}/K$ is therefore a Lie group acting smoothly on $\mathcal{X}$ (and locally isomorphic to the Lie group $\widetilde{\mathcal{G}}$). Because the isomorphism λ transports the smoothness from $\widetilde{\mathcal{G}}/K$ to $\mathcal{G}$, we obtain a smoothness on the group $\mathcal{G}$ that satisfies the conditions of Theorem 10.1.

Exercise 10.2. Show that the smoothness on $\mathcal{G}$ in Theorem 10.1 is unique.

To complete the proof of Theorem 10.1, it remains to prove Lemmas 10.2 and 10.3 (as already noted, Lemma 10.3 implies Lemma 10.1).

Lemma 10.4. *For any complete vector fields $X_1, \dots, X_k \in \mathfrak{g}$ and for any point $p \in \mathcal{X}$, the formula*

$$\gamma_p(\boldsymbol{t}) = (\exp t_1\widetilde{X}_1 \cdot \dots \cdot \exp t_k\widetilde{X}_k, (\varphi_{t_k}^{X_k} \circ \dots \circ \varphi_{t_1}^{X_1})p), \tag{5}$$

where $\boldsymbol{t} = (t_1, \dots, t_k) \in \mathbb{R}^k$, defines a smooth mapping $\gamma_p\colon \mathbb{R}^k \to \mathcal{L}_p$.

Proof. Because the submanifold $\mathcal{L}_p$ is conservative, it suffices to prove that $\gamma_p(\boldsymbol{t}) \in \mathcal{L}_p$ for all $\boldsymbol{t} \in \mathbb{R}^k$.

We first let $k = 1$. In this case, γ_p is a curve $t \mapsto (\beta(t), \varphi(t))$, where $\beta(t) = \exp t\widetilde{X}$, $\varphi(t) = \varphi_t^X(p)$, and $X = X_1$, passing through the point (e, p) for $t = 0$. Moreover, $\dot\gamma_p(t) = (\dot\beta(t), \dot\varphi(t))$, where $\dot\beta(t) = \widetilde{X}_{\beta(t)}$ and $\dot\varphi(t) = X_{\varphi(t)}$, and therefore $\dot\gamma_p(t) \in \mathcal{D}_{\gamma(t)}$ for any $t \in \mathbb{R}$. This means that γ_p is an integral curve of the distribution $\mathcal{D}$. Therefore, $\gamma_p(t) \in \mathcal{L}_p$ for all $t \in \mathbb{R}$.

We suppose that Lemma 10.4 is already proved for the fields $X_1, \dots, X_{k-1}$, $k > 1$, and let

$$a = \exp t_1\widetilde{X}_1 \cdot \dots \cdot \exp t_{k-1}\widetilde{X}_{k-1},$$

$$q = (\varphi_{t_{k-1}}^{X_{k-1}} \circ \dots \circ \varphi_{t_1}^{X_1})p.$$

Then $(a, q) \in \mathcal{L}_p$ and therefore $a\mathcal{L}_q = \mathcal{L}_p$. On the other hand, applying the just proved case where $k = 1$ to the field X_k and the point q, we obtain

$$(\exp t_k \widetilde{X}_k, \varphi_{t_k}^{X_k}(q)) \in \mathcal{L}_q.$$

Therefore, $\gamma_p(t) \in \mathcal{L}_p$. □

Proof (of Lemma 10.3). Mapping (5) constructed for basis (2) (for $k = m$) is connected with mapping (3) by

$$\gamma_p = \varphi_p \circ \beta \circ \chi^{-1} \quad \text{on } U,$$

where $\chi \colon \mathfrak{g} \mapsto \mathbb{R}^m$ is the coordinate isomorphism corresponding to basis (2). □

We note that Lemma 10.2 is used in this proof. Therefore, everything is reduced to the proof of this lemma, for which we need the following elementary algebraic assertion.

Lemma 10.5. *Let a Lie algebra* $\mathfrak{g}$ *be generated by a set* S. *If* S *is closed with respect to multiplication by numbers (i.e.,* $tX \in S$ *for any* $X \in S$ *and* $t \in \mathbb{R}$*) and if*

$$e^{\operatorname{ad} X} Y \in S \quad \textit{for } X, Y \in S, \tag{6}$$

then S *linearly generates* $\mathfrak{g}$.

Proof. Let $\mathcal{S}$ be the linear span of the set S. Because

$$[X, Y] = (\operatorname{ad} X)Y = \lim_{t \to 0} \frac{e^{t \operatorname{ad} X} Y - Y}{t},$$

we have $[X, Y] \in \mathcal{S}$ for any $X, Y \in S$ and therefore for any $X, Y \in \mathcal{S}$ (because of linearity). Therefore, $\mathcal{S}$ is a subalgebra of the Lie algebra $\mathfrak{g}$ that contains S, and then $\mathcal{S} = \mathfrak{g}$. □

Proof (of Lemma 10.2). Lemma 10.2 is equivalent to the assertion that the set S of all complete fields from $\mathfrak{g}$ (which generates the Lie algebra $\mathfrak{g}$ by assumption) linearly generates $\mathfrak{g}$. Because this set is obviously closed with respect to multiplication by numbers, it therefore suffices (by Lemma 10.5) to prove that the set S satisfies condition (6).

Let $X, Y \in S$, $t \in \mathbb{R}$, and let

$$\beta(t) = \exp \widetilde{X} \exp t\widetilde{Y} \exp(-\widetilde{X}),$$
$$\varphi(t) = (\varphi_1^X \circ \varphi_t^Y \circ \varphi_{-1}^X)p.$$

Because $\beta(t) = \operatorname{int}_a(\exp t\widetilde{Y})$, where $a = \exp \widetilde{X}$, and therefore

$$\beta(t) = \exp t(\operatorname{Ad} a)\widetilde{Y} = \exp t\, e^{\operatorname{ad} \widetilde{X}} \widetilde{Y},$$

we have $\dot{\beta}(0) = \widetilde{Z}_e$, where $Z = e^{\operatorname{ad} X} Y$; we recall that $\operatorname{Ad} a = \mathfrak{l}((\operatorname{int}_a))$ by definition. But because the inclusion $(\beta(t), \varphi(t)) \in \mathcal{L}_p$ holds for any $t \in \mathbb{R}$ according to Lemma 10.4, we have $(\dot\beta(0), \dot\varphi(0)) \in \mathcal{D}_{(e,p)}$ and therefore $(\widetilde{Z}_e, \dot\varphi(0)) \in \mathcal{D}_{(e,p)}$, i.e., $\dot\varphi(0) = Z_p$. By the arbitrariness of the point p, this means that the flow $\{\varphi_1^X \circ \varphi_t^Y \circ \varphi_{-1}^X\}$ is generated by the vector field Z. Therefore, this field is complete. Moreover, because $(\widetilde{Z}_e, Z_p) \in \mathcal{D}_{(e,p)}$ for any point $p \in \mathcal{X}$, the field Z belongs to $\mathfrak{g}$ by the definition of the subspaces $\mathcal{D}_{(e,p)}$. Being a complete field from $\mathfrak{g}$, the field Z lies in S. □

Therefore, Theorem 10.1 is completely proved. □

Remark 10.1. In proving Theorem 10.1, we showed that *any subalgebra* $\mathfrak{g}$ *of the Lie algebra* $\mathfrak{a}\mathcal{X}$ *satisfying conditions* 1 *and* 2 *of Theorem* 10.1 *consists of complete vector fields.*

Theorem 10.1 is known as the *Palais theorem.*

§4. Kobayashi Theorem

Exercise 10.3. Let $\mathcal{G}$ be a group, and let $\mathcal{G}_0$ be its invariant subgroup. Show that if

1. the group $\mathcal{G}_0$ is a Lie group and
2. all inner automorphisms

$$\mathrm{int}_a : x \mapsto axa^{-1}, \quad a, x \in \mathcal{G}$$

are smooth mappings on $\mathcal{G}_0$,

then there exists a unique smoothness on $\mathcal{G}$ with respect to which the group $\mathcal{G}$ is a Lie group whose component of the identity is $\mathcal{G}_0$.

The proof of the following variant of the Palais theorem, which is known as the *Kobayashi theorem*, is based on this assertion.

Theorem 10.2. *Let* $\mathcal{G}$ *be a subgroup of the diffeomorphism group of a manifold* $\mathcal{X}$, *and let* S *be the set of all complete fields* $X \in \mathfrak{a}\mathcal{X}$ *such that* $\varphi_t^X \in \mathcal{G}$ *for any* $t \in \mathbb{R}$. *If* S *generates a finite-dimensional Lie subalgebra* $\mathfrak{g}$ *in the Lie algebra* $\mathfrak{a}\mathcal{X}$, *then* $S = \mathfrak{g}$, *and the group* $\mathcal{G}$ *admits a unique smoothness with respect to which it is a Lie group that acts smoothly and effectively on the manifold* $\mathcal{X}$; *moreover, the monomorphism* $X \mapsto X^*$ *corresponding to this action is an isomorphism of the Lie algebra* $\mathfrak{l}\mathcal{G}$ *onto the subalgebra* $\mathfrak{g}$.

Proof. Let $\mathcal{G}_0$ be a subgroup of the group $\mathcal{G}$ generated by all diffeomorphisms of the form φ_t^X, where $t \in \mathbb{R}$ and X is an arbitrary (complete) field from $\mathfrak{g}$. Because the Lie algebra $\mathfrak{g}$ obviously satisfies conditions 1 and 2 of Theorem 10.1, the group $\mathcal{G}_0$ admits the smoothness in this theorem (and according to the assertion in Exercise 10.2, this smoothness is unique). According to Remark 10.1, all fields from $\mathfrak{g}$ are complete; because they generate one-parameter subgroups of the group $\mathcal{G}$, we have $\mathfrak{g} \subset S$ and therefore $S = \mathfrak{g}$.

For any flow $\{\varphi_t^X\}$, $X \in \mathfrak{g}$, and any diffeomorphism $\varphi \in \mathcal{G}$, the diffeomorphism $\varphi^{-1} \circ \varphi_t^X \circ \varphi$ composes a complete flow and is contained in $\mathcal{G}$. Therefore, $\varphi^{-1} \circ \varphi_t^X \circ \varphi \in \mathcal{G}_0$. Because diffeomorphisms of the form φ_t^X generate the subgroup $\mathcal{G}_0$, this implies $\varphi^{-1}\mathcal{G}_0\varphi \subset \mathcal{G}_0$, i.e., the subgroup $\mathcal{G}_0$ is invariant.

Exercise 10.4. Show that condition 2 of Exercise 10.3 holds for the invariant subgroup $\mathcal{G}_0$.

Therefore, there exists a unique smoothness on $\mathcal{G}$ with respect to which the group $\mathcal{G}$ is a Lie group whose component of the identity is $\mathcal{G}_0$. An obvious verification shows that this smoothness satisfies all conditions of Theorem 10.2. □

§5. Affine Automorphism Group

We apply the obtained general theorems to groups that arise in differential geometry.

Proposition 10.1. *The group* Aff $\mathcal{X}$ *of affine automorphisms of an arbirary connected affine connection space* $\mathcal{X}$ *admits a natural smoothness with respect to which it is a Lie group acting smoothly on* $\mathcal{X}$. *In the case where the space* $\mathcal{X}$ *is geodesically complete, the Lie algebra of the group* Aff $\mathcal{X}$ *is the Lie algebra* $\mathfrak{aff}\,\mathcal{X}$ *of affine fields.*

Proof. According to Proposition 8.2, the group Aff $\mathcal{X}$ satisfies the condition of Theorem 10.2. If the space $\mathcal{X}$ is geodesically complete, then according to Proposition 8.3, the corresponding Lie algebra $\mathfrak{g}$ coincides with the algebra $\mathfrak{aff}\,\mathcal{X}$. □

§6. Automorphism Group of a Symmetric Space

Corollary 10.1. *The group* Aut $\mathcal{X}$ *of automorphisms of an arbitrary symmetric space* $\mathcal{X}$ *is a Lie group with the Lie algebra* $\mathfrak{d}\mathcal{X}$.

Proof. According to Proposition 6.1, the group Aut $\mathcal{X}$ coincides with the group Aff $\mathcal{X}$, and the Lie algebra $\mathfrak{d}\mathcal{X}$ coincides with the Lie algebra $\mathfrak{aff}\,\mathcal{X}$. In addition, according to the assertion in Exercise 5.1, the space $\mathcal{X}$ is geodesically complete. □

Exercise 10.5. Let $\beta: \mathbb{R} \to \mathcal{X}$ be a curve in a punctured symmetric space $\mathcal{X}$ that passes through a point p_0 for $t = 0$. Prove that the following assertions are equivalent:

1. The curve β is a morphism of symmetric spaces (where $\mathbb{R}$ is considered as a symmetric space with the multiplication $s \cdot t = 2s - t$; see Exercise 5.6).
2. The curve β is an integral curve of a certain derivation $X \in \mathfrak{s}\mathcal{X}$.

3. For the flow $\{\varphi_t^X\}$ induced by the field X, we have

$$\varphi_{2t}^X = s_{\beta(t)} \circ s_{p_0}, \quad t \in \mathbb{R}. \tag{7}$$

4. The curve β is a geodesic of the canonical connection.
5. Parallel translations along the curve β are differentials of the mappings φ_t.

§7. Translation Group of a Symmetric Space

Proposition 10.2. *The translation group* Trans $\mathcal{X}$ *of an arbitrary symmetric space* $\mathcal{X}$ *is a connected subgroup of the Lie group* Aut $\mathcal{X}$. *Its algebra Lie* trans $\mathcal{X}$ *is the subalgebra of the Lie algebra* $\mathfrak{d}\mathcal{X}$ *generated by all fields* $X \in \mathfrak{s}\mathcal{X}$.

(Compare this with Proposition 9.3.)

Proof. The subalgebra $\mathfrak{g}$ of the Lie algebra $\mathfrak{d}\mathcal{X}$ generated by all fields $X \in \mathfrak{s}\mathcal{X}$ satisfies conditions 1 and 2 of Theorem 10.1, and the corresponding group $\mathcal{G}$ is therefore a connected Lie group acting smoothly on $\mathcal{X}$. In this case, according to (7), the group $\mathcal{G}$ is generated by all possible transformations of the form $s_{\beta(t)} \circ s_{p_0}$, where p_0 is a distinguished point of the punctured symmetric space $\mathcal{X}$ and $\beta\colon \mathbb{R} \to \mathcal{X}$ is an arbitrary geodesic of the space $\mathcal{X}$ passing through the point p_0 for $t = 0$. Therefore, $\mathcal{G} \subset$ Trans $\mathcal{X}$.

To prove the converse inclusion, for any point $p \in \mathcal{X}$, we consider the set W_p of all points $q \in \mathcal{X}$ for which $s_q \circ s_p \in \mathcal{G}$. If $p_0 \in W_p$ and U is an arbitrary normal neighborhood of the point p_0, then the transformation $s_q \circ s_{p_0}$ belongs to $\mathcal{G}$ because any point $q \in U$ has the form $\beta(t)$. Moreover, $s_{p_0} \circ s_p \in \mathcal{G}$ by assumption. Therefore, $s_q \circ s_p = s_p \circ s_{p_0} \circ s_{p_0} \circ s_q \in \mathcal{G}$ and hence $U \subset W_p$. Therefore, an arbitrary normal neighborhood of each point from W_p is contained in W_p, which directly implies (compare with the proof of Proposition 8.3) that the set W_p is open and closed in $\mathcal{X}$. Because the set W_p is not empty (because $p \in W_p$) and the space $\mathcal{X}$ is connected by assumption, this proves that $W_p = \mathcal{X}$, i.e., $s_q \circ s_p \in \mathcal{G}$ for any point $q \in \mathcal{X}$, and therefore Trans $\mathcal{X} \subset \mathcal{G}$.

Therefore, Trans $\mathcal{X} = \mathcal{G}$, which obviously proves Proposition 10.2. □

We note that the Lie algebra trans $\mathcal{X}$ is obviously symmetric and is the minimal symmetric Lie algebra with the socle $\mathfrak{s}\mathcal{X}$.

Chapter 11
Lagrangians in Riemannian Spaces

§1. Riemannian and Pseudo-Riemannian Spaces

Definition 11.1. A *Riemannian structure* on a smooth n-dimensional manifold $\mathcal{X}$ is a smooth Euclidean metric on the tangent bundle $\tau_{\mathcal{X}}$, in other words, a tensor field g of type (2,0) on $\mathcal{X}$ such that for any point $p \in \mathcal{X}$, the bilinear functional g_p on the linear space $\mathbf{T}_p\mathcal{X}$ is an inner product (is symmetric and positive definite). A manifold $\mathcal{X}$ on which a Riemannian structure is given is called a *Riemannian space.* The tensor field g is also called the *metric tensor* of a Riemannian space or merely the *metric* (sometimes, a *Riemannian metric*).

For any vectors $A, B \in \mathbf{T}_p\mathcal{X}$, their inner product $g_p(A, B)$ with respect to the metric g_p is also denoted by $(A, B)_p$ or merely (A, B).

Using the tensor g, we can lower superscripts and raise subscripts of tensor fields, i.e., identify tensor fields of types (a, b) for which the sum $a + b$ is the same. For example, a vector field X with components X^i can be identified with a covector field (linear differential form) that is obtained by contracting the tensor g with the field X (and having the components $g_{ij}X^j$, where g_{ij} are the components of the metric g). Similarly, contracting the differential du of an arbitrary smooth function u on $\mathcal{X}$ (or, more precisely, the corresponding covector field ∇u) with the tensor g^{ij} (satisfying the relation $g^{ij}g_{jk} = \delta^i_k$), we obtain the *gradient vector field* $\operatorname{grad} u$ with the components

$$(\operatorname{grad} u)^i = g^{ij}\frac{\partial u}{\partial x^j}, \quad i = 1, \dots, n. \tag{1}$$

For an arbitrary submanifold $\mathcal{Y}$ of a Riemannian manifold $\mathcal{X}$, the linear space $\mathbf{T}_p\mathcal{Y}$ at each point $p \in \mathcal{Y}$ is a subspace of the space $\mathbf{T}_p\mathcal{X}$ and is therefore equipped with the Euclidean metric $g_p|_{\mathcal{Y}}$. This assigns the structure of a Riemannian space on $\mathcal{Y}$. We say that this structure is *induced* by the Riemannian structure on $\mathcal{X}$. A submanifold $\mathcal{Y}$ equipped with an induced structure is called a *Riemannian subspace.* In what follows, when considering submanifolds of Riemannian spaces, we always assume (if the contrary is not specified) that the induced Riemannian structure is introduced on them.

Of course, any Euclidean space is a Riemannian space. Therefore, *each submanifold of a Euclidean space has a natural Riemannian structure* (induced by the Euclidean structure of the ambient space). For $n = 2$, this structure is just the first quadratic form of a surface in the sense of elementary differential geometry.

A Riemannian structure exists on any Hausdorff paracompact manifold $\mathcal{X}$. (Proof: because a manifold is paracompact iff each of its components satisfies the second countability axiom, we can assume without loss of generality that the manifold $\mathcal{X}$ satisfies the second countability axiom; therefore, it is embeddable into a Euclidean space.)

More generally, we can consider a manifold $\mathcal{X}$ on which we have a tensor field g of type (2,0) that defines a pseudo-Euclidean structure on $\mathbf{T}_p\mathcal{X}$ for each $p \in \mathcal{X}$, i.e., a symmetric nondegenerate, but not positive definite in general, bilinear functional g_p. Such a field is called a *pseudo-Riemannian structure* on $\mathcal{X}$, and a manifold $\mathcal{X}$ equipped with a pseudo-Riemannian structure is called a *pseudo-Riemannian space.*

If a manifold $\mathcal{X}$ is connected, then by obvious continuity arguments, the index (signature) of a pseudo-Euclidean metric g_p does not depend on p. This index (signature) is called the *index* (*signature*) of the pseudo-Riemannian space $\mathcal{X}$.

We note that, in general, a submanifold of a pseudo-Riemannian space has no natural pseudo-Riemannian structure (the restriction to $\mathcal{Y}$ of the pseudo-Riemannian metric can degenerate), and for any k, $0 < k < n$, there exist smooth n-dimensional Hausdorff manifolds satisfying the second countability axiom (and being therefore paracompact) on which it is not possible to introduce a pseudo-Riemannian structure of index k. It turns out that if such a structure exists, some characteristic classes of the tangent bundle of the manifold $\mathcal{X}$ should vanish.

Exercise 11.1. Show that if a pseudo-Riemannian structure of index 1 exists on an orientable manifold $\mathcal{X}$, then the Euler class $e[\mathcal{X}]$ of the bundle $\tau_{\mathcal{X}}$ should vanish. [*Hint*: Eigenvectors of the functionals g_p corresponding to the negative eigenvalue compose an everywhere nonzero field of tangent vectors on $\mathcal{X}$, which is not possible for $e[\mathcal{X}] \neq 0$.]

For any two smooth vector fields X and Y on a pseudo-Riemannian (and, in particular, Riemannian) space $\mathcal{X}$, their contraction $g(X, Y)$ with the tensor g is a smooth function on $\mathcal{X}$. This function is called the *inner product* of the fields X and Y and is denoted by (X, Y) (i.e., $(X, Y) = g(X, Y)$ by definition). The correspondence $X, Y \mapsto (X, Y)$ is a bilinear functional (over the algebra $\mathsf{F}\mathcal{X}$) on the $\mathsf{F}\mathcal{X}$-module $\mathfrak{a}\mathcal{X}$ and is also called the *inner product.*

We note that $(X, Y)(p) = (X_p, Y_p)_p$ for any point $p \in \mathcal{X}$. We also write $(X, Y)_p$ instead of $(X, Y)(p)$. In each chart $(U, x^1, \ldots, x^n)$, the function (X, Y) is expressed by

$$(X, Y) = g_{ij} X^i Y^j \quad \text{on } U,$$

where

$$g_{ij} = g\left(\frac{\partial}{\partial x^i}, \frac{\partial}{\partial x^j}\right)$$

are components of the tensor g and X^i and Y^j are components of the vector fields X and Y.

§2. Riemannian Connections

Definition 11.2. A connection ∇ on a pseudo-Riemannian (in particular, Riemannian) space $\mathcal{X}$ is *compatible with the metric* g if

$$X(Y,Z) = (\nabla_X Y, Z) + (Y, \nabla_X Z) \tag{2}$$

for any three vector fields $X, Y, Z \in \mathfrak{a}\mathcal{X}$.

It is clear that this relation holds for any fields X, Y, and Z iff it holds for the coordinate base fields $\partial/\partial x^i$, $\partial/\partial x^j$, and $\partial/\partial x^k$, $1 \leq i,j,k \leq n$, in each chart $(U, x^1, \dots, x^n)$, i.e., by the relations

$$g_{jk} = \left(\frac{\partial}{\partial x^j}, \frac{\partial}{\partial x^k}\right) \quad \text{and} \quad \Gamma^i_{kj} = \left(\nabla_k \frac{\partial}{\partial x^j}\right)^i,$$

if

$$\frac{\partial g_{jk}}{\partial x^i} = \Gamma_{k,ij} + \Gamma_{j,ik}, \tag{2'}$$

where

$$\Gamma_{k,ij} = g_{kp}\Gamma^p_{ij} \tag{3}$$

are the so-called connection coefficients of the *first kind* (with lowered superscripts). (The coefficients Γ^i_{kj} are called connection coefficients of the *second kind.*)

Formula (2′) can be rewritten as

$$\frac{\partial g_{ij}}{\partial x^k} - \Gamma^p_{ki} g_{pj} - \Gamma^p_{kj} g_{ip} = 0.$$

Because the left-hand expression yields exactly the components of the covariant derivatives $\nabla_k g$ of the tensor g, this proves that *a Riemannian structure is compatible with a connection ∇ iff the metric tensor g is covariantly constant.*

Exercise 11.2. Let $u: I \to \mathcal{X}$ be a curve in a (pseudo-)Riemannian space $\mathcal{X}$, and let $X, Y: t \to X(t), Y(t)$ be two vector fields on the curve u. Show that the formula

$$\frac{d}{dt}(X(t), Y(t)) = \left(\frac{\nabla X}{dt}(t), Y(t)\right) + \left(X(t), \frac{\nabla Y}{dt}(t)\right) \tag{4}$$

holds for the covariant derivatives with respect to a connection that is compatible with the Riemannian structure.

Theorem 11.1. *On a (pseudo-)Riemannian space $\mathcal{X}$, there exists a unique symmetric connection that is compatible with the (pseudo-)Riemannian structure on $\mathcal{X}$.*

Proof. This theorem is proved by direct computation, either in coordinates or in invariant form. As usual for existence and uniqueness theorems, we first prove the uniqueness.

Uniqueness: If a connection ∇ exists, then according to (2′),

$$\frac{\partial g_{ij}}{\partial x^k} = \Gamma_{j,ki} + \Gamma_{i,kj},$$
$$\frac{\partial g_{ik}}{\partial x^j} = \Gamma_{k,ji} + \Gamma_{i,jk},$$
$$\frac{\partial g_{jk}}{\partial x^i} = \Gamma_{k,ij} + \Gamma_{j,ik}.$$

Because the coefficients Γ^i_{kj} (and therefore the coefficients $\Gamma_{i,kj}$) are symmetric in k and j by the symmetry of the connection, the second summands in the right-hand sides of the first two relations are the same, and the first ones coincide with the summands in the right-hand side of the third relation. Therefore, adding the first two relations, subtracting the third, and dividing by 2, we obtain

$$\Gamma_{i,jk} = \frac{1}{2}\left(\frac{\partial g_{ij}}{\partial x^k} + \frac{\partial g_{ik}}{\partial x^j} - \frac{\partial g_{jk}}{\partial x^i}\right), \tag{5}$$

and after raising the subscript i, we therefore have

$$\Gamma^i_{jk} = \frac{1}{2} g^{ip}\left(\frac{\partial g_{pj}}{\partial x^k} + \frac{\partial g_{pk}}{\partial x^j} - \frac{\partial g_{jk}}{\partial x^p}\right). \tag{5'}$$

Therefore, the coefficients of the connection ∇ are expressed through the components of the tensor g, and this connection is therefore unique.

The same argument in invariant form is as follows. Because the connection ∇ is symmetric, we have

$$\nabla_X Z - \nabla_Z X = [X, Z]$$

for any fields $X, Z \in \mathfrak{a}\mathcal{X}$, and relation (2) is therefore equivalent to the relation

$$X(Y,Z) = (\nabla_X Y, Z) + (Y, \nabla_Z X) + (Y, [X, Z]).$$

Cyclically permuting X, Y, and Z here, we obtain two more relations. Adding the first and second relations and subtracting the third, we obtain the identity

$$\begin{aligned} 2(\nabla_X Y, Z) &= X(Y,Z) + Y(Z,X) - Z(X,Y) \\ &\quad + (X, [Z,Y]) - (Y, [X,Z]) - (Z, [Y,X]); \end{aligned} \tag{6}$$

by the arbitrariness of the field Z and the nondegeneracy of the tensor g, this uniquely defines the covariant derivative $\nabla_X Y$.

Existence: We define the connection ∇ by taking functions (5') as its coefficients Γ^i_{jk} in each chart $(U, x^1, \dots, x^n)$. To justify this construction, we must verify that for any chart $(U', x^{1'}, \dots, x^{n'})$, the functions

$$\Gamma^{i'}_{j'k'} = \frac{1}{2} g^{i'p'} \left(\frac{\partial g_{p'j'}}{\partial x^{k'}} + \frac{\partial g_{p'k'}}{\partial x^{j'}} - \frac{\partial g_{j'k'}}{\partial x^{p'}} \right),$$

where $g_{i'j'}$ are components of the metric tensor g in the chart $(U', x^{1'}, \dots, x^{n'})$, are connected with the functions Γ^i_{jk} on $U \cap U'$ by

$$\Gamma^{i'}_{j'k'} = \frac{\partial x^{i'}}{\partial x^i} \frac{\partial x^j}{\partial x^{j'}} \frac{\partial x^k}{\partial x^{k'}} \Gamma^i_{jk} + \frac{\partial x^{i'}}{\partial x^i} \frac{\partial^2 x^i}{\partial x^{j'} \partial x^{k'}}$$

(see (3) in Chap. 1). But because

$$g_{i'j'} = \frac{\partial x^i}{\partial x^{i'}} \frac{\partial x^j}{\partial x^{j'}} g_{ij}$$

and therefore

$$\frac{\partial g_{i'j'}}{\partial x^{k'}} = \frac{\partial^2 x^i}{\partial x^{k'} \partial x^{i'}} \frac{\partial x^j}{\partial x^{j'}} g_{ij} + \frac{\partial x^i}{\partial x^{i'}} \frac{\partial^2 x^j}{\partial x^{k'} \partial x^{j'}} g_{ij} + \frac{\partial x^i}{\partial x^{i'}} \frac{\partial x^j}{\partial x^{j'}} \frac{\partial x^k}{\partial x^{k'}} \frac{\partial g_{ij}}{\partial x^k},$$

we have

$$\begin{aligned} 2\Gamma_{i',j'k'} = {} & \frac{\partial^2 x^i}{\partial x^{k'} \partial x^{i'}} \frac{\partial x^j}{\partial x^{j'}} g_{ij} + \frac{\partial x^i}{\partial x^{i'}} \frac{\partial^2 x^j}{\partial x^{k'} \partial x^{j'}} g_{ij} + \frac{\partial x^i}{\partial x^{i'}} \frac{\partial x^j}{\partial x^{j'}} \frac{\partial x^k}{\partial x^{k'}} \frac{\partial g_{ij}}{\partial x^k} \\ & + \frac{\partial^2 x^i}{\partial x^{j'} \partial x^{i'}} \frac{\partial x^k}{\partial x^{k'}} g_{ik} + \frac{\partial x^i}{\partial x^{i'}} \frac{\partial^2 x^k}{\partial x^{j'} \partial x^{k'}} g_{ik} + \frac{\partial x^i}{\partial x^{i'}} \frac{\partial x^k}{\partial x^{k'}} \frac{\partial x^j}{\partial x^{j'}} \frac{\partial g_{ik}}{\partial x^j} \\ & - \frac{\partial^2 x^j}{\partial x^{i'} \partial x^{j'}} \frac{\partial x^k}{\partial x^{k'}} g_{jk} - \frac{\partial x^j}{\partial x^{j'}} \frac{\partial^2 x^k}{\partial x^{i'} \partial x^{k'}} g_{jk} - \frac{\partial x^j}{\partial x^{j'}} \frac{\partial x^k}{\partial x^{k'}} \frac{\partial x^i}{\partial x^{i'}} \frac{\partial g_{jk}}{\partial x^i}. \end{aligned}$$

In the right-hand side, the second terms equals the fifth, and the first and fourth are canceled by the eighth and seventh. Therefore,

$$\Gamma_{i',j'k'} = \frac{\partial x^i}{\partial x^{i'}} \frac{\partial^2 x^j}{\partial x^{j'} \partial x^{k'}} g_{ij} + \frac{\partial x^i}{\partial x^{i'}} \frac{\partial x^j}{\partial x^{j'}} \frac{\partial x^k}{\partial x^{k'}} \Gamma_{i,jk}.$$

Because

$$g^{i'j'} = \frac{\partial x^{i'}}{\partial x^i} \frac{\partial x^{j'}}{\partial x^j} g^{ij},$$

$$\frac{\partial x^{i'}}{\partial x^l} \frac{\partial x^{p'}}{\partial x^q} \frac{\partial x^p}{\partial x^{p'}} g^{lq} g_{pj} = \frac{\partial x^{i'}}{\partial x^l} g^{lq} g_{qj} = \frac{\partial x^{i'}}{\partial x^j},$$

$$\frac{\partial x^{p'}}{\partial x^q} \frac{\partial x^p}{\partial x^{p'}} g^{lq} \Gamma_{p,jk} = g^{lp} \Gamma_{p,jk} = \Gamma^l_{jk},$$

we obtain

$$\begin{aligned}\Gamma^{i'}_{j'k'} &= \left(\frac{\partial x^{i'}}{\partial x^l}\frac{\partial x^{p'}}{\partial x^q}g^{lq}\right)\left(\frac{\partial x^p}{\partial x^{p'}}\frac{\partial^2 x^j}{\partial x^{j'}\partial x^{k'}}g_{pj} + \frac{\partial x^p}{\partial x^{p'}}\frac{\partial x^j}{\partial x^{j'}}\frac{\partial x^k}{\partial x^{k'}}\Gamma_{p,jk}\right)\\ &= \frac{\partial x^{i'}}{\partial x^j}\frac{\partial^2 x^j}{\partial x^{j'}\partial x^{k'}} + \frac{\partial x^{i'}}{\partial x^l}\frac{\partial x^j}{\partial x^{j'}}\frac{\partial x^k}{\partial x^{k'}}\Gamma^l_{jk}\\ &= \frac{\partial x^{i'}}{\partial x^i}\frac{\partial x^j}{\partial x^{j'}}\frac{\partial x^k}{\partial x^{k'}}\Gamma^i_{jk} + \frac{\partial x^{i'}}{\partial x^i}\frac{\partial^2 x^i}{\partial x^{j'}\partial x^{k'}}.\end{aligned}$$

Because the right-hand side of (5) is symmetric in j and k, the constructed connection is symmetric. Moreover, because

$$\begin{aligned}\Gamma_{j,ik} + \Gamma_{i,jk} &= \frac{1}{2}\left(\frac{\partial g_{ji}}{\partial x^k} + \frac{\partial g_{jk}}{\partial x^i} - \frac{\partial g_{ik}}{\partial x^j}\right) + \frac{1}{2}\left(\frac{\partial g_{ji}}{\partial x^k} + \frac{\partial g_{ik}}{\partial x^j} - \frac{\partial g_{jk}}{\partial x^i}\right)\\ &= \frac{\partial g_{ij}}{\partial x^k},\end{aligned}$$

this connection is compatible with g. □

In invariant form, this argument is as follows. As $\nabla_X Y$, we take a field that is uniquely characterized by (6). Because the right-hand side of this relation is $\mathbf{F}\mathcal{X}$-linear in X and $\mathbb{R}$-linear in Y, the field $\nabla_X Y$ is also $\mathbf{F}\mathcal{X}$-linear in X and $\mathbb{R}$-linear in Y. Moreover, if Y is replaced by fY, where $f \in \mathbf{F}\mathcal{X}$, in the right-hand side of (6), then because X and Z are derivations and $[Z, fY] = f[Z,Y] + Zf \cdot Y$ and $[fY, X] = f[Y,X] - Xf \cdot Y$, the right-hand side of (6) is equal to the sum of the old part multiplied by f and four summands: $Xf \cdot (Y,Z)$ (first term), $-Zf \cdot (X,Y)$ (third term), $Zf \cdot (X,Y)$ (fourth term), and $Xf \cdot (Z,Y)$ (sixth term). Therefore,

$$(\nabla_X(fY), Z) = f(\nabla_X Y, Z) + Xf \cdot (Y,Z) = (f\nabla_X Y + Xf \cdot Y, Z)$$

and hence $\nabla_X(fY) = Xf \cdot Y + f\nabla_X Y$. This proves that the correspondence $\nabla\colon X \mapsto \nabla_X$ is a connection. Subtracting formula (6) with interchanged X and Y from (6) and dividing by 2, we similarly obtain

$$(\nabla_X Y - \nabla_Y X, Z) = ([X,Y], Z),$$

i.e., the connection ∇ is symmetric. Finally, interchanging Z and Y in (6) and adding, we establish that the connection ∇ is compatible with g.

Theorem 11.1 is known as *Levi-Civita theorem*. The connection ∇ is called a *Riemannian connection* or a *Levi-Civita connection* induced by the metric g.

Exercise 11.3. Prove that parallel translation with respect to a Riemannian connection is an isometric mapping of tangent spaces. [*Hint*: Use (4).]

An affine connection ∇ is said to be *metrical* if there exists a pseudo-Riemannian metric inducing ∇, i.e., there exists a covariantly constant (with respect to ∇) symmetric tensor field g_{ij} that is nondegenerate at each point.

Obviously, the covariant constancy condition $\nabla_k g_{ij} = 0$ is linear in g_{ij}. Therefore, if two fields g_{ij} and h_{ij} induce a metric connection ∇, then the fields $g_{ij} + h_{ij}$ and λg_{ij} for any $\lambda \in \mathbb{R}$ induce the same connection, certainly, under the condition that these fields are nondegenerate. This means that the set of all fields g_{ij} induced a given metric connection ∇ is a subset of a certain linear subspace of the linear space of all symmetric fields g_{ij} consisting of nondegenerate fields (and of positive-definite fields in the case where we are only interested in Riemannian metrics).

Hereafter, when considering a (pseudo-)Riemannian space $\mathcal{X}$, we always assume that it is equipped with a Riemannian connection. Because the metric tensor is covariantly constant with respect to this connection, *the operation of contraction with g (lowering and raising superscripts and subscripts) commutes with the operation of covariant differentiation.* For example, the formula

$$g_{ij}(\nabla_X \operatorname{grad} u)^j = (\nabla_X \nabla u)_i, \quad i = 1, \dots, n, \tag{7}$$

where ∇u is the differential du of the function u considered as a covector field (i.e., as the field with the components $\partial u/\partial x^i$), holds in each coordinate neighborhood for any smooth function $u \in \mathbf{F}\mathcal{X}$ and any vector field $X \in \mathfrak{a}\mathcal{X}$.

§3. Geodesics in a Riemannian Space

Geodesics of a (pseudo-)Riemannian space $\mathcal{X}$ are geodesics with respect to its Riemannian connection. In local coordinates, they are given by Eq. (5) in Chap. 1 with the coefficients Γ^i_{jk} from (5′), i.e., by the equations

$$\ddot{x}^i(t) + \frac{1}{2} g^{ip}\left(\frac{\partial g_{pj}}{\partial x^k} + \frac{\partial g_{pk}}{\partial x^j} - \frac{\partial g_{jk}}{\partial x^p}\right)\dot{x}^j(t)\dot{x}^k(t) = 0, \quad i = 1, \dots, n, \tag{8}$$

which can be rewritten in the form

$$\ddot{x}^i(t) + g^{ip}\left(\frac{\partial g_{pj}}{\partial x^k} - \frac{1}{2}\frac{\partial g_{jk}}{\partial x^p}\right)\dot{x}^j(t)\dot{x}^k(t) = 0, \quad i = 1, \dots, n. \tag{8′}$$

It is easy to see that *if a geodesic γ of a Riemannian space $\mathcal{X}$ lies entirely on a submanifold $\mathcal{Y} \subset \mathcal{X}$, then it is also a geodesic as a curve in $\mathcal{Y}$ (with respect to the Riemannian metric induced on $\mathcal{Y}$).*

Because parallel translation with respect to the Riemannian connection is an isometric mapping, the vector $\dot{\gamma}_A(t)$ has the same length for any geodesic $\gamma_A = \gamma_{p,A}$, which is equal to the length of the vector $\dot{\gamma}_A(0) = A$ for all t:

$$|\dot{\gamma}_A(t)| = |A| \quad \text{for all } t \in I_A. \tag{9}$$

(We recall that I_A denotes the interval of the real axis $\mathbb{R}$ on which the maximal geodesic γ_A is defined.) Of course, in a pseudo-Riemannian space, the number $|A|$ can be equal to zero even if $A \neq 0$. The corresponding geodesics are said to be *isotropic*.

§4. Simplest Problem of the Calculus of Variations

For a (pseudo-)Riemannian space $\mathcal{X}$, the results about geodesics obtained in Chap. 1 can be essentially completed and improved. Preparatorily, we present the necessary general notions and results.

Let $\mathcal{X}$ be a connected Hausdorff smooth manifold, and let $\mathbf{T}\mathcal{X}$ be the manifold of its tangent vectors (total space of its tangent bundle $\tau_{\mathcal{X}}$). As we know, each chart (U, h) of the manifold $\mathcal{X}$ defines a chart $(\mathbf{T}U, \mathbf{T}h)$ of the manifold $\mathbf{T}\mathcal{X}$ for which $\mathbf{T}h$ is a fiber diffeomorphism of the open set $\mathbf{T}U = \pi^{-1}U$ onto the product $h(U) \times \mathbb{R}^n$. In contrast to the previous (and subsequent) chapters, we now let $q^1, \dots, q^n$ denote local coordinates of the chart (U, h) and $q^1, \dots, q^n, \dot{q}^1, \dots, \dot{q}^n$ denote local coordinates of the chart $(\mathbf{T}U, \mathbf{T}h)$. (By definition, the vector $A \in \mathbf{T}U$ has the coordinates $q^1, \dots, q^n, \dot{q}^1, \dots, \dot{q}^n$ if $h(\pi A) = (q^1, \dots, q^n)$ and $A = \dot{q}^i(\partial/\partial q^i)_{\pi A}$.)

Definition 11.3. A smooth function L given on the manifold $\mathbf{T}\mathcal{X}$ (or in a certain open subset of it) is called a *Lagrangian* on $\mathcal{X}$.

(It is also possible to consider *Lagrangians depending on time* t, i.e., smooth functions on the product $\mathbf{T}\mathcal{X} \times \mathbb{R}$. But we do not need them in this chapter.)

Any Lagrangian L is given in each chart (U, h) (more precisely, in the chart $(\mathbf{T}U, \mathbf{T}h)$) by a smooth function $L(\boldsymbol{q}, \dot{\boldsymbol{q}}) = L(q^1, \dots, q^n, \dot{q}^1, \dots, \dot{q}^n)$ of $2n$ variables $q^1, \dots, q^n, \dot{q}^1, \dots, \dot{q}^n$. As a rule, we do not make a pedantic distinction between L and $L(\boldsymbol{q}, \dot{\boldsymbol{q}})$.

In mechanics, the role of the manifold $\mathcal{X}$ is played by the configuration space of a mechanical system, and the role of the manifold $\mathbf{T}\mathcal{X}$ is played by the phase space of its velocities. The numbers $q^1, \dots, q^n$ are the generalized coordinates of the system, and the numbers $\dot{q}^1, \dots, \dot{q}^n$ are its generalized velocities.

Let a Lagrangian L and two points p_0 and p_1 be given on a manifold $\mathcal{X}$. For any piecewise smooth curve $\gamma\colon [a, b] \to \mathcal{X}$ connecting the points p_0 and p_1 (and such that its natural lift $\dot{\gamma}\colon [a, b] \to \mathbf{T}\mathcal{X}$ lies in the domain of the Lagrangian L), the integral

$$S = \int_a^b L(\dot{\gamma}(t))\, dt \tag{10}$$

is then well defined. If the curve γ is contained in the chart $(U, q^1, \dots, q^n)$ (i.e., $\gamma(t) \in U$ for any $t \in [a, b]$) and is given in this chart by a vector-valued function $\boldsymbol{q}(t) = (q^1(t), \dots, q^n(t))$ (has the parametric equations $q^1 = q^1(t), \dots, q^n = q^n(t)$), then

$$S = \int_a^b L(\boldsymbol{q}(t), \dot{\boldsymbol{q}}(t))\, dt, \tag{10'}$$

where $\dot{\boldsymbol{q}}(t) = (\dot{q}^1(t), \dots, \dot{q}^n(t))$.

The *simplest problem of the calculus of variations* consists in the search for a curve γ for a given Lagrangian L and given points p_0 and p_1 such that integral (10) assumes the minimum value. (According to the *Lagrange variational principle*, such minimum curves are the equations of motion of a mechanical system with the Lagrangian L.) We restrict ourselves here to the search for necessary conditions that each minimum curve should satisfy (considering conditions ensuring its existence is deferred to Chap. 25 and there only for the particular case of the length functional.)

In elementary textbooks of the calculus of variations, the simplest problem of the calculus of variations is stated as the search for a function $y = y(x)$ on a given closed interval $[a, b]$ of the real axis $\mathbb{R}$ that yields the minimum value for the integral

$$\int_a^b F(x, y, y')\, dx.$$

This is the simplest problem in our sense for a Lagrangian on the manifold $\mathcal{X} = \mathbb{R}$. In this case, the Lagrangian depends on time whose role is played by x.

§5. Euler–Lagrange Equations

First, let the minimum curve γ be smooth and be contained in a chart $(U, h) = (U, q^1, \dots, q^n)$ (integral (10) is therefore given by (10′) for this curve). We *vary* this curve, i.e., include it in a family of smooth curves

$$\gamma_\varepsilon\colon t \mapsto \gamma_\varepsilon(t), \quad t \in [a, b], \quad |\varepsilon| < \varepsilon_0$$

on U such that $\gamma_0 = \gamma$. In this case, we assume that the curves γ_ε *depend smoothly on ε near $\varepsilon = 0$*, i.e., the functions

$$q^i = q^i(t, \varepsilon), \quad t \in [a, b], \quad |\varepsilon| < \varepsilon_0, \tag{11}$$

assigning them in the chart (U, h) have smooth partial derivatives

$$\eta^i(t) = \left.\frac{\partial q^i(t, \varepsilon)}{\partial \varepsilon}\right|_{\varepsilon=0}, \quad i = 1, \dots, n, \tag{12}$$

for $\varepsilon = 0$.

We note that *for arbitrary smooth functions η^i, $1 \le i \le n$, given on the closed interval $[a, b]$, there exists variation* (11) *of the curve γ for which derivatives* (12) *coincide with the functions η^i*. (Such a variation can be given, e.g., by

$$q^i(t, \varepsilon) = q^i(t) + \varepsilon\eta^i(t), \quad t \in [a, b], \quad |\varepsilon| < \varepsilon_0, \tag{13}$$

where ε_0 is a certain sufficiently small positive number.)

Variation (10) is called a *variation with fixed endpoints* if $\gamma_\varepsilon(a) = p_0$ and $\gamma_\varepsilon(b) = p_1$ for all ε, $|\varepsilon| < \varepsilon_0$. In this case, by the minimality of integral (10) on the curve γ, the function

$$S(\varepsilon) = \int_a^b L(\gamma_\varepsilon(t))\, dt = \int_a^b L(\boldsymbol{q}(t, \varepsilon), \dot{\boldsymbol{q}}(t, \varepsilon))\, dt \tag{14}$$

of ε has a minimum for $\varepsilon = 0$, and therefore $S'(0) = 0$. But, by the rule for differentiation of integrals with respect to parameters, we have

$$S'(0) = \int_a^b \left[\frac{\partial L}{\partial q^i} \frac{\partial q^i(t,\varepsilon)}{\partial \varepsilon} + \frac{\partial L}{\partial \dot{q}^i} \frac{\partial \dot{q}^i(t,\varepsilon)}{\partial \varepsilon} \right] \Bigg|_{\varepsilon=0} dt$$

$$= \int_a^b \left[\frac{\partial L}{\partial q^i} \eta^i(t) + \frac{\partial L}{\partial \dot{q}^i} \dot{\eta}^i(t) \right] dt$$

(because $\dot{q}^i(t,\varepsilon) = \partial q^i(t,\varepsilon)/\partial t$ by definition, we have

$$\frac{\partial \dot{q}^i(t,\varepsilon)}{\partial \varepsilon} = \frac{\partial^2 q^i(t,\varepsilon)}{\partial \varepsilon \partial t} = \frac{\partial}{\partial t} \frac{\partial q^i(t,\varepsilon)}{\partial \varepsilon}$$

and therefore

$$\frac{\partial \dot{q}^i(t,\varepsilon)}{\partial \varepsilon} \Bigg|_{\varepsilon=0} = \dot{\eta}^i(t),$$

from which, integrating the second summand by parts, we immediately obtain

$$S'(0) = \frac{\partial L}{\partial \dot{q}^i} \eta^i(t) \Bigg|_a^b + \int_a^b \left[\frac{\partial L}{\partial q^i} - \frac{d}{dt} \frac{\partial L}{\partial \dot{q}^i} \right] \eta^i(t)\, dt, \tag{15}$$

where $q^i = q^i(t)$ and $\dot{q}^i = \dot{q}^i(t)$ are substituted in $\partial L/\partial q^i$ and $\partial L/\partial \dot{q}^i$. For a variation with fixed endpoints, we have $\eta^i(a) = 0$ and $\eta^i(b) = 0$; therefore, the integrated term in the right-hand of (15) vanishes. Because of the above remark, this proves that *a smooth minimum curve contained in a chart* (U, h) *is such that for any smooth functions* $\eta^i(t)$, $t \in [a, b]$, $i = 1, \ldots, n$, *vanishing for* $t = a$ *and* $t = b$, *we have*

$$\int_a^b \left[\frac{d}{dt} \frac{\partial L}{\partial \dot{q}^i} - \frac{\partial L}{\partial q^i} \right] \eta^i(t)\, dt = 0. \tag{16}$$

The following lemma is known from calculus (it is usually called the *basic lemma of the calculus of variations.*

Lemma 11.1. *Let* $A_i(t)$, $i = 1, \ldots, n$, *be smooth functions given on a closed interval* $[a, b]$. *If for any smooth (class* C^∞*) functions* $\eta^i(t)$, $i = 1, \ldots, n$, *given on* $[a, b]$ *and satisfying the relations*

$$\eta^i(a) = 0, \quad \eta^i(b) = 0, \quad i = 1, \ldots, n,$$

we have

$$\int_a^b A_i(t) \eta^i(t)\, dt = 0, \tag{17}$$

then $A_i(t) = 0$ *for all* $t \in [a, b]$ *and for all* $i = 1, \ldots, n$.

For completeness, we prove this lemma.

Proof. If the lemma is not true, then there exist a number i_0, $1 \le i_0 \le n$, and an interval $(\alpha, \beta) \subset [a, b]$ such that $A_{i_0}(t) \ne 0$ for $\alpha < t < \beta$. By the Darboux theorem, the function $A_{i_0}(t)$ preserves the sign on (α, β). For definiteness, let $A_{i_0}(t) > 0$ for $\alpha < t < \beta$.

We know that there exists a function φ of class C^∞ on $\mathbb{R}$ such that $\varphi(t) > 0$ for $\alpha < t < \beta$, and $\varphi(t) = 0$ for $t \le \alpha$ and $t \ge \beta$. For the functions η^i, $1 \le i \le n$, on $[a, b]$ given by

$$\eta^i(t) = \begin{cases} \varphi(t) & \text{if } i = i_0, \\ 0 & \text{if } i \ne i_0, \end{cases}$$

we have

$$\int_a^b A_i(t)\eta^i(t)\,dt = \int_a^b A_{i_0}(t)\varphi(t)\,dt > 0,$$

which contradicts the condition. □

Because (16) has form (17), Lemma 11.1 implies that the functions $q^i(t)$ assigning the smooth minimum curve satisfy the differential equations

$$\frac{d}{dt}\frac{\partial L}{\partial \dot q^i} - \frac{\partial L}{\partial q^i} = 0, \quad i = 1, \dots, n. \tag{18}$$

These equations are called the *Euler–Lagrange equations* (in mechanics, they are called the *Lagrange equations*; in geometry, some call them the *Euler equations*).

§6. Minimum Curves and Extremals

Definition 11.4. A smooth curve $\gamma\colon [a, b] \to \mathcal{X}$ is called an *extremal* of a Lagrangian L if for any closed interval $I \subset [a, b]$ such that the restriction $\gamma|_I$ of the curve γ to I is entirely contained in a certain chart (U, h), the functions $q^i = q^i(t)$, $i = 1, \dots, n$, assigning the curve γ (more precisely, the curve $\gamma|_I$) in this chart satisfy Eqs. (18).

We stress that *any extremal is a smooth curve* by definition. Because the restriction $\gamma|_{[a_1, b_1]}$ of the minimum curve $\gamma\colon [a, b] \to \mathcal{X}$ is also a minimum curve for any closed interval $[a_1, b_1] \subset [a, b]$ (among all curves connecting the points $\gamma(a_1)$ and $\gamma(b_1)$), the above arguments prove that *any smooth minimum curve is an extremal.* If a segment of a piecewise smooth minimum curve (when it exists) is a smooth curve, then it is an extremal.

Remark 11.1. Smooth *local minimum curves* for which integral (10) has the minimum value only as compared with all sufficiently close (in a clear sense) curves are also extremals. This fact is unnecessary for us.

Of course, an extremal in general need not be a minimum curve (even a local minimum curve). The case here is completely similar to the relation between minimum points and critical points (extremum points) of functions.

The left-hand side of Eq. (18) is denoted by $\delta L/\delta q^i$ and is called the *ith partial variational derivative of the Lagrangian L with respect to the curve* $q^i = q^i(t)$. (Therefore, a curve is an extremal iff all variational derivatives of the Lagrangian with respect to this curve vanish. This once more stresses an analogy between the extremals and the extremum points.)

Remark 11.2. In the framework of infinite-dimensional manifold theory (see the beginning of Chap. 10), the set of all curves in $\mathcal{X}$ that connect a point p_0 with a point p_1 is a Banach manifold, integral (10) is a smooth function on this manifold, variational derivatives are components of the differential of this function, and extremals are its critical points.

In opened form, the variational derivatives become

$$\frac{\delta L}{\delta q^i} = \frac{\partial^2 L}{\partial \dot{q}^i \partial \dot{q}^j} \ddot{q}^j(t) + \frac{\partial^2 L}{\partial \dot{q}^i \partial q^j} \dot{q}^j(t) - \frac{\partial L}{\partial q^i}, \quad i = 1, \ldots, n,$$

where it is assumed that $\boldsymbol{q} = \boldsymbol{q}(t)$ and $\dot{\boldsymbol{q}} = \dot{\boldsymbol{q}}(t)$ are substituted in the functions

$$\frac{\partial^2 L}{\partial \dot{q}^i \partial \dot{q}^j}, \qquad \frac{\partial^2 L}{\partial \dot{q}^i \partial q^j}, \qquad \frac{\partial L}{\partial q^i},$$

and the Euler–Lagrange equations therefore become

$$\frac{\partial 2L}{\partial \dot{q}^i \partial \dot{q}^j} \ddot{q}^j(t) + \frac{\partial 2L}{\partial \dot{q}^i \partial q^j} \dot{q}^j(t) - \frac{\partial L}{\partial q^i} = 0. \tag{18'}$$

Therefore, *the Euler–Lagrange equations are second-order differential equations for the functions $q^i(t)$, $i = 1, \ldots, n$, that are linear in the second derivatives $\ddot{q}^j(t)$* (the first derivatives $\dot{q}^j(t)$ generally enter them nonlinearly).

In mechanical problems, as a rule, the Lagrangian L is naturally represented in the form $L = T - U$ (more precisely, in the form $L = T - U \circ \pi$), where T is a function on $\mathbf{T}\mathcal{X}$ called the *kinetic energy* and U is a function on $\mathcal{X}$ called the *potential energy*. According to this, Lagrange equations (18) become

$$\frac{d}{dt}\left(\frac{\partial T}{\partial \dot{q}^i}\right) - \frac{\partial T}{\partial q^i} = -\frac{\partial U}{\partial q^i}, \quad i = 1, \ldots, n. \tag{19}$$

More generally, it is possible to introduce an arbitrary linear horizontal form θ on $\mathbf{T}\mathcal{X}$ (called the *force field*) and to consider equations of the form

$$\frac{d}{dt}\left(\frac{\partial T}{\dot{q}^i}\right) - \frac{\partial T}{\partial q^i} = \theta_i, \quad i = 1, \ldots, n, \tag{19'}$$

instead of Eqs. (19), where $\theta_i = \theta_i(\boldsymbol{q}, \dot{\boldsymbol{q}})$ are coefficients of the form θ. However, the *potential field* ∇U is usually isolated for physical reasons, and Eqs. (19') are written in the form

$$\frac{d}{dt}\left(\frac{\partial T}{\partial \dot{q}^i}\right) - \frac{\partial T}{\partial q^i} = \theta_i - \frac{\partial U}{\partial q^i}, \quad i = 1, \ldots, n, \tag{19''}$$

where θ_i are components of the *field of nonpotential forces*.

§7. Regular Lagrangians

It is important to keep in mind that Eqs. (18′) are not resolved with respect to the second derivatives.

Definition 11.5. A Lagrangian L is called a *regular Lagrangian* if the matrix

$$\left\| \frac{\partial^2 L}{\partial \dot{q}^i \partial \dot{q}^j} \right\| \tag{20}$$

is nonsingular for any chart $(U, q^1, \ldots, q^n)$.

Exercise 11.4. Prove that matrix (20) does not depend on the choice of local coordinates.

For a regular Lagrangian, resolving Eqs. (18′) with respect to the second derivatives, we represent them in the form

$$\ddot{q}^i(t) = F^i(\boldsymbol{q}(t), \dot{\boldsymbol{q}}(t)), \quad i = 1, \ldots, n,$$

where F^i are certain functions. Therefore, the standard theorem of existence and uniqueness of solutions to a system of differential equations can be applied to them. According to this theorem *for any point $p_0 \in \mathcal{X}$ and any vector $A \in \mathbf{T}_{p_0}\mathcal{X}$, there exists a unique maximal (not continued to a large closed interval) extremal γ of the Lagrangian L passing through the point p_0 for $t = 0$ and having the tangent vector A at this point.* (Of course, this theorem is silent regarding the existence of extremals and more so regarding the existence of minimum curves connecting given points p_0 and p_1.) For nonregular Lagrangians, an extremal with a given tangent vector can fail to exist, and when it does exist, it need not be unique.

In mechanical Lagrangians $L = T - U$, the kinetic energy T is usually a positive-definite form in the generalized velocities $\dot{q}^1, \ldots, \dot{q}^n$. In this case, matrix (20) does not depend on the velocities and is just a doubled matrix of the form T. Therefore, *all mechanical Lagrangians $L = T - U$ are regular.*

§8. Extremals of the Energy Lagrangian

From the geometric standpoint, the positive-definite kinetic energy T is just a Riemannian metric on $\mathcal{X}$. This determines a close interaction of classical mechanics and Riemannian geometry, which enriches both these fields. Having no possibility to consider this in detail, we restrict ourselves to inertial motion (where there are no forces, not even potential ones.) This means that we deal with a Riemannian space $\mathcal{X}$ and a Lagrangian L on $\mathbf{T}\mathcal{X}$ defined by

$$L(A) = \frac{1}{2}(A, A), \quad A \in \mathbf{T}\mathcal{X} \tag{21}$$

(the factor 1/2 certainly plays no principal role and is introduced to simplify formulas). Lagrangian (21) is usually called the *energy Lagrangian* of the Riemannian space $\mathcal{X}$, but some prefer to call it the *action Lagrangian.*

In local coordinates, Lagrangian (21) is expressed by the function

$$L(\boldsymbol{q}, \dot{\boldsymbol{q}}) = \frac{1}{2} g_{ij} \dot{q}^i \dot{q}^j. \tag{21'}$$

We note that it has a sense for any pseudo-Riemannian space $\mathcal{X}$. For the energy Lagrangian,

$$\frac{\partial L}{\partial \dot{q}^i} = g_{ik} \dot{q}^k, \qquad \frac{\partial^2 L}{\partial \dot{q}^i \partial \dot{q}^j} = g_{ij}$$

and

$$\frac{\partial L}{\partial q^i} = \frac{1}{2} \frac{\partial g_{jk}}{\partial q^i} \dot{q}^j \dot{q}^k, \qquad \frac{\partial^2 L}{\partial q^i \partial \dot{q}^j} = \frac{\partial g_{jk}}{\partial q^i} \dot{q}^k.$$

Therefore, Eq. (18′) for this Lagrangian becomes

$$g_{ij} \ddot{q}^j(t) + \frac{\partial g_{ik}}{\partial q^j} \dot{q}^j(t) \dot{q}^k(t) - \frac{1}{2} \frac{\partial g_{jk}}{\partial q^i} \dot{q}^j(t) \dot{q}^k(t) = 0.$$

Contracting the left-hand sides of these equations with the tensor g^{il}, we thus resolve them with respect to the second derivatives:

$$\ddot{q}^l(t) + g^{il} \left(\frac{\partial g_{ik}}{\partial q^j} - \frac{1}{2} \frac{\partial g_{jk}}{\partial q^i} \right) \dot{q}^j(t) \dot{q}^k(t) = 0.$$

Because the latter equations only differ from geodesic equations (8') in notation, this proves the following proposition.

Proposition 11.1. *The extremals of the energy Lagrangian are exactly geodesics of a (pseudo-)Riemannian metric.*

Therefore, a geodesic is just a trajectory of inertial motion.

Chapter 12
Metric Properties of Geodesics

§1. Length of a Curve in a Riemannian Space

For a Riemannian (but not a pseudo-Riemannian) space $\mathcal{X}$, along with the energy Lagrangian, we can also consider the Lagrangian

$$L(A) = |A|, \quad A \in \mathbf{T}\mathcal{X}, \tag{1}$$

which is expressed in local coordinates by

$$L(x, \dot{x}) = \sqrt{g_{ij}\dot{x}^i\dot{x}^j}. \tag{1'}$$

(Taking into account that this Lagrangian has no mechanical meaning, we return to the usual notation $x^1, \dots, x^n$ for local coordinates. We note that Lagrangian (1) is a smooth function only outside the zero secant surface of the tangent bundle, i.e., for $A \neq 0$.)

The corresponding integral has the form

$$s_\gamma = \int_a^b |\dot{\gamma}(t)| dt$$

(traditionally, a lowercase, not uppercase, letter is used for its designation) and is called the *length* of a curve γ in the metric g (for $n = 2$, this is exactly the length of a curve on a surface). In accordance with this, Lagrangian (1) is called the *length Lagrangian.*

If a curve γ lies entirely in the chart $(U, h) = (U, x^1, \dots, x^n)$, then

$$s_\gamma = \int_a^b \sqrt{g_{ij}(x(t))\dot{x}^i(t)\dot{x}^j(t)}\, dt, \quad x(t) = (x^1(t), \dots, x^n(t)), \tag{2}$$

where $x^1(t), \dots, x^n(t)$ are the functions that assign the curve γ in the chart (U, h) and

$$g_{ij} = g\left(\frac{\partial}{\partial x^i}, \frac{\partial}{\partial x^j}\right), \quad 1 \le i, j \le n,$$

are the components of the tensor g in this chart. In a conditional mnemonic form, formula (2) can be rewritten as

$$s_\gamma = \int_\gamma \sqrt{g_{ij}dx^i dx^j}$$

or even as

$$ds^2 = g_{ij}dx^i dx^j.$$

If needed, the symbol ds^2 can be considered to denote the *field of quadratic functionals* in which the quadratic functional

$$A \mapsto g_p(A, A), \quad A \in \mathbf{T}_p\mathcal{X},$$

on the space $\mathbf{T}_p\mathcal{X}$ corresponds to each point $p \in U$.

We note that $s_\gamma \geq 0$ and, moreover, $s_\gamma = 0$ iff the curve γ is constant (is a mapping into a point). Of course, integral (2) can also be considered for curves on a pseudo-Riemannian space $\mathcal{X}$. However, in this case, it is generally a complex number (and can be equal to zero for nonconstant curves).

Exercise 12.1. Show that the length of a curve γ (even a noninvertible one) does not change under a reparameterization.

§2. Natural Parameter

For curves in the Euclidean space, the parameter t on a curve γ is said to be *natural* if $|\dot{\gamma}(t)| = 1$. In this case, the length of any segment $\gamma|_{[t_0,t_1]}$ of the curve γ equals $t_1 - t_0$. The natural parameter on an (oriented) curve γ is therefore uniquely defined up to reparameterization of the form $t \mapsto t - t_0$. To fix it completely, it suffices to choose a point p_0 on the curve and assume that the parameter value vanishes at this point, $t_0 = 0$. Under such a choice of the natural parameter, its value for an arbitrary point p of the curve is equal to the length of the segment of the curve from p_0 to p up to the sign. Because of this, the natural parameter is also called the *arc length.*

We note that according to formula (9) in Chap. 11, *the parameter t on a geodesic γ_A is natural iff $|A| = 1$.*

Exercise 12.2. Prove that a geodesic remains a geodesic under a reparameterization iff this reparameterization is linear (has the form $t \mapsto at + b$, where $a \neq 0$).

If the vector $\dot{\gamma}(t)$ is everywhere nonzero for a curve γ (such curves are said to be *regular*), then choosing a point $p_0 = \gamma(t_0)$ on the curve, we can reparameterize the curve to the natural parameter $s = s(t)$, where

$$s(t) = \int_{t_0}^{t} |\dot{\gamma}(t)| \, dt.$$

Such a reparameterization is also possible for nonregular curves, but it is not invertible in general.

§3. Riemannian Distance and Shortest Arcs

Let a manifold $\mathcal{X}$ be connected; therefore, for any two points $p, q \in \mathcal{X}$, there exist piecewise smooth (and even smooth) curves connecting these points.

Definition 12.1. The greatest lower bound

$$\rho(p,q) = \inf \int_\gamma |\dot\gamma(t)|dt \tag{3}$$

of the lengths of all curves γ connecting the point p with the point q is called the *Riemannian* (or *intrinsic*) *distance* between the points p and q. A piecewise smooth curve (if it exists) at which greatest lower bound (3) is attained, i.e., whose length equals the distance $\rho(p,q)$, is called a *shortest arc*. Such a curve does not always exist and, generally speaking, is not unique (even up to a parameterization).

Example 12.1. Let $\mathcal{X}$ be the Euclidean plane $\mathbb{R}^2$ from which the disk $x^2+y^2 \leq 1$ is removed. There is no shortest arc connecting the points $(-2,0)$ and $(0,2)$.

Example 12.2. On a two-dimensional sphere (Earth surface), meridian arcs (large circles) serve as shortest arcs. There are infinitely many meridians connecting diametrally opposite points.

Of course, for any reparameterization, a shortest arc remains a shortest arc.

§4. Extremals of the Length Lagrangian

By definition, each shortest arc is a minimum curve of the length Lagrangian.

Exercise 12.3. Deduce from this that the length Lagrangian is not regular. [*Hint*: Minimum curves (and more so extremals) are not unique.]

Exercise 12.4. Prove by direct computation that for the length Lagrangian, the matrix

$$\left\| \frac{\partial^2 L}{\partial \dot x^i \partial \dot x^j} \right\| \tag{4}$$

is nonsingular (see Definition 11.5). [*Hint*: Taking into account that Lagrangian (1′) is a homogeneous first-degree function of the generalized velocities, use the Euler theorem on homogeneous functions twice.]

Exercise 12.5. Prove that an extremal of the length Lagrangian remains an extremal for any reparameterization.

Exercise 12.6. Prove that the rank of matrix (4) for the length Lagrangian equals $n-1$.

The assertion in Exercise 12.6 implies that for the length Lagrangian, we have exactly $n-1$ linearly independent equations among the Euler–Lagrange equations (Eqs. (18) in Chap. 11). Therefore, we can hope that it is possible to reconstruct the uniqueness by adding one equation to these equations.

As such an additional equation, we choose the equation

$$g_{ij}\dot{x}^i\dot{x}^j = \text{const}, \tag{5}$$

which means that the parameter t on the extremal is proportional to the natural parameter (arc length). This condition allows finding extremals of the length Lagrangians practically without calculation.

Let L_1 be the energy functional of a Riemannian space $\mathcal{X}$. Then $L_1 = L^2/2$, and therefore

$$\frac{\partial L_1}{\partial q^i} = L\frac{\partial L}{\partial q^i}, \qquad \frac{\partial L_1}{\partial \dot{q}^i} = L\frac{\partial L}{\partial \dot{q}^i}$$

(we again return to the q-notation). In particular,

$$\frac{\partial L_1}{\partial \dot{q}^i}(\boldsymbol{q}(t), \dot{\boldsymbol{q}}(t)) = L(\boldsymbol{q}(t), \dot{\boldsymbol{q}}(t))\frac{\partial L}{\partial \dot{q}^i}(\boldsymbol{q}(t), \dot{\boldsymbol{q}}(t)).$$

But condition (5) means exactly that $L(\boldsymbol{q}(t), \dot{\boldsymbol{q}}(t)) = \text{const}$. Therefore (we again omit the arguments),

$$\frac{d}{dt}\frac{\partial L_1}{\partial \dot{q}^i} = L\frac{d}{dt}\frac{\partial L}{\partial \dot{q}^i},$$

and

$$\frac{\delta L_1}{\delta q^i} = L\frac{\delta L}{\delta q^i}, \quad i = 1, \dots, n.$$

Because $L \neq 0$, this proves that under condition (5), the relations $\delta L/\delta q^i = 0$ are exactly equivalent to the relations $\delta L_1/\delta q^i = 0$, which, as we know, characterize geodesics.

Therefore, *geodesics of a Riemannian metric are exactly extremals of the length Lagrangian that satisfy condition* (5) (i.e., are related to a parameter proportional to the natural parameter). We note that by (9) in Chap. 11, any geodesic satisfies condition (5) in advance. In particular, we see that *in any Riemannian space, extremals of energy and extremals of length coincide.* This is a differential-geometric expression of the *Maupertuis least-action principle.*

Applying all these statements to shortest arcs, we now conclude that *each smooth shortest arc related to the natural parameter is a geodesic.* It turns out that the smoothness condition is unnecessary here (any shortest arc is a smooth curve), but to prove this requires a sufficiently long preparation.

§5. Riemannian Coordinates

For any point p_0 of a Riemannian space $\mathcal{X}$, the exponential mapping

$$\exp_{p_0}\colon \mathbf{T}_{p_0}\mathcal{X} \to \mathcal{X}$$

(corresponding to the Riemannian connection ∇) allows transporting the polar (as well as spherical) coordinates from the Euclidean space $\mathbf{T}_{p_0}\mathcal{X}$ to an arbitrary normal neighborhood U_0 of the point p_0. However, to avoid extra conditions, it is convenient to assume a normal neighborhood U_0 to be *spherical* here, i.e., the $\exp_{p_0}$-image of a certain open ball of the space $\mathbf{T}_{p_0}\mathcal{X}$ centered at the point 0. (We note that the radius of this ball is called the *radius* of the normal spherical neighborhood U_0.)

The polar coordinates transported to U_0 are called the *Riemannian* or *semigeodesic* coordinates. By definition, a point $p \in U_0$ has the semigeodesic coordinates $t, \alpha^1, \dots, \alpha^{n-1}$ if $p = \exp_{p_0} tA$, where A is a unit vector of the Euclidean space $\mathbf{T}_{p_0}\mathcal{X}$ having the "geographical coordinates" $\alpha^1, \dots, \alpha^{n-1}$ on the unit sphere of this space. (Therefore, strictly speaking, the coordinates $t, \alpha^1, \dots, \alpha^{n-1}$ are not defined everywhere on U_0; they are defined only on a certain domain that is the $\exp_{p_0}$-image of a cone of the space $T_{p_0}\mathcal{X}$ with the vertex removed. The base of this cone is the domain on the unit sphere where the coordinates $\alpha^1, \dots, \alpha^{n-1}$ are defined.) To simplify the statements, it is conventional to ignore this.]

The coordinate t-lines of the semigeodesic coordinate system are geodesics

$$\gamma_A : t \mapsto \exp_{p_0} tA \tag{6}$$

passing through the point p_0 for $t = 0$, and the coordinate surfaces $t = \text{const}$ are *geodesic spheres* (embedded $(n-1)$-dimensional submanifolds Σ_t, $t \neq 0$, that are images under the diffeomorphism $\exp_{p_0}$ of the spheres $|A| = t$ of the Euclidean space $\mathbf{T}_{p_0}\mathcal{X}$).

§6. Gauss Lemma

For each point $p = \exp_{p_0} tA$ of the sphere Σ_t, the first vector $(\partial/\partial t)_p$ of the coordinate basis

$$\left(\frac{\partial}{\partial t}\right)_p, \quad \left(\frac{\partial}{\partial \alpha^1}\right)_p, \quad \dots, \quad \left(\frac{\partial}{\partial \alpha^{n-1}}\right)_p$$

of the space $\mathbf{T}_p\mathcal{X}$ is just a tangent vector $\dot{\gamma}_A(t)$ of geodesic (6), and the other vectors

$$\left(\frac{\partial}{\partial \alpha^1}\right)_p, \quad \dots, \quad \left(\frac{\partial}{\partial \alpha^{n-1}}\right)_p$$

compose a basis of the tangent space $\mathbf{T}_p\Sigma_t$ of the sphere Σ_t. Because the vector A is a unit vector, according to formula (9) in Chap. 11, we have

$$\left(\frac{\partial}{\partial t}, \frac{\partial}{\partial t}\right) = 1$$

everywhere on U_0.

Proposition 12.1. *At any point $p \in \Sigma_t$, the vector*

$$\left(\frac{\partial}{\partial t}\right)_p = \dot{\gamma}_A(t)$$

is orthogonal to the subspace $\mathbf{T}_p\Sigma_t$.

This important proposition is known as the *Gauss lemma.* In a visual statement, it asserts that as are ordinary Euclidean spheres, *geodesic spheres are orthogonal to their radii.*

Another usual restatement is that in the semigeodesic coordinates, the matrix $\|g_{ij}\|$ of the metric tensor becomes

$$\begin{Vmatrix} 1 & 0 & \cdots & 0 \\ 0 & & & \\ \vdots & & G & \\ 0 & & & \end{Vmatrix}, \tag{7}$$

where G is an $(n-1)\times(n-1)$ matrix.

Proof (of Proposition 12.1). For any t, $|t| < \delta_0$, and for any curve $s \mapsto A(s)$, $|A(s)| = 1$, $|s| < s_0$, on the unit sphere of the space $\mathbf{T}_p\mathcal{X}$ passing through the point A for $s = 0$ (i.e., such that $A(0) = A$), the curve

$$u_t \colon s \mapsto \exp_{p_0} tA(s)$$

is a curve on the geodesic sphere Σ_t, and its tangent vector $\dot{u}_t(0)$ therefore belongs to the subspace $\mathbf{T}_p\Sigma_t$, $p = \exp_{p_0} tA$. Because an arbitrary vector of the latter subspace can obviously be represented in the form $\dot{u}_t(0)$ (for a curve $s \mapsto A(s)$ correspondingly chosen), to prove Proposition 12.1, it therefore suffices to show that for each curve $s \to A(s)$, we have

$$\left(\left(\frac{\partial}{\partial t}\right)_p, \dot{u}_t(0)\right) = 0.$$

For this, we introduce the elementary surface

$$\varphi \colon W \to \mathcal{X}$$

defined on the square $\{(s,t); |s| < s_0, |t| < \delta_0\}$ of the plane $\mathbb{R}^2$ by

$$\varphi(s,t) = \exp_{p_0} tA(s).$$

The coordinate lines $s =$ const of this surface are the geodesics $\gamma_{A(s)}$ (for $s = 0$ in particular, the geodesic γ_A), and the coordinate lines $t =$ const are the curves u_t. Therefore,

$$\left(\frac{\partial \varphi}{\partial t}\right)_p = \dot{\gamma}_A(t) = \left(\frac{\partial}{\partial t}\right)_p \quad \text{and} \quad \left(\frac{\partial \varphi}{\partial s}\right)_p = \dot{u}_t(0),$$

which implies that if we show that the function

$$\left(\frac{\partial\varphi}{\partial t}, \frac{\partial\varphi}{\partial s}\right) \tag{8}$$

of the variables s and t vanishes identically, then Proposition 12.1 is proved. Because function (8) is equal to zero for $t = 0$ in advance (because $u_0(s) = p_0$ for all s and hence $(\partial\varphi/\partial s)_{t=0} = \dot{u}_0(s) = 0$), it in turn suffices to show that

$$\frac{d}{dt}\left(\frac{\partial\varphi}{\partial t}, \frac{\partial\varphi}{\partial s}\right) = 0.$$

But according to (4) in Chap. 11,

$$\frac{d}{dt}\left(\frac{\partial\varphi}{\partial t}, \frac{\partial\varphi}{\partial s}\right) = \left(\frac{\nabla}{\partial t}\frac{\partial\varphi}{\partial t}, \frac{\partial\varphi}{\partial s}\right) + \left(\frac{\partial\varphi}{\partial t}, \frac{\nabla}{\partial t}\frac{\partial\varphi}{\partial s}\right).$$

Moreover, because the curves $s = \text{const}$ are geodesics,

$$\frac{\nabla}{\partial t}\frac{\partial\varphi}{\partial t} = 0,$$

and because the Riemannian connection is symmetric by definition, we have

$$\frac{\nabla}{\partial t}\frac{\partial\varphi}{\partial s} = \frac{\nabla}{\partial s}\frac{\partial\varphi}{\partial t}$$

(see (17) in Chap. 2). On the other hand, because

$$\left(\frac{\partial\varphi}{\partial t}, \frac{\partial\varphi}{\partial t}\right) = \left(\frac{\partial}{\partial t}, \frac{\partial}{\partial t}\right) = 1,$$

we have

$$\left(\frac{\partial\varphi}{\partial t}, \frac{\nabla}{\partial s}\frac{\partial\varphi}{\partial t}\right) = \frac{1}{2}\frac{\partial}{\partial s}\left(\frac{\partial\varphi}{\partial t}, \frac{\partial\varphi}{\partial t}\right) = 0.$$

Therefore, we indeed have

$$\frac{d}{dt}\left(\frac{\partial\varphi}{\partial t}, \frac{\partial\varphi}{\partial s}\right) = 0,$$

and Proposition 12.1 is proved. □

Proposition 12.1 has many important consequences.

§7. Geodesics are Locally Shortest Arcs

We recall (see (13) in Chap. 1) that for any point q belonging to a normal neighborhood U of a point p, the segment $\gamma_{p,q}$ of a geodesic in U connecting the point p with the point q is defined. We additionally assume that the neighborhood U is a normal spherical neighborhood. It turns out that the following statement holds under this assumption.

Proposition 12.2. *The segment $\gamma_{p,q}$ is a shortest arc.*

Proof. Let $(t, \boldsymbol{\alpha}) = (t, \alpha^1, \dots, \alpha^{n-1})$ be semigeodesic coordinates in the neighborhood U. The assertion that the metric tensor has form (7) in these coordinates means that

$$ds^2 = dt^2 + G(d\boldsymbol{\alpha}),$$

where $G(d\boldsymbol{\alpha})$ is a certain positive-definite quadratic form $d\alpha^1, \dots, d\alpha^{n-1}$ (with coefficients depending not only on $\alpha^1, \dots, \alpha^{n-1}$ but also on t). Therefore, for any curve $\gamma: s \mapsto \gamma(s)$ connecting the point p with the point q and lying entirely in U, the estimate

$$\begin{aligned}\int_\gamma |\dot{\gamma}(s)|\,ds &= \int_\gamma \sqrt{(\dot{t}(s))^2 + G(\dot{\boldsymbol{\alpha}}(s))}\,ds \\ &\geq \int_\gamma |\dot{t}(s)|\,ds \\ &\geq \left|\int_\gamma \dot{t}(s)\,ds\right| = t_0,\end{aligned}$$

holds, where $t(s)$ and $\boldsymbol{\alpha}(s)$ are the functions that assign the curve γ in the coordinates $(t, \boldsymbol{\alpha})$ and t_0 is the value of the coordinate t at the point q (i.e., a number such that $q \in \Sigma_{t_0}$). In this case, the equality is attained only for $\dot{\boldsymbol{\alpha}}(s) = 0$ and $\dot{t}(s) \geq 0$, i.e., on the curve that passes into the segment $\gamma_{p,q}$ under a reparameterization $t = t(s)$ (which is noninvertible in general).

We have thus proved the following:

1. The length of the segment $\gamma_{p,q}$ is equal to t_0, where t_0 is a number such that $q \in \Sigma_{t_0}$.
2. The length of any curve connecting the point p with the point q in U is not less than t_0.

It remains to consider curves connecting the points p and q that leave the neighborhood U.

Here, we need the assumption that the neighborhood U is normal and spherical; therefore, it is bounded by a geodesic sphere Σ_δ, where δ is the radius of the neighborhood U. Because a curve connecting the points p and q that is not contained in the neighborhood U should intersect the sphere Σ_δ, assertion 2 (which is already proved) applied to the first point of the intersection of the curve with the sphere Σ_δ implies that the length of such a curve is not less than δ. Because $t_0 < \delta$, such curves do not affect the greatest lower bound, which is therefore equal to t_0 by assertion 1. Therefore, $\rho(p,q) = t_0$, and the segment $\gamma_{p,q}$ is a shortest arc. □

In proving Proposition 12.2, we obtained the following two propositions.

Corollary 12.1. *Up to a reparameterization, the segment $\gamma_{p,q}$ is a unique shortest arc connecting the point p with the point q.*

Corollary 12.2. *The distance from the point p to any point q on the geodesic sphere Σ_{t_0} is equal to its radius t_0.*

In particular, Proposition 12.2 implies that *each sufficiently small segment of a geodesic is a shortest arc*, i.e., for the length Lagrangian on a Riemannian space, each sufficiently small segment of an extremal is a minimum curve. This is no longer true for arbitrary Lagrangians, i.e., there exist Lagrangians with extremals having segments that are not minimum curves.

Example 12.3. For the Lagrangian $L = \dot{x}^2 - \dot{y}^2$ on the (x, y)-plane $\mathbb{R}^2$, the abscissa $x = at$, $y = 0$ is an extremal. Let A be the point $(a, 0)$, $a > 0$, on this extremal. The integral

$$\int_0^1 (\dot{x}^2 - \dot{y}^2)dt$$

over the segment OA of this extremal is obviously equal to a^2. At the same time, this integral over the broken line OBA (correspondingly parameterized), where B is the point $(a/2, b/2)$, $b > 0$, is equal to $a^2 - b^2$. Therefore, the segment OA is not a minimum curve.

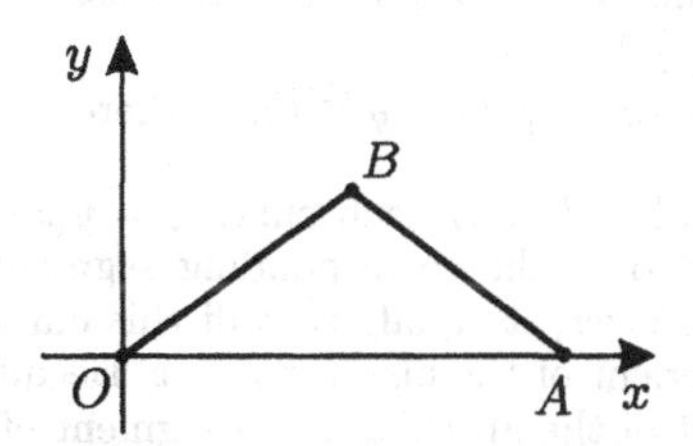

Fig. 12.1.

The above calculation shows that there is no minimum curve connecting the points O and A for the Lagrangian $L = \dot{x}^2 - \dot{y}^2$.

§8. Smoothness of Shortest Arcs

We are now ready to prove that shortest arcs of a Riemannian metric are smooth curves.

Proposition 12.3. *Up to a reparameterization, any shortest arc is a geodesic (and therefore a smooth curve).*

Proof. Each segment of a shortest arc is obviously a shortest arc (otherwise it can be replaced by a shorter segment reducing the length of the whole shortest arc). But if the endpoints p and q of this segment are sufficiently close to each other (the point q is contained in a normal spherical neighborhood of the point p), then according to Corollary 12.1, this segment is a segment of the geodesic $\gamma_{p,q}$. Therefore, locally (in a neighborhood of any point), a shortest arc is a geodesic. To complete the proof, it suffices to note that the property to be a geodesic is expressed by differential equations; therefore, a curve that is geodesic locally is a geodesic globally. □

We note that the converse assertion is not true: a geodesic need not be a shortest arc.

Example 12.4. On a two-dimensional sphere, each curve going around the sphere along the equator several times is a geodesic but certainly not a shortest arc.

We also note that for arbitrary Lagrangians, a minimum curve need not be a smooth curve in general.

Example 12.5. On the (x, y)-plane $\mathbb{R}^2$, we consider the Lagrangian

$$L = \frac{y^2(\dot{x} - \dot{y})^2}{\dot{x}}$$

(defined for $\dot{x} \neq 0$) and the points $Q_0(-1,0)$ and $Q_1(1,1)$. A piecewise smooth curve in $\mathbb{R}^2$ connecting the point Q_0 with the point Q_1 whose natural lift lies in the domain of the Lagrangian L has no vertical tangents and is therefore composed of smooth curves with equations of the form $y = y(x)$. But because

$$\frac{y^2(\dot{x} - \dot{y})^2}{\dot{x}} dt = y^2(1 - y')^2 dx, \quad \text{where } y' = \frac{dy}{dx},$$

for such curves, the integral of L over each curve $y = y(x)$ is equal to the integral of the function $y^2(1 - y')^2$ over the corresponding segment of the abscissa and is therefore nonnegative. Moreover, it equals zero iff this curve is either a segment of the abscissa $y = 0$ or a segment of the bisector $y = x$. Because there exists a broken line consisting of a segment of the abscissa and a segment of the bisector $y = x$ that connects the point Q_0 with the point Q_1, this proves that the minimum of integrals of L exists, is equal to zero, and is attained at the above broken line. Therefore, in this case, a minimum curve (we note that it is unique!) is not a smooth curve in $\mathbb{R}^2$.

§9. Local Existence of Shortest Arcs

Proposition 12.4. *There exists a neighborhood U of a point $p_0 \in \mathcal{X}$ such that any two points $p, q \in U$ are connected by a unique (up to a reparameterization) shortest arc. This shortest arc is a segment of the geodesic $\gamma_{p,q}$. There exists a number $\varepsilon > 0$ such that for any points $p, q \in U$, the length of the shortest arc $\gamma_{p,q}$ (the distance $\rho(p, q)$) is less than ε.*

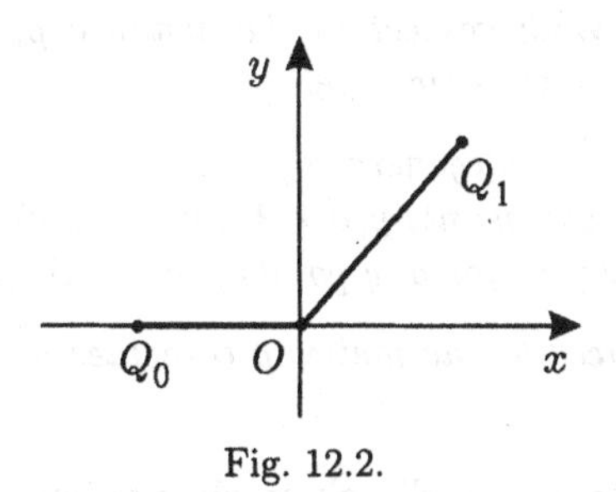

Fig. 12.2.

Proof. At the first glance, it seems that we can take the neighborhood indicated in Proposition 1.1 as U. Indeed, because for any point $p \in U$, we have the inclusion $U \subset U_{\delta,p}$, where $U_{\delta,p}$ is a normal spherical neighborhood of the point p, we see that the conditions of Proposition 12.2 hold for a segment $\gamma_{p,q}$, and its length is less than δ. However, this argument is not correct, because the neighborhood $U_{\delta,p}$ is a spherical neighborhood not with respect to the metric g_p on $\mathbf{T}_p\mathcal{X}$ but with respect to the auxiliary metric introduced in Chap. 1 (which is now denoted by $g_p^{(0)}$). Nevertheless, we can preserve this argument, even by two methods.

First method: Examining the arguments in Chap. 1 once more, where we used the metric $g_p^{(0)}$, we immediately find that a special method for its construction was not used elsewhere and all these arguments are literally repeated if we understand an arbitrary Riemannian structure on U as $g^{(0)}$. In particular, we can take the restriction of the Riemannian structure g to U as $g^{(0)}$. Then the neighborhoods $U_{\delta,p}$ are normal spherical neighborhoodss with respect to the metric g_p, and the above argument is absolutely correct.

Second method: The metrics g_p and $g_p^{(0)}$, being Euclidean metrics on the same finite-dimensional linear space, are equivalent, i.e., there exist numbers $m > 0$ and $M > 0$ such that

$$m|A|^{(0)} \leq |A| \leq M|A|^{(0)} \tag{9}$$

for any vector $A \in \mathbf{T}_p\mathcal{X}$, where $|A|^{(0)}$ is the length of the vector A in the metric $g_p^{(0)}$ and $|A|$ is its length in the metric g_p. In this case, by the compactness of the set $\overline{U}_0$, the numbers m and M can be considered the same for all $p \in U_0$. Therefore, the spherical δ-neighborhood with respect to the metric $g_p^{(0)}$ is contained in the spherical $M\delta$-neighborhood with respect to the metric g_p. This again makes the argument correct if we only replace the number δ with the number $M\delta$ (which certainly plays no role). □

§10. Intrinsic Metric

We now return to the Riemannian distance ρ defined by formula (3).

Proposition 12.5. *With respect to the distance ρ, each connected Riemannian space $\mathcal{X}$ is a metric space, i.e.,*

1. $\rho(p,q) = 0$ *iff* $p = q$ *(nondegeneracy),*
2. $\rho(p,q) = \rho(q,p)$ *for any points* $p, q \in \mathcal{X}$ *(symmetry), and*
3. $\rho(p,r) \leq \rho(p,q) + \rho(q,r)$ *for any points* $p, q, r \in \mathcal{X}$ *(triangle axiom).*

The topology on $\mathcal{X}$ induced by the metric ρ coincides with the topology on the manifold $\mathcal{X}$.

Proof. The symmetry is immediately implied because a curve going in the opposite direction has the same length as the initial curve. To prove the triangle axiom, it suffices to note that for any curve γ_1 connecting a point p with a point q and for any curve γ_2 connecting the point q with a point r, the composition curve $\gamma_1\gamma_2$ connects the point p with the point r, and its length is equal to the sum of the lengths of the curves γ_1 and γ_2. Finally, Corollary 12.2 directly implies that each normal spherical δ-neighborhood U_δ of an arbitrary point $p \in \mathcal{X}$ is also its spherical δ-neighborhood with respect to the distance ρ (i.e., consists of all points $q \in \mathcal{X}$ for which $\rho(p,q) < \delta$). Therefore, the metric ρ is nondegenerate, and the topology it defines coincides with the topology on the manifold $\mathcal{X}$. □

The metric ρ is called the *Riemannian* (or *intrinsic*) *metric* on the Riemannian space $\mathcal{X}$.

Remark 12.1. The nondegeneracy of the metric ρ can be proved in another way without using Proposition 12.2 (the Gauss lemma). Let ρ_0 be a metric in a coordinate neighborhood U_0 that results from transporting the standard Euclidean metric on $\mathbb{R}^n$ (more precisely, its restriction to the ball $h(U_0)$) using the coordinate mapping $h: U_0 \to \mathbb{R}^n$. This metric is certainly nondegenerate. On the other hand, it is easy to see that it is the intrinsic metric on U_0 corresponding to the auxiliary Riemannian structure $g^{(0)}$, which was introduced in Chap. 1. Therefore, estimates (9) imply that similar estimates

$$m\rho_0(p,q) \leq \rho(p,q) \leq M\rho_0(p,q), \quad p, q \in U_0,$$

also hold for the metrics ρ_0 and ρ. Therefore, the metric ρ is also nondegenerate.

We mention two consequences of Proposition 12.5.

Corollary 12.3. *Each paracompact and Hausdorff smooth manifold is metricizable.*

According to the Stone Theorem, *any metricizable space is paracompact.* Therefore, the paracompactness (the fulfillment of the second countability axiom for each component) of a smooth Hausdorff manifold $\mathcal{X}$ is not only sufficient but also necessary for the existence of a Riemannian structure on $\mathcal{X}$.

Corollary 12.4. *For any compact Riemannian space $\mathcal{X}$, there exists a number $\rho_0 > 0$ such that any two points $p, q \in \mathcal{X}$ whose distance from each other is less than ρ_0 are connected by a unique shortest arc.*

Proof. We consider an open covering of the space $\mathcal{X}$ consisting of neighborhoods U of all points $p_0 \in \mathcal{X}$ that have the property indicated in Proposition 12.4. Because the space $\mathcal{X}$ is compact, by the Lebesgue covering lemma, there exists $\rho_0 > 0$ such that any two points $p, q \in \mathcal{X}$ for which $\rho(p, q) < \rho_0$ are contained in the same element of this covering and are therefore connected by a unique shortest arc. □

The number ρ_0 is called the *Morse number* of the Riemannian space $\mathcal{X}$.

A subset U of a Riemannian space $\mathcal{X}$ is said to be *convex* if

1. any two points $p, q \in U$ are connected by a unique shortest arc $\gamma_{p,q}$ and
2. this shortest arc lies entirely in U.

Exercise 12.7. Prove that for any Riemannian space, there exists an open covering consisting of convex sets. [*Hint*: Use the Whitehead theorem from Chap. 1.]

It is clear that any convex set is diffeomorphic to a ball and that the intersection of two convex sets (when it is not empty) is convex. It follows that a covering consisting of convex sets is a Leray covering (see § 1.7).

§11. Hopf–Rinow Theorem

Definition 12.2. A connected Riemannian space $\mathcal{X}$ is said to be

1. *metrically complete* if it is a complete metric space with respect to the metric ρ (each Cauchy sequence converges) and
2. *geodesically complete* if it is geodesically complete with respect to the Riemannian connection, i.e., each maximal geodesic $\gamma_{p,A}$ is defined on the entire real axis $\mathbb{R}$.

Theorem 12.1. *Both completeness concepts coincide, i.e., each of the conditions* 1 *and* 2 *implies the other. Moreover, conditions* 1 *and* 2 *imply that*

3. *any two points of the space $\mathcal{X}$ can be connected by a shortest arc (not necessary unique in general).*

Proof. It suffices to prove that condition 1 implies condition 2, condition 2 implies condition 3, and conditions 2 and 3 together imply condition 1.

Condition 1 implies condition 2: Let a geodesic $\gamma = \gamma_{p,A}$ be defined on an interval with a finite right endpoint b (the case where the left endpoint is finite is reduced to this one by the replacing A with $-A$), let $\{t_k, k \geq 1\}$ be a monotonically increasing sequence of values of the parameter t that converges to b, and let $p_k = \gamma(t_k)$. Then because the distance $\rho(p_k, p_l)$ between the points p_k and p_l for any k and $l > k$ does not exceed the length $(t_l - t_k)|A|$ of the geodesic segment $\gamma|_{[t_k, t_l]}$, the sequence $\{p_k\}$ is a Cauchy sequence. Therefore, because the metric space is complete by condition 1, this sequence converges. Let p_0 be its limit.

Exercise 12.8. Show that the point p_0 is independent of the choice of the sequence $\{t_k\}$.

For an arbitrary chart $(U, x^1, \dots, x^n)$ centered at the point p_0, the segment of the geodesic γ contained in U is given by the functions $x^i = x^i(t)$, $i = 1, \dots, n$, defined on a certain interval of the form (a, b) such that $x^i(t) \to 0$ as $t \to b$. Because the length of the vector $\dot{\gamma}(t)$ equals $|A|$, the first derivatives $\dot{x}^i(t)$ of these functions are bounded on (a, b), and their second derivatives $\ddot{x}^i(t)$ are therefore also bounded on (a, b) because the functions $x^i(t)$ satisfy the equations

$$\ddot{x}^i(t) + \Gamma^i_{jk}(x(t))\dot{x}^j(t)\dot{x}^k(t) = 0$$

on the interval (a, b). Therefore, the first derivatives $\dot{x}^i(t)$ are uniformly continuous on (a, b), and there hence exist the limits

$$a_0^i = \lim_{t \to b} \dot{x}^i(t), \quad i = 1, \dots, n,$$

as $t \to b$. Let A_0 be a vector in $\mathbf{T}_{p_0}\mathcal{X}$ with the coordinates $a_0^1, \dots, a_0^n$ in the chart $(U, x^1, \dots, x^n)$, and let γ_0 be the geodesic γ_{p_0, A_0}.

Exercise 12.9. Let c be a number $c > b$ such that the curve $\gamma_0(t - b)$ is defined on the semiopen interval $[b, c)$. Prove that the curve $\overline{\gamma}$ defined on the interval (a, c) by

$$\overline{\gamma}(t) = \begin{cases} \gamma(t) & \text{if } a < t < b, \\ \gamma_0(t - b) & \text{if } b \le t < c, \end{cases}$$

is a smooth curve (and therefore a geodesic).

Therefore, the maximal geodesic γ is extended beyond the point b, which is impossible. The obtained contradiction proves that condition 1 implies condition 2.

Conditions 2 and 3 together imply condition 3: The metric space $\mathcal{X}$ is said to be *Bolzanian* if any bounded closed subset is compact (the Bolzano–Weierstrass theorem holds in $\mathcal{X}$).

Exercise 12.10. Show that any Bolzanian metric space is complete.

Therefore, it suffices to prove that *a Riemannian space $\mathcal{X}$ satisfying conditions 2 and 3 is Bolzanian.* Let C be an arbitrary closed bounded set of the space $\mathcal{X}$, and let $p_0 \in C$. According to condition 2, the mapping $\exp_{p_0}$ is defined on the whole space $\mathbf{T}_{p_0}\mathcal{X}$, and according to condition 3, its image coincides with the whole manifold $\mathcal{X}$ (in particular, it contains C) because each shortest arc is a geodesic. Moreover, if R is the diameter of the set C, then C is even contained in the set $\exp_{p_0}(\mathbb{B}_R)$, where $\mathbb{B}_R$ is a closed ball of radius R of the space $\mathbf{T}_{p_0}\mathcal{X}$. The latter set is compact because the ball $\mathbb{B}_R$ is compact; therefore, the set C is also compact. Therefore, the space $\mathcal{X}$ is Bolzanian.

Remark 12.2. The latter argument also holds for $p_0 \notin C$ (it suffices to replace C with $\{p_0\} \cup C$). Therefore, *a connected Riemannian space $\mathcal{X}$ is metrically (and therefore geodesically) complete whenever the domain O_{p_0} of the exponential mapping* $\exp_{p_0}$ *exhausts the whole space $\mathbf{T}_{p_0}\mathcal{X}$ for at least one point $p_0 \in \mathcal{X}$.*

We continue the proof of Theorem 12.1.

Condition 2 implies condition 3: This is essentially a single nontrivial fact. Let p_0 be an arbitrary point of a geodesically complete Riemannian manifold $\mathcal{X}$. We must prove that for any other point $q \in \mathcal{X}$, there exists at least one minimal arc connecting the point p_0 with the point q.

For this, we consider a normal spherical δ-neighborhood U_δ of the point p_0 and its boundary sphere Σ_δ. Because there is nothing to prove for $q \in U_\delta$, we can assume without loss of generality that $q \notin U_\delta$, i.e., $\rho(p_0, q) \geq \delta$. Let $p = \exp_{p_0} \delta A$, $|A| = 1$, be a point of the sphere Σ_δ for which the distance to the point q attains its maximum value (by the compactness of the sphere Σ_δ, the point p exists).

We first prove the following lemma.

Lemma 12.1. *The formula*

$$\rho(p_0, q) = \delta + \rho(p, q)$$

holds.

Proof. Because $\delta = \rho(p_0, p)$, we use the triangle axiom to obtain

$$\rho(p_0, q) \leq \delta + \rho(p, q). \tag{10}$$

On the other hand, by the definition of the greatest lower bound, for any $\varepsilon > 0$, there exists a curve γ connecting the points p_0 and q whose length s_γ satisfies the inequalities

$$\rho(p_0, q) \leq s_\gamma \leq \rho(p_0, q) + \varepsilon.$$

Let r be an intersection of the curve γ with the sphere Σ_δ (there exists at least one such point because $q \notin U_\delta$ and $p_0 \in U_\delta$ by condition). Then

$$s_\gamma \geq \rho(p_0, r) + \rho(r, q) = \delta + \rho(r, q) \geq \delta + \rho(p, q)$$

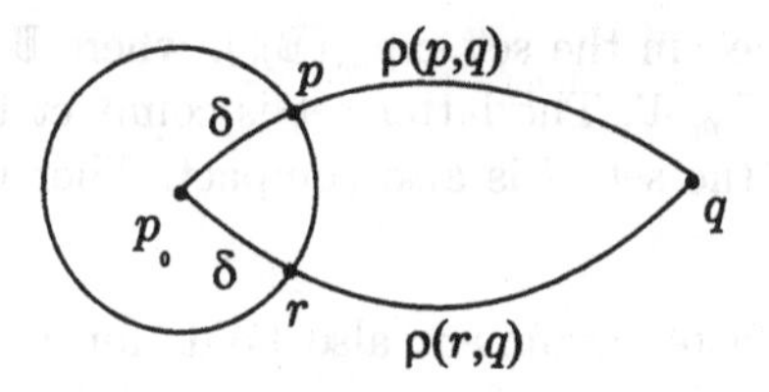

Fig. 12.3.

and therefore $\delta + \rho(p, q) \leq \rho(p_0, q) + \varepsilon$. Because the number $\varepsilon > 0$ is arbitrary, this is possible only if

$$\delta + \rho(p, q) \leq \rho(p_0, q).$$

This, together with (10), proves the lemma. □

Let $\gamma = \gamma_{p_0,A}$ be a geodesic $t \mapsto \exp_{p_0} tA$, and let $\rho_0 = \rho(p_0, q)$.

Lemma 12.2. *For any t, $\delta \leq t \leq \rho_0$, we have*

$$\rho(\gamma(t), q) = \rho_0 - t. \tag{11}$$

Proof. For $t = \delta$, relation (11) passes to the assertion of Lemma 12.1. This means that the set C of all numbers t of the closed interval $[\delta, \rho_0]$ of the real axis $\mathbb{R}$ for which (11) holds is not empty. Let t_0 be its least upper bound. If we show that $t_0 = \rho_0$, then Lemma 12.2 is proved.

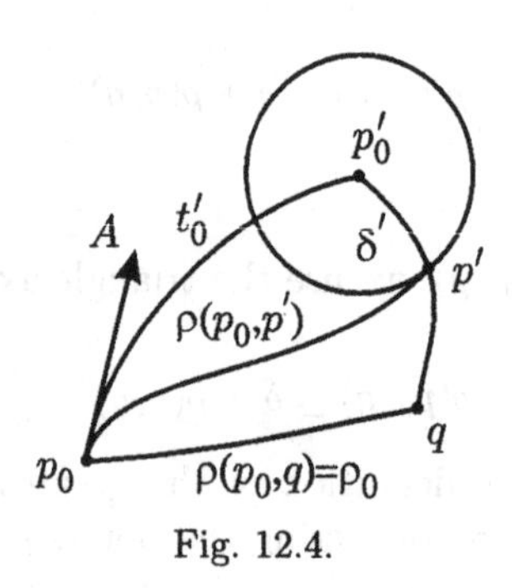

Fig. 12.4.

Let $t_0 < \rho_0$. Then there exists a normal spherical δ'-neighborhood of the point $p_0' = \gamma(t_0)$ that does not contain the point q, and there exists a point p' of the geodesic sphere $\Sigma_{\delta'}$ for which the distance to the point q attains its minimum value. In this case, according to Lemma 12.1, we have

$$\rho(p_0', q) = \delta' + \rho(p', q)$$

and therefore

$$\rho(p', q) = \rho_0 - (t_0 + \delta') \tag{12}$$

(because $\rho(p'_0, q) = \rho_0 - t_0$ by continuity). Therefore, by the triangle axiom (applied to the points p_0, p', and q), we have

$$\rho(p_0, p') \geq \rho(p_0, q) - \rho(p', q) = \rho_0 - (\rho_0 - (t_0 + \delta')) = t_0 + \delta'.$$

On the other hand, we can explicitly give a piecewise smooth curve of length $t_0 + \delta'$ connecting the point p_0 with the point p'; this is the geodesic broken line composed of the segment of the geodesic γ from p_0 to p'_0 (of length t_0) and the normal geodesic segment $\gamma_{p'_0,p'}$ (of length δ'). Therefore, first, $\rho(p_0, p') = t_0 + \delta'$; second, this geodesic broken line is a shortest arc. But because it is a shortest arc and therefore a smooth curve, this broken line cannot have a break (and is therefore a geodesic). Having a common segment with the maximal geodesic $\gamma = \gamma_{p,A}$, it itself should be a segment of this geodesic. Therefore, the point $\gamma(t_0 + \delta')$ of the geodesic γ (which is distant from the point p_0 by $t_0 + \delta'$) should coincide with the point p'. By (12), this means that the inclusion $t_0 + \delta' \in C$ holds, which is impossible for the least upper bound t_0. The obtained contradiction proves that $t_0 = \rho_0$. □

It is now easy to complete the proof of Theorem 12.1. Indeed, for $t = \rho_0$, relation (11) implies $q = \gamma(\rho_0)$, and because the length of the geodesic segment $\gamma|_{[0,\rho_0]}$ equals $\rho_0 = \rho(p_0, q)$, this segment is therefore a shortest arc connecting the point p_0 with the point q. □

Theorem 12.1 is known as the *Hopf–Rinow theorem.*

Connected Riemannian spaces satisfying conditions 1 or 2 in Theorem 12.1 are said to be *complete.*

Remark 12.3. As Example 12.3 shows, the Hopf–Rinow theorem is not directly extended to pseudo-Riemannian spaces. At the same time, we can show (try to do this independently) that the Hopf–Rinow theorem (and also other results in this chapter, for example, the smoothness of minimum curves and the property of extremals to be locally minimum curves) are extended to any Lagrangians L for which the symmetric matrix

$$\left\| \frac{\partial^2 L}{\partial \dot{q}^i \partial \dot{q}^j} \right\|$$

is positive definite in each chart $(U, q^1, \ldots, q^n)$ (at any point of the neighborhood U).

Remark 12.4. In particular, condition 3 means that any two points of the space $\mathcal{X}$ can be connected by a geodesic. In this weakened form, it is meaningful for any affine connection space, and the question whether this property is implied by the geodesic completeness (and, if so, for what spaces it holds) arises. The answer is negative. The Cartan connection on the Lie group $\mathrm{SL}(2; \mathbb{R})$ of unimodular matrices of second order yields a counterexample (as we know, it is geodesically complete; see Chap. 6). Indeed, the Lie algebra of this group consists of real matrices X of the form

$$\begin{Vmatrix} x & y \\ z & -x \end{Vmatrix}.$$

Because $X^2 = \delta E$, where $\delta = x^2 + yz = -\det X$, we have

$$e^X = \cosh\sqrt{\delta}\cdot E + (1/\sqrt{\delta})\sinh\sqrt{\delta}\cdot X$$

(for $\delta \neq 0$; if $\delta = 0$, then $e^X = E + X$); therefore, if the matrix

$$A = \begin{Vmatrix} a & b \\ c & d \end{Vmatrix} \in \mathrm{SL}(2;\mathbb{R})$$

has the form e^X and hence belongs to a certain one-parameter subgroup, then the inequality $a + d \geq -2$ should hold. (We note that the number $\sqrt{\delta}$ is either real or purely imaginary; in fact, the matrix $A \in \mathrm{SL}(2;\mathbb{R})$ has the form e^X iff (prove this) either $a + d > -2$ or $a = d = -1$ and $b = c = 0$, i.e., $A = -E$.) Therefore, not every element of the group $\mathrm{SL}(2;\mathbb{R})$ belongs to at least one one-parameter subgroup, i.e., can be connected with the element E of the Cartan geodesic connection.

Remark 12.5. In Chap. 26, we show (see Proposition 26.2) that on any compact Lie group, the Cartan connection is generated by a Riemannian metric, and the phenomenon indicated in Remark 12.4 is therefore impossible for this connection, i.e., *any element of a compact Lie group is contained in a certain one-parameter subgroup* (the exponential mapping is surjective). However, it is apparently still unknown whether any two points of an arbitrary geodesically complete compact affine connection space can be connected by a geodesic.

Remark 12.6. It is interesting that *not geodesically complete affine connections can exist on a compact manifold.* An example of them is (prove this!) the metric connection on $\mathbb{R}$ induced by the metric $ds^2 = e^x dx^2$ transported to $\mathbb{S}^1$.

Chapter 13
Harmonic Functionals and Related Topics

§1. Riemannian Volume Element

In a (pseudo-)Riemannian space $\mathcal{X}$, we can measure not only lengths but also volumes, i.e., we can define a natural volume density dV on $\mathcal{X}$. Indeed, let (U, h) be an arbitrary chart of an arbitrary (pseudo-)Riemannian space $\mathcal{X}$, let $\|g_{ij}\|$ be the matrix of components of the metric tensor g in the chart (U, h), and let

$$\det g = |g_{ij}|$$

be its determinant. The transformation formula for the matrix of a quadratic form under a change of basis directly implies that under a change of coordinates, the determinant $\det g$ is multiplied by the square of the Jacobian of the transition functions. Therefore, setting

$$dV^U = \sqrt{|\det g|}$$

(as usual, we mean the arithmetical square root), we obtain a certain volume density dV on $\mathcal{X}$. (Of course, for a Riemannian space, it is not necessary to pass from $\det g$ to $|\det g|$.) The density dV is conventionally denoted by

$$\sqrt{|\det g|}\, dx^1 \cdots dx^n \qquad \text{or} \qquad \sqrt{|\det g|}\, dx. \tag{1}$$

Definition 13.1. Density (1) is called the *Riemannian volume element* on the (pseudo-)Riemannian manifold $\mathcal{X}$.

For $n = 2$, this is the already known area element $\sqrt{EG - F^2}\, du\, dv$ on a surface.

If a manifold $\mathcal{X}$ is orientable and oriented, then we can pass from the density dV to the corresponding form of maximal degree. This form is also called the *volume element* (sometimes the *oriented volume element*) and is denoted by the previous symbol dV. By definition,

$$dV = \sqrt{|\det g|}\, dx^1 \wedge \cdots \wedge dx^n$$

in each positively oriented chart $(U, x^1, \ldots, x^n)$.

§2. Discriminant Tensor

The coefficients of the form dV compose a tensor field of type $(n, 0)$ on $\mathcal{X}$. This field is called the *discriminant tensor* of a (pseudo-)Riemannian manifold

$\mathcal{X}$ and is traditionally denoted by e (the symbol ε is also used); its components are denoted by $e_{i_1\cdots i_n}$ (or $\varepsilon_{i_1\cdots i_n}$). Therefore, in each chart (U,h), we have

$$e_{i_1\cdots i_n} = \begin{cases} \varepsilon_\sigma e_{1\cdots n} & \text{if the subscripts } i_1,\ldots,i_n \text{ are distinct,} \\ 0 & \text{otherwise,} \end{cases} \tag{2}$$

where ε_σ is the sign of the permutation

$$\sigma = \begin{pmatrix} 1 & \cdots & n \\ i_1 & \cdots & i_n \end{pmatrix}$$

and

$$e_{1\cdots n} = \sqrt{|\det g|}$$

if the orientation of the chart (U,h) is positive and

$$e_{1\cdots n} = -\sqrt{|\det g|}$$

if the orientation is negative.

For example, for $n=2$,

$$\begin{Vmatrix} e_{11} & e_{12} \\ e_{21} & e_{22} \end{Vmatrix} = \begin{Vmatrix} 0 & \pm\sqrt{EG-F^2} \\ \mp\sqrt{EG-F^2} & 0 \end{Vmatrix},$$

where the signs $\pm$ depend on the orientation of the chart.

§3. Foss–Weyl Formula

The linear differential form

$$\gamma = \Gamma^i_{ik}\, dx^k \tag{3}$$

(see Chap. 2) is closely related to the discriminant tensor.

Proposition 13.1. *The formula*

$$\gamma = d\ln\sqrt{|\det g|} \tag{4}$$

holds.

Proof. Formula (5′) in Chap. 11 implies

$$\Gamma^i_{ik} = \frac{1}{2} g^{ip}\left(\frac{\partial g_{pi}}{\partial x^k} + \frac{\partial g_{pk}}{\partial x^i} - \frac{\partial g_{ik}}{\partial x^p}\right) = \frac{1}{2} g^{ij}\frac{\partial g_{ij}}{\partial x^k},$$

i.e.,

$$\gamma = \frac{1}{2} g^{ij} dg_{ij} = \frac{1}{2}\operatorname{Tr}(g^{-1}dg),$$

where g is the matrix $\|g_{ij}\|$. On the other hand,

$$\begin{aligned}d\ln\sqrt{|\det g|} &= \frac{1}{2}d\ln|\det g| = \frac{1}{2}d\ln e^{\operatorname{Tr}\ln g}\\ &= \frac{1}{2}d\operatorname{Tr}\ln g = \frac{1}{2}\operatorname{Tr}(d\ln g) = \frac{1}{2}\operatorname{Tr}(g^{-1}dg)\end{aligned}$$

by the rules of matrix calculus. □

Formula (4) is known as the *Foss–Weyl formula.*

Corollary 13.1. *The discriminant tensor is covariantly constant,*

$$\nabla_X e = 0$$

for any vector field X on $\mathcal{X}$.

Proof. It suffices to show that

$$(\nabla_j e)_{i_1\cdots i_n} = 0$$

for any subscripts j and $i_1, \ldots, i_n$. Because, in accordance with the general rules for covariant differentiation (see formula (5) in Chap. 2),

$$(\nabla_j e)_{i_1\cdots i_n} = \frac{\partial e_{i_1\cdots i_n}}{\partial x^j} - \sum_{s=1}^{n} \Gamma^p_{ji_s} e_{i_1\cdots i_{s-1}\,p\,i_{s+1}\cdots i_n}, \tag{5}$$

we must show that the right-hand side of this formula vanishes for any subscripts j and $i_1, \ldots, i_n$.

Case 1. We suppose that there are equal subscripts among the subscripts $i_1, \ldots, i_n$. If there exist more that two equal subscripts, then the right-hand side of (5) obviously vanishes. Therefore, we can assume without loss of generality that among the subscripts $i_1, \ldots, i_n$, there are exactly two equal subscripts; therefore, in the series $1, \ldots, n$, we have exactly one element different from all $i_1, \ldots, i_n$. In this case, if $i_a = i_b$ and $k \neq i_1, \ldots, i_n$, the right-hand side of (5) is equal to

$$\Gamma^k_{ji_a} e_{i_1\cdots i_{a-1}\,k\,i_{a+1}\cdots i_n} + \Gamma^k_{ji_b} e_{i_1\cdots i_{b-1}\,k\,i_{b+1}\cdots i_n}$$

(the sum in k is not performed), i.e., it is equal to $(\varepsilon_\sigma + \varepsilon_\tau)e_{1\cdots n}$ up to the factor $\Gamma^k_{ji_a} = \Gamma^k_{ji_b}$, where ε_σ and ε_τ are the signs of the permutations

$$\sigma = \begin{pmatrix} 1 \cdots\cdots\cdots\cdots\cdots n \\ i_1 \cdots i_{a-1}\; k\; i_{a+1} \cdots i_n \end{pmatrix}$$

and

$$\tau = \begin{pmatrix} 1 \cdots\cdots\cdots\cdots\cdots n \\ i_1 \cdots i_{b-1}\; k\; i_{b+1} \cdots i_n \end{pmatrix}.$$

But, as is easily seen, because these permutations differ by the transposition (ik), where $i = i_a = i_b$, we have $\varepsilon_\tau = -\varepsilon_\sigma$. Therefore, the right-hand side of (5) vanishes in this case.

Case 2. We suppose that all the subscripts $i_1, \ldots, i_n$ are different. Then only one summand (corresponding to $p = i_s$) is different from zero in each sum $\Gamma^p_{ji_s} e_{i_1 \cdots i_{s-1} p i_{s+1} \cdots i_n}$. Therefore, the right-hand side of (5) in this case is equal to

$$\frac{\partial e_{i_1 \cdots i_n}}{\partial x^j} - \sum_{s=1}^{n} \Gamma^{i_s}_{ji_s} e_{1 \cdots i_s \cdots n} = \frac{\partial e_{i_1 \cdots i_n}}{\partial x^j} - \Gamma^i_{ji} e_{1 \cdots n}$$

$$= \varepsilon_\sigma \left[\frac{\partial \sqrt{|\det g|}}{\partial x^j} - \Gamma^i_{ji} \sqrt{|\det g|} \right],$$

where ε_σ is the sign of the permutation

$$\sigma = \begin{pmatrix} 1 & \cdots & n \\ i_1 & \cdots & i_n \end{pmatrix}.$$

On the other hand, according to Foss–Weyl formula (4),

$$\Gamma^i_{ji} = \frac{1}{\sqrt{|\det g|}} \frac{\partial}{\partial x^j} \sqrt{|\det g|}$$

(the Riemannian connection is symmetric by definition and therefore $\Gamma^i_{ji} = \Gamma^i_{ij}$). Therefore, the expression in the square brackets vanishes. □

Geometrically, this corollary means that *parallel translation preserves the Riemannian volume.* From this standpoint, it is visually obvious.

§4. Case $n = 2$

For $n = 2$, the discriminant tensor is often used to lower and raise superscripts and subscripts. For example, the contraction of a tensor e with a vector field yields a linear differential form. If the field has the components $X^1 = X$ and $X^2 = Y$, then the form has the coefficients εY and $-\varepsilon X$, where we set $\varepsilon = \sqrt{EG - F^2}$ for simplicity, i.e., it becomes

$$\varepsilon Y \, du - \varepsilon X \, dv, \tag{6}$$

where u and v are local coordinates. Of course, this correspondence of fields and forms differs from the correspondence established by the metric tensor (see Chap. 11), but similarly to the latter, it commutes with covariant differentiation.

It is sometimes convenient to combine these correspondences. For example, taking the contraction of form (6) with the metric tensor g^{ij}, we again obtain a vector field but with the components

$$\frac{FX + GY}{\varepsilon} \quad \text{and} \quad -\frac{EX + FY}{\varepsilon}. \tag{7}$$

(Here, we use the standard notation of surface theory:

$$g_{11} = E, \qquad g_{12} = F, \qquad g_{22} = G.$$

We recall that in this notation,

$$g^{11} = \frac{G}{\varepsilon^2}, \qquad g^{12} = -\frac{F}{\varepsilon^2}, \qquad g^{22} = \frac{E}{\varepsilon^2}, \tag{8}$$

where $\varepsilon^2 = EG - F^2$.)

Exercise 13.1. Show that field (7) is orthogonal to the initial field (at each point). [*Hint*: $g_{ij}X^i g^{jk} e_{kl} X^l = e_{il} X^i X^l = 0$.]

Raising both subscripts of the discriminant tensor via the metric tensor, we obtain a tensor of type $(0, 2)$ with the components

$$e^{ij} = g^{ik} g^{jl} e_{kl}. \tag{9}$$

This tensor is also skew-symmetric (because $e^{ji} = g^{jk} g^{il} e_{kl} = -g^{il} g^{jk} e_{lk} = -e^{ij}$) and its components e^{12} are expressed by

$$e^{12} = g^{1k} g^{2l} e_{kl} = \varepsilon(g^{11} g^{22} - g^{12} g^{21}) = \varepsilon \frac{1}{EG - F^2} = \frac{1}{\varepsilon}.$$

Because

$$\begin{Vmatrix} 0 & \varepsilon \\ -\varepsilon & 0 \end{Vmatrix} \cdot \begin{Vmatrix} 0 & -1/\varepsilon \\ 1/\varepsilon & 0 \end{Vmatrix} = \begin{Vmatrix} 1 & 0 \\ 0 & 1 \end{Vmatrix},$$

this means that *the matrix* $\|e^{ij}\|$ *is inverse to the matrix* $-\|e_{ij}\|$ (similarly to the matrix $\|g^{ij}\|$ being inverse to the matrix $\|g_{ij}\|$).

It is useful to keep the relation

$$g^{ij} e_{jk} = e^{ij} g_{jk} \tag{10}$$

in mind, which means that the identity

$$g^{-1} e = -e^{-1} g$$

holds for the matrices $g = \|g_{ij}\|$ and $e = \|e_{ij}\|$. The easiest way to verify (10) is by direct computation:

$$e^{ij} g_{jk} = g^{ip} g^{jq} g_{jk} e_{pq} = g^{ip} e_{pk} = g^{ij} e_{jk}.$$

In the Gauss notation, this becomes

$$\begin{Vmatrix} G/\varepsilon^2 & -F/\varepsilon^2 \\ -F/\varepsilon^2 & E/\varepsilon^2 \end{Vmatrix} \cdot \begin{Vmatrix} 0 & \varepsilon \\ -\varepsilon & 0 \end{Vmatrix} = \begin{Vmatrix} 0 & 1/\varepsilon \\ -1/\varepsilon & 0 \end{Vmatrix} \cdot \begin{Vmatrix} E & F \\ F & G \end{Vmatrix},$$

which is also verified by direct computation.

§5. Laplace Operator on a Riemannian Space

Again let $n \geq 2$. For any smooth function u on a (pseudo-)Riemannian space $\mathcal{X}$, the components $(\operatorname{grad} u)^i$ of its gradient $\operatorname{grad} u$ in each chart $(U, x^1, \ldots, x^n)$ of the manifold $\mathcal{X}$ by definition (see (1) in Chap. 11) are expressed by the formulas

$$(\operatorname{grad} u)^i = g^{ik}\frac{\partial u}{\partial x^k}, \quad i = 1, \ldots, n.$$

Exercise 13.2. Show that the vector field $\operatorname{grad} u$ is invariantly characterized by

$$Xu = (X, \operatorname{grad} u)$$

for any vector field $X \in \mathfrak{a}\mathcal{X}$.

Because the operation of covariant differentiation ∇_X is $\mathsf{F}\mathcal{X}$-linear in X, the components

$$(\nabla_k \operatorname{grad} u)^i = \frac{\partial(\operatorname{grad} u)^i}{\partial x^k} + \Gamma^i_{kj}(\operatorname{grad} u)^j$$

of partial covariant derivatives of the field $\operatorname{grad} u$ are components of a tensor field of type $(1,1)$ on $\mathcal{X}$. Therefore, the trace

$$\Delta u = (\nabla_i \operatorname{grad} u)^i \tag{11}$$

of this tensor field is a well-defined function on $\mathcal{X}$. This means that formula (11) defines a certain operator

$$\Delta\colon \mathsf{F}\mathcal{X} \to \mathsf{F}\mathcal{X} \tag{12}$$

(obviously linear over $\mathbb{R}$). (Here, we essentially used the assumption that the manifold $\mathcal{X}$ belongs to the class C^∞. For manifolds of class C^r, $r < \infty$, the functions Δu are only functions of class C^{r-2} in general.)

In the case where $\mathcal{X}$ is a Euclidean space and $x^1, \ldots, x^n$ are rectangular coordinates, the function Δu is expressed by the formula

$$\Delta u = \frac{\partial^2 u}{\partial x_1^2} + \cdots + \frac{\partial^2 u}{\partial x_n^2},$$

i.e., the operator Δ is the Laplace operator known from calculus and is therefore also called the *Laplace operator* in this general case. However, the name *Beltrami operator* is also used (see below), and functions $u \in \mathsf{F}\mathcal{X}$ for which $\Delta u = 0$ are said to be *harmonic* (with respect to a given Riemannian metric on $\mathcal{X}$).

By definition, the function Δu is expressed in each chart by the formula

$$\Delta u = \frac{\partial(\operatorname{grad} u)^i}{\partial x^i} + \Gamma^i_{ij}(\operatorname{grad} u)^j.$$

But according to Foss–Weyl formula (4), we have

$$\Gamma^i_{ij} = \frac{\partial}{\partial x^j}(\ln\sqrt{|\det g|}) = \frac{1}{\varepsilon}\frac{\partial\varepsilon}{\partial x^j},$$

where we set $\varepsilon = \sqrt{|\det g|}$ for brevity as in the case $n = 2$. Therefore,

$$\Delta u = \frac{\partial(\operatorname{grad} u)^i}{\partial x^i} + \frac{1}{\varepsilon}\frac{\partial\varepsilon}{\partial x^j}(\operatorname{grad} u)^j = \frac{1}{\varepsilon}\frac{\partial}{\partial x^i}(\varepsilon \operatorname{grad} u)^i,$$

i.e.,

$$\Delta = \frac{1}{\varepsilon}\frac{\partial}{\partial x^i}\left(\varepsilon g^{ij}\frac{\partial}{\partial x^j}\right). \tag{13}$$

For $n = 2$ (for surfaces), operator (13) in the Gauss notation is expressed (see (8)) by the form

$$\Delta = \frac{\dfrac{\partial}{\partial u}\left(\dfrac{G\dfrac{\partial}{\partial u} - F\dfrac{\partial}{\partial v}}{\sqrt{EG-F^2}}\right) + \dfrac{\partial}{\partial v}\left(\dfrac{-F\dfrac{\partial}{\partial u} + E\dfrac{\partial}{\partial v}}{\sqrt{EG-F^2}}\right)}{\sqrt{EG-F^2}} \tag{14}$$

(as usual in the surface theory, the symbols u and v denote local coordinates here).

This formula was first obtained by Beltrami, who represented the result of applying operator (14) to a function f with $\Delta_2 f$ and called it the *second differential parameter* of this function. (The first Beltrami differential parameter $\Delta_1 f$ is just the inner square $(\operatorname{grad} f, \operatorname{grad} f)$ of the gradient). Some authors still use this name (together with the notation $\Delta_2 f$).

§6. The Green Formulas

Exercise 13.3. Assuming that a manifold $\mathcal{X}$ is oriented, define the operator $\operatorname{div}\colon \mathfrak{a}\mathcal{X} \to \mathsf{F}\mathcal{X}$ by

$$(\operatorname{div} X)\omega = \mathfrak{L}_X\omega, \quad X \in \mathfrak{a}\mathcal{X}, \tag{15}$$

where ω is the Riemannian volume element dV and $\mathfrak{L}_X\omega$ is its Lie derivative in X. Show that

$$\operatorname{div} X = \frac{1}{\varepsilon}\frac{\partial}{\partial x^i}(\varepsilon X^i)$$

in each coordinate neighborhood and therefore, in particular,

$$\Delta u = \operatorname{div} \operatorname{grad} u, \tag{16}$$

as in elementary calculus.

The assumption that the manifold $\mathcal{X}$ is orientable is in fact extra here.

Exercise 13.4. Let $\nabla_\bullet X$ be the linear operator $\mathfrak{a}\mathcal{X} \to \mathfrak{a}\mathcal{X}$ defined by $Y \mapsto \nabla_Y X$, and let $\operatorname{Tr}\nabla_\bullet X$ be its trace. Show that

1. if the manifold $\mathcal{X}$ is orientable, then
$$\operatorname{div} X = \operatorname{Tr} \nabla_\bullet X \quad \text{for any field } X \in \mathfrak{a}\mathcal{X}; \tag{17}$$
2. if we take (17) as the definition of the operator div, then (16) is preserved.

We can rewrite formula (15) as
$$(\operatorname{div} X)\,\omega = d\,(X \lrcorner\, \omega).$$
Therefore, if the manifold $\mathcal{X}$ is compact, then
$$\int_{\mathcal{X}} \operatorname{div} X\, dV = 0 \tag{18}$$
for any field $X \in \mathfrak{a}\mathcal{X}$ (dV again denotes the volume form) and, in particular,
$$\int_{\mathcal{X}} \Delta u\, dV = 0. \tag{18'}$$
This is a generalization of the first Green formula or, more precisely, a particular case of it, which is obtained for $\partial D = 0$, and formula (18′) (and sometimes (18)) is therefore usually called the *Green formula.*

Exercise 13.5. Show that for any function $u \in \mathsf{F}\mathcal{X}$, we have
$$\Delta u^2 = 2u \cdot \Delta u + 2|\operatorname{grad} u|^2.$$

By Green formula (18′), this implies
$$\int_{\mathcal{X}} |\operatorname{grad} u|^2 dV = 0$$
for $\Delta u = 0$, and therefore $\operatorname{grad} u = 0$, i.e., $u = \text{const}$. Therefore, *on a compact orientable manifold $\mathcal{X}$, there are no harmonic functions different from constants.*

Exercise 13.6. Prove that this also holds for a nonorientable manifold $\mathcal{X}$. [*Hint*: Pass to a two-sheeted covering orientable manifold.]

§7. Existence of Harmonic Functions with a Nonzero Differential

According to (13), for any function u, we have
$$\Delta u = g^{ij}\frac{\partial^2 u}{\partial x^i \partial x^j} + \dots,$$
where the dots stand for terms depending only on the first derivatives of the function u. Because the matrix $\|g^{ij}\|$ is positive definite for a Riemannian space, this means by definition that on the Riemannian space, the operator Δ is an elliptic second-order differential operator.

We need the following lemma from the theory of elliptic equations.

Lemma 13.1. *Let a linear second-order elliptic differential operator*

$$L = \sum_{i,j=1}^{n} a_{ij}(x)\frac{\partial^2}{\partial x^i \partial x^j} + \sum_{i=1}^{n} b_i(x)\frac{\partial}{\partial x^i} + c(x) \tag{19}$$

(with variable coefficients) be given in a neighborhood of the point $0 \in \mathbb{R}^n$. *If* $c \le 0$, *then in a certain neighborhood* U *of the point* 0, *there exists a smooth function* u *such that* $Lu = 0$ *and* $du \neq 0$ *everywhere in* U.

Although Lemma 13.1 plays a crucial role in surface theory, as is seen in what follows, the author is not familiar with textbooks or monographs devoted to differential geometry or the theory of partial differential equations where it is proved (or even clearly stated). Of course, this lemma is a direct consequence of the theorem on the local solvability of the Cauchy problem, but referring to this theorem is not so satisfactory because the Cauchy problem is usually considered an ill-posed problem for elliptic equations (at least because its solution is not unique in advance), and the theorem on its local solvability is presented only in certain very detailed or nonstandardly oriented monographs (see, e.g., Bers, L., John, F., and Schechter, M., *Partial Differential Equations*, Interscience, New York, 1964) and in any case, is not included to the appropriate university course. Therefore, we here prove Lemma 13.1 (whose idea was told the author by E. M. Landis) using only facts that are almost certainly known to every student of mathematics. (Mainly, we here mean the theorem on the solvability of the Dirichlet problem and the so-called Schauder estimate; see Assertions 13.1 and 13.2 below.)

Let k be a nonnegative integer, $0 < \alpha < 1$, D be a bounded domain of the space $\mathbb{R}^n$, and $\overline{D}$ be its closure. The linear space of functions $u\colon \overline{D} \to \mathbb{R}$ that are differentiable k times in D and whose kth partial derivatives satisfy the Holder condition with the exponent α in D is denoted by $C^{k+\alpha}(D)$. (We usually say that the functions from $C^{k+\alpha}(D)$ belong to the *class* $C^{k+\alpha}$ in D.) It is known that with respect to the norm

$$\|u\|_{C^{k+\alpha}(D)} = \sum_{\nu=0}^{k} (\operatorname{diam} D)^{\nu} \max_{\substack{p_1,\dots,p_n,\\ p_1+\dots+p_n=\nu}} \sup_{x\in D} \left| \frac{\partial^\nu u}{\partial x_1^{p_1}\cdots\partial x_n^{p_n}}(x)\right|$$

$$+ (\operatorname{diam} D)^{k+\alpha} \max_{\substack{p_1,\dots,p_n,\\ p_1+\dots+p_n=k}} \sup_{\substack{x,y\in D,\\ x\neq y}} \frac{\left| \dfrac{\partial^k u}{\partial x_1^{p_1}\cdots\partial x_n^{p_n}}(x) - \dfrac{\partial^k u}{\partial x_1^{p_1}\cdots\partial x_n^{p_n}}(y)\right|}{|x-y|^\alpha}$$

the linear space $C^{k+\alpha}(D)$ is a Banach space. As a rule, we write merely $\|u\|_{k+\alpha}$ instead $\|u\|_{C^{k+\alpha}(D)}$.

Let the domain D be bounded by a thrice differentiable hypersurface ∂D (of class C^3), and let

$$L = \sum_{i,j=1}^{n} a_{ij} \frac{\partial^2}{\partial x^i \partial x^j} + \sum_{i=1}^{n} b_i \frac{\partial}{\partial x^i} + c \tag{19'}$$

be an elliptic operator with coefficients in $C^\alpha(D)$.

Assertion 13.1 (Schauder estimate). *There exists a constant $C > 0$ such that the inequality*

$$\|u\|_{2+\alpha} \le C\|Lu\|_\alpha \tag{20}$$

holds for any function $u \in C^{2+\alpha}(D)$ with $Lu \in C^\alpha(D)$ and $u|_{\partial D} = 0$. The constant C depends only on the domain D, the ellipticity constant of the operator L, and the maximum of the norms of its coefficients in the space $C^\alpha(D)$.

Remark 13.1. In reality, inequality (20) is only the particular case of the classical Schauder estimate where $u|_{\partial D} = 0$, which is enough for our purposes.

Remark 13.2. The Schauder estimate is often presented (see, e.g., the abovementioned book by Bers, John, and Schechter, p. 244 in the Russian edition) in a slightly different form with $\|Lu\|_\alpha$ replaced by $\|Lu\|_\alpha + \|u\|_0$. But it can be can shown (see, e.g., Landis, E.M., *Second-order equations of elliptic and parabolic types* [in Russian], Fizmatgiz, Moscow (1971), pp. 272–273; English translation: *Transl. Math. Monogr.*, Vol. 171, Amer. Math. Soc., Providence, RI (1998)) that $\|u\|_0 \le C_1\|Lu\|_\alpha$, where C_1 is a certain constant, and this estimate therefore implies estimate (20).

Assertion 13.2 (solvability of the Dirichlet problem). *If $c \le 0$ in D, then for any function $f \in C^{2+\alpha}(\partial D)$, there exists a function $u \in C^{2+\alpha}(D)$ such that*

$$Lu = 0 \qquad \textit{and} \qquad u|_{\partial D} = f.$$

(By definition, $f \in C^{2+\alpha}(\partial D)$ if the function f is given by functions of class $C^{2+\alpha}$ in each chart of the smooth manifold ∂D.)

Now, as usual, let $\overline{\mathbb{B}}_r$ be the ball $\|x\| \le r$ of the space $\mathbb{R}^n$ of radius $r > 0$ centered at the point 0, let $\mathbb{B}_r$ be its interior $\|x\| < r$, and let $\mathbb{S}_r$ be the sphere bounding this ball. Further, let L be an elliptic operator of form (19') given in a domain D containing the point 0, and let r_0 be a number such that the ball $\mathbb{B}_{r_0}$ (and therefore each ball $\mathbb{B}_r$, $0 < r \le r_0$) is contained in D.

Restricting the coefficients of the operator L to $\mathbb{B}_r$, $0 < r \le r_0$, we obtain an operator (obviously elliptic) with coefficients in $C^\alpha(\mathbb{B}_r)$. Admitting a certain inaccuracy, we represent this operator with the old symbol L.

Lemma 13.2. *There exists a constant $C > 0$ depending only on the operator L and there exist numbers r_0 such that for any r, $0 < r \le r_0$, and for any function $w \in C^{2+\alpha}(\mathbb{B}_r)$ with $w|_{\mathbb{S}_r} = 0$, the inequality*

$$\left|\frac{\partial w}{\partial x_1}\right| < Cr^2\|Lw\|_\alpha$$

holds on $\overline{\mathbb{B}}_r$.

Proof. Let $h: \overline{\mathbb{B}}_r \to \overline{\mathbb{B}}_1$ be the homothety $\boldsymbol{x} \mapsto \boldsymbol{x}/r$. This homothety transforms the operator L into the operator $\hat{L} = (h^*)^{-1} \circ L \circ h^*$, which acts on $C^{2+\alpha}(\overline{\mathbb{B}}_1)$, where h^* sets the function $h^*v: \boldsymbol{x} \mapsto v(\boldsymbol{x}/r)$ on $\mathbb{B}_r$ in correspondence to a function v on $\mathbb{B}_1$. An obvious computation shows that the operator $L_1 = r^2\hat{L}$ on $C^{2+\alpha}(\mathbb{B}_1)$ is given by

$$L_1 = \sum_{i,j=1}^{n} a_{ij}(r\boldsymbol{x}) \frac{\partial^2}{\partial x^i \partial x^j} + r \sum_{i=1}^{n} b_i(r\boldsymbol{x}) \frac{\partial}{\partial x^i} + r^2 c(r\boldsymbol{x}).$$

If $v = (h^*)^{-1}w$ and $f = (h^*)^{-1}(Lw)$, then $L_1 v = r^2 f$, and according to the Schauder estimate (applied to the operator L_1 in the domain $\mathbb{B}_1$), we therefore have

$$\|v\|_{2+\alpha} \le Cr^2\|f\|_\alpha \quad \text{on } \mathbb{B}_1,$$

where C depends only on the operator L on $\mathbb{B}_1$ and the number r_0. (We note that $v|_{\mathbb{S}_1} = 0$ iff $w|_{\mathbb{S}_r} = 0$.) Because the mapping h^* obviously preserves the norms $\|\cdot\|_{k+\alpha}$, this proves that under the conditions of Lemma 13.2, the inequality

$$\|w\|_{2+\alpha} \le Cr^2\|Lw\|_\alpha$$

holds on $\mathbb{B}_r$. Noting that

$$\left|\frac{\partial w}{\partial x_1}\right| \le \|w\|_{2+\alpha} \quad \text{on } \mathbb{B}_r$$

completes the proof of Lemma 13.2. □

Proof (of Lemma 13.1). Because operator (19) has form (19′) with $c \le 0$, Assertion 13.2 and Lemma 13.2 are applicable to it. As above, let r_0 be a number such that the operator L is defined on $\mathbb{B}_{r_0}$. According to Assertion 2, for any r, $0 < r \le r_0$, there exists a function u on $\overline{\mathbb{B}}_r$ such that $Lu = 0$ in $\mathbb{B}_r$ and $u = -x_1$ on $\mathbb{S}_r$.

Let $w = u + x_1$. Then

$$Lw = Lu + Lx_1 = Lu + b_1 + cx_1 = f \quad \text{on } \mathbb{B}_r,$$

where $f = b_1 + cx_1$ and $w = 0$ on $\mathbb{S}_r$. Therefore, according to Lemma 13.2,

$$\left|\frac{\partial w}{\partial x_1}\right| < C_1 r^2 \quad \text{on } \mathbb{B}_r,$$

where

$$C_1 = C \cdot \max_r \|f\|_{C^\alpha(\mathbb{B}_r)}$$

does not depend on r. Because

$$\frac{\partial w}{\partial x_1} = \frac{\partial u}{\partial x_1} + 1,$$

this proves that for a sufficiently small r, the function $\partial u/\partial x_1$ (and therefore the form du) is different from zero everywhere in $\mathbb{B}_r$. □

Corollary 13.2. *For any point p_0 of a Riemannian space $\mathcal{X}$, there exists a neighborhood of this point U and a harmonic function u on U such that $du \neq 0$ everywhere on U.*

Proof. Operator (14) has form (19) with $c = 0$. □

We need only this corollary.

§8. Conjugate Harmonic Functions

For $n = 2$, the differential $d\omega$ of an arbitrary differential form $\omega = c_1 dx^1 + c_2 dx^2$ on a coordinate neighborhood U (the local coordinates on a surface are not denoted by the usual letters u and v here) is expressed by the formula

$$d\omega = \left(\frac{\partial c_2}{\partial x^1} - \frac{\partial c_1}{\partial x^2}\right) dx^1 \wedge dx^2 = (\nabla_1 c_2 - \nabla_2 c_1)\, dx^1 \wedge dx^2,$$

where

$$\nabla_j c_i = \frac{\partial c_i}{\partial x^j} - \Gamma^k_{ji} c_k, \quad i, j = 1, 2$$

(because $\Gamma^k_{12} = \Gamma^k_{21}$, the terms containing Γ^k_{ji} are canceled). Therefore, *the form ω is closed iff* $\nabla_1 c_2 = \nabla_2 c_1$, i.e., if

$$e^{ij} \nabla_j c_i = 0. \tag{21}$$

Keeping this in mind, we consider the differential form ω_u, which is the result of the convolution of the discriminant tensor with the vector field $\operatorname{grad} u$, where u is an arbitrary smooth function on U. For this form, we have $c_i = e_{ik} (\operatorname{grad} u)^k$, and therefore

$$\nabla_j c_i = e_{ik} (\nabla_j \operatorname{grad} u)^k$$

because the tensor e is covariantly constant. Therefore,

$$e^{ij} \nabla_j c_i = e^{ij} e_{ik} (\nabla_j \operatorname{grad} u)^k = (\nabla_j \operatorname{grad} u)^j = \Delta u,$$

and hence *the form ω_u is closed iff the function u is harmonic.*

But if the form ω_u is closed, then according to the Poincaré lemma, in any circular coordinate neighborhood U of an arbitrary point $p_0 \in \mathcal{X}$, it is an exact differential, i.e., there is a function v on U such that $dv = \omega_u$, and therefore

$$(\operatorname{grad} v)^i = g^{ij} \frac{\partial v}{\partial x^j} = g^{ij} e_{jk} (\operatorname{grad} u)^k.$$

Because the tensors g and e are covariantly constant, this implies that we have the formula

$$\Delta v = (\nabla_i \operatorname{grad} v)^i = g^{ij} e_{jk} (\nabla_i \operatorname{grad} u)^k,$$

i.e. (see (10)),

$$\Delta v = e^{ij} g_{jk} (\nabla_i \operatorname{grad} u)^k = e^{ij} (\nabla_i \nabla u)_j,$$

where ∇u is the form du considered as a covector field (i.e., the field with the components $\partial u / \partial x^j = g_{jk} (\operatorname{grad} u)^k$) and

$$(\nabla_i \nabla u)_j = \frac{\partial^2 u}{\partial x^i \partial x^j} - \Gamma^k_{ji} \frac{\partial u}{\partial x^k} \tag{22}$$

are the components of the ith partial covariant derivative of the field ∇u. Because (as formula (22) shows) the components $(\nabla_i \nabla u)_j$ are symmetric in i and j and the tensor e^{ij} is skew-symmetric, this implies $\Delta v = 0$, i.e., *v is also a harmonic function.* It is called the harmonic function *conjugate* to the harmonic function u.

We note that the function v is defined only locally (on circular coordinate neighborhoods) and only up to a constant summand.

Exercise 13.7. Prove that if $H^1 \mathcal{X} = 0$, then taking constants appropriately, we can choose the local functions v such that they become the restrictions of one function defined on the whole manifold $\mathcal{X}$. (In this case, we conventionally say that the conjugate function v *is globally defined*).

Exercise 13.8. Prove that the function u is conjugate to the function $-v$ (the conjugation relation of harmonic functions is reciprocal up to the sign).

We note that

1. *the gradients of the functions u and v are orthogonal,*

$$(\operatorname{grad} u, \operatorname{grad} v) = 0;$$

2. *the lengths of these gradients are the same,*

$$(\operatorname{grad} u, \operatorname{grad} u) = (\operatorname{grad} v, \operatorname{grad} v).$$

Exercise 13.9. Prove these assertions.

In particular, we see that *if* $\operatorname{grad} u \neq 0$ *in a neighborhood U, then the vectors* $\operatorname{grad} u$ *and* $\operatorname{grad} v$ *form an orthonormal basis of the two-dimensional Euclidean space* $\mathbf{T}_p\mathcal{X}$ at each point $p \in U$.

On the other hand, it is easy to see that (verify this!) up to the factor ε^2, the Jacobian of the functions u and v equals the determinant

$$\begin{vmatrix} (\operatorname{grad} u)^1 & (\operatorname{grad} u)^2 \\ (\operatorname{grad} v)^1 & (\operatorname{grad} v)^2 \end{vmatrix}$$

composed of the coordinates of the vectors $\operatorname{grad} u$ and $\operatorname{grad} v$. Therefore, if $\operatorname{grad} u \neq 0$ and the vectors $\operatorname{grad} u$ and $\operatorname{grad} v$ therefore form a basis, then this Jacobian is different from zero; therefore, in a certain neighborhood of the point p_0 (which is again denoted by U), the functions u and v are local coordinates.

Because the condition $\operatorname{grad} u \neq 0$ (which is obviously equivalent to the condition $du \neq 0$) is always satisfied according to Corollary 13.2 (by choosing a sufficiently small neighborhood U), this proves that *in a neighborhood of each point $p_0 \in \mathcal{X}$, there exist local coordinates u and v that are conjugate harmonic functions.*

Exercise 13.10. Show that in these coordinates, the first quadratic form of a surface becomes

$$ds^2 = \lambda^2(du^2 + dv^2), \tag{23}$$

where $\lambda = (\operatorname{grad} u, \operatorname{grad} u)^{-1/2}$. [*Hint*: Use properties 1 and 2.]

§9. Isothermal Coordinates

Definition 13.2. Local coordinates u and v in which the first qudratic form of a surface has the form (23) are called *isothermal coordinates.*

This name is explained by the fact shown in heat transfer theory that on a thermally isolated heat surface composed of a material with a constant heat conduction, the coordinate lines u = const and v = const are isotherms iff the first quadratic form of this surface is of form (23).

Proposition 13.2. *In a neighborhood of any point of an arbitrary surface, there exist isothermal coordinates.*

Proof. These coordinates are coordinates that are harmonic functions conjugate to u and v. □

Example 13.1. The geographical coordinates u and v on a sphere in the three-dimensional Euclidean space are not isothermal (the first quadratic form of the sphere in these coordinates is

$$ds^2 = du^2 + \cos^2 u\, dv^2.$$

To obtain isothermal coordinates, we must preserve the longitude v and transform the latitude u according to the formula

$$u_1 = \ln \tan\left(\frac{u}{2} + \frac{\pi}{4}\right).$$

Indeed, as an obvious computation shows, in the coordinates u_1 and v (again denoted by u and v), the first quadratic form of the sphere becomes

$$ds^2 = \frac{1}{\cosh^2 u}(du^2 + dv^2),$$

i.e., it has form (23) with $\lambda = 1/\cosh u$. The coordinates

$$x = e^u \cos v, \qquad y = e^u \sin v$$

are also isothermal because

$$ds^2 = \frac{4}{(1+x^2+y^2)^2}(dx^2+dy^2) \tag{24}$$

in these coordinates. (We note that $x + iy = e^{u+iv}$.)

In isothermal coordinates, the Laplace operator is expressed by the formula

$$\Delta = \frac{1}{\lambda^2}\left(\frac{\partial^2}{\partial u^2} + \frac{\partial^2}{\partial v^2}\right),$$

which differs from the formula for the Laplace operator in rectangular coordinates on the plane by only a factor. Therefore, harmonic functions on a surface are exactly the functions that are expressed by the usual harmonic functions of two variables in isothermal coordinates. In particular, this implies that *local coordinates u and v are isothermal iff they are harmonic functions* (they automatically become conjugate).

§10. Semi-Cartesian Coordinates

In some sense, isothermal coordinates are similar to the Euclidean coordinates on the plane. Other coordinates similar to the Euclidean ones are constructed as follows.

Let p_0 be an arbitrary point of the surface $\mathcal{X}$, and let u and v be local coordinates defined in a neighborhood of the point p_0 and equal to zero at p_0. Further, let β_0 be an arbitrary regular curve on $\mathcal{X}$ passing through the point p_0. The parameter on the curve β_0 is denoted by y, and it is assumed that the value $y = 0$ corresponds to the point p_0. For each y (for which the point $\beta_0(y)$ is defined), γ_y denotes a geodesic passing through the point $\beta_0(y)$ that makes a right angle with the curve β_0. (In particular, γ_0 is a geodesic passing

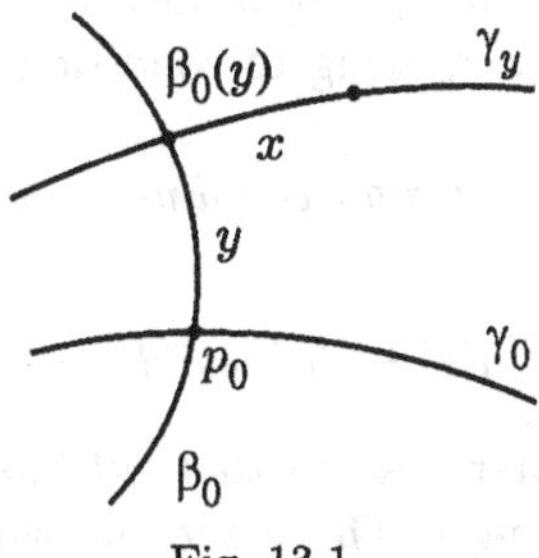

Fig. 13.1.

through the point p_0 that makes a right angle with the curve β_0.) Let x be the natural parameter on the geodesic γ_y counted from the point $\beta_0(y)$, and let

$$u = u(x, y), \quad v = v(x, y) \tag{25}$$

be the equations of the geodesic γ_y in the local coordinates u, v (we assume that the absolute values of the numbers x and y are sufficiently small).

Because $u = u(0, y)$ and $v = v(0, y)$ are obviously just the equations of the curve β_0, we see that for any y_0, the partial derivatives $u_y(0, y_0)$ and $v_y(0, y_0)$ are the components of the vector tangent to the curve β_0 at the point $\beta_0(y_0)$. In particular, the numbers $u_y(0,0)$ and $v_y(0,0)$ are components of this vector at the point p_0.

On the other hand, the numbers $u_x(0,0)$ and $v_x(0,0)$ are components of the vector tangent to the curve γ_0 at the point p_0. Because these two vectors are orthogonal at the point p_0 by assumption and are therefore linearly independent, this implies that the Jacobian $\partial(u, v)/\partial(x, y)$ of functions (25) is nonzero at the point p_0 (and therefore in a certain neighborhood of this point). Therefore, x and y near the point p_0 are local coordinates on $\mathcal{X}$. Again letting u and v denote x and y, we see that we have proved the following proposition.

Proposition 13.3. *In a neighborhood of any point p_0 of a surface $\mathcal{X}$, there exist local coordinates u and v centered at p_0 such that*

1. *the coordinate lines $v = \text{const}$ (related to the parameter u) are geodesics,*
2. *these geodesics are orthogonal to the coordinate line $u = 0$, and*
3. *on each geodesic $v = \text{const}$, the coordinate u is the natural parameter.*

These coordinates are said to be *semi-Cartesian.*

Condition 3 implies that in the semi-Cartesian coordinates u and v, the coefficient E of the first quadratic form of a surface identically equals 1. Therefore, $E_v = 0$ and hence

$$F\Gamma_{11}^1 + G\Gamma_{11}^2 = F_u.$$

Exercise 13.11. Show that condition 1 holds iff $\Gamma_{11}^1 = 0$ and $\Gamma_{11}^2 = 0$.

We therefore see that in the coordinates u and v, the identity $F_u = 0$ holds, which implies $F(u, v) = F(0, v)$ for any u and v. Because, on the other hand, $F(0, v) = 0$ for all v according to condition 2, this proves that $F = 0$ identically.

Therefore, *in the semi-Cartesian coordinates u and v, the first quadratic form of a surface becomes*

$$ds^2 = du^2 + G dv^2. \tag{26}$$

Remark 13.3. In particular, we see that *all lines $u = \text{const}$ intersect the geodesics $v = \text{const}$ orthogonally.* This assertion, which is similar to the Gauss lemma in Chap. 12, is also often called the *Gauss lemma.*

We note that in contrast to semigeodesic coordinates, *semi-Cartesian coordinates u and v are defined at the point p_0.*

§11. Cartesian Coordinates

By construction, the coordinate line $u = 0$ can be arbitrary. In the case where this line is a geodesic and the coordinate v is the natural parameter on this line, the semi-Cartesian coordinates u and v are called the *Cartesian coordinates.*

We stress that the lines $u = c$ for $c \neq 0$ are not geodesics for Cartesian coordinates in general.

Example 13.2. The usual geographical coordinates on the sphere (u is the longitude and v is the latitude) are Cartesian. The lines $v = \text{const}$ are meridians, and the line $u = 0$ is the equator. For $c \neq 0$, the parallels $u = c$ are not geodesics.

Exercise 13.12. Compute the first quadratic form of the Lobachevskii plane in the Cartesian coordinates. Express the Cartesian coordinates on the Lobachevskii plane through the Beltrami coordinates.

Because the Cartesian coordinate v is the natural parameter on the line $u = 0$, we have $G(0, v) = 1$ for any v, and because this line is a geodesic, $\Gamma^1_{22}(0, v) = 0$ and $\Gamma^2_{22}(0, v) = 0$ (compare with the assertion in Exercise 13.11). Therefore, $G_u(0, v) = 0$ (and also $G_v(0, v) = 0$). We therefore see that *for the Cartesian coordinates u and v,*

$$G(0, v) = 1, \qquad G_u(0, v) = 0. \tag{27}$$

The assertion on the existence of local coordinates u and v in which the first quadratic form is like (26) with the coefficient G having property (27) is the content of the "Gauss lemma." We can therefore assume that this lemma is now proved.

Chapter 14
Minimal Surfaces

§1. Conformal Coordinates

We can replace the real coordinates u and v on a surface $\mathcal{X}$ with one complex coordinate $w = u+iv$. In the case where the coordinates u and v are isothermal, the coordinate w is called a *conformal coordinate* on the surface. (Certain authors also apply this name to the coordinates u and v.)

Formula (23) in Chap. 13 means that in the conformal coordinate, a linear surface element has the form

$$ds^2 = \lambda^2\, dw\, d\overline{w} \tag{1}$$

or, in other words,

$$ds = \lambda |dw|. \tag{2}$$

Let $z = x + iy$ and $w = u + iv$ be two conformal coordinates (defined in the respective coordinate neighborhoods U and V). Then, on the intersection $U \cap V$, we have the relations

$$dw = \frac{\partial w}{\partial z} dz + \frac{\partial w}{\partial \overline{z}} d\overline{z} \qquad \text{and} \qquad d\overline{w} = \frac{\overline{\partial w}}{\partial \overline{z}} dz + \frac{\overline{\partial w}}{\partial z} d\overline{z}$$

(here, we use notation in which $\overline{dz} = d\overline{z}$ and $\overline{d\overline{z}} = dz$), and the relation

$$dw\, d\overline{w} = \left(\frac{\partial w}{\partial z}\frac{\overline{\partial w}}{\partial z} + \frac{\partial w}{\partial \overline{z}}\frac{\overline{\partial w}}{\partial \overline{z}} \right) dz\, d\overline{z} + \frac{\partial w}{\partial z}\frac{\overline{\partial w}}{\partial \overline{z}} dz^2 + \frac{\partial w}{\partial \overline{z}}\frac{\overline{\partial w}}{\partial z} d\overline{z}^2.$$

therefore holds. But because both coordinates z and w are conformal and the forms $dz\, d\overline{z}$ and $dw\, d\overline{w}$ are therefore proportional, this relation is possible only if $(\partial w/\partial z)/(\overline{\partial w}/\partial \overline{z}) = 0$, i.e., if either $\partial w/z = 0$ or $\partial w/\overline{z} = 0$ at each point of the intersection $U \cap V$.

But it is easy to see (verify this!) that the Jacobian of the functions u and v with respect to x and y is expressed by the formula

$$\begin{vmatrix} \dfrac{\partial u}{\partial x} & \dfrac{\partial u}{\partial y} \\[2ex] \dfrac{\partial v}{\partial x} & \dfrac{\partial v}{\partial y} \end{vmatrix} = \begin{vmatrix} \dfrac{\partial w}{\partial z} & \dfrac{\partial w}{\partial \overline{z}} \\[2ex] \dfrac{\overline{\partial w}}{\partial \overline{z}} & \dfrac{\overline{\partial w}}{\partial z} \end{vmatrix} = \left|\frac{\partial w}{\partial z}\right|^2 - \left|\frac{\partial w}{\partial \overline{z}}\right|^2. \tag{3}$$

Therefore, there is no point in $U \cap V$ where the functions $\partial w/\partial z$ and $\partial w/\partial \overline{z}$ simultaneously vanish. Therefore, the sets of zeros of these functions (which are closed sets in $U \cap V$ by continuity) are mutually complemented (and therefore open). Therefore, if the set $U \cap V$ is connected, then either $\partial w/\partial \overline{z} = 0$

or $\partial w/\partial z = 0$ everywhere on $U \cap V$. In the first case, the transition function $w = w(z)$ is holomorphic on $U \cap V$, and the charts (U, z) and (V, w) are oriented compatibly; in the second case, the function $w = w(z)$ is antiholomorphic, and the orientations of the charts (U, z) and (V, w) are not compatible.

We note that the function $w = w(z)$ *is antiholomorphic iff the function* $w = w(\overline{z})$ *is holomorphic.*

§2. Conformal Structures

All these arguments motivate the following definition, in which $\mathcal{X}$ is an arbitrary surface, which is not equipped with a Riemannian metric in general.

Definition 14.1. Two complex charts (U, z) and (V, w) on the surface $\mathcal{X}$ are said to be *conformally compatible* if on each component of the set $U \cap V$ (for $U \cap V \neq \emptyset$), one coordinate (w for example) is either a holomorphic or an antiholomorphic function of the other coordinate z. An atlas consisting of conformally compatible charts is called a *conformal atlas.* Two conformal atlases are said to define the same *conformal structure* on $\mathcal{X}$.

Exercise 14.1. Prove that any conformal atlas is contained in a unique maximal conformal atlas.

We can therefore identify conformal structures with the maximal conformal atlases.

A covering $\widetilde{\mathcal{X}} \to \mathcal{X}$ (in particular, a diffeomorphism) of surfaces with conformal structures is said to be *conformal* if it is written in conformal atlases by holomorphic or antiholomorphic functions.

Exercise 14.2. Show that the following assertions hold:

1. For any smooth covering $\widetilde{\mathcal{X}} \to \mathcal{X}$ of a surface $\mathcal{X}$ with a conformal structure, there exists a unique conformal structure on the surface $\widetilde{\mathcal{X}}$ with respect to which this covering is conformal.

2. For any conformal covering $\widetilde{\mathcal{X}} \to \mathcal{X}$, the group of its automorphisms (glidings) Aut $\widetilde{\mathcal{X}}$ consists of conformal diffeomorphisms of the surface $\widetilde{\mathcal{X}}$.

3. If the manifold $\widetilde{\mathcal{X}}$ in the smooth covering $\widetilde{\mathcal{X}} \to \mathcal{X}$ is a surface with a conformal structure and if

a. the covering is regular and
b. its automorphism group consists of conformal diffeomorphisms,

then there exists a unique conformal structure on $\mathcal{X}$ with respect to which the covering is conformal.

(Compare with Exercises 3.4, 3.5, and 3.6.)

Exercise 14.3. Prove that any isometry of surfaces is given in conformal coordinates by holomorphic or antiholomorphic functions (is a conformal diffeomorphism with respect to the conformal structures induced on the surfaces).

Of course, each complex analytic atlas is conformal (and any two of its charts are positively compatible). Conversely, formula (3) directly implies that any conformal atlas consisting of positively compatible charts is a complex analytic atlas. Therefore, *oriented surfaces with conformal structures are exactly one-dimensional complex analytic (Hausdorff and paracompact) manifolds* (which are called *Riemann surfaces* in the terminology of function theory).

We note, however, that the class of conformal mappings of Riemann surfaces is wider than the class of their complex analytic isomorphisms (it also contains diffeomorphisms given by antiholomorphic functions).

In view of Definition 14.1, the arguments preceding it prove the following proposition.

Proposition 14.1. *Each Riemannian metric on a surface uniquely defines a certain conformal structure on $\mathcal{X}$ (which is the complex analytic structure for the orientable surface).*

Corollary 14.1. *On any surface, there exists at least one conformal structure.*

Corollary 14.2. *Each orientable surface can be complexified (is a realification of a one-dimensional complex analytic manifold).*

We note that no Riemannian metric appears in the statement of Corollaries 14.1 and 14.2. It was needed only for their proof.

§3. Minimal Surfaces

One of the most remarkable classes of surfaces, in which isothermal (conformal) coordinates play a crucial role, is introduced in the following definition.

Definition 14.2. A surface $\mathcal{X}$ in the three-dimensional Euclidean space $\mathbb{R}^3$ is called a *minimal surface* if its mean curvature H vanishes identically,

$$H = 0,$$

i.e., if at each of its points, the principal curvatures k_1 and k_2 of the surfaces $\mathcal{X}$ differ from one another by only the sign:

$$k_2 = -k_1$$

(under the assumption $k_1 k_2 \neq 0$, the Dupin indicatrix is an equilateral hyperbola).

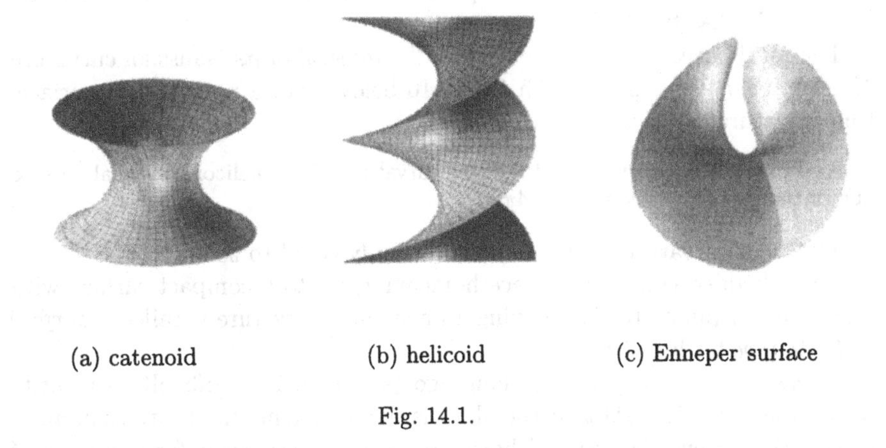

(a) catenoid (b) helicoid (c) Enneper surface

Fig. 14.1.

Ribocur called minimal surfaces *ellasoids* but this name was not used by others.

Examples of minimal surfaces are the *catenoid*

$$\boldsymbol{r}(u, v) = \cosh v \cos u \cdot \boldsymbol{i} + \cosh v \sin u \cdot \boldsymbol{j} + v\boldsymbol{k} \tag{4}$$

and also the *helicoid*

$$\boldsymbol{r}(u, v) = \sinh v \cos u \cdot \boldsymbol{i} + \sinh v \sin u \cdot \boldsymbol{j} + u\boldsymbol{k}. \tag{5}$$

One more example is the so-called Enneper surface

$$\boldsymbol{r}(u, v) = u(3 + 3v^2 - u^2)\boldsymbol{i} - v(3 + 3u^2 - v^2)\boldsymbol{j} - 3(u^2 - v^2)\boldsymbol{k}. \tag{6}$$

Exercise 14.4. Show that surface (6) is minimal.

The Enneper surface has self-intersections, i.e., it is an immersed two-dimensional manifold (see Chap. 3). In contrast, the catenoid and helicoid (and certainly the plane, which is a trivial minimal surface) are embedded surfaces without self-intersections.

We can show (see Remark 14.1 below) that a complete minimal surface cannot be contained in a finite part of the space $\mathbb{R}^3$ (and, in particular, is not compact). From the topological standpoint, the simplest noncompact surfaces are surfaces that are obtained from compact ones by removing finitely many points (punctures) or, equivalently, disjoint disks. For example, the topological plane and also the helicoid are the sphere (compact surface) with one puncture, and the catenoid is the sphere with two punctures. Geometrically, a funnel of the surface going to infinity corresponds to each puncture of the surface. On the catenoid, both funnels are clearly seen, but, for example, the unique funnel of the helicoid is so twisted that it hardly be called by this name. In technical topological languages, funnels of a surface going to infinity are called its *ends*. Therefore, the plane and the helicoid have one end, and the catenoid has two ends.

The *total curvature* of a surface $\mathcal{X}$ is the integral of its Gaussian curvature K taken over $\mathcal{X}$ (compare with Chap. 16 below). For a noncompact surface, this curvature can certainly be infinite.

Exercise 14.5. Prove that the total curvature of the helicoid is equal to $-\infty$ and of the catenoid is equal to -4π.

The total curvature of the plane is certainly equal to zero.

An embedded complete surface homeomorphic to a compact surface with finitely many punctures and having a finite total curvature is called a *surface of finite type* for brevity.

It was shown comparatively long ago (although it is difficult to identify who actually did it first) that the plane is the unique minimal surface of finite type with one end. In 1982, Schoen proved a similar result for a surface of finite type with two ends: the catenoid is the unique such surface. In fact, until 1984, except for the plane and the catenoid, no other minimal surfaces of finite type had been found. A critical step was made in 1984 by the young American mathematicians D. Hoffman and U. Meeks, who (with the help of J. Hoffman) constructed a number of new minimal surfaces of finite type essentially using computer graphics. The Hoffman–Meeks surfaces are beautiful and symmetric. The simplest of them, which is homeomorphic to the torus with three holes, is depicted in Fig. 14.2a.

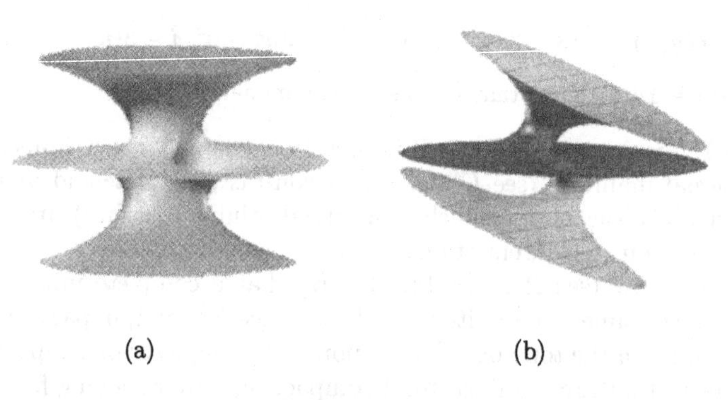

(a) (b)

Fig. 14.2.

In Fig. 14.2b, we depict the so-called *trinoid*, the minimal surface with self-intersections that is homeomorphic to the sphere with three holes. (We note that it is not possible to realize the sphere with three holes as an embedded minimal surface.)

§4. Explanation of Their Name

To explain why surfaces with $H = 0$ are said to be minimal, we consider the *area functional*

$$\text{surface } \mathcal{X} \quad \Rightarrow \quad \text{its area } \sigma\mathcal{X}$$

on the class P_C of all surfaces $\mathcal{X}$ in $\mathbb{R}^3$ with a given boundary C. If a surface $\mathcal{X}$ is subject to a deformation that (in a clear sense) smoothly depends on the parameter t and is fixed on C, then its area $\sigma\mathcal{X}$ (which is assumed to be finite) is a smooth function of t. In the case where for any deformation, the derivative of this function equals zero for the value of the parameter corresponding to the initial surface $\mathcal{X}$, we say (according to an obvious analogy to the length functional considered in Chap. 12) that the surface $\mathcal{X}$ is an *extremal of the area functional on P_C*. In particular, the surface $\mathcal{X}$ is an extremal of the area functional if it has the minimum area among all surfaces in P_C close to it.

A complete surface (without boundary) is called an *extremal of the area functional* if any of its parts bounded by an arbitrary piecewise smooth curve C is an extremal of this functional on P_C.

It turns out (we show this below) that *minimal surfaces are exactly extremals of the area functional.* This explains their name (when it was suggested in the 19th century, a precise distinction was not made between minima and extrema; moreover, it can be shown (but we do not do this) that each sufficiently small part of a minimal surface bounded by a piecewise smooth curve has the minimum area among all surfaces with the same boundary (similar to each sufficiently small part of a geodesic being a shortest arc). To stress the latter fact, the minimal surfaces are sometimes called *locally minimal* surfaces.

§5. Plateau Problem

According to the general physical equilibrium principle, because of the action of surface tension, a soap film spanned on a contour occupies a state whose energy cannot be decreased by small variations. Because this energy (as is shown in the theory of flexible elastic films) is proportional to the area, this implies that *soap films are a physical realization of minimal surfaces.*

Here, we mean surfaces with boundary. For surfaces without a boundary that divide the space into two domains, the exterior and interior ones, the case is controlled by the *Poisson theorem*, which says that the interface of two media at equilibrium is a surface of the constant mean curvature $H = h\Delta p$, where h is the coefficient of surface tension and Δp the difference of pressures inside and outside the surface (for a surface with a boundary that does not divide the space, $\Delta p = 0$ and therefore $H = 0$). In particular, this theorem explains why soap bubbles have the spherical form (when they are not so large and we can ignore the gravitational force).

Soap films were first studied by the Belgian physicist Plateau, in whose honor the problem of constructing minimal surfaces with a given boundary (and also its higher-dimensional analogues) is called the *Plateau problem.* This problem turns out to be very difficult and relates to the purest questions of modern mathematics. We do not deal with it at all.

§6. Free Relativistic Strings

Minimal surfaces (but not in the space $\mathbb{R}^3$) have suddenly also appeared in elementary particle theory, more precisely, in hadron theory. According to the modern view (which is justified by experimental data), hadrons consist of quarks related to each other by Yang–Mills gluon fields. To explain the so-called non-sweeping out of quarks (their absence in a free state), it is assumed that for small distances (of the order of the hadron size), those configurations of gluon fields that do not fill the whole space (as do classical fields) but concentrate along lines connecting quarks are energetically preferable. Then the attraction force between quarks is (as can be shown) constant, i.e., independent of the distance between them; therefore, no external action can separate them and generate a free quark. Visually, we can imagine that two quarks are connected by a thin tube of the gluon field, which degenerates to a line (string) in the limit. The quarks themselves are usually imagined as small, possibly knotted, closed tubes of the gluon field. At large distances, such tubes behave as noninteracting particles, but being sufficiently close, they interlace into one hadron. (The idea of particles as field tubes is very old. As long ago as the beginning of our century, considering an electron as annulus composed of force lines of the electromagnetic field was attempted. This view turned out to be a dead end and was rejected. We hope that this is not the case for quarks and gluons.)

In the classical relativistic approximation, we therefore (abstracting from fields and taking the phenomenological standpoint) come to the problem of studying the dynamics of one-dimensionally stretched objects, the strings. In the course of its motion, a string sweeps out a certain surface in the four-dimensional Minkowski space–time (its "world line"), which is considered an adequate geometric image of the string.

Of course, not every surface in the Minkowski space is a string in this sense. For this, it is first necessary that the tangent plane have the signature (1,1) at each point of the surface; therefore, for this point, it is possible to speak about timelike vectors (in whose direction time develops in the corresponding coordinate system) and spacelike vectors (which assign a momentary position of the string in the section). If

$$r = r(u, v)$$

is a parameterization of the string in a neighborhood of a certain point, then this condition is equivalent (prove this!) to the negativity of the Gram determinant

$$\begin{vmatrix} r_u^2 & r_u r_v \\ r_u r_v & r_v^2 \end{vmatrix} = EG - F^2.$$

Such surfaces are called *surfaces of signature* (1,1).

Exercise 14.6. Prove that an area element of a surface of signature (1,1) equals $\sqrt{|EG - F^2|} = \sqrt{F^2 - EG}$.

The local coordinates u and v on a surface of signature (1,1) are said to be *isothermal* if the first quadratic form of this surface in these coordinates is

$$ds^2 = \lambda(du^2 - dv^2), \quad \lambda > 0.$$

(Compare with Definition 13.2.)

Exercise 14.7. Prove that isothermal coordinates exist in a neighborhood of an arbitrary point of a surface of signature (1,1).

In physics, the isothermal coordinates are called the orthonormal gauge. Therefore, the orthonormal gauge of a string (surface of signature (1,1)) is characterized by

$$r_u^2 + r_v^2 = 0, \qquad r_u r_v = 0$$

or, equivalently,

$$(r_u + r_v)^2 = 0.$$

Certain conditions are also imposed by the dynamics of the string. In general, the equations of motion of a mechanical system are determined by the corresponding variational principle, i.e. (see Chap. 11), are extremals of the Lagrangian corresponding to this system. For example, the Lagrangian of a free (not subject to any external actions) material point is the length of the tangent vector, and according to this, the trajectory of such a particle is a geodesic, i.e., in empty (flat) space, it is a line. Considering a string consisting of material points whose interaction does not contribute to the action (in particular, this means the absence of the usual elasticity forces, which is absurd from the standpoint of classical physics but is probably justified for gluon strings), we (as is easy to understand) obtain from this not the length but the area as the Lagrangian, i.e., we obtain the area functional. Therefore, *from the mathematical standpoint, a free relativistic string is just a minimal surface of signature* $(1, 1)$ *in the Minkowski space.*

§7. Simplest Problem of the Calculus of Variations for Functions of Two Variables

Before solving the problem of extremals of the area functional, we discuss higher-dimensional variational problems (as in Chap. 11 for the length functional). For simplicity, we restrict ourselves to two-dimensional problems of the first order in the space $\mathbb{R}^n$.

Let L be an arbitrary function (called the *Lagrangian* in what follows) on the Euclidean space $\mathbb{R}^{3n}$. Points of the space $\mathbb{R}^{3n}$ are identified with triples of the form $(\boldsymbol{r}, \boldsymbol{r}_u, \boldsymbol{r}_v)$, where $\boldsymbol{r} = (x^1, \dots, x^n)$, $\boldsymbol{r}_u = (x_u^1, \dots, x_u^n)$, and $\boldsymbol{r}_v = (x_v^1, \dots, x_v^n)$ are arbitrary vectors in the space $\mathbb{R}^3$ (which, in contrast to the notation, are totally unrelated).

Further, let W be an arbitrary open set of the plane $\mathbb{R}^2$ (it is convenient to assume that this set is convex and bounded by a piecewise smooth line; these conditions are in fact extra, but without them, the statements and proofs become more complicated, which is not especially needed).

A *parameterization* is a smooth mapping $\boldsymbol{r}: W \to \mathbb{R}^n$, i.e., a smooth vector-valued function

$$\boldsymbol{r} = \boldsymbol{r}(u, v), \quad \boldsymbol{r}(u, v) \in \mathbb{R}^n, \tag{7}$$

of the point $(u, v) \in W$. (No regularity and monomorphicity conditions are imposed here.)

By the formulas

$$\boldsymbol{r} = \boldsymbol{r}(u, v), \qquad \boldsymbol{r}_u = \frac{\partial \boldsymbol{r}}{\partial u}(u, v), \qquad \boldsymbol{r}_v = \frac{\partial \boldsymbol{r}}{\partial v}(u, v,), \tag{8}$$

each parameterization (7) defines a smooth mapping $W \to \mathbb{R}^{3n}$ and therefore a function

$$(u, v) \mapsto L(\boldsymbol{r}(u, v), \boldsymbol{r}_u(u, v), \boldsymbol{r}_v(u, v))$$

on the domain W for a given Lagrangian $L = L(\boldsymbol{r}, \boldsymbol{r}_u, \boldsymbol{r}_v)$. Let

$$S = \iint_W L du\, dv \tag{9}$$

be the integral of this function over the domain W (or, equivalently, over its closure $\overline{W}$).

Proceeding as in Chap. 11, we introduce an arbitrary smooth mapping $\eta: W \to \mathbb{R}^n$, $\boldsymbol{\eta}(u,v) = (\eta^1(u,v), \ldots, \eta^n(u,v))$, vanishing on the boundary ∂W of the domain W and the parameterization $\boldsymbol{r} = \boldsymbol{r}_\varepsilon(u,v)$ defined by

$$\boldsymbol{r}_\varepsilon(u,v) = \boldsymbol{r}(u,v) + \varepsilon\boldsymbol{\eta}(u,v), \quad |\varepsilon| < \varepsilon_0. \tag{10}$$

For parameterization (10), integral (9) is a smooth function of ε, and we can therefore speak about its derivative $S'(0)$ at the point $\varepsilon = 0$.

According to the differentiation rule for integrals with respect to a parameter, we have

$$S'(0) = \iint_W \left(\frac{\partial L}{\partial \boldsymbol{r}}\boldsymbol{\eta} + \frac{\partial L}{\partial \boldsymbol{r}_u}\boldsymbol{\eta}_u + \frac{\partial L}{\partial \boldsymbol{r}_v}\boldsymbol{\eta}_v\right) du\, dv,$$

where

$$\frac{\partial L}{\partial \boldsymbol{r}} = \left(\frac{\partial L}{\partial x^1}, \ldots, \frac{\partial L}{\partial x^n}\right),$$

$$\frac{\partial L}{\partial \boldsymbol{r}_u} = \left(\frac{\partial L}{\partial x_u^1}, \ldots, \frac{\partial L}{\partial x_u^n}\right), \qquad \frac{\partial L}{\partial \boldsymbol{r}_v} = \left(\frac{\partial L}{\partial x_v^1}, \ldots, \frac{\partial L}{\partial x_v^n}\right),$$

$$\boldsymbol{\eta}_u = \left(\frac{\partial \eta^1}{\partial u}, \ldots, \frac{\partial \eta^n}{\partial u}\right), \qquad \boldsymbol{\eta}_v = \left(\frac{\partial \eta^1}{\partial v}, \ldots, \frac{\partial \eta^n}{\partial v}\right),$$

and

$$\frac{\partial L}{\partial \boldsymbol{r}}\boldsymbol{\eta} = \frac{\partial L}{\partial x^i}\eta^i, \qquad \frac{\partial L}{\partial \boldsymbol{r}_u}\boldsymbol{\eta}_u = \frac{\partial L}{\partial x_u^i}\frac{\partial \eta^i}{\partial u}, \qquad \frac{\partial L}{\partial \boldsymbol{r}_v}\boldsymbol{\eta}_v = \frac{\partial L}{\partial x_v^i}\frac{\partial \eta^i}{\partial v}$$

are the inner products of vectors. But

$$\frac{\partial L}{\partial \boldsymbol{r}_u}\boldsymbol{\eta}_u = \frac{\partial}{\partial u}\left(\frac{\partial L}{\partial \boldsymbol{r}_u}\boldsymbol{\eta}\right) - \frac{\partial}{\partial u}\left(\frac{\partial L}{\partial \boldsymbol{r}_u}\right)\boldsymbol{\eta},$$

$$\frac{\partial L}{\partial \boldsymbol{r}_v}\boldsymbol{\eta}_v = \frac{\partial}{\partial v}\left(\frac{\partial L}{\partial \boldsymbol{r}_v}\boldsymbol{\eta}\right) - \frac{\partial}{\partial v}\left(\frac{\partial L}{\partial \boldsymbol{r}_v}\right)\boldsymbol{\eta},$$

and according to the Green formula, we therefore have

$$S'(0) = \iint_W \left[\frac{\partial L}{\partial \boldsymbol{r}} - \frac{\partial}{\partial u}\left(\frac{\partial L}{\partial \boldsymbol{r}_u}\right) - \frac{\partial}{\partial v}\left(\frac{\partial L}{\partial \boldsymbol{r}_v}\right)\right]\boldsymbol{\eta}\, du\, dv$$
$$+ \oint_{\partial W}\left(\frac{\partial L}{\partial \boldsymbol{r}_v}\boldsymbol{\eta}\, du - \frac{\partial L}{\partial \boldsymbol{r}_u}\boldsymbol{\eta}\, dv\right).$$

Because $\boldsymbol{\eta} = 0$ on ∂W, the integral over ∂W vanishes, and therefore

$$S'(0) = \iint_W \left[\frac{\partial L}{\partial \boldsymbol{r}} - \frac{\partial}{\partial u}\left(\frac{\partial L}{\partial \boldsymbol{r}_u}\right) - \frac{\partial}{\partial v}\left(\frac{\partial L}{\partial \boldsymbol{r}_v}\right)\right]\boldsymbol{\eta}\, du\, dv.$$

Exercise 14.8. Deduce from this formula that $S'(0) = 0$ for all $\boldsymbol{\eta}$ iff

$$\frac{\partial}{\partial u}\left(\frac{\partial L}{\partial \boldsymbol{r}_u}\right) + \frac{\partial}{\partial v}\left(\frac{\partial L}{\partial \boldsymbol{r}_v}\right) = \frac{\partial L}{\partial \boldsymbol{r}}. \tag{11}$$

[*Hint*: Prove an analogue of Lemma 1.1.]

Vector differential equation (11) is called the *Euler–Ostrogradskii equation.*

Definition 14.3. Parameterization (7) satisfying Eq. (11) is called an *extremal of the Lagrangian L* (or integral (9)).

(Compare with Definition 11.4.)

Therefore, in particular, *each parameterization* (7) *that yields the minimum value for integral* (9) *in the class of all parameterizations coinciding with each other on ∂W is an extremal of the Lagrangian L.*

Example 14.1 (Dirichlet integral). Let

$$L = \frac{E+G}{2},$$

where $E = \boldsymbol{r}_u^2$ and $G = \boldsymbol{r}_v^2$. (The corresponding integral

$$\frac{1}{2}\iint_W (E+G)du\,dv \tag{12}$$

is called the *Dirichlet integral.*) In this case,

$$\frac{\partial L}{\partial \boldsymbol{r}_u} = \boldsymbol{r}_u, \qquad \frac{\partial L}{\partial \boldsymbol{r}_v} = \boldsymbol{r}_v, \qquad \frac{\partial L}{\partial \boldsymbol{r}} = 0,$$

and Eq. (11) becomes

$$\boldsymbol{r}_{uu} + \boldsymbol{r}_{vv} = 0, \tag{13}$$

which means that *all the components $x^i(u,v)$ of the parameterization $\boldsymbol{r}(u,v)$ are harmonic functions.*

Saying that a parameterization satisfying Eq. (13) is *harmonic*, we therefore obtain that *extremals of the Dirichlet integral are exactly harmonic parameterizations.*

§8. Extremals of the Area Functional

We now consider the area functional

$$\iint \sqrt{EG - F^2}du\,dv, \tag{14}$$

which is of main interest for us and which is an integral of form (9) with the Lagrangian $L = \sqrt{EG - F^2}$, where E and G are the same as above and $F = \boldsymbol{r}_u \boldsymbol{r}_v$.

Integral (14) is not changed when passing to an equivalent parameterization. (The Lagrangian L assigns a density on W.) Therefore (compare with the case of the length Lagrangian in Chap. 12), it is necessary to impose an additional condition on the considered parameterization.

Definition 14.4. Parameterization (7) is said to be *isothermal* if $E = G$ and $F = 0$.

(Compare with Definition 13.2.)

For an isothermal parameterization, integral (14) coincides with Dirichlet integral (12), and an isothermal parameterization is therefore an extremal of the area Lagrangian $\sqrt{EG - F^2}$ iff it is an extremal of the Dirichlet integral, i.e., it is a harmonic parameterization Therefore, *extremals of the area Lagrangian are exactly simultaneously harmonic and isothermal parameterizations* (and all parameterizations equivalent to them).

To apply this result to surfaces, for any surface $\mathcal{X}$ of the space $\mathbb{R}^n$ (in general having self-intersections), we introduce a smooth mapping $i: \mathcal{X} \to \mathbb{R}^n$ that realizes an immersion of the surface $\mathcal{X}$ in $\mathbb{R}^n$. Then, for any chart (U, h) on $\mathcal{X}$ (for which the set $W = h(U)$ is a convex domain with a piecewise smooth boundary), the mapping $i \circ h^{-1}$, which is a certain parameterization $\boldsymbol{r}: W \to \mathbb{R}^n$ by our definitions, is the so-called parameterization of the surface $\mathcal{X}$ in a neighborhood of U. The assertion that this parameterization is isothermal exactly means that the chart (U, h) is isothermal (i.e., the local coordinates u and v of this chart are isothermal), and the assertion that this parameterization is harmonic exactly means that the mapping i is written in the chart (U, h) with harmonic functions, i.e., functions harmonic in U with respect to the Riemannian metric on $\mathcal{X}$ because of the isothermal property of the coordinates u and v.

On the other hand, it is clear that surfaces that are extremals of the area functional in the sense introduced above are exactly the surfaces whose parameterizations are extremals of the area Lagrangian $\sqrt{EG - F^2}$. Therefore, *extremals of the area functional are exactly the surfaces $\mathcal{X}$ whose immersion $\mathcal{X} \to \mathbb{R}^n$ is given by functions harmonic in $\mathcal{X}$*, i.e., using a less-invariant language, the surfaces that are locally given by vector-valued functions

$$\boldsymbol{r} = \boldsymbol{r}(u, v)$$

such that

$$\boldsymbol{r}_u^2 = \boldsymbol{r}_v^2, \qquad \boldsymbol{r}_u \boldsymbol{r}_v = 0 \tag{15}$$

and

$$\boldsymbol{r}_{uu} + \boldsymbol{r}_{vv} = 0. \tag{16}$$

Condition (15) ensures the isothermal property, and condition (16), together with (15), ensures the harmonicity.

Exercise 14.9. Show that for a surface of signature $(1, 1)$ in the Minkowski space, the minimality condition in the isothermal coordinates becomes

$$\boldsymbol{r}_{uu} = \boldsymbol{r}_{vv}. \tag{16'}$$

From the physical standpoint, Eq. (16′) is the equation of motion of a string in the orthonormal gauge.

We stress that in contrast to Eq. (16), which is an elliptic-type equation, Eq. (16′) is a hyperbolic-type equation. This determines the principal distinction between the behavior of strings (minimal surfaces of the Minkowski space) and that of minimal surfaces of the Euclidean space. For example, as is known from the theory of differential equations, the general solution of Eq. (16′) has the form

$$\boldsymbol{r} = \boldsymbol{f}(u+v) + \boldsymbol{f}(u-v),$$

where $\boldsymbol{f}$ is an arbitrary (even nonsmooth if generalized derivatives are used) function, while solutions to Eq. (16), being harmonic functions, should be real analytic.

§9. Case $n = 3$

We now let $n = 3$. In the isothermal coordinates, the known formula for the mean curvature,

$$H = \frac{1}{2}\frac{EN + GL - 2FM}{EG - F^2},$$

becomes

$$H = \frac{1}{2}\frac{L+N}{E},$$

where $L = \boldsymbol{r}_{uu}\boldsymbol{n}$ and $N = \boldsymbol{r}_{vv}\boldsymbol{n}$. Because the vector $\boldsymbol{n}$ is collinear to the vector $\boldsymbol{r}_u \times \boldsymbol{r}_v$, this proves that *in the isothermal coordinates, the relation $H = 0$ is equivalent to*

$$(\boldsymbol{r}_{uu} + \boldsymbol{r}_{vv})\boldsymbol{r}_u\boldsymbol{r}_v = 0. \tag{17}$$

But differentiating relations (15), which characterize the isothermal coordinates, in u and v, we obtain

$$\boldsymbol{r}_u\boldsymbol{r}_{uu} = \boldsymbol{r}_v\boldsymbol{r}_{uv}, \qquad \boldsymbol{r}_{uu}\boldsymbol{r}_v + \boldsymbol{r}_u\boldsymbol{r}_{uv} = 0,$$

$$\boldsymbol{r}_u\boldsymbol{r}_{uv} = \boldsymbol{r}_v\boldsymbol{r}_{vv}, \qquad \boldsymbol{r}_{uv}\boldsymbol{r}_v + \boldsymbol{r}_u\boldsymbol{r}_{vv} = 0,$$

and therefore

$$(\boldsymbol{r}_{uu} + \boldsymbol{r}_{vv})\boldsymbol{r}_u = 0, \qquad (\boldsymbol{r}_{uu} + \boldsymbol{r}_{vv})\boldsymbol{r}_v = 0,$$

i.e., the vector $\boldsymbol{r}_{uu} + \boldsymbol{r}_{vv}$ is orthogonal to both vectors $\boldsymbol{r}_u$ and $\boldsymbol{r}_v$. Therefore, this vector is collinear to the (nonzero!) vector $\boldsymbol{r}_u \times \boldsymbol{r}_v$, and relation (17) is therefore possible *iff the vector $\boldsymbol{r}_{uu} + \boldsymbol{r}_{vv}$ vanishes,* i.e., the isothermal parameterization of the surface $\boldsymbol{r} = \boldsymbol{r}(u, v)$ is harmonic. This proves the above assertion on the coincidence of the class of surfaces in $\mathbb{R}^3$ minimal in sense of Definition 14.2

and the set of surfaces that are extremals of the area functional. Because of this, for any $n \geq 3$, surfaces in $\mathbb{R}^n$ that are extremals of the area functional are called *minimal surfaces.* By what was proved, these are exactly surfaces with the parameterizations $r = \boldsymbol{r}(u, v)$ having properties (15) and (16) or, in a more-invariant language, surfaces whose immersions $\mathcal{X} \to \mathbb{R}^n$ in $\mathbb{R}^n$ are given by functions harmonic in $\mathcal{X}$.

Remark 14.1. Because any harmonic function is constant on a compact surface (see Chap. 13), *there are no compact minimal surfaces in* $\mathbb{R}^n$ (as was already mentioned).

§10. Representation of Minimal Surfaces Via Holomorphic Functions

It is known from calculus that any harmonic function of u and v is (in general, only locally) a real part of a certain holomorphic (single-valued analytic) function of the complex variable $w = u + iv$. Therefore, any harmonic parameterization (satisfying condition (16)) can be represented (in general, in a certain neighborhood of an arbitrary point of its domain) as

$$\boldsymbol{r} = \operatorname{Re} \boldsymbol{f}(w), \tag{18}$$

where $\boldsymbol{f}(w)$ is a certain holomorphic (i.e., having holomorphic coordinates) vector-valued function. In this case, according to the Cauchy–Riemann equations, $\boldsymbol{f}' = \boldsymbol{r}_u - i\boldsymbol{r}_v$, and therefore

$$\boldsymbol{f}'^2 = (\boldsymbol{r}_u^2 - \boldsymbol{r}_v^2) - 2i\boldsymbol{r}_u\boldsymbol{r}_v,$$

which implies that isothermality conditions (15) are equivalent to the relation $\boldsymbol{f}'^2 = 0$.

Therefore, we see that *minimal surfaces of the Euclidean space are exactly surfaces that admit a parameterization of form* (18), *where* $\boldsymbol{f} = \boldsymbol{f}(w)$ *is a holomorphic vector-valued function such that* $\boldsymbol{f}'^2 = 0$.

Remark 14.2. Complex curves $\boldsymbol{r} = \boldsymbol{f}(w)$ in the space $\mathbb{C}^n$ (surfaces from the real standpoint) for which $\boldsymbol{f}'^2 = 0$ are called *isotropic curves.* Their remarkable property is that the distance between any of their points computed according to the usual formula is equal to zero. The proved assertion therefore means that *minimal surfaces are projections of isotropic curves under the mapping* Re: $\mathbb{C}^n \to \mathbb{R}^n$. Isotropic curves are therefore also called *minimal lines.*

§11. Weierstrass Formulas

Again, let $n = 3$, and let $\boldsymbol{f}' = a\boldsymbol{i}+b\boldsymbol{j}+c\boldsymbol{k}$. Assuming that $c \neq 0$ and setting $x = a/c$ and $y = ib/c$, we can write the condition $\boldsymbol{f}'^2 = 0$, i.e., the condition $a^2 + b^2 + c^2 = 0$, in the form of the equation for the hyperbola:

$$x^2 - y^2 = 1.$$

The rational parameterization

$$x = \frac{1}{2}\left(t - \frac{1}{t}\right), \qquad y = \frac{1}{2}\left(t + \frac{1}{t}\right)$$

of the hyperbola suggests passing from the variable w to the variable t (also complex) connected with w by

$$\frac{1}{2}\left(t - \frac{1}{t}\right) = x,$$

i.e., by

$$t = \frac{a + \sqrt{a^2 + c^2}}{c}$$

(we recall that a and c are functions of w). This passage is justified if $x \neq$ const.

Exercise 14.10. Show that $x =$ const iff the considered surface is a plane.

In the new variable, the vector-valued function $\boldsymbol{f}'$ becomes

$$\boldsymbol{f}' = \frac{1}{2}\left(t - \frac{1}{t}\right)c \cdot \boldsymbol{i} + \frac{1}{2i}\left(t + \frac{1}{t}\right)c \cdot \boldsymbol{j} + c \cdot \boldsymbol{k},$$

where $c \not\equiv 0$ is now a function of t (which, in general, does not satisfy any condition except for the holomorphicity). It is convenient to set $c = it\varphi'''$, where φ is a holomorphic function. Then

$$\boldsymbol{f}' = -i\frac{1 - t^2}{2}\varphi''' \cdot \boldsymbol{i} + \frac{1 + t^2}{2}\varphi''' \cdot \boldsymbol{j} + it\varphi''' \cdot \boldsymbol{k}.$$

Integrating and passing to real parts, we obtain a parameterization $\boldsymbol{r} = x\boldsymbol{i} + y\boldsymbol{j} + z\boldsymbol{k}$ for which

$$\begin{aligned} x &= \operatorname{Im}\left(\frac{1 - t^2}{2}\varphi'' + t\varphi' - \varphi\right), \\ y &= \operatorname{Re}\left(\varphi - t\varphi' + \frac{1 + t^2}{2}\varphi''\right), \\ z &= \operatorname{Im}(\varphi' - t\varphi''), \end{aligned} \tag{19}$$

where t is a complex parameter and $\varphi = \varphi(t)$ is a holomorphic function of t different from a quadratic trinomial (i.e., such that $\varphi''' \not\equiv 0$). This proves the following proposition.

Proposition 14.2. *Each minimal surface in $\mathbb{R}^3$ that is not the plane admits a parameterization of form* (19) *near any of its points. Conversely, for any holomorphic function $\varphi = \varphi(t)$ with $\varphi''' \not\equiv 0$, formulas* (19) *assign a parameterization of a certain minimal surface.*

For example, for $\varphi = it^3$, where $\varphi' = 3it^2$ and $\varphi'' = 6it$, we obtain the parameterization

$$x = \operatorname{Re}(3t - t^3), \qquad y = -\operatorname{Im}(3t + t^3), \qquad z = -\operatorname{Re} 3t^2,$$

coinciding (if we set $t = u + iv$) with parameterization (6) of the Enneper surface. Therefore, *for $\varphi = it^3$, we obtain the Enneper surface.*

For $\varphi = t\ln t - t$, where $\varphi' = \ln t$ and $\varphi'' = 1/t$, we obtain the parameterization

$$x = \operatorname{Im}\left(\frac{1+t^2}{2t}\right), \qquad y = \operatorname{Re}\left(\frac{1-t^2}{2t}\right), \qquad z = \operatorname{Im}\ln t,$$

which under the substitution $t = e^w$, passes to the parameterization

$$x = \operatorname{Im}\cosh w, \qquad y = -\operatorname{Re}\sinh w, \qquad z = \operatorname{Im} w, \tag{20}$$

i.e., because of the identities

$$\cosh w = \cosh u \cos v + i \sinh u \sin v,$$

$$\sinh w = \sinh u \cos v + i \cosh u \sin v,$$

where $w = u + iv$, it passes to the parameterization

$$x = \sinh u \sin v, \qquad y = -\sinh u \cos v, \qquad z = v.$$

Replacing u with v and v with u and simultaneously x with y and y with $-x$ in the latter parameterization, we obtain the parameterization of helicoid (5), which we already met. Therefore, *for $\varphi = t\ln t - t$, we obtain the helicoid.*

§12. Adjoined Minimal Surfaces

Of course, if we replace Re with Im in (18) (which corresponds to multiplication of the vector-valued function $\boldsymbol{f}(w)$ by $-i$), then we again obtain a parameterization of a minimal surface that is different from the initial one in general. This surface is said to be *associated with* a given one.

It is easy to see that the replacement of Re with Im and Im with $-$Re in (19) corresponds to the replacement of Re with Im in (18). In particular, under such a replacement, we obtain the parameterization

$$x = -\operatorname{Re}\cosh w, \qquad y = -\operatorname{Im}\sinh w, \qquad z = -\operatorname{Re} w$$

from paramerization (20), i.e., the parameterization

$$x = -\cosh u \cos v, \qquad y = -\cosh u \sin v, \qquad z = -u,$$

passing to the parameterization of catenoid (4) under the replacements $u \mapsto -v$, $v \mapsto -u$, and $x \mapsto -x$. Therefore, *the catenoid is associated with the helicoid* (and vice versa).

Exercise 14.11. Show that minimal surfaces associated with one another are isometric.

Formulas (19) are called the *Weierstrass formulas* (and also the *Enneper–Weierstrass formulas)*. They can be found in the literature in several different forms, which, however, easily transform into one another.

Chapter 15
Curvature in Riemannian Space

§1. Riemannian Curvature Tensor

For a (pseudo-)Riemannian space $\mathcal{X}$, we can use the metric tensor g to lower the superscript of the curvature tensor R, i.e., introduce a tensor of type (4,0) with the components

$$R_{ij,kl} = g_{ip} R^p{}_{j,kl}. \tag{1}$$

We emphasize that the lowered subscript is assumed to be the *first*. Specifically for this reason, the components of the tensor R are denoted by $R^j{}_{i,kl}$.

Definition 15.1. A tensor with components (1) is called the *Riemannian* (or *covariant*) *curvature tensor*. It is denoted by the old symbol R.

We note that in contrast to the contravariant curvature tensor, which is defined only by a connection, *the Riemannian curvature tensor depends on the metric* g, and under the passage from g to λg, $\lambda \neq 0$, which does not change the connection (see Chap. 11), it is multiplied by λ.

The Riemannian tensor identified with the multilinear mapping

$$\mathfrak{a}\mathcal{X} \times \mathfrak{a}\mathcal{X} \times \mathfrak{a}\mathcal{X} \times \mathfrak{a}\mathcal{X} \to \mathsf{F}\mathcal{X},$$

$$(X, Y, Z, W) \mapsto R_{ij,kl} X^i Y^j Z^k W^l,$$

is given by

$$R(X, Y, Z, W) = (R(Z, W)Y, X) \tag{2}$$

for any $X, Y, Z, W \in \mathfrak{a}\mathcal{X}$.

§2. Symmetries of the Riemannian Tensor

All symmetries of the curvature tensor are certainly preserved for the Riemannian curvature tensor. In particular, this tensor satisfies the Bianchi identity

$$R(X, Y, Z, W) + R(X, Z, W, Y) + R(X, W, Y, Z) = 0 \tag{3}$$

and is skew-symmetric in the last two arguments,

$$R(X, Y, W, Z) = -R(X, Y, Z, W), \tag{4}$$

i.e., its components are skew-symmetric in the last two subscripts,

$$R_{ij,lk} = -R_{ij,kl}.$$

It is surprising that it is also skew-symmetric in the first two arguments.

Proposition 15.1. *The identity*

$$R(Y, X, Z, W) = -R(X, Y, Z, W) \tag{5}$$

holds for any vector fields $X, Y, Z, W \in \mathfrak{a}\mathcal{X}$.

We present three proofs of this important proposition.

Proof (First). Identity (5) is equivalent (compare Exercise 6.5) to the identity

$$R(X, X, Z, W) = 0, \tag{5'}$$

which we therefore prove.

First, we prove the identity (5′) under the additional assumption

$$[Z, W] = 0. \tag{6}$$

Under this assumption,

$$\begin{aligned} R(X, X, Z, W) &= (\nabla_Z \nabla_W X - \nabla_W \nabla_Z X, X) \\ &= (\nabla_Z \nabla_W X, X) - (\nabla_W \nabla_Z X, X), \end{aligned}$$

and therefore identity (5′) is equivalent to the assertion that the inner product $(\nabla_W \nabla_Z X, X)$ is symmetric in Z and W.

Keeping this in mind, we first apply the operator Z and then the operator W to the function $|X|^2 = (X, X)$ on $\mathcal{X}$:

$$\begin{aligned} Z(X, X) &= (\nabla_Z X, X) + (X, \nabla_Z X) = 2(\nabla_Z X, X), \\ WZ(X, X) &= 2(\nabla_W \nabla_Z X, X) + 2(\nabla_Z X, \nabla_W X). \end{aligned}$$

Of course, the inner product $(\nabla_Z X, \nabla_W X)$ is symmetric in Z and W. In addition, because $ZW - WZ = [Z, W]$, the expression $WZ(X, X)$ is also symmetric in Z and W by condition (6). Therefore, the inner product $(\nabla_W \nabla_Z X, X)$ is indeed symmetric in Z and W.

This proves identity (5′) (and therefore identity (5)) under condition (6).

Condition (6) holds in particular on an arbitrary coordinate neighborhood U for the base coordinate fields $Z = \partial/\partial x^k$ and $W = \partial/\partial x^l$. Therefore, by what was proved, the components

$$R_{ij,kl} = R\left(\frac{\partial}{\partial x^i}, \frac{\partial}{\partial x^j}, \frac{\partial}{\partial x^k}, \frac{\partial}{\partial x^l}\right)$$

of the Riemannian tensor are skew-symmetric in i and j. Therefore, identity (5) holds for any fields X, Y, Z, and W. □

Proof (Second). It suffices to show that in each chart $(U, x^1, \dots, x^n)$, the components $R_{ij,kl}$ of the Riemannian tensor are skew-symmetric in i and j, i.e., the matrix

$$\widehat{\Omega} = \|\Omega_{ij}\|$$

consisting of the forms

$$\Omega_{ij} = \sum_{k<l} R_{ij,kl} dx^k \wedge dx^l$$

is skew-symmetric.

To prove this, for the module $\mathfrak{a}U = \Gamma(\tau_{\mathcal{X}}|_U)$, we introduce an orthonormal (not holonomic!) basis $X_1, \dots, X_n$ and relate the matrix $\widehat{\Omega}$ with the matrix $\widetilde{\Omega}$ of the Riemannian curvature forms in the basis $X_1, \dots, X_n$, which is a skew-symmetric matrix (as directly follows from the Cartan structural equation; see (1) in Chap. 4). Let C be the transition matrix from the orthonormal basis $X_1, \dots, X_n$ to the holonomic basis

$$\frac{\partial}{\partial x^1}, \quad \dots, \quad \frac{\partial}{\partial x^n}. \tag{7}$$

Then, as is known from linear algebra,

$$\Omega = C^{-1}\widetilde{\Omega}C,$$

where $\Omega = \|\Omega^i_j\|$ is the matrix of curvature forms of the Riemannian connection in basis (7). On the other hand,

$$\Omega^i_j = \sum_{k<l} R^i_{j,kl} dx^k \wedge dx^l,$$

and therefore

$$\Omega_{ij} = g_{ip}\Omega^p_j,$$

i.e.,

$$\widehat{\Omega} = G\Omega,$$

where $G = \|g_{ij}\|$ is the matrix of the metric tensor g in basis (7).

Because $G = C^T C$, this proves that

$$\widehat{\Omega} = C^T\widetilde{\Omega}C. \tag{8}$$

Therefore, the matrix $\widehat{\Omega}$, simultaneously with the matrix $\widetilde{\Omega}$, is skew-symmetric. □

Proof (Third). Because

$$g_{ip}\Gamma^p_{jk} = \Gamma_{i,jk},$$

$$\Gamma_{i,jk} = \frac{1}{2}\left(\frac{\partial g_{ij}}{\partial x^k} + \frac{\partial g_{ik}}{\partial x^j} - \frac{\partial g_{jk}}{\partial x^i}\right),$$

$$\frac{\partial g_{jk}}{\partial x^i} = \Gamma_{k,ij} + \Gamma_{j,ik}$$

(see (2′), (3), and (5) in Chap. 11), we have

$$g_{ip}\left(\frac{\partial \Gamma^p_{lj}}{\partial x^k} + \Gamma^p_{kq}\Gamma^q_{lj}\right) = \frac{\partial}{\partial x^k}(g_{ip}\Gamma^p_{lj}) - \frac{\partial g_{ip}}{\partial x^k}\Gamma^p_{lj} + \Gamma_{i,kq}\Gamma^q_{lj}$$
$$= \frac{1}{2}\frac{\partial}{\partial x^k}\left(\frac{\partial g_{ij}}{\partial x^l} + \frac{\partial g_{il}}{\partial x^j} - \frac{\partial g_{jl}}{\partial x^i}\right)$$
$$- (\Gamma_{i,kp} + \Gamma_{p,ki})\Gamma^p_{lj} + \Gamma_{i,kq}\Gamma^q_{lj}$$
$$= \frac{1}{2}\left(\frac{\partial^2 g_{ij}}{\partial x^k \partial x^l} + \frac{\partial^2 g_{il}}{\partial x^j \partial x^k} - \frac{\partial^2 g_{jl}}{\partial x^i \partial x^k}\right) - \Gamma_{p,ki}\Gamma^p_{lj}.$$

Interchanging k and l and subtracting, we obtain the formula

$$R_{ij,kl} = \frac{1}{2}\left(\frac{\partial^2 g_{il}}{\partial x^j \partial x^k} - \frac{\partial^2 g_{jl}}{\partial x^i \partial x^k} - \frac{\partial^2 g_{ik}}{\partial x^j \partial x^l} + \frac{\partial^2 g_{jk}}{\partial x^i \partial x^l}\right) + g_{pq}(\Gamma^p_{li}\Gamma^q_{kj} - \Gamma^p_{ki}\Gamma^q_{lj}) \quad (9)$$

for the component $R_{ij,kl}$ of the metric tensor after an obvious transformation.

Because both summands of this formula are obviously skew-symmetric in i and j, this proves Proposition 1. □

Remark 15.1. If the local coordinates $x^1, \ldots, x^n$ are normal coordinates centered at a point p_0 and the connection coefficients Γ^i_{kj} therefore vanish at the point p_0 (see Proposition 2.1), then only the first summand

$$R^{(0)}_{ij,kl} = \frac{1}{2}\left(\frac{\partial^2 g_{il}}{\partial x^j \partial x^k} - \frac{\partial^2 g_{jl}}{\partial x^i \partial x^k} - \frac{\partial^2 g_{ik}}{\partial x^j \partial x^l} + \frac{\partial^2 g_{jk}}{\partial x^i \partial x^l}\right)^{(0)} \quad (9')$$

remains for the value $R^{(0)}_{ij,kl}$ of the components $R_{ij,kl}$ at the point p_0 in formula (9).

Remark 15.2. The trick with condition (6), which was used in the first proof, has a general character. For example, it can be used to simplify the proof of Proposition 2.2. On the other hand,we can certainly avoid its use and a certain complexification of calculations.

Exercise 15.1. Transform the first proof of Proposition 15.1 into a proof that does not use assumption (6). Also, prove Proposition 2.2 using this assumption.

The proved symmetries of the Riemannian tensor implies one more remarkable symmetry of this tensor obtained purely algebraically.

Proposition 15.2. *The Riemannian tensor is unchanged under permutation of the first and second pairs of subscripts,*

$$R(Z, W, X, Y) = R(X, Y, Z, W). \quad (10)$$

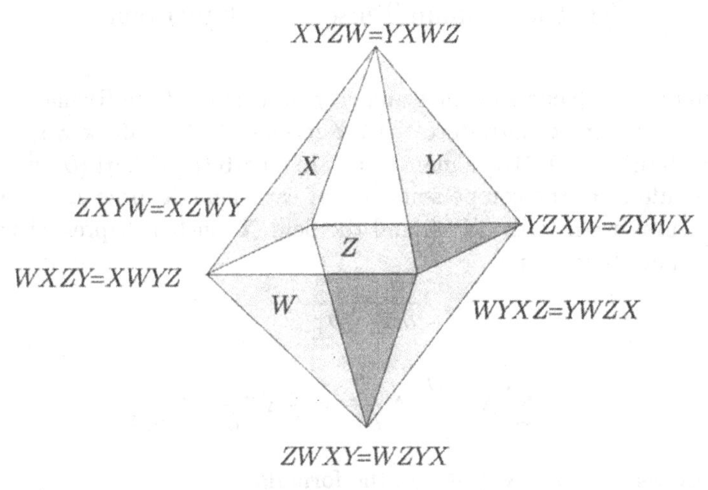

Fig. 15.1.

Proof. We set permutations of the letters X, Y, Z, and W in correspondence to vertices of the octahedron as depicted in Fig. 15.1 (permutations obtained one from another by simultaneous permutations of the first and second pair are assumed to be equal). Then for the four shaded faces of the octahedron, the permutations corresponding to their vertices have the same first component (pointed at the center of the face in the figure), and the other three components are obtained from one another by a cycling. Therefore, Bianchi identity (3) corresponds to each such face. Adding the identities corresponding to the two upper shaded faces, subtracting the identities corresponding to two lower faces, and the dividing by 2, we obtain identity (10) (because all terms corresponding to vertices of the equatorial square are canceled). □

Also, formula (10) is immediately implied by (9).

Because of Proposition 15.2, we can write the values of the Riemannian tensor on vector fields X, Y, Z, and W from $\mathfrak{a}\mathcal{X}$ as

$$R(X,Y,Z,W) = (R(X,Y)W,Z)$$

or

$$R(X,Y,Z,W) = -(R(X,Y)Z,W),$$

where in the left- and right-hand sides, the fields X, Y, Z, and W are located in the same sequence. (By the way, the last formula explains why many authors prefer to define the curvature tensor with the opposite sign).

Therefore, in particular, the Bianchi identity can be written as

$$R(X,Y,Z,W) + R(Z,X,Y,W) + R(Y,Z,X,W) = 0.$$

§3. Riemannian Tensor as a Functional

To study consequences of the symmetry properties of the Riemannian tensor, it is convenient to introduce the $\mathsf{F}\mathcal{X}$-module $\Lambda^2\mathcal{X}$ of all skew-symmetric tensor fields of type $(0,2)$ on a manifold $\mathcal{X}$. In an arbitrary chart $(U, x^1, \ldots, x^n)$ of the manifold $\mathcal{X}$, the components X^{ij} of each field $\boldsymbol{X}$ from $\Lambda^2\mathcal{X}$ compose a skew-symmetric matrix $\|X^{ij}\|$, and the field $\boldsymbol{X}$ on U is expressed through the coordinate bivectors

$$\frac{\partial}{\partial x^i} \wedge \frac{\partial}{\partial x^j}$$

by

$$\boldsymbol{X} = \sum_{i<j} X^{ij} \frac{\partial}{\partial x^i} \wedge \frac{\partial}{\partial x^j} = \frac{1}{2} X^{ij} \frac{\partial}{\partial x^i} \wedge \frac{\partial}{\partial x^j}.$$

For any vector fields $X, Y \in \mathfrak{a}\mathcal{X}$, the formula

$$(X \wedge Y)_p = X_p \wedge Y_p, \quad p \in \mathcal{X},$$

defines the field $X \wedge Y \in \Lambda^2\mathcal{X}$ (the *exterior product* of the fields X and Y), for which

$$(X \wedge Y)^{ij} = X^i Y^j - X^j Y^i$$

in each chart $(U, x^1, \ldots, x^n)$. Fields of the form $X \wedge Y$ are called *bivector fields* on $\mathcal{X}$, and their set is denoted by $\boldsymbol{A}^2\mathcal{X}$. We stress that the set $\boldsymbol{A}^2\mathcal{X}$ in general is not a linear subspace (let alone a submodule).

The Riemannian tensor R is naturally identified with a symmetric $\mathsf{F}\mathcal{X}$-bilinear functional R on $\Lambda^2\mathcal{X}$ that is given by

$$R(\boldsymbol{X}, \boldsymbol{Y}) = \sum_{i<j} \sum_{k<l} R_{ij,kl} X^{ij} Y^{kl} \tag{11}$$

in each chart. Because for $X^{ij} = X^i Y^j - X^j Y^i$ and $Y^{kl} = Z^k W^l - Z^l W^k$,

$$\begin{aligned}\sum_{i<j} \sum_{k<l} R_{ij,kl} X^{ij} Y^{kl} = &\sum_{i<j} \sum_{k<l} R_{ij,kl} X^i Y^j Z^k W^l - \sum_{i<j} \sum_{k<l} R_{ij,kl} X^i Y^j Z^l W^k \\ &- \sum_{i<j} \sum_{k<l} R_{ij,kl} X^j Y^i Z^k W^l + \sum_{i<j} \sum_{k<l} R_{ij,kl} X^j Y^i Z^l W^k\end{aligned}$$

(opening parentheses)

$$\begin{aligned}= &\sum_{i<j} \sum_{k<l} R_{ij,kl} X^i Y^j Z^k W^l - \sum_{i<j} \sum_{l<k} R_{ij,lk} X^i Y^j Z^k W^l \\ &- \sum_{j<i} \sum_{k<l} R_{ji,kl} X^i Y^j Z^k W^l + \sum_{j<i} \sum_{l<k} R_{ji,lk} X^i Y^j Z^k W^l\end{aligned}$$

(renaming the summation indices)

$$=\sum_{i,j}\sum_{k,l} R_{ij,kl}X^iY^jZ^kW^l$$

(interchanging the tensor component indices and uniting the sums), it follows that

$$R(X \wedge Y, Z \wedge W) = R_{ij,kl}X^iY^jZ^kW^l \tag{12}$$

in each chart U for any vector fields $X, Y, Z, W \in \mathfrak{a}\mathcal{X}$ and therefore

$$R(X \wedge Y, Z \wedge W) = R(X, Y, Z, W) \tag{13}$$

on the whole manifold $\mathcal{X}$.

An invariant definition of functional (11) is given on $\boldsymbol{A}^2\mathcal{X}$ by (13) and is then extended by linearity to the whole module $\Lambda^2\mathcal{X}$. Of course, in this approach, the possibility as well as the uniqueness of extension should be proved.

Remark 15.3. We note that in deducing (12) and(13), we used only the skew-symmetry of the components $R_{ij,kl}$ in i, j and k, l.

§4. Walker Identity and Its Consequences

Because the curvature operator $R(X, Y)$ (see (18) in Chap. 2) is skew symmetric in X and Y, we obtain by similar arguments that for any field $\boldsymbol{Z} \in \Lambda^2\mathcal{X}$, the operator $\boldsymbol{R}(\boldsymbol{Z})$ is defined; for $\boldsymbol{Z} = X \wedge Y$, it coincides with $R(X, Y)$ (we write $\boldsymbol{R}(\boldsymbol{Z})$ instead of the natural $R(\boldsymbol{Z})$ because the latter symbol is used below in another sense). This operator is a derivation of tensor fields on $\mathcal{X}$ and is therefore applicable to the Riemannian curvature tensor in particular. Therefore, for any fields $\boldsymbol{X}, \boldsymbol{Y}, \boldsymbol{Z} \in \Lambda^2\mathcal{X}$ on $\mathcal{X}$, the function $(\boldsymbol{R}(\boldsymbol{Z})R)(\boldsymbol{X}, \boldsymbol{Y})$ is defined, and it is clear that Riemannian space $\mathcal{X}$ considered as an affine connection space is semisymmetric (satisfies condition (30) in Chap. 4) iff

$$(\boldsymbol{R}(\boldsymbol{Z})R)(\boldsymbol{X}, \boldsymbol{Y}) = 0 \tag{14}$$

identically for $\boldsymbol{X}, \boldsymbol{Y}, \boldsymbol{Z} \in \Lambda^2\mathcal{X}$.

Exercise 15.2. Prove the following *Walker identity*:

$$(\boldsymbol{R}(\boldsymbol{Z})R)(\boldsymbol{X}, \boldsymbol{Y}) + (\boldsymbol{R}(\boldsymbol{X})R)(\boldsymbol{Y}, \boldsymbol{Z}) + (\boldsymbol{R}(\boldsymbol{Y})R)(\boldsymbol{Z}, \boldsymbol{X}) = 0.$$

Write this identity componentwise in two different ways (in the form of a quadratic relation for the components and in the form of a linear relation for their second partial covariant derivatives).

We assume that on $\Lambda^2\mathcal{X}$, there exist an $\mathsf{F}\mathcal{X}$-linear functional $\boldsymbol{\xi}$ and an $\mathsf{F}\mathcal{X}$-bilinear symmetric functional S such that

$$(\boldsymbol{R}(\boldsymbol{Z})R)(\boldsymbol{X}, \boldsymbol{Y}) = \boldsymbol{\xi}(\boldsymbol{Z})S(\boldsymbol{X}, \boldsymbol{Y}) \tag{15}$$

for any fields $\boldsymbol{X}, \boldsymbol{Y}, \boldsymbol{Z} \in \Lambda^2\mathcal{X}$. It then turns out that at each point $p \in \mathcal{X}$, at least one of the functionals $\boldsymbol{\xi}$ or S vanishes; therefore, the tensor $\boldsymbol{R}(\boldsymbol{Z})R$ also vanishes (*a Riemannian space whose curvature tensor has property* (15) *is semisymmetric*). Indeed, by the Walker identity,

$$\boldsymbol{\xi}(\boldsymbol{Z})S(\boldsymbol{X}, \boldsymbol{Y}) + \boldsymbol{\xi}(\boldsymbol{X})S(\boldsymbol{Y}, \boldsymbol{Z}) + \boldsymbol{\xi}(\boldsymbol{Y})S(\boldsymbol{Z}, \boldsymbol{X}) = 0. \tag{16}$$

We suppose that the functional $\boldsymbol{\xi}$ is different from zero at the point $p \in \mathcal{X}$, i.e., there exists a field $\boldsymbol{Z}_0 \in \Lambda^2\mathcal{X}$ such that $\boldsymbol{\xi}(\boldsymbol{Z}_0)_p = 1$. Then, for $\boldsymbol{Z} = \boldsymbol{Z}_0$, (16) implies

$$S(\boldsymbol{X}, \boldsymbol{Y})_p + \boldsymbol{\xi}(\boldsymbol{X})_p S(\boldsymbol{Y}, \boldsymbol{Z}_0)_p + \boldsymbol{\xi}(\boldsymbol{Y})_p S(\boldsymbol{Z}_0, \boldsymbol{X})_p = 0; \tag{16'}$$

further, for $\boldsymbol{Y} = \boldsymbol{Z}_0$, it follows that

$$2S(\boldsymbol{X}, \boldsymbol{Z}_0)_p + \boldsymbol{\xi}(\boldsymbol{X})_p S(\boldsymbol{Z}_0, \boldsymbol{Z}_0)_p = 0; \tag{16''}$$

finally, for $\boldsymbol{X} = \boldsymbol{Z}_0$, we have

$$3S(\boldsymbol{Z}_0, \boldsymbol{Z}_0)_p = 0,$$

i.e., $S(\boldsymbol{Z}_0, \boldsymbol{Z}_0)_p = 0$. But then, by (16″), we have $S(\boldsymbol{X}, \boldsymbol{Z}_0)_p = 0$, and therefore, by (16′), $S(\boldsymbol{X}, \boldsymbol{Y})_p = 0$ for any $\boldsymbol{X}, \boldsymbol{Y} \in \Lambda^2\mathcal{X}$, i.e., the functional S vanishes at the point $p \in \mathcal{X}$.

§5. Recurrent Spaces

A tensor P on a Riemannian space $\mathcal{X}$ is said to be *recurrent* if there exists an $\mathsf{F}\mathcal{X}$-linear functional ξ on $\mathfrak{a}\mathcal{X}$ such that

$$\nabla_X P = \xi(X)P \tag{17}$$

for any field $X \in \mathfrak{a}\mathcal{X}$. The space $\mathcal{X}$ is said to be *recurrent* if its curvature tensor R is recurrent. *Each recurrent space $\mathcal{X}$ is semisymmetric.* Indeed, if (17) holds for $P = R$, then

$$\nabla_X \nabla_Y R = \nabla_X(\xi(Y)R) = X\xi(Y)R + \xi(Y)\xi(X)R$$

for any fields $X, Y \in \mathfrak{a}\mathcal{X}$, and the relation

$$\begin{aligned} R(X,Y)R &= (\nabla_X\nabla_Y - \nabla_Y\nabla_X - \nabla_{[X,Y]})R \\ &= (X\xi(Y) - Y\xi(X) - \xi([X,Y]))R = (d\xi(X,Y))R \end{aligned}$$

therefore means that the tensor R satisfies condition (15) with $S = R$ and $\boldsymbol{\xi}(\boldsymbol{Z}) = d\xi(X,Y)$ for $\boldsymbol{Z} = X \wedge Y$. Therefore, the space $\mathcal{X}$ is semisymmetric.

Remark 15.4. If $R \neq 0$ (the space $\mathcal{X}$ is not flat), then $d\xi(X, Y) = 0$ for any fields $X, Y \in \mathfrak{a}\mathcal{X}$. This implies (prove this!) that on the space $\mathcal{X}$, we have a function φ (only locally in general) such that

$$\xi(X) = X\varphi$$

for any field $X \in \mathfrak{a}\mathcal{X}$ (the covector ξ is the gradient).

Exercise 15.3. Prove that if we have a nonzero tensor field $\boldsymbol{\xi}$ of type $(m, 0)$, $m \geq 2$, on a Riemannian space $\mathcal{X}$ such that for any fields $X_1, \ldots, X_m \in \mathfrak{a}\mathcal{X}$,

$$\nabla_{X_1} \cdots \nabla_{X_m} R = \boldsymbol{\xi}(X_1, \ldots, X_m) R,$$

then the space $\mathcal{X}$ is locally symmetric.

§6. Virtual Curvature Tensors

The Bianchi identity for the curvature tensor means that on $\boldsymbol{A}^2\mathcal{X}$, the functional R satisfies the identity

$$R(X \wedge Y, Z \wedge W) + R(Y \wedge Z, X \wedge W) + R(Z \wedge X, Y \wedge W) = 0. \tag{18}$$

This identity is also called the *Bianchi identity.*

Definition 15.2. An arbitrary symmetric bilinear functional R on $\Lambda^2\mathcal{X}$ (in other words, a tensor of type (4,0) that is skew-symmetric in the first and second pairs of indices and not changed under a permutation of these pairs) satisfying identity (18) on $\boldsymbol{A}^2\mathcal{X}$ is called a *virtual curvature tensor* (or *Bianchi tensor*) on the manifold $\mathcal{X}$.

We note that in this definition, it is not assumed that the manifold $\mathcal{X}$ is a Riemannian (or pseudo-Riemannian) space.

By the bijective correspondence between symmetric bilinear functionals and quadratic functionals, which is known from linear algebra, each Bianchi tensor is identified with the corresponding quadratic functional $\boldsymbol{X} \mapsto R(\boldsymbol{X})$ on $\Lambda^2\mathcal{X}$, where $R(\boldsymbol{X}) = R(\boldsymbol{X}, \boldsymbol{X})$. All Bianchi tensors on the manifold $\mathcal{X}$ form an $\mathsf{F}\mathcal{X}$-module $\boldsymbol{B}\mathcal{X}$.

For an arbitrary $\mathsf{F}\mathcal{X}$-submodule $\boldsymbol{A}$ of the module of all tensor fields on the manifold $\mathcal{X}$ (of a certain fixed type) and for any coordinate neighborhood $U \subset \mathcal{X}$, we let $\boldsymbol{A}|_U$ denote the $\mathsf{F}U$-module consisting of the restrictions of fields from $\boldsymbol{A}$ to U. In the case where all modules of the form $\boldsymbol{A}|_U$ are free modules of the same dimension m (over $\mathsf{F}U$), we say that *the dimension of the module* $\boldsymbol{A}$ *equals* m and write

$$\dim \boldsymbol{A} = m$$

(bases of the modules $\boldsymbol{A}|_U$ are called *bases of the module* $\boldsymbol{A}$ *over* U, as we have previously done several times). For example, it is clear that

$$\dim \Lambda^2 \mathcal{X} = \frac{n(n-1)}{2},$$

where, as usual, $n = \dim \mathcal{X}$.

Exercise 15.4. Prove that

$$\dim \boldsymbol{B}\mathcal{X} = \frac{n^2(n^2-1)}{12}.$$

[*Hint*: The symmetry of components of an arbitrary Bianchi identity shows that if its nonzero component $R_{ij,kl}$ has only two distinct subscripts (i and j for example), then this component is equal to $R_{ij,ij}$ up to the sign, and if only three subscripts are distinct (i, j, and k for example), then this component is equal to one of the three components $R_{ij,ik}$, $R_{ij,jk}$, and $R_{ik,jk}$ (again up to the sign). In the case where all four subscripts i, j, k, and l are distinct, there also remain only three components $R_{ij,kl}$, $R_{ik,jl}$, and $R_{il,jk}$. However, in the latter case, one of the components is expressed through the other two because of the Bianchi identity; therefore, only two components are linearly independent in this case. Therefore, we have

$$\binom{n}{2} + 3\binom{n}{3} + 2\binom{n}{4} = \frac{n^2(n^2-1)}{12}$$

for the total number of independent components of the Bianchi tensor.]

In particular,

$$\dim \boldsymbol{B}\mathcal{X} = \begin{cases} 1 & \text{for } n = 2, \\ 6 & \text{for } n = 3, \\ 20 & \text{for } n = 4 \end{cases}$$

(we do not need these dimensions for larger n).

§7. Reconstruction of the Bianci Tensor from Its Values on Bivectors

Let $R \in \boldsymbol{B}\mathcal{X}$, and let $p \in \mathcal{X}$.

Exercise 15.5. Show that for any field $\boldsymbol{X} \in \Lambda^2\mathcal{X}$, the value $R(\boldsymbol{X})_p$ of the function $R(\boldsymbol{X})$ at the point p depends on only the value $\boldsymbol{A} = \boldsymbol{X}_p$ of the field $\boldsymbol{X}$ at the point p. [*Hint*: Let A^{ij} be components of the tensor $\boldsymbol{A} \in \Lambda^2(\mathbf{T}_p\mathcal{X})$ in an arbitrary chart $(U, x^1, \dots, x^n)$ (containing the point p). Then the number $R(\boldsymbol{X})_p$ is equal to

$$\sum_{i<j}\sum_{k<l} R^{(0)}_{ij,kl} A^{ij} A^{kl},$$

where $R^{(0)}_{ij,kl}$ are values of the components $R_{ij,kl}$ of the tensor R at the point p.]

This value is denoted by $R(\boldsymbol{A})$. In the case where $\boldsymbol{A}$ is a bivector $A \wedge B$, $A, B \in \mathbf{T}_p\mathcal{X}$, it is expressed by

$$R(A \wedge B) = R^{(0)}_{ij,kl} A^i B^j A^k B^l. \tag{19}$$

(Compare with (12).)

Therefore, by definition, for any vector fields $X, Y \in \mathfrak{a}\mathcal{X}$, we have

$$R(X \wedge Y)_p = R(X_p \wedge Y_p)$$

for the value $R(X \wedge Y)_p$ of the function $R(X \wedge Y)$ at the point $p \in \mathcal{X}$. That the tensor R is completely reconstructed from the functions $R(X \wedge Y)$ determines their special role.

Proposition 15.3. *If two Bianchi tensors (considered as quadratic functionals) coincide on* $\boldsymbol{A}^2\mathcal{X}$*, then they coincide on the whole module* $\Lambda^2\mathcal{X}$*.*

Proof. It suffices to show that if $R(X \wedge Y) = 0$ for any vector fields $X, Y \in \mathfrak{a}\mathcal{X}$, then $R = 0$. On the other hand, because the bivector fields generate the module $\Lambda^2\mathcal{X}$ over each coordinate neighborhood, we see that $R = 0$ iff the function

$$R(X, Y, Z, W) = R(X \wedge Y, Z \wedge W)$$

identically equals zero for any vector fields $X, Y, Z, W \in \mathfrak{a}\mathcal{X}$.

Therefore, we are given that $R(X, Y, X, Y) = 0$, and it is required to prove that $R(X, Y, Z, W) = 0$. To prove this, we substitute the field $X + Z$ for the field X in the identity $R(X, Y, X, Y) = 0$. Because

$$R(X + Z, Y, X + Z, Y) = R(X, Y, X, Y) + 2R(X, Y, Z, Y) + R(Z, Y, Z, Y),$$

we have

$$R(X, Y, Z, Y) = 0.$$

Applying this relation to the field $Y + W$ instead of Y and taking into account that

$$\begin{aligned} R(X, Y + W, Z, Y + W) &= R(X, Y, Z, Y) + R(X, Y, Z, W) \\ &\quad + R(X, W, Z, Y) + R(X, W, Z, W), \end{aligned}$$

we further obtain

$$R(X, Y, Z, W) + R(X, W, Z, Y) = 0,$$

i.e.,

$$R(X, Y, Z, W) = R(Y, Z, X, W).$$

This means that the function $R(X, Y, Z, W)$ is unchanged under a cyclic permutation of the fields X, Y, and Z. Therefore, its cycling in X, Y, and Z only multiplies this function by 3. Because the result of this cycling equals zero according to the Bianchi identity, this is possible only if $R = 0$. □

§8. Sectional Curvatures

To find the geometric sense of the numbers $R(A \wedge B)$ (for the case where R is the curvature tensor of a Riemannian space $\mathcal{X}$), we use the following construction. Let $p_0 \in \mathcal{X}$, and let $\boldsymbol{\xi}_0 \in \mathbf{T}_{p_0}\mathcal{X}$ (it is now convenient to slightly change the standard notation; the point p_0 was denoted by b_0 and the vector $\boldsymbol{\xi}_0$ by p_0). Under the parallel translation along the "parallelogram" with sides sA and sB, where $A, B \in \mathbf{T}_{p_0}\mathcal{X}$, the vector $\boldsymbol{\xi}_0$ passes to the vector

$$\boldsymbol{\xi}_1 = \boldsymbol{\xi}_0 - s^2 R(A,B)\boldsymbol{\xi}_0 + O(s^3).$$

Let the vector $\boldsymbol{\xi}_0$ belong to the plane π of the vectors A and B (i.e., to the tangent plane of the auxiliary two-dimensional surface. Then, in general, the vector $\boldsymbol{\xi}_1$ does not belong to this plane. Let $\boldsymbol{\xi}'_1$ be the orthogonal projection of the vector $\boldsymbol{\xi}_1$ on the plane π. We want to compute the angle φ made by the vectors $\boldsymbol{\xi}_0$ and $\boldsymbol{\xi}'_1$ in the plane π.

Without loss of generality, we can certainly assume that the vector $\boldsymbol{\xi}_0$ is a unit vector. We complete this vector up to an orthonormal basis $\boldsymbol{\xi}_0, \boldsymbol{\eta}_0$ of the plane π that assigns the same orientation as the basis A, B. Then for the vector $\boldsymbol{\xi}'_1 - \boldsymbol{\xi}_0$, we have

$$\boldsymbol{\xi}'_1 - \boldsymbol{\xi}_0 = a\boldsymbol{\xi}_0 + b\boldsymbol{\eta}_0,$$

and the angle φ satisfies

$$\tan\varphi = \frac{b}{1+a}.$$

Moreover,

$$a = (\boldsymbol{\xi}'_1 - \boldsymbol{\xi}_0, \boldsymbol{\xi}_0) = (\boldsymbol{\xi}_1 - \boldsymbol{\xi}_0, \boldsymbol{\xi}_0) = -s^2(R(A,B)\boldsymbol{\xi}_0, \boldsymbol{\xi}_0) + O(s^3) = O(s^3)$$

(because $(R(A,B)\boldsymbol{\xi}_0, \boldsymbol{\xi}_0) = 0$ by the skew-symmetry of the Riemannian tensor with respect to the first pair of arguments) and similarly

$$b = (\boldsymbol{\xi}'_1 - \boldsymbol{\xi}_0, \boldsymbol{\eta}_0) = (\boldsymbol{\xi}_1 - \boldsymbol{\xi}_0, \boldsymbol{\eta}_0) = -s^2(R(A,B)\boldsymbol{\xi}_0, \boldsymbol{\eta}_0) + O(s^3).$$

Because

$$\frac{b}{1+a} = b + O(a)$$

and

$$\varphi = \arctan(\tan\varphi) = \tan\varphi + O(\tan^3\varphi),$$

this implies

$$\varphi = -s^2(R(A,B)\boldsymbol{\xi}_0, \boldsymbol{\eta}_0) + O(s^3).$$

On the other hand,

$$\begin{aligned} -(R(A,B)\boldsymbol{\xi}_0, \boldsymbol{\eta}_0) &= R(A,B,\boldsymbol{\xi}_0,\boldsymbol{\eta}_0) = R(A \wedge B, \boldsymbol{\xi}_0 \wedge \boldsymbol{\eta}_0) \\ &= \frac{R(A \wedge B, A \wedge B)}{|A \wedge B|} = \frac{R(A \wedge B)}{|A \wedge B|}, \end{aligned}$$

where $|A \wedge B|$ is the value (oriented area) of the bivector $A \wedge B$ (for which $A \wedge B = |A \wedge B|(\boldsymbol{\xi}_0 \wedge \boldsymbol{\eta}_0)$); therefore,

$$\varphi = s^2 \frac{R(A \wedge B)}{|A \wedge B|} + O(s^3).$$

Therefore,

$$\lim_{s \to 0} \frac{\varphi}{s^2} = \frac{R(A \wedge B)}{|A \wedge B|}.$$

Because the number $s^2|A \wedge B|$ is equal to the oriented surface σ of the parallelogram of the plane π, which is constructed by the vectors sA and sB, this proves that

$$\lim_{s \to 0} \frac{\varphi}{\sigma} = \frac{R(A \wedge B)}{|A \wedge B|^2}.$$

Setting

$$K_{p_0}(\pi) = \frac{R(A \wedge B)}{|A \wedge B|^2}, \tag{20}$$

we write this formula as

$$K_{p_0}(\pi) = \lim_{s \to 0} \frac{\varphi}{\sigma}. \tag{21}$$

Exercise 15.6. Show that the number $K_{p_0}(\pi)$ depends on only the (nonoriented!) plane π (which justifies its name).

This number is called the *sectional curvature of the Riemannian space $\mathcal{X}$ at the point p_0 with respect to the two-dimensional direction π.*

For a pseudo-Riemannian space $\mathcal{X}$, the curvature $K_{p_0}(\pi)$ is defined in the case where the plane π is not isotropic. A loop which the vector $\boldsymbol{\xi}_0$ traverses to obtain the angle φ should be imagined visually as the result of traversing a certain "curved parallelogram" in the manifold $\mathcal{X}$ in a positive direction, and the number σ as the area of this parallelogram. From this standpoint, formula (21) asserts that *the sectional curvature $K_{p_0}(\pi)$ is the limit value of the angle φ related to the area of the circumscribed domain.*

§9. Formula for the Sectional Curvature

To obtain an analytic formula for $K_{p_0}(\pi)$, we need some facts from linear algebra.

Exercise 15.7. Prove that for any (pseudo-)Euclidean space $\mathcal{V}$, there exists a unique (pseudo-)Euclidean metric on the space $\Lambda^2\mathcal{V}$ of skew-symmetric tensors of type $(0, 2)$ such that

$$(\boldsymbol{a} \wedge \boldsymbol{b}, \boldsymbol{x} \wedge \boldsymbol{y}) = \begin{vmatrix} (\boldsymbol{a}, \boldsymbol{x}) & (\boldsymbol{a}, \boldsymbol{y}) \\ (\boldsymbol{b}, \boldsymbol{y}) & (\boldsymbol{b}, \boldsymbol{y}) \end{vmatrix}$$

for any vectors $\boldsymbol{a}, \boldsymbol{b}, \boldsymbol{x}, \boldsymbol{y} \in \mathcal{V}$.

Exercise 15.8. For $n = 3$, the space $\Lambda^2\mathcal{V} = A^2\mathcal{V}$ is naturally isomorphic to the space $\mathcal{V}$. Show that this isomorphism is an isometry.

Exercise 15.9. Prove that for any vectors $\boldsymbol{a}, \boldsymbol{b} \in \mathcal{V}$, the squared area of the bivector $\boldsymbol{a} \wedge \boldsymbol{b}$ equals its inner square, $|\boldsymbol{a} \wedge \boldsymbol{b}|^2 = (\boldsymbol{a} \wedge \boldsymbol{b}, \boldsymbol{a} \wedge \boldsymbol{b})$.

Let $\boldsymbol{e}_1, \dots, \boldsymbol{e}_n$ be an arbitrary basis of the space $\mathcal{V}$. As we know, bivectors of the form $\boldsymbol{e}_i \wedge \boldsymbol{e}_j$, $i < j$, then compose a basis of the space $\Lambda^2\mathcal{V} = \mathcal{V} \wedge \mathcal{V}$. By definition, *the matrix* $\|g_{ij,kl}\|$, $i < j$, $k < l$, *of metric coefficients of this basis is given by*

$$g_{ij,kl} = \begin{vmatrix} g_{ik} & g_{il} \\ g_{jk} & g_{jl} \end{vmatrix}. \tag{22}$$

This means that the inner product of any tensors

$$\boldsymbol{A} = \sum_{i<j} A^{ij} \boldsymbol{e}_i \wedge \boldsymbol{e}_j, \qquad \text{and} \qquad \boldsymbol{B} = \sum_{k<l} B^{kl} \boldsymbol{e}_k \wedge \boldsymbol{e}_l$$

from $\Lambda^2\mathcal{V}$ is expressed by

$$(\boldsymbol{A}, \boldsymbol{B}) = \sum_{i<j} \sum_{k<l} \begin{vmatrix} g_{ik} & g_{il} \\ g_{jk} & g_{jl} \end{vmatrix} A^{ij} B^{kl}. \tag{23}$$

The numbers $g_{ij,kl}$ are defined by (22) not only for $i < j$ and $k < l$ but also for any i, j, and k, l; moreover, the obtained tuple of numbers $g_{ij,kl}$ is skew-symmetric in i, j and k, l. Therefore (see the deduction of formula (12) from (11) and Remark 15.3), *for any vectors* $\boldsymbol{a}, \boldsymbol{b} \in \mathcal{V}$, *the square area* $|\boldsymbol{a} \wedge \boldsymbol{b}|^2$ *of the bivector* $\boldsymbol{a} \wedge \boldsymbol{b}$ *is expressed by* $|\boldsymbol{a} \wedge \boldsymbol{b}|^2 = g_{ij,kl} a^i b^j a^k b^l$.

We apply these assertions (with corresponding changes in the notation) to the case where $\mathcal{V}$ is the tangent space $\mathbf{T}_{p_0}\mathcal{X}$ of a (pseudo-)Riemannian space $\mathcal{X}$ at a certain point p_0. For example, for any vectors $A, B \in \mathbf{T}_{p_0}\mathcal{X}$, we have

$$|A \wedge B|^2 = g^{(0)}_{ij,kl} A^i B^j A^k B^l, \tag{24}$$

where A^i and B^j are coordinates of the vectors A and B in an arbitrary chart U (containing the point p_0) of the manifold $\mathcal{X}$ and $g^{(0)}_{ij,kl}$ are the values of the functions $g_{ij,kl}$ defined on U by (22) at the point p_0. (As a rule, we omit the superscript (0) for $g_{ij,kl}$ in what follows.)

For the sectional curvature, this yields

$$K_{p_0}(\pi) = \frac{R^{(0)}_{ij,kl} A^i B^j A^k B^l}{g^{(0)}_{ij,kl} A^i B^j A^k B^l}, \tag{25}$$

where A^i and B^j are coordinates of the vectors of the space $\mathbf{T}_{p_0}\mathcal{X}$ that compose an arbitrary basis of the plane π (and $R^{(0)}_{ij,kl}$ are the values of the functions $R_{ij,kl}$ at the point p_0).

Exercise 15.10. Show that the functions $g_{ij,kl}$ defined by (10) are the components of a certain Bianchi tensor.

Chapter 16
Gaussian Curvature

§1. Bianchi Tensors as Operators

The construction of inner product (23) in Chap. 15 is immediately extended to the tensor fields from $\Lambda^2\mathcal{X}$. Let $\boldsymbol{X}, \boldsymbol{Y} \in \Lambda^2\mathcal{X}$, and let

$$\boldsymbol{X} = \sum_{i<j} X^{ij}\frac{\partial}{\partial x^i}\wedge\frac{\partial}{\partial x^j} \quad \text{and} \quad \boldsymbol{Y} = \sum_{k<l} Y^{kl}\frac{\partial}{\partial x^k}\wedge\frac{\partial}{\partial x^l}$$

in a certain chart $(U, x^1, \dots, x^n)$. Then the formula

$$\langle \boldsymbol{X}, \boldsymbol{Y}\rangle = \sum_{i<j}\sum_{k<l} g_{ij,kl}X^{ij}Y^{kl} \tag{1}$$

defines the function $\langle \boldsymbol{X}, \boldsymbol{Y}\rangle$ on U, which does not depend on the choice of the coordinates $x^1, \dots, x^n$. Therefore, this formula correctly defines the function $\langle \boldsymbol{X}, \boldsymbol{Y}\rangle$ on the whole manifold $\mathcal{X}$.

The function $\langle \boldsymbol{X}, \boldsymbol{Y}\rangle$ is given invariantly by

$$\langle \boldsymbol{X}, \boldsymbol{Y}\rangle(p) = (\boldsymbol{X}_p, \boldsymbol{Y}_p), \quad p \in \mathcal{X},$$

where $(\, , \,)$ in the right-hand side is inner product (23) in Chap. 15 on the (pseudo-)Euclidean space $\mathbf{T}_p\mathcal{X}$. This function $\langle \boldsymbol{X}, \boldsymbol{Y}\rangle$ is called the *inner product* of the fields $\boldsymbol{X}$ and $\boldsymbol{Y}$. (Angle brackets are traditionally used here instead of parentheses.)

Using this inner product, we can identify each Bianchi tensor R with a linear operator (over the algebra $\mathbf{F}\mathcal{X}$)

$$R\colon \Lambda^2\mathcal{X} \to \Lambda^2\mathcal{X}. \tag{2}$$

For any fields $\boldsymbol{X}, \boldsymbol{Y} \in \Lambda^2\mathcal{X}$, this operator satisfies the relation

$$\langle R\boldsymbol{X}, \boldsymbol{Y}\rangle = R(\boldsymbol{X}, \boldsymbol{Y}).$$

In an arbitrary chart, the components Z^{ij} of the field $\boldsymbol{Z} = R\boldsymbol{X}$ are given by the formula

$$Z^{ij} = \sum_{k<l} R^{ij}{}_{kl}X^{kl} = \frac{1}{2}R^{ij}{}_{kl}X^{kl},$$

whose coefficients $R^{ij}{}_{kl}$ (entries of the matrix of the operator R in the basis

$$\frac{\partial}{\partial x^i}\wedge\frac{\partial}{\partial x^j}, \quad i<j,$$

of the module $\Lambda^2\mathcal{X}$ over U) are obtained by raising subscripts in the components of the curvature tensor,

$$R^{ij}{}_{kl} = \sum_{a<b} g^{ij,ab} R_{ab,kl}, \tag{3}$$

where

$$g^{ij,kl} = \begin{vmatrix} g^{ik} & g^{il} \\ g^{jk} & g^{jl} \end{vmatrix}$$

are components of the metric tensor with raised subscripts.

Exercise 16.1. Show that for $i < j$ and $k < l$,

$$\sum_{a<b} g_{ij,ab} g^{ab,kl} = \begin{cases} 1 & \text{if } (i,j) = (k,l), \\ 0 & \text{otherwise.} \end{cases} \tag{4}$$

We note that $R^{ij}{}_{kl} = g^{jb} R^i{}_{b,kl}$ in accordance with the rules for raising subscripts.

Because the functional R is symmetric, *the operator R is self-adjoint*, i.e.,

$$\langle R\boldsymbol{X}, \boldsymbol{Y}\rangle = \langle \boldsymbol{X}, R\boldsymbol{Y}\rangle$$

for any fields $\boldsymbol{X}, \boldsymbol{Y} \in \Lambda^2\mathcal{X}$.

§2. Splitting of Trace-Free Tensors

The trace of operator (2) is given by

$$\sum_{i<j} R^{ij}{}_{ij} = \frac{1}{2} R^{ij}{}_{ij} = \frac{1}{2} g^{ik} g^{jl} R_{ij,kl} = \frac{1}{2} g^{jl} R^k_{j,kl}. \tag{5}$$

Trace (5) doubled is denoted by $\mathcal{R}$,

$$\mathcal{R} = g^{ik} g^{jl} R_{ij,kl} = g^{jl} R^k_{j,kl}. \tag{6}$$

In addition, we set

$$K = \frac{\mathcal{R}}{n(n-1)}. \tag{7}$$

Comparison of (3) and (4) immediately reveals that the Bianchi tensor with the components $g_{ij,kl}$ (see Exercise 15.10), interpreted as the operator $\Lambda^2\mathcal{X} \to \Lambda^2\mathcal{X}$, has the components $\delta^i_k \delta^j_l$, i.e., is the identity operator. This tensor is therefore denoted by E.

The trace of the tensor E equals $n(n-1)/2$, and therefore $K = 1$ for this tensor. Therefore, *an arbitrary Bianchi tensor R is uniquely represented in the form*

$$R = KE + R_0, \tag{8}$$

where R_0 is a trace-free tensor (for which $K = 0$).

§3. Gaussian Curvature and the Scalar Curvature

Definition 16.1. When R is the curvature tensor of a (pseudo-)Riemannian space $\mathcal{X}$, the function K is called the *Gaussian curvature* of this space and the function $\mathcal{R}$ its *scalar curvature*.

This terminology is explained by the fact that *in the case where $\mathcal{X}$ is a surface of the Euclidean space, the function K coincides with the Gaussian curvature of this surface*. Indeed, for simplicity of computations, we suppose that the surface is given in the semi-Cartesian coordinates u and v, in which its first quadratic form becomes

$$ds^2 = du^2 + G dv^2 \tag{9}$$

(see Chap. 13). For connection coefficients Γ^i_{jk}, we then have

$$\Gamma^1_{jk} = \begin{cases} 0 & \text{if } (j,k) \neq (2,2), \\ -\dfrac{1}{2}G_u & \text{if } (j,k) = (2,2), \end{cases}$$

$$\Gamma^2_{jk} = \begin{cases} 0 & \text{if } (j,k) = (1,1), \\ \dfrac{1}{2}\dfrac{G_u}{G} & \text{if } (j,k) = (1,2) \text{ or } (2,1), \\ \dfrac{1}{2}\dfrac{G_v}{G} & \text{if } (j,k) = (2,2). \end{cases}$$

Therefore,

$$R^1_{2,12} = \frac{\partial \Gamma^1_{22}}{\partial u} - \frac{\partial \Gamma^1_{12}}{\partial v} + \Gamma^1_{1p}\Gamma^p_{22} - \Gamma^1_{2p}\Gamma^p_{12} = \frac{\partial \Gamma^1_{22}}{\partial u} - \Gamma^1_{22}\Gamma^2_{12} = -\frac{1}{2}G_{uu} + \frac{G_u^2}{4G},$$

and because $g^{21} = 0$ and $g^{22} = G^{-1}$, we therefore have

$$R^{12}{}_{12} = -\frac{2G_{uu}G - G_u^2}{4G^2} = -\frac{(\sqrt{G})_{uu}}{\sqrt{G}}.$$

At the same time, $K = R^{12}{}_{12}$ in the case considered. Therefore,

$$K = -\frac{(\sqrt{G})_{uu}}{\sqrt{G}}.$$

Similar to the Riemannian tensor, the curvatures $\mathcal{R}$ and K are not only determined by the connection: under the passage from g to λg, $\lambda \neq 0$, which does not change the connection, they take the multiplier $1/\lambda$. The choice between $\mathcal{R}$ and K is usually determined by convenience and tradition.

§4. Curvature Tensor for $n = 2$

For $n = 2$ (in the case where the manifold $\mathcal{X}$ is a surface), the curvature tensor is expressed algebraically through the metric tensor and the Gaussian curvature.

Proposition 16.1. *For $n = 2$, any Bianchi tensor R has the form KE, i.e., its components are expressed by*

$$R_{ij,kl} = K(g_{ik}g_{jl} - g_{il}g_{jk}). \tag{10}$$

Proof. Because $\dim \boldsymbol{B}\mathcal{X} = 1$, the term R_0 in decomposition (8) must equal zero. □

In this sense, the curvature tensor of a surface is completely characterized by its Gaussian curvature K.

In formula (10), all nonzero components are equal to $\pm R_{12,12}$. Therefore, this formula is in fact reduced to the one relation

$$R_{12,12} = K \begin{vmatrix} g_{11} & g_{12} \\ g_{21} & g_{22} \end{vmatrix}, \tag{10'}$$

which has the form

$$R_{12,12} = K(EG - F^2) \tag{10''}$$

in the Gauss notation. (We stress that the component $R_{12,12}$ and the determinant $EG - F^2$ depend on the choice of local coordinates. Only their ratio K has an invariant sense.)

In the case where $\mathcal{X}$ is a surface of the three-dimensional Euclidean space, comparison of formula (10″) with the formula for the Gaussian curvature of a surface immediately shows that *the component $R_{12,12}$ of the curvature tensor is equal to the determinant $LN - M^2$ of the second quadratic form of a surface.*

For $n = 2$, there is one and only one two-dimensional plane π in the space $\mathbf{T}_{p_0}\mathcal{X}$, the plane $\mathbf{T}_{p_0}\mathcal{X}$ itself. Therefore, the sectional curvature $K_{p_0}(\pi)$ is also unique.

Exercise 16.2. Prove that the value K_{p_0} of the Gaussian curvature K at the point p_0 is this sectional curvature.

In particular, this explains why $K_{p_0}(\pi)$ is called the curvature.

§5. Geometric Interpretation of the Sectional Curvature

Sectional curvatures of a Riemannian space $\mathcal{X}$ of an arbitrary dimension can also be interpreted as Gaussian curvatures. Two-dimensional submanifolds of the space $\mathcal{X}$ are called *surfaces in $\mathcal{X}$*.

Let $p_0 \in \mathcal{X}$, let $U^{(0)}$ be normal neighborhood of the vector 0 of the linear space $\mathbf{T}_{p_0}\mathcal{X}$, and let $\pi \subset \mathbf{T}_{p_0}\mathcal{X}$ be an arbitrary plane in $\mathbf{T}_{p_0}\mathcal{X}$. Further, let $\mathcal{X}_\pi = \exp_{p_0}(U^{(0)} \cap \pi)$ be the surface in $\mathcal{X}$ that is the image of this plane (more precisely, its part $U^{(0)} \cap \pi$) under the diffeomorphism $\exp_{p_0}|_{U^{(0)}}$). (Visually, the surface $\mathcal{X}_\pi$ is swept by geodesics emanating from the point p_0 and tangent to the plane π at p_0.) Clearly, the plane tangent to the surface $\mathcal{X}_\pi$ at the point p_0 is the plane π.

Proposition 16.2. *The Gauss curvature $K(\mathcal{X}_\pi)_{p_0}$ of the surface $\mathcal{X}_\pi$ at the point p_0 is equal to the sectional curvature $K_{p_0}(\pi)$ of the Riemannian space $\mathcal{X}$ at the point p_0 with respect to the two-dimensional direction π.*

Proof. Choosing a basis in $\mathbf{T}_{p_0}\mathcal{X}$ whose first two vectors generate the plane π, we consider the corresponding normal coordinates $x^1, \dots, x^n$ in a normal neighborhood $U = \exp_{p_0} U^{(0)}$ of the point p_0. According to formula (25) in Chap. 15 for $A = (\partial/\partial x^1)_{p_0}$ and $B = (\partial/\partial x^2)_{p_0}$, the sectional curvature $K_{p_0}(\pi)$ is expressed in these coordinates by

$$K_{p_0}(\pi) = \frac{R^{(0)}_{12,12}}{g^{(0)}_{11} g^{(0)}_{22} - (g^{(0)}_{12})^2}.$$

On the other hand, the surface $\mathcal{X}_\pi$ (lying entirely in U) has the equations $x^3 = 0, \dots, x^n = 0$ in the coordinates $x^1, \dots, x^n$, and the functions $u = x^1$ and $v = x^2$ are local coordinates on $\mathcal{X}_\pi$ (which, however, are defined on the whole surface $\mathcal{X}_\pi$). Therefore, if $\widetilde{R}$ is the Riemannian curvature tensor of this surface, then we have

$$K(\mathcal{X}_\pi)_{p_0} = \frac{\widetilde{R}^{(0)}_{12,12}}{\widetilde{g}^{(0)}_{11} \widetilde{g}^{(0)}_{22} - (\widetilde{g}^{(0)}_{12})^2}$$

according to (10), where $\widetilde{g}^{(0)}_{ij}$ are values of the metric tensor of the surface $\mathcal{X}_\pi$ at the point p_0. Because $\widetilde{g}^{(0)}_{ij} = g^{(0)}_{ij}$ by definition, this implies that to prove Proposition 16.2, it suffices to prove that

$$\widetilde{R}^{(0)}_{12,12} = R^{(0)}_{12,12}. \tag{11}$$

To prove this, we note that because the coordinates $x^1, \dots, x^n$ are normal, we have the following formula for the component $R^{(0)}_{12,12}$ (see Remark 15.1):

$$\begin{aligned}\widetilde{R}^{(0)}_{12,12} &= \frac{1}{2}\left(\frac{\partial^2 g_{12}}{\partial x^2 \partial x^1} - \frac{\partial^2 g_{22}}{\partial x^1 \partial x^1} - \frac{\partial^2 g_{11}}{\partial x^2 \partial x^2} + \frac{\partial^2 g_{21}}{\partial x^1 \partial x^2}\right)^{(0)} \\ &= \frac{1}{2}\left(2\frac{\partial^2 g_{12}}{\partial x^1 \partial x^2} - \frac{\partial^2 g_{11}}{\partial x^2 \partial x^2} - \frac{\partial^2 g_{22}}{\partial x^1 \partial x^1}\right)^{(0)}\end{aligned}$$

(see (9′) in Chap. 15).

But because the coordinates u and v on $\mathcal{X}_\pi$ are also normal (why?), we have a similar formula for $\widetilde{R}^{(0)}_{12,12}$:

$$\widetilde{R}^{(0)}_{12,12} = \frac{1}{2}\left(2\frac{\partial^2 \widetilde{g}_{12}}{\partial u \partial v} - \frac{\partial^2 \widetilde{g}_{11}}{\partial v \partial v} - \frac{\partial^2 \widetilde{g}_{22}}{\partial u \partial u}\right)^{(0)}.$$

Because $\widetilde{g}_{ij} = g_{ij}$ on U and $u = x^1$, $v = x^2$, this proves (11) and thus Proposition 16.2. □

§6. Total Curvature of a Domain on a Surface

Returning to surfaces (we can visualize them as surfaces in the Euclidean space), we note that according formula (21) in Chap. 15 (and the assertion in Exercise 16.2), we have

$$K = \lim_{s \to 0} \frac{\varphi}{\sigma} \tag{12}$$

for the Gauss curvature K of an arbitrary surface $\mathcal{X}$. Here, it should be kept in mind that the case to which (12) refers is essentially simplified for $n = 2$ because there is no necessity to choose an auxiliary two-dimensional surface: its role is now played by the manifold $\mathcal{X}$ itself, and we need not project the encircling tangent vector to its tangent plane. In particular, this means that the angle φ in (12) is just the *angle of turn* of the vector $\boldsymbol{\xi}_0$ under its parallel translation along the loop. This allows ignoring differential relation (12) and finding the turn φ along any loop homotopic to zero in a more geometric and, in essence, equivalent statement: we can find the turn φ along any simple closed curve that bounds a domain diffeomorphic to a disk.

The further reasoning is in fact a realization in the case $n = 2$ of the integration approach for computing the restricted holonomy group. This approach can be performed comparatively simply because the group $\mathrm{SO}(2) \cong \mathbb{S}^1$ is Abelian.

First, let the surface $\mathcal{X}$ be elementary, i.e., covered by one chart (which is assumed to be diffeomorphic to an open disk in the (u, v)-plane $\mathbb{R}^2$). In the case where $\mathcal{X}$ is a surface in the Euclidean space, this means by definition that $\mathcal{X}$ admits a parameterization in the form

$$\boldsymbol{r} = \boldsymbol{r}(u, v), \tag{13}$$

where (u, v) runs over the open disk $\mathbb{B}$ of the plane $\mathbb{R}^2$. By this assumption, we can apply results proved in calculus for a plane domain to domains in $\mathcal{X}$. In addition, we can assume that the surface $\mathcal{X}$ is oriented.

Exercise 16.3. Let G be a domain in an elementary surface that is bounded by a piecewise smooth simple closed curve Γ. Prove that the following assertions hold:

1. For any point $p_0 \in \Gamma$, the turn angle φ of the vector $\boldsymbol{\xi}_0 \in \mathbf{T}_{p_0}\mathcal{X}$ under encircling the domain G in the positive direction (such that the domain remains on the left)

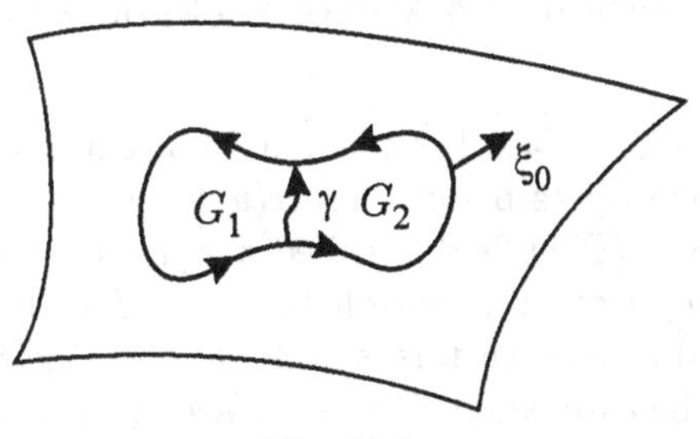

Fig. 16.1.

does not depend on the choice of the vector $\boldsymbol{\xi}_0$ and is the same for all points $p_0 \in \Gamma$. This means that φ depends only on the domain G, i.e., it is a *domain function* in the conventional language of analysis.

2. The domain function φ is additive, i.e., if a simple curve γ cuts the domain G into two domains G_1 and G_2, then

$$\varphi(G) = \varphi(G_1) + \varphi(G_2).$$

[*Hint*: When encircling the domains G_1 and G_2, we go along the curve γ once in each direction.]

3. The domain function φ is smooth, i.e., for any point p of the surface $\mathcal{X}$ and for any sequence of domains $\{G_m\}$ containing the point p_0 and contracting to this point (i.e., such that $\operatorname{diam} G_m \to 0$ as $m \to \infty$), there exists the limit

$$\lim_{m\to\infty} \frac{\varphi(G_m)}{|G_m|}, \tag{14}$$

where $|G_m|$ is the area of the domain G_m (this limit is called the *density* of the additive domain function φ at the point p; see any calculus course).

4. Density (14) equals the Gaussian curvature K_p of the surface $\mathcal{X}$ at the point p. [*Hint*: See (12)].

By the theorem on the reconstruction of a smooth additive domain function from its density known from calculus, this immediately implies the following proposition.

Proposition 16.3. *For any domain G in an elementary surface $\mathcal{X}$ that is bounded by a piecewise smooth simple closed curve, the turn angle φ of tangent vectors encircling the boundary Γ of the domain is equal to the integral of the Gaussian curvature of this surface taken over the domain G:*

$$\varphi = \iint K\, d\sigma. \tag{15}$$

Here, $d\sigma = \sqrt{EG - F^2}\, du\, dv$ is the area element of the surface. The integral in (15) is called the *total curvature* of the domain G in the surface $\mathcal{X}$.

Exercise 16.4. Let a surface $\mathcal{X}$ be nonelementary, and let G be an oriented domain in $\mathcal{X}$ that is diffeomorphic to a plane domain with a piecewise smooth boundary (which, in general, consists of several piecewise smooth simple closed curves). Further, let φ be the sum of the turn angles over all boundary curves equipped with the induced orientation. Prove that formula (15) holds in this more general case.

§7. Rotation of a Vector Field on a Curve

We can usefully reformulate Proposition 16.3. On a surface $\mathcal{X}$ (which is assumed to be elementary as before), let $\boldsymbol{\alpha}(u,v)$ be a given vector field that is nonzero everywhere. (This field is called a *reference field.* The coordinate field $\partial/\partial u$ is an example of a reference field, i.e., under the identification of $\mathcal{X}$ with $\mathbb{B}$, it is the field of unit vectors parallel to the abscissa and directed to its positive side; in the case where $\mathcal{X}$ is a surface of the Euclidean space with parameterization (13), it is the field of coordinate vectors $\boldsymbol{r}_u$.)

Further, let Γ be an arbitrary Jordan curve on the surface $\mathcal{X}$ with the parametric equations $u = u(t)$, $v = v(t)$, $0 \le t \le 1$, and let $\boldsymbol{\zeta} = \boldsymbol{\zeta}(t)$ be a continuous vector field on the curve Γ that is nonzero at each point of this curve. Then, for any $t \in I$, the angle $\widehat{\theta}(t)$ made by the vectors $\boldsymbol{\alpha}(t)$ and $\boldsymbol{\zeta}(t)$ and belonging to $(-\pi, \pi]$ is uniquely defined. In the general case, the function $\widehat{\theta}(t)$ is discontinuous: it can have the jumps of $\pm 2\pi$ at certain points.

Exercise 16.5. Prove that there exists one and only one continuous function $\theta: I \to \mathbb{R}$ with $\theta(0) = \widehat{\theta}(0)$ such that

$$\theta(t) - \widehat{\theta}(t) = 2\pi N, \quad t \in I, \tag{16}$$

where N is a certain integer (depending on t).

The function $\theta(t)$ is called the *angular function* of the field $\boldsymbol{\zeta}$ on the curve Γ (for a given reference field $\boldsymbol{\alpha}$), and the difference

$$\Delta_\Gamma(\boldsymbol{\zeta}) = \theta(1) - \theta(0)$$

of its values at the initial and final points of the curve Γ is called the *turn* of the field $\boldsymbol{\zeta}$ on Γ. If the curve Γ is closed and $\boldsymbol{\zeta}(0) = \boldsymbol{\zeta}(1)$, then, by (16), the ratio

$$V_\Gamma(\boldsymbol{\zeta}) = \frac{1}{2\pi}\Delta_\Gamma(\boldsymbol{\zeta}) \tag{17}$$

is an integer. It is called the *rotation* of the field $\boldsymbol{\zeta}$ along the curve Γ.

Exercise 16.6. For simplicity, let the coefficient F of the first quadratic form of a surface equal zero (the coordinate net is orthogonal). Show that in this case,

$$V_\Gamma(\boldsymbol{\zeta}) = \frac{1}{2\pi}\oint_\Gamma \frac{(\sqrt{E}\xi)d(\sqrt{G}\eta) - (\sqrt{G}\eta)d(\sqrt{E}\xi)}{E\xi^2 + G\eta^2}, \tag{18}$$

where ξ and η are components of the field $\boldsymbol{\zeta}$, is the rotation of an arbitrary smooth field $\boldsymbol{\zeta}$ along a closed curve Γ. We stress that the curve Γ in this formula is not assumed to be smooth.

Let the curve Γ together with the field $\boldsymbol{\zeta}$ be continuously deformed (subjected to a homotopy); moreover, let the curve Γ remain closed and the field $\boldsymbol{\zeta}$ never vanish in the process of deformation. (Such deformations are said to be *admissible.*) Moreover, let the reference field α and even the Riemannian metric ds^2 on $\mathcal{X}$ be continuously varied, i.e., the coefficients E, F, and G of this metric are continuously deformed with preservation of the positive definiteness and the inequalities $E > 0$ and $EG - F^2 > 0$ (for a surface in the Euclidean space, this means that the surface is subjected to an *isotopy*, a continuous deformation that in general changes the length of curves on the surface and angles made by them). Then the rotation $V_\Gamma(\zeta)$ of the field ζ is also continuously varied and because it is an integer, it must remain the same. Therefore, *the rotation $V_\Gamma(\zeta)$ is unchanged under all admissible deformations* (of the curve Γ, field ζ, metric ds^2, and field α).

Exercise 16.7. Show that any two reference fields (and also any two Riemannian metrics ds^2) deform one into another (on a surface diffeomorphic to a disk). Therefore, the rotation $V_\Gamma(\zeta)$ is in fact independent of the field α and the metric (5).

§8. Rotation of the Field of Tangent Vectors

If a curve Γ is smooth, then the *field of tangent vectors* $\boldsymbol{\tau}$ (with the components $\dot{u}$ and $\dot{v}$) is defined on it. If the curve Γ is regular, then this field is nonzero everywhere, and its rotation $V_\Gamma(\boldsymbol{\tau})$ is therefore defined if Γ is closed.

Lemma 16.1. *For an arbitrary closed regular curve Γ, the rotation of the field of tangent vectors equals* 1,

$$V_\Gamma(\boldsymbol{\tau}) = 1. \tag{19}$$

We give two proofs of this lemma.

Proof (First). According to the assertion in Exercise 16.7, we can assume without loss of generality that the surface $\mathcal{X}$ is the Euclidean plane (or an open disk of the plane) and the reference field is a constant field $\partial/\partial u$. Then, for any $t \in I$, the tangent $\tan\theta(t)$ of the value $\theta(t)$ of the angular function of the field $\boldsymbol{\tau}$ at the point t is just the inclination of the tangent line to the curve Γ at its point $\Gamma(t)$ (with the coordinates $u(t)$ and $v(t)$). Therefore, there exists a partition

$$0 = t_0 < t_1 < \cdots < t_n = 1 \tag{20}$$

of the closed interval I such that for any $k = 1, \ldots, n$, the angle (from the semiopen interval $(-\pi, \pi]$) made by the oriented tangent lines to the curve Γ (i.e., between the corresponding tangent vectors) at the points $\Gamma(t_{k-1})$ and $\Gamma(t_k)$ is equal to the increment $\theta(t_k) - \theta(t_{k-1})$ of this function. Moreover, it is visually obvious that for a sufficiently small mesh of partition (20), the tangents at the points $\Gamma(t_k)$ make a simple (without self-intersections) closed polygon (with the exterior angles $\theta(t_k) - \theta(t_{k-1})$) being continued up to the intersection with tangents at the points $\Gamma(t_{k-1}$ and $\Gamma(t_{k+1})$ (we conditionally assume that $t_{n+1} = t_1$).

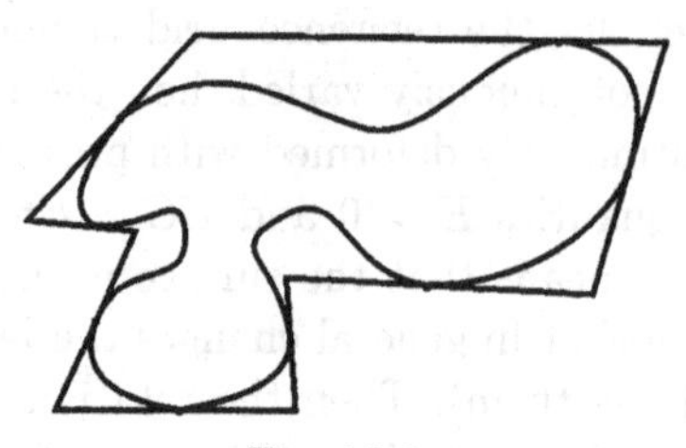

Fig. 16.2.

Because the sum of exterior angles of an arbitrary simple polygon equals 2π (according to the well-known theorem of elementary geometry, the sum of interior angles of a simple n-gon equals $(n-2)\pi$; each inner angle α_k,

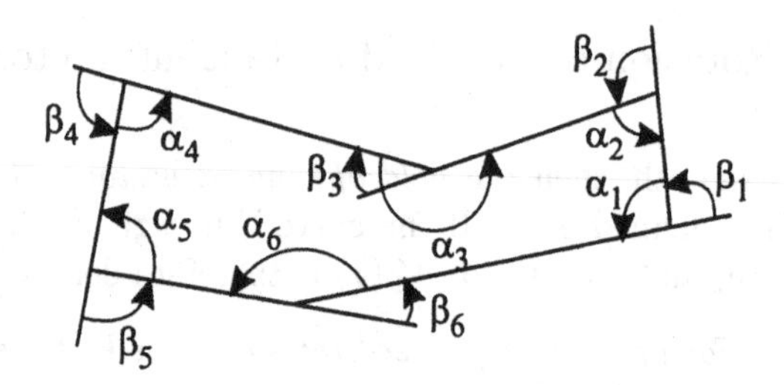

Fig. 16.3.

$1 \leq k \leq n$, is related to the adjacent exterior angle β_k by $\alpha_k + \beta_k = \pi$; then

$$\sum_{k=1}^{n} \beta_k = \sum_{k=1}^{n} (\pi - \alpha_k) = n\pi - \sum_{k=1}^{n} \alpha_k = n\pi - (n-2)\pi = 2\pi$$

obviously follows), we have

$$\theta(1) - \theta(0) = \sum_{k=1}^{n} [\theta(t_k) - \theta(t_{k-1})] = 2\pi,$$

i.e., $V_\Gamma(\tau) = 1$. □

It is difficult to remove the reliance on "visually obvious" from this proof. In the next proof, we demonstrate a general method that shows how this can (and should) be done.

Proof (Second). We introduce a complex coordinate z on the Euclidean plane. Then the curve Γ has the equation $z = z(t)$, $0 \leq t \leq 1$, and each vector field ζ on Γ is just a complex-valued function ζ on Γ.

Exercise 16.8. Show that

$$V_\Gamma(\zeta) = \frac{1}{2\pi} \operatorname{Im} \oint_\Gamma \frac{d\zeta}{\zeta}.$$

(Compare with (18).)

It is known from the theory of functions of a complex variable that the *Riemann theorem* holds, which states that on a domain G bounded by a simple curve Γ, there exists a holomorphic function $w = w(z)$ that conformally maps this domain onto the unit disk $|w| < 1$. Moreover, by the *boundary correspondence theorem*, the function $w(z)$ is continuous in the closed domain $\overline{G}$ and performs its homeomorphism onto the closed disc $|w| \le 1$. Therefore, the equation

$$w = w(z(t))$$

assigns the circle $w\overline{w} = 1$ on the w-plane. We can certainly assume without loss of generality that the parameter t on the curve Γ is chosen such that it is the natural parameter on the circle $w\overline{w} = 1$ and the derivative $\dot{w}(t)$ with respect to t of the function $w(t) = w(z(t))$ therefore equals $iw(t)$ at each point t. On the other hand, by the differentiation rule for a composition, we have

$$\dot{w}(t) = w'(z(t))\dot{z}(t)$$

and therefore

$$\dot{z}(t) = \frac{iw(z(t))}{w'(z(t))}.$$

This means that the vector field $\boldsymbol{\tau}$ on the curve Γ (given by the function $\dot{z}(t)$) is the restriction of the function $\zeta = iw/w'$ to Γ and therefore

$$V_\Gamma(\boldsymbol{\tau}) = \frac{1}{2\pi} \operatorname{Im} \oint_\Gamma \frac{d\zeta}{\zeta} = \frac{1}{2\pi} \operatorname{Im} \oint_\Gamma \left(\frac{dw}{w} - \frac{dw'}{w'} \right)$$

(see Exercise 16.8). But the function w' has no zeros in the domain G; therefore, by the *Cauchy theorem*,

$$\oint_\Gamma \frac{dw'}{w'} = 0.$$

Therefore,

$$V_\Gamma(\tau) = \frac{1}{2\pi} \operatorname{Im} \oint_\Gamma \frac{dw}{w} = \frac{1}{2\pi} \operatorname{Im} \oint_{|w|=1} \frac{dw}{w} = 1. \qquad \square$$

All difficulties are here overcome by the theory of functions of a complex variable.

Lemma 16.1 can also be generalized to curves that are only *piecewise regular*, i.e., consist of finitely many regular arcs. Each such curve has finitely many *breakpoints* where the tangent vector turns by a certain jump angle. The total turn of the tangent vector on the curve (which is the sum of turns on regular arcs and jumps at the breakpoint) is again denoted by $\Delta_\Gamma(\tau) = 2\pi V_\Gamma(\tau)$. Because partitions (20) with the properties above obviously exist for piecewise regular curves (points of a partition should be chosen outside the breakpoints), Lemma 16.1 holds for any piecewise regular curves.

We note that for rectilinear polygons on the Euclidean plane, Lemma 16.1 so generalized is reduced to the theorem on the sum of interior angles of an n-gon.

Lemma 16.1 allows us to not use a reference field (which introduces an undesirable arbitrariness) and to measure all angles using tangent vectors, i.e., for an arbitrary vector field ζ on the curve Γ, we can replace the angular function $\theta(t)$ with the function $\vartheta(t) = \theta(t) - \theta_\tau(t)$, where θ_τ is the angular function of tangent vectors. This changes the rotation V_Γ exactly by 1.

§9. Gauss–Bonnet Formula

Let the field ζ consist of parallel vectors (along the curve Γ). In the case where the surface $\mathcal{X}$ is a plane and all the vectors $\zeta(t)$ are therefore parallel in the usual elementary geometric sense (are equal as free vectors), the derivative $\dot{\vartheta}(t)$ of the function $\vartheta(t)$ is just the curvature of the plane curve Γ (up to the sign). In the general case, this derivative taken with the opposite sign is called the *geodesic curvature* of the curve Γ on the surface $\mathcal{X}$ and is denoted by $k_g(t)$. We note that *the curvature k_g does not depend on the choice of the field ζ* (but does depend on the orientation of the curve Γ).

Obviously *a curve Γ is a geodesic iff its geodesic curvature k_g is identically zero*. Because $\dot{\vartheta}(t) = -k_g(t)$ for the field of parallel vectors ζ, we have

$$\vartheta(1) - \vartheta(0) = -\int_\Gamma k_g\,dt$$

and therefore

$$\begin{aligned}\Delta_\Gamma(\zeta) &= \theta(1) - \theta(0)\\ &= [\theta_\tau(1) - \theta_\tau(0)] + [\vartheta(1) - \vartheta(0)] = \Delta_\Gamma(\tau) - \textstyle\int_\Gamma k_g\,dt,\end{aligned}$$

i.e.,

$$\Delta_\Gamma(\zeta) + \int_\Gamma k_g\,dt = \Delta_\Gamma(\tau)$$

(we do not assume that the curve Γ is closed).

In the case where the curve Γ is piecewise regular, this formula obviously becomes

$$\Delta_\Gamma(\zeta) + \int_\Gamma k_g dt + \sum \theta_i = \Delta_\Gamma(\tau),$$

where $\sum\theta_i$ is the sum of the turn angles θ_i of the tangent vector at the breakpoints. On the other hand, if the curve Γ is closed (and therefore bounds a certain domain G because the surface $\mathcal{X}$ is elementary), the turn $\Delta_\Gamma(\zeta)$ of the field ζ is just the angle φ in (15), and according to this formula, it is therefore equal to the integral of the curvature K over the domain G. By Lemma 16.1, this proves that

$$\iint_G K\,d\sigma + \oint_\Gamma k_g\,dt + \sum \theta_i = 2\pi. \tag{21}$$

Definition 16.2. A domain G on an arbitrary (in general, not elementary) surface $\mathcal{X}$ is called a *curvilinear polygon* if it is bounded by a simple piecewise regular curve Γ and is diffeomorphic to a disk. (For domains on an elementary surface, the second condition is implied by the first, but these conditions are independent in the general case.) Corner points of the boundary Γ are called *vertices* of the polygon G, and the turn angles θ_i of the tangent to the curve Γ at these points (under the assumption that the polygon G and therefore the curve Γ are oriented) are called *exterior angles* of the polygon. The complementary angles $\pi - \theta_i$ are called *interior angles* of the polygon, and the number

$$\delta G = \sum(\pi - \theta_i) - (n-2)\pi = 2\pi - \sum \theta_i,$$

where n is the number of vertices of the polygon G, is called its *angular excess.*

We can now rewrite formula (21) in the form

$$\delta G = \iint_G K\,d\sigma + \oint_\Gamma k_g\,dt; \tag{21'}$$

by what was proved, it holds for any oriented curvilinear polygon G contained in a certain chart of the surface $\mathcal{X}$.

Exercise 16.9. Show that the latter stipulation is extra, i.e., formula (21′) holds for any oriented curvilinear polygon on the surface $\mathcal{X}$. [*Hint*: The polygon G can be decomposed into polygons each of which is contained in a certain chart.]

Formula (21) (or (21′)) is called the *Gauss–Bonnet formula.*

In the particular case of a *geodesic polygon* G (whose boundary consists of geodesic arcs), the curvilinear integral over the boundary equals zero, and formula (21′) becomes

$$\delta G = \iint_G K\,d\sigma \tag{21''}$$

(*the angular excess of the geodesic polygon* G *equals the integral of the Gauss curvature* K *over* G).

For $K = \text{const}$ (for surfaces of constant curvature), this implies that *the angular excess of a geodesic polygon equals its area multiplied by* K. In particular, the angular excess always vanishes for $K = 0$ (on the Euclidean plane); for $K > 0$ (on a sphere), the angular excess is positive; and for $K < 0$ (on a pseudo-sphere), the angular excess is negative (in this case, the quantity $-\delta G$ is usually considered and is called the *angular deficiency*).

Exercise 16.10. Prove the latter assertion in the frameworks of the corresponding geometries (Euclidean, spherical, and Lobachevkii). [*Hint*: First consider the case of a triangle. Of course, the Euclidean case is trivial.]

§10. Triangulated Surfaces

The Gauss–Bonnet formula has interesting and important purely topological consequences.

Definition 16.3. A two-dimensional smooth manifold (surface) is said to be *triangulated* if there exist a finite set of points $p_1, \dots, p_a \in \mathcal{X}$ and a set of simple arcs $\gamma_1, \dots, \gamma_b$ with ends at these points (and having no other common points pairwise) such that any point of the surface $\mathcal{X}$ that does not belong to the arcs $\gamma_1, \dots, \gamma_b$, belongs to one (and only one) curvilinear polygon whose boundary is composed of certain these arcs (briefly speaking, a surface is triangulated if it can be decomposed into the union of finitely many curvilinear polygons $G_1, \dots, G_c$ that touch each other only along their sides).

Exercise 16.11. Show that for any triangulated surface, there exists a decomposition $\{G_1, \dots, G_c\}$ consisting of curvilinear triangles.

This explains the term "triangulated surface." It is clear that a triangulated surface is compact. For smooth surfaces, the converse is true.

Lemma 16.2. *Any compact smooth surface $\mathcal{X}$ is triangulated.*

Proof. We introduce a Riemannian metric ρ on the surface $\mathcal{X}$. The argument proving the existence of the Morse number (see Corollary 12.4) is applicable not only to two-element but also to three-element sets. Therefore, for the surface $\mathcal{X}$, there exists a number ρ_0 such that any three points $p, q, r \in \mathcal{X}$ for which

$$\rho(p, q) < \rho_0, \qquad \rho(p, r) < \rho_0, \qquad \rho(q, r) < \rho_0$$

are contained in one normal convex neighborhood and are therefore vertices of a certain *geodesic triangle* (open set bounded by three geodesic segments $\gamma_{p,q}$, $\gamma_{p,r}$, and $\gamma_{q,r}$).

A simple argument (carry it out!) now shows that each point of the surface $\mathcal{X}$ belongs to a certain geodesic triangle; because of the compactness of the surface $\mathcal{X}$, the surface is therefore covered by a finite set of geodesic triangles (in general, they intersect each other). All possible nonempty intersections of these triangles composes a triangulation of the surface $\mathcal{X}$. □

Remark 16.1. It can be shown (this is the difficult *Rado theorem*) that the assertion on triangulatability holds for any (in general, nonsmooth) compact surface; moreover, if the so-called infinite triangulation is admitted, then any surface (two-dimensional paracompact Hausdorff topological manifold) can be triangulated.

The polygons $G_1, \dots, G_c$ are called *faces*, the arcs $\gamma_1, \dots, \gamma_b$ *edges*, and the points $p_1, \dots, p_a$ *vertices* of the surface $\mathcal{X}$ (or, more precisely, of its decomposition $\{G_1, \dots, G_c\}$). The number

$$\chi(\mathcal{X}) = a - b + c$$

(the number of vertices minus the number of edges plus the number of faces) is called the *Euler characteristic* of the compact surface $\mathcal{X}$.

§11. Gauss–Bonnet Theorem

Proposition 16.4. *For a compact orientable smooth surface $\mathcal{X}$, the Euler characteristic is well defined (does not depend on the choice of the decomposition $\{G_1, \dots, G_c\}$ of the surface $\mathcal{X}$ into polygons) and is a topological invariant (is the same for all diffeomorphic surfaces).*

Proof. Choosing a metric and an orientation on the surface, we apply the Gauss–Bonnet formula to each curvilinear polygon of a certain triangulation $\{G_1, \dots, G_c\}$:

$$\delta G_i = \iint_{G_i} K\,d\sigma + \oint_{\Gamma_i} k_g\,dt.$$

(We assume that the orientation of the polygons is induced by the orientation of the surface.) Adding all these formulas, we obtain

$$\sum_{i=1}^{c} \delta G_i = \sum_{i=1}^{c} \iint_{G_i} K\,d\sigma + \sum_{i=1}^{c} \oint_{\Gamma_i} k_g\,dt.$$

The first sum in the right-hand side equals the total curvature

$$\iint_{\mathcal{X}} K\,d\sigma$$

of the surface $\mathcal{X}$. The second sum (because the two polygons G_i adjacent to the edge γ_k induce opposite orientations on γ_k) equals the sum of integrals of k_g over pairs of oppositely oriented arcs and therefore vanishes.

On the other hand, the sum in the left-hand side is equal to

$$\sum_{i=1}^{c} \left[\sum_{j=1}^{n_i} \beta_{ij} - (n_i - 2)\pi \right], \tag{22}$$

where $\beta_{i1}, \dots, \beta_{in_i}$ are interior angles of the polygon G_i and n_i is the number of its sides. But because the sum of angles at each vertex equals 2π, we have

$$\sum_{i=1}^{c} \sum_{j=1}^{n_i} \beta_{ij} = 2\pi a,$$

and because each edge γ_k is a side of two and only two polygons G_i, we have

$$\sum_{i=1}^{c} n_i = 2b.$$

Therefore,

$$\sum_{i=1}^{c} (n_i - 2)\pi = 2\pi(b - c),$$

and sum (22) equals $2\pi(a - b + c) = 2\pi\chi(\mathcal{X})$.

This proves that

$$\chi(\mathcal{X}) = \frac{1}{2\pi} \iint_{\mathcal{X}} K \, d\sigma. \tag{23}$$

Proposition 16.4 is directly implied by this relation because its right-hand side is independent of the choice of the polygons $G_1, \ldots, G_c$. □

The statement on the fulfillment of (23) is called the *Gauss–Bonnet theorem.*

Remark 16.2. The Euler characteristic is also well defined for nonorientable manifolds, but to prove this requires different methods.

Chapter 17
Some Special Tensors

§1. Characteristic Numbers

The Gauss–Bonnet theorem in the previous chapter directly implies that the *total curvature*

$$\iint_{\mathcal{X}} K\, d\sigma \tag{1}$$

of a surface $\mathcal{X}$ does not depend on the choice of a Riemannian metric on $\mathcal{X}$ (it is the same for all metrics).

Let $\mathcal{X}$ be an arbitrary smooth manifold.

Definition 17.1. The characteristic classes of the tangent bundle of the manifold $\mathcal{X}$ are called the *characteristic classes of the manifold $\mathcal{X}$*.

In the case where the manifold $\mathcal{X}$ is compact and orientable, the group $H^n\mathcal{X}$, $n = \dim \mathcal{X}$, is isomorphic to the group $\mathbb{R}$. The isomorphism depends on the orientation and is given by $c \mapsto c[\mathcal{X}]$ when the orientation is chosen, where

$$c[\mathcal{X}] = \int_{\mathcal{X}} c, \quad c \in H^n\mathcal{X}.$$

(The symbol $\int_{\mathcal{X}} c$ denotes the integral $\int_{\mathcal{X}} \omega$, where ω is an arbitrary closed form of class c; the latter integral is independent of the choice of the form ω.) For each characteristic class c of degree n, the number $c[\mathcal{X}]$ is called the *characteristic c-number* of the oriented manifold $\mathcal{X}$.

The numbers $c[\mathcal{X}]$ depend on the orientation of the manifold $\mathcal{X}$ and change sign if the orientation changes:

$$c[-\mathcal{X}] = -c[\mathcal{X}].$$

§2. Euler Characteristic Number

One exception is the *Euler characteristic number*

$$e[\mathcal{X}] = \int_{\mathcal{X}} e(\tau_{\mathcal{X}}),$$

which does not depend on the choice of orientation of the manifold $\mathcal{X}$ (and is therefore in fact defined for any compact *orientable* manifold $\mathcal{X}$). By definition, to find $e[\mathcal{X}]$, we must choose a certain connection on $\tau_{\mathcal{X}}$. It is natural to take the Levi-Civita connection corresponding to a certain Riemannian metric g on $\mathcal{X}$. Then, for $n = 2l$, we have the following formula for the number $c[\mathcal{X}]$:

$$e[\mathcal{X}] = \left(\frac{1}{2\pi}\right)^l \int_{\mathcal{X}} \operatorname{Pf} R,$$

where, in each coordinate neighborhood U, $\operatorname{Pf} R$ is the differential form that is the Pffafian $\operatorname{Pf} \widetilde{\Omega}$ of the matrix $\widetilde{\Omega}$ of curvature forms of the metric g in a certain positively oriented (but not holonomic in general) basis $X_1, \dots, X_n$ of the module $\mathfrak{a}\mathcal{X}$ over the neighborhood U.

Let C be the transition matrix from the orthonormal basis $X_1, \dots, X_n$ to the holonomic basis

$$\frac{\partial}{\partial x^1}, \quad \dots, \quad \frac{\partial}{\partial x^n} \tag{2}$$

of the same name. According to formula (8) in Chap. 15, we then have

$$C^T \widetilde{\Omega} C = \widehat{\Omega},$$

where $\widehat{\Omega}$ is the matrix with the entries

$$\Omega_{ij} = \sum_{k<l} R_{ij,kl} dx^k \wedge dx^l. \tag{3}$$

Therefore,

$$\operatorname{Pf} \widehat{\Omega} = \det C \cdot \operatorname{Pf} \widetilde{\Omega}.$$

Because $(\det C)^2 = \det g$, where g is the matrix $\|g_{ij}\|$, this proves that *in each chart* $(U, x^1, \dots, x^n)$, *the form* $\operatorname{Pf} R$ *is*

$$\frac{1}{\sqrt{\det g}} \operatorname{Pf} \widehat{\Omega}, \tag{4}$$

where $\widehat{\Omega}$ *is the matrix with entries* (3).

For $n = 2$ (and $l = 1$), the Pffafian $\operatorname{Pf} \widehat{\Omega}$ is the form

$$\widehat{\Omega}_{12} = R_{12,12} du \wedge dv,$$

i.e. (see formula (10″) in Chap. 16), the form

$$K(EG - F^2) du \wedge dv,$$

where $EG - F^2 = \det g$. Therefore, form (4) becomes

$$K\sqrt{EG - F^2} du \wedge dv$$

for $n = 2$. Because $\sqrt{EG - F^2} du \wedge dv$ is exactly the area element of the surface $\mathcal{X}$, *the Euler number* $e[\mathcal{X}]$ *is expressed by*

$$e[\mathcal{X}] = \frac{1}{2\pi} \iint_{\mathcal{X}} K \, d\sigma \tag{5}$$

for $n = 2$.

First, this once more proves that integral (1) is independent of the choice of the metric; second, this proves that *the characteristic Euler number coincides with the Euler characteristic*:

$$e[\mathcal{X}] = \chi(\mathcal{X}). \tag{6}$$

In particular, this explains why the names of the numbers $e[\mathcal{X}]$ and $\chi(\mathcal{X})$ are identical in essence.

§3. Hodge Operator

An analogue of formula (5) holds for any n, but to deduce it, we need to do something. We restrict ourselves to the case $n = 4m$.

We know from linear algebra that for any n-dimensional oriented Euclidean space $\mathcal{V}$ and for any k, $0 \le k \le n$, the Hodge operator $*$ acts on the linear space $\Lambda^k\mathcal{V}$ of skew-symmetric tensors of degree k; this operator transforms each tensor from $\Lambda^k\mathcal{V}$ into a tensor from $\Lambda^{n-k}\mathcal{V}$. In particular, for any n-dimensional oriented Riemannian space $\mathcal{X}$, the operator $*$ acts on the spaces $\Lambda^k\mathbf{T}_p\mathcal{X} = (\Lambda^k\mathcal{X})_p$, $p \in \mathcal{X}$. Therefore, the formula

$$(*\boldsymbol{X})_p = *(\boldsymbol{X}_p), \quad p \in \mathcal{X}, \quad \boldsymbol{X} \in \Lambda^k\mathcal{X},$$

correctly defines the $\mathbf{F}\mathcal{X}$-linear operator

$$*: \Lambda^k\mathcal{X} \to \Lambda^{n-k}\mathcal{X}, \tag{7}$$

which is also called the *Hodge operator*.

In particular, for $n = 2l$ and $k = l$,

$$*: \Lambda^l\mathcal{X} \to \Lambda^l\mathcal{X}. \tag{8}$$

In this case, we have

$$*^2 = (-1)^{(n+1)k} \tag{9}$$

(and, in particular, $*^2 = (-1)^l$ for $k = l$).

The volume element dV of the Riemannian space $\mathcal{X}$, treated as a skew-symmetric tensor of degree n (of type $(0, n)$) on $\mathcal{X}$ is denoted by $\mathfrak{e}$ in this chapter.

Exercise 17.1. Let $X_1, \dots, X_n$ be an orthonormal positively oriented basis of the module $\mathfrak{a}\mathcal{X}$ over a coordinate neighborhood U (i.e., such that at any point $p \in U$, the vectors $(X_1)_p, \dots, (X_n)_p$ form a positively oriented basis of the space $\mathbf{T}_p\mathcal{X}$). Show that

$$\mathfrak{e} = X_1 \wedge \cdots \wedge X_n \quad \text{over } U.$$

The tensor $\mathfrak{e}$ composes the basis of the $\mathbf{F}\mathcal{X}$-module $\Lambda^n\mathcal{X}$ over $\mathcal{X}$, i.e., an arbitrary skew-symmetric tensor of degree n on $\mathcal{X}$ is uniquely represented in the form $f\mathfrak{e}$, where $f \in \mathbf{F}\mathcal{X}$.

In particular, for any tensors $\boldsymbol{X}, \boldsymbol{Y} \in \Lambda^k\mathcal{X}$, $0 \le k \le n$, the tensor $\boldsymbol{X} \wedge *\boldsymbol{Y}$ with the degree n is represented in such a form. The corresponding function f is denoted by $\langle \boldsymbol{X}, \boldsymbol{Y} \rangle$. Therefore, by definition,

$$\boldsymbol{X} \wedge *\boldsymbol{Y} = \langle \boldsymbol{X}, \boldsymbol{Y} \rangle\, \mathfrak{e}. \tag{10}$$

Exercise 17.2. Show that the functional $\boldsymbol{X}, \boldsymbol{Y} \mapsto \langle \boldsymbol{X}, \boldsymbol{Y} \rangle$ is an inner product, i.e., this functional is bilinear, symmetric, and positive definite (in the sense that the number $\langle \boldsymbol{X}, \boldsymbol{X} \rangle$ is positive at every point where $\boldsymbol{X} \neq 0$). Also show that for $k = 2$, this inner product coincides with the product defined by (1) in Chap. 16.

For any subscripts $i_1, \dots, i_k$, $1 \le i_1 < \dots < i_k \le n$, we set

$$\boldsymbol{X}_{i_1 \cdots i_k} = X_{i_1} \wedge \cdots \wedge X_{i_k},$$

where, as above, $X_1, \dots, X_n$ are vector fields composing a positively oriented orthonormal basis of the module $\mathfrak{a}\mathcal{X}$ over a coordinate neighborhood U.

Exercise 17.3. Show that the tensors $\boldsymbol{X}_{i_1 \cdots i_k}$, $1 \le i_1 < \cdots < i_k \le n$, form an orthonormal basis of the module $\Lambda^k \mathcal{X}$ over U.

Exercise 17.4. Show that for $n = 4m$, Hodge operator (8) is self-adjoint (i.e., $\langle *\boldsymbol{X}, \boldsymbol{Y} \rangle = \langle \boldsymbol{X}, *\boldsymbol{Y} \rangle$ for any tensors $\boldsymbol{X}, \boldsymbol{Y} \in \Lambda^{2m} \mathcal{X}$).

§4. Euler Number of a 4m-Dimensional Manifold

Along with these comparatively simple general facts about the Hodge operators, which certainly are of independent interest, we need some sufficiently complicated computations with skew-symmetric matrices of a more special form.

Exercise 17.5. Prove that for the Pffafian $\operatorname{Pf} A$ of a skew-symmetric matrix $A = \|a^i_j\|$ of order $n = 2l$, we have

$$\operatorname{Pf} A = \frac{1}{2^l l!} \sum_\sigma \varepsilon_\sigma a^{\sigma(1)}_{\sigma(2)} \cdots a^{\sigma(n-1)}_{\sigma(n)}, \tag{11}$$

where the sum is taken over all permutations σ of degree n and ε_σ is, as usual, the sign of the permutation σ. [*Hint*: Show that polynomial (11) satisfies the identity $\operatorname{Pf}(C^T AC) = \det C \cdot \operatorname{Pf} A$.]

Exercise 17.6. Show that for any Bianchi tensor R on an n-dimensional ($n = 2l$) Riemannian space $\mathcal{X}$ and for any k, $1 \le k \le l$, there exists a unique $\mathsf{F}\mathcal{X}$-linear operator

$$R_k \colon \Lambda^{2k} \mathcal{X} \to \Lambda^{2k} \mathcal{X}$$

such that for any vector fields $X_1, \dots, X_{2k} \in \mathfrak{a}\mathcal{X}$, we have

$$R_k(X_1 \wedge \cdots \wedge X_{2k}) = \frac{1}{(2k)!} \sum_\rho \varepsilon_\rho R(X_{\rho(1)} \wedge X_{\rho(2)}) \wedge \cdots \wedge R(X_{\rho(2k-1)} \wedge X_{\rho(2k)}), \tag{12}$$

where the sum is taken over all permutations ρ of degree $2k$. [*Hint*: Formula (12) uniquely defines the components of the tensor R_k in each chart.] Show that operator (12) is self-adjoint.

Exercise 17.7. As usual, let $\Omega = \|\Omega^i_j\|$ be the matrix of curvature forms of the metric g in an orthonormal basis $X_1, \dots, X_n$ of the module $\mathfrak{a}\mathcal{X}$ over a coordinate neighborhood U, and let $n = 2l$ as above. Show that

$$\operatorname{Pf} \Omega = \frac{(2l)!}{2^l l!} R_l(X_1 \wedge \cdots \wedge X_n), \tag{13}$$

where R_l is the tensor R_k from Exercise 17.6 for $k = l$ constructed according the Riemannian curvature tensor of the space $\mathcal{X}$. [*Hint*: Because the basis $X_1, \dots, X_n$ is orthonormal, we have $\Omega^i_j(X_p, X_q) = R(X_i, X_j, X_p, X_q)$ for any i, j, p, and q. Therefore, according to (11), for any fields $Y_1, \dots, Y_n$, we have the relation

$$\begin{aligned}\operatorname{Pf}\Omega(Y_1,\dots,Y_n) &= \frac{1}{2^l l!}\sum_\sigma \varepsilon_\sigma(\Omega^{\sigma(1)}_{\sigma(2)}\wedge\dots\wedge\Omega^{\sigma(n-1)}_{\sigma(n)})(Y_1,\dots,Y_n)\\ &= \frac{1}{4^l l!}\sum_{\sigma,\rho}\varepsilon_\sigma\varepsilon_\rho\Omega^{\sigma(1)}_{\sigma(2)}(Y_{\rho(1)},Y_{\rho(2)})\cdots\Omega^{\sigma(n-1)}_{\sigma(n)}(Y_{\rho(n-1)},Y_{\rho(n)})\\ &= \frac{1}{4^l l!}\sum_{\sigma,\rho}\varepsilon_\sigma\varepsilon_\rho R(X_{\sigma(1)},X_{\sigma(2)},Y_{\rho(1)},Y_{\rho(2)})\\ &\qquad\cdots R(X_{\sigma(n-1)},X_{\sigma(n)},Y_{\rho(n-1)},Y_{\rho(n)}).\end{aligned}$$

On the other hand, the relation

$$\begin{aligned}R_l(X_1\wedge\dots\wedge X_n)(Y_1,\dots,Y_n) &= \frac{1}{(2l)!}\Bigg(\sum_\rho \varepsilon_\rho R(X_{\rho(1)}\wedge X_{\rho(2)})\wedge\\ &\qquad\cdots\wedge R(X_{\rho(n-1)}\wedge X_{\rho(n)})\Bigg)(Y_1,\dots,Y_n)\\ &= \frac{1}{2^l(2l)!}\sum_\rho\sum_\sigma \varepsilon_\rho\varepsilon_\sigma R(X_{\rho(1)},X_{\rho(2)},Y_{\sigma(1)},Y_{\sigma(2)})\\ &\qquad\cdots R(X_{\rho(n-1)},X_{\rho(n)},Y_{\sigma(n-1)},Y_{\sigma(n)})\end{aligned}$$

also holds.]

Exercise 17.8. Let $n = 4m$ (and therefore $l = 2m$). Prove that in the notation introduced above,

$$R_l(X_1\wedge\dots\wedge X_n) = \frac{(l!)^2}{(2l)!}\sum\cdots\sum_{1\le i_1<\dots<i_l\le n} R_m(\boldsymbol{X}_{i_1\cdots i_l})\wedge R_m(*\boldsymbol{X}_{i_1\cdots i_l}). \tag{14}$$

[*Hint*: Let ρ be an arbitrary permutation of degree n, and let $i_1 < \dots < i_l$ be the numbers $\rho(1), \dots, \rho(l)$ in increasing order. Similarly, let $j_1 < \dots < j_l$ be the numbers $\rho(l+1), \dots, \rho(n)$ in increasing order. Further, let α and β be permutations of degree l such that

$$i_{\alpha(1)} = \rho(1), \dots, i_{\alpha(l)} = \rho(l), \qquad j_{\beta(1)} = \rho(l+1), \dots, j_{\beta(l)} = \rho(n).$$

The numbers $i_1, \dots, i_l$ and the permutations α and β uniquely define the permutation ρ; moreover, $\varepsilon_\rho = (-1)^w\varepsilon_\alpha\varepsilon_\beta$, where w is the number of inversions in the permutation $(i_1, \dots, i_l, j_1, \dots, j_l)$.]

By formula (10) and because the operators $*$ and R_m are self-adjoint, we have

$$\begin{aligned}R_m(\boldsymbol{X}_{i_1\cdots i_l})\wedge R_m(*\boldsymbol{X}_{i_1\cdots i_l}) &= \langle R_m(\boldsymbol{X}_{i_1\cdots i_l}), *R_m(*\boldsymbol{X}_{i_1\cdots i_l})\rangle\mathfrak{e}\\ &= \langle *R_m * R_m(\boldsymbol{X}_{i_1\cdots i_l}), \boldsymbol{X}_{i_1\cdots i_l}\rangle\mathfrak{e},\end{aligned}$$

and because the basis $\{\boldsymbol{X}_{i_1\cdots i_l},\ 1 \le i_1 < \cdots < i_l \le n\}$ is orthonormal, we have

$$\sum_{1\le i_1<\cdots<i_l\le n} \cdots \sum \langle *R_m * R_m(\boldsymbol{X}_{i_1\cdots i_l}), \boldsymbol{X}_{i_1\cdots i_l}\rangle = \operatorname{Tr}(*R_m * R_m).$$

Therefore, again letting the traditional symbol dV denote the volume element $\mathfrak{e}$, we immediately obtain

$$\operatorname{Pf} R = \frac{l!}{2^l} \operatorname{Tr}(*R_m * R_m) dV$$

from (13) and (14) (on U and therefore on the whole $\mathcal{X}$), i.e.,

$$e[\mathcal{X}] = \frac{l!}{(4\pi)^l} \int_{\mathcal{X}} \operatorname{Tr}(*R_m * R_m) dV. \tag{15}$$

This is an analogue of formula (5) for $n = 4m$.

§5. Euler Characteristic of a Manifold of an Arbitrary Dimension

Does an analogue of formula (6) hold for any n? Here, we first need to define its right-hand side $\chi(\mathcal{X})$ for any n. It can be shown (this is a difficult theorem!) that for any compact triangulated surface $\mathcal{X}$, the Euler characteristic $\chi(\mathcal{X})$ equals the alternated sum $h^0 - h^1 + h^2$ of the numbers $h^i = \dim H^i\mathcal{X}$ (i.e., in the case where the surface $\mathcal{X}$ is connected and compact, it is equal to $2 - h^1$). This suggests *defining* the Euler characteristic $\chi(\mathcal{X})$ for manifolds $\mathcal{X}$ of an arbitrary dimension by

$$\chi(\mathcal{X}) = \sum_{i=1}^{n} (-1)^i h^i, \quad h^i = \dim H^i\mathcal{X}.$$

(If the manifold is compact, then all the linear spaces $H^i\mathcal{X}$ are finite dimensional, and the numbers h^i are therefore well defined.)

It is proved in topology that the numbers h^i and therefore their alternated sum $\chi(\mathcal{X})$ are topological invariants, i.e., they are the same for (even not smoothly!) homeomorphic manifolds.

It turns out (this is again a difficult theorem!) that with this definition of the characteristic $\chi(\mathcal{X})$, *relation* (6) *is preserved for compact and oriented manifolds* $\mathcal{X}$ *of an arbitrary dimension.* In particular, this means that formula (15) (and a similar formula for $n = 4m + 2$) yields a differential-geometric method for computing the Euler characteristic of any compact and oriented even-dimensional manifold. (Because $e(\xi) = 0$ for an odd $\dim \xi$, we see that according to the same formula (6), *the Euler characteristic of any odd-dimensional manifold equals zero.* This is also implied by the *Poincaré duality theorem*, which states that $h^i = h^{n-i}$ for any i, $0 \le i \le n$.)

§6. Signature Theorem

Relation (6) (which was only proved for $n = 2$) is only one of a whole set of remarkable relations that identify one or another topological invariant with characteristic numbers. To give an idea of the theorems existing here, we state the famous *Hirzebruch signature theorem* (unfortunately, without proof).

Let $\mathcal{X}$ be a smooth oriented manifold of dimension $n = 4m$. Then the formula

$$Q(x) = \int_{\mathcal{X}} \omega \wedge \omega, \quad x = [\omega] \in H^{2m}\mathcal{X},$$

correctly defines a quadratic functional Q on the linear space $H^{2m}\mathcal{X}$. The signature of this functional (the difference between the positive and negative inertia index) is denoted by $\operatorname{Sign} \mathcal{X}$ and is called the *signature of the manifold* $\mathcal{X}$.

Exercise 17.9. Show that the functional Q is nondegenerate.

It is proved in topology that similarly to the Euler characteristic, the signature is a topological invariant.

We can construct a nonhomogeneous characteristic class c^F from any formal series

$$F = F_0 + F_1 + \cdots + F_r + \dots,$$

where F_r is a homogeneous symmetric polynomial in n variables $\lambda_1, \dots, \lambda_n$ of degree r. Let $c_n^F[\mathcal{X}]$ be the value of the nth component of this class on the tangent bundle of the manifold $\mathcal{X}$. We define the *characteristic F-number* $\tau^F[\mathcal{X}]$ by

$$\tau^F[\mathcal{X}] = \int_{\mathcal{X}} c_n^F[\mathcal{X}].$$

The particular case where the formal series F is defined by

$$F(\lambda_1, \dots, \lambda_n) = Q(\lambda_1) \cdots Q(\lambda_n),$$

where $Q(\lambda)$ is a certain formal series in the variable λ, is of main interest. In this case, the F-number $\tau^F[\mathcal{X}]$ is called the *Q-genus* of the manifold and is denoted by $Q[\mathcal{X}]$.

Theorem 17.1 (on the signature). *We have*

$$\operatorname{Sign} \mathcal{X} = L[\mathcal{X}],$$

where

$$L(\lambda) = \frac{\sqrt{\lambda}}{\theta\sqrt{\lambda}} = 1 + \sum_{k=1}^{\infty} (-1)^{k-1} \frac{2^{2k}}{(2k)!} B_k \lambda^k.$$

Here, B_k, $k \geq 1$, are the so-called Bernoulli numbers:

$$B_1 = \frac{1}{6}, \quad B_2 = \frac{1}{30}, \quad B_3 = \frac{1}{42}, \quad B_4 = \frac{1}{30}, \quad B_5 = \frac{5}{66},$$
$$B_6 = \frac{691}{2730}, \quad B_7 = \frac{7}{6}, \quad B_8 = \frac{3617}{510}, \quad B_9 = \frac{43867}{798}, \quad \dots .$$

(There exist several concurrent designations of the Bernoulli numbers. We choose the designation in which all Bernoulli numbers are positive and different from 1/2.)

The proof the signature theorem is outside the frame of this work.

Exercise 17.10. Prove that

$$L[\mathcal{X}] = \begin{cases} \frac{1}{3}p_1[\mathcal{X}] & \text{for } n = 4, \\ \frac{1}{45}(7p_2 - p_1^2)[\mathcal{X}] & \text{for } n = 8, \\ \frac{1}{945}(62p_3 - 13p_1p_2 + 2p_1^3)[\mathcal{X}] & \text{for } n = 12, \end{cases}$$

where $p_1, p_2, p_3, \dots$ are the Pontryagin classes of the manifold $\mathcal{X}$.

§7. Ricci Tensor of a Riemannian Space

We now pass to Ricci tensors.

Definition 17.2. The Ricci tensor (see Definition 2.3) of the Riemannian connection of a (pseudo-)Riemannian space $\mathcal{X}$ is called the *Ricci tensor of the space* $\mathcal{X}$ and is denoted by $\operatorname{Ric}\mathcal{X}$ (its components are denoted by R_{ij}).

By definition,

$$R_{ij} = R^k_{i,kj},$$

and therefore

$$R_{ij} = g^{pq}R_{pi,qj}. \tag{16}$$

Exercise 17.11. Show that on each coordinate neighborhood U (or, more generally, on each open set $U \subset \mathcal{X}$ over which the tangent bundle $\tau_{\mathcal{X}}$ is parallelizable), the Ricci tensor is given by

$$\operatorname{Ric}(X,Y) = \sum_{i=1}^{n} R(X_i, Y, X_i, X), \quad X, Y \in \mathfrak{a}\mathcal{X}, \tag{17}$$

where $X_1, \dots, X_n$ is an arbitrary orthonormal basis of the module $\mathfrak{a}\mathcal{X}$ over U.

Proposition 17.1. *The Ricci tensor of each (pseudo-)Riemannian space is symmetric:*

$$R_{ij} = R_{ji}, \quad i, j = 1, \dots, n.$$

We give two proofs of this proposition.

Proof (First). We have

$$R_{ij} = g^{pq}R_{pi,qj} = g^{pq}R_{qj,pi} = R_{ji}$$

by the symmetries of the Riemannian curvature tensor. □

Proof (Second). According to the Foss–Weyl formula (Proposition 13.3), the form γ corresponding to the Riemannian connection is an exact differential. Therefore, it is closed, and (see the corollary from the assertion in Exercise 2.8) the tensor Ric is hence symmetric. □

§8. Ricci Tensor of a Bianchi Tensor

Formula (16) (on a (pseudo-)Riemannian space!) is meaningful for any Bianchi tensor R.

Definition 17.3. The tensor of type (2,0) with component (16) is called the *Ricci tensor of the Bianchi tensor* R and is denoted by $\operatorname{Ric} R$.

Therefore, according to this definition, $\operatorname{Ric} \mathcal{X} = \operatorname{Ric} R_{\mathcal{X}}$, where $R_{\mathcal{X}}$ is the Riemannian curvature tensor of the space $\mathcal{X}$. The first proof of Proposition 17.1 is obviously preserved for each Bianchi tensor R. Therefore, *the Ricci tensor* $\operatorname{Ric} R$ *is symmetric for every tensor* R.

Using the metric, we can raise the first subscript of the tensor $\operatorname{Ric} R$, i.e., pass to the tensor with the components

$$R^i_j = g^{ip}R_{pj}$$

(in the invariant language, the passage from the symmetric bilinear functional $\operatorname{Ric} R$ to the corresponding self-adjoint operator $\mathfrak{a}\mathcal{X} \to \mathfrak{a}\mathcal{X}$) corresponds to this. The trace

$$\operatorname{Tr} \operatorname{Ric} R = R^i_i$$

of this tensor is expressed by the formula

$$g^{ip}R_{pi} = g^{ip}R^k_{p,ki} = g^{jl}R^k_{j,kl}$$

and (see (6) in Chap. 16) is therefore equal to the scalar curvature $\mathcal{R}$,

$$\mathcal{R} = R^i_i.$$

(*The scalar curvature is the trace of the Ricci operator.*)

That the tensor $\operatorname{Ric} R$ is symmetric means that the mapping

$$\operatorname{Ric}: R \mapsto \operatorname{Ric} R \tag{18}$$

is obviously an $\mathsf{F}\mathcal{X}$-linear mapping of the $\mathsf{F}\mathcal{X}$-module $\boldsymbol{B}\mathcal{X}$ into the $\mathsf{F}\mathcal{X}$-module $\boldsymbol{S}^2\mathcal{X}$ of all symmetric $\mathsf{F}\mathcal{X}$-bilinear functionals

$$\mathfrak{a}\mathcal{X} \times \mathfrak{a}\mathcal{X} \to \mathsf{F}\mathcal{X}$$

(or, equivalently, all self-adjoint $\mathsf{F}\mathcal{X}$-linear operators $\mathfrak{a}\mathcal{X} \to \mathfrak{a}\mathcal{X}$).

Exercise 17.12. Show that if $R = KE$, where E is the Bianchi tensor with the components $g_{ij,kl}$ (see (7) and (8) in Chap. 16), then

$$\operatorname{Ric} R = (n-1)Kg = \frac{\mathcal{R}}{n}g,$$

where, as usual, g is the metric tensor.

In particular, we see that for $n = 2$,

$$\operatorname{Ric} R = \frac{\mathcal{R}}{2}g. \tag{19}$$

Therefore, for $n = 2$, mapping (18) is in advance not surjective.

§9. Einstein and Weyl Tensors

For $n \geq 3$, the case is opposite.

Proposition 17.2. *For $n \geq 3$, mapping* (18) *is surjective and moreover admits a section, i.e., there exists its right inverse $\mathsf{F}\mathcal{X}$-linear mapping*

$$Q\colon \boldsymbol{S}^2\mathcal{X} \to \boldsymbol{B}\mathcal{X}. \tag{20}$$

Proof. For an arbitrary tensor $S \in \boldsymbol{S}^2\mathcal{X}$, we consider the tensor P with the components

$$\begin{aligned} P_{ij,kl} &= g_{ik}S_{jl} - g_{il}S_{jk} - g_{jk}S_{il} + g_{jl}S_{ik} \\ &= \begin{vmatrix} g_{ik} & g_{il} \\ S_{jk} & S_{jl} \end{vmatrix} + \begin{vmatrix} S_{ik} & S_{il} \\ g_{jk} & g_{jl} \end{vmatrix}, \end{aligned} \tag{21}$$

where S_{ij} are the components of the tensor S. In an invariant form, the tensor P is defined by the formula

$$\begin{aligned} P(X,Y,Z,W) &= g(X,Z)S(Y,W) - g(X,W)S(Y,Z) \\ &\quad + g(Y,W)S(X,Z) - g(Y,Z)S(X,W) \\ &= \begin{vmatrix} g(X,Z) & g(X,W) \\ S(Y,Z) & S(Y,W) \end{vmatrix} + \begin{vmatrix} S(X,Z) & S(X,W) \\ g(Y,Z) & g(Y,W) \end{vmatrix}, \end{aligned}$$

where X, Y, Z, and W are arbitrary vector fields on $\mathcal{X}$.

Exercise 17.13. Verify that the tensor with components (21) is a Bianchi tensor.

Ricci tensor (21) has the components

$$\begin{aligned} g^{pq}P_{ip,jq} &= g^{pq}g_{ij}S_{pq} - g^{pq}g_{iq}S_{pj} - g^{pq}g_{pj}S_{iq} + g^{pq}g_{pq}S_{ij} \\ &= (\operatorname{Tr} S)g_{ij} - \delta_i^p S_{pj} - \delta_j^q S_{iq} + nS_{ij} = (\operatorname{Tr} S)g_{ij} + (n-2)S_{ij}, \end{aligned}$$

i.e., is expressed by

$$\operatorname{Ric} P = (\operatorname{Tr} S)g + (n-2)S, \tag{22}$$

where $\operatorname{Tr} S = g^{pq}S_{pq}$ is the trace of the tensor S. On the other hand, a direct substitution shows that for $S = g$, tensor (21) is a Bianchi tensor $2E$. Therefore, first,

$$\operatorname{Ric} E = (n-1)g \tag{23}$$

(which incidentally proves the assertion in Exercise 17.12), and, second,

$$\operatorname{Ric}\left(P - \frac{\operatorname{Tr} S}{n-1}E\right) = (n-2)S.$$

Therefore, the formula

$$Q(S) = \frac{P}{n-2} - \frac{\operatorname{Tr} S}{(n-1)(n-2)}E \tag{24}$$

defines an $\mathbf{F}\mathcal{X}$-linear mapping (20) for which $\operatorname{Ric} \circ Q = \mathrm{id}$. □

Corollary 17.1. *For $n \geq 3$, the submodule $\operatorname{Im} Q$ of the module $\boldsymbol{B}\mathcal{X}$ is isomorphic to the module $\boldsymbol{S}^2\mathcal{X}$ and is a direct summand in $\boldsymbol{B}\mathcal{X}$.*

Proof. The complementing direct summand consists of Bianchi tensors R for which $\operatorname{Ric} R = 0$. □

Bianchi tensors R for which $\operatorname{Ric} R = 0$ are called *Weyl tensors* (physicists also call them *Ricci-free tensors*), and tensors of the form $Q(S)$ are called *Einstein tensors.* Therefore, Corollary 17.1 asserts that any Bianchi tensor can be uniquely represented as the sum of an Einstein tensor and a Weyl tensor.

§10. Case $n = 3$

We apply the obtained results to the case $n = 3$.

Corollary 17.2. *For $n = 3$, the submodule* $\operatorname{Im} Q$ *exhausts the whole module $\boldsymbol{B}\mathcal{X}$ (any Bianchi tensor R is a Einstein tensor $Q(\operatorname{Ric} R)$).*

Proof. The dimension of the $\mathbf{F}\mathcal{X}$-module of Weyl tensors is equal to

$$\begin{aligned}\dim \boldsymbol{B}\mathcal{X} - \dim \boldsymbol{S}^2\mathcal{X} &= \frac{n^2(n^2-1)}{12} - \frac{n(n+1)}{2} \\ &= \frac{n(n+1)(n^2-n-6)}{12},\end{aligned}$$

and this number equals zero for $n = 3$. $\square$

In particular, Corollary 17.2 implies that *for $n = 3$, the Riemannian curvature tensor R is expressed through the Ricci tensor and the Gaussian curvature:*

$$R_{ij,kl} = g_{ik}R_{jl} - g_{il}R_{jk} + g_{jl}R_{ik} - g_{jk}R_{il} - 3K(g_{ik}g_{jl} - g_{il}g_{jk}). \qquad (25)$$

Indeed, in the left-hand side, we have the components of tensor (24) for $n = 3$ and $S = \operatorname{Ric} R$.

§11. Einstein Spaces

Definition 17.4. A Riemannian (or pseudo-Riemannian) space $\mathcal{X}$ is called an *Einstein space* if its Ricci tensor $\operatorname{Ric}\mathcal{X}$ is proportional to the metric tensor, i.e., if there exists a function λ such that

$$\operatorname{Ric}\mathcal{X} = \lambda g \qquad (R_{ij} = \lambda g_{ij} \text{ in components}). \qquad (26)$$

It is easy to see, however, that the function λ only differs from the scalar curvature $\mathcal{R}$ by a numerical multiplier. Indeed, passing to the traces in (26) (and taking into account that $\operatorname{Tr} g = g^{ij}g_{ij} = n$), we immediately obtain

$$\lambda = \frac{\mathcal{R}}{n}. \qquad (27)$$

For $n = 2$, condition (26) holds automatically (see (19)); therefore, *any two-dimensional Riemannian space is an Einstein space.* Therefore, condition (26) is interesting only for $n \geq 3$.

Proposition 17.3. *For $n \geq 3$, the scalar curvature of the Einstein space is constant (and the function λ from (26) is therefore also constant).*

Proof. We prove the following lemma below.

Lemma 17.1. *In an arbitrary (pseudo-)Riemannian space $\mathcal{X}$, the relation*

$$\frac{\partial \mathcal{R}}{\partial x^k} = 2\nabla_l R^l_k, \quad k = 1, \dots, n, \tag{28}$$

holds in each chart; here, as above, $R^l_k = g^{lp} R_{pk}$.

On the other hand, if $\mathcal{X}$ is an Einstein space, then

$$R^l_k = \lambda g^{lp} g_{pk} = \lambda \delta^l_k,$$

and therefore (we recall that the tensor δ^l_k is covariantly constant)

$$\nabla_l R^l_k = \frac{\partial \lambda}{\partial x^l} \delta^l_k = \frac{\partial \lambda}{\partial x^k}.$$

Therefore, according to the above lemma,

$$\frac{\partial \mathcal{R}}{\partial x^k} = 2 \frac{\partial \lambda}{\partial x^k},$$

and at the same time (see (27)),

$$\frac{\partial \mathcal{R}}{\partial x^k} = n \frac{\partial \lambda}{\partial x^k}.$$

For $n \geq 3$, this is possible only for $\partial \mathcal{R} / \partial x^k = 0$. □

It is necessary to prove Lemma 17.1.

Proof (of Lemma 17.1). According to the Bianchi–Padov identity (see (23′) in Chap. 2), we have

$$\nabla_i R^i_{j,kl} + \nabla_k R^i_{j,li} + \nabla_l R^i_{j,ik} = 0$$

and therefore

$$\nabla_i R^i_{j,kl} + \nabla_l R_{jk} = \nabla_k R_{jl}. \tag{29}$$

On the other hand, because the metric tensor is covariantly constant, we have

$$\frac{\partial \mathcal{R}}{\partial x^k} = \frac{\partial g^{jl} R_{jl}}{\partial x^k} = \nabla_k (g^{jl} R_{jl}) = g^{jl} \nabla_k R_{jl},$$

$$g^{jl} \nabla_l R_{jk} = \nabla_l (g^{jl} R_{jk}) = \nabla_l R^l_k$$

and

$$\begin{aligned} g^{jl} \nabla_i R^i_{j,kl} &= g^{jl} g^{ip} \nabla_i R_{pj,kl} = g^{jl} g^{ip} \nabla_i R_{jp,lk} \\ &= g^{ip} \nabla_i R^l_{p,lk} = g^{ip} \nabla_i R_{pk} = \nabla_i R^i_k. \end{aligned}$$

Therefore, convoluting (29) with g^{jl}, we obtain (28). □

By the way, Lemma 17.1 serves as a base of arguments that Einstein (simultaneously with Hilbert) used to obtain the equations of the general relativity theory.

Exercise 17.14. Show that formula (28) is equivalent to the relations

$$\nabla_i T^{ij} = 0, \quad j = 1, \dots, n, \tag{30}$$

where

$$T^{ij} = g^{ip} g^{jq} \left(R_{pq} - \frac{\mathcal{R}}{2} g_{pq} \right)$$

is the tensor

$$\mathrm{Ric} - \frac{\mathcal{R}}{2} g$$

with raised subscripts.

On the other hand, in the framework of the four-dimensional formalization of the special relativity theory, the energy and momentum conservation law written in an invariant form is also of form (30), where T^{ij} are components of the so-called energy–momentum tensor T (the left-hand side of (30) is just the components of the divergence of the tensor field T whose vanishing means the absence of sources and sinks for the field).

Because of this, Einstein assumed that these tensors coincide (for the corresponding choice of units of measurement), i.e., in the physical space–time, we have the relations

$$R_{ij} - \frac{\mathcal{R}}{2} g_{ij} = T_{ij}, \quad i, j = 0, 1, 2, 3, \tag{31}$$

where $T_{ij} = g_{ip} g_{jq} T^{pq}$ is the tensor T with lowered superscripts. This assumption is justified by the correspondence of its conclusions with experimental data.

With respect to g_{ij}, relations (31) (for given T^{ij}) are second-order differential equations, and, in principle, they therefore uniquely characterize the metric (for given initial and boundary conditions). We note that these equations are essentially nonlinear.

Proposition 17.3 implies that *in each Einstein space, the Ricci tensor* Ric *is covariantly constant,*

$$\nabla_k R_{ij} = 0,$$

and either is identically zero or is an everywhere nondegenerate symmetric tensor.

Conversely, *if the Ricci tensor is covariantly constant, symmetric, and nondegenerate in an affine connection space $\mathcal{X}$, then the connection on $\mathcal{X}$ is a metric connection and induced by the metric with respect to which $\mathcal{X}$ is an Einstein space.* Indeed, if the tensor Ric is nondegenerate, then it can be taken as a Riemannian metric g on $\mathcal{X}$. Being covariantly constant, this metric induces a given connection on $\mathcal{X}$ (see Theorem 11.1); because $\mathrm{Ric} = g$, the space $\mathcal{X}$ is an Einstein space with respect to it.

§12. Thomas Criterion

Let the space $\mathcal{X}$ be *Riemannian.*

Exercise 17.15. Show that a symmetric tensor $S \in \Lambda^2\mathcal{X}$ in the Riemannian space $\mathcal{X}$ is identically zero iff the function

$$g^{ik}g^{jl}S_{ij}S_{kl} \tag{32}$$

is zero, where S_{ij} are components of this tensor. [*Hint*: Function (32) equals the sum of squares of the components $S_j^k = g^{ik}S_{ij}$.]

For the tensor $R_{ij} - \lambda g_{ij}$, function (32) has the form

$$\begin{aligned} g^{ik}g^{jl}(R_{ij}-\lambda g_{ij})(R_{kl}-\lambda g_{kl}) &= g^{ik}g^{jl}R_{ij}R_{kl}-2\lambda g^{ik}g^{jl}g_{ij}R_{kl}+\lambda^2 g^{ik}g^{jl}g_{ij}g_{kl} \\ &= R_{ij}R^{ij}-2\lambda\delta_i^l R_l^i+\lambda^2 n = R_{ij}R^{ij}-2\lambda\mathcal{R}+\lambda^2 n, \end{aligned}$$

and for $\lambda = \mathcal{R}/n$, it is equal to

$$R_{ij}R^{ij} - \frac{\mathcal{R}^2}{n}.$$

Therefore, *a Riemannian space $\mathcal{X}$ is an Einstein space iff*

$$\mathcal{R}^2 = nR_{ij}R^{ij} \tag{33}$$

(the *Thomas criterion*).

Chapter 18
Surfaces with Conformal Structure

§1. Conformal Transformations of a Metric

The curvature tensor R of a (pseudo-)Riemannian space $\mathcal{X}$, according to Corollary 17.1, admits a decomposition of the form

$$R = Q(\mathrm{Ric}) + W,$$

where W is the Bianchi tensor with the components

$$W_{ij,kl} = R_{ij,kl} - \frac{g_{ik}R_{jl} - g_{il}R_{jk} + g_{jl}R_{ik} - g_{jk}R_{il}}{n-2} + \frac{g_{ik}g_{jl} - g_{jk}g_{il}}{(n-1)(n-2)}\mathcal{R},$$

the *Weyl components of the curvature tensor.* The latter tensor has an interesting geometric sense.

Definition 18.1. A metric $\overline{g}$ on a manifold $\mathcal{X}$ is said to be obtained by a *conformal transformation* of the metric g if $\overline{g}$ differs from g by a positive multiplier. It is convenient to let $e^{2\sigma}$ denote this multiplier, where σ is a certain function on $\mathcal{X}$,

$$\overline{g} = e^{2\sigma} g.$$

Under a conformal transformation, the lengths of curves are changed, but the angles made by curves remain the same.

All objects referred to the metric $\overline{g}$ are denoted by an overline. In particular, $\overline{R}_{ij,kl}$ denotes the components of the Riemannian curvature tensor of this metric (in a certain chart, which is assumed to be fixed here and in what follows).

If $\sigma = \text{const}$, then, as we know (see Chap. 11), the connection ∇ is unchanged under the transformation $g \mapsto \overline{g}$, and the curvature tensor $R^i_{j,kl}$ is therefore also unchanged. We see that the Riemannian curvature tensor obtains the multiplier $e^{2\sigma}$ and the difference $e^{-2\sigma}\overline{R}_{ij,kl} - R_{ij,kl}$ therefore vanishes.

We compute this difference in the general case. This computation is cumbersome if not puzzling. To reduce its volume, for any expression of the form P_{kl} depending on the subscripts k and l (and probably on other subscripts), $P_{kl}[kl]$ denotes the result of its alternation, $P_{[kl]} = P_{kl} - P_{lk}$. For example, the formula for $R^i_{j,kl}$ in Chap. 2 is half as long in this notation,

$$R^i_{j,kl} = \frac{\partial \Gamma^i_{lj}}{\partial x^k} + \Gamma^i_{kp}\Gamma^p_{lj}\ [kl],$$

and the formula for $W_{ij,kl}$ becomes

$$W_{ij,kl} - R_{ij,kl} = -\frac{g_{ik}R_{jl} - g_{jk}R_{il}}{n-2} + \frac{g_{ik}g_{jl} - g_{jk}g_{il}}{2(n-1)(n-2)}\mathcal{R} \quad [kl]. \tag{1}$$

We also write relations like $P_{kl}[kl] = Q_{kl}[kl]$ as

$$P_{kl} \overset{[kl]}{=} Q_{kl}.$$

Moreover, we also use the abbreviated classical notation

$$\sigma_i = \frac{\partial \sigma}{\partial x^i}, \qquad \sigma_{i,l} = \nabla_l \sigma_i.$$

In this notation,

$$\frac{\partial \overline{g}_{ij}}{\partial x^l} = 2e^{2\sigma}\sigma_l g_{ij} + e^{2\sigma}\frac{\partial g_{ij}}{\partial x^l},$$

and therefore

$$\overline{\Gamma}_{i,lj} = \frac{1}{2}\left(\frac{\partial \overline{g}_{ij}}{\partial x^l} + \frac{\partial \overline{g}_{il}}{\partial x^j} - \frac{\partial \overline{g}_{jl}}{\partial x^i}\right) = e^{2\sigma}\Gamma_{i,lj} + e^{2\sigma}(\sigma_l g_{ij} + \sigma_j g_{il} - \sigma_i g_{jl}).$$

Because $\overline{g}^{ij} = e^{-2\sigma}g^{ij}$, this implies $\overline{\Gamma}^i_{lj} = \Gamma^i_{lj} + B^i_{lj}$, where $B^i_{lj} = \delta^i_j\sigma_l + \delta^i_l\sigma_j - g_{jl}g^{ip}\sigma_p$, and therefore

$$\frac{\partial \overline{\Gamma}^i_{lj}}{\partial x^k} + \overline{\Gamma}^i_{kp}\overline{\Gamma}^p_{lj} = \frac{\partial \Gamma^i_{lj}}{\partial x^k} + \frac{\partial B^i_{lj}}{\partial x^k} + \Gamma^i_{kp}\Gamma^p_{lj} + B^i_{kp}\Gamma^p_{lj} + \Gamma^i_{kp}B^p_{lj} + B^i_{kp}B^p_{lj}.$$

But

$$B^i_{kp}\Gamma^p_{lj} \overset{[kl]}{=} -\Gamma^p_{kj}B^i_{lp},$$

$$\frac{\partial B^i_{lj}}{\partial x^k} + \Gamma^i_{kp}B^p_{lj} - \Gamma^p_{kj}B^i_{lp} = \nabla_k B^i_{lj} + \Gamma^p_{kl}B^i_{pj},$$

and $\Gamma^p_{kl}B^i_{pj} \overset{[kl]}{=} 0$. Therefore,

$$\overline{R}^i_{j,kl} - R^i_{j,kl} = \nabla_k B^i_{lj} + B^i_{kp}B^p_{lj} \quad [kl]. \tag{2}$$

On the other hand, because $\nabla_l\delta^i_j = 0$ (the tensor δ^i_j is covariantly constant), we have

$$\nabla_k B^i_{lj} = \delta^i_j\sigma_{l,k} + \delta^i_l\sigma_{j,k} - g_{jl}g^{ip}\sigma_{p,k},$$

and, at the same time,

$$\begin{aligned} B^i_{kp}B^p_{lj} &= (\delta^i_p\sigma_k + \delta^i_k\sigma_p - g_{pk}g^{is}\sigma_s)(\delta^p_j\sigma_l + \delta^p_l\sigma_j - g_{jl}g^{pt}\sigma_t) \\ &= \delta^i_j\sigma_l\sigma_k + \delta^i_k\sigma_l\sigma_j - g_{jk}g^{is}\sigma_l\sigma_s \\ &\quad + \delta^i_l\sigma_j\sigma_k + \delta^i_k\sigma_j\sigma_l - g_{lk}g^{is}\sigma_j\sigma_s \\ &\quad - g_{jl}g^{it}\sigma_t\sigma_k - \delta^i_k g_{jl}g^{pt}\sigma_p\sigma_t + g_{jl}g^{is}\sigma_k\sigma_s. \end{aligned}$$

Because

$$\sigma_{l,k} \overset{[kl]}{=} 0, \qquad \delta^i_l \sigma_j \sigma_k + \delta^i_k \sigma_j \sigma_l \overset{[kl]}{=} 0, \qquad g_{lk} g^{is} \sigma_j \sigma_s \overset{[kl]}{=} 0,$$

and

$$g_{jk} g^{is} \sigma_l \sigma_s + g_{jl} g^{it} \sigma_k \sigma_t \overset{[kl]}{=} 0,$$

this implies

$$\begin{aligned}
\nabla_k B^i_{lj} + B^i_{kp} B^p_{lj} &\overset{[kl]}{=} \delta^i_l(\sigma_{j,k} - \sigma_j \sigma_k) - \delta^i_k g_{jl} g^{st} \sigma_s \sigma_t - g_{jl} g^{ip}(\sigma_{p,k} - \sigma_p \sigma_k) \\
&= \delta^i_l(S_{jk} - \frac{1}{2} g_{jk} g^{st} \sigma_s \sigma_t) - \delta^i_k g_{jl} g^{st} \sigma_s \sigma_t \\
&\quad - g_{jl} g^{ip}(S_{pk} - \frac{1}{2} g_{pk} g^{st} \sigma_s \sigma_t) \\
&= \delta^i_l S_{jk} - g_{jl} g^{ip} S_{pk} - \frac{1}{2} g^{st} \sigma_s \sigma_t (\delta^i_l g_{jk} + \delta^i_k g_{jl}) \\
&\overset{[kl]}{=} \delta^i_l S_{jk} - g_{jl} g^{ip} S_{pk},
\end{aligned}$$

where

$$S_{kl} = \sigma_{k,l} - \sigma_k \sigma_l + \frac{1}{2} g_{kl} g^{st} \sigma_s \sigma_t. \tag{3}$$

Therefore,

$$g_{iq}(\nabla_k B^q_{lj} + B^q_{kp} B^p_{lj}) \overset{[kl]}{=} g_{il} S_{jk} - g_{jl} S_{ik},$$

and hence

$$e^{-2\sigma} \overline{R}_{ij,kl} - R_{ij,kl} = g_{il} S_{jk} - g_{jl} S_{ik} \quad [kl]. \tag{4}$$

In the open form, this formula becomes

$$e^{-2\sigma} \overline{R}_{ij,kl} - R_{ij,kl} = g_{il} S_{jk} - g_{jl} S_{ik} - g_{ik} S_{jl} + g_{jk} S_{il}, \tag{4'}$$

and in the classical bracket notation, it is

$$e^{-2\sigma} \overline{R}_{ij,kl} - R_{ij,kl} = g_{[i|[l} S_{k]|j]} \tag{4''}$$

(here we use the symmetry $S_{kl} = S_{lk}$ of the tensor S_{kl}).

We have thus completely solved our problem.

§2. Conformal Curvature Tensor

For Ricci tensors, the contraction of identity (4′) with the tensor $g^{ik} = e^{2\sigma} \overline{g}^{ik}$ yields the formula

$$\overline{R}_{jl} - R_{jl} = -g_{jl} S - (n-2) S_{jl}, \quad \text{where } S = g^{ik} S_{ik}, \tag{5}$$

which does not contain $e^{-2\sigma}$ in the left-hand side, and a further contraction with g^{jl} yields the formula

$$e^{2\sigma}\overline{\mathcal{R}} - \mathcal{R} = -2(n-1)S \tag{6}$$

for scalar curvatures, in which $e^{2\sigma}$ arises again. Formulas (5) and (6) (for $n > 2$) imply

$$S_{jl} = -\frac{\overline{R}_{jl} - R_{jl}}{n-2} + \frac{e^{2\sigma}\overline{\mathcal{R}} - \mathcal{R}}{2(n-1)(n-2)} g_{jl},$$

and therefore

$$\begin{aligned} g_{jk}S_{il} - g_{ik}S_{jl} = e^{-2\sigma}\frac{\overline{g}_{ik}\overline{R}_{jl} - \overline{g}_{jk}\overline{R}_{il}}{n-2} - \frac{g_{ik}R_{jl} - g_{jk}R_{il}}{n-2} \\ - e^{-2\sigma}\frac{\overline{g}_{ik}\overline{g}_{jl} - \overline{g}_{jk}\overline{g}_{il}}{2(n-1)(n-2)}\overline{\mathcal{R}} + \frac{g_{ik}g_{jl} - g_{jk}g_{il}}{2(n-1)(n-2)}\mathcal{R}. \end{aligned} \tag{7}$$

According to (4), the alternation $[kl]$ of the left-hand side of this relation equals $e^{-2\sigma}\overline{R}_{ij,kl} - R_{ij,kl}$, and according to (1), the alternation of the right-hand side equals $e^{-2\sigma}(-\overline{W}_{ij,kl} + \overline{R}_{ij,kl}) + W_{ij,kl} - R_{ij,kl}$. Therefore, after the alternation, relation (7) becomes

$$e^{-2\sigma}\overline{W}_{ij,kl} = W_{ij,kl}. \tag{8}$$

It is here convenient to pass to the components $W^i_{j,kl}$ of the tensor W with the first subscript raised, which is related to the components $R^i_{j,kl}$ of the curvature tensor by

$$W^i_{j,kl} - R^i_{j,kl} = -\frac{\delta^i_k R_{jl} - g_{jk}R^i_l}{n-2} + \frac{\delta^i_k g_{jl} - \delta^i_l g_{jk}}{2(n-1)(n-2)}\mathcal{R} \quad [kl]. \tag{9}$$

For these components, formula (8) has the form

$$\overline{W}^i_{j,kl} = W^i_{j,kl},$$

which means that *under a conformal transformation of a metric, the tensor W with the components $W^i_{j,kl}$ is unchanged.* This tensor is called the *Weyl conformal curvature tensor*. For an Einstein space, we have

$$W^i_{j,kl} = R^i_{j,kl} - \frac{\mathcal{R}}{n(n-1)}(\delta^i_k g_{jl} - \delta^i_l g_{jk}). \tag{10}$$

§3. Conformal Equivalencies

The assertion on the conformal invariance of the Weyl tensor can be given in another form, which is often more convenient. Let $\mathcal{X}$ and $\mathcal{Y}$ be Riemannian (or pseudo-Riemannian) spaces with the respective metric tensors $g_{\mathcal{X}}$ and $g_{\mathcal{Y}}$. It is clear that for any diffeomorphism $f\colon \mathcal{X} \to \mathcal{Y}$, the tensor $f^*g_{\mathcal{Y}}$ is a (pseudo-)Riemannian metric on $\mathcal{X}$.

Definition 18.2. A diffeomorphism $f: \mathcal{X} \to \mathcal{Y}$ is called a *conformal equivalence* if the metric $f^* g_\mathcal{Y}$ on $\mathcal{X}$ is obtained from the metric $g_\mathcal{X}$ by a conformal transformation, i.e., if there exists an everywhere positive function φ on $\mathcal{X}$ such that

$$f^* g_\mathcal{Y} = \varphi g_\mathcal{X}.$$

In this language, the conformal invariance of the Weyl tensor means that *for any conformal equivalence $f: \mathcal{X} \to \mathcal{Y}$, we have*

$$f^* W_\mathcal{Y} = W_\mathcal{X},$$

where $W_\mathcal{X}$ and $W_\mathcal{Y}$ are the respective Weyl tensors of the spaces $\mathcal{X}$ and $\mathcal{Y}$.

We note that for $n = 2$, conformal equivalencies are exactly conformal mappings with respect to the induced conformal structures on the surfaces $\mathcal{X}$ and $\mathcal{Y}$, i.e. (see Chap. 14), diffeomorphisms that are written by holomorphic or antiholomorphic functions in conformal coordinates.

§4. Conformally Flat Spaces

A Riemannian space $\mathcal{X}$ is said to be *conformally flat* if its metric can be conformally transformed into a flat metric (whose curvature tensor is identically zero). For such a space, the tensor W is identically zero.

The tensor W is also zero in the case where each point of the space $\mathcal{X}$ admits a neighborhood on which the metric can be conformally transformed into a flat metric, i.e., when the space $\mathcal{X}$ is *locally conformally flat.*

Remark 18.1. It turns out that the converse assertion holds for $n \geq 4$: *if $W = 0$, then the space $\mathcal{X}$ is locally conformally flat.* We do not prove this assertion here.

According Corollary 17.2, *for $n = 3$, the tensor W is identically zero.*

Exercise 18.1. Give an example of a three-dimensional Riemannian space that is not locally conformally flat.

For $n = 3$, the role of the tensor W is played by the tensor V with the components

$$V_{ijk} = \frac{\nabla_i R_{jk} - \nabla_j R_{ik}}{n-2} - \frac{\dfrac{\partial \mathcal{R}}{\partial x^i} g_{jk} - \dfrac{\partial \mathcal{R}}{\partial x^j} g_{ik}}{2(n-1)(n-2)}.$$

Exercise 18.2. Show that the following assertions hold:
1. If a three-dimensional Riemanian space is locally conformally flat, then $V = 0$.
2. For $n \geq 4$, the relation $W = 0$ implies $V = 0$.

Remark 18.2. It can be shown that for $n = 3$, the relation $V = 0$ is not only necessary but also sufficient for a Riemannian space to be locally conformally flat. (Compare with Remark 18.1.)

For $n = 2$, the existence of isothermal coordinates (see Proposition 13.2) shows that *for $n = 2$, each metric is locally conformally flat.*

§5. Conformally Equivalent Surfaces

The assertion that any surface is globally conformally flat is not true. Only the following weaker proposition holds.

Proposition 18.1. *Any surface is conformally equivalent to a geodesically complete surface of constant Gaussian curvature.*

By a surface here, we mean an arbitrary two-dimensional Riemannian space $\mathcal{X}$, a two-dimensional (paracompact and Hausdorff) smooth manifold equipped with a Riemannian metric. The Gaussian curvature K of such a space is computed in the usual way.

Corollary 18.1. *On any surface $\mathcal{X}$, there exists a geodesically complete metric of constant Gaussian curvature.*

By a surface here, we mean an arbitrary two-dimensional paracompact Hausdorff smooth manifold having no metric in general.

Of course, we can assume without loss of generality (in Proposition 18.1 as well as in Corollary 18.1) that the surface $\mathcal{X}$ is connected (therefore, it satisfies the second countability axiom).

§6. Classification of Surfaces with a Conformal Structure

The proof of Proposition 18.1 is based on the classification of all possible connected surfaces with a conformal structure (see Definition 14.1). Therefore, we must first consider this classification. Let $\mathcal{X}$ be an arbitrary surface with a conformal structure, and let $\widetilde{\mathcal{X}}$ be a simply connected surface that is a universal covering of $\mathcal{X}$. As we know (see Exercise 14.2), there exists a unique conformal structure on $\widetilde{\mathcal{X}}$ with respect to which the covering mapping $\pi: \widetilde{\mathcal{X}} \to \mathcal{X}$ is conformal. In this case, because the surface $\widetilde{\mathcal{X}}$ being simple connected is oriented, this structure is complex analytic (defines $\mathcal{X}$ as a Riemann surface). On the other hand, according to the *Riemann uniformization theorem* (which is also called the *Koebe theorem*), which is well known in function theory, *any simply connected Riemann surface is complex-analitically equivalent either to the Riemann sphere $\mathbb{C}^+$, to the complex plane $\mathbb{C}$, or to the unit disk $\Delta = \{w \in \mathbb{C} : |w| \leq 1\}$ (or, equivalently, the upper half-plane $\mathbb{P} = \{z \in \mathbb{C} : \operatorname{Im} z > 0\}$).* Because $\mathcal{X} = \widetilde{\mathcal{X}}/\Gamma$, where Γ is the automorphism group of the covering $\widetilde{\mathcal{X}} \to \mathcal{X}$ (consisting of conformal diffeomorphisms $\widetilde{\mathcal{X}} \to \widetilde{\mathcal{X}}$; see assertion 2 in Exercise 14.2), this proves that *any surface $\mathcal{X}$ with a conformal structure has the form $\widetilde{\mathcal{X}}/\Gamma$, where Γ is a certain group of conformal diffeomorphisms $\widetilde{\mathcal{X}} \to \widetilde{\mathcal{X}}$ with a discrete action and $\widetilde{\mathcal{X}}$ is either the sphere $\mathbb{C}^+$, the plane $\mathbb{C}$, or the half-plane $\mathbb{P}$.*

The surface $\mathcal{X}$ is called a surface of *elliptic type* for $\widetilde{\mathcal{X}} = \mathbb{C}^+$, a surface of *parabolic type* for $\widetilde{\mathcal{X}} = \mathbb{C}$, and a surface of *hyperbolic type* for $\widetilde{\mathcal{X}} = \mathbb{P}$.

6.1. Surfaces of Parabolic Type. In the language of function theory, complex-analytic transformations $f: \mathbb{C} \to \mathbb{C}$ are just univalent entire functions. If such a function has an essential singularity at ∞, then by the Weierstrass theorem, it assumes values arbitrary close to $f(0)$ near ∞; therefore, because the set $f(\Delta)$ is open, there exists a point z_0 near ∞ (and therefore outside Δ) such that $f(z_0)\inf(\Delta)$, i.e., $f(z_0) = f(z_1)$, where $z_1 \in \Delta$. Because this contradicts the univalency (the points z_0 and z_1 are distinct in advance), we see that the point ∞ cannot be an essential singularity for the function f and this function, being nonconstant, therefore has a pole at ∞, i.e., is a polynomial (of positive degree). Because (by the main algebra theorem) a polynomial of degree n is an n-sheeted mapping $\mathbb{C} \to \mathbb{C}$, the polynomial f is of degree 1, i.e., $f = az + b$, where $a \neq 0$. Therefore, *complex-analytic diffeomorphisms* $\mathbb{C} \to \mathbb{C}$ *are exactly linear transformations*

$$z' = az + b, \quad a \neq 0.$$

If $a \neq 1$, then such transformation has the fixed point $b/(1-a)$ and therefore cannot belong to the group Γ whose action is discrete. Therefore, *any group* Γ *of complex-analytic transformations* $\mathbb{C} \to \mathbb{C}$ *with a discrete action consists of parallel translations*

$$z' = z + b. \tag{11}$$

Because the transformation $z \mapsto f(\bar{z})$ is complex analytic for a conformal, but not complex analytic, transformation $\mathbb{C} \to \mathbb{C}$, we further obtain that *any group* Γ *of conformal diffeomorphisms* $\mathbb{C} \to \mathbb{C}$ *with a discrete action can contain only transformations of the form*

$$z' = \bar{z} + b \tag{12}$$

(compositions of parallel translations with symmetries with respect to the real axis) *along with parallel translations* (1).

Because transformations (11) and (12) are (proper and improper) motions of the plane $\mathbb{C}$ (with respect to the Euclidean metric on $\mathbb{C} = \mathbb{R}^2$), this proves that *each surface of parabolic type has the form* $\mathbb{R}^2/\Gamma$, *where* Γ *is a certain group of motions of the Euclidean space* $\mathbb{R}^2$ *with a discrete action.*

All these surfaces are easily listed (at least up to a diffeomorphism). For $\Gamma = \{\mathrm{id}\}$, we obtain the plane $\mathbb{R}^2$ itself as the surface $\mathbb{R}^2/\Gamma$. Further, it can be shown that any group of parallel translations (11) with a discrete action on $\mathbb{R}^2$ consists of either translations of the form $z' = z + nb_0$, where b_0 is a fixed nonzero complex number and $n \in \mathbb{Z}$, or translations of the form $z' = z + mb_1 + nb_2$, where b_1 and b_2 are complex numbers that are linearly independent over the field $\mathbb{R}$ and $m, n \in \mathbb{Z}$. In the first case, the quotient space $\mathbb{R}^2/\Gamma$ is diffeomorphic to the cylinder $\mathbb{S}^1 \times (0,1)$ (without boundary circles); in the second case, it is diffeomorphic to the torus $\mathbb{S}^1 \times \mathbb{S}^1$. If the group Γ contains transformations of form (12), then the quotient space $\mathbb{R}^2/\Gamma$ is diffeomorphic to either a Möbius strip or a *Klein bottle*, obtained from the cylinder by gluing together boundary circles with reversed orientation (when the orientation is preserved, we obtain the torus).

Therefore, *there are only five nondiffeomorphic surfaces of parabolic type: the plane, cylinder, torus, Möbius strip, and Klein bottle.*

This assertion is in fact unnecessary for us just now, and we omit its proof. (However, the reader can undoubtedly prove it now). Also, we obtain the classification of parabolic-type surfaces up to not only a diffeomorphism but also a conformal equivalence (and isometry).

6.2. Surfaces of Elliptic Type. An arbitrary linear-fractional transformation

$$\varphi(z) = \frac{az+b}{cz+d}, \quad ad-bc \neq 0, \quad a,b,c,d \in \mathbb{C}, \tag{13}$$

serves as an example of a complex-analytic transformation $\mathbb{C}^+ \to \mathbb{C}^+$. If a complex-analytic transformation $f\colon \mathbb{C}^+ \to \mathbb{C}^+$ transforms a point $z_0 \in \mathbb{C}$ into ∞, then the transformation $g = f \circ \varphi_0$, where

$$\varphi_0(z) = \frac{z_0 z}{z+1},$$

leaves the point ∞ fixed, and its restriction to $\mathbb{C}$ is therefore a diffeomorphism $\mathbb{C} \to \mathbb{C}$, i.e., according to what was proved above, it is a linear transformation of the form $z' = az + b$, $a \neq 0$. Therefore, the transformation $f = g \circ \varphi_0^{-1}$ is linear-fractional. Therefore, *linear-fractional transformations* (13) *exhaust all complex-analytic diffeomorphisms* $\mathbb{C}^+ \to \mathbb{C}^+$. But we know that each nonidentical transformation (13) should have fixed points and therefore cannot belong to a group Γ with a discrete action. This proves that the group of complex-analytic transformations $\mathbb{C}^+ \to \mathbb{C}^+$ with a discrete action should consist of only the identity transformation id; therefore, *the sphere* $\mathbb{C}^+ = \mathbb{S}^2$ *is a unique orientable surface of elliptic type.*

We see that conformal transformations $\mathbb{C}^+ \to \mathbb{C}^+$ that are not complex analytic all have the form

$$z' = \frac{a\bar{z}+b}{c\bar{z}+d}, \quad ad-bc \neq 0, \quad a,b,c,d \in \mathbb{C}. \tag{14}$$

On the other hand, because the composition of any two transformations (14) is obviously of form (13), the group Γ with a discrete action can contain not more than one transformation (14), and this transformation should be involutive (its square is the identity transformation id). An example of such a transformation is the transformation

$$z' = -\frac{1}{\bar{z}} \tag{15}$$

(the composition of the inversion with respect to the unit circle $|z| = 1$ and the central symmetry with respect to the point 0).

Exercise 18.3. Show that the following assertions hold:

1. The surface $\mathbb{C}^+/\Gamma$ corresponding to transformation (15) is just the projective plane $\mathbb{R}P^2$. [*Hint*: The stereographic projection $\mathbb{S}^2 \to \mathbb{C}^+$ transforms the antipodal mapping $\boldsymbol{x} \mapsto -\boldsymbol{x}$, $\boldsymbol{x} \in \mathbb{S}^2$, into transformation (15).]

2. Each involution $\mathbb{C}^+ \to \mathbb{C}^+$ acting on $\mathbb{C}^+$ without fixed points is conjugate to involution (15). [*Hint*: For $ad - bc = 1$, transformation (14) is involutive iff either $b, c \in \mathbb{R}$ and $d = -\overline{a}$ or $ib, ic \in \mathbb{R}$ and $d = \overline{a}$; moreover, the transformation has no fixed points in only the first case.]

This means that *the projective plane* $\mathbb{R}P^2$ *is the unique nonorientable surface of elliptic type.*

We note that *for elliptic-type surfaces, the group* Γ *also consists of motions* (not of the plane here but of the sphere).

6.3. Surfaces of Hyperbolic Type. It turns out that the latter conclusion is preserved for hyperbolic-type surfaces if $\mathbb{P}$ (or Δ) is treated as a model of the Lobachevskii geometry and if the motions $\mathbb{P} \to \mathbb{P}$ are understood in the sense of this geometry. Indeed, according to the so-called boundary correspondence principle, any complex-analytic transformation $f: \mathbb{P} \to \mathbb{P}$ is extended by continuity to a homeomorphism $\overline{\mathbb{P}} \to \overline{\mathbb{P}}$ of the closed half-plane onto itself (transforming the real axis into itself) and according to the *symmetry principle* is therefore extended to a complex-analytic transformation $\mathbb{C}^+ \to \mathbb{C}^+$ (acting by the formulas $z \mapsto f(z)$ if $z \in \overline{\mathbb{P}}$ and $z \mapsto \overline{f(\overline{z})}$ if $z \notin \overline{\mathbb{P}}$), i.e., by what was proved, to linear-fractional transformation (13). On the other hand, transformation (13) is a transformation $\mathbb{P} \to \mathbb{P}$ iff (according to the same boundary correspondence principle) $\operatorname{Im} z' = 0$ for $\operatorname{Im} z = 0$. But if $z = 0$, then $z' = b/d$; if $z = 1$, then $z' = (a+b)/(c+d)$; and if $z = \infty$, then $z' = a/c$. Therefore, if $\operatorname{Im} z' = 0$ for $\operatorname{Im} z = 0$, then $b = dr$, $a = cs$, and $a + b = R(c + d)$, where r, s, and R are real numbers. For the standard normalization $ad - bc = 1$, it is easy to see that this is possible only if all coefficients the a, b, c, and d are real or purely imaginary. But the case where all the coefficients a, b, c, and d are purely imaginary is impossible because we would then have

$$\operatorname{Im} \frac{ai + b}{ci + d} = \operatorname{Im} \frac{(ai + b)\overline{(c + id)}}{|ci + d|^2} = \operatorname{Im} \frac{(ai + b)(c - id)}{|ci + d|^2}$$
$$= \frac{-(ad - bc)}{|ci + d|^2} = \frac{-1}{|ci + d|^2} < 0$$

and the image of the point $i \in \mathbb{P}$ under transformation (13) would therefore not belong to $\mathbb{P}$. Conversely, if these coefficients are real (and $ad - bc = 1$), then

$$\operatorname{Im} z' = \frac{\operatorname{Im}(az + d)(c\overline{z} + d)}{|cz + d|^2} = (ad - bc)\frac{\operatorname{Im} z}{|cz + d|^2} = \frac{\operatorname{Im} z}{|cz + d|^2}, \tag{16}$$

and transformation (13) therefore transforms $\mathbb{P}$ into $\mathbb{P}$.

Therefore, *complex-analytic transformations* $\mathbb{P} \to \mathbb{P}$ *are given by formula* (13) *with real coefficients* (*under the normalization* $ad - bc = 1$). But we know that these are exactly all proper motions of the Lobachevskii plane (and therefore conformal, but not analytic, transformations $\mathbb{P} \to \mathbb{P}$ are improper motions of this plane).

All these prove the following proposition.

Proposition 18.2. *Each surface with a conformal structure has the form* $\mathcal{M}/\Gamma$, *where* $\mathcal{M} = \mathbb{S}^2$, $\mathbb{R}^2$, *or* $\mathbb{P}$, *and* Γ *is the group of motions of the corresponding geometry* (*Euclidean, spherical, or Lobachevskii*) *with a discrete action on* $\mathcal{M}$.

In the next chapter, we deduce Proposition 18.1 from this proposition.

Remark 18.3. The representation of a surface in the form $\mathcal{M}/\Gamma$ allows describing (at least up to a diffeomorphism) all surfaces of elliptic and parabolic types. This is not the case for hyperbolic-type surfaces: there are many such surfaces and their structure is very complicated. However, the number of compact hyperbolic-type surfaces is comparatively small (two countable series), and their topological structure can be described.

Chapter 19
Mappings and Submanifolds I

§1. Locally Isometric Mapping of Riemannian Spaces

Let $\mathcal{X}$ and $\mathcal{Y}$ be Riemannian (or pseudo-Riemannian) spaces with the metric tensors $g_{\mathcal{X}}$ and $g_{\mathcal{Y}}$ and the Riemannian connections $\nabla^{\mathcal{X}}$ and $\nabla^{\mathcal{Y}}$. Moreover, let $n = \dim \mathcal{X}$ and $m = \dim \mathcal{Y}$.

Definition 19.1. A smooth mapping $f: \mathcal{X} \to \mathcal{Y}$ is called a *locally isometric mapping* if $f^* g_{\mathcal{Y}} = g_{\mathcal{X}}$, i.e., for any point $p \in \mathcal{X}$, the mapping

$$(df)_p: \mathbf{T}_p\mathcal{X} \to \mathbf{T}_q\mathcal{Y}, \quad q = f(p),$$

is an isometric mapping of the (pseudo-)Euclidean space $\mathbf{T}_p\mathcal{X}$ onto a subspace of the (pseudo-)Euclidean space $\mathbf{T}_q\mathcal{Y}$.

It is clear that any such mapping is an immersion. Therefore, if there exist locally isometric mappings $\mathcal{X} \to \mathcal{Y}$, then the inequality

$$n \leq m$$

should hold. (Moreover, the signature of the tensors $g_{\mathcal{X}}$ and $g_{\mathcal{Y}}$ should be such that there exist subspaces in the spaces $\mathbf{T}_q\mathcal{Y}$ isometric to the spaces $\mathbf{T}_p\mathcal{X}$; of course, this remark is relevant only for the pseudo-Riemannian spaces $\mathcal{X}$ and $\mathcal{Y}$.)

Because for each locally isometric mapping $f: \mathcal{X} \to \mathcal{Y}$ and for any curve $\gamma: I \to \mathcal{X}$ on $\mathcal{X}$, the length of the curve $f \circ \gamma: I \to \mathcal{Y}$ is obviously equal to the length of the curve γ, we have

$$\rho(f(p), f(q)) \leq \rho(p, q) \tag{1}$$

for any points $p, q \in \mathcal{X}$ (*a locally isometric mapping does not increase the distances*; of course the spaces $\mathcal{X}$ and $\mathcal{Y}$ are here assumed to be Riemannian and connected).

Now let $n = m$ (therefore, each locally isometric immersion is an étale mapping).

Exercise 19.1. Prove that each locally isometric mapping of Riemannian spaces of the same dimension is an affine mapping (with respect to the connections $\nabla^{\mathcal{X}}$ and $\nabla^{\mathcal{Y}}$). [*Hint*: Use the uniqueness of the Riemannian connection.]

Proposition 19.1. *A smooth mapping*

$$f: \mathcal{X} \to \mathcal{Y}$$

of a connected (pseudo-)Riemannian space $\mathcal{X}$ into a (pseudo-)Riemannian space $\mathcal{Y}$ of the same dimension is isometric iff

1. *it is affine and*
2. *there exists a point* $p_0 \in \mathcal{X}$ *such that the mapping*

$$(df)_{p_0}: \mathbf{T}_{p_0}\mathcal{X} \to T_{q_0}\mathcal{Y}, \quad q_0 = f(p_0),$$

is an isometric mapping of the space $\mathbf{T}_{p_0}\mathcal{X}$ *onto the space* $\mathbf{T}_{q_0}\mathcal{Y}$.

Proof. It suffices to note that according to assertion 4 in Proposition 3.1, the diagram

$$\begin{array}{ccc} \mathbf{T}_{p_0}\mathcal{X} & \xrightarrow{\Pi_\gamma} & \mathbf{T}_p\mathcal{X} \\ {\scriptstyle (df)_{p_0}}\downarrow & & \downarrow{\scriptstyle (df)_p} \\ \mathbf{T}_{q_0}\mathcal{Y} & \xrightarrow{\Pi_{f\circ\gamma}} & \mathbf{T}_q\mathcal{Y} \end{array}$$

is commutative for any points $p_0, p \in \mathcal{X}$ and for any curve $\gamma: I \to \mathcal{X}$ connecting the point p_0 with the point p, where $q_0 = f(p_0)$, $q = f(p)$, and Π_γ and $\Pi_{f\circ\gamma}$ are isometries (see Exercise 11.3). □

§2. Metric Coverings

Exercise 19.2. Let $\mathcal{X}$ be a smooth manifold, let $\mathcal{Y}$ be a Riemannian space of the same dimension, and let $f: \mathcal{X} \to \mathcal{Y}$ be an étale mapping. Prove that there exists a unique metric $g_{\mathcal{X}}$ on $\mathcal{X}$ with respect to which the mapping f is locally isometric. [*Hint*: Set $g_{\mathcal{X}} = f^* g_{\mathcal{Y}}$.]

(Compare with Exercise 3.4 and assertion 1 in Exercise 14.2.)

In particular, we see that *for any smooth covering mapping* $\pi: \widetilde{\mathcal{X}} \to \mathcal{X}$ *of a (pseudo-)Riemannian space* $\mathcal{X}$, *we have a unique (pseudo-)Riemannian metric on the manifold* $\widetilde{\mathcal{X}}$ *with respect to which the mapping* π *is locally isometric.* This metric is given by $\widetilde{g} = \pi^* g$, where g is the metric on $\mathcal{X}$.

A smooth covering $(\widetilde{\mathcal{X}}, \pi, \mathcal{X})$ whose projection π is a locally isometric mapping is called a *metric covering*. Of course, it is an affine covering in the sense in Chap. 3. Therefore, *for a metric covering* $(\widetilde{\mathcal{X}}, \pi, \mathcal{X})$, *the space* $\widetilde{\mathcal{X}}$ *is complete iff the space* $\mathcal{X}$ *is complete* (see Proposition 3.4). It is remarkable that in the category of complete Riemannian spaces, metric coverings exhaust all locally isometric mappings.

Theorem 19.1. *Each locally isometric mapping*

$$f: \mathcal{X} \to \mathcal{Y}$$

of complete Riemannian spaces of the same dimension is a covering (and automatically a metric covering).

(We recall (see Chap. 12) that complete Riemannian spaces are connected by definition.)

Proof. Let $p_0 \in \mathcal{X}$ and $q_0 = f(p_0)$. Because the space $\mathcal{Y}$ is complete, we see that according to the Hopf–Rinow theorem, for any point $q \in \mathcal{Y}$, there exists a geodesic $\gamma: t \mapsto \exp_{q_0} tB$, $0 \le t \le 1$, connecting the point q_0 with the point q. Because the mapping f is étale, we can find a vector A in the space $\mathbf{T}_{p_0}\mathcal{X}$ such that $(df)_{p_0}A = B$; because the mapping f is affine, we have $f(\exp_{p_0} A) = \exp_{q_0} B = q$. Therefore, the mapping f is surjective.

To prove Theorem 1, we therefore need only prove that an arbitrary point $q \in \mathcal{Y}$ has a neighborhood V that is exactly covered by the mapping f, i.e., such that its inverse image $f^{-1}V$ is a disjoint union of open sets that f maps diffeomorphically onto V. For any $\epsilon > 0$, we let $V(\epsilon)$ denote the spherical ϵ-neighborhood of the point q and $U_p(\epsilon)$ denote the spherical ϵ-neighborhood of an arbitrary point $p \in f^{-1}(q)$. Let $\delta > 0$ be a number such that the 2δ-neighborhoods $U_p(2\delta)$ and $V(2\delta)$ of the points p and q are normal. We show that as V, we can take the δ-neighborhood $V(\delta)$.

Because in the commutative diagram

$$\begin{array}{ccc} \mathbb{B}_{2\delta}(\mathbf{T}_p\mathcal{X}) & \xrightarrow{(df)_p} & \mathbb{B}_{2\delta}(\mathbf{T}_q\mathcal{Y}) \\ {\scriptstyle \exp_p}\downarrow & & \downarrow{\scriptstyle \exp_q} \\ U_p(2\delta) & \xrightarrow{f} & V(2\delta) \end{array},$$

all the mappings except for the mapping f are diffeomorphisms, the mapping f is also a diffeomorphism. Moreover, if the intersection $U_{p_1}(\delta) \cap U_{p_2}(\delta)$ is not empty for the points $p_1, p_2 \in f^{-1}(q)$, then by the triangle inequality, $p_2 \in U_{p_1}(2\delta)$ and therefore $p_1 = p_2$. Therefore, $U_{p_1}(\delta) \cap U_{p_2}(\delta) \neq \emptyset$ only if $p_1 = p_2$. Finally, let $a \in f^{-1}V$, where $V = V(\delta)$, and let $b = f(a)$. Because $b \in V$, we have $\rho(b,q) = \rho(q,b) < \delta$. Therefore, there exists a vector $B \in \mathbf{T}_b\mathcal{Y}$, $|B| < \delta$, such that $q = \exp_b B$. Let $(df)_a A = B$, $A \in \mathbf{T}_a\mathcal{X}$, and $p = \exp_a A$. Because $f(\exp_a A) = \exp_b B = q$, we have $p \in f^{-1}(q)$ and $\rho(p,a) = \rho(a,p) = \rho(b,q) < \delta$, i.e., $a \in U_p(\delta)$. Therefore,

$$f^{-1}V = \coprod_{p \in f^{-1}(q)} U_p(\delta). \quad \square$$

Exercise 19.3. Show that if Riemannian spaces $\mathcal{X}$ and $\mathcal{Y}$ of the same dimension are connected, if the space $\mathcal{X}$ is complete, and if there exists a locally isometric mapping $f: \mathcal{X} \to \mathcal{Y}$, then the space $\mathcal{Y}$ is also complete.

§3. Theorem on Expanding Mappings

Definition 19.2. A mapping $f: \mathcal{X} \to \mathcal{Y}$ of Riemannian spaces of the same dimension n is said to be *expanding* if

$$|(df)_p A| \geq |A|$$

for any point $p \in \mathcal{X}$ and any vector $A \in \mathbf{T}_p\mathcal{X}$.

Corollary 19.1. *Any expanding mapping $f: \mathcal{X} \to \mathcal{Y}$ of a complete Riemannian space $\mathcal{X}$ into a connected Riemannian space $\mathcal{Y}$ of the same dimension is a covering (which is not metric in general!). If such mapping exists, then the space $\mathcal{Y}$ is complete.*

Proof. Obviously, an expanding mapping is étale. Therefore, the Riemannian metric $f^*g_{\mathcal{Y}}$ is defined on $\mathcal{X}$, and the mapping f is locally isometric with respect to the metrics $f^*g_{\mathcal{Y}}$ and $g_{\mathcal{Y}}$. On the other hand, it is clear that the length of any curve γ in the space $\mathcal{X}$ with respect to the metric $g_{\mathcal{X}}$ is not greater than the length of the curve $f \circ \gamma$ in the space $\mathcal{Y}$ with respect to the metric $g_{\mathcal{Y}}$, i.e., the length of the initial curve γ with respect to the metric $f^*g_{\mathcal{Y}}$. Therefore, for any points $p, q \in \mathcal{X}$, we have

$$\rho'(p,q) \geq \rho(p,q),$$

where ρ' and ρ are Riemannian distances in $\mathcal{X}$ corresponding to the respective metrics $f^*g_{\mathcal{Y}}$ and $g_{\mathcal{X}}$. Therefore, each Cauchy sequence with respect to the metric ρ' is also a Cauchy sequence with respect to the metric ρ and, by the completeness of the space $\mathcal{X}$ with respect to the metric ρ, is therefore a convergent sequence. Therefore, the space $\mathcal{X}$ is also complete with respect to the metric $f^*g_{\mathcal{Y}}$. Therefore, the space $\mathcal{Y}$ is also complete (see Exercise 19.3). Therefore, Theorem 19.1 is applicable to the spaces $\mathcal{X}$ and $\mathcal{Y}$ and to the mapping f. □

§4. Isometric Mappings of Riemannian Spaces

Definition 19.3. If a locally isometric mapping $f: \mathcal{X} \to \mathcal{Y}$ of (pseudo-)Riemannian spaces is bijective (and therefore a diffeomorphism), it is called an *isometric mapping* or merely *isometry.*

An example of an isometry is each parameterization $\boldsymbol{r} = \boldsymbol{r}(u,v)$, $(u,v) \in U$, of an arbitrary elementary surface $\mathcal{X}$. (Here, U plays the role of $\mathcal{X}$, $\mathcal{X}$ plays the role of $\mathcal{Y}$, the metric tensor g of the surface $\mathcal{X}$ plays the role of $g_{\mathcal{Y}}$, and the same tensor considered as a tensor on U plays the role of $g_{\mathcal{X}}$.)

Two Riemannian (or pseudo-Riemannian) spaces $\mathcal{X}$ and $\mathcal{Y}$ are said to be *isometric* if there exists at least one isometry $\mathcal{X} \to \mathcal{Y}$. (We note that another terminology is accepted with respect to locally isometric mappings. Namely, two spaces $\mathcal{X}$ and $\mathcal{Y}$ are said to be *locally isometric* if any point of the space $\mathcal{X}$ has a neighborhood that is isometric to a neighborhood of a certain point of the space $\mathcal{Y}$.)

For isometries, inequality (1) certainly transforms into an equality, i.e., *each isometry $f: \mathcal{X} \to \mathcal{Y}$ of Riemannian spaces $\mathcal{X}$ and $\mathcal{Y}$ preserves the Riemannian distance* (is their isometry as metric spaces).

Exercise 19.4. Show the converse, that each diffeomorphism $f: \mathcal{X} \to \mathcal{Y}$ of Riemannian spaces that preserves the Riemannian distance is an isometry.

§5. Isometry Group of a Riemannian Space

Isometries of a (pseudo-)Riemannian space $\mathcal{X}$ onto itself obviously form a group. We represent this group by $\operatorname{Iso}\mathcal{X}$.

Exercise 19.5. Show that the following assertions hold:

1. For any metric covering $(\widetilde{\mathcal{X}}, \pi, \mathcal{X})$, the group $\operatorname{Aut}\widetilde{\mathcal{X}}$ of its automorphisms (glidings) consists of isometries (is a subgroup of the group $\operatorname{Iso}\widetilde{\mathcal{X}}$).

2. If the space $\widetilde{\mathcal{X}}$ in a smooth covering $(\widetilde{\mathcal{X}}, \pi, \mathcal{X})$ is (pseudo-)Riemannian and if

a. this covering is regular and

b. its automorphism group consists of isometries of the space $\widetilde{\mathcal{X}}$,

then there exists a unique (pseudo-)Riemannian structure on $\mathcal{X}$ with respect to which the covering $(\widetilde{\mathcal{X}}, \pi, \mathcal{X})$ is a metric covering.

In particular, *for any isometry group* Γ *with a discrete action on the (pseudo-)Riemannian space* $\mathcal{X}$, *the quotient manifold* $\mathcal{X}/\Gamma$ *admits a unique (pseudo-)Riemannian structure with respect to which the projection* $\pi: \mathcal{X} \to \mathcal{X}/\Gamma$ *is a locally isometric mapping (metric covering).*

(Compare with Exercises 3.5, 3.6, and 14.2.)

§6. Elliptic Geometry

Example 19.1. Let $\mathbb{S}^n_R$ be a sphere of radius $R > 0$ centered at zero in the $(n+1)$-dimensional space $\mathbb{R}^{n+1}$, and let Γ be a group of the second order generated by the antipodal mapping $\boldsymbol{x} \mapsto -\boldsymbol{x}$. Because the latter mapping is an isometry of the sphere $\mathbb{S}^n_R$ (considered as a Riemannian space with the metric induced by the standard Euclidean metric of the space $\mathbb{R}^{n+1}$), then according to assertion 2 in Exercise 19.5, the n-dimensional projective space

$$\mathbb{R}P^n_R = \mathbb{S}^n_R/\Gamma$$

admits a unique Riemannian metric with respect to which the natural projection $\mathbb{S}^n_R \to \mathbb{R}P^n_R$ is a locally isometric mapping.

The projective space $\mathbb{R}P^n$ equipped with this metric is called the *elliptic space.*

Similar to the space $\mathbb{R}^n$ being the support of the Euclidean geometry (also called the *parabolic geometry*), the ball $\mathbb{B}^n_R$ of radius $R > 0$ being the support of the Lobachevskii geometry (also called the *hyperbolic geometry*), and the sphere $\mathbb{S}^n_R$ being the support of the spherical geometry, the space $\mathbb{R}P^n_R$ is the support of the so-called elliptic geometry (or *Riemann geometry*; do not confuse it with the Riemannian geometry!). In the elliptic geometry, as in the Euclidean and Lobachevskii geometries, any two distinct geodesics (which are also called *lines* here) intersect one another in not more than one point. But in the elliptic geometry in contrast to the Euclidean and Lobachevskii geometries, any two lines necessarily intersect (there are no parallel lines), and each line is a closed line of the finite length πR. (In contrast, in the spherical geometry, geodesics, which are large circles of the sphere, intersect in two points.)

Similar to the Euclidean and Lobachevskii geometries, the elliptic geometry can be synthetically constructed from axioms, but this is far outside the frame of our presentation.

§7. Proof of Proposition 18.1

We are now able to prove Proposition 18.1.

Proof (of Proposition 18.1). Let $\mathcal{X}$ be an arbitrary surface with the Riemannian metric g. According to Proposition 18.2, the surface $\mathcal{X}$ with the conformal structure induced by the metric g is conformally equivalent to a quotient space of the form $\mathcal{M}/\Gamma$, where $\mathcal{M} = \mathbb{S}^2$, $\mathbb{R}^2$, or $\mathbb{P}$, and Γ is the group of motions with a discrete action of the corresponding geometry, i.e., isometries $\mathcal{M} \to \mathcal{M}$. Therefore, the metric of the space $\mathcal{M}$ (which has, as we know, a constant curvature and is complete) induces a certain metric g_0 on the surface $\mathcal{X}$ (of course, of constant curvature and complete). The metric g_0 assigns the same conformal structure as the metric g and is therefore conformally equivalent to it. □

Also, we mention the following proposition, which is a direct consequence of assertions 1 and 2 in Exercise 19.5.

Proposition 19.2. *A connected (pseudo-)Riemannian space is isometric to a space of form $\mathcal{X}/\Gamma$, where $\mathcal{X}$ is a simply connected (pseudo-)Riemannian space and Γ is a certain group of its isometries with a discrete action.*

§8. Dimension of the Isometry Group

According to the assertion in Exercise 19.1, the isometry group $\operatorname{Iso}\mathcal{X}$ is a subgroup of the group $\operatorname{Aff}\mathcal{X}$ of all affine mappings $\mathcal{X} \to \mathcal{X}$ of the space $\mathcal{X}$ onto itself, which, as we know (see Proposition 10.1), is a Lie group.

Exercise 19.6. Prove that the group $\operatorname{Iso}\mathcal{X}$ is closed in the Lie group $\operatorname{Aff}\mathcal{X}$.

According to the Cartan theorem, this proves the first assertion in the following proposition.

Proposition 19.3. *For any connected (pseudo-)Riemannian space $\mathcal{X}$, the group* $\operatorname{Iso}\mathcal{X}$ *of its isometries is a Lie group. Its dimension does not exceed $n(n+1)/2$, where $n = \dim \mathcal{X}$ as usual.*

§9. Killing Fields

This assertion can be proved differently (simultaneously obtaining the proof of the second assertion).

Definition 19.4. A vector field X on a (pseudo-)Riemannian space $\mathcal{X}$ is called a *Killing field* (or *infinitesimal isometry*) if the flow $\{\varphi_t^X\}$ it generates consists of isometries.

Of course, *each Killing field is an affine field* (see Definition 8.1), i.e., the set $\mathfrak{iso}\,\mathcal{X}$ of all Killing fields is contained in the Lie algebra $\mathfrak{aff}\,\mathcal{X}$ of affine fields.

Exercise 19.7. Prove that a field $X \in \mathfrak{a}\mathcal{X}$ is a Killing field iff

$$\mathcal{L}_X g = 0$$

where, as usual, g is a metric tensor of the space $\mathcal{X}$ and $\mathcal{L}_X$ is the Lie derivative.

By the identity $[\mathcal{L}_X, \mathcal{L}_Y] = \mathcal{L}_{[X,Y]}$, this implies that *the set $\mathfrak{iso}\,\mathcal{X}$ is a subalgebra of the Lie algebra $\mathfrak{aff}\,\mathcal{X}$.*

In coordinates, the relation $\mathcal{L}_X g = 0$ becomes

$$\frac{\partial g_{ij}}{\partial x^k} X^k + g_{ik}\frac{\partial X^k}{\partial x^j} + g_{kj}\frac{\partial X^k}{\partial x^i} = 0.$$

If the coordinates $x^1, \ldots, x^n$ are normal at the point p and correspond to an orthonormal basis of the space $\mathbf{T}_p\mathcal{X}$ (for simplicity, we assume that the space $\mathcal{X}$ is Riemannian), then $(g_{ij})_p = \delta_{ij}$ (because the basis is orthonormal) and $(\partial g_{ij}/\partial x^k)_p = 0$ (therefore $(\Gamma^i_{kj})_p = 0$; see Proposition 2.1). Therefore, in the coordinates $xd^1, \ldots, x^n$, the equation $\mathcal{L}_X g = 0$ at the point p has the form

$$\left(\frac{\partial X^j}{\partial x^i}\right)_p + \left(\frac{\partial X^i}{\partial x^j}\right)_p = 0.$$

This means (see Exercise 8.2) that under the injective mapping (on $\mathfrak{aff}\,\mathcal{X}$ and therefore on $\mathfrak{iso}\,\mathcal{X}$)

$$l_p\colon \mathfrak{a}\mathcal{X} \to \operatorname{Hom}[\mathbf{T}_p\mathcal{X}, (\mathbf{F}\mathcal{X})^*]$$

in Chap. 8, each Killing field X transforms into the mapping

$$\left(\frac{\partial}{\partial x^i}\right)_p \mapsto a_i^j\left(\frac{\partial}{\partial x^j}\right)_p + a^j\left(\frac{\partial^2}{\partial x^i \partial x^j}\right)_p \tag{2}$$

with the skew-symmetric matrix $\|a_i^j\|$. Because all mappings of form (2) compose a subspace of the linear space $\mathrm{Hom}[\mathbf{T}_p\mathcal{X}, (\mathbf{F}\mathcal{X})^*]$ of dimension

$$\frac{n(n-1)}{2} + n = \frac{n(n+1)}{2},$$

this proves that $\dim \mathfrak{iso}\,\mathcal{X} \leq n(n+1)/2$. (Verify that this conclusion also holds for a pseudo-Riemannian space $\mathcal{X}$.)

Proof (of Proposition 3). The group $\mathcal{G} = \mathrm{Iso}\,\mathcal{X}$ satisfies the conditions of Theorem 10.2 (with $\mathfrak{g} = \mathfrak{iso}\,\mathcal{X}$). □

We note that *on a complete (pseudo-)Riemannian space* $\mathcal{X}$*, each Killing field* X *is complete* (see Proposition 8.3). Therefore, *for such a space* $\mathcal{X}$*, the Lie algebra of the Lie group* $\mathrm{Iso}\,\mathcal{X}$ *is the Lie algebra* $\mathfrak{iso}\,\mathcal{X}$. (For spaces for which $\dim \mathfrak{iso}\,\mathcal{X} = n(n+1)/2$, see the end of Chap. 23.)

§10. Riemannian Connection on a Submanifold of a Riemannian Space

We now study the case where the dimension n of the manifold $\mathcal{X}$ is *less* than the dimension m of the manifold $\mathcal{Y}$. We restrict ourselves here to the most important case where a locally isometric immersion $f: \mathcal{X} \to \mathcal{Y}$ is injective, in fact, to the case where $\mathcal{X}$ is a submanifold of the manifold $\mathcal{Y}$ and f is an embedding mapping (and the metric $g_\mathcal{X}$ on $\mathcal{X}$ is therefore just the restriction $g|_\mathcal{X}$ to $\mathcal{X}$ of the metric $g = g_\mathcal{Y}$ on $\mathcal{Y}$; see Chap. 11), although up to trivial variations of the terminology, all our results hold for any immersed submanifolds even with self-intersections (compare with Chap. 3). Our main goal is to compare the connections $\nabla^\mathcal{X}$ and $\nabla^\mathcal{Y}$ (in the spirit of Chap. 3) in this case.

As noted in Chap. 11, the metric $g_\mathcal{X}$ is defined for an arbitrary submanifold $\mathcal{X} \subset \mathcal{Y}$ only in the case where the metric $g_\mathcal{Y}$ is Riemannian (positive definite). In the general case, the tensor $g|_\mathcal{X}$ is a metric (is nondegenerate) if at each point $p \in \mathcal{X}$, the subspace $\mathbf{T}_p\mathcal{X} \subset \mathbf{T}_p\mathcal{Y}$ does not touch the *isotropic cone* consisting of the vectors $B \in \mathbf{T}_p\mathcal{Y}$ for which $|B| = 0$ (see Exercise 19.8 below). Such submanifolds $\mathcal{X}$ are called *completely nonisotropic submanifolds.*

Exercise 19.8. Prove that the metric of the pseudo-Euclidean space $\mathcal{V}$ induces a nondegenerate metric on a subspace $\mathcal{P} \subset \mathcal{V}$ iff this subspace is not tangent to the isotropic cone of the space $\mathcal{V}$. [*Hint*: The vector $\boldsymbol{x} \in \mathcal{V}$ is tangent to the isotropic cone at a point c with the radius vector $\boldsymbol{a}$ iff $\boldsymbol{a}\,\boldsymbol{x} = 0$; here, $\boldsymbol{a}\,\boldsymbol{x}$ denotes the inner product of two vectors $\boldsymbol{a}$ and $\boldsymbol{x}$.]

For any point $p \in \mathcal{X}$, we let $\mathbf{N}_p\mathcal{X}$ denote the orthogonal complement in the space $\mathbf{T}_p\mathcal{Y}$ of the subspace $\mathbf{T}_p\mathcal{X}$.

Exercise 19.9. Show that the subspaces $\mathbf{N}_p\mathcal{X}$ are fibers of a certain vector bundle ν (which is a subbundle of the bundle $\tau_{\mathcal{Y}}|_{\mathcal{X}}$).

In the case where the submanifold $\mathcal{X}$ is completely nonisotropic, at each point $p \in \mathcal{X}$, we have

$$\mathbf{T}_p\mathcal{Y} = \mathbf{T}_p\mathcal{X} \oplus \mathbf{N}_p\mathcal{X}.$$

According to Definition 3.2, this means that ν is the normal bundle over the submanifold $\mathcal{X}$. It is called the *Riemannian* (or *metric*) *normal bundle.* When speaking of the normal bundle of a completely nonisotropic submanifold of a pseudo-Riemannian manifold, we always means exactly this bundle.

Therefore, in a pseudo-Riemannian space $\mathcal{Y}$, each completely nonisotropic submanifold $\mathcal{X}$ is naturally normalized, and any connection on $\mathcal{Y}$ therefore uniquely induces a certain connection on $\mathcal{X}$. In particular, the Riemannian connection $\nabla = \nabla^{\mathcal{Y}}$ on $\mathcal{Y}$ induces a certain connection $\nabla^{\mathcal{X}}$ on $\mathcal{X}$.

Proposition 19.4. *The connection $\nabla^{\mathcal{X}}$ induced on $\mathcal{X}$ by the Riemannian connection $\nabla = \nabla^{\mathcal{Y}}$ on $\mathcal{Y}$ is the Riemannian connection corresponding to the metric $g_{\mathcal{X}}$.*

Proof. Because the connection ∇ is symmetric, the connection $\nabla^{\mathcal{X}}$ is also symmetric (see Chap. 3). Therefore, we need only prove that the connection $\nabla^{\mathcal{X}}$ is compatible with the metric $g_{\mathcal{X}}$.

Let X, Y, and Z be arbitrary vector fields on $\mathcal{X}$. On $\mathcal{X}$, we consider the field

$$h(X,Y) = (\nabla|_{\mathcal{X}})_X Y - \nabla^{\mathcal{X}}_X Y \tag{3}$$

(see formula (14) in Chap. 3). As we know, this field is normal (belongs to $\Gamma\nu$), and therefore $g(h(X,Y),Z) = 0$. Therefore,

$$g_{\mathcal{X}}(\nabla^{\mathcal{X}}_X Y, Z) = g((\nabla|_{\mathcal{X}})_X Y, Z) = g(\nabla_X Y, Z)|_{\mathcal{X}},$$

where X, Y, and Z in the right-hand side are extensions of the fields X, Y, and Z to $\mathcal{Y}$ (see formula (9) in Chap. 3). By similar arguments,

$$g_{\mathcal{X}}(Y, \nabla^{\mathcal{X}}_X Z) = g(Y, \nabla_X Z)|_{\mathcal{X}}.$$

On the other hand, it is clear that

$$X g_{\mathcal{X}}(Y,Z) = X g(Y,Z)|_{\mathcal{X}},$$

where X, Y, and Z in the right-hand side are extensions of the fields X, Y, and Z to $\mathcal{Y}$. Therefore,

$$\begin{aligned} X g_{\mathcal{X}}(Y,Z) &= X g(Y,Z)|_{\mathcal{X}} = g(\nabla_X Y, Z)|_{\mathcal{X}} + g(Y, \nabla_X Z)|_{\mathcal{X}} \\ &= g_{\mathcal{X}}(\nabla^{\mathcal{X}}_X Y, Z) + g_{\mathcal{X}}(Y, \nabla^{\mathcal{X}}_X Z). \end{aligned}$$

Therefore, the connection $\nabla^{\mathcal{X}}$ is compatible with the metric $g_{\mathcal{X}}$. $\square$

Exercise 19.10. Show that for the connection D on the bundle ν (see Chap. 3), the formula

$$Xg(s,t) = g(D_X s, t) + g(s, D_X t), \quad s,t \in \Gamma\nu, \tag{4}$$

holds. By definition, this means that *the connection D is compatible with the metric g on ν*.

We write merely g instead of $g_{\mathcal{X}}$ in what follows.

§11. Gauss and Weingarten Formulas for Submanifolds of Riemannian Spaces

Along with the mapping

$$h\colon \mathfrak{a}\mathcal{X} \otimes \mathfrak{a}\mathcal{X} \to \Gamma\nu$$

given by (3) (we recall that it is called the *second fundamental form* of a submanifold $\mathcal{X}$), for the normalized manifold $\mathcal{X}$, the linear operators

$$A_s\colon \mathfrak{a}\mathcal{X} \to \mathfrak{a}\mathcal{X}, \quad s \in \Gamma\nu,$$

satisfying the Weingarten formula

$$(\nabla|_{\mathcal{X}})_X s = -A_s X + D_X s, \quad s \in \Gamma\nu,$$

for any field $X \in \mathfrak{a}\mathcal{X}$ (see (16) in Chap. 3) are also defined.

It turns out that in the case considered, we have

$$g(A_s X, Y) = g(h(X,Y), s), \quad X, Y \in \mathfrak{a}\mathcal{X}, \quad s \in \Gamma\nu, \tag{5}$$

(where $g = g_{\mathcal{X}}$ in the left-hand side and $g = g_{\mathcal{Y}}$ in the right-hand side); therefore, *the form h and the operators A_s define each other* (so that only h plays an independent role). Indeed, because $g(D_X s, Y) = 0$, according to the Weingarten formula, we have

$$g(A_s X, Y) = -g((\nabla|_{\mathcal{X}})_X s, Y) = -g(\nabla_X s, Y)|_{\mathcal{X}}$$

and therefore

$$g(A_s X, Y) = g(s, \nabla_X Y)|_{\mathcal{X}}$$

(because $g(s,Y) = 0$, we have $g(\nabla_X s, Y) + g(s, \nabla_X Y) = 0$). On the other hand, because $g(\nabla^{\mathcal{X}}_X Y, s) = 0$, according to (3), we have

$$g(h(X,Y), s) = g((\nabla|_{\mathcal{X}})_X Y, s) = g(\nabla_X Y, s)|_{\mathcal{X}}.$$

Therefore, $g(A_s X, Y) = g(h(X,Y), s)$.

Because the form h is symmetric (see Chap. 3), formula (5) implies

$$g(A_s X, Y) = g(X, A_s Y)$$

for any fields $X, Y \in \mathfrak{a}\mathcal{X}$. By definition, this means that *the operators A_s are self-adjoint.*

We collect all the facts proved here (and in Chap. 3) into one theorem.

Theorem 19.2. *For any submanifold $\mathcal{X}$ of a Riemannian space $\mathcal{Y}$ (for any completely nonisotropic submanifold $\mathcal{X}$ of a pseudo-Riemannian space $\mathcal{Y}$), the following Gauss and Weingarten formulas hold:*

$$\nabla_X^{\mathcal{Y}} Y = \nabla_X^{\mathcal{X}} Y + h(X, Y), \quad X, Y \in \mathfrak{a}\mathcal{X}, \tag{6}$$

$$\nabla_X^{\mathcal{Y}} s = -A_s X + D_X s, \quad X \in \mathfrak{a}\mathcal{X}, \quad s \in \Gamma\nu, \tag{7}$$

where

h is a symmetric $\mathsf{F}\mathcal{X}$-bilinear mapping $\mathfrak{a}\mathcal{X} \otimes \mathfrak{a}\mathcal{X} \to \Gamma\nu$ (the second fundamental form),

A_s, $s \in \Gamma\nu$, are linear self-adjoint operators $\mathfrak{a}\mathcal{X} \to \mathfrak{a}\mathcal{X}$ connected with h by (5), *and*

D is a connection on the normal bundle ν of the submanifold $\mathcal{X}$ that satisfies relation (4).

Formulas (6) and (7) are obtained from (14) and (16) in Chap. 3 by transformation of their left-hand sides in accordance with formula (9) in Chap. 3. Therefore, by the fields X, Y, and s in the left-hand sides of these formulas, we mean their extensions from $\mathcal{X}$ to $\mathcal{Y}$. Moreover, we presuppose that the left-hand sides are bounded on $\mathcal{X}$.

§12. Normal of the Mean Curvature

Because the trace $\operatorname{Tr} A_s$ of the operator A_s depends on s linearly, the formula

$$s \mapsto \operatorname{Tr} A_s$$

defines a linear functional on $\Gamma\nu$; therefore, there exists a section $t \in \Gamma\nu$ (normal vector field on $\mathcal{X}$) such that

$$(s, t) = \frac{1}{n} \operatorname{Tr} A_s$$

for any $s \in \Gamma\nu$, where $n = \dim \mathcal{X}$ as usual. If $s_1, \dots, s_{m-n}$ is an orthonormal basis of the module $\Gamma\nu$ over a certain coordinate neighborhood U (for definiteness, we assume that the space $\mathcal{Y}$ is Riemannian), then we have the relation

$$t = \frac{1}{n} \sum_{i=1}^{m-n} \operatorname{Tr} A_{s_i} \cdot s_i$$

over U. The section t is called the *field of normals of the mean curvature.*

In the case where it vanishes identically (i.e., $\operatorname{Tr} A_s = 0$ for any $s \in \Gamma\nu$), the submanifold $\mathcal{X}$ (we recall that, generally speaking, it can have self-intersections) is called a *minimal submanifold* of the Riemannian space $\mathcal{Y}$. (We show below that for $n = 2$ and $\mathcal{Y} = \mathbb{R}^3$, this is exactly a minimal surface in the sense of Definition 14.2. It can be proved (we do not do this) that they also have the corresponding extremal property in the general case.)

§13. Gauss, Peterson–Codazzi, and Ricci Relations

Let $R_{\mathcal{X}}$ and $R_{\mathcal{Y}}$ be the curvature tensors of the (pseudo-)Riemannian spaces $\mathcal{X}$ and $\mathcal{Y}$. According to the Gauss–Weingarten formulas (6) and (7), for any fields $X, Y, Z \in \mathfrak{a}\mathcal{X}$, we have the following relation, where X, Y, and Z in the first line are extensions of these fields to $\mathcal{Y}$):

$$\begin{aligned}
R_{\mathcal{Y}}(X,Y)Z &= \nabla_X\nabla_Y Z - \nabla_Y\nabla_X Z - \nabla_{[X,Y]}Z \\
&= \nabla_X((\nabla^{\mathcal{X}})_Y Z + h(Y,Z)) \\
&\quad - \nabla_Y((\nabla^{\mathcal{X}})_X Z + h(X,Z)) - \nabla_{[X,Y]}Z \\
&= (\nabla^{\mathcal{X}})_X(\nabla^{\mathcal{X}})_Y Z + h(X,(\nabla^{\mathcal{X}})_Y Z) - A_{h(Y,Z)}X + D_X h(Y,Z) \\
&\quad - (\nabla^{\mathcal{X}})_Y(\nabla^{\mathcal{X}})_X Z - h(Y,(\nabla^{\mathcal{X}})_X Z) + A_{h(X,Z)}Y \\
&\quad - D_Y h(X,Z) - (\nabla^{\mathcal{X}})_{[\mathcal{X},Y]}Z - h([X,Y],Z) \\
&= R_{\mathcal{X}}(X,Y)Z + h(X,\nabla^{\mathcal{X}})_Y Z) - h(Y,\nabla^{\mathcal{X}})_X Z) - h([X,Y],Z) \\
&\quad - A_{h(Y,Z)}X + A_{h(X,Z)}Y + D_X h(Y,Z) - D_Y h(X,Z). \qquad (8)
\end{aligned}$$

Taking the inner product of this relation and the vector field $W \in \mathfrak{a}\mathcal{X}$ and taking relations (5) (and also that fields of the form $h(X,Y)$ and $D_X s$ are orthogonal to the field W at each point) into account, we immediately obtain *the relation*

$$\begin{aligned}
R_{\mathcal{X}}(X,Y,Z,W) &= R_{\mathcal{Y}}(X,Y,Z,W) \\
&\quad - (h(X,W),h(Y,Z)) + (h(Y,W),h(X,Z)). \qquad (9)
\end{aligned}$$

for any fields $X, Y, Z, W \in \mathfrak{a}\mathcal{X}$. (We write simply (X,Y) instead of $g(X,Y)$; compare with Chap. 11.) This relation is called the *Gauss relation.*

Moreover, because $[X,Y] = \nabla^{\mathcal{X}}_X Y - \nabla^{\mathcal{X}}_Y X$ (see (14) in Chap. 2), it follows from (8) that *the normal component* $R^{\perp}_{\mathcal{Y}}(X,Y)Z$ *of the field* $R_{\mathcal{Y}}(X,Y)Z$ *is expressed by*

$$R^{\perp}_{\mathcal{Y}}(X,Y)Z = (\overline{\nabla}_X h)(Y,Z) - (\overline{\nabla}_Y h)(X,Z), \qquad (10)$$

where

$$(\overline{\nabla}_X h)(Y,Z) = D_X h(Y,Z) - h(\nabla^{\mathcal{X}}_X Y, Z) - h(Y, \nabla^{\mathcal{X}}_X Z)$$

(see (20) in Chap. 3). Relation (10) is called the *Peterson–Codazzi relation.*

Further, if s and t are normal vector fields on $\mathcal{X}$, then (we again omit g)

$$\begin{aligned}
-R_{\mathcal{Y}}(X,Y,s,t) &= (\nabla_X\nabla_Y s,t) - (\nabla_Y\nabla_X s,t) - (\nabla_{[X,Y]}s,t) \\
&= (\nabla_X(-A_s Y + D_Y s),t) - (\nabla_Y(-A_s X + D_X s),t) \\
&\quad - (-A_s[X,Y] + D_{[X,Y]}s,t) = \\
&= -(h(X,A_s Y),t) + (D_X D_Y s,t) + (h(Y,A_s X),t) \\
&\quad - (D_Y D_X s,t) - (D_{[X,Y]}s,t).
\end{aligned}$$

Introducing the analogue

$$-R_D(X, Y, s, t) = (D_X D_Y s, t) - (D_Y D_X s, t) - (D_{[X,Y]} s, t)$$

of the Riemannian curvature tensor for the connection D and taking into account that according to (5), $(h(X, A_t Y), s) = (A_s X, A_t Y)$, we immediately obtain

$$R_D(X, Y, s, t) = R_{\mathcal{Y}}(X, Y, s, t) - (A_t X, A_s Y) + (A_s X, A_t Y),$$

i.e. (because the operators A_s are self-adjoint),

$$R_D(X, Y, s, t) = R_{\mathcal{Y}}(X, Y, s, t) - ([A_s, A_t]X, Y). \tag{11}$$

Relation (11) is called the *Ricci relation.* In particular, relation (11) implies that *the functions $R_D(X, Y, s, t)$ are skew-symmetric functions of s and t (and certainly on X and Y).*

§14. Case of a Flat Ambient Space

In the case where the ambient manifold $\mathcal{Y}$ is flat (its curvature tensor is zero), relations (9), (10), and (11) become

$$R_{\mathcal{X}}(X, Y, Z, W) = (h(Y, W), h(X, Z)) - (h(X, W), h(Y, Z)), \tag{9'}$$

$$(\overline{\nabla}_X h)(Y, Z) = (\overline{\nabla}_Y h)(X, Z), \tag{10'}$$

$$R_D(X, Y, s, t) = -([A_s, A_t]X, Y). \tag{11'}$$

In particular, *these relations hold for any submanifold of the Euclidean space (or any completely nonisotropic submanifold of the pseudo-Euclidean space).*

In each chart $(U, x^1, \ldots, x^n) = (U, h)$ of the submanifold $\mathcal{X}$, Gauss formula (3) for $X = \partial/\partial x^i$ and $Y = \partial/\partial x^j$ becomes

$$(\nabla|_{\mathcal{X}})_i \frac{\partial}{\partial x^j} = \nabla_i^{\mathcal{X}} \frac{\partial}{\partial x^j} + h_{ij}, \quad i, j = 1, \ldots, n, \tag{12}$$

where $h_{ij} = h(\partial/\partial x^i, \partial/\partial x^j)$ are certain normal vectors. In the case where $\mathcal{X}$ is a submanifold of the (pseudo-)Euclidean space $\mathcal{V}$ and is given locally by the vector equation

$$\boldsymbol{r} = \boldsymbol{r}(\boldsymbol{x}), \quad \boldsymbol{x} = (x^1, \ldots, x^n) \in \mathring{U} \subset \mathbb{R}^n, \tag{13}$$

where $\mathring{U} = h(U)$ (i.e., if $\boldsymbol{r}(\boldsymbol{x})$ is the radius vector of a point in U with the coordinates $x^1, \ldots, x^n$), the vectors $\partial/\partial x^i$ are identified with the vectors $\boldsymbol{r}_i = \partial \boldsymbol{r}/\partial x^i$ in the known way. Therefore, the first summand $\nabla_i^{\mathcal{X}} \partial/\partial x^j = \Gamma_{ij}^k \partial/\partial x^k$ in the right-hand side of (12), where Γ_{ij}^k are coefficients of the connection $\nabla^{\mathcal{X}}$

in the chart (U, h), can be written as $\Gamma_{ij}^k \boldsymbol{r}_k$. Moreover, in this case, the covariant derivatives in $\mathcal{Y}$ are usual derivatives (all coefficients of a flat connection are zero); therefore, the left-hand side of (12) is just the second derivative $\boldsymbol{r}_{ij} = \partial^2 \boldsymbol{r}/(\partial x^i \partial x^j)$ of radius vector (13). Therefore, in this case, formulas (12) become

$$\boldsymbol{r}_{ij} = \Gamma_{ij}^k \boldsymbol{r}_k + \boldsymbol{h}_{ij}, \quad i, j = 1, \ldots, n \tag{14}$$

(we now write $\boldsymbol{h}_{ij}$ instead of h_{ij} to stress the vector character).

Chapter 20
Submanifolds II

We apply the general results in the previous chapter to certain remarkable classes of submanifolds in the Euclidean space $\mathcal{V}$.

§1. Locally Symmetric Submanifolds

For any point p of a submanifold $\mathcal{X} \subset \mathcal{V}$, we let σ_p denote the symmetry with respect to the normal at the point p, i.e., the isometry of the space $\mathcal{V}$ that leaves the point p fixed and induces the identical transformation of the vector space $\mathbf{N}_p\mathcal{X}$ and the antipodal transformation (with the matrix $-E$) of the vector space $\mathbf{T}_p\mathcal{X}$.

Definition 20.1. A submanifold $\mathcal{X}$ is called a *(locally) symmetric submanifold* if for each point $p \in \mathcal{X}$, there exists $\varepsilon > 0$ such that

$$\sigma_p(\exp_p sA) = \exp_p(-sA)$$

for any vector $A \in \mathbf{T}_p\mathcal{X}$, $|A| = 1$, and for any s, $|s| < \varepsilon$.

Of course, each such submanifold is a locally symmetric space in the sense in Chap. 4.

Proposition 20.1 (Ferus–Strubing). *A submanifold $\mathcal{X}$ is locally symmetric iff its second fundamental form is covariantly constant with respect to the Van der Waerden–Bortolotti connection. i.e., if*

$$\overline{\nabla}_X h(Y, Z) = 0$$

for any vector fields $X, Y, Z \in \mathfrak{a}\mathcal{X}$.

Proof. Let $p \in \mathcal{X}$, let $A \in \mathbf{T}_p\mathcal{X}$, $|A| = 1$, and let γ be a geodesic $\gamma_{p,A}$ for which $\gamma(0) = p$ and $\dot\gamma(0) = A$. Further, let U be a neighborhood (in $\mathcal{X}$) of the point p, X be a vector field on U, and $\varepsilon > 0$ be a positive number such that $\gamma(s) \in U$ and $\dot\gamma(s) = X_{\gamma(s)}$ for any s, $|s| < \varepsilon$. Finally, let $\boldsymbol{r} = \boldsymbol{r}(s)$ be the equation of the geodesic $\gamma|_{(-\varepsilon,\varepsilon)}$ (which is denoted merely by γ) as a geodesic of the ambient Euclidean space. Then by the usual identifications (and formulas (14″) and (16) in Chap. 3), for any s, $|s| < \varepsilon$, we have

$$\begin{aligned}
\dot{\boldsymbol{r}}(s) &= X_{\gamma(s)}, \\
\ddot{\boldsymbol{r}}(s) &= h(X, X)_{\gamma(s)}, \\
\dddot{\boldsymbol{r}}(s) &= -\left[A_{h(X,X)}X\right]_{\gamma(s)} + [D_X h(X, X)]_{\gamma(s)} \\
&= -\left[A_{h(X,X)}X\right]_{\gamma(s)} + \left[\overline{\nabla}_X h(X, X)\right]_{\gamma(s)}
\end{aligned}$$

because

$$\left(\nabla_X^{\mathcal{X}} X\right)_{\gamma(s)} = \frac{\nabla^{\mathcal{X}} \dot{\gamma}}{dt}(s) = 0,$$

$([D_X h(X,X)]_{\gamma(s)} = \left[\overline{\nabla}_X h(X,X)\right]_{\gamma(s)}$; see (20) in Chap. 3).

If the submanifold $\mathcal{X}$ is locally symmetric, then the curve $\sigma_p \circ \gamma$ is the geodesic $\gamma_{p,-A}$, and therefore

$$\sigma_p(\boldsymbol{r}(s)) = \boldsymbol{r}(-s)$$

for any s, $|s| < \varepsilon$. Therefore,

$$\begin{aligned}
\sigma_p(\dot{\boldsymbol{r}}(s)) &= -\dot{\boldsymbol{r}}(-s),\\
\sigma_p(\ddot{\boldsymbol{r}}(s)) &= \ddot{\boldsymbol{r}}(-s),\\
\sigma_p(\dddot{\boldsymbol{r}}(s)) &= -\dddot{\boldsymbol{r}}(-s),
\end{aligned}$$

and, in particular,

$$\begin{aligned}
\sigma_p(\dot{\boldsymbol{r}}(0)) &= -\dot{\boldsymbol{r}}(0),\\
\sigma_p(\ddot{\boldsymbol{r}}(0)) &= \ddot{\boldsymbol{r}}(0),\\
\sigma_p(\dddot{\boldsymbol{r}}(0)) &= -\dddot{\boldsymbol{r}}(0).
\end{aligned}$$

Because $\sigma_p|_{\mathbf{T}_p\mathcal{X}} = -\operatorname{id}$ and $\sigma_p|_{\mathbf{N}_p\mathcal{X}} = \operatorname{id}$, the first two of these equations are automatically satisfied (because $\dot{\boldsymbol{r}}(0) \in \mathbf{T}_p\mathcal{X}$ and $\ddot{\boldsymbol{r}}(0) \in \mathbf{N}_p\mathcal{X}$), and the third equation implies $\left[\overline{\nabla}_X h(X,X)\right]_p = 0$.

Exercise 20.1. Prove that a trilinear symmetric operator $X, Y, Z \mapsto f(X,Y,Z)$ is identically zero iff it is zero for $X = Y = Z$. [*Hint*: Use the identity

$$\begin{aligned}
6f(X,Y,Z) = {} & f(X+Y+Z, X+Y+Z, X+Y+Z)\\
& - f(X+Y, X+Y, X+Y) - f(X+Z, X+Z, X+Z)\\
& - f(Y+Z, Y+Z, Y+Z) + f(X,X,X)\\
& + f(Y,Y,Y) + f(Z,Z,Z). \quad]
\end{aligned}$$

Applied to the functional $X, Y, Z \mapsto \left[\overline{\nabla}_X h(Y,Z)\right]_p$ (which is symmetric by the Peterson–Codazzi relations; see (10′) in Chap. 19), this (by the arbitrariness of the point p and the unit vector $A \in \mathbf{T}_p\mathcal{X}$) immediately yields $\overline{\nabla}_X h(Y,Z) = 0$ for all $X, Y, Z \in \mathfrak{a}\mathcal{X}$.

To prove the converse, we must study the curve $\boldsymbol{r} = \boldsymbol{r}(s)$, $|s| < \varepsilon$, in more detail. First, let $\boldsymbol{r} = \boldsymbol{r}(s)$ be an arbitrary curve in the Euclidean space that is related to the natural parameter s (and is defined for $|s| < \varepsilon$). We say that this curve *has the rank* m if for any s, $|s| < \varepsilon$, the vectors

$$\dot{\boldsymbol{r}}(s), \quad \ldots, \quad \overset{(m)}{\boldsymbol{r}}(s)$$

are linearly independent and the vector $\overset{(m+1)}{\boldsymbol{r}}(s)$ is linearly expressed through them. (Of course, not every curve has a rank.)

The orthonormal vectors

$$t_1(s), \quad \dots, \quad t_m(s),$$

obtained as a result of applying the Gram–Schmidt orthogonalization process to the vectors $\dot{r}(s), \dots, \overset{(m)}{r}(s)$ compose the *Frenet frame* on the curve $r = r(s)$.

Exercise 20.2. Prove the following:
1. The Frenet formulas

$$\dot{t}_i = -k_{i-1}t_{i-1} + k_i t_{i+1}, \quad i = 1, \dots, m,$$

hold, where $k_1 = k_1(s), \dots, k_{m-1} = k_{m-1}(s)$ are certain positive functions of s, $|s| < \varepsilon$ (and $k_0 = 0, k_m = 0$).
2. The functions $k_1, \dots, k_{m-1}$ together with the point $r_0 = r(0)$ and the vectors $t_1(0), \dots, t_m(0)$ uniquely define the curve $r = r(s)$.

The functions $k_1, \dots, k_m$ are called the *curvatures* of the curve $r = r(s)$. A curve for which all curvatures are constant is called a *helix*.

Proposition 20.2. *If the second fundamental form of a submanifold $\mathcal{X}$ is covariantly constant, then the geodesic $r = r(s)$ as a curve in the space has a rank and is a helix. On a neighborhood U, there exist vector fields $X_1, \dots, X_m$ (where m is the rank of the curve) such that*

1. *all the fields $(X_1)_{\gamma(s)}, \dots, (X_m)_{\gamma(s)}$ on the curve γ are parallel, i.e., $(\nabla^{\mathcal{X}}_X X_i)_{\gamma(s)} = 0$ for any s, $|s| < \varepsilon$, and*
2. *for the vectors $t_i(s)$, $i = 1, \dots, m$, of the Frenet frame on the curve $r = r(s)$, we have*

$$t_i(s) = \begin{cases} (X_i)_{\gamma(s)} & \text{if } i \text{ is odd,} \\ h(X, X_i)_{\gamma(s)} & \text{if } i \text{ is even,} \end{cases}$$

   where X is the vector field on the neighborhood U introduced above (for which $X_{\gamma(s)} = \dot{\gamma}(s)$, $|s| < \varepsilon$).

In particular, $t_i(s) \in \mathbf{T}_{\gamma(s)}\mathcal{X}$ if i is odd, and $t_i(s) \in \mathbf{N}_p\mathcal{X}$ if i is even; therefore,

$$\sigma_p(t_i(0)) = (-1)^i t_i(0)$$

for any $i = 1, \dots, m$.

We prove this proposition below and now continue the proof of Proposition 20.1. Let

$$\overline{r}(s) = \sigma_p(r(-s)), \quad |s| < \varepsilon.$$

It is clear that the curve $r = \overline{r}(s)$, $|s| < \varepsilon$, also has the rank m and the Frenet frame $\overline{t}_1, \dots, \overline{t}_m$ is therefore defined for it.

Exercise 20.3. Prove the following:

1. The curve $\boldsymbol{r} = \overline{\boldsymbol{r}}(s)$ has the same (constant!) curvatures $k_1, \dots, k_{m-1}$ as the curve $\boldsymbol{r} = \boldsymbol{r}(s)$.

2. For any $i = 1, \dots, m$, we have

$$\overline{\boldsymbol{t}}_i(s) = (-1)^i \sigma_p(\boldsymbol{t}_i(-s)), \quad |s| < \varepsilon.$$

[*Hint*: Use induction on i.]

In particular, we see that

$$\overline{\boldsymbol{t}}_i(0) = (-1)^i \sigma_p(\boldsymbol{t}_i(0)) = \boldsymbol{t}_i(0)$$

for any $i = 1, \dots, m$. Because the curves $\boldsymbol{r} = \boldsymbol{r}(s)$ and $\boldsymbol{r} = \overline{\boldsymbol{r}}(s)$ have the same curvatures $k_1, \dots, k_{m-1}$, this implies (see assertion 2 in Exercise 20.2) that $\overline{\boldsymbol{r}}(s) = \boldsymbol{r}(s)$ for all s, i.e.,

$$\sigma_p(\boldsymbol{r}(s)) = \boldsymbol{r}(-s).$$

Because $\boldsymbol{r}(s) = \exp_p sA$, the manifold $\mathcal{X}$ is therefore locally symmetric. □

It remains to prove Proposition 20.2.

Proof (of Proposition 20.2). Because $\boldsymbol{t}_1(s) = \dot{\boldsymbol{r}}(s)$, we can accept the field X as the field X_1. We suppose that for a certain $i \geq 1$, the vectors $\dot{\boldsymbol{r}}, \dots, \overset{(i)}{\boldsymbol{r}}$ are linearly independent at all points of the curve $\boldsymbol{r} = \boldsymbol{r}(s)$ (and the vectors $\boldsymbol{t}_1, \dots, \boldsymbol{t}_i$ are therefore defined), the fields $X_1, \dots, X_i$ are constructed, and the curvatures $k_1, \dots, k_{i-1}$ are constant. If i is odd, then

$$\dot{\boldsymbol{t}}_i(s) = \left(\nabla_X^{\mathcal{X}} X_i\right)_{\gamma(s)} + h\left(X, X_i\right)_{\gamma(s)} = h\left(X, X_i\right)_{\gamma(s)}$$

and therefore

$$\begin{aligned}(\dot{\boldsymbol{t}}_i + k_{i-1}\boldsymbol{t}_{i-1})(s) &= h(X, X_i)_{\gamma(s)} + h(X, k_{i-1}X_{i-1})_{\gamma(s)} \\ &= h(X, X_i + k_{i-1}X_{i-1})_{\gamma(s)}.\end{aligned}$$

Because the form h is covariantly constant (with respect to the connection $\overline{\nabla}$) by assumption and $(\nabla_X^{\mathcal{X}} X_i)_{\gamma(s)} = 0$, we have $[D_X h(X, X_i)]_{\gamma(s)} = 0$ for any i (see (20) in Chap. 3); therefore, the normal vector $\dot{\boldsymbol{t}}_i + k_{i-1}\boldsymbol{t}_{i-1}$ is covariantly constant along the curve γ (with respect to the normal connection D). In particular, its length $k_i = |\dot{\boldsymbol{t}}_i + k_{i-1}\boldsymbol{t}_{i-1}|$ is constant; therefore, it either is zero or is nonzero at all points of the curve γ. In the first case, Proposition 20.2 obviously holds (with $m = i$). In the second case, the vector $\boldsymbol{t}_{i+1} = (\dot{\boldsymbol{t}}_i + k_{i-1}\boldsymbol{t}_{i-1})/k_i$ and the field $X_{i+1} = (X_i + k_{i-1}X_{i-1})/k_i$ connected by the relation $\boldsymbol{t}_{i+1}(s) = h(X, X_{i+1})_{\gamma(s)}$ are defined; moreover, the vectors $(X_{i+1})_{\gamma(s)}$ are parallel along γ.

Similarly, if i is even, then

$$\begin{aligned}\dot{t}_i(s) &= -\left[A_{h(X,X_i)}X\right]_{\gamma(s)} + [D_X h(X,X_i)]_{\gamma(s)}\\ &= -\left[A_{h(X,X_i)}X\right]_{\gamma(s)} + \left[\,\overline{\nabla}_X h(X,X_i)\right]_{\gamma(s)}\\ &= -\left[A_{h(X,X_i)}X\right]_{\gamma(s)}\end{aligned}$$

and therefore

$$(\dot{t}_i + k_{i-1}t_{i-1})(s) = \left[-A_{h(X,X_i)}X + k_{i-1}X_{i-1}\right]_{\gamma(s)}.$$

If for the vector field Y on U, the vectors $Y_{\gamma(s)}$, $|s| < \varepsilon$, are parallel along γ, then

$$\begin{aligned}g_{\gamma(s)}\left(\frac{d}{ds}\left[A_{h(X,X_i)}X\right]_{\gamma(s)}, Y_{\gamma(s)}\right) &= \left[Xg(A_{h(X,X_i)}X, Y)\right]_{\gamma(s)}\\ &= [Xg(h(X,Y), h(X,X_i))]_{\gamma(s)}\\ &= \left[g(\overline{\nabla}_X h(X,Y), h(X,X_i))\right.\\ &\qquad \left.+ g(h(X,Y), \overline{\nabla}_X h(X,X_i))\right]_{\gamma(s)}\\ &= \left[g((\overline{\nabla}_X h)(X,Y), h(X,X_i))\right.\\ &\qquad \left.+ g(h(X,Y), (\overline{\nabla}_X h)(X,X_i))\right]_{\gamma(s)} = 0\end{aligned}$$

(see formula (4) in Chap. 19); therefore,

$$\frac{d}{ds}\left[A_{h(X,X_i)}X\right]_{\gamma(s)} = 0,$$

i.e., the vector field $\left[A_{h(X,X_i)}X\right]_{\gamma(s)}$ is parallel along the curve γ. Therefore, this field $\dot{t}_i + k_{i-1}t_{i-1}$ is also parallel along γ, and the function $k_i = |\dot{t}_i + k_{i-1}t_{i-1}|$ is therefore constant. Therefore, this field either is identically zero (which proves Proposition 20.2 with $m = i$) or is nonzero everywhere. In the latter case, the vector

$$t_{i+1} = \frac{\dot{t}_i + k_{i-1}t_{i-1}}{k_i}$$

and the field

$$X_{i+1} = \frac{-A_{h(X,X_i)}X + k_{i-1}X_{i-1}}{k_i}$$

connected by the relation $t_{i+1}(s) = (X_{i+1})_{\gamma(s)}$ are defined; moreover, the vectors $(X_{i+1})_{\gamma(s)}$ are parallel along γ. Because the curvature k_i is constant in both cases, this completely proves Proposition 20.2 by induction. □

§2. Compact Submanifolds

We now consider the case where the submanifold $\mathcal{X}$ (immersed in the general case) of the Euclidean space is compact.

Lemma 20.1. *In each compact submanifold $\mathcal{X}$ of the Euclidean space $\mathcal{V}$, there exists a point $p_0 \in \mathcal{X}$ such that*

$$h_{p_0}(A, A) \neq 0$$

for any vector $A \neq 0$ of the space $\mathbf{T}_{p_0}\mathcal{X}$.

Proof. Choosing the origin O in the space $\mathcal{V}$, we consider a point p_0 in $\mathcal{X}$ such that the length of its radius vector is maximum. (This point exists because the manifold $\mathcal{X}$ is compact.) In a neighborhood of the point p_0, let the submanifold $\mathcal{X}$ be given by Eq. (13) in Chap. 19. Then for any $i = 1, \dots, n$, the partial derivative $2\boldsymbol{r}_i\boldsymbol{r}$ of the function $\boldsymbol{r}^2$ with respect to x^i vanishes at the point p_0, i.e., all inner products $\boldsymbol{r}_i\boldsymbol{r}$ vanish. Therefore, for the second partial derivatives of this function at the point p_0, we have

$$(\boldsymbol{r}^2)_{ij} = 2(\boldsymbol{r}_{ij}\boldsymbol{r} + \boldsymbol{r}_i\boldsymbol{r}_j) = 2(\boldsymbol{h}_{ij}\boldsymbol{r} + \boldsymbol{r}_i\boldsymbol{r}_j)$$

(here we use formula (14) in Chap. 19). Because the point p_0 is a maximum point of the function $\boldsymbol{r}^2$, the matrix $\|(\boldsymbol{r}^2)_{ij}\|$ of the second derivatives of this function at this point is negative definite. Because the matrix $\|\boldsymbol{r}_i\boldsymbol{r}_j\| = \|g_{ij}\|$ is positive definite, this proves that at the point p_0, the matrix $\|\boldsymbol{h}_{ij}\boldsymbol{r}\|$ is negative definite. Therefore, if $\boldsymbol{h}_{ij}^{(0)}$ and $\boldsymbol{r}^{(0)}$ are values of the vectors $\boldsymbol{h}_{ij}$ and $\boldsymbol{r}$ at the point p_0, then for any nonzero vector $A \in \mathbf{T}_{p_0}\mathcal{X}$, the number

$$(\boldsymbol{h}_{ij}^{(0)}\boldsymbol{r}^{(0)})A^iA^j = \boldsymbol{h}^{(0)}(A, A)\boldsymbol{r}^{(0)}$$

is nonzero (it is even negative). Therefore, the vector $\boldsymbol{h}^{(0)}(A, A) = h_{p_0}(A, A)$ is also nonzero. □

The fact that the numbers $\boldsymbol{h}^{(0)}(A, A)\boldsymbol{r}^{(0)}$ are negative has other interesting consequences.

Because the vector $\boldsymbol{r}^{(0)}$ is orthogonal to all vectors $\boldsymbol{r}_i$, it is normal to the submanifold $\mathcal{X}$ at the point p_0. Therefore, there exists a normal vector field $s \in \Gamma\nu$ on $\mathcal{X}$ such that $s(p_0) = \boldsymbol{r}^{(0)}$. The operator A_s corresponding to this field satisfies the relation

$$(A_s X, X) = (h(X, X), s)$$

for each $X \in \mathfrak{a}\mathcal{X}$. Therefore, for $X_{p_0} \neq 0$, the function $(A_s X, X)$ assumes a negative value at the point p_0 (the functional $X \mapsto (A_s X, X)$ is negative definite at the point p_0). Therefore, the trace $\operatorname{Tr} A_s$ of the operator A_s at the point p_0 is negative and hence nonzero. For the field t of normals of the mean curvature, this means that the field t is nonzero at the point p_0. Therefore, the submanifold $\mathcal{X}$ is not minimal in advance. Therefore, we see that *a minimal submanifold of the Euclidean space cannot be compact.* (Compare with Remark 14.1.)

§3. Chern–Kuiper Theorem

To deduce more interesting consequences, we need one more purely algebraic lemma.

Lemma 20.2. *Let $\mathcal{V}$ and $\mathcal{W}$ be Euclidean spaces, and let*

$$\boldsymbol{h}\colon \mathcal{V} \otimes \mathcal{V} \to \mathcal{W}$$

be a symmetric bilinear mapping such that

$$\boldsymbol{h}(\boldsymbol{x}, \boldsymbol{x})\boldsymbol{h}(\boldsymbol{y}, \boldsymbol{y}) \leq \boldsymbol{h}(\boldsymbol{x}, \boldsymbol{y})^2$$

for any vectors $\boldsymbol{x}, \boldsymbol{y} \in \mathcal{V}$. Then, if $\dim \mathcal{W} < \dim \mathcal{V}$, there exists a nonzero vector $\boldsymbol{x}_0 \in \mathcal{V}$ such that

$$\boldsymbol{h}(\boldsymbol{x}_0, \boldsymbol{x}_0) = 0.$$

Proof. Let $n = \dim \mathcal{V}$ and $m = \dim \mathcal{W}$. The vector equation $\boldsymbol{h}(\boldsymbol{z}, \boldsymbol{z}) = 0$ is equivalent to quadratic numerical equations for n components of the vector $\boldsymbol{z}$. Therefore, for $m < n$, it has a nonzero solution

$$\boldsymbol{z}_0 = \boldsymbol{x}_0 + i\boldsymbol{y}_0,$$

which is complex in general. Because $\boldsymbol{z}_0 \neq 0$, we can assume without loss of generality that $\boldsymbol{x}_0 \neq 0$. On the other hand, because

$$\boldsymbol{h}(\boldsymbol{z}_0, \boldsymbol{z}_0) = \boldsymbol{h}(\boldsymbol{x}_0, \boldsymbol{x}_0) - \boldsymbol{h}(\boldsymbol{y}_0, \boldsymbol{y}_0) + 2i\boldsymbol{h}(\boldsymbol{x}_0, \boldsymbol{y}_0),$$

we have $\boldsymbol{h}(\boldsymbol{x}_0, \boldsymbol{x}_0) = \boldsymbol{h}(\boldsymbol{y}_0, \boldsymbol{y}_0)$ and $\boldsymbol{h}(\boldsymbol{x}_0, \boldsymbol{y}_0) = 0$. But, by assumption,

$$\boldsymbol{h}(\boldsymbol{x}_0, \boldsymbol{x}_0)\boldsymbol{h}(\boldsymbol{y}_0, \boldsymbol{y}_0) \leq \boldsymbol{h}(\boldsymbol{x}_0, \boldsymbol{y}_0)^2$$

and therefore $\boldsymbol{h}(\boldsymbol{x}_0, \boldsymbol{x}_0)^2 \leq 0$, which is possible only if $\boldsymbol{h}(\boldsymbol{x}_0, \boldsymbol{x}_0) = 0$. □

Proposition 20.3 (Chern–Kuiper theorem). *Let $\mathcal{X}$ be a compact n-dimensional submanifold of the m-dimensional Euclidean space $\mathcal{V}$. For any point $p \in \mathcal{X}$, let the linear space $\mathbf{T}_p\mathcal{X}$ contain a k-dimensional subspace $\check{\mathbf{T}}_p\mathcal{X}$ such that for the sectional curvature $K_p(\pi)$ with respect to each two-dimensional direction $\pi \subset \check{\mathbf{T}}_p$, we have*

$$K_p(\pi) \leq 0.$$

Then $m \geq n + k$.

Proof. By the Gauss relation (see (9') in Chap. 19), the condition imposed on the sectional curvature implies that on each linear space $\check{\mathbf{T}}_p\mathcal{X}$ and on the linear space $\check{\mathbf{T}}_{p_0}\mathcal{X}$ in particular, where p_0 is the point in Lemma 20.1, the form h satisfies the conditions of Lemma 20.2 (with $\mathcal{W} = \mathbf{N}_{p_0}$). Therefore, if $m - n < k$, then there would exist a vector $A \in \check{\mathbf{T}}_{p_0}\mathcal{X} \subset \mathbf{T}_{p_0}\mathcal{X}$ such that $h_{p_0}(A, A) = 0$ contrary to Lemma 20.1. Therefore, $m - n \geq k$. □

Corollary 20.1. *A compact n-dimensional Riemannian space $\mathcal{X}$ of nonpositive sectional curvature cannot be isometrically immersed in $\mathbb{R}^{2n-1}$.*

In particular, *a compact n-dimensional flat space cannot be isometrically immersed in* $\mathbb{R}^{2n-1}$. For example, in the three-dimensional space, there is no surface (even with self-intersections) that is isometric to the quotient space $\mathbb{R}^2/\mathbb{Z}^2$ (the torus equipped with the Euclidean structure).

§4. First and Second Quadratic Forms of a Hypersurface

In the case $m = n + 1$, i.e., where $\mathcal{X}$ is a (completely nonisotropic) *hypersurface* of the space $\mathcal{V}$ (which is now assumed to be pseudo-Euclidean in general), the normal subspace $\mathbf{N}_p\mathcal{X}$ at each point $p \in \mathcal{X}$ is one-dimensional and is therefore generated by a certain vector $\boldsymbol{n}$, which is called the *normal vector* to the hypersurface $\mathcal{X}$ at the point p. Because the hypersurface $\mathcal{X}$ is completely nonisotropic, *the vector $\boldsymbol{n}$ is nonisotropic*, i.e., $\boldsymbol{n}^2 \neq 0$. (Otherwise, this vector belongs to a subspace of $\mathbf{T}_p\mathcal{X}$ and is orthogonal to each vector from $\mathbf{T}_p\mathcal{X}$; therefore, the metric on $\mathbf{T}_p\mathcal{X}$ is degenerate.) Therefore, we can assume without loss of generality that *the vector $\boldsymbol{n}$ is normalized*, i.e.,

$$\boldsymbol{n}^2 = \varepsilon,$$

where $\varepsilon = \pm 1$ is the same for all points $p \in \mathcal{X}$.

Exercise 20.4. Let (a, b) be the signature of the pseudo-Euclidean space $\mathcal{V}$. Prove that the signature of the hypersurface $\mathcal{X}$ is equal to $(a - 1, b)$ for $\varepsilon = 1$ and is equal to $(a, b - 1)$ for $\varepsilon = -1$.

In general, it is impossible to choose vectors $\boldsymbol{n}$ at all points $p \in \mathcal{X}$ such that they form a smooth normal field on $\mathcal{X}$. In the case where it is possible, the hypersurface $\mathcal{X}$ is said to be *two-sided*; otherwise, it is said to be *one-sided.* (The standard example of a one-sided surface in the three dimensional case is the Möbius strip.)

A smooth normal field on a two-sided hypersurface is called its *framing*, and a hypersurface on which a framing is chosen is called a *framed hypersurface.* On a two-sided hypersurface, there are two and only two framings, and they differ only by the sign.

Exercise 20.5. Show that a hypersurface of the space $\mathcal{V}$ is two-sided iff it is an oriented manifold.

This implies (it can directly be proved) that a framing exists on any coordinate neighborhood U. Choosing a framing $\boldsymbol{n}$ on U, we can rewrite formulas (14) in Chap. 19 in the form

$$\boldsymbol{r}_{ij} = \Gamma^k_{ij}\boldsymbol{r}_k + h_{ij}\boldsymbol{n}, \quad i,j = 1,\dots,n, \tag{1}$$

where h_{ij} are certain smooth functions on U that compose a symmetric matrix $\|h_{ij}\|$. The quadratic form $h_{ij}dx^i dx^j$ with this matrix is called the *second fundamental form* of the hypersurface $\mathcal{X}$. (Its *first fundamental form* is the form $ds^2 = g_{ij}dx^i dx^j$.) This incidentally explains the use of the term "second fundamental form" for the mapping h in the general case.

For a framed two-sided hypersurface $\mathcal{X}$, the form h is defined on the whole hypersurface (is a field of symmetric tensors of type (2,0) on $\mathcal{X}$). For a surface of the three-dimensional Euclidean space (the case $n = 2$ and $\varepsilon = 1$), formulas (1) are exactly the first three Weingarten formulas.

Formula (7) in Chap. 19 can be interpreted similarly. Because $\boldsymbol{n}^2 = \pm 1$, for any field $X \in \mathfrak{a}\mathcal{X}$, we have

$$\boldsymbol{n} \cdot D_X \boldsymbol{n} = 0.$$

Because the field $\boldsymbol{n}$ forms a basis of the one-dimensional $\mathsf{F}\mathcal{X}$-module $\Gamma\nu$ over U, this implies that $D_X\boldsymbol{n} = 0$ identically. Therefore, for $s = \boldsymbol{n}$, only the first summand remains in the right-hand side of (7) in Chap. 19, and for $X = \partial/\partial x^i$, this formula becomes

$$\boldsymbol{n}_i = -a^j_i \boldsymbol{r}_j,$$

where $\boldsymbol{n}_i = \partial\boldsymbol{n}/\partial x^i$ and a^j_i are entries of the matrix of the operator $A = A_{\boldsymbol{n}}$ in the basis $\boldsymbol{r}_1, \dots, \boldsymbol{r}_n$. Because $a^j_i = \varepsilon h_{ik} g^{kj}$ (see (5) in Chap. 19), this proves that

$$\boldsymbol{n}_i = -\varepsilon h_{ik} g^{kj} \boldsymbol{r}_j. \tag{2}$$

For $n = 2$ and $\varepsilon = 1$, this is exactly the last two Weingarten formulas.

The Gauss relation (formula (9′) in Chap. 19) for a hypersurface becomes

$$\varepsilon R_{ij,kl} = h_{ik}h_{jl} - h_{jk}h_{il}, \quad i,j,k,l = 1,\dots,n, \tag{3}$$

for $X = \partial/\partial x^i$, $Y = \partial/\partial x^j$, $Z = \partial/\partial x^k$, and $W = \partial/\partial x^l$. Therefore, *for a hypersurface of the Euclidean space (a completely nonisotropic hypersurface of the pseudo-Euclidean space), the curvature tensor is expressed through the coefficients of the second fundamental form.*

For a surface of the three-dimensional Euclidean space (the case $n = 2$ and $\varepsilon = 1$), formulas (3) are reduced to the formula

$$R_{12,12} = h_{11}h_{22} - h_{12}^2,$$

which coincides up to the notation with the formula for $R_{12,12}$ proved in Chap. 16.

For the case of a hypersurface, formula (11′) in Chap. 19 (the Ricci relation) is automatically satisfied (both its parts are identically zero). Moreover, because $D_X\boldsymbol{n} = 0$, we have $D_X(h_{ij}\boldsymbol{n}) = Xh_{ij} \cdot \boldsymbol{n}$ and therefore

$$(\overline{\nabla}_k h)\left(\frac{\partial}{\partial x^i}, \frac{\partial}{\partial x^j}\right) = \left(\frac{\partial h_{ij}}{\partial x^k} - \Gamma^p_{ki}h_{pj} - \Gamma^p_{kj}h_{ip}\right)\boldsymbol{n}.$$

Because the coefficient of $\boldsymbol{n}$ in the right-hand side of this formula is just the component $(\nabla_k h)_{ij}$ of the partial derivative $\nabla_k h$ of the tensor h with the components h_{ij}, this implies that for a hypersurface in the (pseudo-)Euclidean space, formula (10′) in Chap. 19 (the Peterson-Codazzi relation) is equivalent to the formula

$$(\nabla_k h)_{ij} = (\nabla_i h)_{kj}, \tag{4}$$

which means that *the tensor* $(\nabla_k h)_{ij}$ *is symmetric* (with respect to all subscripts).

§5. Hypersurfaces Whose Points are All Umbilical

A point of a hypersurface $\mathcal{X}$ is said to be *umbilical* if the tensors g and h are proportional at this point, i.e., there exists a number λ (which is possibly equal to zero) such that in each chart $(U, x^1, \ldots, x^n)$,

$$h_{ij} = \lambda g_{ij} \tag{5}$$

for all $i, j = 1, \ldots, n$. If all points of the hypersurface $\mathcal{X}$ are umbilical, then formula (5) defines a function $\lambda: \mathcal{X} \to \mathbb{R}$ on $\mathcal{X}$ (which is obviously smooth). It turns out that *this function is constant* (certainly, if $\mathcal{X}$ is connected). Indeed (because the tensor g is covariantly constant), formula (5) implies

$$(\nabla_k h)_{ij} = \lambda_k g_{ij}, \quad \text{where } \lambda_k = \frac{\partial \lambda}{\partial x^k}.$$

Therefore, by (4),

$$\lambda_k g_{ij} = \lambda_i g_{kj}.$$

Contracting with the tensor g^{ij}, we immediately obtain the relation $n\lambda_k = \lambda_k$, which is possible (because $n \geq 2$) only if $\lambda_k = 0$. Because this is true for any $k = 1, \ldots, n$, we therefore have $\lambda = \text{const}$. □

It is now easy to see that *a connected hypersurface* $\mathcal{X}$ *whose points are all umbilical is located either on a hypersurface or on a sphere of the space* $\mathcal{V}$ (i.e., is obtained from the sphere or from the hypersurface by removing a certain closed set, which is possibly empty). Indeed, if $\boldsymbol{r}$ is, as above, the radius vector of points of the hypersurface $\mathcal{X}$ and $\boldsymbol{n}$ is the normal vector, then in each chart $(U, x^1, \ldots, x^n)$, we have

$$\boldsymbol{n}_i = -\varepsilon h_{ik}g^{kj}\boldsymbol{r}_j = -\varepsilon\lambda g_{ik}g^{kj}\boldsymbol{r}_j = -\varepsilon\lambda\boldsymbol{r}_i.$$

Therefore, $(\lambda\boldsymbol{r} + \varepsilon\boldsymbol{n})_i = 0$ for any $i = 1, \dots, n$, and the vector $\lambda\boldsymbol{r} + \varepsilon\boldsymbol{n}$ is hence constant (on U and therefore on the whole $\mathcal{X}$). Let $\lambda\boldsymbol{r} + \varepsilon\boldsymbol{n} = \boldsymbol{a}$, where $\boldsymbol{a} = \text{const}$. If $\lambda \neq 0$, then $\boldsymbol{r} - \boldsymbol{a}/\lambda = -(\varepsilon/\lambda)\boldsymbol{n}$, and therefore $|\boldsymbol{r} - \boldsymbol{a}/\lambda|^2 = 1/\lambda^2$. Because the latter equation assigns the sphere of radius $1/|\lambda|$ centered at the point $\boldsymbol{a}/\lambda$, this proves the assertion for $\lambda \neq 0$. If $\lambda = 0$, we consider the function $\boldsymbol{nr}$ on $\mathcal{X}$. Because the vector $\varepsilon\boldsymbol{n} = \boldsymbol{a}$ is constant, the derivatives of this function in each chart are equal to $\boldsymbol{nr}_i$ and therefore vanish. Therefore, $\boldsymbol{nr} = \text{const}$, and to complete the proof, it suffices to note that the equation $\boldsymbol{nr} = \text{const}$ assigns a hyperplane in the space $\mathcal{V}$. □

If the submanifold $\mathcal{X}$ is complete, then $\mathcal{X}$ coincides with the whole hyperplane or sphere (with half the sphere in the pseudo-Euclidean case).

A simple computation (which we prove in Chap. 22 for convenience; see (11) in Chap. 22) shows the converse, that all points of the sphere or hyperplane in the pseudo-)Euclidean space are umbilical points. (For $n = 2$, this is visually obvious because in this case, the umbilicity is equivalent to the the Dupin indicatrix being a circle.)

Therefore, *the property of umbilicity for all points characterizes the spheres and hyperplanes.*

§6. Principal Curvatures of a Hypersurface

Let the ambient space $\mathcal{V}$ be Euclidean again. Being a symmetric bilinear form on the Euclidean space $\mathsf{T}_p\mathcal{X}$ at each point $p \in \mathcal{X}$, the second fundamental form h of a hyperplane $\mathcal{X}$ (which is assumed to be framed) is locally reduced to the normal form, i.e., in a neighborhood U of any point $p \in \mathcal{X}$, there exists an orthonormal (not holonomic!) basis $X_1, \dots, X_n$ of the module $\mathfrak{a}\mathcal{X}$ such that

$$h(X_i, X_j) = 0 \quad \text{for } i \neq j.$$

The diagonal coefficients $h_i = h(X_i, X_i)$ of the form h in this basis are called the *principal curvatures* of the hypersurface X. Being eigenvalues of the self-adjoint operator $A = A_n$ (see Chap. 19), these curvatures do not depend on the choice of the basis $X_1, \dots, X_n$ and are therefore obviously smooth functions defined on the whole hypersurface $\mathcal{X}$.

Exercise 20.6. Prove that for $n = 2$, these are exactly the principal curvatures k_1 and k_2 of the surface $\mathcal{X}$.

A point of a hypersurface $\mathcal{X}$ is umbilical iff all principal curvatures are equal to each other at this point. We note that *the curvatures $h_1, \dots, h_n$ are defined up to the sign*; when the framing changes, they all change sign simultaneously.

The elementary symmetric functions

$$K_i = \sigma_i(h_1, \ldots, h_n), \quad i = 1, \ldots, n,$$

of the principal curvatures, which are expressed through the components of the tensors g and h are of special importance. The function $H = K_1/n$, i.e., the function

$$H = \frac{h_1 + \cdots + h_n}{n}$$

is called the *mean curvature* of the hypersurface $\mathcal{X}$, and the function

$$K_n = h_1 \cdots h_n$$

is called its *total curvature*. For $n = 2$, these are the mean curvature and the Gauss curvature of the surface.

When the framing $\boldsymbol{n}$ is changed to $-\boldsymbol{n}$, the mean curvature changes sign, and the total curvature changes sign for odd n and remains the same for even n.

The components $R_{ij,kl} = R(X_i, X_j, X_k, X_l)$ of the curvature tensor $R = R_{\mathcal{X}}$ of the hypersurface $\mathcal{X}$ in the basis $X_1, \ldots, X_n$ is expressed by

$$R_{ij,kl} = h(X_i, X_k)h(X_j, X_l) - h(X_j, X_k)h(X_i, X_l) = h_i\delta_{ik}h_j\delta_{jl} - h_j\delta_{jk}h_i\delta_{il}$$

according to formula (9′) in Chap. 19, i.e., by

$$R_{ij,kl} = h_i h_j(\delta_{ik}\delta_{jl} - \delta_{jk}\delta_{il}). \tag{6}$$

Therefore, *the products $h_i h_j$ of principal curvatures taken pairwise do not depend on the immersion (and framing) of the hypersurface $\mathcal{X}$ and are determined by only its intrinsic geometry (the metric tensor g)*. Therefore, for even n, the total curvature K_n is also an invariant of the intrinsic geometry. For $n = 2$, this yields the invariance of the Gaussian curvatures under bendings (the Gauss theorema egregium).

§7. Scalar Curvature of a Hypersurface

Formula (6) implies that the components R_{ij} of the Ricci tensor of the hypersurface $\mathcal{X}$ in a basis $X_1, \ldots, X_n$ are expressed by

$$R_{ij} = \delta_{ij}[h_i(h_1 + \cdots + h_n) - h_i^2],$$

and its trace $\mathcal{R}$ (the scalar curvature; see Definition 16.1) is therefore equal to

$$\left(\sum_{i=1}^{n} h_i\right)^2 - \sum_{i=1}^{n} h_i^2 = 2K_2.$$

Therefore, *the scalar curvature of a hypersurface is expressed by the formula*

$$\mathcal{R} = 2K_2.$$

(In particular, we see that *the Gauss curvature* K of the hypersurface equals

$$\frac{2K_2}{n(n-1)}$$

and is, in general, different from its total curvature K_n for $n > 2$.)

§8. Hypersurfaces That are Einstein Spaces

In the case where the hypersurface $\mathcal{X}$ is an Einstein space, the latter formula implies (see Proposition 17.3) that for $n \geq 3$,

$$K_2 = \text{const}\,.$$

Moreover, because the metric tensor has the components δ_{ij} in the orthonormal basis $X_1, \dots, X_n$, the condition

$$\text{Ric} = \frac{\mathcal{R}}{n} g$$

for a space to be Einstein is reduced to the set of n equations

$$h_i(h_1 + \cdots + h_n) - h_i^2 = \frac{2}{n} K_2, \quad i = 1, \dots, n,$$

which mean that the principal curvatures $h_1, \dots, h_n$ are roots of the quadratic equation

$$\lambda^2 - K_1 \lambda + \frac{2}{n} K_2 = 0. \tag{7}$$

Therefore, there exists k, $1 \leq k \leq n$, such that (probably after renumbering the curvatures $h_1, \dots, h_n$) the relations

$$h_1 = \cdots = h_k = \lambda_1, \qquad h_{k+1} = \cdots = h_n = \lambda_2$$

hold, where λ_1 and λ_2 are roots of Eq. (7) (for $k = n$, there are no second relations; in the case $\lambda_1 = \lambda_2$, we conditionally assume that $k = n$). (We note that the roots λ_1 and λ_2 are necessarily real and therefore $K_1^2 \geq 8K_2/n$. This means that *for any hypersurface that is an Einstein space, we have the inequality*

$$n^2 H^2 \geq 4(n-1)K. \tag{8}$$

For $n = 2$, this reduces to the obvious inequality $H^2 \geq K$, which holds for any surface of the three-dimensional space. It is called the *Euler inequality.*)

Exercise 20.7. Prove inequality (8) by a direct computation. [*Hint*: Consider the cases $n = 2$ and $n = 3$ separately.]

Because $\lambda_1 + \lambda_2 = K_1$ by the Vieta formula and, on the other hand,

$$K_1 = h_1 + \cdots + h_n = k\lambda_1 + (n-k)\lambda_2,$$

we have

$$(k-1)\lambda_1 + (n-k-1)\lambda_2 = 0. \tag{9}$$

Moreover, $\lambda_1\lambda_2 = 2K_2/n$.

It can be shown (this is a difficult theorem!) that the function K_2 (we recall that it is constant for $n \geq 3$) cannot be negative, i.e., either $K_2 = 0$ or $K_2 > 0$.

Let $K_2 = 0$ (i.e., $\mathcal{R} = 0$). We can then assume without loss of generality that $\lambda_2 = 0$ and therefore $(k-1)\lambda_1 = 0$. Therefore, either $\lambda_1 = 0$ or $k = 1$. In both cases, $h_i = 0$ for $i > 1$, and according to (6), the tensor R is therefore identically zero. Because we necessarily have $\mathcal{R} = 0$ for $R = 0$, this proves that *a hypersurface of the Euclidean space that is an Einstein space is locally Euclidean* (*is a flat Riemannian space*) *iff its scalar curvature is zero* (and all its principal curvatures are zero except possibly one of them). For $n = 2$, these are exactly developable surfaces.

Let $K_2 > 0$. Then λ_1 and λ_2 are nonzero and have equal signs. Therefore, relation (9) can hold either for $k = 1$ and $n = k+1$ or for $k = n$ and $\lambda_2 = (n-1)\lambda_1$. In the first case, $n = 2$; in the second case, $h_1 = \cdots = h_n$ identically. Therefore, all points of the hypersurface $\mathcal{X}$ are umbilical (and this hypersurface is therefore locally a sphere). This proves that *for $n \geq 3$, a hypersurface of the Euclidean space that is an n-dimensional Einstein space and has a nonzero* (*and therefore positive*) *scalar curvature is locally a sphere.*

In Chap. 22, we prove the converse, that each sphere of the Euclidean space is an Einstein space of positive scalar curvature. Therefore, *for $n \geq 3$, the spheres are the only hypersurfaces of the Euclidean space that are complete Einstein spaces of nonzero scalar curvature.*

§9. Rigidity of the Sphere

In particular, this implies that *for $n \geq 3$, any hypersurface of the $(n+1)$-dimensional Euclidean space that is isometric to the sphere is the sphere itself.* A hypersurface $\mathcal{X}$ of the Euclidean space is said to be *rigid* if any hypersurface isometric to it can be combined with $\mathcal{X}$ by a motion (which is improper in general). In this terminology, the proved assertion means that *the spheres of dimension $n \geq 3$ are rigid hypersurfaces* (*theorem on the rigidity of a sphere*).

Remark 20.1. This theorem also holds for $n = 2$, but its proof is essentially more difficult and requires different methods.

Chapter 21
Fundamental Forms of a Hypersurface

§1. Sufficient Condition for Rigidity of Hypersurfaces

The rigidity property (at least for $n \geq 3$) is very general and holds not only for spheres but also for any two-sided (not necessarily complete!) hypersurface for which at least three principal curvatures are different from zero at each point.

Theorem 21.1. *A two-sided hypersurface of the Euclidean space is rigid if at least three principal curvatures are different from zero at each point.*

We deduce Theorem 21.1 from the following theorem.

Theorem 21.2. *For connected two-sided hypersurfaces $\mathcal{X}$ and $\mathcal{X}^*$ of the Euclidean space, there is an isometry $\mathcal{X} \to \mathcal{X}^*$ that transforms the second fundamental form of the hypersurface $\mathcal{X}$ into the second fundamental form of the hypersurface $\mathcal{X}^*$ iff these hypersurfaces can be made to coincide with one another by a motion.*

In other words, *a connected two-sided hypersurface of the Euclidean space is uniquely defined (up to a motion) by its first and second fundamental forms.* Of course, all hypersurfaces are here assumed to be framed, and the motions transform one framing into another. Without the latter condition, the isometry induced by a motion can change the sign of the second fundamental form. For simplicity, isometries satisfying the conditions of Theorem 21.2 are called *isometries preserving the second fundamental forms* in what follows.

Proof (of Theorem 21.1). To prove Theorem 21.1 using Theorem 21.2, it suffices to show that for isometric two-sided hypersurfaces $\mathcal{X}$ and $\mathcal{X}^*$ satisfying the conditions of this theorem, each isometry $\mathcal{X} \to \mathcal{X}^*$ preserves (up to the sign) the second fundamental forms, i.e., principal curvatures of the hypersurfaces $\mathcal{X}$ and $\mathcal{X}^*$ at the corresponding points are either equal or can made equal by multiplying all curvatures by -1. Because the curvature tensors of isometric surfaces coincide at the corresponding points, everything reduces to the following algebraic lemma by formula (6) in Chap. 20.

Lemma 21.1. *Let $\boldsymbol{h} = (h_1, \dots, h_n)$ and $\boldsymbol{k} = (k_1, \dots, k_n)$ be row vectors of length n such that for $i \neq j$,*

$$h_i h_j = k_i k_j$$

for any $i, j = 1, \dots, n$. If at least three components of the vector $\boldsymbol{h}$ are different from zero, we then have $\boldsymbol{h} = \pm\boldsymbol{k}$.

Proof. For definiteness, let $h_1 \neq 0$, $h_2 \neq 0$, and $h_3 \neq 0$. Then $k_1 = \lambda h_1$ and $k_2 = \lambda^{-1} h_2$, where $\lambda \neq 0$. Therefore, for any $j \neq 1$,

$$h_1 h_j = k_1 k_j = \lambda h_1 k_j$$

and therefore $h_j = \lambda k_j$. Similarly, for any $i \neq 2$,

$$h_i h_2 = k_i k_2 = \lambda^{-1} k_i h_2$$

and therefore $h_i = \lambda^{-1} k_i$. In particular, $\lambda k_3 = \lambda^{-1} k_3$, which is possible for $k_3 \neq 0$ only if $\lambda = \pm 1$. Therefore, $\boldsymbol{h} = \pm \boldsymbol{k}$. □

Therefore, Theorem 21.1 is proved if Theorem 21.2 holds. □

§2. Hypersurfaces with a Given Second Fundamental Form

As for Theorem 21.2, it is clear that the restriction to $\mathcal{X}$ of an arbitrary motion of the space $\mathcal{V}$ that makes $\mathcal{X}$ coincide with $\mathcal{X}^*$ (and transforms the framing of the hypersurface $\mathcal{X}$ into the framing of the hypersurface $\mathcal{X}^*$) is an isometry $\mathcal{X} \to \mathcal{X}^*$ preserving the second fundamental forms.

Conversely, let there exist an isometry $\varphi\colon \mathcal{X} \to \mathcal{X}^*$ preserving the second fundamental forms of two framed hypersurfaces $\mathcal{X}$ and $\mathcal{X}^*$. Choosing a point p_0 in the space $\mathcal{X}$ and a basis $A_1, \dots, A_n$ in the linear space $\mathbf{T}_{p_0}\mathcal{X}$, we consider a motion $\Phi\colon \mathcal{V} \to \mathcal{V}$ transforming the point p_0 into the point $\varphi(p_0)$, the basis $A_1, \dots, A_n$ into the basis $(d\varphi)_{p_0} A_1, \dots, (d\varphi)_{p_0} A_n$ of the linear space $\mathbf{T}_{\varphi(p_0)}\mathcal{X}^*$, and the normal vector $\boldsymbol{n}_0$ into the normal vector $\boldsymbol{n}_0^*$. (Such a motion always exists and is unique.) Let $\mathcal{X}'$ be the image of the hypersurface $\mathcal{X}^*$ under the motion Φ^{-1}. Then $p_0 \in \mathcal{X}'$, $A_1, \dots, A_n \in \mathbf{T}_{p_0}\mathcal{X}'$, and the mapping

$$\psi = (\Phi|_{\mathcal{X}'})^{-1} \circ \varphi \tag{1}$$

is an isometry $\mathcal{X} \to \mathcal{X}'$ preserving the second fundamental forms and leaving the point p_0 and the vectors $A_1, \dots, A_n$ fixed. If an isometry ψ is induced by a motion Ψ, then the isometry φ is induced by the motion $\Phi \circ \Psi$. Therefore, it suffices to prove Theorem 21.2 for isometries of form (1). This reduces Theorem 21.2 to the following particular case (which also asserts that $\Psi = \mathrm{id}$).

Theorem 21.2′. *Let $\mathcal{X}$ and $\mathcal{X}^*$ be two connected framed two-sided hypersurfaces having a common point p_0 such that $\mathbf{T}_{p_0}\mathcal{X} = \mathbf{T}_{p_0}\mathcal{X}^*$. If there exists an isometry $\varphi\colon \mathcal{X} \to \mathcal{X}^*$ that preserves the second fundamental forms, leaves the point p_0 fixed, and $(d\varphi)_{p_0} = \mathrm{id}$, then $\mathcal{X} = \mathcal{X}^*$ and $\varphi = \mathrm{id}$.*

Let f be an arbitrary isometric immersion of an n-dimensional Riemannian space $\mathcal{X}$ in the $(n+1)$-dimensional Euclidean space $\mathcal{V}$. Assuming that f is an embedding, we can consider the space $\mathcal{X}$ as a hypersurface in $\mathcal{V}$, and choosing a framing $\boldsymbol{n}$, we can therefore speak about its second fundamental form h. For brevity, an immersion f for which a framing $\boldsymbol{n}$ is chosen on $\mathcal{X}$ is called a *framed immersion*, and the corresponding second fundamental form on $\mathcal{X}$ is called the *second fundamental form of the framed immersion* f. (We note that if the space $\mathcal{X}$ is connected, then the framing $\boldsymbol{n}$ is uniquely defined by its value $\boldsymbol{n}_0$ at a certain point $p_0 \in \mathcal{X}$, i.e., by the unit vector $\boldsymbol{n}_0$ that is orthogonal to the hyperplane $(df)_{p_0}\mathbf{T}_{p_0}\mathcal{X}$. The vector $\boldsymbol{n}_0$ is therefore called the *framing vector* of the immersion f.) In this terminology, Theorem 21.2′ can be restated as follows.

Theorem 21.2″. *Let f and f^* be two framed isometric immersions of a connected oriented n-dimensional Riemannian space $\mathcal{X}$ in the $(n+1)$-dimensional Euclidean space $\mathcal{V}$. If*

1. *we have $f(p_0) = f^*(p_0)$ and $(df)_{p_0} = (df^*)_{p_0}$ for a certain point $p_0 \in \mathcal{X}$ and*
2. *the second fundamental forms of the immersions f and f^* coincide,*

then $f = f^*$.

This *uniqueness theorem* can be complemented by the corresponding *existence theorem.*

§3. Hypersurfaces with Given First and Second Fundamental Forms

By construction, each hypersurface $\mathcal{X}$ is a Riemannian space with an additional structure, a symmetric tensor field h of type (2,0) that is connected with the metrical tensor by the Gauss relation

$$R_{ij,kl} = h_{ik}h_{jl} - h_{jk}h_{il} \tag{2}$$

and by the Peterson–Codazzi relation

$$(\nabla_k h)_{ij} = (\nabla_i h)_{kj} \tag{3}$$

(see (3) and (4) in Chap. 20; now $\varepsilon = 1$ because the ambient space $\mathcal{V}$ is Euclidean), which should hold in each chart of the manifold $\mathcal{X}$. This means that relations (2) and (3) are *necessary* for the existence of a realization of the Riemannian space $\mathcal{X}$ on which a symmetric tensor field h is given such that it is a hyperplane of the Euclidean space whose second fundamental form is h. Are these relations sufficient? The answer is positive for simply connected manifolds.

Theorem 21.3. *Let $\mathcal{X}$ be a connected and simply connected n-dimensional Riemannian space on which a symmetric tensor field h of type $(2,0)$ connected with the metric tensor g by Gauss relation* (2) *and Peterson–Codazzi relation*

(3) is given. Further, let p_0 be an arbitrary point of the space $\mathcal{X}$, and let $A_1, \dots, A_n$ be a basis of the space $\mathbf{T}_{p_0}\mathcal{X}$. Finally, let $\mathcal{V}$ be a Euclidean $(n+1)$-dimensional space, $\boldsymbol{r}_0$ be a point of the space $\mathcal{V}$, $\boldsymbol{a}_1, \dots, \boldsymbol{a}_n$ be a family of vectors of the space $\mathcal{V}$ such that

$$\boldsymbol{a}_i \boldsymbol{a}_j = g_{ij}(p_0) \quad \textit{for all } i, j = 1, \dots, n,$$

and $\boldsymbol{n}_0$ be a unit vector of the space $\mathcal{V}$ that is orthogonal to the vectors $\boldsymbol{a}_1, \dots, \boldsymbol{a}_n$. Then there exists an isometric immersion

$$f: \mathcal{X} \to \mathcal{V} \tag{4}$$

such that

$$f(p_0) = \boldsymbol{r}_0, \qquad (df)_{p_0} A_1 = \boldsymbol{a}_1, \quad \dots, \quad (df)_{p_0} A_n = \boldsymbol{a}_n; \tag{5}$$

moreover, h is the second fundamental form of the immersion f with respect to the framing vector $\boldsymbol{n}_0$.

According to Theorem 21.2″ *immersion* (4) *is unique.*

Both Theorems 21.2″ and 21.3 have local analogues. We join them into one theorem and simultaneously generalize it to the case of the pseudo-Euclidean space $\mathcal{V}$. (Of course, a similar generalization is also possible for Theorems 21.2″ and 21.3 themselves. [*Question*: Is it true that Theorem 21.1 can be extended to the case of hypersurfaces in the pseudo-Euclidean space?]

Let two symmetric tensors g and h of type (2,0) with the components g_{ij} and h_{ij} be given on a certain connected open set $\overset{\circ}{U}$ of the Euclidean space $\mathbb{R}^n$. At each point of $\overset{\circ}{U}$, let the following conditions hold:

- **a.** The tensor g is nondegenerate (is a pseudo-Riemannian metric).
- **b.** Relations (3) and (4) in Chap. 20 hold, where $R_{ij,kl}$ and ∇_k are constructed according to the Riemannian connection corresponding to the metric g and $\varepsilon = \pm 1$.

Further, let (k, l) be the signature of the tensor g, and let $\mathcal{V}$ be the $(n+1)$-dimensional pseudo-Euclidean space of signature $(k+1, l)$ if $\varepsilon = 1$ and $(k, l-1)$ if $\varepsilon = -1$, where $\varepsilon = \pm 1$ is the number from relation (3) in Chap. 20. Let $\boldsymbol{x}_0 \in \overset{\circ}{U}$, and let

$$\boldsymbol{a}_1, \dots, \boldsymbol{a}_n, \boldsymbol{n}_0$$

be a basis of the space $\mathcal{V}$ such that

1. the relations
$$\boldsymbol{a}_i \boldsymbol{a}_j = g_{ij}(\boldsymbol{x}_0), \quad i, j = 1, \dots, n,$$
hold,
2. the vector $\boldsymbol{n}_0$ is orthogonal to the vectors $\boldsymbol{a}_1, \dots, \boldsymbol{a}_n$, and

3. the relation
$$n_0^2 = \varepsilon$$
holds.

Finally, let r_0 be an arbitrary vector of the space $\mathcal{V}$.

We consider all possible vector-valued functions $r = r(x) \in \mathcal{V}$ defined in a certain neighborhood of the point x_0 such that

$$r(x_0) = r_0, \qquad \frac{\partial r}{\partial x^1}(x_0) = a_1, \quad \ldots, \quad \frac{\partial r}{\partial x^n}(x_0) = a_n. \tag{6}$$

In a sufficiently small neighborhood of the point x_0, each such function satisfies the regularity condition

$$\frac{\partial r}{\partial x^1} \wedge \cdots \wedge \frac{\partial r}{\partial x^n} \neq 0,$$

and the equation $r = r(x)$ therefore defines an elementary (obviously, nonisotropic) hypersurface $\mathcal{X}$ in $\mathcal{V}$. This hypersurface has a framing n that assumes the value n_0 at the point r_0. We assume that the second fundamental form of the hypersurface $\mathcal{X}$ is constructed using namely this framing.

Theorem 21.4. *If two tensors g and h satisfy conditions* **a** *and* **b** *near the point x_0, then there exists a unique vector-valued function $r = r(x)$ satisfying conditions* (6) *such that the first and second fundamental forms of the hypersurface $\mathcal{X}$ respectively coincide with g and hnear x_0.*

In other words, this theorem asserts the existence and uniqueness of a framed immersion in the (pseudo-)Euclidean space $\mathcal{V}$ of a neighborhood of a point of a Riemannian space $\mathcal{X}$ with the metric g whose second fundamental form is h and which satisfies initial conditions (6).

Before proving this theorem, we deduce the global Theorems 21.2″ and 21.3 from it.

§4. Proof of the Uniqueness

Proof (of Theorem 21.2″). The assertion of Theorem 21.4 on the uniqueness of a vector-valued function $r = r(x)$ means that two immersions f and f^* satisfying conditions 1 and 2 in Theorem 21.2″ coincide in a certain neighborhood of the point p_0. Let C be the set of all points $p \in \mathcal{X}$ for which $f(p) = f^*(p)$ and $(df)_p = (df^*)_p$. This set contains the point p_0 and is therefore not empty. Moreover, if $p \in C$, then by the same uniqueness assertion (applied to p instead of p_0), there exists a neighborhood U of the point p such that $U \subset C$. Therefore, the set C is open.

We now prove that the set C is also closed. Let $p \in \overline{C}$, and let Φ be a motion of the space $\mathcal{V}$ that transforms the point $f(p)$ into the point $f^*(p)$, the hyperplane $(df)_p\mathbf{T}_p\mathcal{X}$ into the hyperplane $(df^*)_p\mathbf{T}_p\mathcal{X}$, and the vector normal to $f\mathcal{X}$ at the point $f(p)$ into the vector normal to $f^*\mathcal{X}$ at the point $f^*(p)$. Then the immersions f and $\Phi^{-1}\circ f^*$ satisfy the conditions of Theorem 21.2″ at the point p and (according to what was proved) therefore coincide in a certain neighborhood U of this point. Therefore, $f^* = \Phi\circ f$ on U. Because $p\in\overline{C}$, there exists a point $q\in C$ such that $q\in U$. Then the motion Φ leaves the point $f(q)$ fixed and induces the identical mapping of the linear space associated with $\mathcal{V}$. Therefore, $\Phi = \mathrm{id}$, and hence $f^* = f$ on U. Therefore, $U\subset C$ and, in particular, $p\in C$.

Being an open-closed subset of the connected manifold $\mathcal{X}$, the set C coincides with the whole $\mathcal{X}$. Therefore, $f = f^*$. □

§5. Proof of the Existence

We now prove Theorem 21.3. To simplify the statements, isometric immersions of the form $U\to\mathcal{V}$, where U is an open set in $\mathcal{X}$, whose second fundamental form (for a certain choice of the framing $\boldsymbol{n}$) is the restriction of a given form h to U are called *admissible immersions* in this proof.

Let $J(\mathcal{X},\mathcal{V})$ be a subset of the space $G(\mathcal{X},\mathcal{V})$ of germs of smooth mappings from $\mathcal{X}$ into $\mathcal{V}$ (see Chap. 5) consisting of germs $[f]_p$ of admissible immersions $f\colon U\to\mathcal{V}$, $p\in U$, $U\subset\mathcal{X}$. (Clearly, if $[f]_p = [f_1]_p$ and f is an immersion at p, then f_1 is also an immersion at p. Moreover, we can assume without loss of generality that f_1 is an isometric embedding.) Obviously, the space $J(\mathcal{X},\mathcal{V})$ is not empty.

Lemma 21.2. *On each component of the space $J(\mathcal{X},\mathcal{V})$, the mapping*

$$\alpha\colon J(\mathcal{X},\mathcal{V})\to\mathcal{X},\qquad [f]_p\mapsto p, \tag{7}$$

is a covering.

Proof. According to Theorem 21.2″, which is already proved, if $f\colon U\to\mathcal{V}$ and $f_1\colon U_1\to\mathcal{V}$ are admissible immersions such that for a certain point $p_0\in U\cap U_1$, we have

$$f(p_0) = f_1(p_0),\qquad (df)_{p_0} = (df_1)_{p_0},$$

then $f = f_1$ on the component $(U\cap U_1)_{p_0}$ of the intersection $U\cap U_1$ that contains the point p_0, and therefore $[f]_p = [f_1]_p$ for any point $p\in(U\cap U_1)_{p_0}$. (Compare with Lemma 5.1.)

This means (compare with the proof of Lemma 5.2) that the open sets

$$\sigma_f U = \{[f]_p : p\in U\} \tag{8}$$

(open in $\alpha^{-1}U$ and therefore in $J(\mathcal{X},\mathcal{V})$) constructed for all connected open sets $U\subset\mathcal{X}$ and any admissible immersions $f\colon U\to\mathcal{V}$ either coincide or are disjoint. Moreover, on each of them, the mapping is a homeomorphism onto U (prove this!). Therefore, to prove Lemma 21.2, we need only show that any point $p_0 \in \mathcal{X}$ has a connected neighborhood U such that any point $[f_1]_p \in \alpha^{-1}U$ belongs to a certain set of form (8), i.e., for any point $p \in U$, each admissible mapping f_1 of the form $U_1 \to \mathcal{V}$, where $p \in U_1 \subset U$, is the restriction of a certain admissible mapping $U \to \mathcal{V}$ in a neighborhood of the point p.

As U, we take an arbitrary neighborhood of the point p_0 for which there exists at least one admissible immersion $f: U \to \mathcal{V}$. The existence of such a neighborhood is ensured by Theorem 21.4. Let Φ be a motion of the space $\mathcal{V}$ that transforms the point $f(p)$ into the point $f_1(p)$, the hyperplane $(df)_p\mathbf{T}_p\mathcal{X}$ into the hyperplane $(df_1)_p\mathbf{T}_p\mathcal{X}$, and the vector normal to $f(U)$ at the point $f(p)$ into the corresponding normal vector at the point $f_1(p)$. Then the immersion $\Phi \circ f: U \to \mathcal{V}$ is admissible and, according to Theorem 21.4, coincides with f_1 in a neighborhood of the point p. □

Proof (of Theorem 21.3). According to Theorem 21.4, there exist a neighborhood U of the point p_0 and an admissible immersion $f: U \to \mathcal{V}$ satisfying conditions (5). Let $J_0(\mathcal{X}, \mathcal{V})$ be the component of the space $J(\mathcal{X}, \mathcal{V})$ that contains the germ $[f]_{p_0}$ of this immersion. Because the space $\mathcal{X}$ is connected and simply connected by assumption, mapping (7) on the component $J_0(\mathcal{X}, \mathcal{V})$, being a covering, is a homeomorphism and therefore has an inverse mapping $\alpha^{-1}: \mathcal{X} \to J_0(\mathcal{X}, \mathcal{V})$. As immersion (4), we take the composition $\beta \circ \alpha^{-1}$, where $\beta: J_0(\mathcal{X}, \mathcal{V}) \to \mathcal{V}$ is the mapping $[f]_p \mapsto f(p)$. It is now clear that the composition $\beta \circ \alpha^{-1}$ is an admissible immersion $\mathcal{X} \to \mathcal{V}$ and satisfies conditions (5). □

Remark 21.1. The presented proof of Theorem 21.3 suggests how its analogue for submanifolds of any codimension (not only for hypersurfaces) can be stated and proved. We leave this to the reader.

We now prove Theorem 21.4.

§6. Proof of a Local Variant of the Existence and Uniqueness Theorem

Proof (of Theorem 21.4). We consider the following set of linear differential equations with respect to $n+1$ vector-valued functions $\boldsymbol{r}_1, \ldots, \boldsymbol{r}_n$ and $\boldsymbol{n}$, where the coefficients Γ^k_{ij} are determined according to the tensor g by formulas (5′) in Chap. 11:

$$\begin{aligned} \frac{\partial \boldsymbol{r}_j}{\partial x^i} &= \Gamma^k_{ij}\boldsymbol{r}_k + h_{ij}\boldsymbol{n}, \quad i, j = 1, \ldots, n, \\ \frac{\partial \boldsymbol{n}}{\partial x^i} &= -\varepsilon h_{ik} g^{kj} \boldsymbol{r}_j, \qquad i = 1, \ldots, n. \end{aligned} \tag{9}$$

If the vector-valued function $\boldsymbol{r} = \boldsymbol{r}(\boldsymbol{x})$ from Theorem 21.4 exists, then these equations have a solution for which $\boldsymbol{n}$ is a unit normal vector to the hypersurface $\mathcal{X}$ and the functions $\boldsymbol{r}_1 \dots, \boldsymbol{r}_n$ are partial derivatives $\partial\boldsymbol{r}/\partial x^1, \dots, \partial\boldsymbol{r}/\partial x^n$ of this function (see (1) and (2) in Chap. 20). Conversely, assuming that these equations have a solution $\boldsymbol{r}_1, \dots, \boldsymbol{r}_n, \boldsymbol{n}$ for which

$$\boldsymbol{r}_1(\boldsymbol{x}_0) = \boldsymbol{a}_1, \quad \dots, \quad \boldsymbol{r}_n(\boldsymbol{x}_0) = \boldsymbol{a}_n, \qquad \boldsymbol{n}(\boldsymbol{x}_0) = \boldsymbol{n}_0, \tag{10}$$

we consider the linear differential form $\boldsymbol{r}_j dx^j$ with vector coefficients (more precisely, a family of linear differential forms whose coefficients are the coordinates of the vector-valued functions $\boldsymbol{r}_j$). Because the partial derivatives $\partial\boldsymbol{r}_j/\partial x^i$ are symmetric in i and j by the first equation in (9), this form is closed. Therefore, by the Poincaré lemma, in each spherical neighborhood of the point $\boldsymbol{x}_0$, it is the differential $d\boldsymbol{r}$ of a certain vector-valued function $\boldsymbol{r} = \boldsymbol{r}(\boldsymbol{x})$. Adding a constant, we can always guarantee the relation $\boldsymbol{r}(\boldsymbol{x}_0) = \boldsymbol{r}_0$, and it is clear that a function $\boldsymbol{r} = \boldsymbol{r}(\boldsymbol{x})$ satisfying this condition is uniquely defined.

Let $\mathcal{X}$ be a hypersurface with equation $\boldsymbol{r} = \boldsymbol{r}(\boldsymbol{x})$. Because the vectors $\boldsymbol{r}_i = \partial\boldsymbol{r}/\partial x^i$, $i = 1, \dots, n$, are just the vectors $\partial/\partial x^i$, $i = 1, \dots, n$, for this hypersurface, the inner products $\boldsymbol{r}_i\boldsymbol{r}_j$ are the components of the metric tensor of the hypersurface $\mathcal{X}$. On the other hand, because

$$\frac{\partial\boldsymbol{r}_i\boldsymbol{r}_j}{\partial x^k} = \frac{\partial\boldsymbol{r}_i}{\partial x^k}\boldsymbol{r}_j + \boldsymbol{r}_i\frac{\partial\boldsymbol{r}_j}{\partial x^k} = \Gamma^p_{ki}\boldsymbol{r}_p\boldsymbol{r}_j + \Gamma^p_{kj}\boldsymbol{r}_i\boldsymbol{r}_p$$

and (see (2′) and (3) in Chap. 11)

$$\frac{\partial g_{ij}}{\partial x^k} = \Gamma^p_{ki}g_{pj} + \Gamma^p_{kj}g_{ip}, \tag{11}$$

the functions $\boldsymbol{r}_i\boldsymbol{r}_j$ and g_{ij} are solutions of the same set of differential equations (11) with the same initial conditions $(\boldsymbol{r}_i\boldsymbol{r}_j)(\boldsymbol{x}_0) = \boldsymbol{a}_i\boldsymbol{a}_j = g_{ij}(\boldsymbol{x}_0)$. Therefore, these functions coincide:

$$g_{ij} = \boldsymbol{r}_i\boldsymbol{r}_j, \quad i, j = 1, \dots, n.$$

This proves that *the first fundamental form of the hypersurface $\mathcal{X}$ is the tensor g.*

Further, because

$$\frac{\partial\boldsymbol{r}_j\boldsymbol{n}}{\partial x^i} = \frac{\partial\boldsymbol{r}_j}{\partial x^i}\boldsymbol{n} + \boldsymbol{r}_j\frac{\partial\boldsymbol{n}}{\partial x^i} = h_{ij}\boldsymbol{n}^2 - \varepsilon h_{ik}g^{kp}\boldsymbol{r}_j\boldsymbol{r}_p$$
$$= h_{ij}\boldsymbol{n}^2 - \varepsilon h_{ik}g^{kp}g_{jp} = (\boldsymbol{n}^2 - \varepsilon)h_{ij}$$

and

$$\frac{\partial\boldsymbol{n}^2}{\partial x^i} = 2\frac{\partial\boldsymbol{n}}{\partial x^i}\boldsymbol{n} = -2\varepsilon h_{ik}g^{kj}\boldsymbol{r}_j\boldsymbol{n},$$

the functions $u_0 = \boldsymbol{n}^2$ and $u_j = \boldsymbol{r}_j\boldsymbol{n}$, $j = 1, \ldots, n$, are solutions of the set of differential equations

$$\frac{\partial u_0}{\partial x^i} = -2\varepsilon h_{ik} g^{kj} u_j, \qquad \frac{\partial u_j}{\partial x^i} = (u_0 - \varepsilon) h_{ij} \tag{12}$$

with the initial conditions $u_0(\boldsymbol{x}_0) = \varepsilon$ and $u_j(\boldsymbol{x}_0) = 0$. Because Eqs. (12) are satisfied by the constant functions $u_0 = \varepsilon$ and $u_j = 0$, this proves that the relations

$$\boldsymbol{n}^2 = \varepsilon \quad \text{and} \quad \boldsymbol{r}_j\boldsymbol{n} = 0$$

hold identically. Therefore, $\boldsymbol{n}$ is a unit normal vector of the hypersurface $\mathcal{X}$.

Comparing Gauss formula (1) in Chap. 20 for the hypersurface $\mathcal{X}$ with the first equation in (9), we immediately find that *the second fundamental form of the hypersurface $\mathcal{X}$ is the tensor h*. Therefore, to complete the proof of Theorem 21.4, we need only prove that Eqs. (9) have a unique solution $\boldsymbol{r}_1, \ldots, \boldsymbol{r}_n, \boldsymbol{n}$ satisfying initial conditions (10).

Equations (9) written for the coordinates of the vectors $\boldsymbol{r}_j$ and $\boldsymbol{n}$ belong to the class of equations of the form

$$\frac{\partial u_j}{\partial x^i} = A_{ij}(\boldsymbol{x}, \boldsymbol{u}), \quad 1 \le i \le n, \quad 1 \le j \le m, \tag{13}$$

where $A_{ij} = A_{ij}(\boldsymbol{x}, \boldsymbol{u})$ are smooth functions of the vectors $\boldsymbol{x} = (x^1, \ldots, x^n)$ and $\boldsymbol{u} = (u_1, \ldots, u_m)$, which vary on open connected sets $U \subset \mathbb{R}^n$ and $V \subset \mathbb{R}^m$. Initial conditions (10) have the form $\boldsymbol{u}(\boldsymbol{x}_0) = \boldsymbol{u}_0$, where $\boldsymbol{x}_0 \in U$ and $\boldsymbol{u}_0 \in V$.

We recall the foundations of the theory of Eqs. (13). Let X_i be linear differential operators (vector fields) defined on a domain $W = U \times V$ of the space $\mathbb{R}^{n+m} = \mathbb{R}^n \times \mathbb{R}^m$ by

$$X_i = \frac{\partial}{\partial x^i} + A_{ij}\frac{\partial}{\partial u_j}, \quad i = 1, \ldots, n. \tag{14}$$

Further, let $p = (\boldsymbol{x}, \boldsymbol{u}) \in W$, and let $\mathcal{E}_p$ be the linear span of the vectors $(X_1)_p, \ldots, (X_n)_p$. It is clear that $\mathcal{E}_p$ is an n-dimensional subspace of the tangent space $\mathbf{T}_pW = \mathbb{R}^{n+m}$ that depends smoothly on p, i.e., the correspondence $p \mapsto \mathcal{E}_p$ is a distribution on W. Because each subspace $\mathcal{E}_p$ is projected bijectively on $\mathbf{T}_{\boldsymbol{x}}U = \mathbb{R}^n$, the integral manifolds of the distribution $\mathcal{E}$ are the *graphs* of the vector-valued functions $\boldsymbol{u} = \boldsymbol{u}(\boldsymbol{x})$ on U (i.e., submanifolds of the space $\mathbb{R}^{n+m}$ consisting of all points of the form $(\boldsymbol{x}, \boldsymbol{u}(\boldsymbol{x}))$). The tangent space to the graph of the function $\boldsymbol{u} = \boldsymbol{u}(\boldsymbol{x})$ at the point $p = (\boldsymbol{x}, \boldsymbol{u})$ is obviously generated by the vectors

$$\frac{\partial}{\partial x^i} + \frac{\partial u_j}{\partial x^i}\frac{\partial}{\partial u_j},$$

which directly imply that *the graph of a vector-valued function $\boldsymbol{u} = \boldsymbol{u}(\boldsymbol{x})$ is an integral manifold of the distribution $\mathcal{E}$ iff this vector-valued function is a solution of Eqs.* (13).

By the Frobenius theorem, this implies that Eqs. (13) have a unique solution satisfying the initial condition of the form $\boldsymbol{u}(\boldsymbol{x}_0) = \boldsymbol{u}_0$ iff the distribution $\mathcal{E}$ is involutive, i.e., when for any $i, j = 1, \ldots, n$, the commutator $[X_i, X_j]$ is expressed linearly through operator (14). Because

$$X_i X_j = \left(\frac{\partial}{\partial x^i} + A_{ik}\frac{\partial}{\partial u_k}\right)\left(\frac{\partial}{\partial x^j} + A_{jl}\frac{\partial}{\partial u_l}\right)$$
$$= \left(\frac{\partial A_{jl}}{\partial x^i} + A_{ik}\frac{\partial A_{jl}}{\partial u_k}\right)\frac{\partial}{\partial u^l} + \dots,$$

where the dots denote terms containing derivations of the second order, we have

$$[X_i, X_j] = B_{[ij]l}\frac{\partial}{\partial u_l},$$

where

$$B_{ijl} = \frac{\partial A_{jl}}{\partial x^i} + A_{ik}\frac{\partial A_{jl}}{\partial u_k} \tag{15}$$

(and $B_{[ij]l} = B_{ijl} - B_{jil}$). In particular, we see that the operators $[X_i, X_j]$ are expressed only through $\partial/\partial u_l$. On the other hand, because (obviously) no nontrivial linear combination of operators (14) can have this property, this thus proves that the distribution $\mathcal{E}$ is involutive iff $[X_i, X_j] = 0$ for any $i, j = 1, \dots, n$, i.e., iff

$$B_{ijl} = B_{jil} \tag{16}$$

for any $i, j = 1, \dots, n$ and $l = 1, \dots, m$.

Therefore, *Eqs.* (13) *have a unique solution* $\boldsymbol{u} = \boldsymbol{u}(\boldsymbol{x})$ *satisfying the initial condition of the form* $\boldsymbol{u}(\boldsymbol{x}_0) = \boldsymbol{u}_0$ *iff relations* (16) *hold.* We can rewrite this condition in another form, which is more convenient to remember. Let $F = F(\boldsymbol{x}, \boldsymbol{u})$ be an arbitrary smooth function of $\boldsymbol{x}$ and $\boldsymbol{u}$. If $\boldsymbol{u}$ is a function of $\boldsymbol{x}$, then, as is known, the derivative of F with respect to x^i is equal to

$$\frac{\partial F}{\partial x^i} + \frac{\partial u_k}{\partial x^i}\frac{\partial F}{\partial u_k}.$$

Therefore, if $\boldsymbol{u}$ satisfies Eqs. (13), then this derivative can be written in the form

$$\frac{\partial F}{\partial x^i} + A_{ik}\frac{\partial F}{\partial u_k}. \tag{17}$$

Because of this, function (17) of $\boldsymbol{x}$ and $\boldsymbol{u}$ is called the *derivative of the function F with respect to x^i by Eqs.* (13). (We stress that the arguments $u_1, \dots, u_m$ in (17) are not assumed to be functions of $x^1, \dots, x^n$.)

Because function (17) for $F = A_{jl}$ is just function (15), in this terminology, *condition* (16) *is equivalent to the fact that for each $l = 1, \dots, m$ and for all $i, j = 1, \dots, n$, the derivative with respect to x_i by Eqs.* (13) *of the right-hand side of the jth equation for the function u_l coincides with the derivative with respect to x_j of the right-hand side of the ith equation.* (In this form, the necessity of condition (16) is obvious because of the symmetry of the second partial derivatives $\partial^2 u_l/(\partial x^i \partial x^j)$ in i and j.)

Remark 21.2. Under the assumption that all functions A_{ij} are real analytic (are functions of class C^ω), the sufficiency of condition (16) can be proved using the indeterminate coefficient method. For this, we expand the functions A_{ij} in power series in the variables $x^1, \dots, x^n$ and $u_1, \dots, u_m$, substitute the power series with indeterminate coefficients for the functions $u_1, \dots, u_m$ in Eqs. (13), and equate the coefficients of equal monomials in $x^1, \dots, x^n$. We obtain a set of linear equations for the unknown coefficients whose compatibility condition is exactly equivalent to the symmetry of all coefficients of the power series for function (15) with respect to i

and j. Therefore, if condition (16) holds, then we obtain natural power series for $u_1, \dots, u_m$. It remains to prove (using the majorant method for example) that these series converge (in a neighborhood of the point $\boldsymbol{x}_0$). Of course, this argument also holds in the complex domain.

For Eqs. (9), the derivatives by these equations of their right-hand sides with respect to x^j have the following form (we replace the subscript j in the first group of these equations with l and the superscript j in the second group with p):

$$\frac{\partial \Gamma_{il}^k}{\partial x^j}\boldsymbol{r}_k + \Gamma_{il}^k(\Gamma_{jk}^p \boldsymbol{r}_p + h_{jk}\boldsymbol{n}) + \frac{\partial h_{il}}{\partial x^j}\boldsymbol{n} + h_{il}(-\varepsilon h_{jk}g^{kp}\boldsymbol{r}_p)$$

$$= \left(\frac{\partial \Gamma_{il}^p}{\partial x^j} + \Gamma_{il}^k\Gamma_{jk}^p - \varepsilon g^{kp}h_{il}h_{jk}\right)\boldsymbol{r}_p + \left(\Gamma_{il}^k h_{jk} + \frac{\partial h_{il}}{\partial x^j}\right)\boldsymbol{n},$$

$$-\varepsilon\frac{\partial(h_{ik}g^{kp})}{\partial x^j}\boldsymbol{r}_p - \varepsilon h_{ik}g^{kp}(\Gamma_{jp}^q\boldsymbol{r}_q + h_{jp}\boldsymbol{n})$$

$$= -\varepsilon\left(\frac{\partial(h_{ik}g^{kp})}{\partial x^j} + h_{ik}g^{kq}\Gamma_{jq}^p\right)\boldsymbol{r}_p - \varepsilon h_{ik}h_{jp}g^{kp}\boldsymbol{n}.$$

Therefore, condition (16) for these equations is reduced to the following four relations:

$$\begin{aligned}
&\frac{\partial \Gamma_{il}^p}{\partial x^j} + \Gamma_{il}^k\Gamma_{jk}^p - \varepsilon g^{kp}h_{il}h_{jk} = \frac{\partial \Gamma_{jl}^p}{\partial x^i} + \Gamma_{jl}^k\Gamma_{ik}^p - \varepsilon g^{kp}h_{jl}h_{ik},\\
&\Gamma_{il}^k h_{jk} + \frac{\partial h_{il}}{\partial x^j} = \Gamma_{jl}^k h_{ik} + \frac{\partial h_{jl}}{\partial x^i},\\
&\frac{\partial(h_{ik}g^{kp})}{\partial x^j} + h_{ik}g^{kq}\Gamma_{jq}^p = \frac{\partial(h_{jk}g^{kp})}{\partial x^i} + h_{jk}g^{kq}\Gamma_{iq}^p,\\
&g^{kp}h_{ik}h_{jp} = g^{kp}h_{jk}h_{ip}.
\end{aligned} \tag{18}$$

Therefore, to complete the proof of Theorem 21.4, it remains to verify the fulfillment of these relations. Subtracting the right-hand side of (18) from its left-hand side, we obtain the expression

$$\frac{\partial \Gamma_{il}^p}{\partial x^j} - \frac{\partial \Gamma_{jl}^p}{\partial x^i} + \Gamma_{il}^k\Gamma_{jk}^p - \Gamma_{jl}^k\Gamma_{ik}^p - \varepsilon g^{kp}(h_{il}h_{jk} - h_{jl}h_{ik})$$

$$= R_{l,ji}^p - \varepsilon g^{kp}(h_{il}h_{jk} - h_{jl}h_{ik}) = \varepsilon g^{kp}(\varepsilon R_{kl,ji} - h_{il}h_{jk} + h_{jl}h_{ik}),$$

which is zero by the Gauss relation (formula (3) in Chap. 20). Rewriting the second relation in (18) as

$$\frac{\partial h_{jl}}{\partial x^i} - \Gamma_{il}^k h_{jk} = \frac{\partial h_{il}}{\partial x^j} - \Gamma_{jl}^k h_{ik}$$

and subtracting $\Gamma_{ij}^k h_{kl}$ from both sides, we represent it in the form of the Peterson–Codazzi relation

$$(\nabla_i h)_{jl} = (\nabla_j h)_{il}.$$

Therefore, the second relation also holds.

Similarly, we represent the third relation in (18) as the following equality of covariant partial derivatives of the tensor h with the second subscript raised (having the components $h_i^p = g^{kp} h_{ik}$):

$$(\nabla_i h)_j^p = (\nabla_j h)_i^p. \tag{19}$$

(It suffices to subtract $\Gamma_{ij}^q h_q^p$ from both sides.) Because the metric tensor g is covariantly constant, we have $(\nabla_i h)_j^p = g^{kp}(\nabla_i h)_{jk}$. Therefore, after contraction with the tensor g, relation (19) passes to the Peterson–Codazzi relation.

Because the fourth relation (18) obviously holds, Theorem 21.4 is thus completely proved. □

We need this theorem only in the case where the tensor g is positive definite (has the signature $(n, 0)$). In this case, $\mathcal{X}$ is a Riemannian space embedded in the Euclidean space for $\varepsilon = 1$ and in the pseudo-Euclidean space of signature $(n, 1)$ for $\varepsilon = -1$.

Chapter 22
Spaces of Constant Curvature

§1. Spaces of Constant Curvature

Let a Riemannian space $\mathcal{X}$ have the following property:

★ At each point $p \in \mathcal{X}$, the sectional curvature $K_p(\pi)$ of the space $\mathcal{X}$ is the same for all two-dimensional directions $\pi \subset \mathbf{T}_p\mathcal{X}$.

If this property holds, then the formula

$$K_p = K_p(\pi), \quad p \in \mathcal{X}, \quad \pi \subset \mathbf{T}_p\mathcal{X}, \tag{1}$$

correctly defines a certain function $K: p \mapsto K_p$ on $\mathcal{X}$.

In the language of the Riemannian tensor R of the space $\mathcal{X}$ interpreted as a quadratic functional on $\Lambda^2\mathcal{X}$ (see Chap. 15), property ★ means that for any bivector field $\boldsymbol{P} \in \boldsymbol{A}^2\mathcal{X}$, the value $R(\boldsymbol{P})$ of the tensor R on $\boldsymbol{P}$ is expressed by

$$R(\boldsymbol{P}) = K|\boldsymbol{P}|^2 \tag{2}$$

(compare with formula (20) in Chap. 15). On the other hand, the same formula (2) obviously also holds for the Bianchi tensor KE, where E is the Bianchi tensor, which is the identity operator $\Lambda^2\mathcal{X} \to \Lambda^2\mathcal{X}$ (see Chap. 16). Therefore, $R = KE$ on $\boldsymbol{A}^2\mathcal{X}$ and hence, see Proposition 15.3, on the whole $\Lambda^2\mathcal{X}$. This proves that *a Riemannian space $\mathcal{X}$ has property ★ iff the relation*

$$R = KE \tag{3}$$

holds for its Riemannian curvature tensor, where K is a certain function on $\mathcal{X}$ (connected with the sectional curvature by formula (1)). (We have met condition (3) repeatedly; see, e.g., Proposition 18.3.)

For components $R_{ij,kl}$ of the curvature tensor, condition (3) means that

$$R_{ij,kl} = K(g_{ik}g_{jl} - g_{il}g_{jk}) \tag{3'}$$

(see Chap. 15), and for components $R^i_{j,kl}$, it means that

$$R^i_{j,kl} = K(\delta^i_k g_{jl} - \delta^i_l g_{jk}). \tag{3''}$$

For $n = 2$, this yields a new proof of Proposition 16.1 because property ★ holds automatically in this case. (We note that in this case, K is just the Gaussian curvature of the surface $\mathcal{X}$.)

We consider the tensor W with the components

$$W^i_{j,kl} = R^i_{j,kl} - \frac{1}{n-1}(R_{jl}\delta^i_k - R_{jk}\delta^i_l)$$

or

$$W_{ij,kl} = R_{ij,kl} - \frac{1}{n-1}(R_{jl}g_{ik} - R_{jk}g_{il}).$$

It is easy to see that the tensor W is a *Weyl tensor*. It is called the *projective curvature tensor*. (We have no possibility to explain this name here.)

Proposition 22.1. *The tensor W is zero iff the Riemannian curvature tensor of the space $\mathcal{X}$ has form* (3).

Proof. It is clear that if

$$R_{ij} = \frac{\mathcal{R}}{n}g_{ij}$$

(the space $\mathcal{X}$ is an Einstein space), then identity (3′) is equivalent to the relation $W_{ij,kl} = 0$ (see formulas (10) in Chap. 18 and (7) in Chap. 16). On the other hand, as we know (see Exercise 17.12), if (3′) holds, then the space $\mathcal{X}$ is Einstein, and if $W_{ij,kl} = 0$, i.e., if

$$R_{ij,kl} = \frac{1}{n-1}(R_{jl}g_{ik} - R_{jk}g_{il}),$$

then

$$R_{ik} = g^{jl}R_{ij,kl} = \frac{1}{n-1}(\mathcal{R}g_{ik} - R_{ik}),$$

and $R_{ij} = (\mathcal{R}/n)g_{ij}$ again. □

Proposition 22.2 (Schur theorem). *The function K defined by* (1) *is constant for any connected Riemannian space $\mathcal{X}$ of dimension $n \geq 3$ having property* ★.

Proof. It follows from (3″) that for any i, j, k, l, and s,

$$\nabla_s R^i_{j,kl} = g_{jl}\delta^i_k \partial_s K - g_{jk}\delta^i_l \partial_s K,$$

where $\partial_s K = \partial K/\partial x^s$. (We recall that tensors with the components g_{ij} and δ^i_j are covariantly constant.) Therefore, by the second Bianchi formula (formula (23″) in Chap. 2), the cycling of the right-hand side of this relation with respect k, l, and s yields zero:

$$g_{j(l}\delta^i_k \partial_{s)}K - g_{j(k}\delta^i_l \partial_{s)}K = 0. \tag{4}$$

Because $n \geq 3$ by condition, for any subscript s, there exist subscripts k and l that are different from it and from one another. But if all three subscripts k, l, and s are different, then relation (4) for $i = k$ becomes

$$g_{js}\partial_l K - g_{jl}\partial_s K = 0.$$

Contracting this relation with the tensor g^{pj}, we obtain

$$\delta^p_s \partial_l K - \delta^p_l \partial_s K = 0.$$

In particular, for $p = l$, this implies (because $\delta^l_s = 0$) that $\partial_s K = 0$. Because this is true for any s, we therefore have $K = \text{const}$. □

Exercise 22.1. Present this argument in an invariant form (not using the components).

Proposition 22.2 is the motivation for the following definition.

Definition 22.1. A connected Riemannian space $\mathcal{X}$ is called a *space of constant curvature* K if for each point $p \in \mathcal{X}$ and for any two-dimensional direction $\pi \subset \mathbf{T}_p\mathcal{X}$, its section curvature $K_p(\pi)$ equals K, i.e., in other words, if relation (3) holds for its Riemannian curvature tensor.

For $n \geq 3$, according to Proposition 22.2, these are exactly connected Riemannian spaces for which (2) holds, and for $n = 2$, these are the already known surfaces of constant Gaussian curvature.

§2. Model Spaces of Constant Curvature

As is seen in what follows, the following three examples of constant-curvature spaces play a special role. The spaces in these examples are called *model spaces of constant curvature.*

Example 22.1. Let $\mathcal{X} = \mathbb{R}^n$ with the usual Euclidean metric

$$ds^2 = (dx^1)^2 + \cdots + (dx^n)^2. \tag{5}$$

Because $g_{ij} = \text{const}$, we have $\Gamma^i_{kj} = 0$ and, moreover, $R^i_{j,kl} = 0$. Therefore, relation (3″) holds for $K = 0$. Therefore, *the space* $\mathbb{R}^n$ *is a space of curvature zero* (constant).

Example 22.2. Let $\mathcal{X} = \mathbb{B}^n_R$ be the open ball $|\boldsymbol{x}| < R$ of the space $\mathbb{R}^n$ with the metric

$$ds^2 = 4R^4 \frac{(dx^1)^2 + \cdots + (dx^n)^2}{(R^2 - |\boldsymbol{x}|^2)^2}, \tag{6}$$

where $|\boldsymbol{x}|^2 = (x^1)^2 + \cdots + (x^n)^2$. Although a direct computation of the curvature tensor is automatic in this case, it is a sufficiently tedious problem. Therefore, we choose an indirect approach.

We first note that the ball $\mathbb{B}^n_R$ with metric (6) is exactly the Poincaré model of the n-dimensional Lobachevskii space. Therefore, we can apply Lobachevskii geometry to its study.

First let $n = 2$. Then we can pass to the Poincaré model on the upper half-plane for which

$$ds^2 = R^2 \frac{dx^2 + dy^2}{y^2}.$$

The further transformation

$$u = R \ln \frac{y}{R}, \qquad v = x$$

(with the inverse transformation $x = v$, $y = Re^{u/R}$) transforms this quadratic form into the form

$$ds^2 = du^2 + e^{-2u/R}dv^2,$$

i.e., the form given by formula (9) in Chap. 16 (with $G = e^{-2u/R}$). Therefore, according to the computations in Chap. 16, the Gauss curvature in this case is

$$-\frac{(\sqrt{G})_{uu}}{\sqrt{G}} = -\frac{e^{-u/R}}{R^2 e^{-u/R}} = -\frac{1}{R^2}.$$

This proves that for $n = 2$, our Riemannian space (the Lobachevskii plane) is a space of the constant negative curvature $K = -1/R^2$.

Now let $n > 2$. In the Lobachevskii geometry, lines are minimal arcs and therefore geodesics. Therefore, for any point $p_0 \in \mathcal{X}$ and any two-dimensional plane $\pi \subset \mathbf{T}_{p_0}\mathcal{X}$, the surface in Proposition 16.2 (sweeping out by geodesics emanating from the point p_0) is just a plane in the Lobachevskii space and is therefore itself a Lobachevskii space. By what was proved above, its Gaussian curvature, i.e., its section curvature $K_{p_0}(\pi)$ (by Proposition 16.2), is therefore equal to $-1/R^2$. This proves that *for any $n \geq 2$, the Lobachevskii space $\mathbb{B}^n_R$ is a Riemannian space of the constant negative curvature $K = -1/R^2$.*

We note that metric (6) (after multiplication of all coordinates by 2) can also be written in the form

$$ds^2 = \frac{(dx^1)^2 + \cdots + (dx^n)^2}{\left(1 + (K/4)|\boldsymbol{x}|^2\right)^2}. \tag{7}$$

For $K = 0$, this metric passes into the Euclidean metric (5) in Example 22.1.

Example 22.3. Let $\mathcal{X} = \mathbb{S}^n_R$ be the sphere $|\boldsymbol{t}|^2 = R^2$, $\boldsymbol{t} = (t^0, t^1, \ldots, t^n)$, of radius R in the space $\mathbb{R}^{n+1}$ with the metric induced by the Euclidean metric of the space $\mathbb{R}^{n+1}$, and let $x^1, \ldots, x^n$ be the stereographic coordinates on the sphere $\mathbb{S}^n_R$ connected with the coordinates $t^0, t^1, \ldots, t^n$ by

$$t^0 = R\frac{R^2 - |\boldsymbol{x}|^2}{R^2 + |\boldsymbol{x}|^2}, \qquad |\boldsymbol{x}|^2 = (x^1)^2 + \cdots + (x^n)^2,$$

$$t^i = \frac{2R^2}{R^2 + |\boldsymbol{x}|^2}x^i, \quad i = 1, \ldots, n.$$

Then in a clear notation, we have

$$dt^0 = -\frac{4R^3}{(R^2 + |\boldsymbol{x}|^2)^2}\boldsymbol{x}\,d\boldsymbol{x},$$

$$\begin{aligned} dt^i &= -\frac{4R^2\boldsymbol{x}\,d\boldsymbol{x}}{(R^2 + |\boldsymbol{x}|^2)^2}x^i + \frac{2R^2 dx^i}{R^2 + |\boldsymbol{x}|^2} \\ &= -\frac{2R^2}{(R^2 + |\boldsymbol{x}|^2)^2}[2(\boldsymbol{x}\,d\boldsymbol{x})x^i - (R^2 + |\boldsymbol{x}|^2)dx^i], \end{aligned}$$

and therefore

$$(dt^0)^2 + (dt^1)^2 + \ldots + (dt^n)^2$$
$$= \frac{16R^6}{(R^2+|\boldsymbol{x}|^2)^4}(\boldsymbol{x}\,d\boldsymbol{x})^2 + \frac{4R^4}{(R^2+|\boldsymbol{x}|^2)^4}\Big[4(\boldsymbol{x}d\boldsymbol{x})^2\sum_{i=1}^n (x^i)^2$$
$$- 4(R^2+|\boldsymbol{x}|^2)(\boldsymbol{x}\,d\boldsymbol{x})\sum_{i=1}^n x^i dx^i + (R^2+|\boldsymbol{x}|^2)^2\sum_{i=1}^n (dx^i)^2\Big]$$
$$= \frac{4R^4}{(R^2+|\boldsymbol{x}|^2)^4}[4R^2(\boldsymbol{x}\,d\boldsymbol{x})^2 + 4(\boldsymbol{x}\,d\boldsymbol{x})^2|\boldsymbol{x}|^2$$
$$- 4(R^2+|\boldsymbol{x}|^2)(\boldsymbol{x}d\boldsymbol{x})^2 + (R^2+|\boldsymbol{x}|^2)^2(d\boldsymbol{x})^2] = \frac{4R^4(d\boldsymbol{x})^2}{(R^2+|\boldsymbol{x}|^2)^2}.$$

This means that *the Riemannian metric on* $\mathbb{S}^n_R$ *in the coordinates* $x^1, \ldots, x^n$ *is given by*

$$ds^2 = 4R^4\frac{(dx^1)^2 + \cdots + (dx^n)^2}{(R^2+|\boldsymbol{x}|^2)^2}. \tag{8}$$

To compute the curvature of this metric, we can proceed as in Example 22.2 above (it should only be verified preparatorily that arcs of a great circle are geodesics on the sphere), but we can also proceed more easily by noting that the result obtained in Example 22.2 means that if for $K < 0$, all the calculations needed for computing the curvature tensor are performed for metric (7), then we obtain the tensor with components (3). But it is clear that these calculations cannot depend on the sign of the constant parameter K and yield the same result for K positive (or equal to zero). Because metric (8) (after multiplication of all the coordinates by 2) has form (7) with $K = 1/R^2$, this proves that *the sphere* $\mathbb{S}^n_R$ *with metric* (8) *is a space of the constant curvature* $K = 1/R^2$.

We stress that *all three spaces* $\mathbb{R}^n$, $\mathbb{B}^n_R$, *and* $\mathbb{S}^n_R$ *have a metric of form* (7). The distinction between these spaces is only that K equals zero for $\mathbb{R}^n$, is negative for $\mathbb{B}^n_R$, and is positive for $\mathbb{S}^n_R$. (Moreover, the coordinates $x^1, \ldots, x^n$ are defined everywhere for the spaces $\mathbb{R}^n$ and $\mathbb{B}^n_R$, while they are defined only outside the pole of the stereographic projection for the sphere $\mathbb{S}^n_R$.)

To unify the notation, the spaces $\mathbb{R}^n$, $\mathbb{B}^n_R$, and $\mathbb{S}^n_R$ with metric (7) are denoted by $\mathcal{M}_K$, where $K = 0$ for the space $\mathbb{R}^n$ and $K = \pm 1/R^2$ for the spaces $\mathbb{S}^n_R$ and $\mathbb{B}^n_R$.

§3. Model Spaces as Hypersurfaces

By definition, for $K > 0$, the space $\mathcal{M}_K$ is the hypersurface $\mathbb{S}^n_R$, $R = 1/\sqrt{K}$, of the Euclidean space $\mathbb{R}^{n+1}$, and the Riemannian metric on $\mathcal{M}_K$ is induced by the Euclidean metric on $\mathbb{R}^{n+1}$. Similarly, for $K = 0$, we can identify the space $\mathcal{M}_K = \mathbb{R}^n$ with the hypersurface $x^{n+1} = 0$ of the space $\mathbb{R}^{n+1}$, and the metric on $\mathcal{M}_K$ is then also induced by the metric on $\mathbb{R}^{n+1}$. The space $\mathcal{M}_K$ for $K < 0$ then can be identified with the hyperplane of the pseudo-Euclidean space $\mathbb{R}^{n+1}$ of signature $(1, n)$, which is a half of the sphere of radius $R = 1/\sqrt{-K}$ in this space. Because this hypersurface is obviously completely nonisotropic, the metric of the space $\mathbb{R}^{n+1}$ induces a certain Riemannian metric on $\mathcal{M}_K$, $K < 0$ (more precisely, a Riemannian metric is obtained under an additional change of sign).

Exercise 22.2. Show that the latter metric coincides with metric (7).

Therefore, *each space $\mathcal{M}_K$ is a hyperplane of the space* $\mathbb{R}^{n+1}$ *(which is Euclidean for $K \geq 0$ and pseudo-Euclidean for $K < 0$) equipped with the induced Riemannian metric.*

We compute the second fundamental form h of this hypersurface. Because the radios of the sphere $\mathbb{S}^n_R$ is directed along the normal to this sphere, its unit normal vector at the point with the radius vector $\boldsymbol{a}$ is given by

$$\boldsymbol{n} = \frac{\boldsymbol{a}}{R}. \tag{9}$$

It turns out that *formula* (9) *also holds for the sphere of the pseudo-Euclidean space* (of course, if the orthogonality and the length are understood with respect to the pseudo-Euclidean inner product). Indeed, the sphere of radius R in the Euclidean as well as in the pseudo-Euclidean space has the equation

$$\boldsymbol{t}^2 - R^2 = 0 \tag{10}$$

(but in the pseudo-Euclidean space, $\boldsymbol{t}^2 = (t^0)^2 - (t^1)^2 - \cdots - (t^n)^2$; we assume that the pseudo-Euclidean metric has the signature $(1, n)$ because only this case is needed). On the other hand, it is known from a calculus course that the tangent hyperplane at the point with the radius vector $\boldsymbol{a} = (a^0, a^1, \ldots, a^n)$ to the hyperplane with the equation $F(\boldsymbol{t}) = 0$ is given by the equation

$$\left(\frac{\partial F}{\partial t^0}\right)_{\boldsymbol{a}} (t^0 - a^0) + \left(\frac{\partial F}{\partial t^1}\right)_{\boldsymbol{a}} (t^1 - a^1) + \cdots + \left(\frac{\partial F}{\partial t^n}\right)_{\boldsymbol{a}} (t^n - a^n) = 0.$$

For sphere (10), this equation (in the case of the pseudo-Euclidean space of signature $(1, n)$ and after dividing by 2) becomes

$$a^0(t^0 - a^0) - a^1(t^1 - a^1) - \cdots - a^n(t^n - a^n) = 0,$$

which means that the vector $\boldsymbol{a}$ is orthogonal (with respect to the pseudo-Euclidean metric) to each vector $\boldsymbol{t} - \boldsymbol{a}$ of this hyperplane, i.e., in other words, this vector is the vector normal to sphere (10) at the point considered. Therefore, the unit normal vector is $\lambda\boldsymbol{a}$, where λ is the inverse of the length of the vector $\boldsymbol{a}$ (in the pseudo-Euclidean metric). To complete the proof, it suffices to note that this length is equal to R by assumption.

We note that, in fact, the usual argument that proves formula (9) for the sphere of the Euclidean space is repeated here.

Let g_{ij} be a Riemannian metric on the hypersurface $\mathcal{M}_K$, which is induced by the metric of the space $\mathbb{R}^{n+1}$ in the case $K \geq 0$ and differs from the induced metric by the sign in the case $K < 0$. We prove that for each of the hypersurfaces $\mathcal{M}_K$, we have

$$h_{ij} = -\sigma\sqrt{|K|}g_{ij}, \tag{11}$$

where $\sigma = +1$ for $K \geq 0$ and $\sigma = -1$ for $K < 0$ (and *all points of the hypersurface $\mathcal{M}_K$ are therefore umbilical*; see Chap. 20). Indeed, according to formula (1) in Chap. 20, if $\boldsymbol{a} = \boldsymbol{a}(\boldsymbol{x})$ is the parametric vector equation of the hypersurface $\mathcal{M}_K$, then

$$h_{ij} = \boldsymbol{a}_{ij}\boldsymbol{n}$$

because $\boldsymbol{n}^2 = 1$. On the other hand, because $\boldsymbol{a}_i\boldsymbol{n} = 0$, we have $\boldsymbol{a}_{ij}\boldsymbol{n} = -\boldsymbol{a}_i\boldsymbol{n}_j$, and according to (9),

$$\boldsymbol{n}_j = \frac{\partial}{\partial x^j}\left(\frac{\boldsymbol{a}}{R}\right) = \frac{\boldsymbol{a}_j}{R} = \sqrt{|K|}\boldsymbol{a}_j$$

for $K \neq 0$. Because $\boldsymbol{a}_i\boldsymbol{a}_j = \sigma g_{ij}$, this proves (11). (For $K = 0$, formula (9) is not applicable, but $\boldsymbol{n} = \text{const}$ in this case, and therefore $h_{ij} = 0$.)

§4. Isometries of Model Spaces

The interpretation of the space $\mathcal{M}_K$ as a subspace of the (pseudo-)Euclidean space also allows easily finding its isometry group $\operatorname{Iso}\mathcal{M}_K$. First let $K > 0$. Each orthogonal transformation of the space $\mathbb{R}^{n+1}$ (an element of the group $\mathcal{O}(n+1)$) transforms the sphere $\mathbb{S}^n_R$ into itself and, being an isometry of the space $\mathbb{R}^{n+1}$, therefore induces a certain isometry $\mathbb{S}^n_R \to \mathbb{S}^n_R$. The arising mapping $\mathcal{O}(n+1) \to \operatorname{Iso}\mathbb{S}^n_R$ is obviously an monomorphism. This allows identifying an orthogonal transformation from $\mathcal{O}(n+1)$ with its image in $\operatorname{Iso}\mathbb{S}^n_R$; therefore, we can assume that

$$\mathcal{O}(n+1) \subset \operatorname{Iso}\mathbb{S}^n_R. \tag{12}$$

Similarly for $K < 0$, there arises the inclusion

$$O^{\uparrow}(1,n) \subset \operatorname{Iso}\mathbb{B}^n_R, \tag{13}$$

where $O^{\uparrow}(1,n)$ is the group of all orthochronous (preserving the direction of the time axis) pseudo-orthogonal transformations of the $(n+1)$-dimensional pseudo-Euclidean space of signature $(1,n)$.

Finally, for $K = 0$, we have the inclusion

$$\operatorname{Euc}(n) \subset \operatorname{Iso}\mathbb{R}^n, \tag{14}$$

where $\mathrm{Euc}(n)$ is the group of all motions (proper and improper) of the Euclidean space $\mathbb{R}^n$. (From the algebraic standpoint, the group $\mathrm{Euc}(n)$ is the *semidirect product* $\mathrm{Trans}(n) \ltimes \mathcal{O}(n)$ of the translation group $\mathrm{Trans}(n) \approx \mathbb{R}^n$ and the group $\mathcal{O}(n)$. The geometric expression of this is the representation of any motion $\mathbb{R}^n \to \mathbb{R}^n$ by $\boldsymbol{x} \mapsto \boldsymbol{A}\boldsymbol{x} + \boldsymbol{a}$, where $\boldsymbol{A} \in \mathcal{O}(n)$ and $\boldsymbol{a} \in \mathbb{R}^n$.)

The groups

$$\mathrm{Euc}(n), \qquad \mathcal{O}(n+1), \qquad O^{\uparrow}(1,n) \tag{15}$$

(whose elements are here called *motions* of the corresponding geometries for brevity) are Lie groups consisting of two connected components. Their components of the identity are the corresponding groups

$$\mathrm{Euc}^+(n) = \mathrm{Trans}(n) \ltimes \mathrm{SO}(n),\ \mathrm{SO}(n+1),\ O^{\uparrow}_{+}(1,n) \tag{16}$$

(the groups of *proper motions*). On the other hand, according to Proposition 19.3, the groups $\mathrm{Iso}\,\mathcal{M}_K$ are also Lie groups; moreover, their dimension does not exceed $n(n+1)/2$. But the dimension of groups (16) is equal to this number! Therefore, in fact, we have

$$\dim \mathrm{Iso}\,\mathcal{M}_K = \frac{n(n+1)}{2}, \tag{17}$$

and (because a Lie subgroup of a connected Lie group whose dimension is equal to the dimension of the group coincides with the whole group) *the component of the identity of the group* $\mathrm{Iso}\,\mathcal{M}_K$ *is hence the corresponding group* (16).

This assertion (and even its refinement referring to the whole group $\mathrm{Iso}\,\mathcal{M}_K$) can be proved differently by replacing the reference to Proposition 19.3 with a reference to Proposition 3.2, according to which each isometry $f\colon \mathcal{M}_K \to \mathcal{M}_K$, being an affinity, is uniquely determined by its differential

$$(df)_{p_0}\colon \mathbf{T}_{p_0}\mathcal{M}_K \to \mathbf{T}_{q_0}\mathcal{M}_K, \qquad q_0 = f(p_0),$$

at an arbitrary point $p_0 \in \mathcal{M}_K$, which is chosen and fixed for what follows. On the other hand, it is easy to see that *for any two points* $p_0, q_0 \in \mathcal{M}_K$ *and for any isometric mapping*

$$\varphi\colon \mathbf{T}_{p_0}\mathcal{M}_K \to \mathbf{T}_{q_0}\mathcal{M}_K,$$

there exists a motion f_φ of the space $\mathcal{M}_K$ such that $f_\varphi(p_0) = q_0$ and $(df_\varphi)_{p_0} = \varphi$.

Exercise 22.3. Prove this assertion.

In particular, the motion f_φ constructed according to the differential $\varphi = (df)_{p_0}$ is just the isometry f. This again proves that *each isometry is a motion.* We can therefore see that all three inclusions (12), (13) and (14) are equalities:

$$\mathrm{Iso}\,\mathcal{M}_k = \begin{cases} \mathrm{Euc}(n) & \text{if } K = 0, \\ \mathcal{O}(n+1) & \text{if } K > 0, \\ O^{\uparrow}(1,n) & \text{if } K < 0, \end{cases} \tag{18}$$

i.e., *isometries of the spaces* $\mathcal{M}_K$ *are exactly their motions.* Moreover, we see that *the group* $\mathrm{Iso}\,\mathcal{M}_K$ *acts transitively on the space* $\mathcal{M}_K$.

§5. Fixed Points of Isometries

Fixed points of the Euclidean motion $\gamma: \boldsymbol{x} \to \boldsymbol{A}\,\boldsymbol{x} + \boldsymbol{a}$ are found from the equation

$$(\boldsymbol{A} - \boldsymbol{E})\boldsymbol{x} + \boldsymbol{a} = 0,$$

which assigns a certain plane in the space $\mathbb{R}^n$. Its dimension r can be any integer from -1 up to n; this plane is the empty set for $r = -1$, a point for $r = 0$, and a line for $r = 1$. The relation $r = n$ means that the motion γ is the identity.

Because the motions of the sphere $\mathbb{S}^n_R$ are rotations (which, in general, are improper) of the ambient space $\mathbb{R}^{n+1}$, i.e., elements of the group $\mathcal{O}(n+1)$, and the motions of the Lobachevskii space $\mathbb{B}^n_R$ are elements of the Lorentz group $O^{\uparrow}(1, n)$, we see that this conclusion also holds in the pseudo-Euclidean spaces $\mathbb{S}^n_R$ and $\mathbb{B}^n_R$. For convenience of reference, we state this conclusion in the form of a separate proposition.

Proposition 22.3. *In an arbitrary model space $\mathcal{M}_K$ of constant curvature, fixed points of each motion $\mathcal{M}_K \to \mathcal{M}_K$ compose a certain plane.*

§6. Riemann Theorem

The characteristic of spaces $\mathcal{M}_K$ as model ones is explained by the following *Riemann theorem.*

Theorem 22.1. *Each space $\mathcal{X}$ of constant curvature is locally isometric to the space $\mathcal{M}_K$, i.e., any point of $\mathcal{X}$ has a neighborhood that is isometrically mapped onto a neighborhood of a certain point of the space $\mathcal{M}_K$.*

Because the group $\operatorname{Iso}\mathcal{M}_K$ acts transitively on the space $\mathcal{M}_K$, any two points of this space have isometric neighborhoods. In the coordinate language, Theorem 22.1 asserts that the space $\mathcal{X}$ *admits an atlas in each of whose charts the metric of this space is of form* (7).

Proof. Formula (11) implies that to prove Theorem 22.1, it suffices to show that the tensor σg, where g is the metric tensor of the space $\mathcal{X}$, and the tensor h with the components

$$h_{ij} = -\sigma\sqrt{|K|}g_{ij} \tag{19}$$

satisfy the Gauss relation and the Peterson–Codazzi relation. (Indeed, according to Theorem 21.4, a certain neighborhood of any point of the space $\mathcal{X}$ is then isometrically embedded in $\mathbb{R}^{n+1}$ as an elementary hypersurface with the fundamental tensors σg and h. Therefore, by the uniqueness assertion for this hypersurface, under the corresponding choice of the initial conditions, it is contained in the hypersurface $\mathcal{M}_K$ as an open set.)

But the curvature tensor of the space $\mathcal{X}$ is connected with the metric g_{ij} by (3′), and the curvature tensor for the metric σg_{ij} is therefore expressed by the formula

$$R_{ij,kl} = \sigma K(g_{ik}g_{jl} - g_{il}g_{jk}).$$

By (19), this immediately yields the Gauss relation (formula (3) in Chap. 20) for the tensors σg and h. Further, because the tensor σg is covariantly constant, the tensor with components (19) is also covariantly constant, i.e., $(\nabla_k h)_{ij} = 0$ for all $i, j, k = 1, \ldots, n$. Therefore, the Peterson–Codazzi relation (formula (4) in Chap. 20) also holds. □

Chapter 23
Space Forms

§1. Space Forms

We can obtain other spaces of constant curvature from the space $\mathcal{M}_K$. It is clear that if one of the spaces $\widetilde{\mathcal{X}}$ or $\mathcal{X}$ in a Riemannian covering $(\widetilde{\mathcal{X}}, \pi, \mathcal{X})$ is a Riemannian space of constant curvature K, then the other space is also a space of constant curvature K. In particular, *for any group* Γ *of isometries of the space* $\mathcal{M}_K$ *with a discrete action, the quotient space* $\mathcal{M}_K/\Gamma$ *is a space of constant curvature* K.

Definition 23.1. Quotient spaces of the form $\mathcal{M}_K/\Gamma$ are called *space forms* (they are said to be *Euclidean* or *parabolic* for $K = 0$, *spherical* or *elliptic* for $K > 0$, and *hyperbolic* for $K < 0$). The group Γ is called the *fundamental group* of a space form $\mathcal{M}_K/\Gamma$.

(It is easy to see that *each of the spaces* $\mathcal{M}_K$ *is simply connected* (for $n \geq 2$). Therefore, the group Γ is the fundamental group of the space $\mathcal{X} = \mathcal{M}_K/\Gamma$.)

Proposition 23.1. *Two space forms* $\mathcal{M}_K/\Gamma$ *and* $\mathcal{M}_K/\Delta$ *are isometric iff the groups* Γ *and* Δ *are conjugated in the group* $\operatorname{Iso}\mathcal{M}_K$ *of all isometries of the space* $\mathcal{M}_K$.

Proof. If $\Delta = f\Gamma f^{-1}$, $f \in \operatorname{Iso}\mathcal{M}_K$, then the formula $\overline{f}\colon \Gamma p \mapsto \Delta f(p)$, $p \in \mathcal{M}_K$, correctly defines a mapping $\overline{f}\colon \mathcal{M}_K/\Gamma \to \mathcal{M}_K/\Delta$ that makes the diagram

$$\begin{array}{ccc} \mathcal{M}_K & \xrightarrow{f} & \mathcal{M}_K \\ \downarrow{\scriptstyle\pi} & & \downarrow{\scriptstyle\pi} \\ \mathcal{M}_K/\Gamma & \xrightarrow{\overline{f}} & \mathcal{M}_K/\Delta \end{array} \tag{1}$$

commutative, where the vertical arrows are canonical projections. Because f is an isometry, $\overline{f}$ is also an isometry (and vice versa). Therefore, if the groups Γ and Δ are conjugated, then the space forms $\mathcal{M}_K/\Gamma$ and $\mathcal{M}_K/\Delta$ are isometric.

Conversely, let there exist an isometry $\overline{f}\colon \mathcal{M}_K/\Gamma \to \mathcal{M}_K/\Delta$. Because the space $\mathcal{M}_K$ is simply connected, the mapping

$$\overline{f} \circ \pi\colon \mathcal{M}_K \to \mathcal{M}_K/\Delta$$

is liftable, i.e., there exists a mapping $f\colon \mathcal{M}_K \to \mathcal{M}_K$ that makes diagram (1) commutative. Because $\overline{f}$ is an isometry, f is also an isometry. Moreover, for any element $\gamma \in \Gamma$ and for any point $p_0 \in \mathcal{M}_K$, we have the relation

$$(\pi \circ f)(\gamma p_0) = (\overline{f} \circ \pi)(\gamma p_0) = (\overline{f} \circ \pi)(p_0) = (\pi \circ f)(p_0),$$

which means that $f(\gamma p_0) = \delta f(p_0)$ for a certain $\delta \in \Delta$, i.e., the mappings $\delta^{-1} f \gamma$ and f act identically on the point p_0. Because the mapping $\delta^{-1} f \gamma$ obviously also covers the mapping $\overline{f} \circ \pi$, by the uniqueness assertion, this implies that $\delta^{-1} f \gamma = f$ everywhere, i.e., $f \gamma f^{-1} = \delta$. Therefore, $f \Gamma f^{-1} = \Delta$, i.e., the groups Γ and Δ are conjugated. □

§2. Cartan–Killing Theorem

Being a closed subspace of a complete metric space, each of the model spaces $\mathcal{M}_K$ is complete. (Each is also complete by the Hopf–Rinow theorem; see Theorem 12.1.) Therefore, each space $\mathcal{M}_K/\Gamma$ is also complete (see Exercise 19.3). Therefore, *each space form $\mathcal{M}_K/\Gamma$ is a complete space of constant curvature.* It is noteworthy that the converse is also true.

Theorem 23.1. *Each complete connected space $\mathcal{X}$ of constant curvature K is isometric to a certain space form $\mathcal{M}_K/\Gamma$.*

This theorem is called the *Cartan–Killing theorem.* It once more justifies the representation of the spaces $\mathcal{M}_K$ as model spaces of constant curvature. We mention the following particular case of Theorem 23.1.

Theorem 23.2. *Each connected and simply connected complete space of constant curvature K is isometric to the space $\mathcal{M}_K$.*

This theorem is also called the *Cartan–Killing theorem.*

We note that *Theorem* 23.1 *is easily implied by Theorem* 23.2. Indeed, the space $\widetilde{\mathcal{X}}$ that universally covers the space $\mathcal{X}$ is obviously a complete and simply connected space of constant curvature K. Therefore, according to Theorem 23.2, it is isometric to the space $\mathcal{M}_K$. Therefore, we can assume without loss of generality that $\widetilde{\mathcal{X}} = \mathcal{M}_K$. On the other hand, if Γ is the automorphism group of the covering $\mathcal{M}_K \to \mathcal{X}$ (according to assertion 1 in Exercise 19.5, it consists of isometries), then the quotient space $\mathcal{M}_K/\Gamma$ is homeomorphic (and is also diffeomorphic under our conditions) to the manifold $\mathcal{X}$. To complete the proof, it suffices to note that this diffeomorphism is obviously an isometry. □

Therefore, we need only prove Theorem 23.2. We do this in the more general context of symmetric spaces.

§3. (Pseudo-)Riemannian Symmetric Spaces

Let $\mathcal{X}$ be a (pseudo-)Riemannian space that is simultaneously a symmetric space (in the sense of Definition 5.2).

Definition 23.2. The space $\mathcal{X}$ is called a *(pseudo-)Riemannian symmetric space* if all symmetries

$$s_p: \mathcal{X} \to \mathcal{X}, \quad p \in \mathcal{X},$$

are isometries.

Being isometries, the symmetries s_p are affine mappings with respect to the Levi–Civita connection ∇ on $\mathcal{X}$. Therefore (see Remark 5.2), the space $\mathcal{X}$ is a globally symmetric space with respect to the connection ∇; hence, by the uniqueness of the canonical connection, *the connection ∇ is the canonical connection on $\mathcal{X}$.*

Exercise 23.1. Prove that if a (pseudo-)Riemannian space has a covariantly constant curvature tensor (is a locally symmetric space with respect to the Levi-Civita connection), then for any point $p_0 \in \mathcal{X}$, the local geodesic symmetry

$$\exp_{p_0} A \mapsto \exp_{p_0}(-A), \quad A \in \mathbf{T}_{p_0}\mathcal{X},$$

is an isometry. [*Hint*: The mapping $A \mapsto -A$ of the space $\mathbf{T}_{p_0}\mathcal{X}$ onto itself is isometric.]

Exercise 23.2. Prove that a connected and simply connected (pseudo-)Riemannian space is a (pseudo-)Riemannian symmetric space iff it is complete and its curvature tensor is covariantly constant.

In particular, *any connected and simply connected complete space of constant curvature is a Riemannian symmetric space.*

Exercise 23.3. Prove that connected and simply connected (pseudo-)Riemannian symmetric spaces of the same signature and curvature are isometric. [*Hint*: In the case where the spaces $\mathcal{X}$ and $\mathcal{Y}$ in Theorem 5.2 are (pseudo-)Riemannian spaces of the same signature, we can take an isometry as the isomorphism φ in this theorem.]

We can now prove Theorem 23.2.

Proof. This theorem is directly implied by the assertions in Exercises 23.2 and 23.3. □

We also note that *each connected complete (pseudo-)Riemannian locally symmetric space (in particular, any (pseudo-)Riemannian symmetric space) has the form $\mathcal{X}/\Gamma$, where $\mathcal{X}$ is a simply connected (pseudo-)Riemannian symmetric space and Γ is its isometry group with a discrete action.* (Compare with Proposition 5.2.)

§4. Classification of Space Forms

According to the Cartan–Killing theorem, the problem of enumerating all space forms is reduced to the problem of enumerating (up to a conjugation) all subgroups (15) in Chap. 22 with a discrete action.

The simplest (and completely studied) case arises for $K > 0$, i.e., for *spherical space forms*. The matter is that by the compactness of the group $\mathcal{O}(n+1)$, any discrete isometry group is necessarily finite. Together with the condition that no transformation of the group, except for the identity transformation, has fixed points, this allows finding all subgroups Γ of the group $\mathcal{O}(n+1)$ with a discrete action on the sphere $\mathbb{S}^n_R$ and therefore allows describing all spherical forms. The answer obtained by Vincent in 1966 yields a long and less transparent list. We therefore only consider the even-dimensional case where the answer is surprisingly simple.

The complete classification of the *hyperbolic space form* is only known for $n = 2$ (and then only under the assumption that the group Γ has a finite number of generators). These forms arise as Riemannian surfaces of analytic functions, and their theory is in general referred not to geometry but to function theory. For $n \geq 3$, the theory of the hyperbolic space forms is only in its first stage now and is far from complete.

Euclidean space forms occupy an intermediate position. Although they are still only classified for $n \leq 4$, we know sufficiently more about them, and there are a number of perspective approaches to their complete classification.

§5. Spherical Forms of Even Dimension

The simplest example of a nontrivial spherical form is the already known (see Example 19.1) spherical space

$$\mathbb{R}P^n_R = \mathbb{S}^n_R/\{\mathrm{id}, \sigma\},$$

which is the quotient space of the sphere $\mathbb{S}^n_R$ by the second-order group $\{\mathrm{id}, \sigma\}$ generated by the antipodal mapping

$$\sigma\colon \boldsymbol{x} \mapsto -\boldsymbol{x}, \quad \boldsymbol{x} \in \mathbb{S}^n_R.$$

Proposition 23.2. *Any spherical space form $\mathcal{X}$ of an even dimension $n = 2m$ is isometric to either the sphere $\mathbb{S}^n_R$ or the projective space $\mathbb{R}P^n_R$.*

Proof. Each transformation of the group $\mathrm{SO}\,(2m+1)$ has an eigenvector corresponding to the eigenvalue 1 and, considered as a transformation from $\mathrm{Iso}\,\mathbb{S}^{2m}_R$, therefore has a fixed point. Therefore, for the fundamental group $\Gamma \subset \mathrm{Iso}\,\mathbb{S}^{2m}_R = \mathcal{O}(2m+1)$, the relation $\Gamma \cap \mathrm{SO}(2m+1) = \{\mathrm{id}\}$ should hold. Because the square γ^2 of any element $\gamma \in \mathcal{O}(n+1)$ belongs to $\mathrm{SO}(n+1)$, this implies $\gamma^2 = \mathrm{id}$, i.e., the group Γ consists of only involutions. On the other hand, because each involutive linear operator γ is diagonalizable (and has the eigenvalues ± 1), the operator $\gamma \neq \mathrm{id}$ has no fixed points iff $\gamma\colon \boldsymbol{x} \mapsto -\boldsymbol{x}$. □

In particular, *the spheres $\mathbb{S}^2_R$ and the projective (elliptic) planes $\mathbb{R}P^2_R$ exhaust all two-dimensional elliptic forms.* At the same time, for $n = 3$ for example, along with the spheres and projective spaces, there are six series of spherical space forms.

§6. Orientable Space Forms

It is clear that each model space $\mathcal{M}_K$ is an orientable smooth manifold. Nevertheless, there also exist nonorientable space forms $\mathcal{M}_K/\Gamma$. How can the group Γ be used to recognize whether the manifold $\mathcal{M}_K/\Gamma$ is orientable? We examine this question in the general context of an arbitrary manifold $\mathcal{X}/\Gamma$, where $\mathcal{X}$ is a connected orientable smooth manifold and Γ is the group of diffeomorphisms with a discrete action on $\mathcal{X}$.

Definition 23.3. We say that a diffeomorphism $\gamma: \mathcal{X} \to \mathcal{X}$ of a connected orientable manifold $\mathcal{X}$ onto itself *preserves the orientation* if it is given by functions with a positive Jacobian in the charts of an arbitrary orienting atlas.

Exercise 23.4. Show that this definition is correct, i.e., does not depend on the choice of the orienting atlas.

For $\mathcal{X} = \mathcal{M}_K$, a motion $\gamma \in \operatorname{Iso}\mathcal{M}_K$ preserves the orientation iff it is a proper motion (belongs to the component of the identity of the group $\operatorname{Iso}\mathcal{M}_K$).

Exercise 23.5. Show that for any orienting atlas $\mathcal{A}$ of the manifold $\mathcal{X}$ and for any orientation-preserving diffeomorphism $\gamma: \mathcal{X} \to \mathcal{X}$, each chart (U, h) of the manifold $\mathcal{X}$ with the property that the chart $(\gamma U, h \circ \gamma^{-1})$ belongs to the atlas $\mathcal{A}$ is positively compatible with charts of the atlas $\mathcal{A}$.

Let Γ be a diffeomorphism group of a connected orientable manifold $\mathcal{X}$ with a discrete action.

Proposition 23.3. *The quotient manifold $\mathcal{X}/\Gamma$ of the manifold $\mathcal{X}$ by the group Γ is orientable iff all elements of the group Γ preserve the orientation.*

Proof. An atlas $\{(V, k)\}$ of the manifold $\mathcal{X}/\Gamma$ is said to be *exactly covered* if the support V of each of its charts is exactly covered by the mapping $\pi: \mathcal{X} \to \mathcal{X}/\Gamma$. Clearly, in this case, all charts of the form $(U, k \circ \pi)$, where U is an open set in $\mathcal{X}$, that are diffeomorphically mapped by π onto the support V of a certain chart of the atlas $\{(V, k)\}$ compose an atlas of the manifold $\mathcal{X}$. We say that the atlas $\{(U, k \circ \pi)\}$ is the *inverse image* of the exactly covered atlas $\{(V, k)\}$. Because the transition functions of the atlases $\{(V, k)\}$ and $\{(U, k \circ \pi)\}$ obviously coincide, these atlases either are or are not simultaneously orienting atlases.

On the other hand, each diffeomorphism $\gamma \in \Gamma$, being a gliding transformation, transforms each chart $(U, k \circ \pi)$ into a chart $(\gamma U, k \circ \pi)$ of the same form and therefore acts in these charts according to the equality of coordinates. Because the Jacobian of the identity transformation is positive, this proves that all diffeomorphisms from Γ are given in charts of the atlas $\{(U, k \circ \pi)\}$ by functions with a positive Jacobian and are therefore orientation-preserving diffeomorphisms in the case where this atlas is orienting. This proves that if there exists an exactly covered orienting atlas $\{(V, k)\}$ for $\mathcal{X}/\Gamma$, then the group Γ consists of orientation-preserving diffeomorphisms.

Because such an atlas obviously exists for an orientable manifold $\mathcal{X}/\Gamma$, we see that in the case where the manifold $\mathcal{X}/\Gamma$ is orientable, all diffeomorphisms from the group Γ indeed preserve the orientation. Therefore, it only remains to prove the converse statement.

Exercise 23.6. Prove that for a manifold $\mathcal{X}$, there exists an orienting atlas $\mathcal{A}$ such that

1. the support of each chart of this atlas is diffeomorphically mapped under π onto a certain exactly covered set of the manifold $\mathcal{X}/\Gamma$ and
2. each chart of the manifold $\mathcal{X}$ that is positive compatible with charts of the atlas $\mathcal{A}$ and has property 1 is contained in the atlas $\mathcal{A}$ (the atlas $\mathcal{A}$ is maximal with respect to property 1).

Let V be an exactly covered set of the manifold $\mathcal{X}/\Gamma$ such that there exists a chart (U_0, h_0) in the atlas $\mathcal{A}$ for which $\pi U_0 = V$. Choosing such a chart, we set

$$k = h_0 \circ (\pi|_{U_0})^{-1}.$$

Then the pair (V, k) is a chart of the manifold $\mathcal{X}/\Gamma$, and all the charts of such form compose an exactly covered atlas $\{(V, k)\}$ of this manifold. We consider the inverse image $\{(U, k \circ \pi)\}$ of the atlas $\{(V, k)\}$. For each chart $(U, k \circ \pi)$ of the atlas $\{(U, k \circ \pi)\}$, we have

$$U = \gamma U_0 \qquad \text{and} \qquad k \circ \pi = h_0 \circ \gamma^{-1},$$

where $\gamma \in \Gamma$ and (U_0, h_0) is the chart of the atlas $\mathcal{A}$ chosen for $V = \pi U_0$.

Therefore (see Exercise 23.5), if γ preserves the orientation, then the chart $(U, k \circ \pi)$ is positively compatible with charts of the atlas $\mathcal{A}$ and, by property 2 of this atlas, therefore belongs to the atlas $\mathcal{A}$. Therefore, if all diffeomorphisms of the group Γ preserve the orientation, then the atlas $\{(U, k \circ \pi)\}$ is contained in the orienting atlas $\mathcal{A}$ and is therefore itself an orienting atlas. In this case, the atlas $\{(V, k)\}$ is also orienting; therefore, the manifold $\mathcal{X}/\Gamma$, having an orienting atlas, is orientable. □

Corollary 23.1. *A space form $\mathcal{M}_K/\Gamma$ is an orientable manifold iff its fundamental group Γ consists of proper motions.*

For example, *the projective space $\mathbb{R}P^n$ is orientable iff its dimension n is odd* (because the antipodal mapping $x \mapsto -x$ of the sphere $\mathbb{S}^n$ into itself preserves the orientation only in this case). In general, the following important corollary holds.

Corollary 23.2. *Each odd-dimensional spherical form $\mathbb{S}^n/\Gamma$, $n = 2m-1$, is an orientable manifold.*

Proof. Each improper orthogonal transformation of the even-dimensional space $\mathbb{R}^{2m}$ necessarily has an eigenvector with the eigenvalue 1 and, considered as a transformation of the sphere $\mathbb{S}^{2m-1}$, therefore has a fixed point. Therefore, the isometry group Γ of the sphere $\mathbb{S}^{2m-1}$ with a discrete action is a subgroup of the group $\mathrm{SO}(2m)$. □

Therefore, *only even-dimensional elliptic spaces* $\mathbb{R}P_R^{2m}$ *are nonorientable spherical forms.*

§7. Complex-Analytic and Conformal Quotient Manifolds

An analogue of Proposition 23.3 also holds for complex-analytic manifolds. Let Γ be a diffeomorphism group of a connected complex-analytic manifold $\mathcal{X}$ with a discrete action.

Exercise 23.7. Prove that the quotient manifold $\mathcal{X}/\Gamma$ of the manifold $\mathcal{X}$ by the group Γ has a complex structure with respect to which the natural mapping $\mathcal{X} \to \mathcal{X}/\Gamma$ is complex analytic iff all elements of the group Γ are complex-analytic diffeomorphisms. Also prove that a similar assertion holds for $n = 2$ with respect to conformal structures.

§8. Riemannian Spaces with an Isometry Group of Maximal Dimension

To conclude this chapter, we give one more characterization of model spaces of constant curvature that explains their unique place among all the so-called non-Euclidean spaces (although together with the elliptic space $\mathbb{R}P_R^n$). We recall that $\operatorname{Iso}\mathcal{X}$ denotes the group of all isometries $\mathcal{X} \to \mathcal{X}$ of a Riemannian space $\mathcal{X}$. According to Proposition 19.3, this group is a Lie group whose dimension $\dim \operatorname{Iso}\mathcal{X}$ does not exceed the number $n(n+1)/2$, where $n = \dim \mathcal{X}$.

Proposition 23.4. *If*

$$\dim \operatorname{Iso}\mathcal{X} = \frac{n(n+1)}{2}, \tag{2}$$

then $\mathcal{X}$ is a space of constant curvature.

Proof. Choosing a point $p \in \mathcal{X}$, we consider its stabilizer $\operatorname{Iso}_p \mathcal{X}$ in the group $\operatorname{Iso}\mathcal{X}$. For any element $\gamma \in \operatorname{Iso}_p \mathcal{X}$, its differential $(d\gamma)_p$ is an orthogonal mapping of the Euclidean space $\mathbf{T}_p\mathcal{X}$ onto itself, i.e., is an element of the orthogonal group $\mathcal{O}(n)$. The mapping

$$\operatorname{Iso}_p \mathcal{X} \to \mathcal{O}(n), \qquad \gamma \mapsto (d\gamma)_p, \tag{3}$$

of the Lie group $\operatorname{Iso}_p \mathcal{X}$ into the Lie group $\mathcal{O}(n)$ thus arising is obviously a homomorphism.

Exercise 23.8. Prove that the Lie algebra homomorphism

$$\mathfrak{iso}_p \mathcal{X} \to \mathfrak{so}(n) \tag{4}$$

induced by this homomorphism is a monomorphism. [*Hint*: Homomorphism (4) coincides with the restriction of the monomorphism that sets the linear operator given by formula (8) in Chap. 8 for a vector field X in correspondence to $\mathfrak{iso}_p \mathcal{X}$.]

Exercise 23.9. Prove that when (2) holds, we have

$$\dim \operatorname{Iso}_p \mathcal{X} = \frac{n(n-1)}{2}, \tag{5}$$

and monomorphism (4) is therefore an isomorphism.

Lemma 23.1. *If $\mathcal{G}$ and $\mathcal{H}$ are connected Lie groups of the same dimension, then each homomorphism*

$$\mathcal{G} \to \mathcal{H} \tag{6}$$

inducing a Lie algebra isomorphism is an epimorphism.

Proof. Homomorphism (6) inducing a Lie algebra isomorphism is étale at the identity of the group $\mathcal{G}$, i.e., is a diffeomorphism of a certain neighborhood of the identity onto a certain neighborhood U of the identity of the group $\mathcal{H}$. Therefore, the subgroup $\varphi\mathcal{G}$ of the group $\mathcal{H}$ contains the neighborhood U. Because the group $\mathcal{H}$ is connected, it is generated by this neighborhood. Therefore, $\varphi\mathcal{G} = \mathcal{H}$. □

Remark 23.1. It is useful to keep in mind that homomorphism (6) induces a Lie algebra isomorphism iff its kernel is discrete.

Lemma 23.1 (and the assertion in Exercise 23.9) implies in particular that *the image of homomorphism* (3) *contains the component of the identity* $\mathrm{SO}(n)$ *of the group* $\mathcal{O}(n)$.

Because the group $\mathrm{SO}(n)$ acts transitively on the set of all two-dimensional planes of the n-dimensional Euclidean space, this implies that *for any two planes $\pi, \pi' \subset \mathbf{T}_p\mathcal{X}$, there exists an isometry γ in the group $\operatorname{Iso}_p \mathcal{X}$ such that*

$$(d\gamma)_p \pi = \pi'. \tag{7}$$

Because the sectional curvatures $K_p(\pi)$ and $K_p(\pi')$ with respect to the two-dimensional directions π and π' obviously coincide under condition (7), this directly implies that the Riemannian space $\mathcal{X}$ has the property ★ considered in the beginning of Chap. 22. Therefore, according to the Schur theorem (Proposition 22.2), this space is a space of constant curvature for $n \geq 3$.

Exercise 23.10. Prove the latter assertions also for $n = 2$. [*Hint*: Relations (2) and (5) imply that for any vector $A \in \mathbf{T}_p\mathcal{X}$, there exists a Killing field $X \in \mathfrak{iso}\,\mathcal{X}$ such that $X_p = A$.]

Proposition 23.4 is thus completely proved. □

§9. Their Enumeration

As we know (see formula (17) in Chap. 22), each model space of constant curvature satisfies condition (2). However, one more space satisfies this condition.

Exercise 23.11. Show that condition (2) also holds for $\mathcal{X} = \mathbb{R}P^n_R$.

Therefore, condition (2) holds not only for the model spaces $\mathcal{M} = \mathbb{R}^n$, $\mathbb{S}^n_R$, and $\mathbb{B}^n_R$ but also for the elliptic space $\mathbb{R}P^n_R$. However, this exception is unique.

Theorem 23.3. *A complete connected Riemannian space $\mathcal{X}$ satisfying condition* (2) *either is one of the model spaces $\mathbb{R}^n$, $\mathbb{S}^n_R$, and $\mathbb{B}^n_R$ or is the elliptic space $\mathbb{R}P^n_R$.*

Proof. Let $\mathcal{M}$ be a model space that is a universal covering of the space $\mathcal{X}$, and let $\pi\colon \mathcal{M} \to \mathcal{X}$ be the corresponding projection (covering mapping). Because the mapping $(d\pi)_p\colon \mathsf{T}_p\mathcal{M} \to \mathsf{T}_{\pi(p)}\mathcal{X}$ is an isomorphism for each point $p \in \mathcal{M}$, for any vector field $X \in \mathfrak{a}X$, there exists a unique vector field $Y \in \mathfrak{a}\mathcal{M}$ that is π-related to the field X (the field Y is defined by $Y_p = (d\pi)_p^{-1} X_{\pi(p)}$, $p \in \mathcal{M}$). It is clear that the field Y is a Killing field if the field X is a Killing field. Therefore, the correspondence $X \mapsto Y$ is a Lie algebra homomorphism

$$\mathfrak{iso}\,\mathcal{X} \to \mathfrak{iso}\,\mathcal{M}.$$

This homomorphism is obviously a monomorphism and is therefore an isomorphism because

$$\dim \mathfrak{iso}\,\mathcal{X} = \frac{n(n+1)}{2} = \dim \mathfrak{iso}\,\mathcal{M}$$

by assumption. Therefore, if N is the number of sheets of the covering $\pi\colon \mathcal{M} \to \mathcal{X}$ (which is possibly infinite), then the number of zeros of each field $Y \in \mathfrak{iso}\,\mathcal{M}$ (points at which it vanishes) is divided by N (for $N = \infty$, this number is either zero or infinite). On the other hand, it is easy to see that *on the spaces $\mathbb{R}^n$ and $\mathbb{B}^n_R$, there exists a Killing field that vanishes only at one point, and on the space $\mathbb{S}^n_R$, there exists a Killing field that vanishes only at two diametrically opposite points.*

Exercise 23.12. Prove the latter assertion. [*Hint*: Consider a one-parameter subgroup consisting of rotations centered at a given point.]

Therefore, for $\mathcal{M} = \mathbb{R}^n, \mathbb{B}^n_R$, it should be $N = 1$ (and therefore $\mathcal{X} = \mathcal{M}$, i.e., $\mathcal{X} = \mathbb{R}^n, \mathbb{B}^n_R$), and for $\mathcal{M} = \mathbb{S}^n_R$, either $N = 1$ (and then $\mathcal{X} = \mathbb{S}^n_R$) or $N = 2$ (and then $\mathcal{X} = \mathbb{R}P^n_R$). □

Exercise 23.13. Prove that the completeness condition for the space $\mathcal{X}$ is extra in Theorem 23.3.

§10. Complete Mobility Condition

A *frame* in a Riemannian space $\mathcal{X}$ is a tuple $(p, A_1, \dots, A_n)$ consisting of a point $p \in \mathcal{X}$ and a basis $A_1, \dots, A_n$ of the space $\mathbf{T}_p\mathcal{X}$. The *image* of a frame $(p, A_1, \dots, A_n)$ under an isometry $\gamma: \mathcal{X} \to \mathcal{X}$ is the frame $(\gamma p, (d\gamma)_p A_1, \dots, (d\gamma)_p A_n)$. We say that a connected Riemannian space $\mathcal{X}$ satisfies the *complete mobility condition* if for any two frames, there exists an isometry that transforms one frame into the other (i.e., if the group $\operatorname{Iso} \mathcal{X}$ is transitive on frames).

Exercise 23.14. Prove that a connected Riemannian space $\mathcal{X}$ satisfies the complete mobility condition iff relation (2) holds for it. [*Hint*: For such a space $\mathcal{X}$, the number $\dim \operatorname{Iso} \mathcal{X}$ is not less than $n + n(n-1)/2 = n(n+1)/2$.]

Therefore, *only the spaces* $\mathbb{R}^n$, $\mathbb{S}^n$, $\mathbb{B}^n$, *and* $\mathbb{R}P^n$ *satisfy the complete mobility condition.* This explains why there are no "good" non-Euclidean geometries except for the hyperbolic, spherical, and elliptic ones. A special role of affine spaces is similarly explained.

Exercise 23.15. Prove that each connected n-dimensional affine connection space $\mathcal{X}$ for which $\dim \operatorname{Aff} \mathcal{X} = n + n^2$ is the affine space A^n.

Chapter 24
Four-Dimensional Manifolds

§1. Bianchi Tensors for $n = 4$

In this chapter, we study four-dimensional Riemannian spaces (the case $n = 4$) and, in particular, completely describe their curvature tensors. This was done for $n = 2$ in Chap.16 (see formula (10) there) and for $n = 3$ in Chap. 17 (see formula (25) there). The special attention to the case $n = 4$ is explained not only because an elegant theorem holds only for $n = 4$ but also because according to the Einstein general relativity theory, the physical space–time is a pseudo-Riemannian space of signature (1,3) whose metric is given by the distribution of gravitated masses. (However, we focus on only the case of Riemannian spaces.)

Our main tool in studying the geometry of the four-dimensional Riemannian space $\mathcal{X}$ is the Hodge operator

$$*\colon \Lambda^2\mathcal{X} \to \Lambda^2\mathcal{X}$$

(see formula (7) in Chap. 17). As we know (see the assertion in Exercise 17.3), for any orthonormal basis X_1, X_2, X_3, X_4 of the module $\mathfrak{a}\mathcal{X}$ over a coordinate neighborhood U, the bivectors $\boldsymbol{X}_{ij} = X_i \wedge X_j$, $1 \leq i < j \leq 4$, form an orthonormal basis of the module $\Lambda^2\mathcal{X}$ over U. However, the basis

$$\boldsymbol{X}_{12},\ \boldsymbol{X}_{31} = -\boldsymbol{X}_{13},\ \boldsymbol{X}_{14},\ \boldsymbol{X}_{23},\ \boldsymbol{X}_{24},\ \boldsymbol{X}_{34}, \tag{1}$$

obtained by replacing $\boldsymbol{X}_{13}$ with $\boldsymbol{X}_{31}$, is more convenient for us.

Exercise 24.1. Show that operator $*$ has the matrix

$$\begin{Vmatrix} 0 & 0 & 0 & 0 & 0 & 1 \\ 0 & 0 & 0 & 0 & 1 & 0 \\ 0 & 0 & 0 & 1 & 0 & 0 \\ 0 & 0 & 1 & 0 & 0 & 0 \\ 0 & 1 & 0 & 0 & 0 & 0 \\ 1 & 0 & 0 & 0 & 0 & 0 \end{Vmatrix} \tag{2}$$

in basis (1).

Let R be an arbitrary Bianchi tensor on the manifold $\mathcal{X}$, and let $\|R^{ij}_{kl}\|$ be its matrix as the operator $\Lambda^2\mathcal{X} \to \Lambda^2\mathcal{X}$ (see Chap. 16) in basis (1). Because the operator R is self-adjoint and basis (1) is orthonormal, the matrix $\|R^{ij}_{kl}\|$ is symmetric ($R^{ij}_{kl} = R^{kl}_{ij}$); moreover, lowering the superscripts does not change the numerical values of the components,

$$R^{ij}_{kl} = R_{ij,kl}.$$

Therefore, for $(i, j, k, l) = (1, 2, 3, 4)$, the Bianchi identity for the tensor R has the form

$$R^{12}_{34} + R^{23}_{14} + R^{31}_{24} = 0. \tag{3}$$

Multiplying this identity by 2 and using the symmetry of the matrix $\|R^{ij}_{kl}\|$, we obtain the identity

$$R^{12}_{34} + R^{31}_{24} + R^{14}_{23} + R^{23}_{14} + R^{24}_{31} + R^{34}_{12} = 0, \tag{4}$$

which means that the sum of entries of the secondary diagonal (which consists of unities in matrix (2)) vanishes for the matrix $\|R^{ij}_{kl}\|$. On the other hand, it is easily seen that multiplying matrix (2) by the matrix $\|R^{ij}_{kl}\|$, we obtain a matrix whose principal diagonal is the secondary diagonal of the matrix $\|R^{ij}_{kl}\|$, and its trace is therefore equal to sum (4). This proves that *Bianchi identity* (3) *is equivalent to the relation*

$$\mathrm{Tr}(*R) = 0. \tag{5}$$

Because $\dim \Lambda^2\mathcal{X} = 6$, the dimension of the module of all self-adjoint operators $\Lambda^2\mathcal{X} \to \Lambda^2\mathcal{X}$ is equal to

$$\frac{6(6+1)}{2} = 21,$$

and the dimension of its submodule consisting of operators satisfying relation (5) is therefore equal to $21 - 1 = 20$, i.e. (see Exercise 15.4), is equal to the dimension of the module $B\mathcal{X}$. This proves that *for* $n = 4$, *condition* (5) *characterizes Bianchi tensors* (each Bianchi identity is a consequence of identity (5)).

§2. Matrix Representation of Bianchi Tensors for $n = 4$

To avoid writing six-order matrices, it is convenient to introduce the eigenspaces

$$\begin{aligned} \Lambda^2_+\mathcal{X} &= \{\boldsymbol{X} \in \Lambda^2\mathcal{X} : *\boldsymbol{X} = \boldsymbol{X}\}, \\ \Lambda^2_-\mathcal{X} &= \{\boldsymbol{X} \in \Lambda^2\mathcal{X} : *\boldsymbol{X} = -\boldsymbol{X}\} \end{aligned} \tag{6}$$

of the operator $*$ corresponding to the eigenvalues $+1$ and -1. These subspaces are obviously submodules of the $\mathsf{F}\mathcal{X}$-module $\Lambda^2\mathcal{X}$. Moreover, because the operator $*$ is involutive (see (9) in Chap. 17), the module $\Lambda^2\mathcal{X}$ is decomposed into the direct sum

$$\Lambda^2\mathcal{X} = \Lambda^2_+\mathcal{X} + \Lambda^2_-\mathcal{X} \tag{7}$$

of submodules (6).

Exercise 24.2. Show that submodules (6) are orthogonal, i.e., $\langle \boldsymbol{X}, \boldsymbol{Y} \rangle = 0$ for any tensors $\boldsymbol{X} \in \Lambda^2_+\mathcal{X}$ and $\boldsymbol{Y} \in \Lambda^2_-\mathcal{X}$. [*Hint*: The operator $*$ is self-adjoint.]

Exercise 24.3. Show that

$$\dim \Lambda^2_+ \mathcal{X} = 3 \qquad \text{and} \qquad \dim \Lambda^2_- \mathcal{X} = 3.$$

[*Hint*: Over each coordinate neighborhood U, the tensors

$$\frac{\boldsymbol{X}_{12} + \boldsymbol{X}_{34}}{\sqrt{2}}, \qquad \frac{\boldsymbol{X}_{31} + \boldsymbol{X}_{24}}{\sqrt{2}}, \qquad \text{and} \qquad \frac{\boldsymbol{X}_{14} + \boldsymbol{X}_{23}}{\sqrt{2}} \tag{8}$$

form an orthonormal basis of the module $\Lambda^2_+ \mathcal{X}$, and the tensors

$$\frac{\boldsymbol{X}_{12} - \boldsymbol{X}_{34}}{\sqrt{2}}, \qquad \frac{\boldsymbol{X}_{31} - \boldsymbol{X}_{24}}{\sqrt{2}}, \qquad \text{and} \qquad \frac{\boldsymbol{X}_{14} - \boldsymbol{X}_{23}}{\sqrt{2}} \tag{9}$$

form an orthonormal basis of the module $\Lambda^2_- \mathcal{X}$.]

In accordance with decomposition (7), each self-adjoint operator $R\colon \Lambda^2 \mathcal{X} \to \Lambda^2 \mathcal{X}$ can be written in the form of the matrix

$$\begin{Vmatrix} A & B \\ B' & C \end{Vmatrix}, \tag{10}$$

where the operators

$$A\colon \Lambda^2_+ \mathcal{X} \to \Lambda^2_+ \mathcal{X} \qquad \text{and} \qquad C\colon \Lambda^2_- \mathcal{X} \to \Lambda^2_- \mathcal{X}$$

are self-adjoint, the operator

$$B\colon \Lambda^2_- \mathcal{X} \to \Lambda^2_+ \mathcal{X}$$

is arbitrary, and the operator $B'\colon \Lambda^2_+ \mathcal{X} \to \Lambda^2_- \mathcal{X}$ is adjoint to the operator B (i.e., it satisfies the relation $\langle B'\boldsymbol{X}, \boldsymbol{Y} \rangle = \langle \boldsymbol{X}, B\boldsymbol{Y} \rangle$ for any tensors $\boldsymbol{X} \in \Lambda^2_+ \mathcal{X}$ and $\boldsymbol{Y} \in \Lambda^2_- \mathcal{X}$).

Matrix (10) becomes

$$\begin{Vmatrix} \mathrm{id} & 0 \\ 0 & \mathrm{id} \end{Vmatrix}$$

for a Bianchi tensor E and

$$\begin{Vmatrix} \mathrm{id} & 0 \\ 0 & -\mathrm{id} \end{Vmatrix}$$

for the Hodge operator $*$. Therefore, for any operator R, the operator $*R$ has the matrix

$$\begin{Vmatrix} A & B \\ -B' & -C \end{Vmatrix},$$

which implies that condition (5) is equivalent to the relation

$$\operatorname{Tr} A = \operatorname{Tr} C. \tag{11}$$

Therefore, *operator* (10) *is a Bianchi tensor iff it satisfies condition* (11). In this case, we have

$$K = \frac{1}{3} \operatorname{Tr} A$$

for its invariant Gaussian curvature (see (7) in Chap. 16).

§3. Explicit Form of Bianchi Tensors for $n = 4$

To find an explicit form of the Bianchi tensors (and therefore the Riemannian curvature tensor of the space $\mathcal{X}$), we first describe all Einstein tensors $Q(S)$ (see Chap. 17). Let S be an arbitrary functional from $S^2\mathcal{X}$ (see Chap. 17). Because this functional, considered as an operator $\mathfrak{a}\mathcal{X} \to \mathfrak{a}\mathcal{X}$ is self-adjoint, over an arbitrary coordinate neighborhood U, the module $\mathfrak{a}\mathcal{X}$ admits an orthonormal basis X_1, X_2, X_3, X_4 consisting of eigenvectors of the operator S, i.e., such that S has a diagonal matrix of the form

$$\begin{Vmatrix} \lambda_1 & 0 & 0 & 0 \\ 0 & \lambda_2 & 0 & 0 \\ 0 & 0 & \lambda_3 & 0 \\ 0 & 0 & 0 & \lambda_4 \end{Vmatrix} \tag{12}$$

in this basis. The Bianchi tensor P (defined by (21) in Chap. 17) corresponding to the functional S has the following components in the corresponding basis (1):

$$P_{ij,kl} = \delta_{ik}\delta_{jl}\lambda_j - \delta_{il}\delta_{jk}\lambda_j + \delta_{jl}\delta_{ik}\lambda_i - \delta_{jk}\delta_{il}\lambda_i = (\delta_{ik}\delta_{jl} - \delta_{il}\delta_{jk})(\lambda_i + \lambda_j).$$

For $i < j$ and $k < l$, this means that

$$P_{ij,kl} = \begin{cases} \lambda_{ij} & \text{if } (i,j) = (k,l), \\ 0 & \text{otherwise,} \end{cases}$$

where we set $\lambda_{ij} = \lambda_i + \lambda_j$ to simplify the formulas.

The operators A and C corresponding to this tensor therefore have the same matrix

$$\frac{1}{2}\begin{Vmatrix} \lambda & 0 & 0 \\ 0 & \lambda & 0 \\ 0 & 0 & \lambda \end{Vmatrix} = \frac{\lambda}{2}\,\mathrm{id}$$

in basis (8), where $\lambda = \lambda_1 + \lambda_2 + \lambda_3 + \lambda_4 = \operatorname{Tr} S$, and the operator B has the matrix

$$\frac{1}{2}\begin{Vmatrix} \lambda_{12} - \lambda_{34} & 0 & 0 \\ 0 & \lambda_{13} - \lambda_{24} & 0 \\ 0 & 0 & \lambda_{14} - \lambda_{23} \end{Vmatrix}. \tag{13}$$

Therefore, the operator

$$Q(S) = \frac{P}{2} - \frac{\operatorname{Tr} S}{6} E$$

(see (24) in Chap. 17) has the matrix

$$\begin{Vmatrix} \dfrac{\lambda}{12}\,\mathrm{id} & \dfrac{1}{2}B \\ \dfrac{1}{2}B' & \dfrac{\lambda}{12}\,\mathrm{id} \end{Vmatrix}.$$

This proves that *for $n = 4$, any Einstein tensor has the form*

$$\begin{Vmatrix} K & B \\ B' & K \end{Vmatrix}, \tag{14}$$

where K is the operator $K\,\mathrm{id}$ (the operator of multiplication by K) and B is a certain operator $\Lambda^2_-\mathcal{X} \to \Lambda^2_+\mathcal{X}$. In this case, because the dimension of the module of all operators B is $9 = 3 \times 3$ and the dimension of the module of Einstein tensors is 10, *operator* (14) *is an Einstein tensor for any operator B and any function K.*

Exercise 24.4. Prove that a Bianchi tensor R is

1. a trace-free Einstein tensor when

$$*R = -R*;$$

2. a Weyl tensor when its trace is zero and

$$*R = R*.$$

Therefore, *for $n = 4$, Weyl tensors are exactly operators of the form*

$$\begin{Vmatrix} A & 0 \\ 0 & C \end{Vmatrix}, \quad \text{where } \operatorname{Tr} A = \operatorname{Tr} C = 0.$$

Summarizing, we see that *the decomposition of matrix* (10) *into the sum of matrices*

$$K\begin{Vmatrix} \mathrm{id} & 0 \\ 0 & \mathrm{id} \end{Vmatrix} + \begin{Vmatrix} 0 & B \\ B' & 0 \end{Vmatrix} + \begin{Vmatrix} A & 0 \\ 0 & C \end{Vmatrix}, \quad \operatorname{Tr} A = \operatorname{Tr} C = 0,$$

exactly reflects the decomposition of an arbitrary Bianchi tensor into the sum of a tensor that is a multiple of the tensor E, a trace-free Einstein tensor, and a Weyl tensor. In particular, this yields the general form of the Riemannian curvature tensor of an arbitrary four-dimensional Riemannian space $\mathcal{X}$.

Moreover, *the Ricci tensor $\operatorname{Ric} R$ of a Bianchi tensor R with matrix* (10) *is expressed through the operator B and the trace $\operatorname{Tr} A = \operatorname{Tr} C$ of the operators A and C.* Namely, if X_1, X_2, X_3, X_4 is an orthonormal basis of the module $\mathfrak{a}\mathcal{X}$ over a coordinate neighborhood U such that in the corresponding bases (8) and (9) of the modules $\Lambda^2_+\mathcal{X}$ and $\Lambda^2_-\mathcal{X}$, the matrix of the operator B is diagonal (such bases necessarily exist), then the Ricci tensor $\operatorname{Ric} R$ is just the operator $\mathfrak{a}\mathcal{X} \to \mathfrak{a}\mathcal{X}$ having matrix (12) in the basis X_1, X_2, X_3, X_4 whose diagonal entries are

$$\begin{aligned} \lambda_1 &= b_1 + b_2 + b_3 + \operatorname{Tr} A, \\ \lambda_2 &= b_1 - b_2 - b_3 + \operatorname{Tr} A, \\ \lambda_3 &= -b_1 + b_2 - b_3 + \operatorname{Tr} A, \\ \lambda_4 &= -b_1 - b_2 + b_3 + \operatorname{Tr} A, \end{aligned} \tag{15}$$

where b_1, b_2, and b_3 are diagonal entries of the matrix of the operator B in bases (8) and (9). (To prove this, it suffices to equate the doubled matrix $\operatorname{diag}(b_1, b_2, b_3)$ and matrix (13) and take into account that $\lambda_1+\lambda_2+\lambda_3+\lambda_4 = \operatorname{Tr}\operatorname{Ric} R = 2\operatorname{Tr} R = 4\operatorname{Tr} A$ (see (5) in Chap. 16).) $\square$

§4. Euler Numbers for $n = 4$

As is implied by the Gauss–Bonnet theorem, the topology of the manifold $\mathcal{X}$ imposes certain, sometimes sufficiently strong, conditions that a possible differential-geometric structure on $\mathcal{X}$ should satisfy. We now illustrate this general principle by examining compact connected oriented four-dimensional manifolds.

According to formula (15) in Chap. 17 (for $n = 4$), we have

$$e[\mathcal{X}] = \frac{1}{8\pi^2}\int_{\mathcal{X}} \operatorname{Tr}(*R*R)\,dV, \tag{16}$$

where R is the Riemannian curvature tensor of the manifold $\mathcal{X}$. On the other hand, if R has form (10), then

$$*R*R = \begin{Vmatrix} A^2 - BB' & AB - BC \\ CB' - B'A & C^2 - B'B \end{Vmatrix}$$

and therefore

$$\operatorname{Tr}(*R*R) = \operatorname{Tr}(A^2 + C^2 - 2BB').$$

Therefore, *the Euler characteristic number of a compact connected oriented Riemannian four-dimensional manifold $\mathcal{X}$ with curvature tensor* (10) *is expressed by the formula*

$$e[\mathcal{X}] = \frac{1}{8\pi^2}\int_{\mathcal{X}} \operatorname{Tr}(A^2 + C^2 - 2BB')\,dV. \tag{17}$$

(Of course, this formula can be proved by a direct computation without reference to general formula (15) in Chap. 17 similar to the proof of formula (5) in Chap. 17. But although this computation is completely automatic, it is sufficiently cumbersome.)

Exercise 24.5. Prove the d'Ave formula

$$e[\mathcal{X}] = \frac{1}{8\pi^2}\int_{\mathcal{X}} [\operatorname{Tr} R^2 - \operatorname{Tr}(\operatorname{Ric} R)^2 + (\operatorname{Tr} R)^2]\,dV.$$

[*Hint*: $\operatorname{Tr} R^2 = \operatorname{Tr}(A^2 + C^2 + 2BB')$ and

$$\operatorname{Tr}(\operatorname{Ric} R)^2 = \lambda_1^2 + \lambda_2^2 + \lambda_3^2 + \lambda_4^2 = 4(b_1^2 + b_2^2 + b_3^2) + 4(\operatorname{Tr} A)^2 = 4\operatorname{Tr} BB' + (\operatorname{Tr} R)^2$$

in the notation of formula (15).]

§5. Chern–Milnor Theorem

If we decompose the curvature tensor R of the space $\mathcal{X}$ into the sum

$$R = W + Z \tag{18}$$

of the tensor W, the matrix

$$\begin{Vmatrix} A & 0 \\ 0 & C \end{Vmatrix},$$

and the trace-free Einstein tensor on Z with the matrix

$$\begin{Vmatrix} 0 & B \\ B' & 0 \end{Vmatrix},$$

then we have the formula for the trace

$$\operatorname{Tr}(A^2 + C^2 - 2BB') = \operatorname{Tr}(W^2 - Z^2).$$

Therefore, (17) can be written in the form

$$e[\mathcal{X}] = \frac{1}{8\pi^2}\int_{\mathcal{X}} \operatorname{Tr}(W^2 - Z^2)\, dV. \tag{17'}$$

Because

$$*R* = W - Z$$

(see Exercise 24.4), we have the formula

$$\langle R(*\boldsymbol{P}), *\boldsymbol{P}\rangle = \langle (*R*)(\boldsymbol{P}), \boldsymbol{P}\rangle = \langle (W-Z)(\boldsymbol{P}), \boldsymbol{P}\rangle = \langle W(\boldsymbol{P}), \boldsymbol{P}\rangle - \langle Z(\boldsymbol{P}), \boldsymbol{P}\rangle,$$

for any bivector $\boldsymbol{P}$ (at the arbitrary point $p \in \mathcal{X}$), while

$$\langle R(\boldsymbol{P}), \boldsymbol{P}\rangle = \langle W(\boldsymbol{P}), \boldsymbol{P}\rangle + \langle Z(\boldsymbol{P}), \boldsymbol{P}\rangle.$$

Therefore,

$$\begin{aligned} \langle R(\boldsymbol{P}), \boldsymbol{P}\rangle + \langle R(*\boldsymbol{P}), *\boldsymbol{P}\rangle &= 2\langle W(\boldsymbol{P}), \boldsymbol{P}\rangle, \\ \langle R(\boldsymbol{P}), \boldsymbol{P}\rangle - \langle R(*\boldsymbol{P}), *\boldsymbol{P}\rangle &= 2\langle Z(\boldsymbol{P}), \boldsymbol{P}\rangle. \end{aligned} \tag{19}$$

By definition, if the area $|\boldsymbol{P}| = \sqrt{\langle \boldsymbol{P}, \boldsymbol{P}\rangle}$ of the bivector $\boldsymbol{P}$ equals 1, then the number $\langle R(\boldsymbol{P}), \boldsymbol{P}\rangle$ is just the sectional curvature $K_p(\pi)$ of the Riemannian space $\mathcal{X}$, where π is the two-dimensional direction at the point p given by the bivector $\boldsymbol{P}$ (the plane of the space $\mathsf{T}_p\mathcal{X}$ with the directing bivector $\boldsymbol{P}$) and the number $\langle R(*\boldsymbol{P}), *\boldsymbol{P}\rangle$ is the sectional curvature $K_p(\pi^\perp)$ with respect to orthogonal direction $\pi^\perp$ (because $n = 4$, we have $\dim \pi^\perp = \dim \pi = 2$). Therefore, formula (19) can be rewritten in the form

$$\begin{aligned} K_p(\pi) + K_p(\pi^\perp) &= 2\langle W(\boldsymbol{P}), \boldsymbol{P}\rangle, \\ K_p(\pi) - K_p(\pi^\perp) &= 2\langle Z(\boldsymbol{P}), \boldsymbol{P}\rangle. \end{aligned} \tag{19'}$$

Being an Einstein tensor, the tensor Z has the form $Q(S)$, where S is the self-adjoint operator $\mathfrak{a}\mathcal{X} \to \mathfrak{a}\mathcal{X}$ with a zero trace. As above, let X_1, X_2, X_3, X_4 be an orthonormal basis of the module $\mathfrak{a}\mathcal{X}$ over a coordinate neighborhood U in which the operator S has diagonal matrix (12).

Exercise 24.6. Show that in orthonormal basis (1) of the linear space $\Lambda^2\mathcal{X}$ corresponding to the basis X_1, X_2, X_3, X_4, the matrix of the operator $Z = Q(S)$ is also diagonal. [*Hint*: The operator Z has form (14), where $K = 0$, and the operator B is represented in bases (8) and (9) by a diagonal matrix.]

The diagonal entries of the matrix of the operator Z are the numbers $\langle Z(\boldsymbol{X}_{ij}), \boldsymbol{X}_{ij}\rangle$, $1 \le i < j \le 4$, and the matrix of the operator Z^2 is formed by their squares $\langle Z(\boldsymbol{X}_{ij}), \boldsymbol{X}_{ij}\rangle^2$. Therefore, according to the second formula in (19′), we have

$$\operatorname{Tr} Z^2 = \sum_{i<j} \langle Z(\boldsymbol{X}_{ij}), \boldsymbol{X}_{ij}\rangle^2 = \frac{1}{4}\sum_{i<j}(K_p(\pi_{ij}^{\perp}) - K_p(\pi_{ij}))^2,$$

where π_{ij} is the plane of the bivector $\boldsymbol{X}_{ij} = X_i \wedge X_j$ (at a given point $p \in U$). (We note that $|\boldsymbol{X}_{ij}| = 1$.)

On the other hand, if all curvatures $K_p(\pi)$ have the same sign, then

$$(K_p(\pi^{\perp}) - K_p(\pi))^2 \le (K_p(\pi^{\perp}) + K_p(\pi))^2$$

for an arbitrary plane $\pi \subset \mathbf{T}_p\mathcal{X}$. Therefore, in this case,

$$\operatorname{Tr} Z^2 \le \frac{1}{4}\sum_{i<j}(K_p(\pi_{ij}^{\perp}) + K_p(\pi_{ij}))^2 = \sum_{i<j}\langle W(\boldsymbol{X}_{ij}), \boldsymbol{X}_{ij}\rangle^2 \le \operatorname{Tr} W^2.$$

(For any symmetric matrix $A = \|a_{ij}\|$, the trace $\operatorname{Tr} A^2$ of its square is equal to the sum of the squares of its entries,

$$\operatorname{Tr} A^2 = \sum_{i,j=1}^{n} a_{ij}a_{ji} = \sum_{i,j=1}^{n} a_{ij}^2,$$

and is therefore estimated from below by the sum of the squares of its diagonal entries, $\operatorname{Tr} A^2 \ge \sum_{i=1}^{n} a_{ii}^2$.) By formula (17′), this proves the following proposition.

Proposition 24.1 (Chern–Milnor theorem). *If the sectional curvature has the same sign at all points and in all two-dimensional directions in a connected compact orientable four-dimensional Riemannian space $\mathcal{X}$, then the Euler number of this space is nonnegative,*

$$e[\mathcal{X}] \ge 0.$$

Remark 24.1. It is interesting that the inequality $e[\mathcal{X}] < 0$ (and even the inequality $e[\mathcal{X}] < 2$) is possible (for $n = 4$) only for non-simply connected (compact and orientable) manifolds $\mathcal{X}$. (If the manifold $\mathcal{X}$ is simply connected, then, as is sufficiently easy to understand, its first Betti number h^1 is equal to zero. Therefore, by the Poincaré duality theorem, the number h^3 is also equal to zero, and the Euler characteristic $\chi(\mathcal{X})$ is therefore equal to $h^0 + h^2 + h^4 = 2 + h^2 \ge 2$ (we recall that $h^0 = h^n = 1$ for a compact connected orientable manifold).)

§6. Sectional Curvatures of Four-Dimensional Einstein Spaces

Four-dimensional Einstein spaces are certainly of special interest. The above calculation shows that *a four-dimensional Riemannian space is an Einstein space iff its curvature tensor has the form*

$$\left\| \begin{matrix} A & 0 \\ 0 & C \end{matrix} \right\|, \quad \operatorname{Tr} A = \operatorname{Tr} C, \tag{20}$$

i.e., is a Weyl tensor (in decomposition (18), its first Einstein component Z vanishes).

On the other hand, the second formula in (19′) directly implies that in the space $\mathcal{X}$, we have $K_p(\pi^\perp) = K_p(\pi)$ for all two-dimensional directions $\pi \subset \mathsf{T}_p\mathcal{X}$ at all points $p \in \mathcal{X}$ iff $\langle Z(\boldsymbol{P}), \boldsymbol{P}\rangle = 0$ for all bivectors $\boldsymbol{P}$, i.e. (see Proposition 15.3), iff $Z = 0$. Therefore, *a four-dimensional Riemannian space $\mathcal{X}$ is an Einstein space iff for any two-dimensional direction $\pi \subset \mathsf{T}_p\mathcal{X}$ at each point $p \in \mathcal{X}$, we have*

$$K_p(\pi^\perp) = K_p(\pi), \tag{21}$$

where $\pi^\perp$ is the orthogonal direction.

§7. Berger Theorem

For an Einstein space $\mathcal{X}$, formula (17) becomes

$$\mathrm{e}[\mathcal{X}] = \frac{1}{8\pi^2} \int_{\mathcal{X}} \operatorname{Tr}(A^2 + C^2)\, dV. \tag{22}$$

Proposition 24.2 (Berger Theorem). *The Euler number (Euler characteristic) of a connected four-dimensional compact orientable Einstein manifold $\mathcal{X}$ is nonnegative,*

$$\mathrm{e}[\mathcal{X}] \geq 0.$$

Moreover, if $\mathrm{e}[\mathcal{X}] = 0$, then the space $\mathcal{X}$ is flat (its curvature tensor is identically zero).

Proof. For self-adjoint operators A and C, the trace $\operatorname{Tr}(A^2 + C^2)$ is nonnegative and is equal to zero iff $A = C = 0$. □

It is surprising that two absolutely different differential-geometric properties lead to the same topological condition.

The Berger theorem can be strengthened.

§8. Pontryagin Number of a Four-Dimensional Riemannian Space

To strengthen the Berger theorem, we need to orient the manifold $\mathcal{X}$ and consider its Pontryagin number $p_1[\mathcal{X}]$, which is equal to the tripled signature of the manifold $\mathcal{X}$, $3\,\mathrm{Sign}\,\mathcal{X}$ (see Exercise 17.10). Let X_1, X_2, X_3, X_4 be a positively oriented orthonormal basis of the module $\mathfrak{a}\mathcal{X}$ over a coordinate neighborhood U, and let $\boldsymbol{X}_{ij} = X_i \wedge X_j$, $1 \le i < j \le 4$, as usual.

We first note that by the definition of the exterior product, for any differential form Ω of degree two, we have

$$(\Omega \wedge \Omega)(X_1, X_2, X_3, X_4) = \sum_{i_1 < i_2} (-1)^w \Omega(X_{i_1}, X_{i_2}) \Omega(X_{j_1}, X_{j_2}),$$

where $j_1 < j_2$ are subscripts such that the sequence (i_1, i_2, j_1, j_2) is a permutation of the sequence $(1, 2, 3, 4)$ and w is the number of inversions in this permutation. Comparing this relation with the definition of the Hodge operator and taking into account that

$$\Omega(X, Y) = \Omega(X \wedge Y),$$

we immediately obtain

$$(\Omega \wedge \Omega)(X_1, X_2, X_3, X_4) = \sum_{i_1 < i_2} \Omega(\boldsymbol{X}_{i_1 i_2}) \Omega(*\boldsymbol{X}_{i_1 i_2}).$$

On the other hand, because the number

$$\sigma_2(A) = \sum_{i<j} \begin{vmatrix} a_i^i & a_j^i \\ a_i^j & a_j^j \end{vmatrix}$$

is equal to

$$\sum_{i<j} \begin{vmatrix} 0 & a_j^i \\ -a_j^i & 0 \end{vmatrix} = \sum_{i<j} (a_j^i)^2,$$

for any skew-symmetric matrix $A = \|a_j^i\|$, the Pontryagin class $p_1 = p_1[\tau_{\mathcal{X}}]$ of the manifold $\mathcal{X}$ is given by a differential form that has the form

$$\frac{1}{4\pi^2} \sum_{i<j} \Omega_j^i \wedge \Omega_j^i$$

on U, where Ω_j^i is the curvature form of the manifold $\mathcal{X}$ on U.

Therefore, the value $p_1[\mathfrak{e}]$ of this form on the volume element $\mathfrak{e} = X_1 \wedge X_2 \wedge X_3 \wedge X_4$ is expressed by

$$p_1[\mathfrak{e}] = \frac{1}{4\pi^2} \sum_{i<j} \sum_{i_1 < i_2} \Omega_j^i(\boldsymbol{X}_{i_1 i_2}) \Omega_j^i(*\boldsymbol{X}_{i_1 i_2}).$$

Moreover, because the basis X_1, X_2, X_3, X_4 is orthonormal and therefore

$$\Omega^i_j(X_{i_1}, X_{i_2}) = R(X_i, X_j, X_{i_1}, X_{i_2}),$$

we have

$$\Omega^i_j(\boldsymbol{X}_{i_1 i_2}) = R(\boldsymbol{X}_{ij}, \boldsymbol{X}_{i_1 i_2}) = \langle R(\boldsymbol{X}_{ij}), \boldsymbol{X}_{i_1 i_2}\rangle.$$

Similarly,

$$\Omega^i_j(*\boldsymbol{X}_{i_1 i_2}) = \langle R(\boldsymbol{X}_{ij}), *\boldsymbol{X}_{i_1 i_2}\rangle.$$

Moreover, because the numbers $\langle R(\boldsymbol{X}_{ij}), \boldsymbol{X}_{i_1 i_2}\rangle$, $i < j$, $i_1 < i_2$, are just the entries of the matrix of the curvature operator R in the (orthonormal!) basis $\{\boldsymbol{X}_{ij},\ i < j\}$, we have

$$\begin{aligned}\sum_{i_1<i_2} \Omega^i_j(\boldsymbol{X}_{i_1 i_2})\Omega^i_j(*\boldsymbol{X}_{i_1 i_2}) &= \sum_{i_1<i_2} \langle R(\boldsymbol{X}_{ij}), \boldsymbol{X}_{i_1 i_2}\rangle\langle *R(\boldsymbol{X}_{ij}), \boldsymbol{X}_{i_1 i_2}\rangle \\ &= \langle (R*R)(\boldsymbol{X}_{ij}), \boldsymbol{X}_{ij}\rangle\end{aligned}$$

and

$$\sum_{i<j}\sum_{i_1<i_2} \Omega^i_j(\boldsymbol{X}_{i_1 i_2})\Omega^i_j(*\boldsymbol{X}_{i_1 i_2}) = \sum_{i<j}\langle (R*R)(\boldsymbol{X}_{ij}), \boldsymbol{X}_{ij}\rangle = \mathrm{Tr}(R*R).$$

Therefore,

$$p_1[\mathfrak{e}] = \frac{1}{4\pi^2}\,\mathrm{Tr}(R*R),$$

and the cohomology class p_1 is hence given by the form

$$\frac{1}{4\pi^2}\,\mathrm{Tr}(R*R)\,dV$$

(on U and therefore on the whole $\mathcal{X}$).

This proves that *the Pontryagin number $p_1[\mathcal{X}]$ of a connected four-dimensional compact oriented Riemannian space $\mathcal{X}$ is expressed by*

$$p_1[\mathcal{X}] = \frac{1}{4\pi^2}\int_{\mathcal{X}} \mathrm{Tr}(R*R)\,dV, \tag{23}$$

where R is the Riemannian curvature tensor of the manifold $\mathcal{X}$.

Exercise 24.7. Using a similar computation, show that the formula

$$p_k[\mathcal{X}] = \left(\frac{(2k)!}{(4\pi)^k k!}\right)^2 \int_{\mathcal{X}} \mathrm{Tr}(R_k * R_k)\,dV$$

holds for $n = 4k$, where R_k is given by formula (12) in Chap. 17.

§9. Thorp Theorem

If

$$R = \begin{Vmatrix} A & B \\ B' & C \end{Vmatrix},$$

then

$$R * R = \begin{Vmatrix} A^2 - BB' & AB - BC \\ B'A - CB' & B'B - C^2 \end{Vmatrix}$$

and therefore $\operatorname{Tr}(R * R) = \operatorname{Tr}(A^2 - C^2)$. This shows that formula (23) can be rewritten in the form

$$p_1[\mathcal{X}] = \frac{1}{4\pi^2} \int_{\mathcal{X}} (\operatorname{Tr} A^2 - \operatorname{Tr} C^2)\, dV. \tag{23'}$$

Comparing (22) and (23′), we immediately obtain that in the case where $\mathcal{X}$ is an Einstein space,

$$2e[\mathcal{X}] - p_1[\mathcal{X}] = \frac{1}{2\pi^2} \int_{\mathcal{X}} \operatorname{Tr} C^2\, dV \geq 0 \tag{24}$$

and therefore

$$e[\mathcal{X}] \geq \frac{p_1[\mathcal{X}]}{2}.$$

When the orientation of the manifold $\mathcal{X}$ changes, the number $e[\mathcal{X}]$ remains the same, while the number $p_1[\mathcal{X}]$ changes its sign. Therefore, the obtained inequality in fact has the form

$$e[\mathcal{X}] \geq \left| \frac{p_1[\mathcal{X}]}{2} \right|. \tag{25}$$

This proves the following proposition, which strengthens the Berger theorem.

Proposition 24.3 (Thorp theorem). *Inequality* (25) *holds for any connected four-dimensional compact oriented Einstein space.*

By the relations $e[\mathcal{X}] = \chi(\mathcal{X})$ and $p_1[\mathcal{X}] = 3 \operatorname{Sign} \mathcal{X}$, we can also rewrite inequality (25) in the form

$$\frac{3}{2} |\operatorname{Sign} \mathcal{X}| \leq \chi(\mathcal{X}). \tag{25'}$$

Here, the equality can also be attained for nonflat metrics, but in this case, as Hitchin showed, the manifold $\mathcal{X}$ should be one of three specific complex algebraic varieties (which have the dimension two over the field $\mathbb{C}$ and are called the *Enriques surfaces* and the *K*3 *surface*).

Exercise 24.8. Show that if the equality holds in (25), then the Ricci tensor of the manifold $\mathcal{X}$ equals zero.

There exist simply connected four-dimensional manifolds that do not satisfy condition (25), and there is therefore no Einstein metric on them.

§10. Sentenac Theorem

The Chern–Milnor theorem admits a similar improvement for Einstein spaces.

Proposition 24.4. *If sectional curvatures have the same sign at all points and in all directions in a connected four-dimensional compact orientable Einstein space* $\mathcal{X}$, *then*

$$\left(\frac{3}{2}\right)^{3/2} |\operatorname{Sign}\mathcal{X}| \leq \chi(\mathcal{X}). \tag{26}$$

Because the number $(3/2)^{3/2}$ is irrational, the equality is possible here only for $\operatorname{Sign}\mathcal{X} = 0$ and $\chi(\mathcal{X}) = 0$. But if $\chi(\mathcal{X}) = e[\mathcal{X}] = 0$, then $R = 0$ (see (22)), i.e., the metric on $\mathcal{X}$ is flat. Therefore, *the equality in* (26) *is only attained for flat Einstein metrics.*

Sentenac first proved Proposition 24.4, based on the following lemma.

Lemma 24.1. *Over each coordinate neighborhood* U *of an arbitrary four-dimensional Einstein space* $\mathcal{X}$, *the module of vector fields* $\mathfrak{a}\mathcal{X}$ *has an orthonormal basis* X_1, X_2, X_3, X_4 *such that in the corresponding basis* $X_i \wedge X_j$, $1 \leq i < j \leq 4$, *of the module* $\Lambda^2\mathcal{X}$, *the curvature tensor* R *of the space* $\mathcal{X}$, *considered as an operator* $\Lambda^2\mathcal{X} \to \Lambda^2\mathcal{X}$, *has a matrix of the following form for the corresponding location of the bivectors* $X_i \wedge X_j$:

$$\begin{Vmatrix} L & M \\ M & L \end{Vmatrix}, \tag{27}$$

where

$$L = \begin{Vmatrix} \lambda_1 & & 0 \\ & \lambda_2 & \\ 0 & & \lambda_3 \end{Vmatrix}, \qquad M = \begin{Vmatrix} \mu_1 & & 0 \\ & \mu_2 & \\ 0 & & \mu_3 \end{Vmatrix},$$

and $\mu_1 + \mu_2 + \mu_3 = 0$.

Proof. By assumption, the tensor R has form (20), where A and C are self-adjoint and therefore orthogonally diagonalizable operators with $\operatorname{Tr} A = \operatorname{Tr} C$. The submodules $\Lambda^2_+\mathcal{X}$ and $\Lambda^2_-\mathcal{X}$ therefore admit orthonormal bases $\boldsymbol{\xi}_1, \boldsymbol{\xi}_2, \boldsymbol{\xi}_3$ and $\boldsymbol{\eta}_1, \boldsymbol{\eta}_2, \boldsymbol{\eta}_3$ over U such that

$$\begin{array}{lll} A\boldsymbol{\xi}_1 = a_1\boldsymbol{\xi}_1, & A\boldsymbol{\xi}_2 = a_2\boldsymbol{\xi}_2, & A\boldsymbol{\xi}_3 = a_3\boldsymbol{\xi}_3, \\ C\boldsymbol{\eta}_1 = c_1\boldsymbol{\eta}_1, & C\boldsymbol{\eta}_2 = c_2\boldsymbol{\eta}_2, & C\boldsymbol{\eta}_3 = c_3\boldsymbol{\eta}_3, \end{array}$$

where $a_1 + a_2 + a_3 = c_1 + c_2 + c_3$. We set

$$\begin{array}{lll} \boldsymbol{P}_1 = \dfrac{\boldsymbol{\xi}_1 + \boldsymbol{\eta}_1}{\sqrt{2}}, & \boldsymbol{P}_2 = \dfrac{\boldsymbol{\xi}_2 + \boldsymbol{\eta}_2}{\sqrt{2}}, & \boldsymbol{P}_3 = \dfrac{\boldsymbol{\xi}_3 + \boldsymbol{\eta}_3}{\sqrt{2}}, \\ \boldsymbol{P}_1^{\perp} = \dfrac{\boldsymbol{\xi}_1 - \boldsymbol{\eta}_1}{\sqrt{2}}, & \boldsymbol{P}_2^{\perp} = \dfrac{\boldsymbol{\xi}_2 - \boldsymbol{\eta}_2}{\sqrt{2}}, & \boldsymbol{P}_3^{\perp} = \dfrac{\boldsymbol{\xi}_3 - \boldsymbol{\eta}_3}{\sqrt{2}}. \end{array} \tag{28}$$

Then in the basis $\boldsymbol{P}_1, \boldsymbol{P}_2, \boldsymbol{P}_3, \boldsymbol{P}_1^\perp, \boldsymbol{P}_2^\perp, \boldsymbol{P}_3^\perp$, the operator R has a matrix of form (27) with

$$\lambda_i = \frac{a_i + c_i}{2}, \qquad \mu_i = \frac{a_i - c_i}{2}, \quad i = 1, 2, 3,$$

for which

$$\mu_1 + \mu_2 + \mu_3 = \frac{(a_1 + a_2 + a_3) - (c_1 + c_2 + c_3)}{2} = 0.$$

To complete the proof of Lemma 24.1, we therefore need only prove that this basis consists of bivectors of the form $X_i \wedge X_j$, $1 \le i < j \le 4$. To prove this, we note that according to formula (10) in Chap. 17, we have

$$\boldsymbol{\xi}_i \wedge \boldsymbol{\eta}_j = -\boldsymbol{\xi}_i \wedge *\boldsymbol{\eta}_j = -\langle \boldsymbol{\xi}_i, \boldsymbol{\eta}_j \rangle \mathfrak{e} = 0$$

and similarly

$$\boldsymbol{\xi}_i \wedge \boldsymbol{\xi}_j = \delta_{ij}\mathfrak{e}, \qquad \boldsymbol{\eta}_i \wedge \boldsymbol{\eta}_j = -\delta_{ij}\mathfrak{e}.$$

Therefore,

$$\boldsymbol{P}_i \wedge \boldsymbol{P}_j = \frac{\boldsymbol{\xi}_i + \boldsymbol{\eta}_i}{\sqrt{2}} \wedge \frac{\boldsymbol{\xi}_j + \boldsymbol{\eta}_j}{\sqrt{2}} = 0 \tag{29}$$

for any $i, j = 1, 2, 3$, and hence $\langle \boldsymbol{P}_i, *\boldsymbol{P}_j \rangle = 0$. Similarly, we show that

$$\langle \boldsymbol{P}_i, \boldsymbol{P}_j \rangle = \delta_{ij}, \quad i, j = 1, 2, 3.$$

Exercise 24.9. Show that for a skew-symmetric tensor $\boldsymbol{P}$ of degree two in the four-dimensional space, the relation $\langle \boldsymbol{P}, *\boldsymbol{P} \rangle = 0$ holds iff the tensor $\boldsymbol{P}$ is a bivector. [*Hint*: The number $\langle \boldsymbol{P}, *\boldsymbol{P} \rangle$ is equal to the left-hand side of the Plücker relation.]

In particular, this implies that all the tensors $\boldsymbol{P}_i$, $i = 1, 2, 3$, are bivectors (more precisely, bivector fields).

Exercise 24.10. Prove that the relation $\boldsymbol{P} \wedge \boldsymbol{Q} = 0$ holds for the bivectors $\boldsymbol{P}$ and $\boldsymbol{Q}$ in the four-dimensional space iff there exists a vector that is parallel to both bivectors.

Because $\boldsymbol{P}_1 \wedge \boldsymbol{P}_2 = 0$ (and $\langle \boldsymbol{P}_1, \boldsymbol{P}_1 \rangle = \langle \boldsymbol{P}_2, \boldsymbol{P}_2 \rangle = 1$), this implies that there exist orthonormal vector fields X_1, X_2, X_3 on U such that

$$\boldsymbol{P}_1 = X_1 \wedge X_2 \qquad \text{and} \qquad \boldsymbol{P}_2 = X_1 \wedge X_3.$$

We add a field X_4 to them such that the basis X_1, X_2, X_3, X_4 of the module $\mathfrak{a}\mathcal{X}$ over U becomes orthonormal. Because $\boldsymbol{P}_1 \wedge \boldsymbol{P}_3 = 0$ and $\boldsymbol{P}_2 \wedge \boldsymbol{P}_3 = 0$, the coefficients of the bivectors $X_3 \wedge X_4$ and $X_2 \wedge X_4$ vanish in the expansion of the bivector $\boldsymbol{P}_3$ with respect to the basis $X_i \wedge X_j$, $1 \le i < j \le 4$. Moreover, because $\langle \boldsymbol{P}_1, \boldsymbol{P}_3 \rangle = 0$ and $\langle \boldsymbol{P}_2, \boldsymbol{P}_3 \rangle = 0$, the same is true for the coefficients of the bivectors $X_1 \wedge X_2 = \boldsymbol{P}_1$ and $X_1 \wedge X_3 = \boldsymbol{P}_2$. Therefore, $\boldsymbol{P}_3 = a(X_1 \wedge X_4) + b(X_2 \wedge X_3)$, where $a^2 + b^2 = 1$. Because $\boldsymbol{P}_3$ is a bivector, this relation is possible only for $a = \pm 1$ or $b = \pm 1$. Because, obviously, $*\boldsymbol{P}_i = \boldsymbol{P}_i^\perp$, $i = 1, 2, 3$, replacing X_4 with $-X_4$ if necessary, we obtain the coincidence of the bivectors $X_i \wedge X_j$ with bivectors (28) up to the order. □

Proof (Proposition 24.4). Because the operator R has matrix (27) in the basis $\boldsymbol{P}_i, \boldsymbol{P}_i^{\perp}$, $i = 1, 2, 3$, we have

$$\lambda_i = \langle R(\boldsymbol{P}_i), \boldsymbol{P}_i \rangle, \quad i = 1, 2, 3,$$

and the elements λ_i are therefore sectional curvatures of the space $\mathcal{X}$. Therefore, under the conditions of Proposition 24.4, they have the same sign.

On the other hand, because the Hodge operator in this basis has the matrix

$$\begin{Vmatrix} 0 & \mathrm{id} \\ \mathrm{id} & 0 \end{Vmatrix},$$

the operators $R * R$ and $*R * R$ have the matrices

$$\begin{Vmatrix} LM + ML & L^2 + M^2 \\ M^2 + L^2 & ML + LM \end{Vmatrix} \quad \text{and} \quad \begin{Vmatrix} M^2 + L^2 & ML + LM \\ LM + ML & L^2 + M^2 \end{Vmatrix}.$$

Therefore,

$$\begin{aligned} \operatorname{Tr}(R * R) &= 2 \operatorname{Tr}(LM + ML) = 4(\lambda_1\mu_1 + \lambda_2\mu_2 + \lambda_3\mu_3) = 4\boldsymbol{\lambda\mu} \\ \operatorname{Tr}(*R * R) &= 2 \operatorname{Tr}(L^2 + M^2) \\ &= 2(\lambda_1^2 + \lambda_2^2 + \lambda_3^2 + \mu_1^2 + \mu_2^2 + \mu_3^2) = 2(\boldsymbol{\lambda}^2 + \boldsymbol{\mu}^2), \end{aligned}$$

where $\boldsymbol{\lambda} = (\lambda_1, \lambda_2, \lambda_3)$ and $\boldsymbol{\mu} = (\mu_1, \mu_2, \mu_3)$.

Lemma 24.2. *If the numbers λ_1, λ_2, and λ_3 have the same sign and the numbers μ_1, μ_2, and μ_3 satisfy the relation $\mu_1 + \mu_2 + \mu_3 = 0$, then*

$$\boldsymbol{\lambda\mu} \le \frac{1}{\sqrt{6}} (\boldsymbol{\lambda}^2 + \boldsymbol{\mu}^2), \tag{30}$$

where $\boldsymbol{\lambda} = (\lambda_1, \lambda_2, \lambda_3)$ and $\boldsymbol{\mu} = (\mu_1, \mu_2, \mu_3)$.

According to this lemma,

$$\operatorname{Tr}(R * R) \le \frac{2}{\sqrt{6}} \operatorname{Tr}(*R * R),$$

and therefore (see (16) and (23))

$$p_1[\mathcal{X}] \le \frac{4}{\sqrt{6}} e[\mathcal{X}].$$

Changing, the orientation of the manifold $\mathcal{X}$ if necessary, we immediately obtain the inequality

$$|p_1[\mathcal{X}]| \le \frac{4}{\sqrt{6}} e[\mathcal{X}],$$

which is equivalent to inequality (26). □

It remains to prove Lemma 2.

Proof (of Lemma 2). If $\boldsymbol{\lambda} = 0$ or $\boldsymbol{\mu} = 0$, then inequality (30) is obvious, and the case $\lambda_i < 0$, $i = 1, 2, 3$, is reduced to the case $\lambda_i > 0$, $i = 1, 2, 3$, by changing the sign of $\boldsymbol{\mu}$. We can therefore assume without loss of generality that $\lambda_i > 0$, $i = 1, 2, 3$, and $\boldsymbol{\mu} \neq 0$.

Let θ be the angle made by two vectors $\boldsymbol{\lambda}$ and $\boldsymbol{\mu}$ of the space $\mathbb{R}^3$. Because the minimum value of this angle for a fixed vector $\boldsymbol{\lambda}$ and a variable vector $\boldsymbol{\mu}$ equals the angle made by the vector $\boldsymbol{\lambda}$ and the plane $\mu_1 + \mu_2 + \mu_3 = 0$, we have

$$\sin\theta \geq \frac{\boldsymbol{\lambda e}}{|\boldsymbol{\lambda}| \cdot |\boldsymbol{e}|} = \frac{1}{\sqrt{3}} \frac{\lambda_1 + \lambda_2 + \lambda_3}{\sqrt{\lambda_1^2 + \lambda_2^2 + \lambda_3^2}},$$

where $\boldsymbol{e} = (1, 1, 1)$ is the vector orthogonal to the plane $\mu_1 + \mu_2 + \mu_3 = 0$. Because, as is easily seen,

$$\min_{\lambda_i \geq 0} \frac{\lambda_1 + \lambda_2 + \lambda_3}{\sqrt{\lambda_1^2 + \lambda_2^2 + \lambda_3^2}} = 1,$$

this proves that $\sin\theta \geq 1/\sqrt{3}$ and therefore $\cos\theta \leq \sqrt{2/3}$. Because $\boldsymbol{\lambda}^2 + \boldsymbol{\mu}^2 = |\boldsymbol{\lambda}|^2 + |\boldsymbol{\mu}|^2 \geq 2|\boldsymbol{\lambda}| \cdot |\boldsymbol{\mu}|$, we have

$$2\frac{\boldsymbol{\lambda\mu}}{\boldsymbol{\lambda}^2 + \boldsymbol{\mu}^2} \leq \frac{\boldsymbol{\lambda\mu}}{|\boldsymbol{\lambda}| \cdot |\boldsymbol{\mu}|} = \cos\theta \leq \sqrt{\frac{2}{3}},$$

which is equivalent to inequality (30). □

Chapter 25
Metrics on a Lie Group I

§1. Left-Invariant Metrics on a Lie Group

Let $\mathcal{G}$ be an arbitrary connected Lie group, and let $\mathfrak{g}$ be its Lie algebra. We recall that L_a, $a \in \mathcal{G}$ denotes the *left translation*

$$L_a: \mathcal{G} \to \mathcal{G}, \quad p \mapsto ap, \quad p \in \mathcal{G}.$$

Definition 25.1. A Riemannian (or pseudo-Riemannian) metric g on the Lie group $\mathcal{G}$ is said to be *left-invariant* if $L_a^* g = g$ for any element $a \in \mathcal{G}$, i.e., if for any point $p \in \mathcal{G}$ and any vectors $A, B \in \mathbf{T}_p\mathcal{G}$,

$$g_p(A, B) = g_{ap}((dL_a)_p A, (dL_a)_p B). \tag{1}$$

In other words, *a metric is left-invariant if all left translations L_a are isometries.*

Therefore (see Exercises 6.3 and 19.1), *the Levi-Civita connection corresponding to a left-invariant metric is left-invariant.*

In the interpretation of the field g as a $\mathbf{F}\mathcal{X}$-morphism $\mathfrak{a}\mathfrak{g} \otimes \mathfrak{a}\mathfrak{g} \to \mathbf{F}\mathcal{X}$, condition (1) is equivalent to $g(X, Y) = \text{const}$ for any left-invariant fields $X, Y \in \mathfrak{g}$, i.e., to the functions

$$g_{ij} = g(X_i, X_j), \quad i, j = 1, \dots, n,$$

being constants for any basis

$$X_1, \dots, X_n \tag{2}$$

of the Lie algebra $\mathfrak{g}$. This means that *the left-invariant (pseudo-)Riemannian metrics on the Lie group $\mathcal{G}$ are in a natural bijective correspondence with the (pseudo-)Euclidean structures on the linear space $\mathfrak{g}$ and can therefore be identified with them.* As a rule, we write merely (X, Y) instead of $g(X, Y)$ (compare with Chap. 11).

§2. Invariant Metrics on a Lie Group

Of course, metrics on a Lie group for which geodesics passing through the point e coincide with one-parameter subgroups, i.e., for which the corresponding Levi-Civita connection is a Cartan connection (see Definition 6.2), are of main interest. Because the symmetric Cartan connection is unique, this connection should be given by formula (10) in Chap. 6; therefore, a left-invariant connection has this property iff

$$([X,Y],Z)+(Y,[X,Z])=0 \tag{3}$$

for any fields $X, Y, Z \in \mathfrak{g}$ (see formula (2) in Chap. 11; because the metric g is left-invariant, we have $(Y, Z) = \text{const}$, and the left-hand side of this formula therefore equals zero.)

Exercise 25.1. Show that condition (3) holds iff

$$(X,[X,Y])=0 \tag{3'}$$

for any fields $X, Y \in \mathfrak{g}$.

A (pseudo-)Euclidean metric on the Lie algebra $\mathfrak{g}$ that satisfies condition (3) (or condition (3′)) is called an *invariant metric.*

This terminology is based on the following. For simplicity, we suppose that a metric g on the Lie algebra $\mathfrak{g}$ is Euclidean. Then condition (3) means exactly that the operator

$$\operatorname{ad} X\colon Y \mapsto [X,Y] \tag{4}$$

on the Euclidean space $\mathfrak{g}$ is skew-symmetric. Therefore, the linear operator

$$e^{\operatorname{ad} X}\colon \mathfrak{g} \to \mathfrak{g}$$

is orthogonal (isometric). Because $e^{\operatorname{ad} X} = (d\operatorname{int}_a)_e$, where $a = \exp X$ and

$$R_a = L_{ba} \circ \operatorname{int}_{a^{-1}} \circ L_b^{-1}, \quad a, b \in \mathcal{G},$$

we have

$$(dR_a)_b = (dL_{ba})_e \circ (d\operatorname{int}_{a^{-1}})_e \circ (dL_b^{-1})_b.$$

This proves that *the mapping $R_a\colon \mathcal{G} \to \mathcal{G}$ is an isometry* (for points a from a certain neighborhood of the identity and therefore, because the group $\mathcal{G}$ is connected, for all points $a \in \mathcal{G}$). Conversely, if all mappings R_a, $a \in \mathcal{G}$, are isometries, then all mappings int_a are also isometries. In particular, for any field $X \in \mathfrak{g}$, the operator $e^{\operatorname{ad} X}$ is an isometry, i.e., the operator $\operatorname{ad} X$ is skew-symmetric. When we say that a metric g on the Lie group is *invariant* if all left and right translations L_a and R_a with respect to this metric are isometries, we see therefore that *condition* (3) *for a left-invariant metric g on the Lie group $\mathcal{G}$ is equivalent to this metric being invariant.*

Not on every Lie group has an invariant metric.

Example 25.1. Let $\mathcal{G}$ be the group of all 2×2 matrices of the form

$$\begin{Vmatrix} a & b \\ 0 & 1 \end{Vmatrix}, \quad a \neq 0$$

(this group is isomorphic to the group of all affine transformations $y = ax + b$ of the real line). The Lie algebra $\mathfrak{g}$ of the group $\mathcal{G}$ consists of matrices of the form

$$\begin{Vmatrix} x & y \\ 0 & 0 \end{Vmatrix}$$

(prove this!) and is generated by the matrices

$$E_1 = \begin{Vmatrix} 1 & 0 \\ 0 & 0 \end{Vmatrix}, \qquad E_2 = \begin{Vmatrix} 0 & 1 \\ 0 & 0 \end{Vmatrix},$$

connected by the relation $[E_1, E_2] = E_2$. Therefore, relation (3′) yields $(E_1, E_2) = 0$ for $X = E_1$ and $Y = E_2$ and $(E_2, E_2) = 0$ for $X = E_2$ and $Y = E_1$. Therefore, the element E_2 is orthogonal to all elements from $\mathfrak{g}$, which contradicts the nondegeneracy of the metric. Therefore, no metric on the Lie algebra $\mathfrak{g}$ satisfies condition (3).

§3. Semisimple Lie Groups and Algebras

Setting

$$\mathrm{Kil}(X, Y) = \mathrm{Tr}(\mathrm{ad}\, X \circ \mathrm{ad}\, Y), \quad X, Y \in \mathfrak{g},$$

we obviously obtain a bilinear symmetric functional (form) Kil on the Lie algebra $\mathfrak{g}$. This functional is called the *Killing form* on $\mathfrak{g}$. (Certain authors call it the *Cartan–Killing form.*)

Because for any automorphism $\varphi: \mathfrak{g} \to \mathfrak{g}$ of the Lie algebra $\mathfrak{g}$,

$$\varphi[X, \varphi^{-1} Y] = [\varphi X, Y], \quad X, Y \in \mathfrak{g},$$

we have

$$\mathrm{ad}(\varphi X) = \varphi \circ \mathrm{ad}\, X \circ \varphi^{-1}, \quad X \in \mathfrak{g}, \tag{5}$$

which directly implies

$$\mathrm{Kil}(\varphi X, \varphi Y) = \mathrm{Kil}(X, Y), \quad X, Y \in \mathfrak{g}$$

(*the form* Kil *is invariant with respect to automorphisms*).

Moreover, because $\mathrm{ad}[X, Y] = [\mathrm{ad}\, X, \mathrm{ad}\, Y]$ and $\mathrm{Tr}\, AB = \mathrm{Tr}\, BA$, the form Kil satisfies condition (3′):

$$\begin{aligned} \mathrm{Kil}(X, [X, Y]) &= \mathrm{Tr}(\mathrm{ad}\, X \circ \mathrm{ad}[X, Y]) = \\ &= \mathrm{Tr}(\mathrm{ad}\, X \circ \mathrm{ad}\, X \circ \mathrm{ad}\, Y - \mathrm{ad}\, X \circ \mathrm{ad}\, Y \circ \mathrm{ad}\, X) = 0. \end{aligned}$$

However, this form is degenerate in general.

Definition 25.2. A Lie algebra $\mathfrak{g}$ (and also the corresponding Lie group $\mathcal{G}$) is said to be *semisimple* if the Killing form Kil is nondegenerate.

Remark 25.1. This definition is meaningful for Lie algebras over an arbitrary field $\mathbb{K}$ (and, in particular, over the field $\mathbb{C}$).

Therefore, on a semisimple Lie algebra (over the field $\mathbb{R}$), the Killing form is an invariant metric (and therefore assigns an invariant (pseudo-)Riemannian metric on the Lie group $\mathcal{G}$). In what follows, we always assume that a semisimple Lie algebra $\mathfrak{g}$ is equipped with this metric and, as a rule, merely write (X, Y) instead of $\mathrm{Kil}(X, Y)$.

In accordance with general algebraic definitions, a subalgebra $\mathfrak{h}$ of a Lie algebra $\mathfrak{g}$ is called an *ideal* if $[X, Y] \in \mathfrak{h}$ for any $X \in \mathfrak{g}$ and $Y \in \mathfrak{h}$, i.e., if this subalgebra is an invariant subspace of the operator $\operatorname{ad} X: \mathfrak{g} \to \mathfrak{g}$ for any $X \in \mathfrak{g}$.

Exercise 25.2. Show that a subalgebra $\mathfrak{h}$ is an ideal of a Lie algebra $\mathfrak{g}$ iff the connected subgroup $\mathcal{H}$ of the Lie group $\mathcal{G}$ corresponding to it is invariant.

If $\mathfrak{b}$ is a subspace complemented to an ideal $\mathfrak{h}$, then for each $X \in \mathfrak{h}$, the operator $\operatorname{ad} X$ maps $\mathfrak{h}$ and $\mathfrak{b}$ into $\mathfrak{h}$. Schematically, we have

	$\mathfrak{h}$	$\mathfrak{b}$
$\mathfrak{h}$	$\operatorname{ad}_{\mathfrak{h}} X$	$*$
$\mathfrak{b}$	0	0

,

where $\operatorname{ad}_{\mathfrak{h}} X$ is the operator $\operatorname{ad} X$ for the Lie algebra $\mathfrak{h}$ and $*$ is a certain linear operator, which is outside our interest. Therefore, the operator $\operatorname{ad} X \circ \operatorname{ad} Y$ for $X, Y \in \mathfrak{h}$ (or even for $X \in \mathfrak{h}$ and $Y \in \mathfrak{g}$) has the form

	$\mathfrak{h}$	$\mathfrak{b}$
$\mathfrak{h}$	$\operatorname{ad}_{\mathfrak{h}} X \circ \operatorname{ad}_{\mathfrak{h}} Y$	$*$
$\mathfrak{b}$	0	0

,

which directly implies that *the restriction of the Killing form of the Lie algebra* $\mathfrak{g}$ *to the ideal* $\mathfrak{h}$ *is the Killing form of the Lie algebra* $\mathfrak{h}$. In a conditional, but clear, notation,

$$\mathrm{Kil}^{\mathfrak{g}}|_{\mathfrak{h}\times\mathfrak{h}} = \mathrm{Kil}^{\mathfrak{h}}.$$

We stress that this is true only under the assumption that $\mathfrak{h}$ is an ideal.

Furthermore, it is easy to see that *the orthogonal complement* $\mathfrak{h}^{\perp}$ *of any ideal* $\mathfrak{h} \subset \mathfrak{g}$ *is also an ideal.* Indeed, if $X \in \mathfrak{g}$, $Y \in \mathfrak{h}^{\perp}$, and $Z \in \mathfrak{h}$, then $[X, Z] \in \mathfrak{h}$, and therefore

$$([X, Y], Z) = -(Y, [X, Z]) = 0. \qquad \square$$

Moreover, if $X, Y \in \mathfrak{h} \cap \mathfrak{h}^\perp$, then

$$([X,Y],Z) = -(Y,[X,Z]) = 0$$

for any $Z \in \mathfrak{g}$ because $Y \in \mathfrak{h}^\perp$ and $[X,Z] \in \mathfrak{h}$. Therefore, if the metric is nondegenerate (the Lie algebra is semisimple), then $[X,Y] = 0$.

A Lie algebra $\mathfrak{g}$ with zero multiplication, i.e., such that $[X,Y] = 0$ for any elements $X, Y \in \mathfrak{g}$ is said to be *Abelian.* The Killing form on such an algebra vanishes identically. We have thus proved that *for any ideal $\mathfrak{h} \subset \mathfrak{g}$ of a semisimple Lie algebra $\mathfrak{g}$, the ideal $\mathfrak{h} \cap \mathfrak{h}^\perp$ is Abelian.* □

Exercise 25.3. Prove that a connected Lie group $\mathcal{G}$ is Abelian (commutative) iff its Lie algebra $\mathfrak{g} = \mathfrak{l}\mathcal{G}$ is Abelian. (This explains the terminology.)

On the other hand, it is easy to see that *a semisimple algebra $\mathfrak{g}$ does not contain nonzero Abelian ideals.* Indeed, if $\mathfrak{a} \subset \mathfrak{g}$ is an Abelian ideal, then for any elements $X \in \mathfrak{a}$ and $Y \in \mathfrak{g}$, the operator $\operatorname{ad} X \circ \operatorname{ad} Y$ has the form

	$\mathfrak{a}$	$\mathfrak{b}$
$\mathfrak{a}$	0	*
$\mathfrak{b}$	0	0

,

where, as above, $\mathfrak{b}$ is the subspace complemented to the ideal $\mathfrak{a} = \mathfrak{h}$; therefore, $\operatorname{Kil}(X,Y) = 0$, which is possible only for $X = 0$ because the form Kil is nondegenerate. □

Remark 25.2. The converse can be proved (it is a difficult theorem!): *a Lie algebra is semisimple if it does not contain nonzero Abelian ideals.* We do not need this assertion (which belongs to H. Cartan).

Combining the last two assertions, we immediately find that in a semisimple Lie algebra $\mathfrak{g}$, we have the relation $\mathfrak{h} \cap \mathfrak{h}^\perp = 0$ for any ideal $\mathfrak{h}$. Because $\dim \mathfrak{h} + \dim \mathfrak{h}^\perp = \dim \mathfrak{g}$ by the nondegeneracy of the metric, this implies that *a semisimple Lie algebra $\mathfrak{g}$ is the direct sum of the ideals $\mathfrak{h}$ and $\mathfrak{h}^\perp$:*

$$\mathfrak{g} = \mathfrak{h} \oplus \mathfrak{h}^\perp. \tag{6}$$

But then the Killing form of the algebra $\mathfrak{g}$ is the direct sum (in a clear sense) of the Killing forms of the Lie algebras $\mathfrak{h}$ and $\mathfrak{h}^\perp$, and both of the latter forms are therefore nondegenerate. Therefore, *each ideal $\mathfrak{h}$ of a semisimple Lie algebra $\mathfrak{g}$ is itself a semisimple Lie algebra.* □

§4. Simple Lie Groups and Algebras

A Lie algebra $\mathfrak{g}$ (and also the corresponding Lie group $\mathcal{G}$) is said to be *simple* if it is semisimple and does not contain any nontrivial ideals (different from the zero ideal 0 and the whole algebra $\mathfrak{g}$).

Remark 25.3. This definition was proposed by Helgason. By the Cartan theorem mentioned in Remark 25.2, it is equivalent to the classical definition in which only the non-Abelian property is required instead of the semisimplicity.

Remark 25.4. In the general theory of algebras, an algebra is said to be *simple* if it has no nontrivial ideals. From this standpoint, simple algebras should also contain a one-dimensional Lie algebra (which is automatically Abelian), but this is inconvenient in Lie algebra theory. In exactly the same way, the general group theory definition of simple groups as groups that do not contain nontrivial invariant subgroups is inconvenient in Lie group theory (according to our definition, a simple Lie group can contain nontrivial invariant subgroups, but these groups should be discrete (see Exercise 25.2)).

Because both summands in decomposition (6) are ideals, we have $[X, Y] = 0$ for $X \in \mathfrak{h}$ and $Y \in \mathfrak{h}^{\perp}$. Therefore, *any ideal of the Lie algebra* $\mathfrak{h}$ *is also an ideal of the Lie algebra* $\mathfrak{g}$. □

By an obvious induction, this implies that *each semisimple Lie algebra* $\mathfrak{g}$ *is an orthogonal direct sum of ideals that are simple Lie algebras,*

$$\mathfrak{g} = \mathfrak{g}_1 \oplus \cdots \oplus \mathfrak{g}_r. \tag{7}$$

Exercise 25.4. Show that each ideal $\mathfrak{h}$ of a semisimple Lie algebra $\mathfrak{g}$ is a sum of simple ideals from decomposition (7).

§5. Inner Derivations of Lie Algebras

General results concerning automorphism groups of finite-dimensional algebras can be applied to an arbitrary Lie algebra $\mathfrak{g}$ in particular. Therefore, for each (finite-dimensional) Lie algebra $\mathfrak{g}$, its automorphism group $\operatorname{Aut}\mathfrak{g}$ is a Lie group (which is not connected in general); moreover, the Lie algebra of this group is the Lie algebra $\operatorname{Der}\mathfrak{g}$ of all derivations of the algebra $\mathfrak{g}$, i.e., linear mappings $D: \mathfrak{g} \to \mathfrak{g}$ such that

$$D[X, Y] = [DX, Y] + [X, DY] \tag{8}$$

for any elements $X, Y \in \mathfrak{g}$.

The Jacobi identity directly implies that *each mapping* $\operatorname{ad} X$, $X \in \mathfrak{g}$, *is a derivation of the Lie algebra* $\mathfrak{g}$, *and the mapping*

$$\operatorname{ad}: \mathfrak{g} \to \operatorname{Der}\mathfrak{g}, \qquad X \mapsto \operatorname{ad} X, \quad X \in \mathfrak{g}, \tag{9}$$

is a Lie algebra homomorphism. □

In particular, we see that the mappings $\operatorname{ad} X$, $X \in \mathfrak{g}$, compose a subalgebra of the Lie algebra $\operatorname{Der}\mathfrak{g}$. This subalgebra is denoted by $\operatorname{ad}\mathfrak{g}$, and its elements are called *inner derivations* of the Lie algebra $\mathfrak{g}$.

Because

$$\operatorname{ad} DX = [D, \operatorname{ad} X]$$

for any $D \in \operatorname{Der}\mathfrak{g}$ and $X \in \mathfrak{g}$ (this is only another form of identity (8)), we see that *the subalgebra* $\operatorname{ad}\mathfrak{g}$ *is an ideal of the Lie algebra* $\operatorname{Der}\mathfrak{g}$. □

The kernel of mapping (9) is the *center* $\mathfrak{z}$ of the Lie algebra $\mathfrak{g}$, which is its Abelian ideal consisting of elements $X \in \mathfrak{g}$ for which

$$[X, Y] = 0$$

for any $Y \in \mathfrak{g}$. Therefore, *if* $\mathfrak{z} = 0$, *then this subalgebra* $\operatorname{ad}\mathfrak{g}$ *is isomorphic to the algebra* $\mathfrak{g}$. In particular, $\operatorname{ad}\mathfrak{g} = \mathfrak{g}$ *for any semisimple Lie algebra* $\mathfrak{g}$.

Therefore, if a Lie algebra $\mathfrak{g}$ is semisimple, then the algebra $\operatorname{ad}\mathfrak{g}$ is also semisimple, and the restriction of the Killing form of the Lie algebra $\operatorname{Der}\mathfrak{g}$ to its ideal $\operatorname{ad}\mathfrak{g}$ is hence nondegenerate. Therefore, if $(\operatorname{ad}\mathfrak{g})^{\perp}$ is the orthogonal complement of the algebra $\operatorname{ad}\mathfrak{g}$ in the algebra $\operatorname{Der}\mathfrak{g}$ with respect to the Killing form of the algebra $\operatorname{Der}\mathfrak{g}$, then the ideal $\operatorname{ad}\mathfrak{g} \cap (\operatorname{ad}\mathfrak{g})^{\perp}$ of the algebra $\operatorname{Der}\mathfrak{g}$ (which is just the null space of the Killing form of the algebra $\operatorname{ad}\mathfrak{g}$) is equal to zero. Because the inclusion

$$\operatorname{ad}(DX) = [D, \operatorname{ad} X] \in \operatorname{ad}\mathfrak{g} \cap (\operatorname{ad}\mathfrak{g})^{\perp}$$

holds for any $D \in (\operatorname{ad}\mathfrak{g})^{\perp}$ and $X \in \mathfrak{g}$, this proves that $\operatorname{ad}(DX) = 0$ and therefore (because the kernel of the mapping ad equals zero) $DX = 0$, i.e., $D = 0$. Therefore, $(\operatorname{ad}\mathfrak{g})^{\perp} = 0$ and hence $\operatorname{ad}\mathfrak{g} = \operatorname{Der}\mathfrak{g}$. This means that *any derivation of a semisimple Lie algebra* $\mathfrak{g}$ *is inner.* □

A connected subgroup of the Lie group $\operatorname{Aut}\mathfrak{g}$ that corresponds to the subalgebra (ideal) $\operatorname{ad}\mathfrak{g}$ is denoted by $\operatorname{Int}\mathfrak{g}$. Being connected, it is contained in the component of the identity $(\operatorname{Aut}\mathfrak{g})_0$ of the group $\operatorname{Aut}\mathfrak{g}$ and coincides with this component iff $\operatorname{ad}\mathfrak{g} = \operatorname{Der}\mathfrak{g}$. In particular, we see that *for any semisimple Lie algebra* $\mathfrak{g}$, *the group* $\operatorname{Int}\mathfrak{g}$ *is the component of the identity of the group* $\operatorname{Aut}\mathfrak{g}$ (*and is therefore closed*). □

There exist Lie algebras (which are not semisimple in advance) for which this is no longer true.

Exercise 25.5 (according to van Est and Hochschild). Fixing a certain irrational number h, we define the operation $[\,,\,]$ in the linear space $\mathfrak{g} = \mathbb{C} \oplus \mathbb{C} \oplus \mathbb{R}$ over the field $\mathbb{R}$ by

$$[(z_1, z_2, r), (w_1, w_2, s)] = (2\pi i(rw_1 - sz_1), 2h\pi i(rw_2 - sz_2), 0),$$

where $z_1, z_2, w_1, w_2 \in \mathbb{C}$ and $r, s \in \mathbb{R}$. Show that

1. the linear space $\mathfrak{g}$ is a Lie algebra with respect to this operation;
2. the formula
$$\alpha_{s,t}(z_1, z_2, r) = (e^{2\pi is} z_1, e^{2\pi it} z_2, r)$$
defines an automorphism of the Lie algebra $\mathfrak{g}$ for any $s, t \in \mathbb{R}$;
3. if $t = (s+n)h$, where n is an integer, then $\alpha_{s,t} = e^{\operatorname{ad} X}$, where $X = (0, 0, s+n)$;
4. the automorphism $\alpha_{0,1/3}$ of the Lie algebra $\mathfrak{g}$ does not belong to the group $\operatorname{Int} \mathfrak{g}$;
5. if $s_n \to s$ and $t_n \to t$, then $\alpha_{s_n,t_n} \to \alpha_{s,t}$ in the group $\operatorname{Aut} \mathfrak{g}$.

It follows from 3, 4, and 5 that the group $\operatorname{Int} \mathfrak{g}$ is not closed in the group $\operatorname{Aut} \mathfrak{g}$.

§6. Adjoint Group

According to the general results in Chap. 7, for any Lie group $\mathcal{G}$, the Lie functor defines a homomorphism

$$\mathfrak{l}\colon \operatorname{Aut} \mathcal{G} \to \operatorname{Aut} \mathfrak{g} \tag{10}$$

of the automorphism group $\operatorname{Aut} \mathcal{G}$ of the group $\mathcal{G}$ into the automorphism group $\operatorname{Aut} \mathfrak{g}$ of its Lie algebra $\mathfrak{g}$, which is a monomorphism for a connected group $\mathcal{G}$ (and even an isomorphism for a connected and simply connected group). (Therefore, the automorphism group $\operatorname{Aut} \mathcal{G}$ is automatically a Lie group in the case where the group $\mathcal{G}$ is connected and simply connected.) By definition, the composition of the homomorphism

$$\operatorname{int}\colon \mathcal{G} \to \operatorname{Aut} \mathcal{G}, \qquad a \mapsto \text{rm int}_a, \quad a \in \mathcal{G}, \tag{11}$$

with this mapping is just the adjoint representation Ad,

$$\mathfrak{l} \circ \operatorname{int} = \operatorname{Ad}\colon \mathcal{G} \to \operatorname{Aut} \mathfrak{g}. \tag{12}$$

The image $\operatorname{Ad} \mathcal{G}$ of the group $\mathcal{G}$ under homomorphism (12) is called the *adjoint group*. Because the kernel of homomorphism (11) (and therefore homomorphism (12) in the case where the group $\mathcal{G}$ is connected) is the center $\mathcal{Z}$ of the group $\mathcal{G}$, we see that *for a connected Lie group $\mathcal{G}$, the Lie group $\operatorname{Ad} \mathcal{G}$ is isomorphic to the quotient group $\mathcal{G}/\mathcal{Z}$* (see Exercise 7.4); because $\mathfrak{l}(\operatorname{Ad}) = \operatorname{ad}$ by definition, *the Lie algebra of the group $\operatorname{Ad} \mathcal{G}$ is the algebra* $\operatorname{ad} \mathfrak{g}$, and the group $\operatorname{Ad} \mathcal{G}$ is therefore the group $\operatorname{Int} \mathfrak{g}$, which was introduced above,

$$\operatorname{Ad} \mathcal{G} = \operatorname{Int} \mathfrak{g}.$$

(We stress that the Lie group $\mathcal{G}$ is here assumed to be connected.)

Exercise 25.6. Show that the group $\operatorname{Ad} \mathcal{G}$ (and also the group $\operatorname{Aut} \mathcal{G}$) has a natural structure of a Lie group also for *any* Lie group $\mathcal{G}$ (not necessarily connected and simply connected). In this case, $\operatorname{Int} \mathfrak{g} = (\operatorname{Ad} \mathcal{G})_e$.

§7. Lie Groups and Algebras Without Center

A group $\mathcal{G}$ is called a *group without center* if its center $\mathcal{Z}$ consists of only the identity. Similarly, a Lie algebra $\mathfrak{g}$ is called an *algebra without center* if its center $\mathfrak{z}$ consists of only zero. Of course, any semisimple Lie algebra is an algebra without center. According to what was said above, *a connected Lie group $\mathcal{G}$ without center is isomorphic to the adjoint group* $\operatorname{Ad}\mathcal{G}$.

Exercise 25.7. Prove that *the center $\mathfrak{z}$ of the Lie algebra $\mathfrak{g}$ is the Lie algebra of the center $\mathcal{Z}$ of a Lie group $\mathcal{G}$*:

$$\mathfrak{l}\mathcal{Z} = \mathfrak{z}.$$

Therefore, *the Lie algebra $\mathfrak{g}$ of a group $\mathcal{G}$ is an algebra without center iff the center of the group $\mathcal{G}$ is discrete* (and the homomorphism $\operatorname{Ad}: \mathcal{G} \to \operatorname{Ad}\mathcal{G}$ in this case is a group covering for the connected group $\mathcal{G}$). In particular, *the center of a semisimple Lie group $\mathcal{G}$ is discrete.*

Proposition 25.1. *For a Lie algebra $\mathfrak{g}$ without center, the Lie group* $\operatorname{Int}\mathfrak{g}$ *is a group without center.*

Proof. Because for any automorphism $\varphi: \mathfrak{g} \to \mathfrak{g}$,

$$\operatorname{int}_\varphi(e^{\operatorname{ad} X}) = \varphi \circ e^{\operatorname{ad} X} \circ \varphi^{-1} = e^{\operatorname{ad} \varphi X}, \quad X \in \mathfrak{g}$$

in the group $\operatorname{Aut}\mathfrak{g}$ (see (5)), we have

$$(d\operatorname{int}_\varphi)_e(\operatorname{ad} X) = \operatorname{ad}\varphi X,$$

i.e.,

$$(d\operatorname{int}_\varphi)_e = \operatorname{ad} \circ \varphi \circ \operatorname{ad}^{-1}$$

(because the Lie algebra $\mathfrak{g}$ has no center by assumption, the mapping $\operatorname{ad}: \mathfrak{g} \to \operatorname{ad}\mathfrak{g}$ is an isomorphism, and the mapping ad^{-1} is therefore well defined). Applied to $\varphi \in \operatorname{Int}\mathfrak{g}$, this means that the adjoint representation $\operatorname{Int}\mathfrak{g} \to \operatorname{Aut}(\operatorname{ad}\mathfrak{g})$ of the group $\operatorname{Int}\mathfrak{g}$ acts according to the formula

$$\varphi \mapsto \operatorname{ad} \circ \varphi \circ \operatorname{ad}^{-1}.$$

Therefore, it is a monomorphism, and the group $\operatorname{Int}\mathfrak{g}$ hence has no center. □

For a Lie algebra whose center is not trivial, the center of the group $\operatorname{Int}\mathfrak{g}$ can also be nontrivial.

Exercise 25.8. Let $\mathfrak{g}$ be a three-dimensional Lie algebra with a basis X_1, X_2, X_3 for which

$$[X_1, X_2] = X_3, \qquad [X_1, X_3] = 0, \qquad [X_2, X_3] = 0.$$

Show that the group $\operatorname{Int}\mathfrak{g}$ is Abelian (and two-dimensional).

Chapter 26
Metrics on a Lie Group II

To study semisimple Lie groups and algebras in more detail, we need some additional facts from Lie group theory, which, however, are of independent interest.

§1. Maurer–Cartan Forms

As directly follows from the assertion in Exercise 6.4, each linear differential form ω on a Lie group $\mathcal{G}$ is uniquely characterized by its values $\omega(X)$ on fields $X \in \mathfrak{g}$. In particular, for any basis

$$X_1, \dots, X_n \tag{1}$$

of the Lie algebra $\mathfrak{g}$ there exist uniquely defined forms

$$\omega^1, \dots, \omega^n \tag{2}$$

on the group $\mathcal{G}$ for which $\omega^i(X_j) = \delta^i_j$, $i, j = 1, \dots, n$.

Definition 26.1. Forms (2) are called the *Maurer–Cartan forms* on the Lie group $\mathcal{G}$ that correspond to basis (1) of the Lie algebra $\mathfrak{g}$.

The exterior differential $d\omega^i$ of each form in (2) is uniquely characterized by its values $d\omega^i(X_p, X_q)$, $p, q = 1, \dots, n$, on elements of basis (1). But according to the Cartan formula,

$$(d\omega^i)(X_p, X_q) = X_p\omega^i(X_q) - X_q\omega^i(X_p) - \omega^i([X_p, X_q]) = -\omega^i([X_p, X_q])$$

(because the functions $\omega^i(X_q) = \delta^i_q$ and $\omega^i(X_p) = \delta^i_p$ are constant, we have $X_p\omega^i(X_q) = 0$ and $X_q\omega^i(X_p) = 0$). Therefore, if

$$[X_p, X_q] = c^r_{pq}X_r, \quad p, q, r = 1, \dots, n,$$

then

$$\begin{aligned}(d\omega^i)(X_p, X_q) &= -c^i_{pq} = -c^i_{jk}\delta^j_p\delta^k_q = -c^i_{jk}\omega^j(X_p)\omega^k(X_q) \\ &= -\frac{1}{2}c^i_{jk}(\omega^j(X_p)\omega^k(X_q) - \omega^k(X_p)\omega^j(X_q)) \\ &= -\frac{1}{2}c^i_{jk}(\omega^j \wedge \omega^k)(X_p, X_q)\end{aligned}$$

and hence

$$d\omega^i = -\frac{1}{2}\, c^i_{jk}\omega^j \wedge \omega^k, \quad i, j, k = 1, \dots, n. \tag{3}$$

Formulas (3) are known as *Maurer–Cartan formulas.*

These formulas can be written in an elegant form without subscripts and superscripts if we introduce $\mathfrak{g}$-valued linear differential forms on a Lie group $\mathcal{G}$. By definition, such forms look like $\omega = \omega^i \otimes X_i$ and are naturally identified with $\mathsf{F}\mathcal{G}$-morphisms of the form

$$\mathfrak{a}\mathcal{G} \to \mathsf{F}_{\mathfrak{g}}\mathcal{G},$$

where, as usual, $\mathfrak{a}\mathcal{G}$ is the linear space (more precisely, the $\mathsf{F}\mathcal{G}$-module) of smooth vector fields on the Lie group $\mathcal{G}$ and $\mathsf{F}_{\mathfrak{g}}\mathcal{G}$ is the linear space ($\mathsf{F}\mathcal{G}$-module) of smooth $\mathfrak{g}$-valued functions on $\mathcal{G}$.

Exercise 26.1. Show that the $\mathsf{F}\mathcal{G}$-module $\mathsf{F}_{\mathfrak{g}}\mathcal{G}$ is naturally isomorphic to the $\mathsf{F}\mathcal{G}$-module $\mathfrak{a}\mathcal{G}$. [*Hint*: The function $p \mapsto X(p)$, $p \in \mathcal{G}$, where $X(p)$ is a left-invariant field on the Lie group $\mathcal{G}$ that coincides with the field X at the point p, corresponds to the field $X \in \mathfrak{a}\mathcal{G}$.]

By this isomorphism, $\mathfrak{g}$-valued forms on $\mathcal{G}$ are just $\mathsf{F}\mathcal{G}$-morphisms of the form

$$\mathfrak{a}\mathcal{G} \to \mathfrak{a}\mathcal{G}. \tag{4}$$

In particular, we can consider a form ω on $\mathcal{G}$ to which identical morphism (4) corresponds. The coefficients ω^i of this form are characterized by the property that

$$\omega^i(X) = f^i, \quad i = 1, \dots, n,$$

for any field $X = f^i X_i$ from $\mathfrak{a}\mathcal{G}$. Therefore, in particular, the relations $\omega^i(X_j) = \delta^i_j$ hold for the forms ω^i. Because these relations characterize form (2), this proves that *Maurer–Cartan forms* (2) *are just the coefficients in basis* (1) *of the* $\mathfrak{g}$*-valued form on* $\mathcal{G}$ *corresponding to identical morphism* (4).

On the other hand, because $\mathfrak{g}$ is a Lie algebra, the form

$$[\omega, \omega] = (\omega^j \wedge \omega^k) \otimes [X_j, X_k] = c^i_{jk}(\omega^j \wedge \omega^k) \otimes X_i$$

is well defined. Therefore, formulas (3) are equivalent to the relation

$$d\omega = -\frac{1}{2}[\omega, \omega]. \tag{5}$$

The Maurer–Cartan formulas are usually written in this form now.

§2. Left-Invariant Differential Forms

A differential (not necessarily linear) form ω on a Lie group $\mathcal{G}$ is said to be *left invariant* if $L_a^*\omega = \omega$ for any element $a \in \mathcal{G}$.

Proposition 26.1. *A linear differential form* ω *on a Lie group* $\mathcal{G}$ *is left invariant iff* $\omega(X) = \text{const}$ *for any left-invariant vector field* $X \in \mathfrak{g}$.

Proof. The operation of transport of a tensor field by a diffeomorphism preserves all algebraic operations on fields and the contraction in particular. Applied to the vector fields X and the linear differential forms ω, this means that for any diffeomorphism φ, we have

$$(\varphi^*\omega)(X) = \omega(\varphi_* X) \circ \varphi.$$

In particular, for $\varphi = L_a$ and $X \in \mathfrak{g}$, we have

$$(L_a^*\omega)(X) = \omega(X) \circ L_a,$$

i.e.,

$$(L_a^*\omega)(X)(p) = \omega(X)(ap)$$

for any point $p \in \mathcal{G}$. Therefore, if $L_a^*\omega = \omega$, then $\omega(X)(a^{-1}) = \omega(X)(e)$ for any $a \in \mathcal{G}$, and hence $\omega(X) = \text{const}$. Conversely, if $\omega(X) = \text{const}$, then $(L_a^*\omega)(X)(p) = \omega(X)(p)$, and therefore $L_a^*\omega = \omega$. □

Corollary 26.1. *Left-invariant linear differential forms on a Lie group $\mathcal{G}$ compose a linear space $\mathfrak{g}^*$ of dimension $n = \dim \mathcal{G}$. The Maurer–Cartan forms are exactly forms composing a certain basis of this space.*

Proof. Let $\omega^1, \ldots, \omega^n$ be forms (2). If $X = c^i X_i$ for $c^i \in \mathbb{R}$, then $\omega^i(X) = c^i$. Therefore, forms (2) are left-invariant. If $\omega(X_i) = c_i$ for $c_i \in \mathbb{R}$, then $(\omega - c_i\omega^i)(X_k) = 0$ for any $k = 1, \ldots, n$, and therefore $\omega = c_i\omega^i$ (we recall that any form on $\mathcal{G}$ is uniquely characterized by its restriction to $\mathfrak{g}$). Therefore, forms (2) compose a basis of the space of left-invariant linear differential forms. Therefore, this space is n-dimensional, and Maurer–Cartan forms (2) compose its basis.

Conversely, let $\omega^1, \ldots, \omega^n$ be an arbitrary basis of the space $\mathfrak{g}^*$. Choosing a basis $X_{1'}, \ldots, X_{n'}$ of the linear space $\mathfrak{g}$, we consider the corresponding Maurer–Cartan forms $\omega^{1'}, \ldots, \omega^{n'}$. By what was proved, these forms compose a basis of the space $\mathfrak{g}^*$; therefore, the formulas

$$\omega^{i'} = c_i^{i'}\omega^i, \quad i, i' = 1, \ldots, n,$$

hold. Then the fields $X_i = c_i^{i'} X_{i'}$ obviously compose a basis of the space $\mathfrak{g}$, and the given forms $\omega^1, \ldots, \omega^n$ are Maurer–Cartan forms corresponding to this basis. □

On the basis of this consequence, family (2) is also called the *Maurer–Cartan basis.*

Example 26.1 (Left-invariant forms on matrix Lie groups). On the general linear group $\mathrm{GL}(n;\mathbb{R})$, local coordinates are the entries x_j^i of an arbitrary matrix $X = \|x_j^i\| \in \mathrm{GL}(n;\mathbb{R})$, and each linear differential form ω on $\mathrm{GL}(n;\mathbb{R})$ therefore has the form

$$\omega = f_i^j dx_j^i, \quad i, j = 1, \ldots, n,$$

where f_i^j are certain functions on $\mathrm{GL}(n;\mathbb{R})$. Introducing the matrix of differentials $dX = \|dx_j^i\|$ and the matrix of functions $F = \|f_i^j\|$, the form ω can be written as the trace of the differential matrix FdX,

$$\omega = \mathrm{Tr}(F\,dX). \tag{6}$$

On the other hand, as we know, the Lie algebra $\mathrm{gl}(n;\mathbb{R})$ of the Lie group $\mathrm{GL}(n;\mathbb{R})$ is the commutator algebra $[\mathrm{Mat}_n\mathbb{R}]$.

Exercise 26.2. Show that for any matrix $A \in \mathrm{GL}(n;\mathbb{R})$ and any form (6), the equality

$$L_A^*\omega = \mathrm{Tr}(F_A\,dX)$$

holds, where $F_A(X) = F(AX)A$, $X \in \mathrm{GL}(n;\mathbb{R})$ (i.e., $F_A = R_A \circ F \circ L_A$). [*Hint*: Tangent vectors at each point $X \in \mathrm{GL}(n;\mathbb{R})$ are naturally identified with matrices $C \in \mathrm{Mat}_n\,\mathbb{R}$. Because of this identification, each mapping $(dL_A)_X$ is the left translation $L_A\colon C \mapsto AC$, and the value $\omega_X(C)$ of form (6) on a vector C is equal to $\mathrm{Tr}(F(X)C)$.]

Therefore, form (6) is left invariant iff $F(AX)A = F(X)$ for any matrices $A, X \in \mathrm{GL}(n;\mathbb{R})$. Because the general solution to this equation obviously has the form $F(X) = DX^{-1}$, where $D = F(E)$ is an arbitrary fixed matrix (it suffices to set $X = E$), we therefore conclude that *left-invariant linear differential forms on a Lie group* $\mathrm{GL}(n;\mathbb{R})$ *are exactly the forms of the type*

$$\omega = \mathrm{Tr}(DX^{-1}\,dX), \tag{7}$$

where D is an arbitrary matrix.

Introducing the matrix $X^{-1}dX$ of differential forms, we immediately have from this that *the entries of the matrix $X^{-1}dX$ compose a basis of the linear space* $\mathrm{gl}(n;\mathbb{R})^*$ (are the Maurer–Cartan forms). We can also compose the matrix $X^{-1}dX$ for an arbitrary matrix Lie group $\mathcal{G}$, but its entries are linearly dependent in general.

Exercise 26.3. Show that the entries of the matrix $X^{-1}dX$ generate the linear space $\mathfrak{g}^*$ (and we can therefore choose the Maurer–Cartan basis).

In this sense, the matrix $X^{-1}dX$ yields all Maurer–Cartan forms on the matrix Lie group $\mathcal{G}$.

§3. Haar Measure on a Lie group

Because forms (2) are left invariant, their exterior product

$$\omega_0 = \omega^1 \wedge \ldots \wedge \omega^n \tag{8}$$

(which is a differential form of maximum degree on the Lie group $\mathcal{G}$) is also left invariant. In this case, because forms (2) compose a basis, the form ω_0 does not vanish at any point; therefore, any other form ω of maximum degree on the Lie group $\mathcal{G}$ can be represented as $f\omega_0$, where f is a certain function on $\mathcal{G}$. If the form ω is left invariant, then $f \circ L_a = f$ for any element $a \in \mathcal{G}$, which is only possible for $f = \mathrm{const}$. Therefore, *all left-invariant forms of maximum* *degree on a Lie group $\mathcal{G}$ are exhausted by forms of the type $c\omega_0$, where $c \in \mathbb{R}$ and ω_0 is form* (8).

As we know, forms of maximum degree on an oriented manifold are in a natural bijective correspondence with densities. On the other hand, each basis (1) of a Lie algebra $\mathfrak{g}$, being a basis of the space $\mathbf{T}_p\mathcal{G}$ at each point $p \in \mathcal{G}$, assigns a certain orientation of the manifold $\mathcal{G}$. Let ρ_0 be the density on $\mathcal{G}$ corresponding to form (8) in this orientation. This density depends on only basis (1).

Exercise 26.4. Show that

1. the density ρ_0 is positive (is a volume density) and
2. the density corresponding to any other basis of the Lie algebra $\mathfrak{g}$ has the form $c\rho_0$, where $c > 0$, $c \in \mathbb{R}$.

Therefore, we see that *on any Lie group, there exists a unique left-invariant density* (up to a positive multiplier). □

This volume density is called the *Haar measure* on the group $\mathcal{G}$.

Fixing a certain Haar measure ρ_0, for any finitely supported, locally bounded, and almost continuous function f on the group $\mathcal{G}$, we can consider the integral

$$\int f\rho_0.$$

This integral is called the *Haar integral* of the function f, which is traditionally denoted by

$$\int f(p)\,dp \qquad \left(\text{or } \int_{\mathcal{G}} f(p)\,dp\right). \tag{9}$$

We stress that the integrand $f(p)\,dp$ has no special sense here. However, roughly speaking, it is sometimes convenient to consider it a designation of the density $f\rho_0$ (and, in accordance with this, use the symbol dp to designate the Haar measure ρ_0).

Exercise 26.5. Prove that for any element $a \in \mathcal{G}$, the Haar integrals of the functions $f \circ L_a$ and f coincide, i.e., in notation (9),

$$\int f(ap)\,dp = \int f(p)\,dp. \tag{10}$$

[*Hint*: $L_a^*(f\omega_0) = (f \circ L_a)L_a^*\omega_0$.]

Property (10) is called the *left-invariance* of the Haar measure (integral).

In each chart $(U, h) = (U, x^1, \dots, x^n)$, the Haar measure ρ_0 is given by a certain function $\rho_0 = \rho_0(\boldsymbol{x})$, $\boldsymbol{x} = (x^1, \dots, x^n)$, on the open set $h(U) \subset \mathbb{R}^n$. This function is called the *Haar kernel.* If $U = \mathcal{G}$ (or at least $\overline{U} = \mathcal{G}$), then for any function f, the Haar integral (9) is equal to the Riemann integral

$$\int_{h(U)} f(\boldsymbol{x})\rho_0(\boldsymbol{x})\,d\boldsymbol{x}, \quad d\boldsymbol{x} = dx^1 \cdots dx^n. \tag{11}$$

In the general case, integral (9) is the sum of integrals of form (11).

Exercise 26.6. Show that the Haar kernel on the group $GL(n;\mathbb{R})$ is expressed (in the coordinates x^i_j) by

$$\rho_0(X) = \frac{\text{const}}{|\det X|^n}, \quad X = \|x^i_j\|.$$

The Haar kernel written in the canonical coordinates can be considered as a function on the Lie algebra $\mathfrak{g}$. It turns out that *the formula*

$$\rho_0(X) = \text{const} \cdot \det[f_0(-\operatorname{ad} X)], \quad X \in \mathfrak{g}, \tag{12}$$

holds for this function, where

$$f_0(z) = \frac{e^z - 1}{z} = \sum_{n=0}^{\infty} \frac{z^n}{(n+1)!}. \tag{13}$$

The proof is based on the formula

$$\exp(X + tY)\exp(-X) = \exp(tf_0(\operatorname{ad} X)Y) + o(t), \quad X, Y \in \mathfrak{g}. \tag{14}$$

Exercise 26.7. Deduce (12) from (14).

Exercise 26.8. Prove formula (14) for matrix Lie groups (when $\exp X = e^X$).

Because right translations as well as left translations commute, the form $R^*_a\omega_0$ for any element $a \in \mathcal{G}$ is also left invariant and therefore has the form $\Delta(a)\omega_0$, where $\Delta(a)$ is a certain constant that is different from zero and depends on only the element a (and not on the choice of the form ω_0).

Exercise 26.9. Show that the constant $\Delta(a)$ depends smoothly on a, i.e., the correspondence $a \mapsto \Delta(a)$ is a smooth function on the Lie group $\mathcal{G}$.

Being a smooth function, the function Δ is continuous and, not vanishing everywhere, therefore preserves a constant sign on each component of the group $\mathcal{G}$. Because $\Delta(e) = 1$, this proves that *on a connected Lie group $\mathcal{G}$, the function Δ is positive.* □

The function Δ is called the *module* of the Lie group $\mathcal{G}$. Because $R_b R_a = R_{ab}$ and because the product of real numbers is commutative, we have

$$\Delta(ab) = \Delta(a)\,\Delta(b) \tag{15}$$

for any elements $a, b \in \mathcal{G}$ (*the module is a homomorphism of the group $\mathcal{G}$ into the multiplicative group of nonzero real numbers*).

Exercise 26.10. Prove that

$$\Delta(\exp X) = e^{\operatorname{Tr} \operatorname{ad} X}$$

for any $X \in \mathfrak{g}$. [*Hint*: Use formula (12).]

§4. Unimodular Lie Groups

A Lie group $\mathcal{G}$ is said to be *unimodular* if each left-invariant form ω of maximum degree on the group $\mathcal{G}$ is right invariant ($R_a^*\omega = \omega$ for any $a \in \mathcal{G}$), i.e., if the module Δ of this group identically equals unity.

The assertion in Exercise 26.10 implies that *a connected Lie group $\mathcal{G}$ is unimodular iff*

$$\operatorname{Tr}\operatorname{ad} X = 0 \tag{16}$$

for any $X \in \mathfrak{g}$. (If the homomorphism Δ identically equals unity on a neighborhood of the point $e \in \mathcal{G}$, then it equals unity everywhere.)

Exercise 26.11. Show that the Haar measure on a unimodular group $\mathcal{G}$ is right-invariant, i.e.,

$$\int f(pa)\,dp = \int f(p)\,dp \tag{17}$$

for any element $a \in \mathcal{G}$ and any function f.

Exercise 26.12. Show that the group $\mathrm{GL}(n;\mathbb{R})$ is unimodular. [*Hint*: See Exercise 26.6.]

Exercise 26.13. Deduce the unimodularity of the group $\mathrm{GL}(n;\mathbb{R})$ (or, more precisely, that of its component of the identity) from criterion (16). [*Hint*: For any matrix $X = \|x_j^i\| = x_j^i E_i^j$ from $\mathrm{gl}(n;\mathbb{R})$, we have the relation

$$(\operatorname{ad} X)E_i^j = x_i^p E_p^j - x_p^j E_i^p, \quad i,j = 1,\dots,n, \tag{18}$$

where E_i^j are matrix units (with the entries $(E_i^j)_p^q = \delta_p^j \delta_i^q$).]

If a Lie group $\mathcal{G}$ is compact, then the integral

$$\int \Delta(p)\,dp = \int \Delta(ap)\,dp = \Delta(a)\int \Delta(p)\,dp \tag{19}$$

is meaningful. On the other hand, if the group $\mathcal{G}$ is connected, then

$$\int \Delta(p)\,dp \neq 0$$

(because $\Delta(p) > 0$). Therefore, in this case, relation (19) is only possible for $\Delta(a) = 1$. This proves that each *compact connected Lie group $\mathcal{G}$ is unimodular.*

For any Haar measure dp on a compact group $\mathcal{G}$, the integral

$$\int_{\mathcal{G}} dp \tag{20}$$

is also meaningful. A Haar measure dp is said to be *normalized* if this integral equals unity. To obtain a normalized measure, it suffices to divide an arbitrary measure dp by integral (20). A normalized Haar measure is unique, of course.

§5. Invariant Riemannian Metrics on a Compact Lie Group

Proposition 26.2. *There exists an invariant Riemannian metric on any compact Lie group $\mathcal{G}$.*

Proof. Let $g^{(0)}$ be an arbitrary left-invariant Riemannian metric on $\mathcal{G}$ (Euclidean metric on $\mathfrak{g}$). For any point $b \in \mathcal{G}$, we define a positive-definite bilinear symmetric tensor g_b of type $(2,0)$ (inner product) on the space $\mathbf{T}_b\mathcal{G}$ by setting

$$g_b(A,B) = \int g^{(0)}_{bp}((dR_p)_b A, (dR_p)_b B)\,dp$$

for any vectors $A, B \in \mathbf{T}_b\mathcal{G}$, where dp is an arbitrary Haar measure on $\mathcal{G}$. Clearly, the tensor g_b depends smoothly on b, and the tensor field $b \mapsto g_b$ is therefore a Riemannian metric on $\mathcal{G}$.

For any elements $a, b \in \mathcal{G}$ and any vectors $A, B \in \mathbf{T}_b\mathcal{G}$, we have

$$\begin{aligned}
(L_a^* g)_b(A,B) &= g_{ab}((dL_a)_b A, (dL_a)_b B) \\
&= \int g^{(0)}_{(ab)p}((dR_p)_{ab}(dL_a)_b A, (dR_p)_{ab}(dL_a)_b B)\,dp \\
&= \int g^{(0)}_{a(bp)}((dL_a)_{bp}(dR_p)_b A, (dL_a)_{bp}(dR_p)_b B)\,dp \\
&= \int (L_a^* g^{(0)})_{bp}((dR_p)_b A, (dR_p)_b B)\,dp \\
&= \int g^{(0)}_{bp}((dR_p)_b A, (dR_p)_b B)\,dp = g_b(A,B).
\end{aligned}$$

(We recall that $L_a^* g^{(0)} = g^{(0)}$ by assumption.) This means that $L_a^* g = g$, i.e., the metric g is left invariant.

On the other hand, because the measure dp is left invariant, we have

$$\begin{aligned}
(R_a^* g)_b(A,B) &= g_{ba}((dR_a)_b A, (dR_a)_b B) \\
&= \int g^{(0)}_{(ba)p}((dR_p)_{ba}(dR_a)_b A, (dR_p)_{ba}(dR_a)_b B)\,dp \\
&= \int g^{(0)}_{b(ap)}((dR_{ap})_b A, (dR_{ap})_b B)\,dp \\
&= \int g^{(0)}_{bp}((dR_p)_b A, (dR_p)_b B)\,dp = g_b(A,B),
\end{aligned}$$

and the metric g is therefore right invariant. □

Control question: Where is the compactness of the group $\mathcal{G}$ used in this proof?

The method used in the proof of Proposition 2 is called *averaging over the group*. It was first suggested by A. Hurwitz.

§6. Lie Groups with a Compact Lie Algebra

Proposition 26.2 suggests the following definition.

Definition 26.2. A Lie algebra on which an invariant Euclidean metric exists is called a *compact Lie algebra.*

In other words, a Lie algebra $\mathfrak{g}$ is compact if there exists a positive-definite inner product on $\mathfrak{g}$ with respect to which all operators

$$\operatorname{ad} X: \mathfrak{g} \to \mathfrak{g}, \qquad Y \mapsto [X, Y], \quad Y \in \mathfrak{g},$$

are skew-symmetric. According to Proposition 2, *the Lie algebra of a compact Lie group is compact.* The converse statement has the following form.

Proposition 26.3. *For any compact Lie algebra $\mathfrak{g}$, there exists a compact Lie group $\mathcal{G}$ whose Lie algebra is isomorphic to the algebra $\mathfrak{g}$.*

Proof. We consider the Lie algebra $\operatorname{ad} \mathfrak{g}$. With respect to the invariant Euclidean metric on $\mathfrak{g}$, each element $\operatorname{ad} X$ of this algebra is a skew-symmetric operator on the Euclidean space $\mathfrak{g}$ and therefore has the skew-symmetric matrix $a_{ij}(X)$ in an arbitrary basis. But then

$$\operatorname{Tr}(\operatorname{ad} X)^2 = \sum_{i,j} a_{ij}(X) a_{ji}(X) = -\sum_{i,j} a_{ij}(X)^2$$

and therefore $\operatorname{Tr}(\operatorname{ad} X)^2 \le 0$; moreover, the equality is attained here iff $\operatorname{ad} X = 0$, i.e., when X belongs to the center $\mathfrak{z}$ of the Lie algebra $\mathfrak{g}$. Therefore, if $\mathfrak{z} = \mathfrak{o}$, then $\operatorname{Tr}(\operatorname{ad} X)^2 < 0$ for $X \neq 0$, i.e., *the Killing form* Kil *of a compact Lie algebra without center is negative definite.* In particular, this form is nondegenerate; therefore, *a compact Lie algebra without center is semisimple.*

Moreover, we see that for $\mathfrak{z} = \mathfrak{o}$, the form $-\operatorname{Kil}$ assigns an invariant metric on $\mathfrak{g}$. We can therefore assume without loss of generality in this case that the metric g on $\mathfrak{g}$ is the metric $-\operatorname{Kil}$ and all automorphisms of the Lie algebra $\mathfrak{g}$ are therefore its orthogonal transformations, i.e., this group $\operatorname{Aut} \mathfrak{g}$ is a subgroup of the group of all orthogonal transformations $\mathfrak{g} \to \mathfrak{g}$ of the Euclidean space $\mathfrak{g}$. Because the latter group is compact and the group $\operatorname{Aut} \mathfrak{g}$ is closed in it (prove this!), *for $\mathfrak{z} = \mathfrak{o}$, the group* $\operatorname{Aut} \mathfrak{g}$ *is therefore compact.*

On the other hand, because the Lie algebra $\mathfrak{g}$ is semisimple in this case, the group $\operatorname{Int} \mathfrak{g}$ is closed in the group $\operatorname{Aut} \mathfrak{g}$ (is its component of the identity). Therefore, this group is also compact. Because the Lie algebra $\mathfrak{g}$ for $\mathfrak{z} = \mathfrak{o}$ is isomorphic to the Lie algebra $\operatorname{ad} \mathfrak{g}$ of the group $\operatorname{Int} \mathfrak{g}$, this proves Proposition 26.3 for $\mathfrak{z} = \mathfrak{o}$.

Now let the center $\mathfrak{z}$ of the Lie algebra $\mathfrak{g}$ be arbitrary. We consider its orthogonal complement $\mathfrak{z}^\perp$. Because the linear space $\mathfrak{g}$ is Euclidean, we have

$$\mathfrak{g} = \mathfrak{z} \oplus \mathfrak{z}^\perp.$$

Moreover, $\mathfrak{z}^\perp$ is an ideal that has no center and is a compact Lie algebra with respect to the restriction of the invariant metric on $\mathfrak{g}$ to $\mathfrak{z}^\perp$. Therefore, by what was proved, the ideal $\mathfrak{z}^\perp$ is the Lie algebra of a certain compact Lie group $\mathcal{H} = \operatorname{Int}\mathfrak{z}^\perp$. The center $\mathfrak{z}$ is obviously the Lie algebra of the torus

$$\mathbb{T}^k = \underbrace{\mathbb{S}^1 \times \cdots \times \mathbb{S}^1}_{k},$$

where $k = \dim \mathfrak{z}$.

Exercise 26.14. Show that a Lie algebra $\mathfrak{g}$ that is a direct sum of ideals is isomorphic to the Lie algebra of the product of Lie groups whose Lie algebras are these ideals:

$$\text{if } \mathfrak{g} = \mathfrak{g}_1 \oplus \mathfrak{g}_2, \text{ where } \mathfrak{g}_1 = \mathfrak{l}\mathcal{G}_1 \text{ and } \mathfrak{g}_2 = \mathfrak{l}\mathcal{G}_2, \text{ then } \mathfrak{g} = \mathfrak{l}(\mathcal{G}_1 \times \mathcal{G}_2).$$

Therefore, in particular, the Lie algebra $\mathfrak{g}$ under consideration is isomorphic to the Lie algebra of the direct product $\mathbb{T}^k \times \mathcal{H}$. To complete the proof, it suffices to note that the group $\mathbb{T}^k \times \mathcal{H}$ is compact. □

In the process of the proof, we also proved that the Killing form is negative definite for a semisimple compact Lie algebra. Because the converse statement is obvious, we see that the following proposition holds.

Proposition 26.4. *A Lie algebra is semisimple iff its Killing form is negative definite.*

According to this proposition, on any Lie group $\mathcal{G}$ with a semisimple compact Lie algebra $\mathfrak{g}$, the formula $g = -\operatorname{Kil}$ defines an invariant Riemannian metric on this group. Because this netric induces a symmetric Cartan connection on $\mathcal{G}$, we see (see (11) in Chap. 6) that for its curvature tensor R, the identity

$$R(X,Y)Z = -\frac{1}{4}[[X,Y],Z], \quad X,Y,Z \in \mathfrak{g}$$

holds. But then

$$\begin{aligned}\operatorname{Tr}(Z \mapsto R(Z,X)Y) &= \operatorname{Tr}(Z \mapsto -\frac{1}{4}[[Z,X],Y]) \\ &= \operatorname{Tr}\left(-\frac{1}{4}\operatorname{ad}Y \circ \operatorname{ad}X\right) \\ &= -\frac{1}{4}\operatorname{Tr}\left(\operatorname{ad}X \circ \operatorname{ad}Y\right) = \frac{1}{4}g(X,Y)\end{aligned}$$

for any $X, Y, Z \in \mathfrak{g}$. Therefore, $\operatorname{Ric} = g/4$ (on $\mathfrak{g}$ and therefore on $\mathfrak{a}\mathcal{G}$), i.e., the group $\mathcal{G}$ is a Riemannian Einstein space with the positive scalar curvature $n/4$, where $n = \dim\mathcal{G}$. For convenience of reference, we state this result as a separate proposition.

Proposition 26.5. *With respect to the metric* $-\operatorname{Kil}$, *each Lie group with a semisimple compact Lie algebra is an Einstein space of constant positive scalar curvature.*

§7. Weyl Theorem

Of course, a Lie group with a compact Lie algebra can be noncompact: the additive group $\mathbb{R}^n$ with the Abelian Lie algebra is an example. However, to a certain extent, this example is the only exception.

Theorem 26.1 (Weyl theorem). *A Lie group with a semisimple compact Lie algebra is compact. Its fundamental group $\pi_1\mathcal{G}$ is finite.*

Corollary 26.2. *Each semisimple compact Lie group is the quotient group $\mathcal{G}/\Gamma$ of a compact simply connected Lie group $\mathcal{G}$ by a certain finite Abelian invariant subgroup Γ of its center.*

Corollary 26.3. *Each simply connected Lie group with a compact Lie algebra is the product $\mathbb{R}^n \times \mathcal{G}$ of the additive group $\mathbb{R}^n$ and a compact simply connected semisimple Lie group $\mathcal{G}$.*

An arbitrary Lie group with a compact Lie algebra is locally isomorphic to the group $\mathbb{R}^n \times \mathcal{G}$, i.e., is the quotient group $(\mathbb{R}^n \times \mathcal{G})/\Gamma$ of the product $\mathbb{R}^n \times \mathcal{G}$ by a certain discrete Abelian invariant subgroup Γ of its center. Therefore, similar to how the description (up to a local isomorphism) of all connected semisimple Lie groups is reduced to the description of all semisimple Lie algebras and therefore (see decomposition (7) in Chap. 25) to the description of all simple Lie algebras, the description of all connected Lie groups with a compact Lie algebra is reduced to the description of all simple compact Lie algebras.

The Weyl theorem is proved in Chap. 28 after we present facts from the general theory of Riemannian spaces that are necessary for this reduction.

Chapter 27
Jacobi Theory

In this chapter, returning to the general theory of Riemannian spaces and examining geodesics, we consider the question of the necessary conditions for an extremal to be a minimum curve (and a geodesic to be a shortest arc). This question is directly related to the presentation in Chap. 12, but only now do we have all the necessary facts for its successful consideration. It turns out that the answer to this question is closely related to critical points (more precisely, critical values) of the exponential mapping.

§1. Conjugate Points

Let $\mathcal{X}$ be a Riemannian space, $p_0 \in \mathcal{X}$ be an arbitrary point, and $\gamma: t \mapsto \exp_{p_0} tA$, $A = \dot{\gamma}(0)$, be an arbitrary geodesic passing through the point p_0 for $t = 0$.

Definition 27.1. A point $q = \exp_{p_0} t_0 A$, $t_0 > 0$, of the geodesic γ *is conjugate to the point* p_0 *along the geodesic* γ if it is a critical value of the exponential mapping, i.e., if the vector $t_0 A$ is a critical point of this mapping (the linear mapping

$$(d\exp_{p_0})_{t_0 A}: \mathbf{T}_{p_0}\mathcal{X} \to \mathbf{T}_q\mathcal{X}$$

has a nontrivial kernel; as usual, we identify $\mathbf{T}_{t_0 A}(\mathbf{T}_{p_0}\mathcal{X})$ with $\mathbf{T}_{p_0}\mathcal{X}$).

We note that this definition is meaningful for an arbitrary affine connection space $\mathcal{X}$.

Because the set of all critical points of the exponential mapping is obviously closed (and does not contain the vector 0), if the set of all conjugate points is not empty, then among all conjugate points on the geodesic γ, there is a point with the minimal $t_0 > 0$. It is called the *first conjugate point.*

The main purpose of this chapter is to prove the following theorem, known as the *Jacobi Theorem.*

Theorem 27.1. *If a geodesic* $\gamma: I \to \mathcal{X}$, $\gamma(0) = p_0$, *is a shortest arc, then none of its points except possibly the endpoint* $p_1 = \gamma(1)$ *is conjugate to the point* p_0.

To prove this theorem, we need to relate conjugate points to variations of geodesics.

§2. Second Variation of Length

In accordance with the general definitions in Chap. 11, a *variation* of a curve $\gamma: I \to \mathcal{X}$ is a smooth mapping

$$\varphi: [0,1] \times (-\varepsilon, \varepsilon) \to \mathcal{X} \tag{1}$$

such that $\varphi(t,0) = \gamma(t)$ for any $t \in I$. A variation φ is naturally identified with the family $\{\gamma_\tau\}$ of curves

$$\gamma_\tau: t \mapsto \varphi(t,\tau), \quad -\varepsilon < \tau < \varepsilon, \quad 0 \le t \le 1. \tag{2}$$

A variation φ is called a *variation with fixed endpoints* if $\varphi(0,\tau) = \gamma(0)$ and $\varphi(1,\tau) = \gamma(1)$ for any τ, $|\tau| < \varepsilon$ (all curves (2) start and end at the same points).

It is convenient for us to slightly extend the concept of variation and introduce *piecewise smooth variations*. By definition, each such variation is a continuous mapping of form (1) (satisfying the condition $\varphi(t,0) = \gamma(t)$, $0 \le t \le 1$) for which

1. there exists a partition

   $$0 = t_0 < t_1 < \cdots < t_r = 1$$

   of the closed interval $I = [0,1]$ (each variation φ has its own partition) such that on each rectangle $[t_{i-1}, t_i] \times (-\varepsilon, \varepsilon)$, $i = 1, \dots, r$, the mapping φ is smooth and
2. the vector field $X: t \mapsto X(t)$ defined by

   $$X(t) = \frac{\partial \varphi}{\partial \tau}(t,0), \quad 0 \le t \le 1, \tag{3}$$

   is a piecewise smooth continuous vector field on the curve γ (and at all points of the curve γ except the points $\gamma(t_i)$, $i = 0, \dots, r$, its covariant derivatives $\nabla X/dt$ are therefore defined; this derivative is a smooth vector field on γ outside these points).

We note that for a variation with fixed endpoints, the field X equals zero for $t = 0$ and $t = 1$.

For a piecewise smooth variation φ, each curve γ_τ is piecewise smooth, and its length

$$s_\varphi(\tau) = \int_0^1 |\dot{\gamma}_\tau(t)|\, dt, \quad -\varepsilon < \tau < \varepsilon,$$

is therefore defined. Moreover, by condition 2, the function $s = s_\varphi(\tau)$ is differentiable at the point $\tau = 0$ on the interval $(-\varepsilon, \varepsilon)$.

On the other hand, if φ is a variation with fixed endpoints, then in the case where the curve γ is a shortest arc, the function $s = s_\varphi(\tau)$ has a minimum at the point $\tau = 0$, and in the case where the curve γ is a geodesic, the relation

$s'_\varphi(0) = 0$ holds at the point $\tau = 0$ (the prime denotes differentiation with respect to τ). As is known from an elementary calculus course, for $s'_\varphi(0) = 0$ and $s''_\varphi(0) < 0$, the point $\tau = 0$ cannot be a minimum point of the function $s = s_\varphi(\tau)$ (it is a local maximum point); we therefore see that to prove Theorem 27.1, it suffices to prove the following proposition (which is also often called the Jacobi theorem).

Proposition 27.1. *If the point $q = \gamma(t_0)$ is conjugate to the point $p_0 = \gamma(0)$ along a geodesic $\gamma: I \to \mathcal{X}$ for a certain t_0, $0 < t_0 < 1$, then there exists a piecewise smooth variation φ of the curve γ with fixed endpoints such that $s''_\varphi(0) < 0$.*

The number $s''_\varphi(0)$ is called the *second variation of the length.*

§3. Formula for the Second Variation

To prove this proposition, we find an explicit formula for $s''_\varphi(0)$. To obtain this formula, because

$$s''_\varphi(0) = \int_0^1 (|\dot{\gamma}_\tau(t)|'')|_{\tau=0}\, dt$$

by the rule for differentiating integrals with respect to a parameter, we must first compute the second derivative

$$|\dot{\gamma}_\tau(t)|'' = \frac{\partial^2}{\partial \tau^2} |\dot{\gamma}_\tau(t)|$$

for $\tau = 0$ (which exists everywhere except finitely many points; these points are ignored in what follows).

By definition,

$$|\dot{\gamma}_\tau(t)|^2 = (\dot{\gamma}_\tau(t), \dot{\gamma}_\tau(t)) = \left(\frac{\partial \varphi}{\partial t}, \frac{\partial \varphi}{\partial t}\right)$$

(to simplify the formulas here and in what follows, we omit arguments of functions as a rule). Therefore, by the rule for differentiating the inner product of vector fields (see formula (4) in Chap. 11), we have

$$|\dot{\gamma}_\tau(t)|' = \frac{1}{|\dot{\gamma}_\tau(t)|}\left(\frac{\nabla}{\partial \tau}\frac{\partial \varphi}{\partial t}, \frac{\partial \varphi}{\partial t}\right)$$

and further

$$\begin{aligned} |\dot{\gamma}_\tau(t)|'' = &-\frac{1}{|\dot{\gamma}_\tau(t)|^3}\left(\frac{\nabla}{\partial \tau}\frac{\partial \varphi}{\partial t}, \frac{\partial \varphi}{\partial t}\right)^2 \\ &+ \frac{1}{|\dot{\gamma}_\tau(t)|}\left[\left(\frac{\nabla}{\partial \tau}\frac{\nabla}{\partial \tau}\frac{\partial \varphi}{\partial t}, \frac{\partial \varphi}{\partial t}\right) + \left(\frac{\nabla}{\partial \tau}\frac{\partial \varphi}{\partial t}, \frac{\nabla}{\partial \tau}\frac{\partial \varphi}{\partial t}\right)\right]. \end{aligned}$$

Because (see formula (17) in Chap. 2)

$$\frac{\nabla}{\partial \tau}\frac{\partial \varphi}{\partial t} = \frac{\nabla}{\partial t}\frac{\partial \varphi}{\partial \tau},$$

where $\partial\varphi/\partial\tau$ is the field of vectors tangent to the curve $\tau \mapsto \gamma_\tau(t)$, $t = \text{const}$ and (see Exercise 2.5)

$$\frac{\nabla}{\partial \tau}\frac{\nabla}{\partial \tau}\frac{\partial \varphi}{\partial t} = \frac{\nabla}{\partial \tau}\frac{\nabla}{\partial t}\frac{\partial \varphi}{\partial \tau} = \frac{\nabla}{\partial t}\frac{\nabla}{\partial \tau}\frac{\partial \varphi}{\partial \tau} + R_{\varphi(t,\tau)}\left(\frac{\partial \varphi}{\partial \tau}, \frac{\partial \varphi}{\partial t}\right)\frac{\partial \varphi}{\partial \tau},$$

we have (see formula (3))

$$\left(\frac{\nabla}{\partial \tau}\frac{\partial \varphi}{\partial t}\right)\bigg|_{\tau=0} = \frac{\nabla X}{dt}$$

and

$$\left(\frac{\nabla}{\partial \tau}\frac{\nabla}{\partial \tau}\frac{\partial \varphi}{\partial t}\right)\bigg|_{\tau=0} = \frac{\nabla}{dt}\left(\left(\frac{\nabla}{\partial \tau}\frac{\partial \varphi}{\partial \tau}\right)\bigg|_{\tau=0}\right) + R(X,\dot\gamma)X,$$

where we write merely R instead of $R_{\gamma(t)}$ for simplicity. (Here,

$$\dot\gamma = \frac{\partial \varphi}{\partial t}\bigg|_{\tau=0}$$

is the field $t \mapsto \dot\gamma(t)$ of vectors tangent to the curve γ.)

Moreover, because (see formula (4) in Chap. 11)

$$\left(\frac{\nabla Y}{dt}, \dot\gamma\right) = \frac{d}{dt}(Y, \dot\gamma)$$

for any field Y on the curve γ (we recall that the field $\dot\gamma$ is covariantly constant), we have

$$\left(\frac{\nabla}{dt}\left(\left(\frac{\nabla}{\partial \tau}\frac{\partial \varphi}{\partial \tau}\right)\bigg|_{\tau=0}\right), \dot\gamma\right) = \frac{d}{dt}\left(\left(\frac{\nabla}{\partial \tau}\frac{\partial \varphi}{\partial \tau}\right)\bigg|_{\tau=0}, \dot\gamma\right).$$

Assuming for simplicity that the geodesic γ is parameterized by the natural parameter, i.e., $|\dot\gamma(t)| = 1$, we immediately obtain

$$\begin{aligned}(|\dot\gamma_\tau(t)|'')|_{\tau=0} = &-\left(\frac{\nabla X}{dt}, \dot\gamma\right)^2 + \frac{d}{dt}\left(\left(\frac{\nabla}{\partial \tau}\frac{\partial \varphi}{\partial \tau}\right)\bigg|_{\tau=0}, \dot\gamma\right) \\ &- R(X,\dot\gamma,X,\dot\gamma) + \left(\frac{\nabla X}{dt}, \frac{\nabla X}{dt}\right);\end{aligned}$$

we recall (see formula (2) in Chap. 15) that

$$(R(X,\dot\gamma)X,\dot\gamma) = R(\dot\gamma,X,X,\dot\gamma) = -R(X,\dot\gamma,X,\dot\gamma).$$

Because

$$\int_0^1 \frac{d}{dt}\left(\left(\frac{\nabla}{\partial \tau}\frac{\partial \varphi}{\partial \tau}\right)\bigg|_{\tau=0}, \dot{\gamma}\right) dt = \left(\left(\frac{\nabla}{\partial \tau}\frac{\partial \varphi}{\partial \tau}\right)\bigg|_{\tau=0}, \dot{\gamma}\right)\bigg|_{t=0}^{t=1} = 0$$

(because the variation φ has fixed endpoints,

$$\frac{\partial \varphi}{\partial \tau}\bigg|_{t=0} = 0 \quad \text{and} \quad \frac{\partial \varphi}{\partial \tau}\bigg|_{t=1} = 0$$

identically with respect to τ), this proves that

$$s''_{\varphi}(0) = \int_0^1 \left[\left(\frac{\nabla X}{dt}, \frac{\nabla X}{dt}\right) - R(X, \dot{\gamma}, X, \dot{\gamma}) - \left(\frac{\nabla X}{dt}, \dot{\gamma}\right)^2\right] dt. \tag{4}$$

Therefore, to prove Proposition 27.1 (and Theorem 27.1), it suffices to find a variation φ of the geodesic γ for which the right-hand side of formula (4) is negative.

§4. Reduction of the Problem

We first note that the number $s''_{\varphi}(0)$ depends on only the vector field X (and the curve γ). Moreover, it is easy to see that for any piecewise smooth vector field X on the curve γ that is equal to zero for $t = 0$ and $t = 1$, there exists a variation φ with fixed endpoints that induces this field by formula (3). (On each segment of the curve γ that is contained in one chart and does not contain breakpoints of the field X, the variation φ is constructed in an obvious way; these partial variations can be easily glued into one piecewise smooth variation of the whole curve.) Therefore, Proposition 27.1 is proved if we can construct a piecewise smooth field X on the curve γ that is equal to zero for $t = 0$ and $t = 1$ and for which the integral in the right-hand side of formula (4) is negative.

A field X is called a *field of normal vectors* (in brief, a *normal field*) if

$$(X(t), \dot{\gamma}(t)) = 0 \quad \text{for any } t, \quad 0 \le t \le 1.$$

Because

$$\left(\frac{\nabla X}{dt}, \dot{\gamma}\right) = \frac{d}{dt}(X, \dot{\gamma}),$$

the integral from formula (4) for such a field is simplified and becomes

$$\int_0^1 \left[\left(\frac{\nabla X}{dt}, \frac{\nabla X}{dt}\right) - R(X, \dot{\gamma}, X, \dot{\gamma})\right] dt. \tag{5}$$

For any a and b, $0 \le a < b \le 1$, we set

$$I_a^b(X) = \int_a^b \left[\left(\frac{\nabla X}{dt}, \frac{\nabla X}{dt}\right) - R(X, \dot{\gamma}, X, \dot{\gamma})\right] dt; \tag{6}$$

this means that we have proved the following proposition.

Proposition 27.2. *On a geodesic $\gamma: I \to \mathcal{X}$, let there exist a piecewise smooth vector field X such that*

1. *the field X is zero for $t = 0$ and $t = 1$,*
$$X(0) = 0, \qquad X(1) = 0;$$
2. *the field X is normal;*
3. *the inequality*
$$I_0^1(X) < 0$$
holds.

Then there exists a variation φ of the geodesic γ such that $s''_\varphi(0) < 0$ (and the geodesic γ is therefore not a shortest arc).

§5. Minimal Fields and Jacobi Fields

Definition 27.2. A field $X^{(0)}$ on the segment $\gamma|_{[a,b]}$ of a geodesic γ, $0 \le a < b \le 1$, is called a *minimal field* if it yields the strict minimum to integral (6) in the class of all piecewise smooth fields X on the curve $\gamma|_{[a,b]}$ for which
$$X(a) = X^{(0)}(a), \quad X(b) = X^{(0)}(b),$$
i.e., $I_a^b(X) \ge I_a^b(X^{(0)})$ for any such field X and, moreover, the equality is attained only for $X = X^{(0)}$.

Choosing an orthonormal basis
$$A_1, \quad \dots, \quad A_n \tag{7}$$
in the space $\mathbf{T}_{p_0}\mathcal{X}$ and translating it in a parallel way to all points of the geodesic γ, we obtain the vector fields
$$X_1, \quad \dots, \quad X_n \tag{8}$$
on γ such that any field X on γ is uniquely represented as a linear combination of them,
$$X = q^1 X_1 + \ldots + q^n X_n = q^i X_i,$$
where $q^1 = q^1(t), \dots, q^n = q^n(t)$ are functions on I (which are smooth if the field X is smooth and piecewise smooth if the field X is piecewise smooth). Moreover,
$$\frac{\nabla X}{dt} = \dot{q}^i X_i, \tag{9}$$
where $\dot{q}^i(t) = dq^i(t)/dt$, $i = 1, \dots, n$, and the field of parallel vectors is therefore characterized by the condition $q^i = \text{const}$, $i = 1, \dots, n$. In addition, because fields (8) are orthonormal at any point of the curve γ, we have

$$\left(\frac{\nabla X}{dt}, \frac{\nabla X}{dt}\right) = \sum_{i=1}^{n} (\dot{q}^i(t))^2.$$

Finally,

$$R(X, \dot{\gamma}, X, \dot{\gamma}) = R_{ij,kl} q^i(t) a^j q^k(t) a^l,$$

where $R_{ij,kl} = R^i_{j,kl}$ and a^j are components of the tensor R_{p_0} and the vector $A = \dot{\gamma}(0)$ in basis (7). (We note that $R_{ij,kl} = (R(X_k, X_l)X_j, X_i)$, $a^i = (\dot{\gamma}, X_i)$, and $\dot{\gamma} = a^i X_i$.)

We see that integral (6) belongs to the class of integrals considered in Chap. 11. The configuration space of this integral is the space $\mathbb{R}^n$, and the Lagrangian is given by

$$L(q, \dot{q}) = \sum_{i=1}^{n} (\dot{q}^i)^2 - R_{ij,kl} q^i a^j q^k a^l.$$

Because $\partial L / \partial \dot{q}^i = 2\dot{q}^i$ and

$$\frac{\partial L}{\partial q^i} = -R_{ij,kl} a^j q^k a^l - R_{kj,il} a^j q^k a^l = -(R_{ij,kl} + R_{kj,il}) a^j a^l q^k,$$

the Euler–Lagrange equations for integral (6) are

$$2\ddot{q}^i(t) + (R_{ij,kl} + R_{kj,il}) a^j a^l q^k(t) = 0.$$

But according to the Bianchi identity (formula (3) in Chap. 15),

$$R_{ij,kl} + R_{jk,il} + R_{ki,jl} = 0$$

and therefore

$$R_{ij,kl} + R_{kj,il} = 2R_{ij,kl} + R_{ki,jl}.$$

Moreover, because the tensor $R_{ki,jl}$ is skew-symmetric with respect to the subscripts j and l, we have

$$R_{ki,jl} a^j a^l = 0.$$

Therefore, we can rewrite the Euler–Lagrange equations for integral (6) (after dividing by 2) in the form

$$\ddot{q}^i(t) + R_{ij,kl} a^j a^l q^k(t) = 0, \quad i = 1, \dots, n. \tag{10}$$

Multiplying these equations by X_i and summing them, we obtain the equation

$$\ddot{q}^i(t) X_i + \sum_{i=1}^{n} R_{ij,kl} a^j a^l q^k(t) X_i = 0,$$

where

$$\ddot{q}^i(t) X_i = \frac{\nabla}{dt}\frac{\nabla}{dt} X$$

(compare with (9)), and

$$\sum_{i=1}^{n} R_{ij,kl}a^j a^l q^k(t)X_i = \sum_{i=1}^{n}(R(X_k, X_l)X_j, X_i)a^j a^l q^k(t)X$$
$$= \sum_{i=1}^{n}(R(X, \dot\gamma)\dot\gamma, X_i)X_i = R(X, \dot\gamma)\dot\gamma.$$

(We recall that basis (8) is orthonormal and therefore $\sum_{i=1}^{n}(Y, X_i)X_i = Y$ for any field Y on γ.)

This proves that the Euler–Lagrange equations for integral (6) in the invariant coordinate-free form are

$$\frac{\nabla}{dt}\frac{\nabla}{dt}X + R(X, \dot\gamma)\dot\gamma = 0. \tag{11}$$

Definition 27.3. Equation (11) is called the *Jacobi equation*, and its smooth solutions are called the *Jacobi fields* (on the geodesic γ).

Therefore, Jacobi fields are exactly extremals of the integral I_0^1, and their restrictions to $[a, b]$ are extremals of the integral I_a^b (with respect to variations constant at the ends of the closed interval). In particular, we see that *any smooth minimal field is a Jacobi field.*

Because Eq. (11) in coordinates $q^1, \dots, q^n$ is written in the form of the set of second-order ordinary differential equations (10), we see that according to the standard theorems of the theory of differential equations *all Jacobi fields on a geodesic γ compose a $2n$-dimensional linear space J_γ, and each Jacobi field X is uniquely defined by its initial value $X(0)$ and the initial value $\nabla X(0)/dt$ of its covariant derivative* (which are vectors of the space $\mathbf{T}_{p_0}\mathcal{X}$).

We are first interested in Jacobi fields with $X(0) = 0$. They compose an n-dimensional linear space $J_\gamma^{(0)}$, and each field $X \in J_\gamma^{(0)}$ is uniquely determined by the vector $\nabla X(0)/dt \in \mathbf{T}_{p_0}\mathcal{X}$.

§6. Jacobi Variation

A smooth variation $\varphi = \{\gamma_\tau\}$ of the geodesic γ is called a *Jacobi variation* if it is composed of geodesics and is fixed at the point p_0, i.e., if

$$\gamma_\tau(t) = \exp_{p_0} tA(\tau), \quad -\varepsilon < \tau < \varepsilon, \tag{12}$$

where $\tau \mapsto A(\tau)$ is a certain smooth curve of the space $\mathbf{T}_{p_0}\mathcal{X}$ passing through the point $A = \dot\gamma(0)$ for $\tau = 0$.

Because the relation

$$\frac{\nabla}{\partial t}\frac{\nabla}{\partial \tau}\dot\gamma_\tau(t) - \frac{\nabla}{\partial \tau}\frac{\nabla}{\partial t}\dot\gamma_\tau(t) = R_{\gamma_\tau(t)}\left(\dot\gamma_\tau(t), \frac{\partial\gamma_\tau(t)}{\partial\tau}\right)\dot\gamma_\tau(t)$$

holds for the field $(t,\tau) \mapsto \dot{\gamma}_\tau(t)$ on the surface φ (see Exercise 2.5) and

$$\frac{\nabla}{\partial t}\dot{\gamma}_\tau(t) = 0,$$

we have

$$\frac{\nabla}{\partial t}\frac{\nabla}{\partial \tau}\dot{\gamma}_\tau(t) + R_{\gamma_\tau(t)}\left(\frac{\partial \gamma_\tau(t)}{\partial \tau}, \dot{\gamma}_\tau(t)\right)\dot{\gamma}_\tau(t) = 0.$$

Setting $\tau = 0$ (and omitting the arguments as always), we directly conclude from the above that field (3), which is induced by variation (12), satisfies relation (11), i.e., it is a Jacobi field.

Because variation (12) is fixed for $t = 0$, we have $X(0) = 0$, i.e., $X \in J_\gamma^{(0)}$. In addition, according to (17) in Chap. 2, we have

$$\frac{\nabla X}{dt}(t) = \frac{\nabla}{dt}\left(\left.\frac{\partial \gamma_\tau(t)}{\partial \tau}\right|_{\tau=0}\right) = \left.\left(\frac{\nabla}{\partial \tau}\dot{\gamma}_\tau(t)\right)\right|_{\tau=0},$$

where $\tau \mapsto \dot{\gamma}_\tau(t)$ in the right-hand side is considered a vector field on the curve $\tau \mapsto \gamma_\tau(t)$, $t =$ const. Because the latter curve is constant for $t = 0$ and the covariant field derivatives of this curve hence coincide with ordinary derivatives with respect to τ and every vector $\dot{\gamma}_\tau(0)$ is just the vector $A(\tau)$, this proves that

$$\frac{\nabla X}{dt}(0) = A'(0). \tag{13}$$

Therefore, *for any Jacobi variation* (12), *the induced vector field* (3) *is a Jacobi field from* $J_\gamma^{(0)}$, for which (13) holds.

In the explicit form, this field is given by

$$X(t) = (d\exp_{p_0})_{tA}(tA'(0)). \tag{14}$$

Now let X' be an arbitrary Jacobi field from $J_\gamma^{(0)}$, and let the relation

$$B = \frac{\nabla X'}{dt}(0)$$

hold. We consider variation (12) with $A(\tau) = A + \tau B$ and the Jacobi field X that is induced by this variation. Because $A'(0) = B$, we have

$$\frac{\nabla X}{dt}(0) = \frac{\nabla X'}{dt}(0)$$

and therefore $X = X'$. Therefore, *any Jacobi field from* $J_\gamma^{(0)}$ *is induced by a certain Jacobi variation* (12).

Summarizing what was said above, we see that Jacobi fields $J_\gamma^{(0)}$ are exactly vector fields induced by Jacobi variations.

§7. Jacobi Fields and Conjugate Points

We can now return to conjugate points.

Proposition 27.3. *A point $q = \gamma(t_0)$, $0 < t_0 \le 1$, of the geodesic $\gamma: I \to \mathcal{X}$ is conjugate to the point $p_0 = \gamma(0)$ iff there exists $X \neq 0$ on γ such that $X(0) = 0$ and $X(t_0) = 0$.*

Proof. It suffices to note that the relation $X(t_0) = 0$ holds for field (14) iff $A'(0) \in \operatorname{Ker}(d\exp_{p_0})_{t_0 A}$ and that $X \neq 0$ iff $A'(0) \neq 0$. □

Corollary 27.1. *The conjugacy relation is symmetric, i.e., if the point q is conjugate to the point p_0, then the point p_0 is conjugate to the point q (on the geodesic γ taken in the opposite direction).*

Proof. The Jacobi relation (11) is invariant with respect to the change of variables $t \mapsto -t$. □

Corollary 27.2. *Let $0 \le t_0 < t_1 \le 1$. If the point $\gamma(t_1)$ is not conjugate to the point $\gamma(t_0)$ along the geodesic $\gamma|_{[t_0,t_1]}$, then for any vectors $A_0 \in \mathbf{T}_{\gamma(t_0)}\mathcal{X}$ and $A_1 \in \mathbf{T}_{\gamma(t_1)}\mathcal{X}$, there exists a unique Jacobi field X on the geodesic $\gamma|_{[t_0,t_1]}$ for which*

$$X(t_0) = A_0 \qquad \text{and} \qquad X(t_1) = A_1.$$

Proof. We can assume without loss of generality that $t_0 = 0$ and $t_1 = 1$. We consider the mapping $\alpha: X \mapsto (X(0), X(1))$ of the linear space J_γ into the direct sum $\mathbf{T}_{p_0}\mathcal{X} \oplus \mathbf{T}_{p_1}\mathcal{X}$, $p_0 = \gamma(t_0)$, $p_1 = \gamma(t_1)$. By Proposition 3, this (obviously linear) mapping is injective. But the linear spaces J_γ and $\mathbf{T}_{p_0}\mathcal{X} \oplus \mathbf{T}_{p_1}\mathcal{X}$ have the same dimension $2n$. Therefore, the mapping α is bijective. □

§8. Properties of Jacobi Fields

It is of interest that, first, the Jacobi field X, which is allowed by Proposition 27.3, is normal and, second, the integral $I_0^{t_0}(X)$ for it is equal to zero. This is implied by the following general proposition.

Proposition 27.4. *If there exist points t_0 and t_1 for the Jacobi field X such that $0 \le t_0 < t_1 \le 1$ and*

$$(X(t_0), \dot\gamma(t_0)) = 0, \qquad (X(t_1), \dot\gamma(t_1)) = 0,$$

then $(X(t), \dot\gamma(t)) = 0$ is identical with respect to t (the field X is normal). If, on the other hand, $X(t_0) = 0$ and $X(t_1) = 0$, then

$$I_{t_0}^{t_1}(X) = 0.$$

Proof. Because the relation

$$\frac{d}{dt}(X, \dot\gamma) = \left(\frac{\nabla X}{dt}, \dot\gamma\right)$$

holds for any field X on γ (we recall that $\nabla\dot\gamma/dt = 0$), we have

$$\frac{d^2}{dt^2}(X, \dot\gamma) = \left(\frac{\nabla}{dt}\frac{\nabla}{dt}X, \dot\gamma\right).$$

Therefore, if X is a Jacobi field, then

$$\frac{d^2}{dt^2}(X, \dot\gamma) = -(R(X, \dot\gamma)\dot\gamma, \dot\gamma) = -R(\dot\gamma, \dot\gamma, X, \dot\gamma),$$

and by virtue of the skew-symmetry of the tensor R with respect to the first two subscripts, the above expression is equal to zero. This means that for any Jacobi field $X \in J_\gamma$, we have

$$(X(t), \dot\gamma(t)) = at + b, \quad \text{where } a, b \in \mathbb{R}. \tag{15}$$

If now $(X(t_0), \dot\gamma(t_0)) = 0$ and $(X(t_1), \dot\gamma(t_1)) = 0$, then $at_0 + b = 0$ and $at_1 + b = 0$; for $t_0 < t_1$, this is only possible when $a = b = 0$. Therefore, $(X, \dot\gamma) = 0$.

Furthermore, because

$$\frac{d}{dt}\left(\frac{\nabla X}{dt}, X\right) = \left(\frac{\nabla}{dt}\frac{\nabla X}{dt}, X\right) + \left(\frac{\nabla X}{dt}, \frac{\nabla X}{dt}\right)$$

and $(R(X, \dot\gamma)\dot\gamma, X) = R(X, \dot\gamma, X, \dot\gamma)$, we have

$$\frac{d}{dt}\left(\frac{\nabla X}{dt}, X\right) = \left(\frac{\nabla X}{dt}, \frac{\nabla X}{dt}\right) - R(X, \dot\gamma, X, \dot\gamma)$$

for any Jacobi field X, and therefore

$$I_{t_0}^{t_1}(X) = \left(\frac{\nabla X}{dt}, X\right)\Bigg|_{t_0}^{t_1}. \tag{16}$$

Hence, if $X(t_0) = 0$ and $X(t_1) = 0$, then $I_{t_0}^{t_1}(X) = 0$. □

Because $\nabla\dot\gamma/dt = 0$ and $R(\dot\gamma, \dot\gamma) = 0$, the field $\dot\gamma\colon t \mapsto \dot\gamma(t)$ is a Jacobi field. By the same reasoning, the field $X_0\colon t \mapsto t\dot\gamma(t)$ is also a Jacobi field. The field $\dot\gamma$ is nonzero everywhere, and the field X_0 is zero only at the point p_0 (for $t = 0$).

Proposition 27.5. *Any Jacobi field $X \in J_\gamma$ can be uniquely represented in the form*

$$X = aX_0 + b\dot\gamma + X^{(0)},$$

where $X^{(0)}$ is a normal Jacobi field. Moreover, $X \in J_\gamma^{(0)}$ iff $b = 0$ and $X^{(0)} \in J_\gamma^{(0)}$.

Proof. We can assume without loss of generality that the parameter on the geodesic γ is natural ($|\dot\gamma| = 1$). Then the relation

$$(X - aX_0 - b\dot\gamma, \dot\gamma) = 0$$

holds for each Jacobi field X, where a and b are the numbers from (15); this relation means that the field $X^{(0)} = X - aX_0 - b\dot\gamma$ is normal. □

Corollary 27.3. *The subspace $J_\gamma^{(00)}$ of the space $J_\gamma^{(0)}$ consisting of normal Jacobi fields is of dimension $n-1$.*

§9. Minimality of Normal Jacobi Fields

Just as geodesics are locally shortest, so Jacobi fields, at least normal ones, are locally minimal (on sufficiently small segments of the curve γ). The precise statement has the following form.

Proposition 27.6. *If there are no points that are conjugate to the point $\gamma(a)$ on the geodesic $\gamma|_{[a,b]}$, $0 \le a < b \le 1$, then every normal Jacobi field $X^{(0)}$ on γ that is equal to zero for $t = a$ is minimal on $\gamma|_{[a,b]}$ (its restriction to $[a,b]$ is the minimal field on $\gamma|_{[a,b]}$).*

Before proving this proposition, we give some auxiliary considerations. Let

$$X_1, \quad \dots, \quad X_{n-1} \tag{17}$$

be an arbitrary basis of the $(n-1)$-dimensional linear space $J_\gamma^{(00)}$. Because there are no conjugate points on the geodesic and hence none of the fields from $J_\gamma^{(00)}$ is zero for $t > 0$, we see that all fields (1) are linearly independent for $t > 0$. Therefore, each normal piecewise smooth field X on γ that is equal to zero for $t = 0$ can be uniquely represented in the form

$$X = f^i X_i, \tag{18}$$

where f^i, $i = 1, \dots, n-1$, are certain piecewise smooth functions on the closed interval $I = [0,1]$.

Lemma 27.1. *The formula*

$$I_0^1(X) = \int_0^1 |\dot f^i X_i|^2 dt + b^i b^j \left(\frac{\nabla X_i}{dt}, X_j\right)\bigg|_{t=1} \tag{19}$$

holds, where $b^i = f^i(1)$.

Proof. We have

$$\left(\frac{\nabla X}{dt}, \frac{\nabla X}{dt}\right) = \left(\dot{f}^i X_i + f^i \frac{\nabla X_i}{dt}, \dot{f}^i X_i + f^i \frac{\nabla X_i}{dt}\right)$$

$$= |\dot{f}^i X_i|^2 + 2\dot{f}^i f^j \left(X_i, \frac{\nabla X_j}{dt}\right) + f^i f^j \left(\frac{\nabla X_i}{dt}, \frac{\nabla X_j}{dt}\right),$$

where $|\dot{f}^i X_i|^2 = (\dot{f}^i X_i, \dot{f}^i X_i)$, and similarly

$$R(X, \dot{\gamma}, X, \dot{\gamma}) = (R(X, \dot{\gamma})\dot{\gamma}, X)$$

$$= f^i f^j (R(X_i, \dot{\gamma})\dot{\gamma}, X_j) = -f^i f^j \left(\frac{\nabla}{dt}\frac{\nabla}{dt} X_i, X_j\right).$$

Because

$$\frac{d}{dt}\left(\frac{\nabla X_i}{dt}, X_j\right) = \left(\frac{\nabla}{dt}\frac{\nabla}{dt} X_i, X_j\right) + \left(\frac{\nabla X_i}{dt}, \frac{\nabla X_j}{dt}\right),$$

it follows that

$$\left(\frac{\nabla X}{dt}, \frac{\nabla X}{dt}\right) - R(X, \dot{\gamma}, X, \dot{\gamma})$$

$$= |\dot{f}^i X_i|^2 + 2\dot{f}^i f^j \left(X_i, \frac{\nabla X_j}{dt}\right) + f^i f^j \frac{d}{dt}\left(\frac{\nabla X_i}{dt}, X_j\right).$$

On the other hand, we have

$$\frac{d}{dt}\left(f^i \frac{\nabla X_i}{dt}, f^j X_j\right) = \frac{d(f^i f^j)}{dt}\left(\frac{\nabla X_i}{dt}, X_j\right) + f^i f^j \frac{d}{dt}\left(\frac{\nabla X_i}{dt}, X_j\right)$$

and

$$\frac{d(f^i f^j)}{dt}\left(\frac{\nabla X_i}{dt}, X_j\right) = (\dot{f}^i f^j + f^i \dot{f}^j)\left(\frac{\nabla X_i}{dt}, X_j\right)$$

$$= \dot{f}^i f^j \left(\frac{\nabla X_i}{dt}, X_j\right) + f^j \dot{f}^i \left(\frac{\nabla X_j}{dt}, X_i\right).$$

Therefore,

$$\left(\frac{\nabla X}{dt}, \frac{\nabla X}{dt}\right) - R(X, \dot{\gamma}, X, \dot{\gamma})$$

$$= |\dot{f}^i X_i|^2 + \dot{f}^i f^j \left[\left(X_i, \frac{\nabla X_j}{dt}\right) - \left(\frac{\nabla X_i}{dt}, X_j\right)\right] + \frac{d}{dt}\left(f^i \frac{\nabla X_i}{dt}, f^j X_j\right).$$

We now note that

$$\begin{aligned}\frac{d}{dt}\left(\frac{\nabla X_i}{dt}, X_j\right) &= \left(\frac{\nabla}{dt}\frac{\nabla}{dt}X_i, X_j\right) + \left(\frac{\nabla X_i}{dt}, \frac{\nabla X_j}{dt}\right) \\ &= \left(\frac{\nabla X_i}{dt}, \frac{\nabla X_j}{dt}\right) - (R(X_i, \dot\gamma)\dot\gamma, X_j) \\ &= \left(\frac{\nabla X_i}{dt}, \frac{\nabla X_j}{dt}\right) - R(X_j, \dot\gamma, X_i, \dot\gamma)\end{aligned}$$

and similarly

$$\frac{d}{dt}\left(X_i, \frac{\nabla X_j}{dt}\right) = \left(\frac{\nabla X_i}{dt}, \frac{\nabla X_j}{dt}\right) - R(X_i, \dot\gamma, X_j, \dot\gamma).$$

Because the right-hand sides of the above identities coincide (see Proposition 15.2), it follows that

$$\frac{d}{dt}\left[\left(X_i, \frac{\nabla X_j}{dt}\right) - \left(\frac{\nabla X_i}{dt}, X_j\right)\right] = 0,$$

i.e.,

$$\left(X_i, \frac{\nabla X_j}{dt}\right) - \left(\frac{\nabla X_i}{dt}, X_j\right) = \text{const}.$$

Because $X_i(0) = 0$ and $X_j(0) = 0$, this constant is zero, i.e.,

$$\left(X_i, \frac{\nabla X_j}{dt}\right) - \left(\frac{\nabla X_i}{dt}, X_j\right) = 0.$$

This implies

$$\left(\frac{\nabla X}{dt}, \frac{\nabla X}{dt}\right) - R(X, \dot\gamma, X, \dot\gamma) = |\dot f^i X_i|^2 + \frac{d}{dt}\left(f^i \frac{\nabla X_i}{dt}, f^j X_j\right).$$

Integrating (and taking $X_i(0) = 0$ into account), we immediately obtain formula (19). □

Proof (of Proposition 27.6). We can assume without loss of generality that $a = 0$ and $b = 1$, i.e., that none of the points of the geodesic $\gamma: I \to \mathcal{X}$ is conjugate to the point $p_0 = \gamma(0)$ (and hence $X^{(0)} \in J_\gamma^{(00)}$).

We apply formula (19) to the normal Jacobi field $X^{(0)}$. This field is of form (18) with $f^i = \text{const}$. Therefore,

$$I_0^1(X^{(0)}) = b^i b^j \left(\frac{\nabla X_i}{dt}, X_j\right)\bigg|_{t=1};$$

hence, for any normal piecewise smooth field X coinciding with $X^{(0)}$ for $t = 0$ and $t = 1$, the inequality

$$I_0^1(X) - I_0^1(X^{(0)}) = \int_0^1 |\dot f^i X_i|^2\, dt \geq 0$$

holds. Therefore,

$$I_0^1(X^{(0)}) \le I_0^1(X) \tag{20}$$

for any normal field X; moreover, the equality is only attained for $X = X^{(0)}$.

Exercise 27.1. Show that for any piecewise smooth field X on the geodesic γ, the inequality

$$I_0^1(X^\perp) \le I_0^1(X)$$

holds, where $X^\perp$ is the *normal component* of the field X (i.e., a field γ such that the vector $X(t) - X^\perp(t)$ is collinear to the vector $\dot\gamma(t)$ for every t).

Taken together with inequality (20), this result obviously proves Proposition 27.6. □

Corollary 27.4. *If there are no points that are conjugate to the point $p_0 = \gamma(0)$ on the segment $\gamma|_{[0,t_0]}$ of the geodesic γ, then $I_0^{t_0}(X) \ge 0$ for every piecewise smooth field X on γ that is equal to zero for $t = 0$ and $t = t_0$.*

§10. Proof of the Jacobi Theorem

We can now prove Proposition 27.1 (and Theorem 27.1 along with it).

Proof (of Proposition 27.1). By Proposition 27.3, there exists a Jacobi field $X^{(0)} \ne 0$ on $\gamma|_{[0,t_0]}$ (a normal one by Proposition 27.4) such that $X^{(0)}(0) = 0$ and $X^{(0)}(t_0) = 0$. According to Proposition 1.1, there exists a neighborhood U of the point $q = \gamma(t_0)$ that is contained in a normal neighborhood of each of its points and is hence such that for any point $p \in U$, the mapping $\exp_p$ has no critical values in U. Therefore, there exists $\delta > 0$, $\delta \le t_0$, $t_0 + \delta \le 1$ (we recall that $0 < t_0 < 1$ by assumption) such that none of the points on the segment $\gamma|_{[t_0-\delta, t_0+\delta]}$ of the geodesic γ is conjugate to the point $\gamma(t_0 - \delta)$. In particular, the point $\gamma(t_0+\delta)$ is not conjugate to the point $\gamma(t_0-\delta)$; hence (see Corollary 27.1), there exists a Jacobi field $X^{(1)}$ on the interval $[t_0 - \delta, t_0 + \delta]$ (a normal one by Proposition 27.4) such that

$$X^{(1)}(t_0 - \delta) = X^{(0)}(t_0 - \delta), \qquad X^{(1)}(t_0 + \delta) = 0.$$

We define the vector field X on γ by the formulas

$$X(t) = \begin{cases} X^{(0)}(t) & \text{if } 0 \le t \le t_0 - \delta, \\ X^{(1)}(t) & \text{if } t_0 - \delta \le t \le t_0 + \delta, \\ 0 & \text{if } t_0 + \delta \le t \le 1. \end{cases}$$

The field X is piecewise smooth and $X(0) = 0$ and $X(1) = 0$. In addition, because both fields $X^{(0)}$ and $X^{(1)}$ are normal, the field X is also normal. Therefore (see Proposition 27.2), to prove Proposition 27.1 it is suffices to show that $I_0^1(X) < 0$.

Because $X^{(0)}(0) = 0$ and $X^{(0)}(t_0) = 0$, we have $I_0^{t_0}(X^{(0)}) = 0$ (see Proposition 27.4). On the other hand,

$$\begin{aligned} I_0^1(X) = I_0^{t_0+\delta}(X) &= I_0^{t_0-\delta}(X^{(0)}) + I_{t_0-\delta}^{t_0+\delta}(X^{(1)}) \\ &= I_0^{t_0}(X^{(0)}) - I_{t_0-\delta}^{t_0}(X^{(0)}) + I_{t_0-\delta}^{t_0+\delta}(X^{(1)}) \end{aligned}$$

and therefore

$$I_0^1(X) = I_{t_0-\delta}^{t_0+\delta}(X^{(1)}) - I_{t_0-\delta}^{t_0}(X^{(0)}).$$

Because the field $X^{(1)}$ is minimal on the segment $[t_0 - \delta, t_0 + \delta]$ (by Proposition 27.6), we have

$$I_{t_0-\delta}^{t_0+\delta}(X^{(1)}) < I_{t_0-\delta}^{t_0+\delta}(Y)$$

for any piecewise smooth field Y on $\gamma|_{[t_0-\delta,t_0+\delta]}$ that is different from the field $X^{(1)}$ and is such that

$$Y(t_0 - \delta) = X^{(0)}(t_0 - \delta), \qquad Y(t_0 + \delta) = 0.$$

In particular, this is true for the field Y defined by

$$Y(t) = \begin{cases} X^{(0)} & \text{if } t_0 - \delta \le t \le t_0, \\ 0 & \text{if } t_0 \le t \le t_0 + \delta. \end{cases}$$

Because $I_{t_0-\delta}^{t_0}(X^{(0)}) = I_{t_0-\delta}^{t_0+\delta}(Y)$, this implies $I_0^1(X) < 0$. □

Chapter 28
Some Additional Theorems I

§1. Cut Points

Proposition 27.3 and Eq. (11) in Chap. 27 show that conjugate points are related to the curvature tensor via Jacobi fields; this tensor largely defines the topology of the manifold $\mathcal{X}$ via the characteristic classes. Therefore, the topology and conjugate points interact with each other rather intimately. We first illustrate this general remark in connection with the notion of a normal neighborhood.

Let $\mathcal{X}$ be a connected complete Riemannian space, $p_0 \in \mathcal{X}$ be an arbitrary point, and γ be a geodesic $t \mapsto \exp_{p_0} tA$, $A = \dot{\gamma}(0)$, passing through the point p_0 for $t = 0$ and parameterized by the natural parameter (i.e., such that $|A| = 1$). We consider the set T_γ of all numbers $t > 0$ for which the segment $\gamma|_{[0,t]}$ of the geodesic γ is a shortest arc.

Exercise 28.1. Prove that either $T_\gamma = (0, \infty)$ or $T_\gamma = (0, \mu_0]$, where μ_0 is a certain positive number. [*Hint*: If $t \in T_\gamma$ and $0 < s < t$, then $s \in T_\gamma$. If $s \in T_\gamma$ for any $s < t$, then $t \in T_\gamma$.]

Definition 28.1. For $T_\gamma = (0, \mu_0]$, the point $\gamma(\mu_0)$ is called a *cut point* of the geodesic γ.

Proposition 28.1. *A cut point $q = \gamma(\mu_0)$ of the geodesic γ is the first point on γ for which at least one of the following conditions holds:*

1. *The point q is conjugate to the point p_0 on the geodesic γ.*
2. *In $\mathcal{X}$, along with the shortest arc $\gamma|_{[0,\mu_0]}$, there exists at least one more shortest arc joining the points p_0 and q.*

Proof. Choosing the monotonically decreasing sequence $t_1 > t_2 > \dots > t_i > \dots$ converging to the point μ_0 and setting $\rho_i = \rho(p_0, p_i)$ and $p_i = \gamma(t_i)$, we consider the shortest arc

$$\gamma_i \colon [0, \rho_i] \to \mathcal{X}, \qquad t \mapsto \exp_{p_0} tA_i, \quad |A_i| = 1,$$

parameterized by the natural parameter and joining the points p_0 and p_i. (The existence of the shortest arc γ_i is guaranteed by the Hopf–Rinow theorem; see Chap. 12.) Because $t_i > \mu_0$ and $\gamma(\mu_0)$ is a cut point, we have $A_i \neq A$ and $t_i > \rho_i$. On the other hand, because $\rho_i = \rho(p_0, p_i)$ and $p_i \to q$, we have $\rho_i \to \mu_0$. Hence, the sequence $\{\rho_i\}$ is bounded. Therefore, by the Bolzano–Weierstrass theorem, we can assume without loss of generality that the vectors $\rho_i A_i$ converge to a certain vector of the form $\mu_0 B$, where $|B| = 1$. Because

$$\exp_{p_0} \mu_0 B = \lim_{i\to\infty} \exp_{p_0} \rho_i A_i = \lim_{i\to\infty} p_i = q,$$

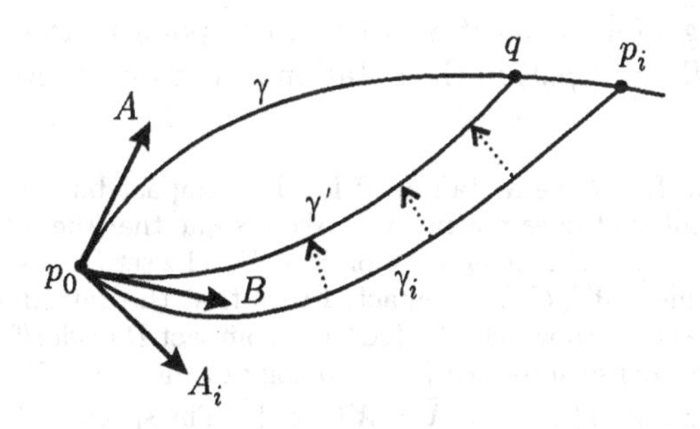

Fig. 28.1.

the geodesic $\gamma': t \mapsto \exp_{p_0} tB$, $0 \le t \le \mu_0$, joins the points p_0 and q, and its length is equal to μ_0: this geodesic is a shortest arc. Therefore, property 2 holds for $\gamma' \neq \gamma$. To prove Proposition 28.1, it hence suffices to prove that property 1 holds for $\gamma = \gamma'$, i.e., for $B = A$. We suppose the contrary.

Let $B = A$, and let the point q be not conjugate to the point p_0 along the geodesic γ. Then, by definition, the mapping

$$\exp_{p_0}: \mathbf{T}_{p_0}\mathcal{X} \to \mathcal{X}$$

is étale at the point $\mu_0 A$ and is hence a diffeomorphism of some neighborhood of the point $\mu_0 A$ onto some neighborhood of the point q. Therefore, if i is sufficiently large, the relation

$$\exp_{p_0} t_i A = p_i = \exp_{p_0} \rho_i A_i$$

implies the relation $t_i A = \rho_i A_i$, which only holds for $t_i = \rho_i$ and $A = A_i$. By assumption, we have $t_i > \rho_i$; this proves that for $B = A$, the point q is conjugate to the point p_0. Moreover, because the geodesic $\gamma|_{[0,\mu_0]}$ is a shortest arc, this point is the first point of the geodesic γ that is conjugate to the point p_0 (by the Jacobi Theorem). Finally, for any positive $\mu < \mu_0$, the shortest arc $\gamma|_{[0,\mu_0]}$ is a unique shortest arc joining the points p and $\gamma(\mu)$ because any other shortest arc joining the points p_0 and $\gamma(\mu)$, together with the segment $\gamma|_{[\mu,\mu']}$, $\mu < \mu' < \mu_0$, of the geodesic γ, forms a broken shortest arc joining the points p_0 and $\gamma(\mu')$, whereas we know (see Proposition 12.3) that each shortest arc is a smooth curve. □

§2. Lemma on Continuity

Let $\mathbb{S}_{p_0}$ be the unit sphere of the space $\mathbf{T}_{p_0}\mathcal{X}$ (the set of all vectors $A \in \mathbf{T}_{p_0}\mathcal{X}$ for which $|A| = 1$). We define the function

$$\mu: \mathbb{S}_{p_0} \to \mathbb{R} \cup \{+\infty\} \tag{1}$$

on $\mathbb{S}_{p_0}$ by setting $\mu(A) = \mu_0$ if $\gamma(\mu_0)$ is a cut point of the geodesic $\gamma: t \mapsto \exp_{p_0} tA$ (i.e., if $T_\gamma = (0, \mu_0])$ in the notation introduced above and $\mu(A) = \infty$ if $T_\gamma = (0, \infty)$.

Exercise 28.2. Let $\mathcal{X}$ be a Hausdorff locally compact but not compact space, and let ∞ be a point that does not lie in $\mathcal{X}$. We assume that the set $U \subset \mathcal{X} \cup \{\infty\}$ is open if either $U \subset \mathcal{X}$ and U is open in $\mathcal{X}$ or $\infty \in U$. The set $U' = U \setminus \{\infty\}$ is open in $\mathcal{X}$, and its complement $\mathcal{X} \setminus U'$ is compact. Prove that this introduces the topology on $\mathcal{X} \cup \{\infty\}$ with respect to which $\mathcal{X} \cup \{\infty\}$ is a compact Hausdorff space containing $\mathcal{X}$ as an everywhere dense subspace (the topology of the space $\mathcal{X}$ is induced by the topology of the space $\mathcal{X} \cup \{\infty\}$ and $\overline{\mathcal{X}} = \mathcal{X} \cup \{\infty\}$). The space $\mathcal{X} \cup \{\infty\}$ is called the *one-point compactification* (or the *Aleksandrov compactification*) of the space $\mathcal{X}$.

In particular, the above construction is applicable to $\mathcal{X} = \mathbb{R}$, and it yields a compact Hausdorff space $\mathbb{R} \cup \{\infty\}$.

Exercise 28.3. Prove that the space $\mathbb{R} \cup \{\infty\}$ is homeomorphic to the sphere $\mathbb{S}^1$.

Because $\mathbb{R} \cup \{\infty\}$ is a topological space, it is reasonable to ask whether the function defined by (1) is continuous. The answer to this question is affirmative.

Lemma 28.1. *The function defined by* (1) *is continuous.*

We prove this lemma later in order not to interrupt the presentation now.

§3. Cut Loci and Maximal Normal Neighborhoods

Let $C_0(p_0)$ be the set of all vectors of the linear space $\mathbf{T}_{p_0}\mathcal{X}$ of the form $\mu(A)A$, where $A \in \mathbb{S}_{p_0}$ and $\mu(A) < \infty$, and $C(p_0)$ be its image in $\mathcal{X}$ under the exponential mapping, i.e.,

$$C(p_0) = \exp_{p_0} C_0(p_0).$$

Therefore, $C(p_0)$ is just the set of all cut points on the geodesics emanating from the point p_0.

Definition 28.2. The set $C(p_0)$ or the set $C_0(p_0)$) is called the *cut locus* of the point p_0 in the respective space $\mathcal{X}$ or tangent space $\mathbf{T}_{p_0}\mathcal{X}$.

Example 28.1. The cut locus of an arbitrary point p_0 of the sphere $\mathbb{S}^n$ in the tangent space $\mathbf{T}_{p_0}\mathbb{S}^n$ is the sphere of radius π; in the sphere $\mathbb{S}^n$ itself, it is the diametrically opposite point (the south pole for the north pole).

Example 28.2. The cut locus of an arbitrary point p_0 of the elliptic space $\mathbb{R}P^n$ (see Example 19.1) in $\mathbf{T}_{p_0}(\mathbb{R}P^n)$ is the sphere of radius $\pi/2$; in $\mathbb{R}P^n$, it is the hyperplane $\mathbb{R}P^{n-1}$.

Let $U_0(p_0)$ be the set of all vectors of the linear space $\mathbf{T}_{p_0}\mathcal{X}$ of the form tA, where $A \in \mathbb{S}_{p_0}$ and $0 \le t < \mu(A)$, and let $U(p_0)$ be its image in $\mathcal{X}$ under the exponential mapping, i.e., let

$$U(p_0) = \exp_{p_0} U_0(p_0).$$

By Proposition 28.1, the set $U(p_0)$ consists of all points $p \in \mathcal{X}$ that

1. are joined with the point p_0 by a unique shortest arc and
2. are not conjugate to the point p_0 along this shortest arc.

It is clear that the set $U_0(p_0)$ is star-shaped (if $A \in U_0(p_0)$, then $\lambda A \in U_0(p_0)$ for any λ, $0 \le \lambda \le 1$).

Exercise 28.4. Prove that

1. the set $U_0(p_0)$ is open and
2. the cut locus $C_0(p_0)$ is the boundary of this set, i.e.,

$$C_0(p_0) = \overline{U_0(p_0)} \setminus U_0(p_0)$$

(and is hence a closed set).

[*Hint*: Apply Lemma 28.1.]

The *set* $U_0(p_0)$, *being open and star-shaped, is homeomorphic to the open ball* $\mathbb{B}^n$ *of radius* 1 *in the space* $\mathbf{T}_{p_0}\mathcal{X}$. The homeomorphism $U_0(p_0) \to \mathbb{B}^n$ can be given explicitly, for example, by the formula

$$tA \mapsto \frac{\mu(A)+1}{\mu(A)}\frac{t}{1+t}A, \quad A \in \mathbb{S}_{p_0}, \quad 0 \le t < \mu(A)$$

(it is assumed that $(\infty + 1)/\infty = 1$).

By virtue of properties 1 and 2 of the points $p \in U_0(p_0)$, the mapping

$$\exp_{p_0}\colon U_0(p_0) \to U(p_0)$$

is bijective and étale, i.e., it is a diffeomorphism. By definition (see Definition 1.4), this means that *open sets* $U_0(p_0)$ *and* $U(p_0)$ *are normal neighborhoods* (of the respective points $0 \in \mathbf{T}_{p_0}\mathcal{X}$ and $p_0 \in \mathcal{X}$). The following proposition, in particular, shows that *these neighborhoods are maximal*, i.e., they are not contained in any larger normal neighborhoods.

Proposition 28.2. *Every complete Riemannian space* $\mathcal{X}$ *is a disjoint union of the maximal normal neighborhood* $U(p_0)$ *of an arbitrary point* p_0 *and the corresponding cut locus* $C(p_0)$:

$$\mathcal{X} = U(p_0) \bigsqcup C(p_0).$$

Proof. Let $p \in \mathcal{X}$, and let

$$\gamma: t \mapsto \exp_{p_0} tA, \quad |A| = 1,$$

be a geodesic parameterized by the natural parameter and having the property that the segment $\gamma|_{[0,t_0]}$ is a shortest arc joining the points p_0 and p. Then $t_0 \in T_\gamma$ and hence $t_0 \le \mu(A)$. If $t_0 < \mu(A)$, then $p \in U(p_0)$, and if $t_0 = \mu(A)$, then $p \in C(p_0)$. □

Corollary 28.1. *Every cut locus $C(p_0)$ is a closed set.*

Corollary 28.2. *The maximal normal neighborhood $U(p_0)$ is everywhere dense in $\mathcal{X}$, i.e.,*

$$\overline{U(p_0)} = \mathcal{X}.$$

Corollary 28.3. *If the function defined by* (1) *is finite (does not take the value ∞) for at least one point $p_0 \in \mathcal{X}$, then the space $\mathcal{X}$ is compact.*

Proof. The function defined by (1), being a continuous real-valued function on the compact space $\mathbb{S}_{p_0}$, is bounded, i.e., there exists a number $d \ge 0$ such that $\mu(A) \le d$ for any vector $A \in \mathbb{S}_{p_0}$. Then the closed ball $\overline{\mathbb{B}}_d$ of radius d in the space $\mathbf{T}_{p_0}\mathcal{X}$ contains the closure $\overline{U_0(p_0)}$ of the neighborhood $U_0(p_0)$ and is therefore mapped onto the whole space $\mathcal{X}$ under the exponential mapping $\exp_{p_0}$. Therefore, the space $\mathcal{X}$, being a continuous image of the compact set $\overline{\mathbb{B}}_d$, is compact. □

We note that the converse is true in the following strengthened statement. *If the Riemannian space $\mathcal{X}$ is compact, then for any point $p_0 \in \mathcal{X}$, the function defined by* (1) *is bounded.* Indeed, because a compact metric space has a finite diameter d, the segment $\gamma|_{[0,t_0]}$, $t_0 > d$, of an arbitrary geodesic $\gamma: t \mapsto \exp_{p_0} tA$, $|A| = 1$, cannot be a shortest arc. Therefore, $\mu(A) \le d$. □

Summarizing, we see that *every smooth connected compact manifold is obtained from the Euclidean ball by a certain identification of its boundary points.*

§4. Proof of Lemma 28.1

We now prove Lemma 28.1.

Proof (of Lemma 28.1). If the function defined by (1) is not continuous, then there exists a convergent sequence $\{A_k,\ k \ge 1\}$ of vectors $A_k \in \mathbb{S}_{p_0}$ such that

$$\mu(\lim_{k\to\infty} A_k) \ne \lim_{k\to\infty} \mu(A_k)$$

(because the space $\mathbb{R} \cup \{\infty\}$ is compact, we can assume without loss of generality that the limit on the right exists). Let

$$A = \lim_{k\to\infty} A_k, \quad \mu_k = \mu(A_k), \quad \mu_0 = \lim_{k\to\infty} \mu_k.$$

Case 1. We suppose that $\mu(A) > \mu_0$. According to Proposition 28.1, the point $q = \exp_{p_0}\mu_0 A$ of the geodesic $\gamma: t \mapsto \exp_{p_0} tA$ is not conjugate to the point p_0 and the mapping $\exp_{p_0}$ is therefore étale at the point $\mu_0 A$. Therefore, the point $\mu_0 A$ has a neighborhood V on which the mapping $\exp_{p_0}$ is a diffeo-

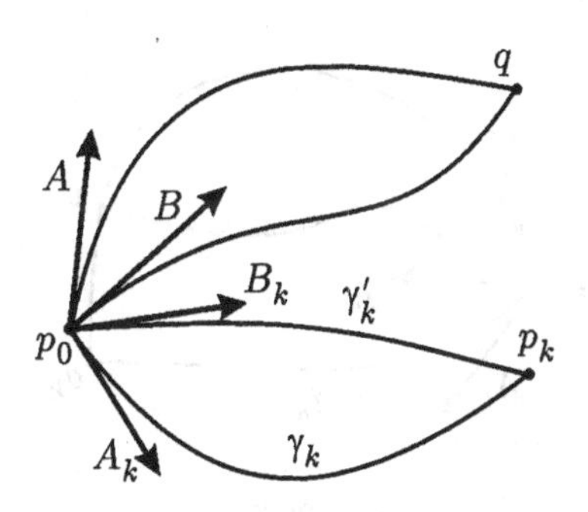

Fig. 28.2.

morphism. On the other hand, because $\mu_k A_k \to \mu_0 A$, there exists a $k_0 \geq 1$ such that $\mu_k A_k \in V$ for $k \geq k_0$; therefore, the mapping $\exp_{p_0}$ is étale at the point $\mu_k A_k$, i.e., the point $p_k = \exp_{p_0}\mu_k A_k$ of the geodesic $\gamma_k: t \mapsto \exp_{p_0} tA_k$ is not conjugate to the point p_0 along this geodesic. Therefore, for the point p_k and the geodesic γ_k, property 1 in Proposition 28.1 does not hold. We therefore have property 2, i.e., along with the shortest arc $\gamma_k|_{[0,\mu_k]}$, there exists another shortest arc

$$\gamma'_k: t \mapsto \exp_{p_0} tB_k, \quad 0 \leq t \leq \mu_k, \quad |B_k| = 1,$$

joining the points p_0 and p_k. Moreover, because $A_k \neq B_k$ and

$$\exp_{p_0}\mu_k A_k = p_k = \exp_{p_0}\mu_k B_k,$$

we have $\mu_k B_k \notin V$. Passing, if necessary, to a subsequence, we can assume without loss of generality that there exists the limit

$$B = \lim_{k\to\infty} B_k.$$

Moreover, because $\mu_k B_k \to \mu_0 B$, we have $\mu_0 B \notin V$ (and hence $B \neq A$). At the same time

$$\begin{aligned}\exp_{p_0}\mu_0 B &= \exp_{p_0}(\lim_{k\to\infty} \mu_k B_k) = \lim_{k\to\infty} (\exp_{p_0}\mu_k B_k) \\ &= \lim_{k\to\infty} (\exp_{p_0}\mu_k A_k) = \exp_{p_0}(\lim_{k\to\infty} \mu_k A_k) = \exp_{p_0}\mu_0 A.\end{aligned}$$

Therefore, the points $\exp_{p_0}\mu_0 A$ and p_0 are joined by two distinct shortest arcs $t \mapsto \exp_{p_0} tA$ and $t \mapsto \exp_{p_0} tB$, $0 \le t \le \mu_0$. Because this is impossible for $\mu_0 < \mu(A)$, we have thus proved that Case 1 cannot hold.

Case 2. We suppose that $\mu(A) < \mu_0$. Choosing the number t_0 on the interval $(\mu(A), \mu_0)$, we can assume without loss of generality that $t_0 < \mu_k$ for

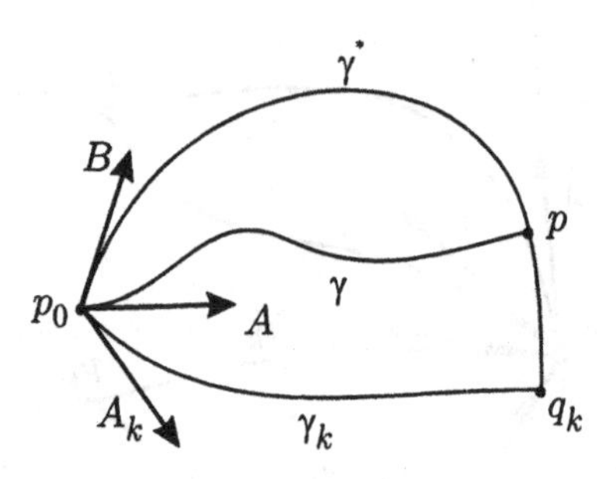

Fig. 28.3.

all $k \ge 1$. Because $t_0 > \mu(A)$, the distance ρ_0 from the point p_0 to the point $p = \exp_{p_0} t_0 A$ is less than t_0, and the shortest arc γ^* joining the points p_0 and p has the form

$$\gamma^*: [0, \rho_0] \to \mathcal{X}, \qquad t \mapsto \exp_{p_0} tB,$$

where $|B| = 1$. On the other hand, because $\exp_{p_0} t_0 A_k \to \exp_{p_0} t_0 A = p$, we can assume without loss of generality that

$$\rho(p, q_k) < \frac{t_0 - \rho_0}{2}, \quad \text{where } q_k = \exp_{p_0} t_0 A_k.$$

Then the length of the curve composed of the shortest arc γ^* and the shortest arc joining the points p and q_k is equal to

$$\rho_0 + \rho(p, q_k) < \frac{t_0 + \rho_0}{2} < t_0,$$

i.e., it is less then the length t_0 of the geodesic

$$\gamma_k: t \mapsto \exp_{p_0} tA_k, \quad 0 \le t \le t_0,$$

which (because of the inequality $t_0 < \mu_k$) is the shortest arc joining the points p_0 and q_k. Therefore, Case 2 is also impossible.

Therefore, we must have $\mu(A) = \mu_0$, and Lemma 28.1 is proved. □

We now discuss more immediate relations between the curvature and the topology of the space $\mathcal{X}$.

§5. Spaces of Strictly Positive Ricci Curvature

We say that a Riemmanian space $\mathcal{X}$ has a *Ricci curvature* $\geq k_0$, where $k_0 \in \mathbb{R}$, if the tensor $\operatorname{Ric} - k_0 g$ is positive semidefinite, i.e., if for any field $X \in \mathfrak{a}\mathcal{X}$, we have

$$\operatorname{Ric}(X, X) \geq k_0(X, X) \quad \text{everywhere on } \mathcal{X}.$$

Definition 28.3. We say that a Riemannian space $\mathcal{X}$ has a *strictly positive Ricci curvature* if there exists a number $k_0 > 0$ such that the Ricci curvature of this space is $\geq k_0$.

Proposition 28.3. *In a Riemannian space $\mathcal{X}$ having a Ricci curvature $\geq k_0$, where $k_0 > 0$, the length of the segment $\gamma|_{[a,b]}$ of a geodesic γ that does not contain points conjugate to the point $\gamma(a)$ does not exceed $\pi\sqrt{n/k_0}$.*

Proof. This proposition is equivalent to the assertion that *if each of the points $\gamma(t)$, $0 < t < 1$, on the geodesic $\gamma\colon t \mapsto \exp_{p_0} tA$ is not conjugate to the point $p_0 = \gamma(0)$, then*

$$|A| \leq \pi\sqrt{n/k_0}. \tag{2}$$

We prove the proposition in this equivalent form.

Once again (see Chapter 27), we consider the fields $X_1, \dots, X_n$ on the curve γ that are obtained by the parallel translation of the orthonormal basis $A_1, \dots, A_n$ of the linear space $\mathbf{T}_{p_0}\mathcal{X}$. Then for any two fields X and Y on γ, the value $\operatorname{Ric}(X, Y)$ of the Ricci tensor is expressed by

$$\operatorname{Ric}(X, Y) = \sum_{i=1}^{n} R(X_i, X, X_i, Y)$$

(see (17) in Chap. 17), and, in particular, we have

$$\operatorname{Ric}(\dot\gamma, \dot\gamma) = \sum_{i=1}^{n} R(X_i, \dot\gamma, X_i, \dot\gamma).$$

Bearing this in mind and choosing an arbitrary number t_0, $0 < t_0 < 1$, we consider the function

$$f(t) = \sin\frac{\pi t}{t_0}.$$

This function vanishes for $t = 0$ and $t = t_0$, and we can therefore apply Corollary 27.1 to the field fX_i (for each i) on the segment $\gamma|_{[0,t_0]}$. Therefore,

$$\sum_{i=1}^{n} I_0^{t_0}(fX_i) \geq 0,$$

and hence

$$0 \le \sum_{i=1}^{n} I_0^{t_0}(fX_i) = \sum_{i=1}^{n} \int_0^{t_0} [(\dot{f}X_i, \dot{f}X_i) - f^2 R(X_i, \dot{\gamma}, X_i, \dot{\gamma})]\, dt$$

$$= \int_0^{t_0} [n\dot{f}^2 - f^2 \operatorname{Ric}(\dot{\gamma}, \dot{\gamma})] dt \le \int_0^{t_0} [n\dot{f}^2 - f^2 k_0(\dot{\gamma}, \dot{\gamma})]\, dt$$

(we recall that $\nabla X_i/dt = 0$). Because $(\dot{\gamma}, \dot{\gamma}) = |A|^2$, this proves that

$$\int_0^{t_0} [n\dot{f}^2 - k_0|A|^2 f^2]\, dt \ge 0,$$

i.e.,

$$k_0|A|^2 \le n\Big(\int_0^{t_0} \dot{f}^2\, dt\Big) \Big/ \Big(\int_0^{t_0} f^2\, dt\Big).$$

But

$$\int_0^{t_0} f^2\, dt = \frac{t_0}{\pi} \int_0^{\pi} \sin^2 x\, dx = \frac{t_0}{2}$$

and

$$\int_0^{t_0} \dot{f}^2 dt = \frac{\pi}{t_0} \int_0^{\pi} \cos^2 x\, dx = \frac{\pi^2}{2t_0}.$$

Therefore,

$$k_0|A|^2 \le \frac{n\pi^2}{t_0^2}.$$

To complete the proof, it remains to pass to the limit as $t_0 \to 1$ and take the square root. □

Exercise 28.5. Show that in the estimate defined by (2), the multiplier n can be replaced by $n - 1$. [*Hint*: We can assume that the field X_n is tangent at every point of the curve γ.]

§6. Mayers Theorem

Theorem 28.1. *Every connected complete Riemannian space $\mathcal{X}$ of strictly positive Ricci curvature*

1. *is a metric space of finite diameter with respect to the Riemannian metric,*
2. *is compact, and*
3. *has a finite fundamental group.*

Proof. Let the Ricci curvature be $\ge k_0$, where $k_0 > 0$. According to the Hopf–Rinow theorem (Theorem 12.1), any two points $p, q \in \mathcal{X}$ can be joined by the shortest arc γ. According to the Jacobi Theorem (Theorem 27.1), none of the points of this shortest arc is conjugate to the point p. Therefore, by Proposition 28.3, the length of the curve γ, i.e., the distance $\rho(p, q)$ from the point p to the point q, is not greater than $\pi\sqrt{n/k_0}$. This proves assertion 1.

Because the metric space $\mathcal{X}$ is Bolzanian (see the proof of the Hopf–Rinow theorem), assertion 2 is a direct consequence of assertion 1.

To prove assertion 3, we consider the universal covering space $\widetilde{\mathcal{X}}$ of the space $\mathcal{X}$. Because the condition of the strict positiveness of the Ricci curvature is of local nature, the space $\widetilde{\mathcal{X}}$ is always a space of strictly positive Ricci curvature with respect to the induced Riemannian metric. According to assertion 2, this space is therefore compact and is therefore a finite-sheeted covering of the space $\mathcal{X}$. Therefore, the fundamental group of the space $\mathcal{X}$, which is in a bijective correspondence with the inverse image in $\widetilde{\mathcal{X}}$ of an arbitrary point of the space $\mathcal{X}$, consists of finitely many elements. □

Theorem 28.1 is known as the *Mayers theorem.*

Corollary 28.4. *Every connected and complete Einstein space that has a scalar curvature bounded from below by a positive constant has properties* 1, 2, *and* 3.

The Weyl theorem (Theorem 26.1) directly follows from comparing this consequence with Proposition 26.5.

§7. Spaces of Strictly Positive Sectional Curvature

The Mayers theorem has other interesting corollaries.

Definition 28.4. We say that a Riemannian space $\mathcal{X}$ is a space of *strictly positive sectional curvature* if there exists a number $k_0 > 0$ such that $K_p(\pi) \geq k_0$ for any point $p \in \mathcal{X}$ and for any two-dimensional subspace $\pi \subset \mathbf{T}_p\mathcal{X}$.

Every vector field X on any coordinate neighborhood U consisting of vectors of length 1 can be obviously completed up to an orthonormal basis

$$X = X_1, X_2, \ldots, X_n$$

of the module $\mathfrak{a}\mathcal{X}$ over U. Let $p \in U$, and let π_{ij} be a two-dimensional direction at the point p given by the bivector $(X_i \wedge X_j)_p$. By definition (see (20) in Chap. 15), the sectional curvature $K_p(\pi_{ij})$ of the space $\mathcal{X}$ at the point p in the direction π_{ij} is equal to the value of the function

$$R(X_i \wedge X_j) = R(X_i, X_j, X_i, X_j)$$

at the point p. Therefore, if $K_p(\pi) \geq k_0$ for all p and π, then for any $j = 1, \ldots, n$, we have

$$\operatorname{Ric}(X_j, X_j) = \sum_i R(X_i, X_j, X_i, X_j) = \sum_{j \neq i} R(X_i, X_j, X_i, X_j) \geq (n-1)k_0$$

on U (we recall that $R(X_i, X_i, X_i, X_i) = 0$), and therefore (we recall that $X = X_1$)

$$\operatorname{Ric}(X, X) \geq (n-1)k_0.$$

This means that on U and hence on the whole $\mathcal{X}$ (by virtue of arbitrariness of the neighborhood U), the Ricci curvature of the space $\mathcal{X}$ is not less than $(n-1)k_0$.

Therefore, *every Riemannian space of strictly positive sectional curvature has a strictly positive Ricci curvature and hence has properties* 1, 2, *and* 3 *in Theorem* 28.1.

§8. Spaces of Nonpositive Sectional Curvature

Spaces of nonpositive sectional curvature have the opposite properties: in these spaces

$$K_p(\pi) \leq 0$$

for any point $p \in \mathcal{X}$ and for any two-dimensional subspace $\pi \in \mathbf{T}_p\mathcal{X}$.

Proposition 28.4. *In a space of the nonpositive sectional curvature, no two points of any geodesic are conjugate.*

Proof. Let X be a Jacobi field on a geodesic γ equal to zero for $t = 0$. It is required to show that either $X = 0$ identically or $X(t) \neq 0$ for every $t > 0$. For this, we note that

$$\begin{aligned}\frac{d}{dt}\left(\frac{\nabla X}{dt}, X\right) &= \left(\frac{\nabla}{dt}\frac{\nabla}{dt}X, X\right) + \left(\frac{\nabla X}{dt}, \frac{\nabla X}{dt}\right)\\ &= -(R(X,\dot\gamma)\dot\gamma, X) + \left|\frac{\nabla X}{dt}\right|^2 = -R(X\wedge\dot\gamma) + \left|\frac{\nabla X}{dt}\right|^2\\ &= -K_p(\pi) + \left|\frac{\nabla X}{dt}\right|^2 \geq 0,\end{aligned}$$

where π is a two-dimensional plane with the directing bivector $(X \wedge \dot\gamma)_p$; moreover, the equality is only attained for $X = 0$. Therefore, for $X \neq 0$, the function $(\nabla X/dt, X)$ is monotonically increasing and hence is nonzero for $t > 0$. Therefore, $X(t) \neq 0$ for $t > 0$. □

Chapter 29
Some Additional Theorems II

§1. Cartan–Hadamard Theorem

Proposition 28.4 means that for each point p_0 of a Riemannian space $\mathcal{X}$ of nonpositive sectional curvature (which is assumed to be complete), the exponential mapping

$$\exp_{p_0}: \mathbf{T}_{p_0}\mathcal{X} \to \mathcal{X} \tag{1}$$

has no critical points and is therefore an étale mapping. In fact, we show now (under the natural assumption of connectedness of the space $\mathcal{X}$) that this mapping is even a covering.

Theorem 29.1 (Cartan–Hadamard). *Exponential mapping* (1) *for any connected complete Riemannian space $\mathcal{X}$ of nonpositive sectional curvature is a covering mapping.*

Proof. According to Corollary 19.1, it suffices to prove that under the conditions of Theorem 29.1, exponential mapping (1) is an expanding mapping, i.e., for any vectors $A, B \in \mathbf{T}_{p_0}\mathcal{X}$, we have the inequality

$$|B| \le |(d\exp_{p_0})_A B| \tag{2}$$

(we assume here, by the standard identification $\mathbf{T}_A(\mathbf{T}_{p_0}\mathcal{X}) = \mathbf{T}_{p_0}\mathcal{X}$, that the mapping $(d\exp_{p_0})_A$ is a mapping $\mathbf{T}_{p_0}\mathcal{X} \to \mathbf{T}_p\mathcal{X}$, $p = \exp_{p_0} A$). In this case, we certainly can assume that $A \ne 0$.

Because

$$|(d\exp_{p_0})_A A| = |\dot\gamma(1)| = |\dot\gamma(0)| = |A|, \tag{3}$$

where γ is the geodesic $t \mapsto \exp_{p_0} tA$, inequality (2) holds in advance for $B = A$ and therefore when the vector B is collinear with the vector A. Therefore, *it suffices to prove inequality* (2) *only for vectors B that are orthogonal to the vector A.* (Indeed, if $B = B_1 + B_2$, where the vector B_1 is collinear to the vector A and the vector B_2 is orthogonal to it (and therefore orthogonal to the vector B_1), then by the Gauss Lemma (Proposition 12.1), the vectors $C_1 = (d\exp_{p_0})_A B_1$ and $C_2 = (d\exp_{p_0})_A B_2$ are orthogonal, and therefore $|(d\exp_{p_0})_A B|^2 = |C_1 + C_2|^2 = |C_1|^2 + |C_2|^2$. On the other hand, if $|B_2| \le |C_2|$, then $|B|^2 = |B_1|^2 + |B_2|^2 \le |C_1|^2 + |C_2|^2$ because $|B_1| = |C_1|$ by what was proved.)

With this purpose, we consider the normal Jacobi field

$$X(t) = (d\exp_{p_0})_{tA} tB \tag{4}$$

on the geodesic $\gamma: t \mapsto \exp_{p_0} tA$ (see formula (14) in Chap. 27). Let $|B| = 1$ (we do not lose generality here of course), and let

$$f(t) = |X(t)|^2.$$

Because inequality (3) (for $|B| = 1$) is equivalent to the inequality $1 \leq f(1)$, Theorem 1 is proved if we prove that the estimate $t^2 \leq f(t)$ holds for any t; for this, in turn, it suffices to prove that for $t > 0$, the similar estimate

$$\frac{2}{t} \leq \frac{f'(t)}{f(t)} \tag{5}$$

holds for logarithmic derivatives and

$$\lim_{t\to 0} \frac{f(t)}{t^2} = 1. \tag{6}$$

(If the function $u(t) = \ln f(t) - \ln t^2$ is continuous for $t \geq 0$, is equal to zero for $t = 0$, and $u'(t) \geq 0$ for $t > 0$, then $f(t) \geq t^2$ for all $t \geq 0$.) Because

$$f'(t) = 2\left(\frac{\nabla X}{dt}(t), X(t)\right),$$

$$f''(t) = 2\left(\frac{\nabla^2 X}{dt^2}(t), X(t)\right) + 2\left(\frac{\nabla X}{dt}(t), \frac{\nabla X}{dt}(t)\right)$$

and

$$\lim_{t\to 0} \frac{f(t)}{t^2} = \lim_{t\to 0} \frac{f'(t)}{2t} = \lim_{t\to 0} \frac{f''(t)}{2}$$

by the L'Hospital rule, we have

$$\lim_{t\to 0} \frac{f(t)}{t^2} = \left|\frac{\nabla X}{dt}(0)\right|^2 = |B|^2 = 1.$$

(In the normal coordinates centered at p_0, the vector $X(t)$ has the same coordinates as the vector tB and the covariant derivatives coincide with ordinary ones because $(\Gamma^i_{kj})_{p_0} = 0$ (see Proposition 2.1) and the mapping $(d\exp_{p_0})_0$ acts with respect to the equality of coordinates.)

Therefore, to prove Theorem 1, we need only prove inequality (5) (in fact, this is the only difficulty). The idea of the following proof probably belongs to Rauch.

Let $t_0 > 0$. For any t, letting $\Pi_t^{t_0}$ denote the parallel translation $\mathbf{T}_{\gamma(t)}\mathcal{X} \to \mathbf{T}_{\gamma(t_0)}\mathcal{X}$ along the geodesic γ, we set

$$Y(t) = \frac{t_0}{\sqrt{f(t_0)}} \left(\varphi \circ \Pi_t^{t_0}\right) X(t),$$

where $\varphi\colon \mathbf{T}_{\gamma(t_0)}\mathcal{X} \to \mathbf{T}_{p_0}\mathcal{X}$ is a certain fixed isometry transforming the unit vector $X(t_0)/|X(t_0)|$ into the unit vector B. Similar to the function $Y_0\colon t \mapsto tB$, we can consider the function $Y\colon t \mapsto Y(t)$ as a vector field on the ray $t \mapsto tA$ of the Euclidean linear space $\mathbf{T}_{p_0}\mathcal{X}$. Both these fields vanish for $t = 0$ and assume the same value $t_0 B$ for $t = t_0$. Moreover, the field Y_0 is a normal Jacobi field (in the Euclidean metric of the linear space $\mathbf{T}_{p_0}\mathcal{X}$) and is therefore minimal (see Proposition 27.6; obviously, the ray $t \mapsto tA$ does not contain conjugate points). Therefore, in particular,

$$I_0^{t_0}(Y_0) \le I_0^{t_0}(Y),$$

where

$$I_0^{t_0}(Y_0) = \int_0^{t_0} \left| \frac{dY_0}{dt}(t) \right|^2 dt = \int_0^{t_0} |B|^2 dt = t_0$$

and

$$I_0^{t_0}(Y) = \int_0^{t_0} \left| \frac{dY}{dt}(t) \right|^2 dt = \frac{t_0^2}{f(t_0)} \int_0^{t_0} \left| \frac{d\Pi_t^{t_0} X}{dt}(t) \right|^2 dt$$

(obviously, the isometry φ commutes with the operation of differentiation of vector-valued functions). On the other hand,

$$\begin{aligned} \frac{d\Pi_t^{t_0} X}{dt}(t) &= \lim_{h\to 0} \frac{\Pi_{t+h}^{t_0} X(t+h) - \Pi_t^{t_0} X(t)}{h} \\ &= \Pi_t^{t_0} \left(\lim_{h\to 0} \frac{\Pi_{t+h}^{t} X(t+h) - X(t)}{h} \right) = \Pi_t^{t_0} \frac{\nabla X}{dt}(t) \end{aligned}$$

(see Exercise 1.1) and therefore

$$\left| \frac{d\Pi_t^{t_0} X}{dt}(t) \right| = \left| \frac{\nabla X}{dt}(t) \right|$$

(the parallel translation is an isometry).

Therefore,

$$\begin{aligned} I_0^{t_0}(Y) &= \frac{t_0^2}{f(t_0)} \int_0^{t_0} \left| \frac{\nabla X}{dt} \right|^2 dt \\ &\le \frac{t_0^2}{f(t_0)} \int_0^{t_0} \left(\left| \frac{\nabla X}{dt} \right|^2 - R(X, \dot\gamma, X, \dot\gamma) \right) dt = \frac{t_0^2}{f(t_0)} I_0^{t_0}(X) \end{aligned}$$

(because $K(\pi) \le 0$ for the plane π with the directing bivector $X \wedge \dot\gamma$, we have $-R(X, \dot\gamma, X, \dot\gamma) = -K(\pi)|X \wedge \dot\gamma|^2 \ge 0$), where

$$I_0^{t_0}(X) = \left(\frac{\nabla X}{dt}(t_0), X(t_0) \right) = \frac{1}{2} f'(t_0)$$

(see formula (16) in Chap. 27; we recall that $X(t)$ is a normal Jacobi field that vanishes for $t = 0$).

Therefore,

$$t_0 \le I_0^{t_0}(Y) \le \frac{t_0^2}{2} \frac{f'(t_0)}{f(t_0)},$$

which was required to be proved. □

Corollary 29.1. *An n-dimensional connected and simply connected complete Riemannian space of positive sectional curvature is diffeomorphic to the space $\mathbb{R}^n$. Any two points of such a space are joined by a unique geodesic.*

§2. Consequence of the Cartan–Hadamard Theorem

A connected topological space $\mathcal{X}$ is called a *space of type* $K(\pi, 1)$ if $\pi_m\mathcal{X} = 0$ for $m \geq 2$ and $\pi_1\mathcal{X} = \pi$.

Corollary 29.2. *Each connected complete Riemannian space $\mathcal{X}$ of nonpositive sectional curvature is a space of type $K(\pi, 1)$.*

Proof. As we know, each covering mapping induces an isomorphism of homotopy groups in dimensions $m \geq 2$. On the other hand, all homotopy groups of the space $\mathbb{R}^n$ are obviously equal to zero. □

Exercise 29.1. Prove that the fundamental group $\pi_1\mathcal{X}$ of a connected complete Riemannian space $\mathcal{X}$ of nonpositive sectional curvature has no elements of finite order. [*Hint*: It suffices to show that any isometry $f: \mathcal{X} \to \mathcal{X}$ of a finite order of a simply connected complete Riemannian space $\mathcal{X}$ of nonpositive sectional curvature has a fixed point. Let Γ be a cyclic subgroup of the group $\operatorname{Iso}\mathcal{X}$ generated by an

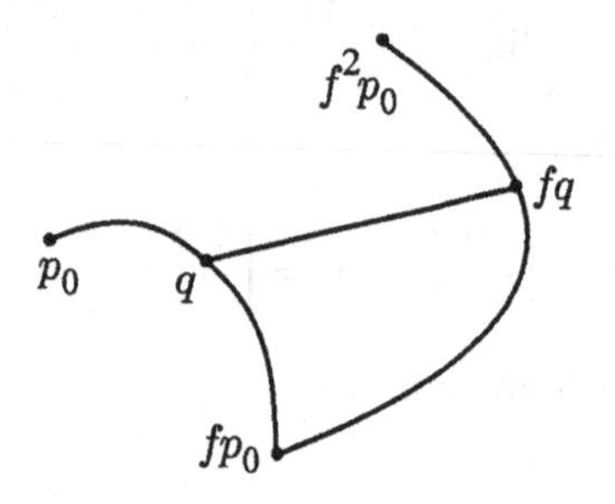

Fig. 29.1.

isometry f, let U be a normal convex neighborhood with a compact closure that contains at least one orbit of the group Γ (it is easily seen that such a neighborhood exists), and let K be the set of all points whose Γ-orbits are contained in $\overline{U}$. This set is convex, compact, and nonempty. Therefore, it contains a point p_0 such that

$$\rho(p_0, fp_0) \leq \rho(p, fp)$$

for any point $p \in K$. If $fp_0 \neq p_0$, then the points p_0, fp_0, and f^2p_0 do not lie on the same geodesic (why?), and therefore

$$\rho(q, fq) < \rho(q, fp_0) + \rho(fp_0, fq) = \rho(q, fp_0) + \rho(p_0, q) = \rho(p_0, fp_0)$$

for any point q belonging to a shortest arc joining the points p_0 and fp_0 (see Fig. 29.1), which contradicts the choice of the point p_0.]

Also, Theorem 29.1 implies that *each connected complete Riemannian space $\mathcal{X}$ of nonpositive sectional curvature is diffeomorphic to a space of the form $\mathbb{R}^n/\Gamma$*, where Γ is a certain diffeomorphism group with a discrete action on $\mathbb{R}^n$ (which is isomorphic to the fundamental group $\pi_1\mathcal{X}$ of the space $\mathcal{X}$). In particular, this is true for any connected complete space of constant curvature $K \leq 0$.

§3. Cartan–Killing Theorem for $K = 0$

For the case where the space $\mathcal{X}$ is flat (is a space of constant curvature $K = 0$), we can obtain more precise results. Indeed, it follows from the Jacobi equation (and the relation $R = 0$) that for a flat space $\mathcal{X}$, the function $f(t) = (X(t), X(t))$ is such that

$$\frac{d^2 f}{dt^2} = 2\left(\frac{\nabla X}{dt}, \frac{\nabla X}{dt}\right)$$

and $d^3 f/dt^3 = 0$, where, as above, X is field (4) (constructed for a certain vector B). Consequently, $d^2 f/dt^2 = \text{const}$ and therefore

$$\left(\frac{\nabla X}{dt}, \frac{\nabla X}{dt}\right) = |B|^2.$$

Moreover, because $X(0) = 0$, we have $f(0) = (X(0), X(0)) = 0$ and $\dot{f}(0) = 2(B, X(0)) = 0$. Therefore, $f(t) = |B|^2 t^2$ and hence $f(1) = |B|^2$, i.e., $|X(1)| = |B|$. Therefore, the mapping $\exp_{p_0}$ is locally isometric in this case. In particular, if the space is simply connected, then this mapping is an isometry. This proves the following proposition.

Proposition 29.1. *Each connected and simply connected complete flat space is isometric to the Euclidean space* $\mathbb{R}^n$.

This is exactly the Cartan–Killing Theorem (Theorem 23.2) for $K = 0$.

As we know, it implies that each connected complete space of constant curvature $K = 0$ is not only diffeomorphic but also isomorphic to a space of the form $\mathbb{R}^n/\Gamma$, where Γ is an isometry group with a discrete action.

§4. Bochner Theorem

Riemannian spaces $\mathcal{X}$ whose Ricci tensor is negative definite also have interesting properties. Restricting ourselves to the case where the space $\mathcal{X}$ is compact, as an illustration, we prove the following proposition, which is known as the *Bochner Theorem.*

Proposition 29.2. *The isometry group* $\operatorname{Iso}\mathcal{X}$ *of an arbitrary compact Riemannian space* $\mathcal{X}$ *with a negative definite Ricci tensor is a finite group.*

Before proving this theorem, we deduce several necessary formulas.

§5. Operators A_X

For any vector field $X \in \mathfrak{a}\mathcal{X}$, we set

$$A_X = \mathcal{L}_X - \nabla_X, \tag{7}$$

where $\mathcal{L}_X$ is the Lie derivative with respect to the field X. Because both the operators $\mathcal{L}_X$ and ∇_X commute with contractions of derivations of the algebra of tensor fields on the manifold $\mathcal{X}$, the operator A_X is also a derivation commuting with contractions.

Moreover, because $\mathcal{L}_X f = Xf = \nabla_X f$ for any function $f \in \mathbf{F}\mathcal{X}$, *the operator A_X equals zero on functions* and is therefore an $\mathbf{F}\mathcal{X}$-linear mapping $\mathfrak{a}\mathcal{X} \to \mathfrak{a}\mathcal{X}$ on vector fields. In explicit form, this mapping is given by

$$A_X Y = -\nabla_Y X, \quad Y \in \mathfrak{a}\mathcal{X}. \tag{8}$$

(Indeed, we have

$$A_X Y = [X, Y] - \nabla_X Y,$$

and because the connection ∇ is symmetric, the right-hand side of this formula is equal to the right-hand side of formula (8) (see formula (14) in Chap. 2.)

By definition, for any derivation A, any tensor field S of type $(2,0)$, and any vector fields $X_1, X_2 \in \mathfrak{a}\mathcal{X}$, we have the relation

$$A(S \otimes X_1 \otimes X_2) = AS \otimes X_1 \otimes X_2 + S \otimes AX_1 \otimes X_2 + S \otimes X_1 \otimes AX_2.$$

Therefore, if the derivation A commutes with contractions, we also have the relation

$$A(S(X_1, X_2)) = (AS)(X_1, X_2) + S(AX_1, X_2) + S(X_1, AX_2)$$

(we recall that the value $S(X_1, X_2)$ of the field S on vector fields X_1 and X_2 is just the result of the complete contraction of the tensor $S \otimes X_1 \otimes X_2$ of type $(2,2)$). Therefore, if the derivation A, moreover, is equal to zero on functions, then

$$(AS)(X_1, X_2) = -S(AX_1, X_2) - S(X_1, AX_2). \tag{9}$$

In particular, *formula* (9) *holds for each derivation of the form A_X*.

Of course, an analogue of formula (9) also holds for any tensor fields S of type $(r, 0)$ and for differential forms in particular. For example, if the manifold $\mathcal{X}$ is oriented and ω is its Riemannian volume element, then for any fields $X_1, \dots, X_n \in \mathfrak{a}\mathcal{X}$, we have

$$(A\omega)(X_1, \dots, X_n) = -\sum_{i=1}^{n} \omega(X_1, \dots, AX_i, \dots, X_n).$$

If the fields $X_1, \dots, X_n$ form a basis of the module $\mathfrak{a}\mathcal{X}$ (over a certain coordinate neighborhood U or, more generally, over a neighborhood U that trivializes the tangent bundle $\tau\mathcal{X}$) and if $\|A^i_j\|$ is the matrix of the operator A in this basis, then because the form ω is skew-symmetric, we have

$$\omega(X_1,\dots,AX_i,\dots,X_n) = A^i_i\omega(X_1,\dots,X_i,\dots,X_n)$$

(the summation with respect to i is not performed!) and therefore

$$(A\omega)(X_1,\dots,X_n) = -\left(\sum_{i=1}^n A^i_i\right)\omega(X_1,\dots,X_n),$$

i.e.,

$$A\omega = -(\operatorname{Tr} A)\omega \tag{10}$$

(on U and therefore on the whole $\mathcal{X}$).

Because $\nabla_X\omega = 0$ (see Corollary 13.1), formula (10) for $A = A_X$ becomes

$$\pounds_X\omega = -(\operatorname{Tr} A_X)\omega.$$

By definition (see formula (15) in Chap. 13), this means that

$$\operatorname{div} X = -\operatorname{Tr} A_X. \tag{11}$$

Because $A_XY = -\nabla_Y X$ (see formula (8)), we can rewrite this formula as

$$\operatorname{div} X = \operatorname{Tr}[Y \mapsto \nabla_Y X]. \tag{12}$$

Now let the field X be affine. Then (see Proposition 8.1), we have

$$[X, \nabla_Y Z] = \nabla_Y[X,Z] + \nabla_{[X,Y]}Z$$

for any fields $Y, Z \in \mathfrak{a}\mathcal{X}$, i.e., we have the relation (we recall that $\pounds_X Y = [X,Y]$)

$$\pounds_X\nabla_Y Z = \nabla_Y\pounds_X Z + \nabla_{[X,Y]}Z \tag{13}$$

(in the operator form, this is $[\pounds_X, \nabla_Y] = \nabla_{[X,Y]}$). Then

$$\begin{aligned} R(X,Y)Z &= \nabla_X\nabla_Y Z - \nabla_Y\nabla_X Z - \nabla_{[X,Y]}Z \\ &= \nabla_X\nabla_Y Z - \nabla_Y\nabla_X Z - \pounds_X\nabla_Y Z + \nabla_Y\pounds_X Z \\ &= -A_X\nabla_Y Z + \nabla_Y A_X Z = A_X A_Z Y + \nabla_Y A_X Z \end{aligned}$$

(see formula (8)). Changing the notation and passing to the traces, we obtain

$$\operatorname{Tr}[Z \mapsto R(Y,Z)X] = \operatorname{Tr} A_X A_Y + \operatorname{Tr}[Z \mapsto \nabla_Z A_Y X].$$

The left hand-side of this formula is equal to $-\operatorname{Ric}(X,Y)$ (see formula (26) in Chap. 2), and the second summand is equal to $\operatorname{div}(A_Y X)$. Therefore,

$$\operatorname{div}(A_Y X) = -\operatorname{Ric}(X,Y) - \operatorname{Tr} A_X A_Y.$$

For $Y = X$, we obtain

$$\operatorname{div}(A_X X) = -\operatorname{Ric}(X,X) - \operatorname{Tr} A_X^2.$$

In the case where the manifold $\mathcal{X}$ is compact and orientable, by the Green formula (see formula (18) in Chap. 13), this implies

$$\int_{\mathcal{X}} (\operatorname{Ric}(X,X) + \operatorname{Tr} A_X^2)\, dV = 0 \tag{14}$$

for any affine field X.

§6. Infinitesimal Variant of the Bochner Theorem

In particular, formula (14) holds for any Killing field $X \in \mathfrak{iso}\,\mathcal{X}$, i.e. (see Exercise 19.7), for a vector field X on $\mathcal{X}$ such that $\mathcal{L}_X g = 0$. But if $\mathcal{L}_X g = 0$, then also $A_X g = 0$ (because $\nabla_X g = 0$); therefore (see formula (9) for $S = g$),

$$(A_X X_1, X_2) + (X_1, A_X X_2) = 0 \tag{15}$$

for any fields $X_1, X_2 \in \mathfrak{a}\mathcal{X}$, i.e., the operator A_X is skew-symmetric. (We note that this conclusion is completely reversible; therefore, *the operator A_X is skew-symmetric iff X is a Killing field.*)

It follows from (15) that in an orthonormal basis $X_1, \dots, X_n$, the matrix $\|A^i_j\|$ of the operator A_X is skew-symmetric, and the trace $\operatorname{Tr} A^2_X$ of the operator A^2_X is therefore expressed by

$$\operatorname{Tr} A^2_X = A^i_j A^j_i = -\sum_{i,j=1}^{n} (A^i_j)^2$$

and is therefore not positive, $\operatorname{Tr} A^2_X \le 0$ (moreover, $\operatorname{Tr} A^2_X = 0$ only for $A_X = 0$).

Therefore, if the Ricci tensor is negative definite ($\operatorname{Ric}(X, X) \le 0$ and, moreover, the equality holds only for $X = 0$), then (14) is possible only for $\operatorname{Ric}(X, X) = 0$ (and $\operatorname{Tr} A^2_X = 0$), i.e., only for $X = 0$. This proves the following proposition.

Proposition 29.3 (infinitesimal variant of the Bochner theorem). *There are no nonzero Killing fields on a compact oriented space $\mathcal{X}$ with a negative-definite Ricci tensor:*

$$\mathfrak{iso}\,\mathcal{X} = 0.$$

Exercise 29.2. Show that this is also true for a nonorientable space $\mathcal{X}$. [*Hint*: Pass to an oriented two-sheeted covering]

Because the Lie algebra $\mathfrak{iso}\,\mathcal{X}$ of all Killing fields is the Lie algebra of the isometry group $\operatorname{Iso}\mathcal{X}$ (see Chap. 19), Proposition 29.3 means that *the isometry group* $\operatorname{Iso}\mathcal{X}$ *of the space $\mathcal{X}$ is discrete.* Therefore, to prove the Bochner theorem, it suffices to prove that *for any compact Riemannian space $\mathcal{X}$, the group* $\operatorname{Iso}\mathcal{X}$ *is compact.*

§7. Isometry Group of a Compact Space

It turns out that the assertion that the group $\operatorname{Iso}\mathcal{X}$ is compact for any compact Riemannian space is purely topological.

Proposition 29.4. *The isometry group* $\operatorname{Iso}\mathcal{X}$ *of an arbitrary compact metric space $\mathcal{X}$ is a compact topological space with respect to the topology of pointwise convergence.*

Proof. We need the following three theorems of general topology, which are stated as exercises.

Exercise 29.3. Prove that each compact metric space $\mathcal{X}$ contains a countable everywhere dense set, i.e., a countable set C such that $\overline{C} = \mathcal{X}$. [*Hint*: For any $n \geq 1$, there exists a finite covering of the space $\mathcal{X}$ that is composed of spherical $(1/n)$-neighborhoods. Letting C_n denote the set of all centers of these neighborhoods, set $C = \bigcup C_n$.]

Exercise 29.4. Prove the following:

1. For any compact metric space $\mathcal{X}$, the formula

$$\breve{\rho}(\varphi, \psi) = \sup_{p \in \mathcal{X}} \rho(\varphi(p), \psi(p)), \qquad \varphi, \psi \colon \mathcal{X} \to \mathcal{X}$$

defines a metric on the set of all continuous mappings $\mathcal{X} \to \mathcal{X}$.
2. The topology corresponding to this metric coincides with the topology of pointwise convergence.

[*Hint*: This means that a sequence of mappings $\varphi_n \colon \mathcal{X} \to \mathcal{X}$ is uniformly convergent iff it converges at each point $p \in \mathcal{X}$.]

Exercise 29.5. Prove that a metric space is compact iff each sequence of its points contains a convergent subsequence. [*Hint*: See the Heine–Borel theorem in a calculus course.]

It follows from the assertions in Exercises 29.4 and 29.5 that to prove Proposition 29.4, it suffices to prove the following two assertions:

1. Any sequence $\{\varphi_n\}$ of isometries of a compact metric space $\mathcal{X}$ contains a convergent (in the space of continuous mappings $\mathcal{X} \to \mathcal{X}$) subsequence.
2. The limit φ of any convergent subsequence $\{\varphi_n\}$ of isometries of the space $\mathcal{X}$ is an isometry.

Proof (of Assertion 1). Let $C = \{c_1, c_2, \dots\}$ be a countable everywhere dense subset of the space $\mathcal{X}$. Because the space $\mathcal{X}$ is compact, the sequence $\{\varphi_n\}$ contains a subsequence $\{\varphi_{n_i}\}$ such that the sequence of points $\{\varphi_{n_i}(c_1)\}$ converges. Setting $\varphi_{n_i} = \varphi_{1,i}$, we similarly find that the sequence $\{\varphi_n\}$ contains a subsequence $\{\varphi_{2,i}\}$ for which the sequence of points $\{\varphi_{2,i}(c_2)\}$ converges, and so on. Continuing the process, for any $k \geq 1$, we obtain a subsequence $\{\varphi_{k,i}\}$ of the sequence $\{\varphi_n\}$ such that the sequence of points $\{\varphi_{k,i}(c_k)\}$ converges. Then the subsequence $\{\psi_n\}$, where $\psi_n = \varphi_{n,n}$, is such that $\{\psi_n(c_k)\}$ converges for any point c_k, $k \geq 1$. (This is the well-known *Cantor diagonal process*).

Now if $\varepsilon > 0$, $p \in \mathcal{X}$, c is a point in C such that $\rho(p, c) < \varepsilon$, and N is a number such that $\rho(\psi_n(c), \psi_m(c)) < \varepsilon$ for $n, m > N$, then

$$\rho(\psi_n(p), \psi_m(p)) \leq \rho(\psi_n(p), \psi_n(c)) + \rho(\psi_n(c), \psi_m(c)) + \rho(\psi_m(c), \psi_m(p)) \leq 3\varepsilon$$

for $n, m > N$ (we have $\rho(\psi_n(p), \psi_n(c)) = \rho(p, c) < \varepsilon$ because ψ_n and ψ_m are isometries; similarly, $\rho(\psi_m(c), \psi_m(p)) < \varepsilon$). This means that the subsequence $\{\psi_n(p)\}$ is a Cauchy subsequence. Therefore, this sequence converges (each compact metric space is complete). Therefore, the sequence $\{\psi_n\}$ converges everywhere on $\mathcal{X}$. □

Proof (of Assertion 2). According to assertion 1, the sequence of isometries $\{\varphi_n^{-1}\}$ contains a convergent subsequence $\{\varphi_{n_i}^{-1}\}$. Let φ' be its limit. By continuity, it is clear that

$$\rho(\varphi(p), \varphi(q)) = \rho(p, q) = \rho(\varphi'(p), \varphi'(q)) \tag{16}$$

for any points $p, q \in \mathcal{X}$. In particular, for any point $p \in \mathcal{X}$, we have

$$\begin{aligned}\rho(\varphi'(\varphi(p)), p) &= \lim_{i\to\infty} \rho(\varphi_{n_i}^{-1}(\varphi(p)), p) \\ &= \lim_{i\to\infty} \rho(\varphi(p), \varphi_{n_i}(p)) = \rho(\varphi(p), \varphi(p)) = 0\end{aligned}$$

(we recall that the sequence $\{\varphi_n\}$ converges by assumption) and therefore $\varphi'(\varphi(p)) = p$, i.e., $\varphi' \circ \varphi = \mathrm{id}$. Similarly, we prove that $\varphi \circ \varphi' = \mathrm{id}$. Therefore, the mapping φ is bijective. Because it satisfies (16), it is an isometry. □

(We note that in the case where $\mathcal{X}$ is a Riemannian space, assertion 2, which is the subject of Exercise 19.6, is trivial because the limit of isometries is an affinity in this case (belongs to the Lie group $\operatorname{Aff} \mathcal{X}$) and is therefore bijective in advance.)

Exercise 29.6. Prove that $\operatorname{Iso} \mathcal{X}$ is a topological group. [*Hint*: It is necessary to prove not only the continuity of the multiplication but also the continuity of the inversion mapping $\varphi \mapsto \varphi^{-1}$.]

Addendum

In the following supplementary chapters, we briefly present the major facts used in the main text from the theory of smooth manifolds and of connections in vector bundles. As a rule, we omit proofs.

Chapter 30
Smooth Manifolds

§1. Introductory Remarks

The concept of a smooth manifold is one of the main concepts in modern mathematics. It arises as a result of explicating the intuitive concept of a surface, considered independently of its location in the space, and simultaneously generalizing it to higher dimensions. The main principles of this explication come from cartography.

We can adequately describe separate parts of the Earth's surface using maps, drawing them in a plane. Each point of the Earth can be drawn in a map, but it is not possible to cover the whole Earth using one chart; we need an atlas for this, i.e., a set of several maps. Any map allows transporting rectangular coordinates on the plane to the corresponding domain on the surface, thus obtaining *local coordinates* on it. (In reality, we proceed in the reverse order in mathematical cartography: geographical coordinates on the Earth's surface are transported to a curvilinear coordinate net on the plane, but this distinction is not of principal importance.) Local coordinates in different charts are related by transition functions that allow expressing one set of coordinates through another (in a common part of two maps).

In the corresponding general definitions, we replace the plane with the standard Euclidean space $\mathbb{R}^n$, where n is an integer, assumed to be fixed here and in what follows.

§2. Open Sets in the Space $\mathbb{R}^n$ and Their Diffeomorphisms

We first recall some facts and definitions from calculus about the space $\mathbb{R}^n$.

A point x of a subset $U \subset \mathbb{R}^n$ is called an *interior point* of it if there exists $\varepsilon > 0$ such that a ball of radius ε centered at the point x lies entirely in U. A set U is said to be *open* (in $\mathbb{R}^n$) if all its points are interior points.

An arbitrary mapping $\varphi: U \to V$ of two subsets of the space $\mathbb{R}^n$ is given by n functions

$$y^1 = \varphi^1(x^1, ..., x^n), \quad \dots, \quad y^n = \varphi^n(x^1, ..., x^n) \tag{1}$$

of n variables that express the coordinates $y^1, \dots, y^n$ of the point $y = \varphi(x)$ through the coordinates $x^1, \dots, x^n$ of an arbitrary point x. In the case where the set U is open, we can speak about derivatives of an arbitrary order of functions (1) in $x^1, \dots, x^n$ at any point in U.

A mapping φ of an open set U is called a *mapping of class* C^r, where r is a positive integer or the symbol ∞, if functions (1) have all continuous partial derivatives of order $\leq r$ for $r \neq \infty$ and all continuous partial derivatives for $r = \infty$ at all points of the set U. In the case where functions (1) are real analytic at any point of U (i.e., they admit expansions in power series with a nonzero radius of convergence), the mapping φ is called a *mapping of class* C^ω.

In what follows, we assume that a certain class of smoothness C^r, where $r \leq \infty$ or $r = \omega$, is fixed, and the mappings of this class are merely called *smooth mappings*. As a rule, any $r \geq 2$, and even $r = 1$, is appropriate for us; however, if not explicitly specified otherwise, we assume without stipulation (to avoid controlling the appearance of higher-order derivatives) that $r = \infty$ or $r = \omega$. However, in some cases, we include a specific case where $r = 0$, in which case smooth mappings are merely continuous ones. The case where $r = \infty$ is assumed to be the main one, and all cases where it differs from the case $r = \omega$ are specified.

Each smooth mapping $\varphi: U \to V$ defines the smooth function

$$D\varphi = \det \left\| \frac{\partial \varphi^i}{\partial x^j} \right\|, \quad i, j = 1, \dots n,$$

on U, which is called the *Jacobian* of this mapping (and is also denoted by the symbol J_φ). As shown in a calculus course, for any smooth mappings $\varphi: U \to V$ and $\psi: V \to W$, where U, V, and W are open subsets of the space $\mathbb{R}^n$, the composition mapping $\psi \circ \varphi: U \to W$ (acting according to the formula $(\psi \circ \varphi)(x) = \psi(\varphi(x))$, $x \in U$) is also smooth and

$$D(\psi \circ \varphi)(x) = (D\psi)(y) \cdot (D\varphi)(x), \quad y = \varphi(x), \tag{2}$$

for any point $x \in U$. Formula (2) is usually called the *chain rule* and the mapping $\psi \circ \varphi$ the *composition* of the mappings φ and ψ).

A mapping $\varphi: U \to V$ of open sets of the space $\mathbb{R}^n$ is called a *diffeomorphic mapping* (or *diffeomorphism*) if it is smooth, it is bijective, and its inverse mapping $\varphi^{-1}: V \to U$ is also smooth. (For $r = 0$, a bijective continuous mapping whose inverse mapping is also continuous is called a *homeomorphism.*)

Formula (2) directly implies that *the Jacobian $D\varphi$ of an arbitrary diffeomorphism $\varphi: U \to V$ is nonzero at all points of the set U* (moreover, $(D\varphi^{-1})(y) = (D\varphi)(x)^{-1}$ for any point $y = \varphi(x)$ of the set V).

The *inverse mapping theorem* states that *if the Jacobian of a smooth mapping* $\varphi: U \to V$ *is nonzero at a point* $x_0 \in U$, *then there exists an open set* $U' \subset U$ *containing the point* x_0 *such that the restriction* $\varphi|_{U'}$ *of the mapping* φ *to* U' *is a diffeomorphism of the set* U' *onto a certain open set* $V' \subset V$ *containing the point* $y_0 = \varphi(x_0)$.

In particular, this implies that *a smooth bijective mapping* $\varphi: U \to V$ *whose Jacobian is nonzero everywhere is a diffeomorphism.*

Remark 30.1. There exist smooth nonbijective mappings $\varphi: U \to V$ whose Jacobian is nonzero everywhere. The mapping φ of the plane annulus $1 < x^2 + y^2 < 2$ onto the annulus $1 < x^2 + y^2 < 4$ given by the formula $\varphi(x, y) = (x^2 - y^2, 2xy)$ is an example of such.

§3. Charts and Atlases

We can now give the main definitions of the "abstract cartography." Let $\mathcal{X}$ be an arbitrary set.

Definition 30.1. A *chart on* $\mathcal{X}$ is a pair (U, h), where U is an arbitrary subset in $\mathcal{X}$ and h is a mapping of the set U into $\mathbb{R}^n$ that bijectively maps U onto a certain open set of the space $\mathbb{R}^n$. The set U is called the *domain* (or *support*) of the chart (U, h), and the mapping h the *charting mapping.* For any point $p \in U$, the point $h(p)$ has the form $(x^1(p), \dots, x^n(p))$, where $x^1(p), \dots, x^n(p)) \in \mathbb{R}$. This yields n numerical functions

$$x^1: p \mapsto x^1(p), \quad \dots, \quad x^n: p \mapsto x^n(p), \quad p \in U, \tag{3}$$

on U, which are called *local coordinates* of the chart (U, h).

Because coordinates (3) uniquely define the mapping h, we often write $(U, x^1, \dots, x^n)$ instead of (U, h), and the mapping h is called the *coordinate mapping.*

A Note on Notation. The function $\varphi\circ: p \mapsto \varphi(x^1(p), \dots, x^n(p))$ on U is denoted by the symbol $\varphi(x^1, \dots, x^n)$, where φ is a certain function on $h(U)$. This is in full correspondence with the traditional notation for a composition function. Therefore, the formula $y = \varphi(x^1, \dots, x^n)$ means that y is another symbol for the function $\varphi(x^1, \dots, x^n)$. To reduce the number of letters used, the function φ is also sometimes denoted by the symbol y; therefore, we write $y = y(x^1, \dots, x^n)$. We note that in this formula, the letter y in the left- and right-hand sides denote *different functions.* In the left-hand side, it is a function on U, and in the right-hand side, it is a function on $h(U)$, which expresses a function on U through the functions $x^1, \dots, x^n$. In the cases where it is necessary to clearly distinguish these functions, the first of them is denoted by the symbol $y \circ h$.

We stress that in contrast to some, we never use the symbols $\varphi(x^1, \dots, x^n)$ (or $y(x^1, \dots, y^n)$) to designate the function φ (or the function y) on $h(U)$ (and only in some cases do we use them to designate the value of these functions at a point $(x^1, \dots, x^n) \in h(U)$.

Two charts (U, h) and (V, k) are said to be *intersecting* if $U \cap V \neq \emptyset$. (This notation reflects the general tendency, which we sometimes follow, of not pedantically distinguishing (U, h) and U.) Let (U, h) and (V, k) be two intersecting charts, and let $W = U \cap V$. Then two sets $h(W)$ and $k(W)$ and the mapping

$$(k|_W) \circ (h|_W)^{-1}: h(W) \to k(W) \tag{4}$$

of one set into the other are defined on $\mathbb{R}^n$. Admitting a certain inaccuracy, we let $k \circ h^{-1}$ denote mapping (4).

Definition 30.2. Two charts (U, h) and (V, k) in $\mathcal{X}$ are said to be *compatible* if either they are disjoint ($U \cap V = \emptyset$) or

1. both the sets $h(W)$ and $k(W)$, where $W = U \cap V$, are open in $\mathbb{R}^n$ and
2. mapping (4) is a diffeomorphism (homeomorphism for $r = 0$).

If $\varphi^1, \dots, \varphi^n$ are functions that define diffeomorphism (4), then for the restrictions $x^1|_W, \dots, x^n|_W$ and $y^1|_W, \dots, y^n|_W$ of local coordinates of the charts (U, h) and (V, h) to W, we have

$$y^1|_W = \varphi^1(x^1|_W, \dots, x^n|_W), \quad \dots, \quad y^n|_W = \varphi(x^1|_W, \dots, x^n|_W). \tag{5}$$

In this connection, the functions $\varphi^1, \dots, \varphi^n$ are called the *transition function* (to W) from the coordinates $x^1, \dots, x^n$ to the coordinates $y^1, \dots, y^n$; formulas (5) are called the *transition formulas*. The diffeomorphism $k \circ h$ is called the *transition mapping*.

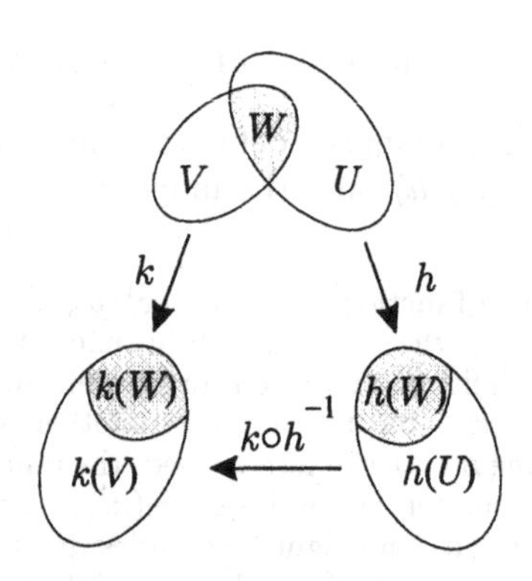

Fig. 30.1. For simplicity, we assume here that the sets $h(U)$ and $k(V)$ are disjoint, although this can be untrue in general. In exactly the same way, the sets $h(U), h(W)$, and $k(V)$ can be not connected.

As a rule, the subscript W is omitted in formulas (5), and they are written as

$$y^1 = \varphi^1(x^1, \ldots, y^n), \quad \ldots, \quad y^n = \varphi^n(x^1, \ldots, x^n) \tag{6}$$

(sometimes, with addition of the reference "to W"). We write them in a shorter ("vector") form as

$$\boldsymbol{y} = \varphi(\boldsymbol{x}) \quad \text{or} \quad \boldsymbol{y} = (k \circ h^{-1}\boldsymbol{x}). \tag{7}$$

The conditional character of formulas (7) should be understood, and it should be remembered every time that they are only an abbreviation of formulas (6) and (5). In particular, although formulas (7) have the form of relations between points of certain subspaces of the space $\mathbb{R}^n$, they in fact relate not the points of this space but functions given on a subset of the space $\mathcal{X}$ (in this sense, they are relations in $\mathcal{X}$).

Definition 30.3. A set of charts $\{(U_\alpha, h_\alpha)\}$ is called an *atlas* on $\mathcal{X}$ if

1. any two charts in this atlas are compatible and
2. the relation

$$\bigcup_\alpha U_\alpha = \mathcal{X}$$

holds (the charts (U_α, h_α) cover the whole $\mathcal{X}$).

§4. Maximal Atlases

For any atlas $\boldsymbol{A}$, $\boldsymbol{A}_{\max}$ denotes the set of all charts that are compatible with each chart of the atlas $\boldsymbol{A}$.

Proposition 30.1. *The set $\boldsymbol{A}_{\max}$ is an atlas.*

Corollary 30.1. *Each atlas $\boldsymbol{A}$ is contained in a unique maximal atlas $\boldsymbol{A}_{\max}$.*

We can now give our main definition.

Definition 30.4. Maximal atlases are also called *smooth structures.* A given set $\mathcal{X}$ with a smooth structure $\boldsymbol{A}_{\max}$ is called a *smooth manifold.* (Therefore, smooth manifolds are in fact pairs of the form $(\mathcal{X}, \boldsymbol{A}_{\max})$, but to simplify the notation and statements, we use this complete notation only when it cannot be avoided.) Charts of the atlas $\boldsymbol{A}_{\max}$ are called *charts* of the manifold $\mathcal{X}$, and even its *smooth charts* for a more complete expression.

By definition, two manifolds $(\mathcal{X}, \boldsymbol{A}_{\max})$ and $(\mathcal{Y}, \boldsymbol{A}^*_{\max})$ are the same iff $\mathcal{X} = \mathcal{Y}$ and $\boldsymbol{A}_{\max} = \boldsymbol{A}^*_{\max}$. Two atlases are said to be *equivalent* if they are contained in the same maximal atlas. Clearly, two atlases $\boldsymbol{A}$ and $\boldsymbol{A}^*$ are equivalent iff their union $\boldsymbol{A} \cup \boldsymbol{A}^*$ is an atlas (i.e., each chart of either of these atlases is compatible with each chart of the other atlas).

§5. Smooth Manifolds

Of course, to assign a smooth structure $\boldsymbol{A}_{\max}$ on $\mathcal{X}$, it suffices to assign an arbitrary atlas $\boldsymbol{A} \subset \boldsymbol{A}_{\max}$. Therefore, as smooth manifolds, we can consider pairs of the form $(\mathcal{X}, \boldsymbol{A})$, where $\boldsymbol{A}$ is an arbitrary atlas on $\mathcal{X}$. In this case two manifolds $(\mathcal{X}, \boldsymbol{A})$ and $(\mathcal{Y}, \boldsymbol{A}^*)$ are equivalent iff $\mathcal{X} = \mathcal{Y}$ and the atlases $\boldsymbol{A}$ and $\boldsymbol{A}^*$ are equivalent (i.e., when $\boldsymbol{A} \cup \boldsymbol{A}^*$ is an atlas).

We stress that the number n, the dimension of the space $\mathbb{R}^n$ containing the images $h(U)$ of supports of charts, enters the definition of a smooth manifold.

Definition 30.5. This dimension of the space $\mathbb{R}^n$ is called the *dimension* of the smooth manifold $\mathcal{X}$ and is denoted $\dim \mathcal{X}$.

The class of the smoothness of the transition mappings $k \circ h^{-1}$ (a number $r \geq 0$ or the symbol ∞ or ω) also enters the definition of the smooth manifold $\mathcal{X}$. It is called the *class of smoothness of the manifold* $\mathcal{X}$ (also, we say that $\mathcal{X}$ is a *manifold of class* C^r). Manifolds of class C^ω are also called *real-analytic manifolds.* For $r = 0$, the term "smooth manifold" is not used and is replaced with the term "topological manifold." Of course, each manifold of class C^r is automatically a manifold of class $C^{r'}$ for any $r' < r$ (and is a topological manifold in particular). In accordance with the convention stated above, as a rule, we assume a manifold to be of class C^∞.

§6. Smooth Manifold Topology

Let $\mathcal{X}$ be an arbitrary smooth (or topological) manifold.

Definition 30.6. A subset $O \subset \mathcal{X}$ is said to be *open* (in $\mathcal{X}$) if for any chart (U, h) of the manifold $\mathcal{X}$, the set $h(O \cap U) \subset \mathbb{R}^n$ is open (in $\mathbb{R}^n$).

This introduces a *topology* on $\mathcal{X}$, i.e., defines $\mathcal{X}$ as a *topological space.* In what follows, when speaking about a topology on a smooth (or topological) manifold, we always keep in mind the topology introduced by Definition 30.6.

We assume that the reader is familiar with the main concepts of general topology: open and closed sets, interiors and closures of sets, neighborhoods of points and sets, bases and subbases, Hausdorff (separable) and normal spaces, open coverings, compact and paracompact spaces, countability axioms, direct products, continuous mappings and homeomorphisms, connected, locally connected, and arcwise connected spaces, connected and arcwise connected components, paths, homotopies, and fundamental groups.

According to Definition 30.6, to verify whether a given subset $O \subset \mathcal{X}$ is open or not, we must consider the sets $h(O \cap U)$ for all charts (U, h) of the manifold $\mathcal{X}$. Of course, this is not possible in practice, and there arises the question: Is it possible to restrict these charts in a certain way? It turns out that it suffices to consider only charts of one arbitrary atlas and, moreover, only charts that cover the set O. Namely, the following statement holds.

Proposition 30.2. *Let $\{(U_\alpha, h_\alpha)\}$ be a family of charts of a manifold $\mathcal{X}$ such that*

$$O \subset \bigcup_\alpha U_\alpha$$

Then a set O is open in $\mathcal{X}$ iff for any α, the set $h_\alpha(O \cap U_\alpha)$ is open in $\mathbb{R}^n$.

Corollary 30.2. *For any chart (U, h) of a manifold $\mathcal{X}$, a set $V \subset U$ is open in $\mathcal{X}$ iff the set $h(V)$ is open in $\mathbb{R}^n$. In particular, the set U itself is open in $\mathbb{R}^n$.*

For any open set O of an arbitrary manifold $\mathcal{X}$ (of class C^r), all charts (O, h) for which $U \subset O$ obviously compose an atlas on O. This atlas is maximal, i.e., it is a smooth structure on O (of the same class C^r). We say that this smooth structure is *induced* by the smooth structure of the manifold $\mathcal{X}$; the set O with this smooth structure is called an *open submanifold* of the manifold $\mathcal{X}$. By definition, $\dim O = \dim \mathcal{X}$. In what follows, we always regard each open set O of an arbitrary manifold $\mathcal{X}$ as a manifold with the induced smooth structure.

We note that each atlas $\boldsymbol{A} = \{(U_\alpha, h_\alpha)\}$ of a manifold $\mathcal{X}$ defines the atlas $O \cap \boldsymbol{A}$ of the manifold O, which consists of the charts $(O \cap U_\alpha, h_\alpha|_{O \cap U_\alpha})$. Therefore, to obtain the induced smooth structure on O, it is not necessary to consider all charts on $\mathcal{X}$: it suffices to consider charts of only one atlas.

For any chart (U, h) of an arbitrary manifold $\mathcal{X}$, the set U (the support of the chart) is a neighborhood of each point $p \in U$. For this reason, supports of charts of $\mathcal{X}$ are also called *coordinate neighborhoods.*

It is clear that any neighborhood V of a point $p \in U$ that is contained in the coordinate neighborhood U is also a coordinate neighborhood (with the coordinate mapping $h|_V$). Therefore, *any neighborhood O of the point p contains a certain coordinate neighborhood* (for example, the intersection $O \cap U$ is such a neighborhood), i.e., *coordinate neighborhoods of each point of an arbitrary manifold form a basis of its neighborhoods.*

Corollary 1 directly implies that *for any chart (U, h) of an arbitrary manifold $\mathcal{X}$, the mapping $h\colon U \to h(U)$ is a homeomorphism* (*moreover, the set $h(U)$ is open in $\mathbb{R}^n$*). (For this reason, coordinate mappings h considered as mappings onto $h(U)$ are usually called *coordinate homeomorphisms.*)

Therefore, because coordinate neighborhoods compose an open covering of a manifold $\mathcal{X}$ (even coordinate neighborhoods that are domains of charts of a certain atlas have this property), *any manifold is a locally Euclidean space* and hence is arcwise connected.

Proposition 30.3. *A compact Hausdorff manifold is normal. A connected manifold is paracompact iff it satisfies the second countability axiom. Homeo-*

morphic smooth manifolds are of the same dimension. There exist connected Hausdorff manifolds that are not paracompact.

Example 30.1 (from Calabi and Rosenlicht). Let $\mathcal{X}$ be a subset of the space $\mathbb{R}^3$ consisting of points (x, y, z) for which either $x = 0$ or $z = 0$ (the union of the coordinate planes Oyz and Oxy), and let U_a, $a \in \mathbb{R}$, be its subset consisting of points (x, y, z) for which either $x \neq 0$ or $y = a$ (the union of the plane Oxy with the axis $x = 0$ removed and the line $x = 0$, $y = a$). Further, let $h_a: U_a \to \mathbb{R}^2$ be the mapping of the set U_a onto the plane $\mathbb{R}^2$ with the coordinates (u_a, v_a) given by

$$u_a = x, \qquad v_a = \begin{cases} \dfrac{y-a}{x} & \text{if } x \neq 0, \\ z & \text{if } x = 0. \end{cases}$$

This mapping is obviously bijective (the inverse mapping maps a point $(u_a, v_a) \in \mathbb{R}^2$ to the point $(u_a, a + u_a v_a, v_a) \in U_a$), and the pair (U_a, h_a) is therefore a chart on $\mathcal{X}$. For any two such charts (U_a, h_a) and (U_b, h_b), the set $U_a \cap U_b$ is the plane Oxy with the axis $x = 0$ removed, the sets $h_a(U_a \cap U_b)$ and $h_b(U_a \cap U_b)$ are the plane $\mathbb{R}^2$ with the axis $u = 0$ removed, and the mapping

$$h_a \circ h_b^{-1}: h_a(U_a \cap U_b) \to h_b(U_a \cap U_b)$$

is given by

$$u_b = u_a, \qquad v_b = v_a + \frac{a-b}{u_a}$$

and is therefore real analytic. This means that the charts (U_a, h_a), $a \in \mathbb{R}$, are compatible in a real analytic way and therefore define the structure of a smooth two-dimensional manifold of class C^ω on $\mathcal{X}$. This manifold is connected and Hausdorff but not paracompact.

Let $\mathcal{X}$ and $\mathcal{Y}$ be two manifolds (of the respective dimensions n and m), and let $\mathcal{X} \times \mathcal{Y}$ be the set of all pairs (p, q), where $p \in \mathcal{X}$ and $q \in \mathcal{Y}$. For any sets $U \subset \mathcal{X}$ and $V \subset \mathcal{Y}$, the set $U \times V$ is a subset of the set $\mathcal{X} \times \mathcal{Y}$, and for any mappings $h: U \to \mathbb{R}^n$ and $k: V \to \mathbb{R}^n$, the formula

$$(h \times k)(p, q) = (h(p), k(q))$$

defines a certain mapping

$$h \times k: U \times V \to \mathbb{R}^{n+m},$$

where we identify $\mathbb{R}^n \times \mathbb{R}^m$ with $\mathbb{R}^{n+m}$. Moreover, if the mappings h and k are injective, then the mapping $h \times k$ is also injective; if the sets $h(U)$ and $k(U)$ are open (in $\mathbb{R}^n$ and $\mathbb{R}^m$ respectively), then the set

$$(h \times k)(U \times V) = h(U) \times k(V)$$

is open in $\mathbb{R}^{n+m}$. This means that if (U, h) and (V, k) are two charts, then $(U \times V, h \times k)$ is also a chart. In addition, it is easy to see that *if a chart*

(U,h) is compatible with a chart (U',h') and a chart (V,k) is compatible with a chart (V',k'), then the chart $(U\times V, h\times k)$ is compatible with the chart $(U'\times V', h'\times k')$. Indeed, it is clear that

$$(U\times V)\cap(U'\times V') = (U\cap U')\times(V\cap V')$$

and, similarly,

$$(h(U)\times k(V))\cap(h'(U')\times k'(V')) = (h(U)\cap h'(U'))\times(k(V)\cap k'(V'))$$

(we assume conditionally that $A\times B=\emptyset$ if $A=\emptyset$ or $B=\emptyset$). Moreover,

$$(h\times k|_{(U\times V)\cap(U'\times V')})\circ(h'\times k'|_{(U\times V)\cap(U'\times V')})^{-1} = [(h|_{U\cap U'})\circ(h'|_{U\cap U'})^{-1}]\times[(k|_{V\cap V'})\circ(k'|_{V\cap V'})^{-1}].$$

To complete the proof, it remains to note that for any diffeomorphisms $\varphi\colon W\to W_1$ and $\varphi'\colon W'\to W_1'$ of open sets of the spaces $\mathbb{R}^n$ and $\mathbb{R}^m$, the mapping

$$\varphi\times\varphi'\colon W\times W'\to W_1\times W_1'$$

is also a diffeomorphism of open sets (of the space $R^{n+m}=\mathbb{R}^n\times\mathbb{R}^m$).

Therefore, the charts $(U\times V, h\times k)$ constructed for all possible charts (U,h) and (V,k) of the manifolds $\mathcal{X}$ and $\mathcal{Y}$ compose an atlas on $\mathcal{X}\times\mathcal{Y}$.

Definition 30.7. The corresponding smooth structure on $\mathcal{X}\times\mathcal{Y}$ is called the *direct product of the smooth structures* of the manifolds $\mathcal{X}$ and $\mathcal{Y}$, and the set $\mathcal{X}\times\mathcal{Y}$ is called the *direct product of the manifolds* $\mathcal{X}$ and $\mathcal{Y}$. Its dimension is equal to the sum of the dimensions of the factors:

$$\dim(\mathcal{X}\times\mathcal{Y}) = \dim\mathcal{X}+\dim\mathcal{Y}.$$

The topology of the manifold $\mathcal{X}\times\mathcal{Y}$ is obviously the direct product of the topologies of the manifolds $\mathcal{X}$ and $\mathcal{Y}$.

Remark 30.2. For an arbitrary group $\mathcal{G}$, we define the mapping

$$\mathcal{G}\times\mathcal{G}\to\mathcal{G},\qquad (a,b)\mapsto ab,\quad a,b\in\mathcal{G} \tag{8}$$

A group $\mathcal{G}$ that is a smooth manifold and for which mapping (8) is smooth is called a *Lie group* (the term *smooth group* is also used).

We assume that the reader is familiar with the main facts of Lie group theory and with the concept of the Lie algebra of a Lie group in particular (see Remark 33.4).

§7. Smooth Structures on a Topological Space

Very often, the structure of a smooth manifold should be introduced on a set $\mathcal{X}$ which always has a topology, i.e., on a topological space. In this case, we always implicitly assume that this structure should define namely this topology on $\mathcal{X}$, i.e., it should be *compatible* with this topology as we usually say.

Of course, for this purpose, it is necessary that an atlas defining a smooth structure on $\mathcal{X}$ with this property consists of charts (U, h) whose supports U are open and whose mappings $h: U \to h(U)$ are homeomorphisms. It turns out that this condition is also sufficient.

Proposition 30.4. *Let $A = \{(U_\alpha, h_\alpha)\}$ be an atlas on a topological space $\mathcal{X}$ such that for any α, the set U_α is open in $\mathcal{X}$ and the mapping $h_\alpha: U_\alpha \to h_\alpha(U_\alpha)$ is a homeomorphism. Then the smooth structure defined by the atlas A is compatible with the topology of the space $\mathcal{X}$.*

Remark 30.3. By Proposition 30.4, a smooth manifold can be *defined* as a topological space $\mathcal{X}$ on which we have an atlas such that the supports U_α of the charts (U_α, h_α) are open and the coordinate mappings h_α are homeomorphisms. In this case, the requirement that for any α and β, the sets $h_\alpha(U_\alpha \cap U_\beta)$ and $h_\alpha(U_\alpha \cap U_\beta)$ are open in $\mathbb{R}^n$ holds automatically. Namely this definition is usually found in the literature (sometimes, in a slightly different environment of words) despite its obvious methodological deficiency (a similar definition of metric space, in which the deficiency arises most effectively, would say that a metric space is a topological space $\mathcal{X}$ on which a continuous function $\rho: \mathcal{X} \times \mathcal{X}$ satisfying the usual axioms of a metric space is given).

Remark 30.4. If on a topological space $\mathcal{X}$, we have the structure of a topological manifold (of class C^0) that is compatible with the topology (this holds iff the space is locally Euclidean), then this structure is unique, i.e., it is uniquely determined by the topology. For $r > 0$ the case is different: if we have the structure of a smooth manifold of class C^r, $r > 0$, on a topological space $\mathcal{X}$, then there are infinitely many such structures (there is even a structure of class C^ω among them). This phenomenon arises even for $\mathcal{X} = \mathbb{R}$. Necessary and sufficient conditions for the existence a smooth structure of class C^r, $r > 0$, on a topological space are known but are very complicated. In what follows, if not explicitly stated otherwise, we exclude topological manifolds from our consideration.

If $\mathcal{X}$ and $\mathcal{Y}$ are smooth manifolds (of the respective dimensions n and m), then for any point $p_0 \in \mathcal{X}$, any continuous mapping $f: \mathcal{X} \to \mathcal{Y}$, and any coordinate neighborhood V of the point $f(p)$ in $\mathcal{Y}$, a neighborhood U of the point p_0 in $\mathcal{X}$ such that $fU \subset V$ can be considered a coordinate neighborhood. Therefore, if

$$h: U \to h(U) \in \mathbb{R}^n \qquad \text{and} \qquad k: V \to k(V) \in \mathbb{R}^m$$

are coordinate mappings, then the formula

$$\overset{\circ}{f} = k \circ (f|_U) \circ h^{-1}$$

defines a certain mapping

$$\overset{\circ}{f}: h(U) \to k(V) \tag{9}$$

of the open set $h(U)$ of the space $\mathbb{R}^n$ into the open set $k(V)$ of the space $\mathbb{R}^m$. This mapping is given by m functions

$$y^j = f^j(x^1, \dots, x^n), \quad j = 1, \dots, m, \tag{10}$$

of n variables that express the coordinates $y^1, \dots, y^m$ of the point $\boldsymbol{y} = \overset{\circ}{f}(\boldsymbol{x}) \in k(V) \subset \mathbb{R}^m$ through the coordinates $x^1, \dots, x^n$ of the point $\boldsymbol{x} \in h(U) \in \mathbb{R}^n$ (in other words, the local coordinates of the point $q = f(p) \in V$ through the local coordinates of the point $p \in U$). We say that functions (10) *express* (or *assign*) *the mapping f in the charts* (U, h) *and* (V, k) (in the local coordinates $x^1, \dots, x^n$ and $y^1, \dots, y^m$).

If (U', h') and (V', k') are another pair of charts having the property that $p_0 \in U'$ and $fU' \in V'$, then the functions f'^j that express the mapping f in the charts (U', h') and (V', k') are smooth at the point $h'(p_0)$ iff functions (10) are smooth at the point $h(p_0)$. In this sense, the smoothness property of functions (10) does not depend on the choice of the charts (U, h) and (V, k).

§8. DIFF Category

Definition 30.8. A continuous mapping $f: \mathcal{X} \to \mathcal{Y}$ is said to be *smooth* at a point $p_0 \in \mathcal{X}$ if in certain (and therefore in all) charts (U, h) and (V, k) that have the properties $p_0 \in U$ and $fU \in V$, functions (3) expressing the mapping f are smooth functions (of a given class C^r) at the point p_0. A mapping $f: \mathcal{X} \to \mathcal{Y}$ that is smooth at all points $p \in \mathcal{X}$ is said to be *smooth.*

Proposition 30.5. *The set* DIFF *of all smooth manifolds and their smooth mappings is a category.*

A path $u: \mathrm{I} \to \mathcal{X}$ in a smooth manifold $\mathcal{X}$ is said to be *smooth* if it is the restriction to I of a certain smooth mapping $u': (-\varepsilon, 1+\varepsilon) \to \mathcal{X}$, where $\varepsilon > 0$. (We recall that the interval $(-\varepsilon, 1+\varepsilon)$ is a smooth manifold). If the mapping u' is smooth everywhere except finitely many points, then the path u is said to be *piecewise smooth.*

Proposition 30.6. *Any two points of a connected smooth manifold can be connected by a piecewise smooth (and even smooth) path.*

Definition 30.9. A mapping $f: \mathcal{X} \to \mathcal{Y}$ of smooth manifolds is called a *diffeomorphism* if

1. it is bijective,
2. it is smooth, and
3. the inverse mapping $f^{-1}: \mathcal{X} \to \mathcal{Y}$ is smooth (and is therefore also a diffeomorphism).

An arbitrary coordinate mapping $h: U \to h(U)$ is an obvious example of a diffeomorphism. Conversely, it is easy to see that *for any open set U of a smooth manifold $\mathcal{X}$ and any diffeomorphism $h: U \to h(U)$ onto the open set $h(U)$, the pair (U, h) is a chart of the manifold $\mathcal{X}$* (belongs to its maximal atlas). Two manifolds $\mathcal{X}$ and $\mathcal{Y}$ are said to be *diffeomorphic* if there exists at least one diffeomorphism $\mathcal{X} \to \mathcal{Y}$. Such manifolds have the same *differential properties* (properties that are expressed in terms of smooth structures) and are the same in this sense.

Remark 30.5. It can be shown that any one-dimensional noncompact manifold satisfying the second countability axiom is diffeomorphic to the line $\mathbb{R}$ with the standard smooth structure (and a compact one is diffeomorphic to the circle $\mathbb{S}^1$). It is interesting that for $n = 4$ (and only for $n = 4$), there exist smooth structures on $\mathbb{R}^n$ (whose construction is very complicated) that are compatible with a topology on $\mathbb{R}^n$ with respect to which $\mathbb{R}^n$ is not diffeomorphic to $\mathbb{R}^n$ in the standard smooth structure.

§9. Transport of Smooth Structures

Let $\mathcal{X}$ be a certain set, $\mathcal{Y}$ be a smooth manifold, and $f: \mathcal{X} \to \mathcal{Y}$ be a bijective mapping. Then *there exists a unique smooth structure on $\mathcal{X}$ such that f is a diffeomorphism with respect to it.* All possible pairs of the form $(f^{-1}U, h \circ f)$, where (U, h) is an arbitrary chart of the manifold $\mathcal{Y}$, are charts of this smooth structure. We say that this smooth structure *is transported* from $\mathcal{Y}$ to $\mathcal{X}$ via f. It is clear that *smooth structures transported from $\mathcal{Y}$ to $\mathcal{X}$ via bijective mappings $f: \mathcal{X} \to \mathcal{Y}$ and $g: \mathcal{X} \to \mathcal{Y}$ coincide iff the mapping $g \circ f^{-1}: \mathcal{Y} \to \mathcal{Y}$ is a diffeomorphism.*

Let

$$\pi: \mathcal{E} \to \mathcal{B} \tag{11}$$

be a smooth mapping. We say that an open set is *exactly* (*uniformly*) *covered* by the mapping π if the full inverse image $\mathcal{E}_U = \pi^{-1}$ of this set (when it is not empty) is a disjoint union of open sets $V_\nu \in \mathcal{E}$,

$$\pi^{-1}U = \coprod_\nu V_\nu,$$

and these sets are such that for any ν, the mapping

$$\pi|_{V_\nu}: V_\nu \to U$$

is a homeomorphism. Let the space $\mathcal{B}$ be connected.

Definition 30.10. Mapping (11) is called a *covering* or a *covering mapping* if

1. the space $\mathcal{E}$ is connected and
2. there exists an open covering $\mathfrak{A} = \{U_\alpha\}$ of the space $\mathcal{B}$ consisting of sets exactly covered by mapping (11).

Also, a *covering* is defined as a triple $\xi = (\mathcal{E}, \pi, \mathcal{B})$ consisting of two spaces $\mathcal{E}$ and $\mathcal{B}$ and a covering mapping $\pi\mathcal{E} \to \mathcal{B}$. It is clear that *any covering* (11) *is a surjective mapping.* The total space $\mathcal{E}$ of the covering ξ is also called the *covering space.*

In the case where all sheets of a covering are finite (and the number of points in each of the sheets are therefore the same), this covering is said to be *finite sheeted,* and the number of points in a sheet is called the *number of sheets* of this covering. (We note that the notion of a *sheet of a covering* is not defined.)

Chapter 31
Tangent Vectors

§1. Vectors Tangent to a Smooth Manifold

Now let $\mathcal{X}$ be an arbitrary smooth n-dimensional manifold (of class C^r, $r \geq 1$). For any point $p_0 \in \mathcal{X}$, we let $\boldsymbol{A}(p_0)$ denote the set of all charts (U, h) of this manifold for which $p_0 \in U$ (we say in this case that these charts are *centered* at p_0).

Definition 31.1. A *vector tangent to the manifold* $\mathcal{X}$ (or merely a *vector of the manifold* $\mathcal{X}$) at the point p_0 is the mapping

$$A\colon \boldsymbol{A}(p_0) \to \mathbb{R}^n \tag{1}$$

such that for any charts $(U, h)=(U, x^1, \ldots, x^n)$ and $(U', h')=(U', x'^1, \ldots, x'^n)$ in $\boldsymbol{A}(p_0)$, the vectors $A(U, h) = (a^1, \ldots, a^n)$ and $A(U', h') = (a'^1, \ldots, a'^n)$ of the linear space $\mathbb{R}^n$ are related by formula (1), i.e., by the formula

$$A(U', h') = \left(\frac{\partial h'}{\partial h}\right)_0 A(U, h), \tag{2}$$

where $(\partial h'/\partial h)_0$ is a linear operator $\mathbb{R}^n \to \mathbb{R}^n$ with the matrix $\|(\partial x'^i/\partial x^i)_0\|$. The components $a^1, \ldots, a^n$ of the vector $A(U, h) \in \mathbb{R}^n$ are called the *coordinates of the vector A in the chart* (U, h) (or in the *local coordinates* $x^1, \ldots, x^n$). To simplify the formulas, the relation $A(U, h) = (a^1, \ldots, a^n)$ is usually written as

$$A = (a^1, \ldots, a^n) \quad \text{in } (U, h).$$

In the case where the chart (U, h) is fixed, the indication "in (U, h)" is omitted as a rule.

The set of all vectors of a manifold $\mathcal{X}$ at a point p_0 is denoted by $\mathbf{T}_{p_0}\mathcal{X}$ and is called the *tangent space* of the manifold $\mathcal{X}$ at the point p_0. It is a linear space over the field $\mathbb{R}$ with respect to linear operations defined by

$$(A + B)(U, h) = A(U, h) + B(U, h),$$
$$(\lambda A(U, h) = \lambda A(U, h),$$

where $A, B \in \mathbf{T}_{p_0}\mathcal{X}$, $\lambda \in \mathbb{R}$, and $(U, h) \in \boldsymbol{A}(p)$. (The inclusions $A+B \in \mathbf{T}_{p_0}\mathcal{X}$ and $\lambda A \in \mathbf{T}_{p_0}\mathcal{X}$ are ensured by the linearity of the operators $(\partial h'/\partial h)_0$.)

Therefore, by definition, if $A = (a^1, \ldots, a^n)$ and $B = (b^1, \ldots, b^n)$ are in (U, h), then $A + B = (a^1 + b^1, \ldots, a^n + b^n)$ and $\lambda A = (\lambda a^1, \ldots, \lambda a^n)$ are in (U, h). Therefore, for any charts (U, h), the correspondence

$$A \mapsto A(U, h)$$

defines the linear mapping

$$\mathbf{T}_{p_0}\mathcal{X} \to \mathbb{R}^n. \tag{3}$$

It is easy to see that *mapping (3) is an isomorphism.* Therefore, $\mathbf{T}_{p_0}\mathcal{X}$ *is a linear space of dimension n*:

$$\dim \mathbf{T}_{p_0}\mathcal{X} = \dim \mathcal{X}.$$

The vectors that pass to the standard basis $e_1, \ldots, e_n$ of the space $\mathbb{R}^n$ under isomorphism (3) are denoted by the symbols

$$\left(\frac{\partial}{\partial x^1}\right)_{p_0}, \quad \ldots, \quad \left(\frac{\partial}{\partial x^n}\right)_{p_0}. \tag{4}$$

They compose a basis of the space $\mathbf{T}_{p_0}\mathcal{X}$; moreover, the coordinates of a vector $A \in \mathbf{T}_{p_0}\mathcal{X}$ with respect to this basis are exactly its coordinates in the chart (U, h): if $A = (a^1, \ldots, a^n)$ in (U, h), then

$$A = a^i \left(\frac{\partial}{\partial x^i}\right)_{p_0}, \tag{5}$$

and vice versa.

In the case where $\mathcal{X}$ is a subspace of $\mathbb{R}^n$, among all isomorphisms (3), we find one distinguished isomorphism that corresponds to $(\mathbb{R}^n, \mathrm{id})$, and we can identify the space $\mathbf{T}_{p_0}\mathbb{R}^n$ with the space $\mathbb{R}^n$ using this isomorphism. Therefore,

$$\mathbf{T}_{p_0}\mathbb{R}^n = \mathbb{R}^n \tag{6}$$

for any point $p_0 \in \mathbb{R}^n$, which is full correspondence with what we began with.

Because the space $\mathbf{T}_{p_0}U$ is identified in a natural way with $\mathbf{T}_{p_0}\mathcal{X}$,

$$\mathbf{T}_{p_0}U = \mathbf{T}_{p_0}\mathcal{X}$$

for any point p_0 of any open submanifold U of an arbitrary manifold $\mathcal{X}$, we obtain, in particular,

$$\mathbf{T}_{p_0}U = \mathbb{R}^n \tag{7}$$

for any point $p_0 \in U$ of an arbitrary open set $U \subset \mathbb{R}^n$. We constantly use these identifications, not always explicitly indicating them.

More generally, if $\mathcal{X}$ is a linear space $\mathcal{V}$ (or an open set in $\mathcal{V}$), then among isomorphisms (3), we distinguish the isomorphisms corresponding to the charts of the form $(\mathcal{V}, h)$, where h is the coordinate isomorphism $\mathcal{V} \to \mathbb{R}^n$ corresponding to a certain basis $e_1, \ldots, e_n$ in $\mathcal{V}$. Because the linear operator $(\partial h'/\partial h)_0$, $p_0 \in \mathcal{V}$, obviously coincides with the operator $h' \circ h$ (and does not depend on p_0 in particular) for any two such isomorphisms h and h', the composition $\mathbf{T}_{p_0}\mathcal{X} \to \mathcal{V}$ of isomorphism (3) corresponding to the chart $(\mathcal{V}, h)$ and the isomorphism $h^{-1}\colon \mathbb{R}^n \to \mathcal{V}$ is the same for all h. Therefore, $\mathbf{T}_{p_0}\mathcal{V}$ *is naturally identified with* $\mathcal{V}$:

$$\mathsf{T}_{p_0}\mathcal{V} = \mathcal{V} \quad \text{for any } p_0 \in \mathcal{V}.$$

We note that the basis $e_1, \ldots, e_n$ corresponds to basis (4) in this identification.

If $\mathcal{X}$ is an affine space $\mathcal{A}$, then the tangent space $\mathsf{T}_{p_0}\mathcal{A}$ for any point $p_0 \in \mathcal{A}$ is similarly identified with the associated linear space $\mathcal{V}$ (and is therefore again the same for all p_0). By the way, it is convenient here to artificially introduce a dependence on p_0 and identify $\mathsf{T}_{p_0}\mathcal{A}$ with the space $\mathcal{A}$ in which we choose the point p_0 as the origin, i.e., to consider all vectors from $\mathsf{T}_{p_0}\mathcal{A}$ to be related to the point p_0.

In the case where $\mathcal{X}$ is an elementary surface in the affine space $\mathcal{A}$, the tangent space $\mathsf{T}_{p_0}\mathcal{X}$ is naturally identified with its tangent space (which is a subspace of the associated linear space $\mathcal{V}$). Namely, if u, v are local coordinates on $\mathcal{X}$ that correspond to a certain parameterization $\boldsymbol{r} = \boldsymbol{r}(u, v)$ of the surface $\mathcal{X}$, then the corresponding base vectors $(\partial/\partial u)|_{p_0}$ and $(\partial/\partial v)|_{p_0}$ of the space $\mathsf{T}_{p_0}\mathcal{X}$ are identified with the vectors r_{u_0} and r_{v_0} of the space $\mathcal{V}$. Because the vectors $(\partial/\partial u)|_{p_0}$ and $(\partial/\partial v)|_{p_0}$ are transformed according to the same formulas as the vectors r_u and r_v under a change of variables, this identification of the space $\mathsf{T}_{p_0}\mathcal{X}$ with the subspace of the space $\mathcal{V}$ does not depend on the choice of the parameterization $\boldsymbol{r} = \boldsymbol{r}(u, v)$.

§2. Oriented Manifolds

Definition 31.2. We say that two charts $(U, h) = (U, x^1, \ldots, x^n)$ and $(U', h') = (U', x'^1, \ldots, x'^n)$ of an n-dimensional manifold $\mathcal{X}$ are *positively compatible* if either $U \cap U' = \emptyset$ or $U \cap U' \neq \emptyset$ and

$$\det \frac{\partial h'}{\partial h} > 0 \quad \text{on } U \cap U',$$

i.e., at each point $p \in U \cap U'$, the bases

$$\left(\frac{\partial}{\partial x^1}\right)_p, \quad \ldots, \quad \left(\frac{\partial}{\partial x^n}\right)_p \quad \text{and} \quad \left(\frac{\partial}{\partial x'^1}\right)_p, \quad \ldots, \quad \left(\frac{\partial}{\partial x'^n}\right)_p$$

of the tangent space $\mathsf{T}_p\mathcal{X}$ are co-oriented. An atlas consisting of positively compatible charts is said to be *orienting*. A manifold $\mathcal{X}$ for which there exists at least one orienting atlas is said to be *orientable*.

It is clear that *a manifold is orientable iff all its components are orientable.*

It is easy to see (compare with Corollary 30.1) that for any orienting atlas $\boldsymbol{A}$ of an orientable manifold $\mathcal{X}$, the set $\boldsymbol{A}^+_{\max}$ of all charts that are positively compatible with each chart of the atlas $\boldsymbol{A}$ is an orienting atlas and this atlas is maximal (i.e., has the property that if an orienting atlas $\boldsymbol{A}^*$ contains the atlas $\boldsymbol{A}$, then $\boldsymbol{A}^* \subset \boldsymbol{A}^+_{\max}$).

It is known that the following manifolds are orientable:

1. the spheres $\mathbb{S}^n$, $n \geq 0$;
2. the projective spaces $\mathbb{R}P^{2n+1}$ of odd dimension;
3. the realifications of complex-analytic manifolds;
4. Lie groups.

(It can be shown that projective spaces of even dimension, the projective plane $\mathbb{R}P$ in particular, are not orientable.)

Definition 31.3. An orientable manifold in which the maximal orienting atlas is chosen is said to be *oriented,* and this atlas is called its *orientation.* Charts belonging to the orientation of an oriented manifold are said to be *positively oriented* (or merely *positive*).

By definition, for each point $p \in \mathcal{X}$, the orientation of the tangent space $\mathbf{T}_p\mathcal{X}$ defined by the basis

$$\left(\frac{\partial}{\partial x^1}\right)_p, \quad \ldots, \quad \left(\frac{\partial}{\partial x^n}\right)_p$$

is the same for all positively oriented charts $(U, x^1, \ldots, x^n)$. We say that this orientation is *induced* by the given orientation of the manifold $\mathcal{X}$. Therefore, roughly speaking, the orientation of a manifold is the choice of compatible orientations of its tangent spaces. It is easy to see that *for each connected chart* $(U, x^1, x^2, \ldots, x^n)$ *of an oriented manifold* $\mathcal{X}$, *one (and only one) of the charts* $(U, x^1, \ldots, x^n)$ *and* $(U, -x^1, x^2, \ldots x^n)$ *is positively oriented.*

Exercise 31.1. Deduce from this that *on a connected oriented manifold of dimension* $n > 0$, *we have two and only two orientations.* These orientations are said to be *opposite.* If $\mathcal{X}$ is a manifold with one orientation, then this manifold equipped with the opposite orientation is usually denoted by $-\mathcal{X}$.)

Clearly, we can assign orientations on distinct components of an oriented manifold $\mathcal{X}$ independently of each other. Therefore, on a manifold $\mathcal{X}$ with N components, we have exactly 2^N distinct orientations.

§3. Differential of a Smooth Mapping

Let $\mathcal{X}$ and $\mathcal{Y}$ be two smooth manifolds (of the respective dimensions n and m), and let $f\colon \mathcal{X} \to \mathcal{Y}$ be an arbitrary smooth mapping. Further, let p be an arbitrary point of the manifold $\mathcal{X}$, let $q = f(p)$ be its image in the manifold $\mathcal{Y}$, and let

$$(U, h) = (U, x^1, \ldots, x^n) \qquad \text{and} \qquad (V, k) = (V, y^1, \ldots, y^m)$$

be charts of the respective manifolds $\mathcal{X}$ and Y such that $p \in U$ and $fV \subset V$.

As we already know, the mapping f is written in the charts (U, h) and (V, k) by

$$y^j = f^j(x^1, \dots, x^n), \quad j = 1, \dots, m,$$

where f^j are certain smooth functions. The $n \times m$ matrix

$$\left\| \left(\frac{\partial f^j}{\partial x^i} \right)_p \right\|, \quad i = 1, \dots, n, \quad j = 1, \dots, m, \tag{8}$$

whose entries are the values $(\partial f^j / \partial x^i)_p$ of partial derivatives of the functions f^j with respect to x^i at the point p (more precisely, at the point $h(p) \in \mathbb{R}^n$) is called the *Jacobi matrix* of the mapping f in the charts (U, h) and (V, k). This matrix assigns the linear mapping $\mathbb{R}^n \to \mathbb{R}^m$ that maps a vector $(a^1, \dots, a^n)$ into the vector $(b^1, \dots, b^m)$, where

$$b^j = \left(\frac{\partial f^j}{\partial x^i} \right)_p a^i, \quad j = 1, \dots, m. \tag{9}$$

The coordinate isomorphisms $A \mapsto A(U, h)$ and $A \mapsto A(V, k)$ allow interpreting this mapping as the linear mapping $\mathbf{T}_p \mathcal{X} \to \mathbf{T}_q \mathcal{Y}$ of tangent spaces.

Definition 31.4. The constructed mapping $\mathbf{T}_p \mathcal{X} \to \mathbf{T}_q \mathcal{Y}$ is called the *differential of the smooth mapping* f at the point p and is denoted by the symbol $(df)_p$ or merely df_p.

Therefore, if $A = (a^1, \dots, a^n)$ in a chart (U, h), then $df_p(A) = (b^1, \dots, b^m)$, where $b^j, j = 1, \dots, m$, are defined by (9). In the language of linear algebra, this means that *the mapping* df_p *is just the linear mapping* $\mathbf{T}_p \mathcal{X} \to \mathbf{T}_q \mathcal{Y}$ *with matrix* (8) *in the basis* $(\partial/\partial x^1)_p, \dots, (\partial/\partial x^n)_p$ *and* $(\partial/\partial y^1)_q, \dots, (\partial/\partial y^m)_q$.

Of course, it is necessary to verify that this definition is *correct,* i.e., verify that the mapping df_p does not depend on the choice of the charts (U, h) and (V, k). We note that the mapping df_p depends only on the local behavior of the mapping f in a neighborhood of the point p, i.e., *if for mappings* $f, g: \mathcal{X} \to \mathcal{Y}$, *there exists a neighborhood* U *of the point* p *such that* $f = g$ *on* U, *then* $df_p = dg_p$.

§4. Chain Rule

Let $f: \mathcal{X} \to \mathcal{Y}$ and $g: \mathcal{Y} \to \mathcal{Z}$ be smooth mappings. If $(U, h) = (U, x^1, \dots, x^n)$, $(V, k) = (V, y^1, \dots, y^m)$, and $(W, l) = (W, z^1, \dots, z^s)$ are charts of the manifolds $\mathcal{X}$, $\mathcal{Y}$, and $\mathcal{Z}$ such that $fU \subset V$ and $gV \subset W$ and if $\boldsymbol{y} = \boldsymbol{f}(\boldsymbol{x})$ and $\boldsymbol{z} = \boldsymbol{g}(\boldsymbol{y})$ are functions that define the mappings f and g in these charts (we use the abbreviated notation introduced above), then the smooth mapping $f \circ g: \mathcal{X} \to \mathcal{Z}$ is obviously given by the function $\boldsymbol{z} = \boldsymbol{g}(\boldsymbol{f}(\boldsymbol{x}))$ in the charts (U, h) and (W, l). Therefore, by the differentiation formula for a composition of functions, we have the relations

$$\left(\frac{\partial z^k}{\partial x^i}\right)_p = \left(\frac{\partial z^k}{\partial y^i}\right)_q \left(\frac{\partial y^j}{\partial x^i}\right)_p, \quad \text{where } q = f(p),$$

for any point $p \in U$, which means that the linear mapping $d(f \circ g)_p$ is a composition of the linear mappings dF_p and dg_p:

$$d(f \circ g) = dg_q \circ df_p. \tag{10}$$

This formula is called the *chain rule.*

§5. Gradient of a Smooth Function

Two cases are of special interest: where $\mathcal{Y} = \mathbb{R}$ and where $m = n$. In the case where $\mathcal{Y}$ is the real axis $\mathbb{R}$ (and the mapping f is therefore a smooth function on $\mathcal{X}$), the differential df_p is usually called the *gradient* of the function f.

By identification (6), the gradient is a linear mapping $\mathbf{T}_p\mathcal{X} \to \mathbb{R}$, i.e., it is a covector of the space $\mathbf{T}_p\mathcal{X}$ (a vector of the dual space $\mathbf{T}_p^*\mathcal{X}$, which, by the way, is called the *cotangent space* of the manifold $\mathcal{X}$ at the point p). By definition, the covector df_p assumes the value

$$df_p(A) = \left(\frac{\partial f}{\partial x^i}\right)_p a^i$$

on any vector $A \in \mathbf{T}_p\mathcal{X}$. This value is called the *derivative of the function f with respect to the vector A* and is denoted by Af. Therefore,

$$Af = df_p(A)$$

and

$$Af = \left(\frac{\partial f}{\partial x^i}\right)_p a^i \quad \text{in } (U, x^1, \ldots, x^n). \tag{11}$$

In particular, $(\partial/\partial x^i)_p f = (\partial f/\partial x^i)_p$ for any $i = 1, \ldots, n$, which explains the choice of notation for vectors of basis (4).

Formula (11) means that in the basis of the space $\mathbf{T}_p^*\mathcal{X}$ dual to basis (4) of the space $\mathbf{T}_p\mathcal{X}$, the covector df_p has the coordinates

$$\left(\frac{\partial f}{\partial x^i}\right)_p, \quad \ldots, \quad \left(\frac{\partial f}{\partial x^n}\right)_p.$$

Therefore, first, this basis consists of the covectors $dx_p^1, \ldots, dx_p^n$ (clearly, the covector df_p is also defined for functions f that are given only in a certain neighborhood of the point p), and, second,

$$df_p = \left(\frac{\partial f}{\partial x^1}\right)_p dx_p^1 + \cdots + \left(\frac{\partial f}{\partial x^n}\right)_p dx_p^n = \left(\frac{\partial f}{\partial x^i}\right)_p dx_p^i.$$

Remark 31.1. We call attention to the fact that *the gradient is a covector.* The familiar notion that the gradient of a smooth function is a vector, which comes from calculus, is based on an implicit identification of vectors and covectors via the standard Euclidean structure on $\mathbb{R}^n$.

§6. Étale Mapping Theorem

Now let $n = m$.

Definition 31.5. A smooth mapping $f: \mathcal{X} \to \mathcal{Y}$ of manifolds of the same dimension is said to be *étale at a point* $p \in \mathcal{X}$ (or a *local diffeomorphism*) if it is a diffeomerphism of a certain neighborhood U of this point onto the neighborhood $V = fU$ of the point $q = f(p)$.

Of course, any diffeomorphism $\mathcal{X} \to \mathcal{Y}$ is an étale mapping (at each point $p \in \mathcal{X}$). Conversely, according to the calculus theorem on the differentiability of the inverse function, *each étale bijective mapping* $f: \mathcal{X} \to \mathcal{Y}$ *is a diffeomorphism*; moreover, the bijectivity condition here is necessary in general (i.e., it is not implied by the étality) as simple examples show (see Remark 31.1).

The following assertion is known as the *étale mapping theorem.*

Proposition 31.1. *A smooth mapping* $f: \mathcal{X} \to \mathcal{Y}$ *is étale at a point* $p \in \mathcal{X}$ *iff its differential*

$$df_p: \mathbf{T}_p\mathcal{X} \to \mathbf{T}_q\mathcal{Y}, \quad q = f(p),$$

is an isomorphism.

If a mapping $f: \mathcal{X} \to \mathcal{Y}$ that is étale at a point p is a diffeomorphism on the support U of a chart (U, h), then the pair $(V, k) = (fU, h \circ (f|_U)^{-1})$ is obviously a chart in $\mathcal{Y}$. Moreover, the mapping $\varphi: h(U) \to k(V)$ corresponding to the mapping f is the identity mapping, i.e., in the corresponding local coordinates $x^1, \ldots, x^n$ and $y^1, \ldots, y^n$, the mapping f is given by

$$y^i = x^i, \quad i = 1, \ldots, n. \tag{12}$$

Therefore, *if a mapping* $f: \mathcal{X} \to \mathcal{Y}$ *is étale at a point* $p \in \mathcal{X}$, *then there are local coordinates* $x^1, \ldots, x^n$ *and* $y^1, \ldots, y^n$ *in the manifolds* $\mathcal{X}$ *and* $\mathcal{Y}$ *in which the mapping* f *is written by formulas* (12) (i.e., it is a mapping defined by the equality of coordinates).

Proposition 31.1 is, in essence, only a restatement of the inverse mapping theorem. It is interesting that the latter theorem admits a principally different interpretation.

§7. Theorem on a Local Coordinate Change

Let $(U, h) = (U, x^1, \ldots, x^n)$ be an arbitrary chart of a smooth manifold $\mathcal{X}$ centered at a point $p_0 \in \mathcal{X}$ (i.e., such that $p_0 \in U$), and let

$$x^{i'} = x^{i'}(x^1, \ldots, x^n), \quad i' = 1', \ldots, n', \tag{13}$$

be smooth functions given in a certain neighborhood $U' \subset U$ of p_0.

We recall that the symbol $x^{i'}$ in (13) has a twofold meaning: in the left-hand side, it denotes a function on U'; in the right-hand side, it denotes a function on the neighborhood $h(U')$ of the point $x_0 = h(p_0)$ in the space $\mathbb{R}^n$. In accordance with this, we can consider the Jacobian

$$\det \left\| \frac{\partial x^{i'}}{\partial x} \right\|, \quad i = 1, \dots, n, \quad i' = 1, \dots, n, \tag{14}$$

of functions (13) either as a function on U' or as a function on $h(U')$. Every time, it should be clear from the context which of these two meanings applies.

If functions (13) are local coordinates in the neighborhood U', then at any point of this neighborhood and at the point p_0 in particular, Jacobian (14) is nonzero. The converse statement holds in the following form.

Proposition 31.2. *If Jacobian* (14) *is nonzero at the point* p_0, *then functions* (13) *are local coordinates in a certain neighborhood of this point.*

Proposition 31.2 is known as the *theorem on a local coordinate change.*

§8. Locally Flat Mappings

We now return to arbitrary smooth mappings $f\colon \mathcal{X} \to \mathcal{Y}$, where $\dim \mathcal{X} = n$ and $\dim \mathcal{Y} = m$.

Definition 31.6. The *rank* of a smooth mapping $f\colon \mathcal{X} \to \mathcal{Y}$ at a point $p \in \mathcal{X}$ is the rank r of the linear mapping $df_p\colon \mathbf{T}_p\mathcal{X} \to \mathbf{T}_q\mathcal{Y}$, $q = f(p)$, i.e., the rank of the Jacobi matrix of the mapping f at the point p.

It is clear that

$$0 \le r \le \min(n, m).$$

Because the rank of a matrix can only increase under a small variation of its entries, the rank of the mapping f at an arbitrary point of a sufficiently small neighborhood of p is not less than its rank at the point p. However, it can be strictly greater than the rank of f at p.

Definition 31.7. A mapping $f\colon \mathcal{X} \to \mathcal{Y}$ is said to be *locally flat* at a point $p \in \mathcal{X}$ if there exists a neighborhood U of p where the rank of f is constant (equals the rank r at the point p).

Proposition 31.3. *If the mapping* $f\colon \mathcal{X} \to \mathcal{Y}$ *is locally flat at a point* $p \in \mathcal{X}$, *then there exist charts* $(U, x^1, \dots, x^n)$ *and* $(V, y^1, \dots, y^m)$ *in the respective manifolds* $\mathcal{X}$ *and* $\mathcal{Y}$ *such that* $p \in U$, $fU \subset V$, *and the mapping* f *is written in the local coordinates* $x^1, \dots, x^n$ *and* $y^1, \dots, y^m$ *as*

$$y^i = \begin{cases} x^j & \text{if } j = 1, \dots, r, \\ 0 & \text{if } j = r+1, \dots, m. \end{cases} \tag{15}$$

(We note that the coordinates $x^{r+1}, \dots, x^n$ do not enter formula (15).)

Proposition 31.3 clearly asserts that near the point p, a locally flat mapping at p looks like the projection $\mathbb{R}^n \to \mathbb{R}^r \subset \mathbb{R}^m$ of the space $\mathbb{R}^n$ along the last $n-r$ coordinate axes on the coordinate subspace $\mathbb{R}^r$ of the space $\mathbb{R}^m$ consisting of points whose last $m-r$ coordinates are equal to zero.

§9. Immersions and Submersions

Definition 31.8. A smooth mapping $f: \mathcal{X} \to \mathcal{Y}$ of a smooth n-dimensional manifold $\mathcal{X}$ into a smooth m-dimensional manifold $\mathcal{Y}$ is called an *immersion* at a point if its rank at this point equals n (of course, this is possible only for $n \leq m$), i.e., the mapping

$$df_p: \mathsf{T}_p\mathcal{X} \to \mathsf{T}_q\mathcal{Y}, \quad q = f(p), \tag{16}$$

is a monomorphism.

Similarly, a mapping $f: \mathcal{X} \to \mathcal{Y}$ is called a *submersion* at a point $p \in \mathcal{X}$ if its rank at this point equals m (and therefore $n \geq m$), i.e., if mapping (16) is an epimorphism.

Therefore, a mapping f is an immersion or submersion if its rank assumes the maximum possible value (for given n and m). Therefore, immersions and submersions are also called *mappings of maximal rank.*

According to Proposition 31.2, a mapping is simultaneously a submersion and immersion (for $n = m$) iff it is étale.

A mapping $f: \mathcal{X} \to \mathcal{Y}$ that is an immersion or a submersion at each point $p \in \mathcal{X}$ is merely called a respective *immersion* or *submersion.*

It is clear that submersions and immersions, being mappings of maximal rank, are locally flat at p. Therefore, according to Proposition 31.2, *for any submersion $f: \mathcal{X} \to \mathcal{Y}$ at a point p, there exist charts $(U, x^1, \dots, x^n)$ and $(V, y^1, \dots, y^m)$ of the respective manifolds $\mathcal{X}$ and $\mathcal{Y}$ such that $p \in U$, $fU \subset V$, and the mapping f is written in the coordinates $x^1, \dots, x^n$ and $y^1, \dots, y^m$ as*

$$y^1 = x^1, \quad \dots, \quad y^m = x^m. \tag{17}$$

This clearly means that in the corresponding coordinates, any submersion is locally represented by the projection $\mathbb{R}^n \to \mathbb{R}^m$ transforming a point $(x^1, \dots, x^m, \dots, x^n) \in \mathbb{R}^n$ into the point $(x^1, \dots, x^m) \in \mathbb{R}^m$.

For immersions, we interchange the designations and assume that f is a mapping $\mathcal{Y} \to \mathcal{X}$. Then *for any immersion $f: \mathcal{Y} \to \mathcal{X}$ at a point $q \in \mathcal{Y}$, there exist charts $(V, y^1, \dots, y^m)$ and $(U, x^1, \dots, x^n)$ of the respective manifolds $\mathcal{Y}$ and $\mathcal{X}$ such that $q \in V$, $fV \in U$, and the mapping f is written in the coordinates $y^1, \dots, y^m$ and $x^1, \dots, x^n$ as*

$$x^1 = y^1, \quad \dots, \quad x^m = y^m, \quad x^{m+1} = 0, \quad \dots, \quad x^n = 0. \tag{18}$$

This clearly means that in the corresponding coordinates, any immersion is locally represented by an embedding $\mathbb{R}^m \to \mathbb{R}^n$ that transforms a point $(y^1, \dots, y^m) \in \mathbb{R}^m$ into the point $(y^1, \dots, y^m, 0, \dots, 0) \in \mathbb{R}^n$.

Chapter 32
Submanifolds of a Smooth Manifold

§1. Submanifolds of a Smooth Manifold

Definition 32.1. A smooth manifold $\mathcal{Y}$ is called a *submanifold* of a smooth manifold $\mathcal{X}$ if it is contained in $\mathcal{X}$ and the corresponding embedding mapping

$$\imath\colon \mathcal{Y} \to \mathcal{X}, \quad \imath(p) = p, \tag{1}$$

is an immersion at each point $p \in \mathcal{Y}$ (and is a smooth mapping in particular).

Because $\mathcal{Y} \subset \mathcal{X}$, we have the topology $\mathbb{T}_{\mathcal{Y}/\mathcal{X}}$ on $\mathcal{Y}$ that is induced by the topology $\mathbb{T}_{\mathcal{X}}$ of the manifold $\mathcal{X}$. Generally speaking (see below), this topology *differs* from the topology $\mathbb{T}_{\mathcal{Y}}$ of the manifold $\mathcal{Y}$. We can only assert that $\mathbb{T}_{\mathcal{X}/\mathcal{Y}} \subset \mathbb{T}_{\mathcal{Y}}$ (this inclusion is equivalent to the continuity of the mapping $\imath$).

Definition 32.2. If $\imath$ is a homeomorphism onto its image (i.e., if $\mathbb{T}_{\mathcal{X}/\mathcal{Y}} = \mathbb{T}_{\mathcal{Y}}$), the manifold $\mathcal{Y}$ is called an *embedded submanifold.* Otherwise, $\mathcal{Y}$ is called an *immersed submanifold.* (The latter term is also used as a synonym for the term "submanifold" when it must be stressed that a submanifold $\mathcal{Y}$ is not an embedded submanifold.)

A submanifold $\mathcal{Y}$ of the manifold $\mathcal{X}$ is said to be *conservative* if for any smooth submanifold $\mathcal{Z}$, each mapping $\varphi\colon \mathcal{Z} \to \mathcal{Y}$ is smooth iff it is smooth as a mapping into $\mathcal{X}$ (i.e., the mapping $\imath \circ \varphi\colon \mathcal{Z} \to \mathcal{X}$, where $\imath\colon \mathcal{Y} \to \mathcal{X}$ is an embedding, is smooth). It is clear that *each embedded manifold is conservative,* but immersed conservative manifolds also exist.

Proposition 32.1. *If a submanifold $\mathcal{Y}$ is connected and the manifold $\mathcal{X}$ satisfies the second countability axiom, then the submanifold $\mathcal{Y}$ also satisfies this axiom.*

Of course, this proposition, which was probably first proved by Chevalley, refers to the case of an immersed submanifold Y because it is trivial for embedded submanifolds.

Immersion (1) is certainly an injective mapping. Conversely, let $f\colon \mathcal{X} \to \mathcal{Y}$ be an arbitrary immersion that is an injective mapping, and let $f = \imath \circ f'$ be its decomposition into the composition of a bijective mapping $f'\colon \mathcal{Y} \to f(\mathcal{Y})$ and an embedding $\imath\colon f(\mathcal{Y}) \to \mathcal{X}$. Because the mapping f' is bijective, we can transport the smooth structure from $\mathcal{Y}$ to $\mathcal{Y}' = f(\mathcal{Y})$ via this mapping. Then $\mathcal{Y}'$ is a smooth manifold, f' is a diffeomorphic mapping, and $\imath = f \circ (f')^{-1}$ is an immersion (because it is a composition of an immersion and a diffeomorphism), i.e., $\mathcal{Y}$ is a submanifold of the manifold $\mathcal{X}$.

Therefore, submanifolds of a manifold X are exactly the images in $\mathcal{X}$ of arbitrary immersions $\mathcal{Y} \to \mathcal{X}$ that are injective mappings. Moreover, embedded manifolds are the images of immersions that are monomorphisms (homeomorphisms onto their images).

Simple regular arcs and elementary surfaces are examples of embedded submanifolds. The parameterizations are the corresponding immersions. In this case, the regularity condition exactly means that a parameterization is an immersion.

§2. Subspace Tangent to a Submanifold

Because embedding (1) is an immersion for any submanifold $\mathcal{Y} \to \mathcal{X}$, we see that for any point $p \in \mathcal{Y}$, the mapping

$$d\imath_p \colon \mathbf{T}_p\mathcal{Y} \to \mathbf{T}_p\mathcal{X}$$

is an isomorphism of the linear space $\mathbf{T}_p\mathcal{Y}$ onto the linear subspace $\operatorname{Im} d\imath_p = (d\imath_p)(\mathbf{T}_p\mathcal{Y})$ of the space $\mathbf{T}_p\mathcal{X}$. This subspace is called the *tangent space of the submanifold $\mathcal{Y}$ at the point p*. It is usually identified with $\mathbf{T}_p\mathcal{Y}$ (via the isomorphism $d\imath_p$).

A certain specific feature arises in the case where $\mathcal{X}$ is an affine space $\mathcal{A}$. Then, for any point $p \in \mathcal{X}$ (and for any point $p \in \mathcal{Y}$ in particular), the linear space $\mathbf{T}_p\mathcal{X}$ is identified with the associated linear space $\mathcal{V}$. In this case, it is usual to identify the subspace $\mathbf{T}_pY$ with the linear subvariety $p + \mathbf{T}_p\mathcal{Y}$ of the affine space $\mathcal{A}$ (i.e., in a clear interpretation, to consider its vectors as vectors taken at the point p).

§3. Local Representation of a Submanifold

Being an immersion, mapping (1) can be written in local coordinates by formulas (18) in Chap. 31. This means that for any point p of the manifold $\mathcal{Y}$, there exists a chart $(U, x^1, \ldots, x^n)$, $p \in U$, of the manifold $\mathcal{X}$ such that, first, on a certain set $V \subset U \cap \mathcal{Y}$ (open in $\mathcal{Y}$), the restrictions

$$y^1 = x^1|_V, \quad \ldots, \quad y^m = x^m|_V$$

of the first m coordinates $x^1, \ldots, x^m$ are local coordinates on V and, second, a point $q \in U$ belongs to V iff

$$x^{m+1}(q) = 0, \quad \ldots, \quad x^n(q) = 0. \tag{2}$$

Such local coordinates $x^1, \ldots, x^m$ are said to be *compatible* with the submanifold $\mathcal{Y}$.

The coordinates $y^1, \dots, y^m$ define the basis

$$\left(\frac{\partial}{\partial y^1}\right)_p, \quad \dots, \quad \left(\frac{\partial}{\partial y^m}\right)_p, \tag{3}$$

in $\mathbf{T}_p\mathcal{Y}$, and the coordinates $x^1, \dots, x^n$ define the basis

$$\left(\frac{\partial}{\partial x^1}\right)_p, \quad \dots, \quad \left(\frac{\partial}{\partial x^n}\right)_p \tag{4}$$

in $\mathbf{T}_p\mathcal{X}$. Moreover, because the embedding $\imath\colon \mathcal{Y} \to \mathcal{X}$ is written in these coordinates by the functions $y^1 = x^1, \dots, y^m = x^m$, its differential $d\imath_p$ transforms basis (3) into the first m vectors of basis (4). By the identification of $\mathbf{T}_p\mathcal{Y}$ with $\operatorname{Im} d\imath_p$, this means that

$$\left(\frac{\partial}{\partial y^1}\right)_p = \left(\frac{\partial}{\partial x^1}\right)_p, \quad \dots, \quad \left(\frac{\partial}{\partial y^m}\right)_p = \left(\frac{\partial}{\partial x^m}\right)_p,$$

i.e., *the vectors* $(\partial/\partial x^1)_p, \dots, (\partial/\partial x^m)_p$ *form a basis of the subspace* $\mathbf{T}_p\mathcal{Y}$ *of the space* $\mathbf{T}_p\mathcal{X}$.

If the submanifold $\mathcal{Y}$ is embedded in $\mathcal{X}$, then we can assume without loss of generality that

$$V = U \cap \mathcal{Y}. \tag{5}$$

Indeed, because V is open in $\mathcal{Y}$ and the topology in $\mathcal{Y}$ is induced by the topology of $\mathcal{X}$, there exists an open set W in $\mathcal{X}$ such that $V = W \cap \mathcal{Y}$. Replacing U by $U \cap V$, we can assume without loss of generality that (5) holds.

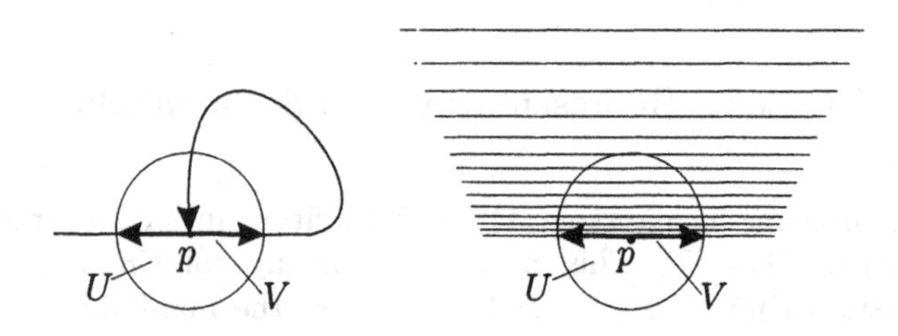

Fig. 32.1. Two typical cases where the relation $V = U \cap \mathcal{Y}$ is impossible.

Relation (5) means that a point $q \in U$ belongs to $\mathcal{Y}$ iff relations (2) hold for it. In other words, *locally* (i.e., in the neighborhood U), the submanifold $\mathcal{Y}$ is given by $n-m$ equations

$$x^{m+1} = 0, \quad \dots, \quad x^n = 0. \tag{6}$$

In the general case for a nonembedded (immersed) submanifold $\mathcal{Y}$, it is generally impossible to attain the fulfillment of (6), and the relation between V and $U \cap \mathcal{Y}$ becomes more complicated.

Because the functions $x^{m+1}, \ldots, x^n$ are continuous and conditions (2) therefore define a closed subset in U, the set V is closed in U (with respect to the topology induced in U by the topology $\mathbb{T}_{\mathcal{Y}/\mathcal{X}}$). Therefore, V is also closed in $U \cap \mathcal{Y}$ (with respect to the topology induced by $\mathbb{T}_{\mathcal{Y}/\mathcal{X}}$ and therefore by the topology induced by the topology $\mathbb{T}_{\mathcal{Y}}$). On the other hand, the set V is open in $\mathcal{Y}$ and therefore in $U \cap \mathcal{Y}$ (with respect to the topology induced by $\mathbb{T}_{\mathcal{Y}}$). Therefore, the set V is open and closed in $U \cap \mathcal{Y}$ (with respect to the topology induced by the topology $\mathbb{T}_{\mathcal{Y}}$). Therefore, *if the coordinate neighborhood V is connected, then it is a connected component of the set $U \cap \mathcal{Y}$ that contains the point p.* (We recall that each point of a smooth manifold has a fundamental system of connected coordinate neighborhoods, for example, neighborhoods that are homeomorphic to open balls of the Euclidean space of the corresponding dimension.)

§4. Uniqueness of a Submanifold Structure

We now suppose that on a subset $\mathcal{Y}$ of a smooth manifold $\mathcal{X}$, we have two smooth structures with respect to which $\mathcal{Y}$ is a submanifold of $\mathcal{X}$ and which define *the same topology* on $\mathcal{Y}$. Then, for both these smooth structures, charts of the form $(V, y^1, \ldots, y^m)$ having the above properties are the same (because the functions $y^1, \ldots, y^m$ are characterized as the restriction of the local coordinates $x^1, \ldots, x^m$ and the sets V are characterized as connected components of the sets $U \cap \mathcal{Y}$). On the other hand, in each of these smooth structures, the charts obviously compose a subatlas of the maximal atlas. Therefore, these smooth structures coincide. Therefore, *for a given topology on a set $\mathcal{Y}$ of a smooth manifold $\mathcal{X}$, we have not more than one smooth manifold structure on $\mathcal{Y}$ with respect to which $\mathcal{Y}$ becomes a submanifold of $\mathcal{X}$.*

Of course, varying the topology, we can obtain many distinct submanifold structures on $\mathcal{Y}$. For example, any subset $\mathcal{Y} \subset \mathcal{X}$ can be equipped with the discrete topology and can be thus transformed into a zero-dimensional submanifold.

§5. Case of Embedded Submanifolds

We obtain a more interesting example by considering the set depicted in Fig. 32.2(A) ("figure-eight curve"). It cannot be a submanifold in the induced topology, because of a singular point at the center. However, in a weaker topology that is conditionally depicted in Fig. 32.2(B), it is an immersed submanifold, which is diffeomorphic to the real line $\mathbb{R}$; in another topology that is conditionally depicted in Fig. 32.2(C), it is also an immersed submanifold diffeomorphic to $\mathbb{R}$. The corresponding immersions are nonequivalent regular curves without double points having the same support.

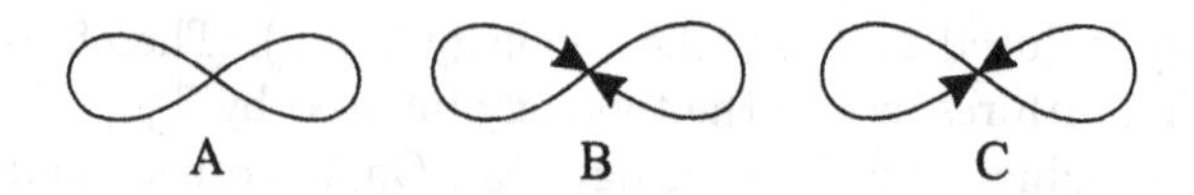

Fig. 32.2.

Considering the topology on $\mathcal{Y}$ induced by the topology of the manifold $\mathcal{X}$, we find in particular that *on a subset* $\mathcal{Y}$ *of a smooth manifold* $\mathcal{X}$, *there cannot be more than one smooth manifold structure with respect to which* $\mathcal{Y}$ *is an embedded manifold.* Therefore, it is legitimate to say that the *subset* $\mathcal{Y}$ is an embedded submanifold.

§6. Tangent Space of a Direct Product

Let $\mathcal{X}$ and $\mathcal{Y}$ be two smooth manifolds, and let $\mathcal{X} \times \mathbf{T}$ be their direct product. Then, for any point $(p_0, q_0) \in \mathcal{X} \times \mathcal{Y}$, the formulas

$$i_{q_0}(p) = (p, q_0), \qquad j_{p_0}(q) = (p_0, q_0), \quad p \in \mathcal{X}, \quad q \in \mathcal{Y},$$

define two smooth mappings

$$i_{q_0}: \mathcal{X} \to \mathcal{X} \times \mathcal{Y} \qquad \text{and} \qquad j_{p_0}: \mathcal{Y} \to \mathcal{X} \times \mathcal{Y} \tag{7}$$

that are related to the projections

$$\pi_1: \mathcal{X} \times \mathcal{Y} \to \mathcal{X} \qquad \text{and} \qquad \pi_2: \mathcal{X} \times \mathcal{Y}$$

by the relations

$$\pi \circ i_{q_0} = \mathrm{id} \qquad \text{and} \qquad \pi \circ j_{p_0} = \mathrm{id}.$$

These mappings are injective monomorphic mappings and are immersions. Their images $i_{q_0}\mathcal{X} = \{(p, q_0) \in X \times \mathcal{Y};\ p \in X\}$ and $j_{p_0}\mathcal{X} = \{(p_0, q) \in X \times \mathcal{Y};\ q \in Y\}$ are embedded submanifolds of the manifold $X \times \mathcal{Y}$ that are diffeomorphic to the respective manifolds $\mathcal{X}$ and $\mathcal{Y}$. The differentials

$$\begin{aligned} (di_{q_0})_{p_0}&: \mathbf{T}_{p_0}\mathcal{X} \to \mathbf{T}_{(p_0,q_0)}(\mathcal{X} \times \mathcal{Y}), \\ (dj_{p_0})_{q_0}&: \mathbf{T}_{q_0}\mathcal{Y} \to \mathbf{T}_{(p_0,q_0)}(\mathcal{X} \times \mathcal{Y}) \end{aligned} \tag{8}$$

of mappings (7) are monomorphisms, and we assume that the linear spaces $\mathbf{T}_{p_0}\mathcal{X}$ and $\mathbf{T}_{q_0}\mathcal{Y}$ are embedded in the linear space $\mathbf{T}_{(p_0,q_0)}(\mathcal{X} \times \mathcal{Y})$ by these monomorphisms.

Lemma 32.1. *The linear space* $\mathsf{T}_{(p_0,q_0)}(\mathcal{X}\times\mathcal{Y})$ *is a direct sum of the linear spaces* $\mathsf{T}_{p_0}\mathcal{X}$ *and* $\mathsf{T}_{q_0}\mathcal{Y}$,

$$\mathsf{T}_{(p_0,q_0)}(\mathcal{X}\times\mathcal{Y}) = \mathsf{T}_{p_0}\mathcal{X} \oplus \mathsf{T}_{q_0}\mathcal{Y}.$$

For any vectors $A \in \mathsf{T}_{p_0}\mathcal{X}$ and $B \in \mathsf{T}_{q_0}\mathcal{Y}$, the vector

$$(di_{q_0})_{p_0}A + (dj_{q_0})_{q_0}B$$

is denoted by (A, B).

§7. Theorem on the Inverse Image of a Regular Value

Definition 32.3. A point $q_0 \in \mathcal{Z}$ is called a *regular value* of a mapping $f\colon \mathcal{X} \to \mathcal{Z}$ if f is a submersion at each point $p \in f_{q_0}^{-1}$.

Proposition 32.2 (theorem on the inverse image of a regular value). *The inverse image* $Y = f^{-1}(q_0)$ *of an arbitrary regular value* q_0 *is (if it is not empty) an embedded submanifold of the manifold* $\mathcal{X}$. *The dimension of this submanifold is* $n - r$, *where* $n = \dim\mathcal{X}$ *and* $r = \dim\mathcal{Z}$.

For any point $p \in \mathcal{Y}$, there exists a chart $(U, x^1, \ldots, x^n) = (U, h)$ in the manifold $\mathcal{X}$ and a chart $(W, z^1, \ldots, z^r)$ in the manifold $\mathcal{Z}$ such that $p \in U$, $fU \subset W$, and our mapping is written in these charts by

$$z^1 = x^{m+1}, \quad \ldots, \quad z^r = x^n, \tag{9}$$

where $m = n-r$. In this case, we can assume without loss of generality that all the coordinates $z^1, \ldots, z^r$ are equal to zero at the point q_0 and a point $u \in U$ therefore belongs to the set $\mathcal{Y}$ (more precisely, to the intersection $V = U \cap \mathcal{Y}$) iff

$$x^{m+1}(u) = 0, \quad \ldots, \quad x^n(u) = 0.$$

Formulas (9), which assign the mapping f in local coordinates, directly imply that the differential

$$df_p\colon \mathsf{T}_p\mathcal{X} \to \mathsf{T}_{q_0}\mathcal{Z}$$

of this mapping at the point $p \in \mathcal{Y}$ acts on the basis vectors of the space $\mathsf{T}_p\mathcal{X}$ by

$$df_p\left(\frac{\partial}{\partial x^i}\right)_p = \begin{cases} 0 & \text{if } i = 1, \ldots, m, \\ \left(\dfrac{\partial}{\partial z^{i-m}}\right)_{q_0} & \text{if } i = m+1, \ldots, n. \end{cases} \tag{10}$$

Therefore, *the vectors*

$$\left(\frac{\partial}{\partial x^1}\right)_p, \quad \ldots, \quad \left(\frac{\partial}{\partial x^n}\right)_p \tag{11}$$

generate the kernel $\operatorname{Ker} df_p$ *of the mapping* df_p*, and the vectors*

$$\left(\frac{\partial}{\partial x^{m+1}}\right)_p, \quad \dots \quad \left(\frac{\partial}{\partial x^n}\right)_p \tag{12}$$

generate a subspace that is isomorphically mapped onto the tangent space $\mathbf{T}_{q_0}\mathcal{Z}$ *of the manifold* $\mathcal{Z}$. Therefore,

$$\mathbf{T}_p\mathcal{X} = \operatorname{Ker} df_p \oplus \widehat{\mathbf{T}}_{q_0}\mathcal{Z},$$

where $\widehat{\mathbf{T}}_{q_0}\mathcal{Z}$ is the subspace spanned by vectors (12).

On the other hand, because $y^1 = x^1|_V, \dots, y^m = x^m|_V$, vectors (11) generate the subspace $\mathbf{T}_p\mathcal{Y}$ of the space $\mathbf{T}_p\mathcal{X}$. Therefore, *for any point* $p \in \mathcal{Y}$, *the kernel* $\operatorname{Ker} df_p$ *of the mapping* df_p *coincides with the tangent space* $\mathbf{T}_p\mathcal{Y}$ *of the submanifold* $\mathcal{Y}$,

$$\mathbf{T}_p\mathcal{Y} = \operatorname{Ker} df_p, \tag{13}$$

and therefore

$$\mathbf{T}_p\mathcal{X} = \mathbf{T}_p\mathcal{Y} \oplus \widehat{\mathbf{T}}_{q_0}\mathcal{Z}. \tag{14}$$

We stress that this decomposition holds for any point $p \in \mathcal{Y}$.

Remark 32.1. It should be kept in mind that not every embedded submanifold $\mathcal{Y} \subset \mathcal{X}$ is the inverse image of a regular value under a certain mapping $f\colon \mathcal{X} \to \mathcal{Z}$.

According to (14), if an embedded submanifold $\mathcal{Y} \in \mathcal{X}$ is the inverse image of a regular value, then a decomposition of the form

$$\mathbf{T}_p\mathcal{X} = \mathbf{T}_p\mathcal{Y} \oplus N_p$$

should hold for any point $p \in \mathcal{Z}$, where N_p is a subspace such that there is an isomorphism $N_p \to \mathbb{R}^{n-m}$ smoothly depending on the point p (in an obvious sense) and given *for all* p. It can be proved that this condition is also sufficient.

§8. Solution of Sets of Equations

Let $f^1, \dots, f^r$ be smooth functions on a smooth manifold $\mathcal{X}$, and let $a^1, \dots, a^r$ be real numbers. A point $p \in \mathcal{X}$ is called a solution of the set of equations

$$f^1 = a^1, \quad \dots, \quad f^r = a^r \tag{15}$$

if $f^i(p) = a^i$ for each $i = 1, \dots, r$. Let $\mathcal{Y}$ be the set of solutions of system (15).

We say that Eqs. (15) are *functionally independent* if the covectors

$$df_p^1, \quad \dots, \quad df_p^r$$

are linearly independent for any point $p \in \mathcal{Y}$ and the equations

$$df_p^1 = o, \quad \dots, \quad df_p^r = 0 \tag{16}$$

therefore define a linear subspace of dimension $m = n - r$ in $\mathbf{T}_p\mathcal{X}$.

Proposition 32.3. *For each system* (15) *of functionally independent equations, the set $\mathcal{Y}$ of its solutions is an embedded submanifold of the manifold $\mathcal{X}$ of dimension $m = n - r$. For each point $p \in \mathcal{Y}$, the subspace $\mathbf{T}_p\mathcal{Y}$ is a subspace of solutions to the system of linear equations* (16).

The first assertion in Proposition 31.3 is a reformulation of Proposition 31.2 for the case where $\mathcal{Z} = \mathbb{R}^n$, and the second one is a restatement of (13).

Corollary 32.1. *Let f be a smooth function on a smooth manifold $\mathcal{X}$, and let $\mathcal{Y}$ be the set of points $p \in \mathcal{X}$ for which $f(p) = a$, where $a \in \mathbb{R}$ is fixed. If $df_p \neq 0$ for each point $p \in \mathcal{Y}$, then $\mathcal{Y}$ is an embedded submanifold of dimension $n - 1$ whose tangent space at each point $y \in Y$ is the hyperplane $df_p = 0$.*

For a function $f = f(x, y)$ on $\mathbb{R}^2$, the condition $df_p \neq 0$ means that either $\partial f/\partial x \neq 0$ or $\partial f/\partial y \neq 0$. Therefore, the condition that a subset on the plane with the equation $f(x, y) = 0$ is an embedded one-dimensional manifold means exactly that it has no singular points.

Remark 32.2. In general, it is not possible to remove the condition $df_p \neq 0$ in Corollary 32.1. However, this condition is not necessary for $\mathcal{Y}$ to be an embedded (not necessarily $(n-1)$-dimensional) submanifold. For example, the equation

$$x^2 + y^2 = 0$$

in $\mathbb{R}^3$ defines a line (embedded submanifold), but the differential df_p of the function $f = x^2 + y^2$ is identically zero at all points of this line.

Moreover, according to Remark 32.1, not every submanifold can be given by a set of functionally independent equations. For example, on the projective plane $\mathbb{R}P^2$, there are no smooth functions whose set of zeros is the projective line $\mathbb{R}P^1 \subset \mathbb{R}P^2$.

Nevertheless, in practice, that one or another subset of a smooth manifold is an embedded manifold is usually proved using Proposition 32.3 or its Corollary 32.1.

§9. Embedding Theorem

The simplest and most easily visualized class of manifolds consists of manifolds that are diffeomorphic to embedded submanifolds of $\mathbb{R}^N$ for a certain $N > 0$ or, as we say for brevity, the manifolds *embeddable* in $\mathbb{R}^N$.

It is clear that any subspace of a Hausdorff topological space satisfying the second countability axiom is also Hausdorff and satisfies this axiom. Therefore, *any manifold embeddable in* $\mathbb{R}^N$ *is Hausdorff and satisfies the second countability axiom.* It turns out that this necessary condition is also sufficient.

Theorem 32.1. *For any Hausdorff smooth manifold* $\mathcal{X}$ *satisfying the second countability axiom, there exists* N *such that the manifold* $\mathcal{X}$ *is embeddable in* $\mathbb{R}^N$.

What can we say about N?

Proposition 32.4. *If a smooth manifold of dimension* n *is embeddable in* $\mathbb{R}^N$ *with* $N > 2n+1$, *then it is also embeddable in* $\mathbb{R}^{N-1}$.

Corollary 32.2 (Whitney embedding theorem). *Any smooth Hausdorff manifold* $\mathcal{X}$ *of dimension* n *satisfying the second countability axiom is embeddable in* $\mathbb{R}^{2n+1}$.

Chapter 33
Vector and Tensor Fields. Differential Forms

§1. Tensor Fields

We recall that a *tensor* S *of type* (a, b), where $a, b \geq 0$, on a linear space $\mathcal{V}$ is a mapping that sets a tuple of n^{a+b} numbers $S^{j_1 \cdots j_b}_{i_1 \cdots i_a}$ in correspondence to an arbitrary basis $e_1, \ldots, e_n$ of the space $\mathcal{V}$ (these numbers $S^{j_1 \cdots j_b}_{i_1 \cdots i_a}$ are called *components* of the tensor S in this basis) such that for any two bases $e_1, \ldots, e_n$ and $e_{1'}, \ldots, e_{n'}$ of the space $\mathcal{V}$, the components $S^{j_1 \cdots jb}_{i_1 \cdots i_a}$ and $S^{j'_1 \cdots j'_b}_{i_1 \cdots i_a}$ are related by the formula

$$Sj'_1 \cdots j_{b i'_1 \cdots i'_a} = c^{i_1}_{i'_1} \cdots c^{i_a}_{i'_a} c^{i'_1}_{j_1} \cdots c^{i'_b}_{j_b} S^{j_1 \cdots j_b}_{i_1 \cdots i_a},$$

where $c^{i'}_i$ and $c^i_{i'}$ are entries of mutually inverse transition matrices, i.e., the numbers such that $e_{i'} = c^i_{i'} e_i$ and $e_i = c^{i'}_i e_{i'}$. Any tensor correctly defines the multilinear functional

$$S(x_1, \ldots, x_a, \xi^1, \ldots, \xi^b) = S^{j_1 \cdots j_b}_{i_1 \cdots i_a} x^{i_1}_1 \cdots x^{i_a}_a \xi^1_{j_1} \cdots \xi^b_{j_b}$$

of a vector and b covector arguments and is identified with this functional as a rule.

With respect to the operations of addition and multiplication by numbers, which are defined in a natural way, all tensors of a given type (a, b) form a linear space $\mathsf{T}^b_a(\mathcal{V})$. For any two tensors S and R of the respective types (a, b) and (c, d), the formula

$$(S \otimes R)^{j_1 \cdots j_{a+b}}_{i_1 \cdots i_{a+c}} = S^{j_1 \cdots j_b}_{i_1 \cdots i_a} R^{j_{b+1} \cdots j_{a+b}}_{i_{a+1} \cdots i_{a+c}}$$

defines the tensor $S \otimes R$ of type $(a + c, b + d)$, which is called the *tensor product* of the tensors S and R. This product is associative and distributive with respect to addition. Moreover, there is a special *contraction* operation for tensors.

Any vector is naturally interpreted as a tensor of type (0,1), and any covector is naturally interpreted as a tensor of type (1,0). Therefore, for any basis $e_1, \ldots, e_n$ of the space $\mathcal{V}$ and for any subscripts $i_1, \ldots, i_a$ and superscripts $j_1, \ldots, j_b$ that vary from 1 to n, the tensor

$$e^{i_1} \otimes \cdots \otimes e^{i_a} \otimes e_{j_1} \otimes \cdots \otimes e_{j_b},$$

where $e^{i_1} \otimes \cdots \otimes e^{i_a}$ are the dual basis vectors of the dual space $\mathcal{V}^*$, is well defined. All such tensors form a basis of the space $\mathsf{T}^b_a(\mathcal{V})$, and in this case, exactly the components of a tensor serve as its components in this basis, i.e.,

$$S = S^{j_1 \cdots j_b}_{i_1 \cdots i_a} e^{j_1} \otimes \cdots \otimes e^{i_a} \otimes e_{j_1} \cdots \otimes e_{j_b}$$

for any tensor S.

We now apply these general concepts of linear algebra to the case where $\mathcal{V}$ is the tangent space $\mathbf{T}_p\mathcal{X}$ of a smooth manifold $\mathcal{X}$ at a point p. Let $(U, h) = (U, x^1, \dots, x^n)$ be an arbitrary chart of $\mathcal{X}$ containing p. This chart defines the basis

$$\left(\frac{\partial}{\partial x^1}\right)_p, \quad \dots, \quad \left(\frac{\partial}{\partial x^n}\right)_p \tag{1}$$

in the space $\mathbf{T}_p\mathcal{X}$ and the dual basis

$$(dx^1)_p, \quad \dots, \quad (dx^n)_p$$

in the dual space $\mathbf{T}_p^*\mathcal{X}$. Therefore, for each tensor S_p of type (a, b) on the space $\mathbf{T}_p\mathcal{X}$, we have the representation

$$S_p = S_{i_1\cdots i_a}^{j_1\cdots j_b}(dx^{i_1})_p \otimes \cdots \otimes (dx^{i_a})_p \otimes \left(\frac{\partial}{\partial x^{j_1}}\right)_p \otimes \left(\frac{\partial}{\partial x^{j_b}}\right)_p, \tag{2}$$

whose components $S_{i_1\cdots i_a}^{j_1\cdots j_b}$ (i.e., the components of the tensor S_p in basis (1)) are called the *components of the tensor* S_p *in the chart* (U, h). (We omit the subscript p in the notation of these components to avoid typographic complexity.)

Any other chart $(U', h') = (U', x^{i'_1}, \dots, x^{i'_n})$ (with $p \in U'$) defines a basis

$$\left(\frac{\partial}{\partial x^{1'}}\right)_p, \quad \dots, \quad \left(\frac{\partial}{\partial x^{n'}}\right)_p \tag{3}$$

of the space $\mathbf{T}_p\mathcal{X}$, which is related to basis (1) by the transition matrix

$$\left\| \left(\frac{\partial x^{i'}}{\partial x^i}\right)_p \right\|, \quad i, i' = 1, \dots, n. \tag{4}$$

Therefore, the components of the tensor S_p in the charts (U, h) and (U', h') are related by

$$S_{i'_1\cdots i'_a}^{j'_1\cdots j'_b} = \left(\frac{\partial x^{i_1}}{\partial x^{i'_1}}\right)_p \cdots \left(\frac{\partial x^{i_a}}{\partial x^{i'_a}}\right)_p \left(\frac{\partial x^{j'_1}}{\partial x^{j_1}}\right)_p \cdots \left(\frac{\partial x^{j'_b}}{\partial x^{j_b}}\right)_p S_{i_1\cdots i_a}^{j_1\cdots j_b}. \tag{5}$$

If a tensor S_p is given for any point $p \in \mathcal{X}$, then the components $S_{i_1\cdots i_a}^{j_1\cdots j_b}$ in representation (2) are functions of p. If these functions are smooth, i.e., if they are smoothly expressed through the coordinates $x^1, \dots, x^n$ in the chart (U, h), then the correspondence $p \mapsto S_p$ is called a (*smooth*) *tensor field* (or simply, a *tensor*) *of type* (a, b) *on the manifold* $\mathcal{X}$. Correspondence (5) for the *functions* $S_{i_1\cdots i_a}^{j_1\cdots j_b}$ has the form

$$S_{i'_1\cdots i'_a}^{j'_1\cdots j'_b} = \frac{\partial x^{i_1}}{\partial x^{i'_1}} \cdots \frac{\partial x^{i_a}}{\partial x^{i'_a}} \frac{\partial x^{j'_1}}{\partial x^{j_1}} \cdots \frac{\partial x^{j'_b}}{\partial x^{j_b}} S_{i_1\cdots i_a}^{j_1\cdots j_b}, \tag{6}$$

which implies that the smoothness condition for a tensor field does not depend on the choice of a chart.

Remark 33.1. For manifolds of a finite class of smoothness C^r, $r \geq 1$, a characteristic difficulty arises here because, in general, the entries of matrix (4) are functions of only class C^{r-1}. Therefore, the smoothness of tensor fields should be understood only in the C^{r-1} sense. To avoid this stipulation, we agree to restrict ourselves to only manifolds of class C^∞ (or C^ω), where no such difficulties arise.

For any open covering $\{U_\alpha\}$ of a manifold $\mathcal{X}$, each tensor field S defines the family of fields

$$S_\alpha = S|_{U_\alpha}$$

such that for any subscripts α and β, we have

$$S_\alpha = S_\beta \quad \text{on } U_\alpha \cap U_\beta. \tag{7}$$

Conversely, for given fields S_α on U_α satisfying relation (7) (we say that such fields are *compatible on intersections*), the formula

$$S_p = (S_\alpha)_p \quad \text{if } p \in U_\alpha$$

correctly defines a tensor field S on $\mathcal{X}$ with the property

$$S|_{U_\alpha} = S_\alpha$$

for any α (this field is therefore smooth). We say that the fields S_α *compose* the field S.

Remark 33.2. A tensor field on a manifold can be considered a correspondence that sets a tuple $S^{j_1 \cdots j_b}_{i_1 \cdots i_a}$ on U in correspondence to each chart (U, h) of a manifold $\mathcal{X}$ and is such that for any two charts (U, h) and (U', h'), relation (6) holds on the intersection $U \cap U'$. This can be accepted as the *definition* of a tensor field. One advantage of this definition is that it can be stated directly after introducting the concept of a smooth manifold without any intermediate definitions; one deficiency is the absence of a formal connection (relating by analogy) with the concept of a tensor on a linear space.

All algebraic operations on tensors (including the contraction operation) are automatically extended to tensor fields. For example, the tensor product $S \otimes R$ of two tensor fields S and R is defined by

$$(S \otimes R)_p = S_p \otimes R_p. \tag{8}$$

Clearly, in this case, we always obtain smooth vector fields from smooth ones.

In particular, we see that *the set* $\mathsf{T}^b_a\mathcal{X}$ *of all tensor fields of type* (a, b) *on a manifold* $\mathcal{X}$ *is a linear space.* This space is infinite dimensional (for $n > 0$). For $(a, b) = (0, 0)$, tensor fields are smooth functions on $\mathcal{X}$, and the linear space $\mathsf{T}^0_0\mathcal{X}$ is the linear space $\mathbf{F}\mathcal{X}$ of all smooth functions on $\mathcal{X}$. The linear space $\mathbf{F}\mathcal{X}$ is an algebra with respect to multiplication of functions; moreover, the formula

$$(fS)_p = f(p)S_p, \quad f \in \mathsf{F}\mathcal{X}, \quad S \in \mathsf{T}_a^b\mathcal{X}$$

(which is a particular case of (8)) defines the operation of multiplication

$$\mathsf{F}\mathcal{X} \times \mathsf{T}_a^b\mathcal{X} \to \mathsf{T}_a^b\mathcal{X}$$

with respect to which *the linear space* $\mathsf{T}_a^b\mathcal{X}$ *is a module over the algebra* $\mathsf{F}\mathcal{X}$.

§2. Vector Fields and Derivations

For $(a, b) = (0, 1)$, tensor fields are called *vector fields.* An example of a vector field on a coordinate neighborhood U (which is considered as a manifold) is given by the field

$$\frac{\partial}{\partial x^i}: p \mapsto \left(\frac{\partial}{\partial x^i}\right)_p, \quad i = 1, \ldots, n. \tag{9}$$

This field is called the ith *coordinate vector field* on U.

For $(a, b) = (1, 0)$, tensor fields are called *covector fields.* An example of a covector field is given by the ith *coordinate covector field*

$$dx: p \mapsto (dx^i)_p \tag{10}$$

on the coordinate neighborhood U.

Formula (2) asserts that each tensor field S on U is uniquely expanded in tensor products of vector and covector coordinate fields:

$$S = S_{i_1 \cdots i_a}^{j_1 \cdots j_b} (dx^{i_1})_p \otimes \cdots \otimes dx^{i_a} \otimes \left(\frac{\partial}{\partial x^{j_1}}\right) \otimes \cdots \otimes \left(\frac{\partial}{\partial x^{j_b}}\right) \quad \text{on } U. \tag{11}$$

In particular, each vector field X on U has the form

$$X = X^i \frac{\partial}{\partial x^i}, \tag{12}$$

and each covector field α has the form

$$\alpha = \alpha_i dx^i, \tag{13}$$

where X^i and α_i, $i = 1, \ldots, n$, are certain smooth functions on U. (Lowercase Greek letters are traditionally used to designate covector fields, and capital Latin letters from the end of the alphabet are used to designate vector fields.)

By definition, the existence of expansion (11) means that *for any coordinate neighborhood* U, *the linear space* $\mathsf{T}_a^b U$ *is a free module over the algebra* $\mathsf{F}U$ *with the basis*

$$(dx^{i_1})_p \otimes \cdots \otimes (dx^{i_a}) \otimes \left(\frac{\partial}{\partial x^{j_1}}\right) \otimes \cdots \otimes \left(\frac{\partial}{\partial x^{j_b}}\right).$$

For arbitrary manifolds $\mathcal{X}$, the module $\mathsf{T}_a^b\mathcal{X}$ is not a free module (over the algebra $\mathbf{F}\mathcal{X}$) in general, and its algebraic structure can be very complicated. Manifolds $\mathcal{X}$ for which all the modules $\mathsf{T}_a^b\mathcal{X}$ are free are said to be *parallelizable.*

As already stated, each vector $A \in \mathbf{T}_p\mathcal{X}$ allows setting a certain number Af, the derivative of f in the direction of the vector A, in correspondence to each function f (that is defined and smooth in a neighborhood of the point p). This implies that for any vector field on the manifold $\mathcal{X}$ and any function $f \in \mathbf{F}\mathcal{X}$, the formula

$$(Xf)(p) = X_p f, \quad p \in \mathcal{X}, \tag{14}$$

defines a certain function Xf on $\mathcal{X}$. The above formulas for Af imply that in an arbitrary chart $(U, h) = (U, x^1, \dots, x^n)$ of the manifold $\mathcal{X}$, the restriction of the function Xf to U is defined by

$$Xf = X^i \frac{\partial f}{\partial x^i} \quad \text{on } U, \tag{15}$$

where X^i, $i = 1, \dots, n$, are components of the vector field X in the chart (U, h). Therefore, the function Xf is smooth on U and hence on the whole $\mathcal{X}$ because U is arbitrary.

Therefore, formula (14) defines a certain (obviously linear) mapping X of the algebra $\mathbf{F}\mathcal{X}$ of smooth functions on $\mathcal{X}$ into itself. This mapping is called the *first-order linear differential operator* on the manifold $\mathcal{X}$ generated by the vector field X. (The motivation for this terminology is formula (15), whose comparison with formula (12) also explains the choice of the notation $\partial/\partial x^i$ for coordinate vector fields.)

Let $\mathcal{A}$ be an arbitrary algebra (not necessary finite-dimensional and associative).

Definition 33.1. A linear mapping

$$D\colon \mathcal{A} \to \mathcal{A}$$

of the algebra $\mathcal{A}$ into itself is called a *derivation* if

$$D(ab) = Da \cdot b + a \cdot Db$$

for any elements $a, b \in \mathcal{A}$.

In particular, derivations of the algebra $\mathbf{F}\mathcal{X}$ (which are merely called *derivations on $\mathcal{X}$*) are linear mappings

$$D\colon \mathbf{F}\mathcal{X} \to \mathbf{F}\mathcal{X}$$

such that

$$D(fg) = Df \cdot g + f \cdot Dg \tag{16}$$

for any two smooth functions f and g on $\mathcal{X}$.

For each derivation D of an arbitrary algebra $\mathcal{A}$ with unity 1, we have the relation

$$D1 = D(1 \cdot 1) = D1 \cdot 1 + 1 \cdot 1 = D1 + D1 = 2D1,$$

and therefore

$$D1 = 0.$$

By the linearity of the mapping D, this implies $D\alpha = 0$ *for any* $\alpha \in \mathbb{R}$. For derivations on a smooth manifold $\mathcal{X}$ (i.e., for derivations of the algebra $\mathsf{F}\mathcal{X}$), this means that *each derivation on* $\mathcal{X}$ *transforms any constant function into zero.*

It is easy to see that the *linear differential operator* $X: \mathsf{F}\mathcal{X} \to \mathsf{F}\mathcal{X}$ *generated by a vector field* X *is a derivation on* X. It turns out that if $\mathcal{X}$ is a Hausdorff manifold, then all the derivations on it are exhausted by the derivations indicated above.

Theorem 33.1. *Each derivation D on a Hausdorff smooth (of class C^∞) manifold $\mathcal{X}$ is generated by a vector field. This field is unique.*

In accordance with Theorem 33.1, a vector field on $\mathcal{X}$ is usually identified with its derivation. (In particular, this a posteriori justifies the notation Xf for the result of applying the derivation generated by the vector field X to the function f.)

The following lemma, which itself is of independent interest, is used in the proof of Theorem 33.1.

Lemma 33.1. *Let X be a Hausdorff smooth manifold, let U be an open submanifold of it, and let f be a smooth function on U. Then for any point $p_0 \in U$, there exist a smooth function f_1 on $\mathcal{X}$ and a neighborhood W of the point p_0 such that $\overline{W} \in U$ and*

$$f = f_1 \quad \text{on } W.$$

Moreover, it can also be assumed that $f_1 = 0$ outside U.

Corollary 33.1. *For each neighborhood U of an arbitrary point p_0 of a smooth Hausdorff manifold $\mathcal{X}$, there exist a neighborhood W of the point p_0 and a smooth function φ such that*

$$\varphi(p) = \begin{cases} 1 & \text{if } p \in W, \\ 0 & \text{if } p \notin W. \end{cases}$$

We note that $\varphi(p_0) = 1$ and $\overline{W} \subset U$.

Now let D be an arbitrary derivation on a manifold $\mathcal{X}$. We say that two smooth functions f and g on $\mathcal{X}$ *are equal near a point* $p_0 \in \mathcal{X}$ if they assume equal values in a certain neighborhood of this point.

Corollary 33.2. *If two functions f and g are equal near a point of $\mathcal{X}$, then the functions Df and Dg are also equal near this point.*

Remark 33.3. Lemma 33.1 also implies that *for any point $p_0 \in U$ and any tensor field S on U, there exist a neighborhood $W \subset U$ and a tensor field S_1 such that*

$$S = S_1 \quad \textit{on } W.$$

To prove this, it suffices to apply Lemma 33.1 to each component of the field S (and take the intersection of the corresponding neighborhoods in W).

The property expressed by Corollary 33.2 is called the *localization property* of the mapping D. This implies the following important statement.

Proposition 33.1. *For any open set $U \subset \mathcal{X}$, there exists a unique derivation*

$$D_U\colon \mathsf{F}U \to \mathsf{F}U$$

that is compatible with the derivation D, i.e.,

$$D_U(f|_U) = Df|_U \tag{17}$$

for any smooth function f on $\mathcal{X}$.

Clearly, this means that in the diagram

$$\begin{array}{ccc} \mathsf{F}\mathcal{X} & \longrightarrow & \mathsf{F}U \\ {\scriptstyle D}\downarrow & & \downarrow{\scriptstyle D_U} \\ \mathsf{F}\mathcal{X} & \longrightarrow & \mathsf{F}U \end{array}$$

whose horizontal mappings are restriction mappings, when moving from the upper left corner to the lower right corner in two possible ways, we obtain the same mapping $\mathsf{F}\mathcal{X} \to \mathsf{F}U$. Diagrams with this property are said to be *commutative*.

§3. Lie Algebra of Vector Fields

It is clear that for any algebra $\mathcal{A}$, the sum of two of its derivations and the product of a derivation and a number are also derivations and *the set* $\operatorname{Der}\mathcal{A}$ *of all derivations of the algebra $\mathcal{A}$ is therefore a linear space* (a subspace of the linear space $\operatorname{End}_{\text{lin}}\mathcal{A}$ of all linear operators $\mathcal{A} \to \mathcal{A}$).

For any two linear operators $D_1, D_2\colon \mathcal{A} \to \mathcal{A}$, the operator

$$[D_1, D_2] = D_1D_2 - D_2D_1$$

is called its *commutator* (or *Lie bracket*). It is easy to see that *for any algebra $\mathcal{A}$, the commutator $[D_1, D_2]$ of any two derivations $D_1, D_2\colon \mathcal{A} \to \mathcal{A}$ is also a derivation.* Because the mapping $(D_1, D_2) \mapsto [D_1, D_2]$ is obviously linear in

D_1 and D_2, *the linear space* Der $\mathcal{A}$ *itself is an algebra with respect to the Lie bracket.* It is clear that *the Lie bracket is anticommutative*, i.e.,

$$[D_1, D_2] = -[D_2, D_1] \tag{18}$$

for any two linear operators D_1 and D_2 (which even are not derivations). In addition, *for any three operators* D_1, D_2, *and* D_3, *we have the identity*

$$[[D_1, D_2], D_3] + [[D_2, D_3], D_1] + [[D_3, D_1], D_2] = 0, \tag{19}$$

which is called the Jacobi identity.

All these facts provide the motivation for the following general definition.

Definition 33.2. An algebra whose multiplication is anticommutative and satisfies the Jacobi identity is called a *Lie algebra.*

We have thus proved that *the algebra* Der $\mathcal{A}$ *is a Lie algebra.*

Because derivations of the algebra of smooth functions $\mathbf{F}\mathcal{X}$ are vector fields on $\mathcal{X}$ by our identifications, we find, in particular, that *vector fields on a smooth Hausdorff manifold* $\mathcal{X}$ *of class* C^∞ *form a Lie algebra.* This algebra is usually denoted by $\mathfrak{a}\mathcal{X}$. By definition,

$$[X, Y]f = X(Yf) - Y(Xf)$$

for any vector fields X and Y and any function $f \in \mathbf{F}\mathcal{X}$. This implies

$$[gX, Y] = g[X, Y] - Yg \cdot X$$

for any function $g \in \mathbf{F}\mathcal{X}$.

Remark 33.4. For any Lie group $\mathfrak{G}$, the tangent space $\mathbf{T}_e\mathfrak{G}$ at the identity $e \in \mathfrak{G}$ is naturally identified with a certain Lie subalgebra of the algebra $\mathfrak{a}\mathcal{X}$ and is therefore a Lie algebra. This algebra is called the *Lie algebra of the group* $\mathfrak{G}$ and is usually denoted by $\mathfrak{g}$.

For arbitrary vector fields X and Y, the components $[X, Y]^i$ of the vector field $[X, Y]$ in each chart $(U, x^1, \dots, x^n)$ are expressed by

$$[X, Y]^i = X^j \frac{\partial Y^i}{\partial x^i} - Y^j \frac{\partial X^i}{\partial x^j}, \quad i = 1, \dots, n, \tag{20}$$

where $X^1, \dots, X^n$ and $Y^1, \dots, Y^n$ are components of the respective vector fields X and Y.

§4. Integral Curves of Vector Fields

By a *curve* on a smooth manifold $\mathcal{X}$, we mean an arbitrary smooth mapping

$$\gamma\colon (a,b) \to \mathcal{X} \tag{21}$$

of a certain interval (a, b) of the real axis $\mathbb{R}$ into $\mathcal{X}$.

The differential $(d\gamma)_t$ of a curve γ at a point $t \in (a, b)$ is a linear mapping of the one-dimensional space $\mathsf{T}_t(a, b) = \mathbb{R}$ into the space $\mathsf{T}_{\gamma(t)}\mathcal{X}$ and is uniquely characterized by the vector

$$\dot{\gamma}(t) = (d\gamma)_t \left(\frac{\partial}{\partial t}\right)_t \tag{22}$$

of the space $\mathsf{T}_{\gamma(t)}\mathcal{X}$ into which it transforms the basis vector $(\partial/\partial t)_t$ of the space $\mathsf{T}_t(a, b)$.

Definition 33.3. The vector $\dot{\gamma}(t)$ is called a *tangent vector* of the curve γ at the point t. (Admitting a slightly inaccuracy, we also call the vector $\dot{\gamma}(t)$ a vector tangent to the curve γ at the point $p = \gamma(t)$.)

Now let X be a vector field on a manifold $\mathcal{X}$. By definition, the field X sets a certain vector $X_p \in \mathsf{T}_p\mathcal{X}$ in correspondence to each point $p \in \mathcal{X}$.

Definition 33.4. A curve γ is called an *integral curve* (or *trajectory*) of the vector field X if

$$\dot{\gamma}(t) = X_{\gamma(t)} \quad \text{for any } t, \quad a < t < b. \tag{23}$$

We say that a curve γ *is contained* in a chart (U, h) if $\gamma(t) \in U$ for any t, $a < t < b$. Such a curve is given by n smooth functions

$$x^i = x^i(t), \quad a < t < b, \quad i = 1, \dots, n, \tag{24}$$

and for its tangent vector $\dot{\gamma}(t)$, we have

$$\dot{\gamma}(t) = \dot{x}^i(t) \left(\frac{\partial}{\partial x^i}\right)_{\gamma(t)}.$$

Therefore, for a curve in U, Eq. (23) is equivalent to the set of n first-order differential equations

$$\dot{x}^i(t) = X^i(x^1(t), \dots, x^n(t)), \quad i = 1, \dots, n, \tag{25}$$

where X^i, $i = 1, \dots, n$, are the components of the vector field X in the chart (U, h) (more precisely, their expressions through the coordinates $x^1, \dots, x^n$. Equations of form (23) are therefore called *differential equations on the manifold* $\mathcal{X}$.

It is clear that for each subinterval $I' \subset I$ of the interval $I = (a, b)$, the restriction $\gamma|_{I'}$ of the integral curve γ on I' is also an integral curve of the vector field X. Integral curve (22) is said to be *maximal* if it is not the restriction of any other integral curve defined on a larger interval.

Let $t_0 \in \mathbb{R}$. We say that curve (21) *passes through a point* $p \in \mathcal{X}$ *for* $t = t_0$ if, first, this curve is defined on an interval (a, b) of the real axis R such that $a < t_0 < b$ and, second, $\gamma(t_0) = p$.

Theorem 33.2. *If $\mathcal{X}$ is a Hausdorff manifold, then for any point $p_0 \in \mathcal{X}$ and any vector field X on $\mathcal{X}$, there exists a unique maximal integral curve $\gamma: I \to \mathcal{X}$ of the field X that passes through the point p for $t = t_0$.*

The condition that our manifold is a Hausdorff manifold is essential here.

§5. Vector Fields and Flows

For an arbitrary point $p \in \mathcal{X}$, we let γ_p^X denote the maximal integral curve of a field X that passes through a point $p \in X$ for $t = 0$, and we set

$$\varphi_t^X(p) = \gamma_p^X(t)$$

for any $t \in \mathbb{R}$ for which the point $\gamma_p^X(t)$ is defined. Therefore, φ_t^X is a mapping of the subset D_t consisting of points $p \in \mathcal{X}$ for which the point $\gamma_p^X(t)$ is defined into the manifold $\mathcal{X}$. It is easy to see that the set D_t is open and the mappings $\varphi_t = \varphi_t^X$ have the following properties:

1. There exists a continuous function $\varepsilon: \mathcal{X} \to \mathbb{R}$ assuming positive values such that $p \in D_t$ for $|t| < \varepsilon(p)$.
2. The mapping φ_0 is defined on the whole $\mathcal{X}$ (i.e., $D_0 = \mathcal{X}$) and is the identity mapping id of the manifold $\mathcal{X}$.
3. If $\varphi_t(p) \in D_s$ (in particular, if $|s| < \varepsilon(\varphi_t(p))$), then $p \in D_{s+t}$ and

$$\varphi_s(\varphi_t(p)) = \varphi_{s+t}(p). \tag{26}$$

Admitting a certain inaccuracy, the latter property is usually written as

$$\varphi_s \circ \varphi_t = \varphi_{s+t}.$$

Definition 33.5. A family of smooth mappings $\varphi_t: D_t \to \mathcal{X}$ having properties 1, 2, and 3 is called a *flow* on a manifold $\mathcal{X}$.

Therefore, we see that *each vector field $X \in \mathfrak{a}\mathcal{X}$ induces a certain flow $\{\varphi_t^X\}$ on $\mathcal{X}$*. Conversely, each flow $\{\varphi_t^X\}$ defines a certain vector field X on the manifold $\mathcal{X}$ by the formula

$$X_p = \dot{\gamma}_p(0), \quad p \in \mathcal{X},$$

where γ_p is the curve $t \mapsto \varphi_t(p)$, $|t| < \varepsilon(p)$.

A flow $\{\varphi'_t : D'_t \to \mathcal{X}\}$ is called a *part* of a flow $\{\varphi_t : D_t \to \mathcal{X}\}$ if $D'_t \subset D_t$ and $\varphi_t|_{D'_t} = \varphi'_t$ for any $t \in \mathbb{R}$. A flow $\{\varphi_t\}$ is said to be *maximal* if it is not a part of any other flow.

It is clear that the following assertions hold:

1. A flow $\{\varphi_t\}$ and any part $\{\varphi'_t\}$ of it generate the same vector field X.
2. The flow $\{\varphi^X_t\}$ induced by the vector field X is maximal.

Therefore, *the formula*

$$\textit{field } X \Longrightarrow \textit{flow } \{\varphi^X_t\}$$

establishes a one-to-one correspondence between vector fields and maximal flows on $\mathcal{X}$.

Because the function ε is continuous and therefore

$$\lim_{t\to 0} \varepsilon(\varphi_t(p)) = \varepsilon(p) > 0$$

for any point $p \in \mathcal{X}$, we see that there exists a continuous function $\delta : \mathcal{X} \to \mathbb{R}$ such that

$$|t| < \varepsilon(\varphi_t(p)) \quad \text{for } |t| < \delta(p).$$

Let O_t be an open subset of the manifold $\mathcal{X}$ that consists of all those points $p \in \mathcal{X}$ for which $|t| < \delta(p)$. Then there exists $\delta_0 > 0$, namely, $\delta_0 = \delta(p_0)$, such that $p_0 \in O_t$ for $|t| < \delta_0$ (therefore, the set O_t is not empty in advance for a sufficiently small t).

Because the point $\varphi_t^{-1}(\varphi_t(p))$ is defined for $p \in O_t$, i.e., for $|t| < \delta(p)$, and because this point coincides with the point p because of property 3, the restriction of φ_t to the set O_t is a bijective mapping of this set to the (obviously open) set $O'_t = \varphi_t O_t$ (with the inverse mapping $\varphi_{-t}|_{O'_t}$. Because both mappings φ_t and φ_{-t} are smooth by definition, this proves that *for any $t \in \mathbb{R}$ (for which the set O_t is not empty), the mapping φ_t is a diffeomorphism $O_t \to O'_t$.*

§6. Transport of Vector Fields via Diffeomorphisms

Now let $\varphi : \mathcal{X} \to \mathcal{Y}$ be an arbitrary diffeomorphism of smooth manifolds, and let S be a tensor field of type (a, b) on the manifold $\mathcal{Y}$. By definition, in each chart (V, k), the field S has components $S^{j_1\cdots j_b}_{i_1\cdots i_a}$ that are smooth functions on V. On the other hand, for any chart (U, h) on $\mathcal{X}$, the pair (V, k), where $V = \varphi U$ and $k = h \circ \varphi^{-1}$, is obviously a chart on $\mathcal{Y}$. Using this fact, we define the tensor field φ^* on $\mathcal{X}$ assuming that in the chart (U, h), it has the components

$$(\varphi^* S)^{j_1\cdots j_b}_{i_1\cdots i_a} = S^{j_1\cdots j_b}_{i_1\cdots i_a} \circ (\varphi|_U), \tag{27}$$

where $S^{j_1\cdots j_b}_{i_1\cdots i_a}$ are components of the field S in the chart $(\varphi U, h \circ \varphi^{-1})$. Because for any two charts (U, h) and (U', h') on $\mathcal{X}$, the transition mapping $h' \circ h^{-1}$

coincides with the transition mapping $(h' \circ \varphi^{-1}) \circ (h \circ \varphi^{-1})^{-1}$ for the charts $(\varphi U, h \circ \varphi^{-1})$ and $(\varphi U', h' \circ \varphi^{-1})$, the functions $(\varphi^* S)^{j_1 \cdots j_b}_{i_1 \cdots i_a}$ are related in different charts (on their intersection) by the same tensor transformation law as the components $S^{j_1 \cdots j_b}_{i_1 \cdots i_a}$. Therefore, these functions are in fact the components of a certain tensor field $\varphi^* S$. We say that the field $\varphi^* S$ is obtained as a result of *transport* of the field S from $\mathcal{Y}$ to $\mathcal{X}$ via the diffeomorphism φ.

Example 33.1. If a field S is of type (0,0) (is a smooth function f on $\mathcal{Y}$), then

$$\varphi^* f = f \circ \varphi. \tag{28}$$

Example 33.2. If a field S is of type (0,1) (is a vector field X), then

$$(\varphi^* X)_p = (d\varphi_p)^{-1} X_{\varphi(p)} \tag{29}$$

for any point $p \in \mathcal{X}$, where $(d\varphi_p)^{-1} \colon \mathbf{T}_{\varphi(p)}\mathcal{Y} \to \mathbf{T}_p\mathcal{X}$ is the mapping inverse to the isomorphism $d\varphi_p \colon \mathbf{T}_p\mathcal{X} \to \mathbf{T}_{\varphi(p)}\mathcal{Y}$.

Example 33.3. If a field S is of type (1,0) (is a covector field α), then

$$(\varphi^* \alpha)_p = (d\varphi_p)^* \alpha_{\varphi(p)} \tag{30}$$

for any point $p \in \mathcal{X}$, where $(d\varphi_p)^*$ is the mapping $\mathbf{T}^*_{\varphi(p)}\mathcal{Y} \to \mathbf{T}^*_p\mathcal{X}$ adjoint to the mapping $d\varphi_p \colon \mathbf{T}_p\mathcal{X} \to \mathbf{T}_{\varphi(p)}\mathcal{Y}$.

Similarly, the tensors $(\varphi^* S)_p$ are described for a tensor field S of an arbitrary type (a, b) because the correspondence $S \mapsto \varphi^* S$ preserves all algebraic operations on tensor fields. In particular,

$$\varphi^*(S \otimes T) = \varphi^* S \otimes \varphi^* T \tag{31}$$

for any tensor fields S and T on $\mathcal{X}$.

Proposition 33.2. *For any vector fields $X, Y \in \mathfrak{a}\mathcal{Y}$ and any diffeomorphism $\varphi \colon \mathcal{X} \to \mathcal{Y}$, we have*

$$[\varphi^* X, \varphi^* Y] = \varphi^* [X, Y],$$

i.e., the mapping $\varphi^* \colon \mathfrak{a}\mathcal{Y} \to \mathfrak{a}\mathcal{X}$ (which is obviously linear) is an isomorphism of the Lie algebra $\mathfrak{a}\mathcal{Y}$ onto the Lie algebra $\mathfrak{a}\mathcal{X}$.

Remark 33.5. We note the fact easily seen from formula (29) that the transport of a vector field is in general possible only using a diffeomorphism. In contrast, formula (30) is meaningful for any smooth mapping $\varphi \colon \mathcal{X} \to \mathcal{Y}$. Therefore, covector fields can be transported via any mappings.

Remark 33.6. Letting Y denote the field X and X denote the field $\varphi^* X$, we can rewrite formula (29) as

$$Y_{\varphi(p)} = (d\varphi)_p X_p. \tag{32}$$

This form is meaningful for any smooth mapping $\varphi \colon \mathcal{X} \to \mathcal{Y}$.

Definition 33.6. If relation (32) holds for fields $X \in \mathfrak{a}\mathcal{X}$ and $Y \in \mathfrak{a}\mathcal{Y}$ at each point $p \in \mathcal{X}$, then the fields X and Y are said to be *φ-related.*

Proposition 33.3. *Two fields $X \in \mathfrak{a}\mathcal{X}$ and $Y \in \mathfrak{a}\mathcal{Y}$ are φ-related iff $X(f \circ \varphi) = Yf \circ \varphi$ for any function $f \in \mathsf{F}\mathcal{Y}$. If two fields $X_1, X_2 \in \mathfrak{a}\mathcal{X}$ are φ-related to the fields $Y_1, Y_2 \in \mathfrak{a}\mathcal{Y}$, then the field $[X_1, X_2]$ is φ-related to the field $[Y_1, Y_2]$.*

§7. Lie Derivative of a Tensor Field

Let S be an arbitrary tensor field of type (a, b) on a manifold $\mathcal{X}$, and let $p \in \mathcal{X}$. Because $p \in O_t$ by definition for any t such that $|t| < \delta(p)$, the tensor $(\varphi_t^* S)_p$ and hence the tensor $(\varphi_t^* S)_p - S_p$ are well defined at p (where S certainly denotes the restriction of the tensor field S to O_t). We set

$$(\pounds_X S)_p = \lim_{t \to 0} \frac{(\varphi_t^* S)_p - S_p}{t}, \tag{33}$$

where X is the vector field generated by the flow $\{\varphi_t\}$. Because the point p is an arbitrary point of our manifold, tensors (33) compose a tensor field $\pounds_X S$ of type (a, b) on the manifold $\mathcal{X}$. Below, computing its elements in an arbitrary chart (U, h), we show that $\pounds_X S$ is smooth.

Definition 33.7. The field $\pounds_X S$ is called the *Lie derivative* of a tensor field S with respect to a vector field X.

It is easy to see that *for each vector field X, the mapping $\pounds_X$ is a derivation of the algebra of tensor fields on the manifold $\mathcal{X}$*, i.e., it is linear, and for any two tensor fields S and T, we have

$$\pounds_X(S \otimes T) = \pounds_X S \otimes T + S \otimes \pounds_X T. \tag{34}$$

Moreover, *the operation $\pounds_X$ commutes with the operation of contraction of tensor fields* (with respect to any pair of subscript and superscript). If the field S is a smooth function f, then

$$\pounds_X f = Xf \quad \text{on } \mathcal{X}. \tag{35}$$

Therefore, *the operation $\pounds_X$ is a generalization of the operation X from functions to arbitrary tensor fields.* By (34), this in particular implies that

$$\pounds_X(fS) = Xf \cdot S + f\pounds_X S$$

for any function f and any tensor field S.

§8. Linear Differential Forms

Now let the field S be a covector field α. If $\alpha = \alpha_j dx^j$ in the chart $(U, x^1, \dots, x^n)$, then

$$\pounds_X \alpha = \left(X\alpha_i + \alpha_k \frac{\partial X^k}{\partial x^i} \right) dx^i \quad \text{on } U, \tag{36}$$

which can be rewritten in a more symmetric form as

$$\pounds_X \alpha = \left(X^k \frac{\partial \alpha_i}{\partial x^k} + \alpha_k \frac{\partial X^k}{\partial x^i} \right) dx^i \quad \text{on } U.$$

In particular, we see that the covector field $\pounds_X$ is indeed smooth.

The action of the operation $\pounds_X$ on an arbitrary vector field Y is given by

$$\pounds_X Y = [X, Y]. \tag{37}$$

In an arbitrary Lie algebra $\mathfrak{g}$, the mapping

$$x \mapsto [a, x], \quad x, a \in \mathfrak{g},$$

is denoted by $\operatorname{ad} a$. Using this notation, we can rewrite formula (37) as

$$\pounds_X = \operatorname{ad} X \quad \text{on } \mathfrak{a}\mathcal{X}.$$

For an arbitrary tensor field S with the components $S^{j_1 \cdots j_b}_{i_1 \cdots i_a}$, we have

$$\begin{aligned}
(\pounds_X S)^{j_1 \cdots j_b}_{i_1 \cdots i_a} &= \frac{\partial S^{j_1 \cdots j_b}_{i_1 \cdots i_a}}{\partial x^k} X^k \\
&\quad + S^{j_1 j_2 \cdots j_b}_{k\, i_2 \cdots i_a} \frac{\partial X^k}{\partial x^{i_1}} + S^{j_1 j_2 \cdots j_b}_{i_1 k \cdots i_a} \frac{\partial X^k}{\partial x^{i_2}} + \cdots + S^{j_1 \cdots j_{b-1} j_b}_{i_1 \cdots i_{a-1} k} \frac{\partial X^k}{\partial x^{i_a}} \\
&\quad - S^{k\, j_2 \cdots j_b}_{i_1 i_2 \cdots i_a} \frac{\partial X^{j_1}}{\partial x^k} - S^{j_1 k \cdots j_b}_{i_1 i_2 \cdots i_a} \frac{\partial X^{j_2}}{\partial x^k} - \cdots - S^{j_1 \cdots j_{b-1} k}_{i_1 \cdots i_{a-1} i_a} \frac{\partial X^{j_b}}{\partial x^k}.
\end{aligned}$$

Proposition 33.4. *For any vector fields X and Y, we have*

$$[\pounds_X, \pounds_Y] = \pounds_{[X,Y]}.$$

Each such field α on $\mathcal{X}$ is written in an arbitrary chart $(U, x^1, \dots, x^n)$ as

$$\alpha = \alpha_i dx^i \quad \text{on } U,$$

i.e., as a linear form in the differentials $dx^1, \dots, dx^n$. Therefore, covector fields are usually called *linear differential forms* (and the functions α_i, $i = 1, \dots, n$, are called their *coefficients* in the chart (U, h)). The linear space $\mathsf{T}_1\mathcal{X}$ of linear differential forms is also denoted by $\Omega^1\mathcal{X}$.

For any linear differential form α and any vector field X, the formula

$$\alpha(X)(p) = \alpha_p(X_p), \quad p \in \mathcal{X},$$

defines a function $\alpha(X)$ on $\mathcal{X}$. This function is also denoted by the symbols $\langle \alpha, X \rangle$, $i_X \alpha$, or $X \lrcorner \alpha$ and is called the *inner product* of the form α and the field X. In each chart $(U, h) = (U, x^1, \dots, x^n)$, the function $\alpha(X)$ is expressed by the formula

$$\alpha(X) = \alpha_i X^i, \tag{38}$$

where X^i are the components of the field X and α_i are the coefficients of the form α in the chart (U, h); this directly implies that *the function* $\alpha(X)$ *is smooth on* $\mathcal{X}$. (This function was considered ad hoc when computing the Lie derivative of a vector field.)

Therefore, for any differential form $\alpha \in \Omega^1\mathcal{X}$, the formula

$$X \mapsto \alpha(X)$$

defines a certain mapping

$$\alpha: \mathfrak{a}\mathcal{X} \to \mathbf{F}\mathcal{X}, \tag{39}$$

which is obviously a morphism of $\mathbf{F}\mathcal{X}$*-modules*, i.e., it satisfies the relation

$$\alpha(fX) = f\alpha(X)$$

for any function $f \in \mathbf{F}\mathcal{X}$ and any field $X \in \mathfrak{a}\mathcal{X}$. We say that morphism (39) is generated by the differential form α.

We say that two vector fields X and Y *coincide near a point* $p \in \mathcal{X}$ if they coincide in a certain neighborhood of this point. (Compare this with a similar definition for functions given above.)

Lemma 33.2 (localization property of morphisms $\mathfrak{a}\mathcal{X} \to \mathbf{F}\mathcal{X}$). *If two vector fields X and Y coincide near a point $p \in \mathcal{X}$, then for any morphism* (39), *the functions $\alpha(X)$ and $\alpha(Y)$ also coincide near the point p.*

(Compare with Corollary 33.2.)

Proposition 33.5. *For any Hausdorff smooth manifold $\mathcal{X}$, the correspondence*

$$\textit{form } \alpha \Longrightarrow \textit{morphism } (39)$$

is an isomorphism of the linear space $\Omega^1\mathcal{X}$ onto the linear space of morphisms (39), $\operatorname{Hom}_{\mathbf{F}\mathcal{X}}(\mathfrak{a}\mathcal{X}, \mathbf{F}\mathcal{X})$.

As a rule, we identify linear differential forms (covector fields) on $\mathcal{X}$ and morphisms (39) generated by them in what follows.

Remark 33.7. We note that *mapping* (39) *itself is a morphism of* $\mathbf{F}\mathcal{X}$*-modules*, i.e., for any form $\alpha \in \Omega^1\mathcal{X}$ and for any function $f \in \mathbf{F}\mathcal{X}$, we have

$$\widehat{f\alpha} = f\widehat{\alpha},$$

where $\widehat{\alpha}$ is morphism (39) generated by the form α and the morphism $f\widehat{\alpha}$ is defined by $(\widehat{f\alpha})(X) = f\widehat{\alpha}(X)$ as is conventional in algebra.

§9. Differential Forms of an Arbitrary Degree

Tensor fields ω that set a *skew-symmetric tensor* ω_p, i.e., a tensor of type $(r,0)$ whose components change their signs under any permutation of indices, in correspondence to each point $p \in \mathcal{X}$ are of special significance. (The number r here is called the *degree* of the field ω.) For any two fields of skew-symmetric tensors θ and ω, the formula

$$(\theta \wedge \omega)_p = \theta_p \wedge \omega_p,$$

where $\theta_p \wedge \omega_p$ is the exterior product of the tensors θ_p and ω_p, defines the field $\theta \wedge \omega$ of skew-symmetric tensors whose degree is equal to the sum of the degrees of the fields θ and ω. Because the components of the exterior product of two tensors are algebraically expressed (in the well-known way) through the components of the factors, *the field $\theta \wedge \omega$ is smooth for smooth θ and ω.*

All algebraic properties of the exterior product of skew-symmetric tensors (for example, the associativity and skew-commutativity) are certainly preserved for their fields. Therefore, in an exterior product of arbitrarily many fields of skew-symmetric tensors, we can omit the parentheses, and the formula

$$\omega \wedge \theta = (-1)^{rs} \theta \wedge \omega$$

holds for any two fields of skew-symmetric tensors θ and ω, where r and s are the degrees of the respective fields ω and θ. For $r = 0$, the field ω is an ordinary function f, and the exterior product $\omega \wedge \theta$ is the ordinary product $f\theta$ of the function f and the field θ.

The known expressions of skew-symmetric tensors of type $(r,0)$ through the exterior product of covectors of the dual basis show that each field of skew-symmetric tensors on an arbitrary coordinate neighborhood U is expressed by

$$\begin{aligned} \omega &= \frac{1}{r!} \omega_{i_1 \cdots i_r} dx^{i_1} \wedge \cdots \wedge dx^{i_r} \\ &= \sum_{1 \le i_1 < \cdots < i_n \le n} \omega_{i_1 \cdots i_r} dx^{i_1} \wedge \cdots \wedge dx^{i_r}, \end{aligned} \tag{40}$$

where $\omega_{i_1 \cdots i_r}$ are components of the field ω in the chart $(U, x^1, \dots, x^n)$ (and are smooth functions on U). Therefore, fields of skew-symmetric tensors are also called *differential forms on the manifold $\mathcal{X}$*. For $r = 1$, we obtain the linear differential forms considered above, and for $r = 0$, we obtain smooth functions on $\mathcal{X}$.

All differential forms of degree $r \ge 0$ form a linear subspace $\Omega^r \mathcal{X}$ of the space $\mathsf{T}_r \mathcal{X}$ of all tensor fields of type $(r,0)$. Therefore, $\Omega^0 \mathcal{X} = \mathsf{F} \mathcal{X}$ and $\Omega^1 \mathcal{X} = \mathsf{T}_1 \mathcal{X}$ (while the inclusion $\Omega^r \mathcal{X} \subset \mathsf{T}_r \mathcal{X}$ is strict in advance for $r > 1$). The symbol $\Omega^1 \mathcal{X}$ was already used above.

We note that

$$\Omega^r \mathcal{X} = 0$$

for any $r > n$. *For any vector field $X \in \mathfrak{a}\mathcal{X}$, the operator $\mathcal{L}_X$ is a derivation of the algebra of forms on $\mathcal{X}$.*

§10. Differential Forms as Functionals on Vector Fields

Interpreting tensors of type $(r, 0)$ as multilinear functionals in vectors, we can set the function $\omega(X_1, \ldots, X_r)$ on $\mathcal{X}$, whose value at a point $p \in \mathcal{X}$ is given by

$$\omega(X_1, \ldots, X_r)(p) = \omega_p((X_1)_p, \ldots, (X_r)_p),$$

in correspondence to each differential form ω of degree r and to any vector fields $X_1, \ldots, X_r$. If we have

$$\omega = \sum_{1 \le i_1 < \cdots < i_n \le n} \omega_{i_1 \cdots i_n} dx^{i-1} \wedge \cdots \wedge dx^{i_r}$$

and

$$X_1 = X_1^{i_1} \frac{\partial}{\partial x^{i_1}}, \quad \ldots, \quad X_r = X_r^{i_r} \frac{\partial}{\partial x^{i_r}}$$

in a chart $(u, x^1, \ldots, x^n)$, then

$$\omega(X_1, \ldots, X_r) = \sum_{1 \le i_1 < \cdots < i_r \le n} \omega_{i_1 \cdots i_r} \begin{vmatrix} X_1^{i_1} & \cdots & X_1^{i_r} \\ \vdots & \ddots & \vdots \\ X_r^{i_r} & \cdots & X_r^{i_r} \end{vmatrix} \quad \text{on } U.$$

Therefore, *the function* $\omega(X_1, \ldots, X_r)$ *is smooth.*

The obtained mapping

$$\omega\colon \underbrace{\mathfrak{a}\mathcal{X} \times \cdots \times \mathfrak{a}\mathcal{X}}_{r \text{ times}} \to \mathsf{F}\mathcal{X}, \qquad (X_1, \ldots, X_r) \mapsto \omega(X_1, \ldots, X_r) \tag{41}$$

is obviously $\mathsf{F}\mathcal{X}$*-multilinear*, i.e., it is a morphism of $\mathsf{F}\mathcal{X}$-modules with respect to each argument. Moreover, *it is skew-symmetric*, i.e., changes its sign under any permutation of subscripts. As in the case $r = 1$ (see Proposition 33.5 and Remark 33.7), *if the manifold* $\mathcal{X}$ *is Hausdorff, then for any* $r \ge 1$, *the correspondence*

$$\textit{form of degree } r \Longrightarrow \textit{mapping } (40) \tag{42}$$

assigns an isomorphic mapping of the $\mathsf{F}\mathcal{X}$*-module* $\Omega^r\mathcal{X}$ *onto the* $\mathsf{F}\mathcal{X}$*-module of all skew-symmetric* $\mathsf{F}\mathcal{X}$*-multilinear mappings* (41).

As a rule, we identify differential forms with the corresponding mappings (41) in what follows.

Remark 33.8. Similarly, arbitrary tensor fields of type $(r, 0)$, $r > 0$, on $\mathcal{X}$ are identified with (not necessarily skew-symmetric) $\mathsf{F}\mathcal{X}$-multilinear mappings

$$S\colon \underbrace{\mathfrak{a}\mathcal{X} \times \cdots \mathfrak{a}\mathcal{X}}_{r \text{ times}} \to \mathsf{F}\mathcal{X},$$

and the tensor fields of type (r, s), $r, s \ge 0$, are identified with $\mathsf{F}\mathcal{X}$-multilinear mappings

$$S: \underbrace{\mathfrak{a}\mathcal{X} \times \cdots \to \mathfrak{a}\mathcal{X}}_{r \text{ times}} \times \underbrace{\Omega^1\mathcal{X} \times \cdots \times \Omega^1\mathcal{X}}_{s \text{ times}} \to \mathbf{F}\mathcal{X}. \tag{43}$$

In particular,

$$\mathfrak{a}\mathcal{X} = \mathrm{Hom}_{\mathbf{F}\mathcal{X}}(\Omega^1\mathcal{X}, \mathbf{F}\mathcal{X})$$

(the $\mathbf{F}\mathcal{X}$-linear mapping $i_X: \alpha \to X \lrcorner \alpha$ corresponds to a field X). Therefore, for $s = 1$, each field (43) can be interpreted as an $\mathbf{F}\mathcal{X}$-multilinear mapping

$$S: \underbrace{\mathfrak{a}\mathcal{X} \times \cdots \times \mathfrak{a}\mathcal{X}}_{r \text{ times}} \to \mathfrak{a}\mathcal{X}$$

that sets a vector field $S(X_1, \dots, X_r)$ such that

$$S(X_1, \dots, X_r) \lrcorner \alpha = S(X_1, \dots, X_r, \alpha)$$

for any differential form $\alpha \in \Omega^1\mathcal{X}$ in correspondence to vector fields $X_1, \dots,$ X_r. A similar interpretation of (43) is also possible for $s > 1$.

In the interpretation of differential forms as mappings (41), the exterior product $\theta \wedge \omega$ of a form θ of degree r and a form ω of degree s is given by the formula

$$(\theta \wedge \omega)(X_1, \dots, X_{r+s}) = \sum_{\sigma} \varepsilon_\sigma \theta(X_{\sigma(1)}, \dots, X_{\sigma(r)}) \omega(X_{\sigma(r+1)}, \dots, X_{\sigma(r+s)}), \tag{44}$$

where the summation is carried out over all shuffles of type (r, s).

§11. Inner Product of Vector Fields and Differential Forms

By identification (42), each vector field X allows setting the form $i_X\omega = X \lrcorner \omega$ of degree $r > 0$ in correspondence to an arbitrary form ω; its value on vector fields $X_1, \dots, X_{r-1}$ is given by

$$(X \lrcorner \omega)(X_1, \dots, X_{r-1} = \omega(X, X_1, \dots, X_{r-1}).$$

The form $X \lrcorner \omega$ is called the *inner product* of a field X and a form ω. (For $r = 1$, it was already considered above.) For $r = 0$ (i.e., in the case where the form ω is a function), we assume that $X \lrcorner \omega = 0$ for any field X by definition.

Proposition 33.6. *For the inner product $x \lrcorner (\theta \wedge \omega)$ of a vector field X and the form $\theta \wedge \omega$, we have*

$$X \lrcorner (\theta \wedge \omega) = (X \lrcorner \theta) \wedge \omega + (-1)^r \theta \wedge (X \lrcorner \omega), \tag{45}$$

where r is the degree of the form θ.

A linear operator D that transforms forms into forms and satisfies the relations

$$D(\theta \wedge \omega) = D\theta \wedge \omega + (-1)^r \theta \wedge \omega$$

for any forms θ and ω, where r is the degree of the form θ, is called an *antiderivation.* Therefore, *for any vector field X, the inner product operator $i_X = X \lrcorner$ (by X) is an antiderivation. In addition,*

$$[\pounds_X, Y] = i_{[X,Y]}.$$

§12. Transport of a Differential Form via a Smooth Mapping

Let $f\colon \mathcal{X} \to \mathcal{Y}$ be an arbitrary smooth mapping, and let ω be an arbitrary differential form of degree $r \geq 0$ on the manifold $\mathcal{Y}$. In correspondence to each point $p \in \mathcal{X}$, we set the skew-symmetric tensor $(f^*\omega)_p$ of the space $\mathbf{T}_p\mathcal{X}$ that assumes the value

$$(f^*\omega)_p(A_1, \dots, A_r) = \omega_q((df)_p A_1, \dots, (df)_p A_r)$$

at vectors $A_1, \dots, A_r \in \mathbf{T}_p\mathcal{X}$, where, as usual, $q = f(p)$ and

$$(df)_p\colon \mathbf{T}_p\mathcal{X} \to \mathbf{T}_q\mathcal{Y}$$

is the differential of the mapping f at the point p. If

$$(U, h) = (U, x^1, \dots, x^n) \quad \text{and} \quad (V, k) = (V, y^1, \dots, y^m)$$

are charts of the manifolds $\mathcal{X}$ and $\mathcal{Y}$ such that $fU \subset V$ and if

$$y^j = f^j(x^1, \dots, x^n), \quad j = 1, \dots, m,$$

are the functions that express the mapping f in the charts (U, h) and (V, k), then

$$(f^*\omega)_{i_1 \cdots i_r} = \frac{\partial f^{j_r}}{\partial x^{i_1}} \cdots \frac{\partial f^{j_r}}{\partial x^{i_r}} (\omega_{j_1 \cdots j_r} \circ f), \tag{46}$$

and, in particular, the functions $(f^*\omega)_{i_1 \cdots i_r}$ are smooth on U. Therefore, the formula

$$f^*\omega = \sum_{1 \leq i_1 < \cdots < i_r \leq n} (f^*\omega)_{i_1 \cdots i_r} dx^{i_1} \wedge \cdots \wedge dx^{i_r} \tag{47}$$

defines a differential form f^* on U.

Comparing the definitions shows that at each point $p \in U$, this form assumes the value $(f^*\omega)_p$; therefore, *the correspondence*

$$p \mapsto (f^*\omega)_p$$

defines a differential form $f^*\omega$ on $\mathcal{X}$. In each chart $(U, x^1, \dots, x^n)$, the form $f^*\omega$ is expressed by formula (47) (in other words, functions (46) are coefficients of this form in the chart $(U, x^1, \dots, x^n)$).

Definition 33.8. We say that the form $f^*\omega$ is obtained by a *transport* via a smooth mapping f.

It is clear that the mapping

$$f^*: \Omega^r\mathcal{Y} \to \Omega^r\mathcal{X}, \qquad \omega \mapsto f^*\omega,$$

is linear and commutes with the exterior product,

$$f^*(\theta \wedge \omega) = f^*\theta \wedge f^*\omega$$

for any forms θ and ω on $\mathcal{Y}$. Moreover, if $f: \mathcal{X} \to \mathcal{Y}$ and $g: \mathcal{Y} \to Z$, then

$$(g \circ f)^* = f^* \circ g^*,$$

and if $f = \mathrm{id}: \mathcal{X} \to \mathcal{X}$ is the identity mapping, then $f^*: \Omega^r\mathcal{X} \to \Omega^r\mathcal{X}$ is also the identity mapping. For $r = 0$, the form ω is a smooth function $g: \mathcal{Y} \to \mathbb{R}$, and we have

$$f^*g = g \circ f.$$

In the case where f is a diffeomorphism φ, the construction of the form $f^*\omega$ is a particular case of the general construction of φ^*S (see Remark 33.5).

For an arbitrary submanifold $\mathcal{Y}$ of a manifold $\mathcal{X}$ and for an embedding $\imath: \mathcal{Y} \to \mathcal{X}$ corresponding to it, the mapping

$$\imath^*: \Omega^r\mathcal{X} \to \Omega^r\mathcal{Y}$$

is the *restriction mapping* transforming the form ω on $\mathcal{X}$ into the form $\omega|_\mathcal{Y}$ on $\mathcal{Y}$ for which

$$(\omega|_\mathcal{Y})_p(A_1, \dots, A_r) = \omega_p(A_1, \dots, A_r)$$

for any vectors $A_1, \dots, A_r \in \mathbf{T}_p\mathcal{Y}$ at any point $p \in \mathcal{Y}$ (where the space $\mathbf{T}_p\mathcal{Y}$ is certainly considered as a subspace of the space $\mathbf{T}_p\mathcal{X}$).

Remark 33.9. The construction of the form $f^*\omega$ is immediately extended to arbitrary tensor fields of type $(r, 0)$, $r \geq 0$. For each such field S on $\mathcal{Y}$, the field f^*S on $\mathcal{X}$ is given by the formula

$$(f^*S)_p(A_1, \dots, A_r) = S_q((df)_pA_1, \dots, (df)_pA_r),$$

where $p \in \mathcal{X}$, $q = f(p)$, and $A_1, \dots, A_r \in \mathbf{T}_p\mathcal{X}$; the components $(f^*S)_{i_1 \cdots i_r}$ of the field f^*S are expressed through the components $S_{i_1 \cdots i_r}$ of the field S by the formula

$$(f^*S)_{i_1 \cdots i_r} = \frac{\partial f^{j_1}}{\partial x^{i_1}} \cdots \frac{\partial f^{j_r}}{\partial x^{i_r}} (S_{i_1 \cdots i_r} \circ f).$$

This construction cannot be extended to fields of type (r, s) with $s > 0$ (of course, if f is not a diffeomorphism).

§13. Exterior Differential

As we know, each smooth function f on a manifold $\mathcal{X}$ defines the covector $(df)_p$ at each point $p \in \mathcal{X}$, which acts according to the formula

$$(df)_p A = Af, \quad A \in \mathbf{T}_p\mathcal{X},$$

and therefore defines a linear differential form

$$df : p \mapsto (df)_p,$$

which is called the *differential* of the function f. In each chart $(U, x^1, \dots, x^n)$, this form is expressed by the formula

$$df = \frac{\partial f}{\partial x^i} dx^i$$

and is therefore a smooth form. As a morphism of $\mathbf{F}\mathcal{X}$-modules $\mathfrak{a}\mathcal{X} \to \mathbf{F}\mathcal{X}$, the form df acts according to the formula

$$df(X) = Xf, \quad X \in \mathfrak{a}\mathcal{X}. \tag{48}$$

It is clear that the mapping

$$d : \Omega^0\mathcal{X} \to \Omega^1\mathcal{X}, \qquad f \mapsto df,$$

is linear and has the property

$$d(fg) = df \cdot g + f \cdot dg$$

for any two functions f and g. It turns out that d is naturally extended to differential forms of arbitrary degree.

Proposition 33.7. *For any smooth manifold $\mathcal{X}$ and for any $r \geq 0$, there exists a unique mapping*

$$d : \Omega^r\mathcal{X} \to \Omega^{r+1}\mathcal{X}$$

that has the following properties:

1. *The mapping d is linear.*
2. *The mapping d is an antiderivation, i.e., for any two differential forms θ and ω, we have*

$$d(\theta \wedge \omega) = d\theta \wedge \omega + (-1)^r \theta \wedge d\omega,$$

   where r is the degree of the form θ.
3. *For any smooth mapping $f : \mathcal{X} \to \mathcal{Y}$ and any form ω on $\mathcal{Y}$, we have*

$$df^*\omega = f^*d\omega.$$

4. *For each function $f \in \Omega^0\mathcal{X}$, the form df is its differential* (48).

5. *If* $\omega = df$, *where* $f \in \Omega^0\mathcal{X}$, *then*

$$d\omega = 0.$$

Proposition 33.8. *For any differential form* ω, *we have*

$$dd\omega = 0.$$

Definition 33.9. The form $d\omega$ is called the *exterior differential of the form* ω.

The coefficients $(d\omega)_{j_1\cdots j_{r+1}}$ of the form $d\omega$ in a chart $(U, x^1, \dots, x^n)$ are expressed through the coefficients $\omega_{i_1\cdots i_r}$ by

$$(d\omega)_{j_1\cdots j_{r+1}} = \sum_{a=1}^{r+1}(-1)^{a+1}\frac{\partial\omega_{j_1\cdots\widehat{j_a}\cdots j_{r+1}}}{\partial x^{j_a}}. \tag{49}$$

For example,

$$(d\alpha)_{ij} = \frac{\partial\alpha_j}{\partial x^i} - \frac{\partial\alpha_i}{\partial x^j}, \quad i < j,$$

for a linear form $\alpha = \alpha_i dx^i$.

Proposition 33.9. *The form* $d\omega$ *considered as an* $\mathbf{F}\mathcal{X}$-*multilinear functional in vector fields is given by the formula*

$$(d\omega)(X_1, \dots, X_{r+1}) = \sum_{a=1}^{r+1}(-1)^{a+1}X_a\omega(X_1, \dots, \widehat{X}_a, \dots, X_{r+1})$$
$$+\sum_{a=1}^{r}\sum_{b=a+1}^{r+1}(-1)^{a+b}\omega([X_a, X_b], X_1, \dots, \widehat{X}_a, \dots, \widehat{X}_b, \dots, X_{r+1}), \tag{50}$$

where the sign $\widehat{\ }$ *indicates that the corresponding field should be omitted.*

For $r = 1$, formula (50) becomes

$$d\omega(X, Y) = X\omega(Y) - Y\omega(X) - \omega([X, Y]). \tag{51}$$

Further, property 3 of the operator d immediately implies that *the operator* d *commutes with the operator* $\pounds_X$, i.e.,

$$d\pounds_X\omega = \pounds_X d\omega \tag{52}$$

for any form ω and any vector field X. In particular,

$$\pounds_X df = d\pounds_X f = d(Xf)$$

for any function f. However, for differential forms, the Lie derivative operator $\pounds_X$ is expressed through the inner product of the field X and the operator d.

Proposition 33.10. *For any vector field X and for any differential form ω, we have*

$$\pounds_X \omega = X \,\lrcorner\, d\omega + d(X \,\lrcorner\, \omega). \tag{53}$$

In the operator form, relation (53) becomes

$$\pounds_X = i_X \circ d + d \circ i_X.$$

Chapter 34
Vector Bundles

§1. Bundles and Their Morphisms

The word "bundle" recalls an association with a certain set $\mathcal{E}$ divided into nonempty disjoint sets, the fibers. Let $\mathcal{B}$ be the set of all fibers, and let $\pi: \mathcal{E} \to \mathcal{B}$ be the mapping that sets the fiber containing a point $p \in \mathcal{E}$ in correspondence to this point. The mapping π is uniquely defined by the bundle and in turn uniquely defines this bundle. Moreover, any surjective mapping of the form $\pi: \mathcal{E} \to \mathcal{B}$ yields a certain bundle of the set $\mathcal{E}$ (consisting of the inverse images $\pi^{-1}(b)$ of points $b \in \mathcal{B}$). Of course, in the topological case (where $\mathcal{E}$ is a topological space), it is natural to assume that the mapping π is continuous. All this is an explanation and motivation for the following definition.

Definition 34.1. A *bundle* is an arbitrary triple of the form

$$\xi = (\mathcal{E}, \pi, \mathcal{B}),$$

where $\mathcal{E}$ and $\mathcal{B}$ are topological spaces and $\pi: \mathcal{E} \to \mathcal{B}$ is a continuous mapping (it is surjective as a rule). The space $\mathcal{E}$ is called the *total space* of the bundle ξ. (Also, the terms *bundle space* and *fibered space* are used.) The space $\mathcal{B}$ is called the *base* of the bundle ξ. A bundle with the base $\mathcal{B}$ is also called the *bundle over* $\mathcal{B}$. The mapping $\pi: \mathcal{E} \to \mathcal{B}$ is called the *projection* of the bundle ξ. For any point $b \in \mathcal{B}$, its inverse image $\mathcal{F}_b = \pi^{-1}(b)$ is called the *fiber* of the bundle ξ over the point b.

Remark 34.1. Because the assignment of the mapping $\pi: \mathcal{E} \to \mathcal{B}$ assumes that the spaces $\mathcal{E}$ and $\mathcal{B}$ are given, the middle component of the triple $(\mathcal{E}, \pi, \mathcal{B})$ uniquely defines the other two. Therefore, formally, the bundle $(\mathcal{E}, \pi, \mathcal{B})$ coincides with the mapping π. The distinction consists only in the point of view (a mapping is a rule that sets the point $\pi(p) \in B$ in correspondence to each point $p \in \mathcal{E}$, while a bundle sets the subspace $\mathcal{F}_b$ in correspondence to each point $b \in \mathcal{B}$).

Sometimes, it is convenient to not distinguish ξ and π and to let *bundle* mean the mapping $\pi: \mathcal{E} \to \mathcal{B}$.

When it is necessary to indicate the dependence of $\mathcal{E}$, $\mathcal{B}$, and π on ξ, we write $\mathcal{E}^\xi$, $\mathcal{B}^\xi$, and π^ξ (and sometimes $\mathcal{E}(\xi)$, $\mathcal{B}(\xi)$, and $\pi(\xi)$, according to the requirement of a publisher).

A bundle $\xi = (\mathcal{E}, \pi, \mathcal{B})$ is called a *bundle with a typical fiber* if for any two points $b_1, b_2 \in \mathcal{B}$, the fibers $\mathcal{F}_{b_1}$ and $\mathcal{F}_{b_2}$ are homeomorphic, i.e., if there exists a topological space $\mathcal{F}$ such that for any point $b \in \mathcal{B}$, the fiber $\mathcal{F}_b$

is homeomorphic to the space $\mathcal{F}$. The space $\mathcal{F}$ is uniquely defined up to a homeomorphism and is called the *typical fiber* of the bundle. It is clear that the projection of each bundle with a typical fiber is a surjective mapping.

Let $\xi = (\mathcal{E}, \pi, \mathcal{B})$ and $\xi' = (\mathcal{E}', \pi', \mathcal{B}')$ be two bundles. A continuous mapping $\varphi: \mathcal{E} \to \mathcal{E}'$ is said to be *fiberwise* if it transforms each fiber $\mathcal{F}_b = \pi^{-1}(b)$, $b \in \mathcal{B}$, of the bundle ξ into a certain fiber $\mathcal{F}_{b'}$ of the bundle ξ'. By the formula $\psi(b) = b'$, such a mapping defines the mapping $\psi: \mathcal{B} \to \mathcal{B}'$ that satisfies the relation $\pi' \circ \varphi = \psi \circ \pi$, i.e., it closes the commutative diagram

$$\begin{array}{ccc} \mathcal{E} & \xrightarrow{\varphi} & \mathcal{E}' \\ {\scriptstyle\pi}\downarrow & & \downarrow{\scriptstyle\pi'} \\ \mathcal{B} & \xrightarrow{\psi} & \mathcal{B}' \end{array} \tag{1}$$

but is not a continuous mapping in general. If the mapping ψ is continuous, then the fiberwise mapping φ is called a *morphism* or *bundle mapping* of the bundle ξ into the bundle ξ', and we write $\varphi: \xi \to \xi'$ in this case.

Because the mapping ψ is uniquely characterized as a mapping that closes diagram (1), we can also say that the mapping $\varphi: \mathcal{E} \to \mathcal{E}'$ is a morphism $\varphi: \xi \to \xi'$ iff there exists a continuous mapping $\psi: \mathcal{B} \to \mathcal{B}'$ that closes diagram (1). Sometimes the pair (φ, ψ) is called the morphism $\xi \to \xi'$.

A morphism $\varphi: \xi \to \xi'$ is called an *isomorphism* if both mappings φ and ψ are homeomorphisms, i.e., if a continuous mapping $\varphi^{-1}: \mathcal{E}' \to \mathcal{E}$ is defined and this mapping is a morphism $\xi' \to \xi$. Two bundles ξ and ξ' are said to be *isomorphic* if there exists at least one isomorphism $\xi \to \xi'$.

In the case where $\mathcal{B} = \mathcal{B}'$ and $\psi = \mathrm{id}$, the fiberwise mapping φ (which is automatically a morphism) is called a *morphism over* $\mathcal{B}$. We have the commutative diagram

$$\begin{array}{ccc} \mathcal{E} & \xrightarrow{\varphi} & \mathcal{E}' \\ & {\scriptstyle\pi}\searrow \quad \swarrow{\scriptstyle\pi'} & \\ & \mathcal{B} & \end{array}$$

for it. A morphism over $\mathcal{B}$ that is a homeomorphism (and therefore an isomorphism) is called an *isomorphism over* $\mathcal{B}$. Two bundles ξ and ξ' with the same base $\mathcal{B}$ are said to be *isomorphic* if they are isomorphic over $\mathcal{B}$.

If the mapping π is open (transforms open sets into open sets), then any fiberwise mapping $\mathcal{E} \to \mathcal{E}'$ is a morphism $\xi \to \xi'$. Therefore, for bundles whose projections are open mappings, morphisms are exactly fiberwise mappings (and isomorphisms are fiberwise homeomorphisms).

§2. Vector Bundles

All these concepts are immediately generalized to the case where all spaces are smooth manifolds and all mappings are smooth mappings. In what follows, as a rule, we consider namely this case.

Definition 34.2. A triple $\xi = (\mathcal{E}, \pi, \mathcal{B})$ consisting of smooth manifolds (topological spaces) $\mathcal{E}$ and $\mathcal{B}$ and a submersion

$$\pi: \mathcal{E} \to \mathcal{B} \tag{2}$$

is called a real vector bundle of rank n if

1. for any point $b \in \mathcal{B}$, the set
$$\mathcal{F}_b = \pi^{-1}(b)$$
is a linear (vector) space over the field $\mathbb{R}$ of real numbers and
2. (*local triviality condition*) there exist an open covering $\{U_\alpha\}$ of the manifold $\mathcal{B}$ and diffeomorphisms $\varphi_\alpha : U_\alpha \times \mathbb{R}^n \to \mathcal{E}_{U_\alpha}$, where $\mathcal{E}_{U_\alpha} = \pi^{-1}U_\alpha$, such that
   a. for any point $(b, \boldsymbol{x}) \in U_\alpha \times \mathbb{R}^n$, we have the inclusion $\varphi_\alpha(b, \boldsymbol{x}) \in \mathcal{F}_b$ (i.e., the diagram
$$\begin{array}{ccc} U_\alpha \times \mathbb{R}^n & \xrightarrow{\varphi_\alpha} & \mathcal{E}_{U_\alpha} \\ & \searrow \quad \swarrow & \\ & U_\alpha & \end{array} \tag{3}$$
   is commutative, where the left inclined arrow is the natural projection $(b, \boldsymbol{x}) \to b$ of the direct product $U_\alpha \times \mathbb{R}^n$ on its first factor and the right inclined arrow is the restriction of mapping (2) to $\mathcal{E}_{U_\alpha}$) and
   b. for each point $b \in \mathcal{B}$, the mapping
$$\varphi_{\alpha,b}: \mathbb{R}^n \to \mathcal{F}_b$$
   defined by
$$\varphi_{\alpha,b}(\boldsymbol{x}) = \varphi_\alpha(b, \boldsymbol{x}), \quad \boldsymbol{x} \in \mathbb{R}^n,$$
   is an isomorphism of linear spaces.

In accordance with the general terminology introduced above, the space $\mathcal{B}$ (also denoted by $\mathcal{B}^\xi$ or $B(\xi)$) is called the *base* of the vector bundle ξ, the space $\mathcal{E}$ (also denoted by $\mathcal{E}^\xi$ or $\mathcal{E}(\xi)$) is called its *total space*, and the mapping π (also denoted by π^ξ or $\pi(\xi)$) is called the *projection*. Very often, ξ and π are not distinguished.

The linear space $\mathcal{F}_b$ (also denoted by $\mathcal{F}_b^\xi$ or $\mathcal{F}_b(\xi)$) is called the *fiber* of the bundle ξ (projection π) over the point $b \in \mathcal{B}$. It is a smooth submanifold of the manifold $\mathcal{B}$. The rank n of the vector bundle ξ is also called the *dimension* of this bundle and is denoted by $\dim \xi$.

The diffeomorphism φ_α in diagram (3) is called a *trivialization* of the bundle ξ over the open set U_α, and the open set U_α is called the *trivializing neighborhood.* Sometimes, a pair $(U_\alpha, \varphi_\alpha)$ is also called a *trivialization.* A covering $\{U_\alpha\}$ consisting of trivializing neighborhoods is called a *trivializing covering.* A family $\{(U_\alpha, \varphi_\alpha)\}$ of trivializations $(U_\alpha, \varphi_\alpha)$ is called a *trivializing atlas.*

§3. Sections of Vector Bundles

Definition 34.3. A smooth mapping $s: \mathcal{B} \to \mathcal{E}$ satisfying the relation $\pi \circ s = \mathrm{id}$ is called a *section* of the bundle ξ. A mapping $s: \mathcal{B} \to \mathcal{E}$ is a section iff $s(b) \in \mathcal{F}_b$ for any point $b \in \mathcal{B}$, i.e., if it selects the vector $s(b)$ in each fiber $\mathcal{F}_b$. Therefore, sections of the bundle ξ are also called *ξ-vector fields* on $\mathcal{B}$.

It is easy to see that

1. for any sections s, s_1, and s_2 and any element $\lambda \in \mathbb{R}$, the formulas

$$(s_1 + s_2)(b) = s_1(b) + s_2(b), \qquad (\lambda s)(b) = \lambda s(b), \quad b \in \mathcal{B},$$

   define the sections $s_1 + s_2$ and λs of the bundle ξ, and with respect to the operations $(s_1, s_2) \mapsto s_1 + s_2$ and $(\lambda, s) \mapsto \lambda s$, the *set $\Gamma\xi$ of all sections of the vector bundle ξ is a linear space over the field $\mathbb{R}$ whose zero is the zero section* 0 (this section sets the zero of the linear space $\mathcal{F}_b$ in correspondence to each point $b \in \mathcal{B}$);
2. for any smooth function f on $\mathcal{B}$ and any section $s \in \Gamma\xi$, the formula

$$(fs)(b) = f(b)s(b), \quad b \in \mathcal{B},$$

   defines a section $fs \in \Gamma\xi$, and with respect to the operation $(f, s) \mapsto fs$, *the linear space $\Gamma\xi$ is a module over the algebra $\mathsf{F}\mathcal{B}$ of all smooth functions on $\mathcal{B}$.*

For each submanifold $U \in \mathcal{B}$, the triple $(\mathcal{E}_U, \pi_U, U)$, where $\mathcal{E}_U = \pi^{-1}U$ and $\pi_U = \pi|_U$, is obviously a vector bundle. It is called the *part* of the bundle ξ over U and is denoted by ξ_U.

If U is a trivializing neighborhood, then each trivialization $\varphi: U \times \mathbb{R}^n \to \mathcal{E}_U$ defines the sections

$$s_1, \ldots, s_n \tag{4}$$

in $\Gamma(\xi|_U)$ that act according to the formulas

$$s_1(b) = \varphi(b, \mathbf{e}_1), \quad \ldots, \quad s_n(b) = \varphi(b, \mathbf{e}_n),$$

where $\mathbf{e}_1, \ldots, \mathbf{e}_n$ is the standard basis of the space $\mathbb{R}^n$. Because the vectors $s_1(b), \ldots, s_n(b)$ compose a basis of the linear space $\mathcal{F}_b$ for each point $b \in U$, each section $s\colon U \to \mathcal{E}_U$ assigns smooth functions $s^1, \dots, s^n\colon U \to \mathbb{R}$ on U that satisfy the relation

$$s(b) = s^1(b)s_1 + \cdots + s^n(b)s_n$$

for any point $b \in \mathcal{B}$. Moreover, on the $\mathsf{F}(U)$-module $\Gamma(\xi|_U)$, we have the relation

$$s = s^1 s_1 + \cdots + s^n s_n, \quad s^1, \dots, s^n \in \mathsf{F}(U).$$

By definition, this means that the $\mathsf{F}(U)$-module $\Gamma(\xi|_U)$ *is a free module of rank n with basis* (4).

Conversely, let U be an open set of the space $\mathcal{B}$ such that the module $\Gamma(\xi|_U)$ is a free module of rank n, and let (4) be its arbitrary basis. We define the mapping $\varphi\colon U \times \mathbb{R}^n \to \mathcal{E}_U$ by

$$\varphi(b, \boldsymbol{x}) = x^i s_i(b), \quad b \in U, \quad \boldsymbol{x} = (x^1, \dots, x^n) \in \mathbb{R}^n.$$

An automatic verification shows that the mapping φ is a trivialization of the bundle ξ over U.

This proves the following proposition.

Proposition 34.1. *An open set $U \subset \mathcal{B}$ is a trivializating neighborhood of a bundle ξ iff the $\mathsf{F}(U)$-module $\Gamma(\xi|_U)$ is a free module of rank n. Moreover, bases* (4) *of this module are in a natural bijective correspondence with the trivializations φ of the bundle ξ over U.*

Because of this proposition, bases (4) of the module $\Gamma(\xi|_U)$ are also called trivializations of the bundle ξ over U. Trivializations (4) are also called *bases of the module $\Gamma(\xi|_U)$ over U*. This shortened terminology is often convenient.

§4. Morphisms of Vector Bundles

Let $\xi = (\mathcal{E}, \pi, \mathcal{B})$ and $\xi' = (\mathcal{E}', \pi', \mathcal{B}')$ be two vector bundles. A smooth mapping $\varphi\colon \mathcal{E} \to \mathcal{E}'$ is said to be *fiberwise* if the condition that the points $p_1, p_2 \in \mathcal{E}$ belong to the same fiber (i.e., $\pi(p_1) = \pi(p_2)$) implies that the points $p_1' = \varphi(p_1)$ and $p_2' = \varphi(p_2)$ also belong to the same fiber (i.e., $\pi'(p_1') = \pi'(p_2')$). For each fiberwise mapping $\varphi\colon \mathcal{E} \to \mathcal{E}'$, the formula

$$\widehat{\varphi}(b) = \pi'(\varphi(p)), \quad p \in \pi^{-1}(b),$$

correctly defines a smooth mapping

$$\widehat{\varphi}\colon \mathcal{B} \to \mathcal{B}'$$

that closes the commutative diagram

$$\begin{array}{ccc} \mathcal{E} & \xrightarrow{\varphi} & \mathcal{E}' \\ {\scriptstyle \pi}\downarrow & & \downarrow{\scriptstyle \pi'} \\ \mathcal{B} & \xrightarrow{\widehat{\varphi}} & \mathcal{B}' \end{array}.$$

For any point $b \in \mathcal{B}$, the mapping φ induces the mapping

$$\varphi_b: \mathcal{F}_b \to \mathcal{F}_{b'}, \quad b' = \widehat{\varphi}(b).$$

Definition 34.4. A fiberwise mapping φ for which all mappings φ_b, $b \in \mathcal{B}$, are linear is called a *morphism* of the vector bundle ξ into the vector bundle ξ'. This is written as $\varphi: \xi \to \xi'$.

Morphisms $\varphi: \eta \to \xi$ for which all the mappings φ_b, $b \in \mathcal{B}^\eta$, are isomorphisms are of special importance. Unfortunately, these morphisms have no good conventional name. For lack of a better term, we call them *regular morphisms.*

For $\mathcal{B} = \mathcal{B}'$, a morphism φ for which $\widehat{\varphi} = \text{id}$ is called a *morphism over* $\mathcal{B}$. A morphism over $\mathcal{B}$ that is a diffeomorphism is called an *isomorphism.* (For such a morphism φ, the inverse diffeomorphism $\varphi^{-1}: \mathcal{E}' \to \mathcal{E}$ is also a morphism over $\mathcal{B}$ and is therefore an isomorphism.) Bundles ξ and ξ' over $\mathcal{B}$ for which there exists at least one isomorphism $\xi \to \xi'$ are said to be *isomorphic.* Isomorphisms of vector bundles over $\mathcal{B}$ are exactly regular morphisms that are simultaneously morphisms over $\mathcal{B}$. Isomorphisms of the form $\xi \to \xi$ are called *automorphisms.*

Remark 34.2. If we replace smooth manifolds with topological spaces and smooth mappings with continuous ones in Definition 34.1, we obtain a formally more general concept of a vector bundle. However, in the framework of smooth manifold theory, this generalization is not rich in content, because each such vector bundle over a smooth (Hausdorff and paracompact) manifold is isomorphic to a bundle in the sense of Definition 34.2.

The set of all morphisms $\xi \to \xi'$ over $\mathcal{B}$ is denoted by $\text{Mor}(\xi, \xi')$. It is a linear space over the field $\mathbb{R}$ and a module over the algebra $\mathsf{F}\mathcal{B}$ with respect to obviously defined operations. Each morphism $\varphi: \xi \to \xi'$ of vector bundles over $\mathcal{B}$ defines the $\mathsf{F}\mathcal{B}$-linear mapping

$$\varphi\circ: \Gamma\xi \to \Gamma\xi'$$

of linear spaces of sections by

$$s \mapsto \varphi \circ s, \quad s \in \Gamma\xi.$$

Similarly, the correspondence $\varphi \mapsto \varphi\circ$ defines the $\mathsf{F}\mathcal{B}$-linear mapping of the $\mathsf{F}\mathcal{B}$-module $\text{Mor}(\xi, \xi')$ into the $\mathsf{F}\mathcal{B}$-module $\text{Hom}_{\mathsf{F}\mathcal{B}}(\Gamma\xi, \Gamma\xi')$ of all $\mathsf{F}\mathcal{B}$-linear mappings $\Gamma\xi \to \Gamma\xi'$. Moreover, any $\mathsf{F}\mathcal{B}$-linear mapping $\Gamma\xi \to \Gamma\xi'$ has the form $\varphi\circ$ for a certain uniquely defined morphism $\varphi: \xi \to \xi'$. This means that *the modules* $\operatorname{Mor}(\xi,\xi')$ *and* $\operatorname{Hom}_{\mathsf{F}\mathcal{B}}(\Gamma\xi,\Gamma\xi')$ *are naturally isomorphic* and can therefore be identified:

$$\operatorname{Mor}(\xi,\xi') = \operatorname{Hom}_{\mathsf{F}\mathcal{B}}(\Gamma\xi,\Gamma\xi').$$

§5. Trivial Vector Bundles

For any manifold (topological space) $\mathcal{B}$ and any linear space $\mathcal{V}$ over the field $\mathbb{R}$, the triple $(\mathcal{B}\times\mathcal{V},\pi,\mathcal{B})$, where $\pi\colon\mathcal{B}\times\mathcal{V}\to\mathcal{B}$ is the projection of the direct product on the first factor, is a vector bundle. For this bundle, the trivializing covering $\{U_\alpha\}$ consists of one element $U=\mathcal{B}$, and the trivialization $\varphi\colon\mathcal{B}\times\mathbb{R}^n\to\mathcal{B}\times V$ is defined by the choice of a basis $e_1,\dots,e_n$ in $\mathcal{V}$ and is given by

$$\varphi(b,x) = (b,\alpha^{-1}(x)),\quad b\in\mathcal{B},\quad x\in\mathbb{R}^n,$$

where $\alpha\colon\mathcal{V}\to\mathbb{R}^n$ is the coordinate isomorphism corresponding to the basis $e_1,\dots,e_n$. This vector bundle is denoted by $\theta^n_{\mathcal{B}}$. This bundle (and each bundle that is isomorphic to it) is called a *trivial vector bundle of rank n*. According to Proposition 34.1, *a vector bundle* ξ *is trivial iff the* $\mathsf{F}(\mathcal{B})$*-module* $\Gamma\xi$ *is free and its rank is equal to the rank n of the bundle* ξ.

Trivializations φ of the vector bundle ξ over an open set $U\subset\mathcal{B}$ are isomorphisms of the bundle θ^n_U onto the bundle $\xi|_U$, and the assertion that U is a trivializing neighborhood means that the bundle $\xi|_U$ is trivial.

§6. Tangent Bundles

Let $\mathcal{X}$ be a smooth n-dimensional manifold, and let

$$\mathsf{T}\mathcal{X} = \bigsqcup_{p\in\mathcal{X}} \mathsf{T}_p\mathcal{X}$$

be the disjoint union of all tangent spaces $\mathsf{T}_p\mathcal{X}$, $p\in\mathcal{X}$. (All possible tangent vectors A of the manifold $\mathcal{X}$ are therefore points of the set $\mathsf{T}\mathcal{X}$.) For each vector $A\in\mathsf{T}\mathcal{X}$, a (unique!) point $p\in\mathcal{X}$ for which $A\in\mathsf{T}_p\mathcal{X}$ is denoted by πA. There thus arises the mapping

$$\pi\colon\mathsf{T}\mathcal{X}\to\mathcal{X}$$

such that $\pi^{-1}(p)=\mathsf{T}_p\mathcal{X}$ for each point $p\in\mathcal{X}$.

For an arbitrary open set $U\subset\mathcal{X}$, the subset $\pi^{-1}U\subset\mathsf{T}\mathcal{X}$ is naturally identified with the set $\mathsf{T}U$. In the case where U is the support of the chart $(U,h)=(U,x^1,\dots,x^n)$, the mapping

$$\mathsf{T}h\colon\mathsf{T}U\to\mathbb{R}^{2n}$$

is well defined and transforms an arbitrary tangent vector $A \in \mathbf{T}U$ into the vector

$$(\mathbf{T}h)A = (x^1, \dots, x^n, a^1, \dots, a^n) \in \mathbb{R}^{2n},$$

where $x^1, \dots, x^n$ are the coordinates of the point $p = \pi A$ in the chart (U, h) and $a^1, \dots, a^n$ are the coordinates of the vector A in the basis

$$\left(\frac{\partial}{\partial x^1}\right)_p, \quad \dots, \quad \left(\frac{\partial}{\partial x^n}\right)_p$$

of the space $\mathbf{T}_p\mathcal{X}$. Therefore,

$$A = a^1 \left(\frac{\partial}{\partial x^1}\right)_p + \dots + a^n \left(\frac{\partial}{\partial x^n}\right)_p;$$

we note that the numbers $a^1, \dots, a^n$ are uniquely determined by the vector A. It is clear that the mapping $\mathbf{T}h$ is bijective and the pair $(\mathbf{T}U, \mathbf{T}h)$ is therefore a chart in $\mathbf{T}\mathcal{X}$.

Let (U, h) and (U', h') be two charts of the manifold $\mathcal{X}$ (for definiteness, we assume that they intersect), and let

$$x^{i'} = x^{i'}(x^1, \dots, x^n), \quad i = 1, \dots, n, \tag{5}$$

be transition formulas of the corresponding local coordinates (in the intersection $U \cap U'$). By definition, for each point $p \in U \cap U'$, the coordinates $a^1, \dots, a^n$ and $a^{1'}, \dots, a^{n'}$ of an arbitrary vector $A \in \mathbf{T}_p\mathcal{X}$ in the charts (U, h) and (U', h') are related by

$$a^{i'} = \left(\frac{\partial x^{i'}}{\partial x^i}\right)_p a^i, \tag{6}$$

where $(\partial x^{i'}/\partial x^i)_p$ are the values of the partial derivatives of function (5) at the point p.

On the other hand, it is clear that $\mathbf{T}h$ and $\mathbf{T}h'$ map the set

$$\mathbf{T}(U \cap U') = \mathbf{T}U \cap \mathbf{T}U'$$

onto the respective open sets $h(U \cap U') \times \mathbb{R}^n$ and $h'(U \cap U') \times \mathbb{R}^n$ of the space $\mathbb{R}^{2n} = \mathbb{R}^n \times \mathbb{R}^n$ and, moreover, formulas (5) and (6) taken together define the mapping $\mathbf{T}h' \circ (\mathbf{T}h)^{-1}$ of the first set onto the second one. Because the Jacobian $\|(\partial x^{i'}/\partial x^i)_p\|$ is nonzero at all points $p \in U \cap U'$, this implies that this mapping is a diffeomorphism.

Therefore, *the charts* $(\mathbf{T}U, \mathbf{T}h)$ *and* $(\mathbf{T}U', \mathbf{T}h')$ *are compatible.* Obviously, this conclusion also holds for $U \cap U' = \emptyset$ (because $\mathbf{T}U \cap \mathbf{T}U' = \emptyset$ if $U \cap U' = \emptyset$).

Therefore, charts of the form $(\mathbf{T}U, \mathbf{T}h)$ compose an atlas on $\mathbf{T}\mathcal{X}$ and therefore define a certain smooth structure on $\mathbf{T}\mathcal{X}$.

Definition 34.5. The constructed smooth manifold $\mathbf{T}\mathcal{X}$ is called the *manifold of tangent vectors* of the manifold $\mathcal{X}$. Its dimension is equal to $2n$, where $n = \dim \mathcal{X}$.

We note that because the derivatives are in formula (6), the class of smoothness of the manifold $\mathbf{T}\mathcal{X}$ is one less than the class r of smoothness of the manifold $\mathcal{X}$ (for $r = \infty$ or $r = \omega$, the class of smoothness is obviously unchanged).

By definition, the local coordinates corresponding to the chart $(\mathbf{T}U, \mathbf{T}h)$ are the numbers $x^1, \dots, x^n, a^1, \dots, a^n$. (Therefore, $x^1, \dots, x^n$ simultaneously denote the local coordinates in U as well as a part of the local coordinates in $\mathbf{T}U$. When we are sufficiently careful, this does not lead to confusion).

In the local coordinates $x^1, \dots, x^n, a^1, \dots, a^n$ (on $\mathbf{T}U$) and $x^1, \dots, x^n$ (on U), the mapping π is written by

$$x^i = x^i, \quad i = 1, \dots, n,$$

where x^i are coordinates on U in the left-hand side and coordinates on $\mathbf{T}U$ in the right-hand side. Therefore, we see that *the mapping* π *is smooth and is a submersion.* By Proposition 32.2, *each tangent space* $\mathbf{T}_p\mathcal{X} = \pi^{-1}(p)$ *is therefore an embedded submanifold of the manifold* $\mathbf{T}\mathcal{X}$. The numbers $a^1, \dots, a^n$ are coordinates on this submanifold (defined on the whole $\mathbf{T}_p\mathcal{X}$).

In the manifold $\mathbf{T}\mathcal{X}$, we isolate a subset $\mathcal{X}_0$ consisting of zero vectors of the spaces $\mathbf{T}_p\mathcal{X}$. It is easy to see that $\mathcal{X}_0$ *is a closed submanifold that is diffeomorphic to the manifold* $\mathcal{X}$ (the diffeomorphism $\mathcal{X}_0 \to \mathcal{X}$ is induced by the mapping $\pi\colon \mathbf{T}\mathcal{X} \to \mathcal{X}$).

It is convenient to replace the mapping $\mathbf{T}h$ by the mapping $(h^{-1} \times \mathrm{id}) \circ \mathbf{T}h\colon \mathbf{T}U \to U \times \mathbb{R}^n$ that acts according to the formula

$$A \mapsto (p, \boldsymbol{a}), \quad p = \pi(A), \quad \boldsymbol{a} = (a^1, \dots, a^n).$$

Let

$$\varphi_h\colon U \times \mathbb{R}^n \to \mathbf{T}U$$

be the inverse mapping,

$$\varphi_h(p, \boldsymbol{a}) = a^i \left(\frac{\partial}{\partial x^i}\right)_p, \quad p \in U, \quad \boldsymbol{a} \in \mathbb{R}^n.$$

The mapping φ_h is a diffeomorphism that closes the commutative diagram

$$\begin{array}{ccc} U \times \mathbb{R}^n & \xrightarrow{\varphi_h} & \mathbf{T}U \\ & \searrow \quad \swarrow & \\ & U & \end{array},$$

i.e., is a trivialization of the bundle $(\mathbf{T}\mathcal{X}, \pi, \mathcal{X})$ over the neighborhood U. Therefore, *the triple* $\tau_{\mathcal{X}} = (\mathbf{T}\mathcal{X}, \pi, \mathcal{X})$ *is a vector bundle of rank* n. (The bundle $\tau_{\mathcal{X}}$ is also denoted by $\tau\mathcal{X}$ or $\tau(\mathcal{X})$.)

Definition 34.6. This bundle is called the *tangent bundle of the manifold* $\mathcal{X}$.

We stress that *the fibers* $\pi^{-1}(p)$ *of the tangent bundle* $\tau_{\mathcal{X}}$ *are tangent spaces* $\mathbf{T}_p\mathcal{X}$ *of the manifold* $\mathcal{X}$. The bundle $\tau_{\mathcal{X}}$ serves as a directing example to which all general constructions of vector bundles can be set in correspondence.

A manifold $\mathcal{X}$ for which the bundle $\tau\mathcal{X}$ is trivial is said to be *parallelizable.*

§7. Frame Bundles

Definition 34.7. A vector bundle $\xi = (\mathcal{E}, \pi, \mathcal{X})$ is called a *subbundle* of a smooth vector bundle $\xi' = (\mathcal{E}', \pi', \mathcal{X}')$ if $\mathcal{E} \subset \mathcal{E}'$ and $\pi = \pi'|_{\mathcal{E}}$. A subbundle is uniquely determined by the manifold $\mathcal{E}$ and is identified with it as a rule.

Subbundles of the tangent bundles $\tau_{\mathcal{X}} = (\tau_{\mathcal{X}}, \pi, \mathcal{X})$ are of particular importance. They are called *distributions on* $\mathcal{X}$. The fiber over a point $p \in \mathcal{X}$ of an arbitrary distribution $\mathcal{E}$ is the subspace $\mathbf{T}_p\mathcal{X} \cap \mathcal{E}$ of the tangent space $\mathbf{T}_p\mathcal{X}$. It is denoted by $\mathcal{E}_p$. In accordance with the general terminology of vector bundle theory, the dimension of subspaces $\mathcal{E}_p$ (which is the same for all p) is called the *rank* of the distribution $\mathcal{E}$.

Definition 34.8. A submanifold $\mathcal{Y}$ (which is in general only immersed) of the manifold $\mathcal{X}$ is called an *integral manifold* of the distribution $\mathcal{E}$ if $\mathbf{T}_p\mathcal{Y} \subset \mathcal{E}_p$ for any point $p \in \mathcal{Y}$. A connected integral manifold is said to be *maximal* if it is not contained in any larger integral manifold. A distribution $\mathcal{E}$ of rank n is said to be *completely integrable* if a unique maximal integral manifold $\mathcal{Y}$ of dimension n passes through each point of the manifold $\mathcal{X}$ (i.e., $\mathbf{T}_p\mathcal{Y} = \mathcal{E}_p$).

Definition 34.9. Let $\mathfrak{a}[\mathcal{E}]$ be the set of all vector fields $X \in \mathfrak{a}\mathcal{X}$ such that $X_p \in \mathcal{E}_p$ for any point $p \in \mathcal{X}$. It is clear that $\mathfrak{a}[\mathcal{E}]$ is a submodule of the Lie algebra $\mathfrak{a}\mathcal{X}$ of all vector fields on $\mathcal{X}$. A distribution $\mathcal{E}$ is said to be *involutive* if $\mathfrak{a}[\mathcal{E}]$ is a subalgebra of the Lie algebra $\mathfrak{a}\mathcal{X}$, i.e., if $[X, Y] \in \mathfrak{a}[\mathcal{E}]$ for any fields $X, Y \in \mathfrak{a}[\mathcal{E}]$.

Exercise 34.1. Prove that for any distribution $\mathcal{E}$ of rank n, the module $\mathfrak{a}[\mathcal{E}]$ is a locally free module of rank n, i.e., there is an open covering $\{U\}$ of the manifold $\mathcal{X}$ such that over any element U, the $\mathbf{F}U$-module $\mathfrak{a}[\mathcal{E}|_U]$ of vector fields on U, where $\mathcal{E}|_U$ is the restriction of the distribution $\mathcal{E}$ to U, is a free module of rank n over the algebra $\mathbf{F}U$. [*Hint*: Take trivializing neighborhoods of the bundle $\mathcal{E}$ as U.]

The bases of the $\mathbf{F}U$-module $\mathfrak{a}[\mathcal{E}|_U]$ are called *bases over* U *of the module* $\mathfrak{a}[\mathcal{E}]$.

Each distribution $\mathcal{E}$ defines the field $p \mapsto \mathcal{E}_p$ of subspaces $\mathcal{E}_p \subset \mathbf{T}_p\mathcal{X}$; this is a bijective correspondence between distributions of rank n and smooth fields of n-dimensional subspaces. Therefore, *distributions and smooth fields of subspaces are in fact the same objects.*

Exercise 34.2. Prove that the following properties of a distribution $\mathcal{E}$ are equivalent:

1. The distribution $\mathcal{E}$ is completely integrable.
2. There is an atlas of the manifold X consisting of those charts $(U, x^1, \dots, x^m)$, $m = \dim \mathcal{X}$, for which each submanifold in U given by the equations
$$x^{n+1} = \text{const}, \quad \dots, \quad x^m = \text{const} \tag{7}$$
is an integral manifold of the distribution $\mathcal{E}$.
3. There is an atlas of the manifold $\mathcal{X}$ that consists of those charts $(U, x^1, \dots, x^m)$ for which the vector fields
$$\frac{\partial}{\partial x^1}, \quad \dots, \quad \frac{\partial}{\partial x^m}$$
form a basis over U of the module $\mathfrak{a}[\mathcal{E}]$.

[*Hint*: Assertions 2 and 3 are obvious restatements of one another. The implication $1 \Rightarrow 2$ easily follows from the general properties of submanifolds. To prove the implication $2 \Rightarrow 1$, introduce a topology on $\mathcal{X}$ whose basis of open sets consists of integral manifolds of the distribution $\mathcal{E}$ and consider components of the manifold $\mathcal{X}$ in this topology.]

Exercise 34.3. Prove that a maximal integral manifold $\mathcal{Y}$ of a completely integrable distribution satisfying the second countability axiom is conservative. [*Hint*: Each point $p \in \mathcal{Y}$ admits a coordinate neighborhood U in $\mathcal{X}$ such that components of the intersection $U \cap \mathcal{Y}$ (in the topology of the manifold $\mathcal{Y}$) are given by equations of form (7) and, moreover, the number of these components is no more than countable. Therefore, they are also the components of the intersection $U \cap \mathcal{X}$ in the topology of the manifold $\mathcal{X}$.]

Exercise 34.4. Show that a distribution $\mathcal{E}$ is involutive iff there exists an open covering $\{U\}$ of the manifold $\mathcal{X}$ over each element of which the module $\mathfrak{a}[\mathcal{E}]$ admits a basis $X_1, \dots, X_n$ such that
$$[X_i, X_j] = f_{ij}^k X_k \quad \text{on } U,$$
where f_{ij}^k are certain smooth functions on U.

The assertions in Exercises 34.2 and 34.4 show that the complete integrability property as well as the involutiveness property are *local properties* (i.e., they hold everywhere if they hold in a neighborhood of each point).

Exercise 34.5. Prove that any involutive distribution is completely integrable. [*Hint*: If two fields X and Y are tangent to a submanifold, then the field $[X, Y]$ also has this property.]

It turns out that the converse statement also holds,

Theorem 34.1. *Each involutive distribution is completely integrable.*

This theorem is known as the *Frobenius theorem*.

For any vector bundle $\xi = (\mathcal{E}, \pi, \mathcal{B})$, we consider the submanifold $\mathcal{E}$ of the direct product

$$\underbrace{\mathcal{E} \times \cdots \times \mathcal{E}}_{n \text{ times}}$$

consisting of the points $\boldsymbol{p} = (p_1, \ldots, p_n)$ all of whose components $p_1, \ldots, p_n \in \mathcal{E}$ belong to the same fiber of the bundle ξ (i.e., satisfy the relations $\pi(p_1) = \cdots = \pi(p_n)$) and compose a basis in this fiber. Setting $\pi(\boldsymbol{p}) = \pi(p_1)$, we obtain the mapping

$$\pi: \mathcal{E} \to \mathcal{B},$$

which is obviously continuous and surjective.

Definition 34.10. The corresponding triple $\xi = (\mathcal{E}, \pi, \mathcal{B})$ is called the *frame bundle* of the vector bundle ξ.

Remark 34.3. The frame bundle is obviously the so-called locally trivial principal bundle, but, in essence, we do not need the general theory of bundles of such type.

§8. Metricizable Bundles

We note that for any neighborhood U that trivializes the bundle ξ, each basis (4) of the module $\Gamma(\xi|_U)$ (the trivialization φ of the bundle ξ over U) defines a certain section of the bundle ξ over U by

$$s(b) = (s_1(b), \ldots, s_n(b)), \quad b \in U.$$

Trivializations of the bundle ξ over U are thus identified with sections of the bundle ξ over U.

Definition 34.11. Each smooth function $Q: \mathcal{E} \to \mathbb{R}$ whose restriction to each fiber $\mathcal{F}_b$, $b \in \mathcal{B}$, is a positive-definite quadratic functional is called a *metric* on ξ.

A vector bundle on which a metric exists is said to be *metricizable*, and a metricizable bundle with a metric given on it is called a *metric bundle*. Metric bundles are characterized by each of their fibers being an Euclidean space.

Proposition 34.2. *Any vector bundle over a paracompact space $\mathcal{B}$ is metricizable.*

§9. ξ-Tensor Fields

As usual, let $\xi = (\mathcal{E}, \pi, \mathcal{B})$ be a smooth $\mathbb{R}$-vector bundle of rank n over an m-dimensional Hausdorff manifold $\mathcal{B}$, and for any point $b \in \mathcal{B}$, let a tensor S_p be given in the linear space $\mathcal{F}_b = \mathcal{F}_b^\xi$. Moreover, let the tensors S_b have the same type (r, s) at all points $b \in \mathcal{B}$. Further, let $s_1, \dots, s_n$ be a smooth trivialization of the bundle ξ over a neighborhood $U \subset \mathcal{B}$. Because the vectors $s_1(b), \dots, s_n(b)$ form a basis of the linear space $\mathcal{F}_b$ for any point $b \in U$, we have the formula

$$S_b = S^{j_1 \cdots j_s}_{i_1 \cdots i_r}(b) s^{i_1}(b) \otimes \cdots \otimes s^{i_r}(b) \otimes s_{j_1}(b) \otimes \cdots \otimes_{j_s}(b) \tag{8}$$

for the tensor S_b, where $s^1(b), \dots, s^n(b)$ is the dual basis of the dual space $\mathcal{F}^*_b$ and $S^{j_1 \cdots j_s}_{i_1 \cdots i_r}$ are certain functions defined on U.

Definition 34.12. If the functions $S^{j_1 \cdots j_s}_{i_1 \cdots i_r}$ are smooth for any smooth trivialization $(U, s_1, \dots, s_n)$, then the correspondence

$$S \colon b \mapsto S_b$$

is called a *ξ-tensor field* (or merely *ξ-tensor*) on $\mathcal{B}$, and the functions $S^{j_1 \cdots j_s}_{i_1 \cdots i_r}$ are called its *components* (in the given trivialization).

In particular, for $(r, s) = (0, 1)$, we obtain *ξ-vector fields* (sections of the bundle ξ), which are already known. For $(r, s) = (1, 0)$, the field S is called a *ξ-covector field.* The field $s^i \colon b \mapsto s^i(b), 1 \le i \le n$, is an example of a ξ-covector field (defined on a neighborhood U).

For $\xi = \tau_{\mathcal{X}}$, where $\tau_{\mathcal{X}}$ is the tangent bundle of a smooth manifold $\mathcal{X}$, ξ-tensor fields are tensor fields on $\mathcal{X}$.

Remark 34.4. Traditionally, instead of the tensor S, we often speak about the *tensor* $S^{j_1 \cdots j_s}_{i_1 \cdots i_r}$; of course, this terminology (which we also use) assumes that the trivialization $(U, s_1, \dots, s_n)$ is chosen and fixed for what follows.

It is clear that on the intersection $U \cap U'$ of two trivializing neighborhoods U and U', the components $S^{j_1 \cdots j_s}_{i_1 \cdots i_r}$ and $S^{j'_1 \cdots j'_s}_{i'_1 \cdots i'_r}$ of the same tensor S are related by

$$S^{j'_1 \cdots j'_s}_{i'_1 \cdots i'_r} = \varphi^{i_1}_{i'_1} \cdots \varphi^{i_r}_{i'_r} \varphi^{j'_1}_{j_1} \cdots \varphi^{j'_s}_{j_s} S^{j_1 \dots j_s}_{i_1 \dots i_r}, \tag{9}$$

where $\|\varphi^{j'}_j\|$ and $\|\varphi^i_{i'}\|$ are mutually inverse transition matrices. We are already familiar with formula (9) for the case of a tangent bundle.

Obviously, all ξ-tensor fields of a given type (r, s) form a module over the algebra $\mathbf{F}\mathcal{B}$ of smooth $\mathbb{R}$-valued functions on $\mathcal{B}$. This module is denoted by $\Gamma^s_r\xi$.

For any ξ-tensor fields S and T (of the respective types (r, s) and (r_1, s_1)), the formula

$$S \otimes T \colon b \mapsto S_b \otimes T_b, \quad b \in \mathcal{B},$$

correctly defines the ξ-vector field $S \otimes T$ of type $(r + r_1, s + s_1)$ for which

$$(S \otimes T)^{j_1 \cdots j_s j_{s+1} \cdots j_{s+s_1}}_{i_1 \cdots i_r i_{r+1} \cdots i_{r+r_1}} = S^{j_1 \cdots j_s}_{i_1 \cdots i_r} T^{j_{s+1} \cdots j_{s+s_1}}_{i_{r+1} \cdots i_{r+r_1}}.$$

This field is called the tensor product of the fields S and T. The tensor product $S \otimes T \otimes \cdots \otimes R$ of any number of ξ-tensor fields is defined similarly.

Using the tensor products

$$s^{i_1} \otimes \cdots \otimes s^{j_r} \otimes s_{j_1} \otimes \cdots \otimes s_{j_s} \tag{10}$$

(which are ξ-tensor fields over U), we can rewrite (8) as

$$S = S^{j_1 \cdots j_s}_{i_1 \cdots i_r} s^{i_1} \otimes \cdots \otimes s^{i_r} \otimes s_{j_1} \otimes \cdots \otimes s_{j_s}.$$

This means that *tensors* (10) *form a basis of the* $\mathsf{F}U$*-module* $\Gamma^s_r(\xi|_U)$ (or, as we say, *a basis of the* $\mathsf{F}\mathcal{B}$*-module* $\Gamma^s_r\xi$ *over* U). For example, for $(r, s) = (1, 0)$, a basis of the module $\Gamma^0_1\xi$ (which is also denoted by $\Gamma\xi^*$) over U consists of the ξ-covector fields $s^1, \dots, s^n$.

§10. Multilinear Functions and ξ-Tensor Fields

Tensor fields of type $(r, 0)$ are of special importance. Let S be such a field, and let $t_1, \dots, t_r \in \Gamma\xi$. Then for any trivialization $(U, s_1, \dots, s_n)$ of the bundle ξ on an open set U, the function

$$S(t_1, \cdots, t_n) = S_{i_1 \cdots i_r} t_1^{i_1} \cdots t_r^{i_r} \tag{11}$$

is well defined, where $t_1^{i_1}, \dots, t_r^{i_r}$ are components of the ξ-vector fields $t_1, \dots, t_r$ on U and $S_{i_1 \cdots i_r}$ are components of the ξ-covector fields S. It is easy to see that formula (11) correctly defines the function $S(t_1, \dots, t_r)$ on the whole manifold $\mathcal{B}$. The function $S(t_1, \dots, t_r)$ is called the contraction of the tensor S by the fields $t_1, \dots, t_r$.

Similar to the case of linear spaces, the mapping

$$S: \underbrace{\Gamma\xi \times \cdots \times \Gamma\xi}_{r \text{ times}} \mapsto \mathsf{F}\mathcal{B} \tag{12}$$

is called an $\mathsf{F}\mathcal{B}$*-multilinear functional* if it is $\mathsf{F}\mathcal{B}$-linear in each of the arguments. It is clear that mapping (12) defined by formula (11) is an $\mathsf{F}\mathcal{B}$-multilinear functional, and *the correspondence*

$$\text{tensor } S \Longrightarrow \text{functional (12)}$$

is an isomorphism of the $\mathsf{F}\mathcal{B}$*-module* $\Gamma_r\xi = \Gamma^0_r\xi$ *onto the* $\mathsf{F}\mathcal{B}$*-module of all* $\mathsf{F}\mathcal{B}$*-multilinear functionals* (12).

In what follows, we use this isomorphism to identify ξ-tensor fields of type $(r, 0)$ on $\mathcal{B}$ with $\mathsf{F}\mathcal{B}$-multilinear functionals (12). In particular, because of this identification, ξ-covector fields $c \in \Gamma_1\xi = \Gamma\xi^*$ are $\mathsf{F}\mathcal{B}$-linear functionals of the form

$$c: \Gamma\xi \to \mathsf{F}\mathcal{B}.$$

The value $c(s)$ of such a field on a section $s \in \Gamma\xi$ (we stress that it is a function on $\mathcal{B}$) is also denoted by $\langle s, c \rangle$.

§11. Tensor Product of Vector Bundles

Let $\xi = (\mathcal{E}^\xi, \pi^\xi, \mathcal{B})$ and $\eta = (\mathcal{E}^\eta, \pi^\eta, \mathcal{B})$ be vector bundles over the field $\mathbb{R}$ of the respective ranks n and m with the same base $\mathcal{B}$, and let $\{U_\alpha\}$ be an open covering of the space $\mathcal{B}$ consisting of coordinate neighborhoods that trivialize each of these bundles (obviously, such a covering always exists). Further, let

$$\varphi^\xi_\alpha : U_\alpha \times \mathbb{R}^n \to \mathcal{E}^\xi_{U_\alpha} \qquad \text{and} \qquad \varphi^\eta_\alpha : U_\alpha \times \mathbb{R}^m \to \mathcal{E}^\eta_{U_\alpha}$$

be trivializations of the bundles ξ and η over U_α.

We consider the disjoint union

$$\mathcal{E} = \bigsqcup_{b \in \mathcal{B}} \mathcal{F}^\xi_b \otimes \mathcal{F}^\eta_b$$

and its projection

$$\pi : \mathcal{E} \to \mathcal{B}$$

on $\mathcal{B}$ that maps each product $\mathcal{F}^\xi_b \otimes \mathcal{F}^\eta_b$ into the point $b \in \mathcal{B}$. It is easily verified that *the triple* $(\mathcal{E}, \pi, \mathcal{B})$ *is a vector bundle of rank* nm. The corresponding trivializing mappings

$$\varphi_\alpha : U_\alpha \times \mathbb{R}^{nm} \to \mathcal{E}_{U_\alpha} = \bigsqcup_{b \in U_\alpha} \mathcal{F}^\xi_b \otimes \mathcal{F}^\eta_b$$

are given by the formula

$$\varphi_\alpha(b, \boldsymbol{x}) = (\varphi^\xi_{b,\alpha} \otimes \varphi^\eta_{b,\alpha})(\boldsymbol{x}), \quad \text{where } b \in U_\alpha, \quad \boldsymbol{x} \in \mathbb{R}^{nm},$$

because of the standard identification $\mathbb{R}^{nm} = \mathbb{R}^n \otimes \mathbb{R}^m$.

Definition 34.13. The constructed bundle $(\mathcal{E}, \pi, \mathcal{B})$ is called the *tensor product* of the bundles ξ and η and is denoted by $\xi \otimes \eta$.

§12. Generalization

The direct sum $\xi \oplus \eta$ of bundles ξ and η (which is called the *Whitney sum* of these bundles) with the fibers $\mathcal{F}^\xi_b \oplus \mathcal{F}^\eta_b$ and the morphism bundle $\operatorname{Hom}(\xi, \eta)$, whose fibers are the spaces $\operatorname{Hom}(\mathcal{F}^\xi_b, \mathcal{F}^\eta_b)$ of linear mappings $\mathcal{F}^\xi_b \to \mathcal{F}^\eta_b$, are defined similarly.

For a single bundle ξ, the bundle ξ^* is also defined; the fibers of this bundle are dual spaces $(\mathcal{F}^\xi_b)^*$, and for each $p \geq 0$, the bundle Λ^p with the fibers $\Lambda^p \mathcal{F}^\xi_b$ is also defined.

In particular, for an arbitrary smooth manifold $\mathcal{X}$, the smooth vector bundle

$$\tau^*_{\mathcal{X}} = \tau^* \mathcal{X} = (\mathbf{T}^* \mathcal{X}, \pi, \mathcal{X})$$

is defined; its fibers are the cotangent spaces $\mathsf{T}_p^*\mathcal{X}$ of the manifold $\mathcal{X}$. This bundle is called the *cotangent bundle over* $\mathcal{X}$. Each chart $(U,h) = (U, x^1, \dots, x^n)$ of the manifold $\mathcal{X}$ defines a trivialization $\varphi: U \times \mathbb{R}^n \to \mathsf{T}^*U$ of the bundle $\tau_{\mathcal{X}}^*$ over U for which

$$\varphi(p, \boldsymbol{a}) = a_i (dx^i)_p,$$

where $p \in U$ and $\boldsymbol{a} = (a_1, \dots, a_n) \in \mathbb{R}^n$. Sections of this bundle are linear differential forms on $\mathcal{X}$.

More generally, for each $r \geq 0$, we can consider the bundle $\Lambda^r \tau^* \mathcal{X}$ over $\mathcal{X}$. Sections of the bundle $\Lambda^r \tau^* \mathcal{X}$ over $\mathcal{X}$ are exactly differential forms of degree r on $\mathcal{X}$. We note that the bundles $\Lambda^r \tau^* \mathcal{X}$ and $(\Lambda^r \tau \mathcal{X})^*$ are naturally isomorphic.

§13. Tensor Product of Sections

Let r and s be nonnegative integers. The bundle

$$\underbrace{\tau_{\mathcal{X}}^* \otimes \cdots \otimes \tau_{\mathcal{X}}^*}_{r \text{ times}} \otimes \underbrace{\tau_{\mathcal{X}} \otimes \cdots \otimes \tau_{\mathcal{X}}}_{s \text{ times}}$$

is denoted by $\tau_r^s \mathcal{X}$ and is called the *tensor bundle of type* (r,s) *over* $\mathcal{X}$. Its sections are tensor fields of type (r,s) on $\mathcal{X}$. More generally, for any vector bundle ξ, we set

$$T_r^s \xi = \underbrace{\xi^* \otimes \cdots \otimes \xi^*}_{r \text{ times}} \otimes \underbrace{\xi \otimes \cdots \otimes \xi}_{s \text{ times}}.$$

Therefore, $\tau_r^s \mathcal{X} = T_r^s \tau \mathcal{X}$.

Sections of the bundle $T_r^s \xi$ are exactly the ξ-tensor fields of type (r,s) on the manifold $\mathcal{B}$ introduced above. Therefore, the modules $\Gamma_r^s \xi$ introduced there are exactly the modules of sections $\Gamma T_r^s \xi$ of the bundle $T_r^s \xi$:

$$\Gamma_s^r \xi = \Gamma T_r^s \xi. \tag{13}$$

Each section s of the bundle $\operatorname{Hom}(\xi, \eta)$ sets the linear mapping $s(b): \mathcal{F}_b^\xi \to \mathcal{F}_b^\eta$, in correspondence to an arbitrary point $b \in \mathcal{B}$, and the formula

$$s^\sharp(p) = s(b)p, \quad p \in \mathcal{E}^\xi,$$

where $b = \pi(p)$, therefore defines the mapping $s^\sharp: \mathcal{E}^\xi \to \mathcal{E}^\eta$ that closes the commutative diagram

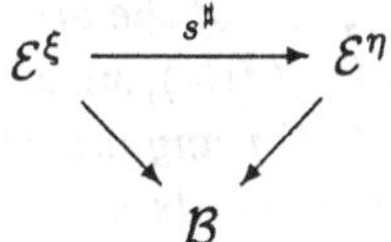

and is a morphism $\xi \to \eta$. It turns out that the correspondence

$$s \mapsto s^\sharp$$

defines an isomorphism of the $\mathsf{F}\mathcal{B}$-module $\Gamma(\operatorname{Hom}(\xi,\eta))$ of all sections of the bundle $\operatorname{Hom}(\xi,\eta)$ onto the $\mathsf{F}\mathcal{B}$-module $\operatorname{Mor}(\xi,\eta)$ of all morphisms $\xi \to \eta$. Therefore,

$$\operatorname{Mor}(\xi,\eta) = \Gamma(\operatorname{Hom}(\xi,\eta)), \tag{14}$$

and morphisms $\xi \to \eta$ are sections of the bundle $\operatorname{Hom}(\xi,\eta)$.

For any two sections $s^\xi: \mathcal{B} \to \mathcal{E}^\xi$ and $s^\eta: \mathcal{B} \to \mathcal{E}^\eta$ of the bundles ξ and η, the formula

$$(S^\xi \otimes s^\eta)(b) = s^\xi(b) \otimes s^\eta(b), \quad b \in \mathcal{B},$$

defines the section

$$s^\xi \otimes s^\eta: \mathcal{B} \to \mathcal{E}^{\xi\otimes\eta}$$

of the bundle $\xi \otimes \eta$, which is called the *tensor product of the sections* s^ξ and s^η.

If bundles ξ and η are trivial over an open set U, then for any bases

$$s_1^\xi, \dots, s_n^\xi \qquad \text{and} \qquad s_1^\eta, \dots, s_n^\eta$$

of the $\mathsf{F}\mathcal{B}$-modules $\Gamma\xi$ and $\Gamma\eta$ over U, the sections

$$s_i^\xi \otimes s_k^\eta, \quad 1 \le i \le n, \quad 1 \le k \le m,$$

obviously form a basis of the $\mathsf{F}\mathcal{B}$-module $\Gamma(\xi \otimes \eta)$ over U. Therefore, any section of the bundle $\xi \otimes \eta$ over U is uniquely represented in the form

$$s^k \otimes s_k^\eta, \quad k = 1, \dots, m,$$

where s^k are certain sections of the bundle ξ over U.

§14. Inverse Image of a Vector Bundle

Proposition 34.3. *For any vector bundle $\xi = (\mathcal{E}, \pi, \mathcal{B})$ and any continuous mapping $f: \mathcal{B}' \to \mathcal{B}$, there exist a vector bundle $\xi' = (\mathcal{E}', \pi', \mathcal{B}')$ over $\mathcal{B}'$ and a regular morphism $\varphi: \xi' \to \xi$ that induces the mapping f:*

$$\begin{array}{ccc} \mathcal{E}' & \xrightarrow{\varphi} & \mathcal{E} \\ {\scriptstyle \pi'}\downarrow & & \downarrow{\scriptstyle \pi} \\ \mathcal{B}' & \xrightarrow{f} & \mathcal{B} \end{array} . \tag{15}$$

For any trivializing atlas $\{(U_\alpha, \varphi_\alpha)\}$ of the bundle ξ, the pairs $(U'_\alpha, \varphi'_\alpha)$, where $U'_\alpha = f^{-1}U_\alpha$ and $\varphi'_\alpha(b', \boldsymbol{x}) = (\varphi_\alpha(f(b'), \boldsymbol{x}), \boldsymbol{x})$, $b' \in U'_\alpha$, $\boldsymbol{x} \in \mathbb{R}^n$, compose a trivializing atlas of the bundle ξ'. For any morphism $\psi: \eta \to \xi$ of vector bundles that induces the mapping f, there exists a unique morphism $\chi: \eta \to f^*\xi$ over $\mathcal{B}'$ satisfying the relation

$$\psi = \varphi \circ \chi. \tag{16}$$

Therefore, the bundle ξ' is uniquely defined up to an isomorphism.

Definition 34.14. The bundle ξ' is called the *bundle induced by f from* ξ or the *inverse image* of the bundle ξ. It is denoted by f^* (sometimes by $f^{-1}\xi$).

In the language of category theory, commutative square (15) is a *universal square*; in other terminology, the bundle $f^*\xi$ is the *coamalgam* of the diagram $\mathcal{B}' \xrightarrow{f} \mathcal{B} \xleftarrow{\pi} \mathcal{E}$.

For any mappings $g: \mathcal{B}'' \to \mathcal{B}'$ and $f: \mathcal{B}' \to \mathcal{B}$, the bundle $(f \circ g)^*\xi$ is naturally isomorphic to the bundle $g^*(f^*\xi)$ (and can therefore be identified with it).

In diagram (3), we can assume without loss of generality that the neighborhood U_α is a coordinate neighborhood of the manifold $\mathcal{B}$. Then the mapping

$$g_\alpha = (\mathrm{id} \times h_\alpha^{-1}) \circ \varphi_\alpha^{-1}: \mathcal{E}_{U_\alpha} \to \mathbb{R}^n \times \mathbb{R}^n = \mathbb{R}^{n+m},$$

where $h_\alpha: U\alpha \to \mathbb{R}^m$ is a coordinate mapping, is the coordinate mapping of the chart $(\mathcal{E}_{U_\alpha}, g_\alpha)$ of the manifold $\mathcal{E}$. Moreover, the local coordinates

$$a^1, \dots, a^n, x^1, \dots, x^m \tag{17}$$

of a point $p \in \mathcal{E}_{U_\alpha}$ are the coordinates $a^1, \dots, a^n$ of the vector $\boldsymbol{a} = \varphi_{\alpha,b}^{-1}(p)$, $b = \pi(p)$, of the space $\mathbb{R}^n$ and the coordinates $x^1, \dots, x^m$ of the vector $\boldsymbol{x} = h_\alpha(b)$ of the space $\mathbb{R}^m$.

Chapter 35
Connections on Vector Bundles

§1. Vertical Subspaces

We note that symbols $x^1, \dots, x^m$ denote not only a part of the local coordinates in formula (17) in Chap. 34 in the chart $(\mathcal{E}_{U_\alpha}, g_\alpha)$ but also the local coordinates in the chart (U_α, h_α) of the manifold $\mathcal{B}$. Because of this convention, which does not lead to confusion if we are sufficiently careful, the projection π is written in local coordinates by the tautological relations

$$x^k = x^k, \quad k = 1, \dots, m.$$

In particular, this implies that *for any point* $p \in \mathcal{F}_b$, *the tangent space* $\mathsf{T}_p\mathcal{F}_p$ *of the fiber* $\mathcal{F}_b$ *coincides with the kernel* $\operatorname{Ker}(d\pi)_p$ *of the linear surjective mapping*

$$(d\pi)_p : \mathsf{T}_p\mathcal{E} \to \mathsf{T}_b\mathcal{B}, \quad b = \pi(p),$$

i.e., *for any point* $p \in \mathcal{F}_b$, *we have the exact sequence*

$$0 \longrightarrow \mathsf{T}_p\mathcal{F}_b \xrightarrow{(d\imath_b)_p} \mathsf{T}_p\mathcal{E} \xrightarrow{(d\pi)_p} \mathsf{T}_b\mathcal{B} \longrightarrow 0, \tag{1}$$

where $\imath$ *is an embedding* $\mathsf{F}_b \to \mathcal{E}$. We note that the term $\mathsf{T}_p\mathcal{F}_b$ in this sequence, which is usually called the *vertical space* of the linear space $\mathsf{T}_p\mathcal{E}$, is naturally identified with the linear space $\mathcal{F}_b$.

We say that a *field of* m*-dimensional subspaces* H is given on an $(m+n)$-dimensional manifold $\mathcal{E}$ if a certain m-dimensional subspace H_p of the space $\mathsf{T}_p\mathcal{E}$ corresponds to each point $p \in \mathcal{E}$.

Let V be an open set of the manifold $\mathcal{E}$.

Proposition 35.1. *The following conditions are equivalent for a field* H *of* m*-dimensional subspaces:*

1. *There exist smooth linear differential forms* $\theta^1, \dots, \theta^n$ *on* V *such that for any point* $p \in V$, *the subspace* H_p *is the annihilator of the covectors* $\theta^1_p, \dots, \theta^n_p$,
$$H_p = \operatorname{Ann}(\theta^1_p, \dots, \theta^n_p). \tag{2}$$
2. *There exist smooth vector fields* $X^{(1)}, \dots, X^{(m)}$ *on* V *such that for any point* $p \in V$, *the space* H_p *is generated by the vectors* $X^{(1)}, \dots, X^{(m)}_p$,
$$H_p = [X^{(1)}_p, \dots, X^{(m)}_p]. \tag{3}$$

Definition 35.1. A field of subspaces

$$H : p \mapsto H_p, \quad p \in \mathcal{E},$$

is said to be *smooth* if there exists an open covering $\{V_\alpha\}$ of the manifold $\mathcal{E}$ on each element of which the field H satisfies condition 1 or 2.

§2. Fields of Horizontal Subspaces

Definition 35.2. In the case where $\mathcal{E}$ is the total space of a (vector) bundle ξ (and the vertical space $\mathbf{T}_p\mathcal{F}_b$, $b = \pi(p)$, is therefore defined at each point $p \in \mathcal{E}$) and for any point $p \in \mathcal{E}$, we have the relation

$$\mathbf{T}_p\mathcal{E} = \mathbf{T}_pF_b \oplus H_p \tag{5}$$

(the subspace H_p is complementary to the subspace $\mathbf{T}_p\mathcal{F}_b$), field (4) is called a *field of horizontal subspaces.*

Because $\operatorname{Ker} d\pi_p = \mathbf{T}_p\mathcal{F}_b$, the linear mapping $d\pi_p$ is an isomorphism on each such subspace H_p. We stress that there is a considerable arbitrariness in the choice of horizontal subspaces, i.e., there exist many (even smooth) fields (4) on $\mathcal{E}$ that satisfy condition (5), while the vertical subspaces $\mathbf{T}_p\mathcal{F}_b$ are uniquely defined.

When a field H of horizontal subspaces is given, every vector field X on $\mathcal{E}$ is uniquely represented in the form

$$X = X^V + X^H,$$

where X^V and X^H are fields on $\mathcal{E}$ such that for any point $p \in \mathcal{E}$, the vector X_p^V is vertical and the vector X_p^H is horizontal.

Proposition 35.2.

1. *A field H of horizontal subspaces is smooth iff for any smooth vector field X on $\mathcal{E}$, the fields X^V and X^H are also smooth.*
2. *If the field H is smooth, then a field X is smooth iff the fields X^V and X^H are smooth.*

As a rule, we define fields of horizontal subspaces by formulas of form (2) taking open sets of the form $\mathcal{E}_U$, where $U \in \mathcal{B}$, as V. Admitting a certain incorrectness of terminology, we call such a field the *annihilator of the forms $\theta^1, \dots, \theta^n$ over U* and write

$$H = \operatorname{Ann}(\theta^1, \dots, \theta^n) \quad \text{over } U.$$

By definition, this means that for each point $p \in \mathcal{E}_U$, the subspace H_p consists of those vectors $A \in \mathbf{T}_p\mathcal{F}_b$ for which

$$\theta_p^i(A) = 0 \quad \text{for any } i = 1, \dots, n. \tag{7}$$

Such a field is a field of horizontal subspaces iff for any point $p \in \mathcal{E}_U$, the restrictions of the covectors $\theta_p^1, \dots, \theta_p^n$ to the vertical space $\mathbf{T}_p\mathcal{F}_b$, $b = \pi(p)$, are linearly independent, i.e., for a vertical vector $A \in \mathbf{T}_p\mathcal{F}_b$, relations (7) only hold for $A = 0$. Moreover, the forms $\theta^1, \dots, \theta^n$ and $\bar{\theta}^1, \dots, \bar{\theta}^n$ assign the same field H on U, i.e., at any point $p \in \mathcal{E}_U$, satisfy the relation

$$\mathrm{Ann}(\theta_p^1, \ldots, \theta_p^n) = \mathrm{Ann}(\bar{\theta}^1, \ldots, \bar{\theta}^n),$$

iff the covectors $\theta_p^1, \ldots, \theta_p^n$ are linearly equivalent to the covectors $\bar{\theta}_p^1, \ldots, \bar{\theta}_p^n$, i.e., there exist functions

$$c_i^j : p \mapsto c_i^j(p), \quad p \in \mathcal{E}_U,$$

where $i, j = 1, \ldots, n$, such that

$$\bar{\theta}^i = c_i^j \theta^j \quad \text{on } \mathcal{E}_U.$$

For the considered vector bundle $\xi = (\mathcal{E}, \pi, \mathcal{B})$, we now choose and fix a trivializing atlas $\{(U_\alpha, \varphi_\alpha)\}$ such that all trivializing neighborhoods U_α are simultaneously coordinate neighborhoods in the manifold $\mathcal{B}$. The trivializations $\varphi_\alpha : U_\alpha \times \mathbb{R}^n \to \mathcal{E}_{U_\alpha}$ and the coordinate mappings $h_\alpha : U_\alpha \to \mathbb{R}^m$ define new local coordinates (see (17) in Chap. 34) on $\mathcal{E}_{U_\alpha}$ and thus the corresponding coordinate vector and covector fields

$$\frac{\partial}{\partial a^1}, \quad \ldots, \quad \frac{\partial}{\partial a^n}, \quad \frac{\partial}{\partial x^1}, \quad \ldots, \quad \frac{\partial}{\partial x^m},$$
$$da^1, \quad \ldots, \quad da^n, \quad dx^1, \quad \ldots, \quad dx^m.$$

Moreover, at each point $p \in \mathcal{E}_{U_\alpha}$, we have the relations

$$(d\pi_p)\left(\frac{\partial}{\partial a^i}\right)_p = 0, \qquad i = 1, \ldots, n,$$
$$(d\pi_p)\left(\frac{\partial}{\partial x^k}\right)_p = \left(\frac{\partial}{\partial x^k}\right)_b, \quad k = 1, \ldots, m,$$

where $b = \pi(p)$. Therefore, the vectors

$$\left(\frac{\partial}{\partial a^1}\right)_p, \quad \ldots, \quad \left(\frac{\partial}{\partial a^n}\right)_p$$

form a basis of the vertical space $\operatorname{Ker} d\pi_p = \mathbf{T}_p\mathcal{F}_b$, and the covectors $da_p^1, \ldots, da_p^n$ (more precisely, their restrictions to $\mathbf{T}_p\mathcal{F}_b$) form a basis of the dual space $\mathbf{T}_p^*\mathcal{F}_b$.

Therefore, if

$$\theta^i = f_j^i da^j + g_k^i dx^k \quad \text{on } \mathcal{E}_{U_\alpha}, \tag{8}$$

where $i, j = 1, \ldots, n$ and $k = 1, \ldots, m$, then the restrictions of the covectors θ_p^i to $\mathbf{T}_p\mathcal{F}_b$ are linearly independent for any point $p \in \mathcal{E}_{U_\alpha}$ iff the matrix $\|f_j^i\|$ is nonsingular on $\mathcal{E}_{U_\alpha}$.

Proposition 35.3. *For any field H of horizontal subspaces, there exist forms*

$$\overset{(\alpha)}{\theta}{}^i = da^i + e^i_k dx^k, \quad i = 1, \dots, n, \quad k = 1, \dots, m, \tag{9}$$

on the coordinate neighborhood $\mathcal{E}_{U_\alpha}$, where e^i_k are certain functions on $\mathcal{E}_{U_\alpha}$, such that

$$H = \operatorname{Ann}\left(\overset{(\alpha)}{\theta}{}^1, \dots, \overset{(\alpha)}{\theta}{}^i\right) \quad \text{over } U_\alpha.$$

Forms (9) are uniquely characterized by these conditions.

Obviously, a field H of horizontal subspaces is smooth iff for each neighborhood U_α, forms (9), i.e., the functions e^i_k, are smooth on $\mathcal{E}_{U_\alpha}$.

§3. Connections and Their Forms

Among all smooth fields H of horizontal spaces, a special role is played by the fields for which the coefficients e^i_k of forms (9) depend linearly on $a^1, \dots, a^n$, i.e.,

$$e^i_k = \overset{(\alpha)}{\Gamma}{}^i_{kj}\, a^j,$$

where $\overset{(\alpha)}{\Gamma}{}^i_{kj}$ are certain smooth functions of $x^1, \dots, x^m$ (i.e., in other words, smooth functions on the coordinate neighborhood U_α).

Definition 35.3. A smooth field

$$H: p \mapsto H_p, \quad p \in \mathcal{E},$$

of horizontal subspaces is called a *connection* on the vector bundle ξ if for each trivializing coordinate neighborhood U_α, forms (9) defining this field are

$$\overset{(\alpha)}{\theta}{}^i = da^i + \overset{(\alpha)}{\Gamma}{}^i_{kj}\, a^j dx^k, \quad i, j = 1, \dots, n, \quad k = 1, \dots, m,$$

where $\overset{(\alpha)}{\Gamma}{}^i_{kj}$ are certain smooth functions on U_α.

The functions $\overset{(\alpha)}{\Gamma}{}^i_{kj}$ are called the *connection coefficients* of H in the neighborhood U_α. (The order of the subscripts k and j in the symbol for the connection coefficients is still not conventional. Certain authors write them in the reverse order.)

It is appropriate to introduce the linear differential forms

$$\overset{(\alpha)}{\omega}{}^i_j = \overset{(\alpha)}{\Gamma}{}^i_{kj}\, dx^k \quad \text{on } U_\alpha.$$

These forms uniquely determine the connection H on U_α (and are uniquely determined by it). They are called the *connection forms* of H on U_α. It is convenient to compose the matrix of these forms:

$$\overset{(\alpha)}{\omega} = \| \overset{(\alpha)}{\omega}{}^i_j \|, \quad i, j = 1, \ldots n.$$

If we naturally consider the forms $\overset{(\alpha)}{\omega}{}^i_j$ as forms on $\mathcal{E}_{U_\alpha}$, then the relation

$$\overset{(\alpha)}{\theta}{}^i = da^i + \overset{(\alpha)}{\omega}{}^i_j a^j \quad \text{on } \mathcal{E}_{U_\alpha} \tag{10}$$

holds. Forms (10) are called *connection forms* on $\mathcal{E}_{U_\alpha}$ by certain authors. The forms $\overset{(\alpha)}{\omega}{}^i_j$ and $\overset{(\alpha)}{\theta}{}^i$ thus define one another. The advantage of the form $\overset{(\alpha)}{\omega}{}^i_j$ is that they are defined on open sets of the manifold $\mathcal{B}$.

To simplify formulas in what follows, we omit the superposed (α) as a rule and write merely ω^i_j, for example, instead of $\overset{(\alpha)}{\omega}{}^i_j$. When it is necessary to consider two neighborhoods U_α and U_β simultaneously, the superposed (β) is replaced by primed coordinates; for example, we write $\omega^{i'}_{j'}$ instead of $\overset{(\beta)}{\omega}{}^i_j$. This agrees with the conventional designation above (see, e.g., formulas (9) in Chap. 34) of local coordinates in $\mathcal{E}_{U_\beta}$ by $a^{i'}$ and $x^{k'}$.

We let $\varphi^{i'}_i(\boldsymbol{x}) = \varphi^{i'}_i(x^1, \ldots, x^n)$ or merely $\varphi^{i'}_i$ denote the entries of the matrix $\varphi_{\beta\alpha}(b)$, $b \in U_\alpha \cap U_\beta$, which are considered as functions of the local coordinates $x^1, \ldots, x^n$ of the chart (U_α, h_α) and $\varphi^i_{i'}(\boldsymbol{x}) = \varphi^i_{i'}(x^1, \ldots, x^n)$ or merely $\varphi^i_{i'}$ denote the entries of the inverse matrix $\varphi_{\beta\alpha}(b)^{-1} = \varphi_{\alpha\beta}(b)$, $b \in U_\alpha \cap U_\beta$, which are considered as functions of the local coordinates $x^{1'}, \ldots, x^{n'}$ of the chart (U_β, h_β).

Proposition 35.4. *For each α, let n^2m smooth functions Γ^i_{kj} be given on the neighborhood U_α. These functions are coefficients of a certain connection H iff for any α and β, we have the relations*

$$\Gamma^{i'}_{k'j'} = \varphi^{i'}_i \varphi^j_{j'} \frac{\partial x^k}{\partial x^{k'}} \Gamma^i_{kj} + \varphi^{i'}_i \frac{\partial \varphi^i_{j'}}{\partial x^{k'}} \tag{11}$$

on the intersection $U_\alpha \cap U_\beta$.

Because $(\partial x^k / \partial x^{k'}) dx^{k'} = dx^k$ and $(\partial \varphi^i_{j'} / \partial x^{k'}) dx^{k'} = d\varphi^i_{j'}$, we can rewrite formula (11) in the simpler form

$$\omega^{i'}_{j'} = \varphi^{i'}_i \varphi^j_{j'} \omega^i_j + \varphi^{i'}_i d\varphi^i_{j'}. \tag{12}$$

In the matrix notation, these relations become

$$\omega' = \varphi^{-1} \omega \varphi + \varphi^{-1} d\varphi, \tag{13}$$

i.e., in the initial systematic notation, they become

$$\overset{(\beta)}{\omega} = \varphi^{-1}_{\beta\alpha} \overset{(\alpha)}{\omega} \varphi_{\beta\alpha} + \varphi^{-1}_{\beta\alpha} d\varphi_{\beta\alpha}. \tag{14}$$

§4. Inverse Image of a Connection

Let $\varphi: \mathcal{E}' \to \mathcal{E}$ be a smooth morphism of an $\mathbb{R}$-vector bundle $\xi' = (\mathcal{E}', \pi', \mathcal{B}')$ into an $\mathbb{R}$-vector bundle $\xi = (\mathcal{E}, \pi, \mathcal{B})$. Then for any point $b' \in \mathcal{B}'$, we have the commutative diagram

$$\begin{array}{ccccc} \mathcal{F}_{\beta'} & \longrightarrow & \mathcal{E}' & \longrightarrow & \mathcal{B}' \\ \Big\downarrow{\scriptstyle \varphi_{\beta'}} & & \Big\downarrow{\scriptstyle \varphi} & & \Big\downarrow{\scriptstyle f} \\ \mathcal{F}_b & \longrightarrow & \mathcal{E} & \longrightarrow & \mathcal{B} \end{array} \quad , \quad b = f(b'),$$

of smooth mappings, where $f = \hat{\varphi}$ is the mapping $\mathcal{B}' \to \mathcal{B}$ induced by the morphism φ. Therefore, for any point $p' \in \mathcal{F}'_{b'}$, we have the commutative diagram

$$\begin{array}{ccccccccc} 0 & \longrightarrow & \mathbf{T}_{p'}\mathcal{F}_{b'} & \longrightarrow & \mathbf{T}_{p'}\mathcal{E}' & \longrightarrow & \mathbf{T}_{b'}\mathcal{B}' & \longrightarrow & 0 \\ & & \Big\downarrow{\scriptstyle (t\varphi_{b'})_{p'}} & & \Big\downarrow{\scriptstyle (d\varphi)_p} & & \Big\downarrow{\scriptstyle (df)_p} & & \\ 0 & \longrightarrow & \mathbf{T}_p\mathcal{F}_b & \longrightarrow & \mathbf{T}_p\mathcal{E} & \longrightarrow & \mathbf{T}_b\mathcal{B} & \longrightarrow & 0 \end{array} \quad , \quad p = \varphi(p'),$$

of linear spaces and their linear mappings such that both its rows are exact sequences (left mappings are monomorphic, right mappings are epimorphic, and the image of the left mapping coincides with the kernel of the right mapping).

Let a connection H be given in the bundle ξ.

Proposition 35.5.

1. *For any point $p' \in \mathcal{E}'$, there exists one and only one subspace $H'_{p'}$ in the space $\mathbf{T}_p\mathcal{E}'$ such that*
$$\mathbf{T}_{p'}\mathcal{E}' = \mathbf{T}_{p'}\mathcal{F}_{b'} \oplus H'_{p'},$$
and $(d\varphi)_{p'}$ isomorphically maps this space onto the subspace $H_p \subset \mathbf{T}_p\mathcal{E}$.
2. *The subspaces $H'_{p'}$ compose a connection H' on $\mathcal{E}'$.*
3. *For any trivializing neighborhood $U \subset \mathcal{B}$, the connection forms of H' over the neighborhood $f^{-1}U \subset \mathcal{B}'$ are the forms $f^*\omega^i_j$, where ω^i_j are connection forms of H over the neighborhood U.*

The connection H' on the bundle ξ' is said to be *induced by* φ or is called the *inverse image* of the connection H under the morphism $\varphi: \xi' \to \xi$ and is denoted by φ^*H. In the case where $\xi' = f^*\xi$ and $\varphi = f$, the connection φ^*H is denoted by f^*H and is called the *inverse image of the connection H under the mapping $f: \mathcal{B}' \to \mathcal{B}$.*

We note that for any smooth mappings $g: \mathcal{B}'' \to \mathcal{B}$ and $f: \mathcal{B}' \to \mathcal{B}$, the connection $(f \circ g)^*H$ on the bundle $(f \circ g)^*\xi = g^*(f^*\xi)$ coincides with the connection $g^*(f^*H)$.

§5. Horizontal Curves

Let $\xi = (\mathcal{E}, \pi, \mathcal{B})$ be a vector bundle with the connection H.

Definition 35.4. A smooth curve $v: I \to \mathcal{E}$ on the manifold $\mathcal{E}$, where I is a certain closed interval of the real line $\mathbb{R}$ (for definiteness, we can assume, for example, that $I = \mathrm{I}$), is said to be *horizontal* if for any $t \in I$, its tangent vector $\dot{v}(t)$ belongs to the subspace $H_{v(t)}$. (Here and in what follows, the dot denotes differentiation with respect to t.)

In the local coordinates

$$a^1, \dots, a^n, x^1, \dots, x^m \tag{15}$$

of the chart $(\mathcal{E}_{U_\alpha}, g_\alpha)$ (see (17) in Chap. 34), each curve v (under the condition that $v(t) \in \mathcal{E}_{U_\alpha}$ for all $t \in I$) has the parametric equations of the form

$$a^i = a^i(t), \qquad x^k = x^k(t), \quad 1 \le i \le n, \quad 1 \le k \le m, \tag{16}$$

where $a^i(t)$ and $x^k(t)$ are certain smooth function on I. Its tangent vector $\dot{v}(t)$ has the coordinates

$$\dot{a}^1(t), \dots, \dot{a}^n(t), \dot{x}^1(t), \dots, \dot{x}^n(t),$$

and the horizontality condition $\dot{v}(t) \in H_{v(t)}$ (which means that $\theta^i_{v(t)}(\dot{v}(t)) = 0$, $1 \le i \le n$, where $\theta^1, \dots, \theta^n$ are connection forms of H on $\mathcal{E}_{U_\alpha}$) becomes

$$\dot{a}^i(t) + \Gamma^i_{kj}(\boldsymbol{x}(t))\dot{x}^k(t)a^j(t) = 0, \quad i = 1, \dots, n, \tag{17}$$

where $\boldsymbol{x}(t) = (x^1(t), \dots, x^m(t))$. Therefore, *a curve with parametric equations* (16) *is horizontal iff relations* (17) *hold identically with respect to* t.

In accordance with the general definition of bundle theory, the curve $u = \pi \circ v$ on the manifold $\mathcal{B}$ is called the *projection* of the curve v, and the curve v on the manifold $\mathcal{E}$ is called the *lift* of the curve u. We also say that the curve v *covers* the curve u. Sometimes, it is useful to imagine the curve v as a vector field on the curve u that sets the vector $v(t)$ of the linear space $\mathcal{F}_{v(t)}$ in correspondence to each point $u(t)$ of this curve, i.e., briefly, as a *ξ-vector field* on u. The horizontal curve v considered as a field on u is called the *field of parallel vectors on* u (with respect to the connection H).

In local coordinates $x^1, \dots, x^m$ on U_α, the curve u has the parametric equations

$$x^k = x^k(t), \quad 1 \le k \le m; \tag{18}$$

therefore, from the analytic standpoint, passing to the curve u consists in rejecting the first n equations (16), and vice versa, passing to the covering curve v consists in adding the additional n equations

$$a^i = a^i(t), \quad 1 \le i \le n, \tag{19}$$

where $a^i(t)$, $1 \leq i \leq n$, are, in general, arbitrary smooth $\mathbb{R}$-valued functions on I, to the m equations (18). In the interpretation of the curve v as a ξ-vector field on u, the functions $a^i(t)$ are called its *components* (in a given local coordinate system).

Proposition 35.6. *If the manifold $\mathcal{B}$ is Hausdorff, then for any smooth curve $u: I \to \mathcal{B}$, any point $t_0 \in I$, and any point $p_0 \in \mathcal{F}_{b_0}$, $b_0 = u(t_0)$, there exists a unique horizontal curve $v: I \to \mathcal{E}$ that covers the curve u and is such that $v(t_0) = p_0$.*

§6. Covariant Derivatives of Sections

We return to the study of the geometric properties of horizontal covering curves below and now apply these curves to the problem of invariantly defining connection without using trivializing coordinate neighborhoods. Let $s: \mathcal{B} \to \mathcal{E}$ be an arbitrary smooth section of the bundle ξ, and let $v: I \to \mathcal{E}$ be a horizontal curve that covers the integral curve u of the field X and satisfies the relation

$$v(0) = s(b).$$

Then for any $t \in I$, two vectors $s(u(t))$ and $v(t)$ are defined in the fiber $\mathcal{F}_{u(t)}$, and for $t \neq 0$, the vector

$$\frac{s(u(t)) - v(t)}{t} \tag{20}$$

is therefore also defined. Although vectors (20) in general belong to different linear spaces $\mathcal{F}_{u(t)}$ for distinct t, we can speak about their limit

$$\lim_{t \to 0} \frac{s(u(t)) - v(t)}{t}. \tag{21}$$

This limit belongs to the fiber $\mathcal{F}_{u(0)} = \mathcal{F}_b$ and is denoted by $(\nabla_X s)(b)$. Because the point b is an arbitrary point of the manifold $\mathcal{B}$, this construction yields the mapping

$$\nabla_X s: b \mapsto (\nabla_X s)(b)$$

acting from $\mathcal{B}$ to $\mathcal{E}$, which is a section of the bundle ξ by construction.

Definition 35.5. The section $\nabla_X s$ is called the *covariant derivative of the section s with respect to the vector field X* (in the given connection H).

Let U be an arbitrary coordinate and simultaneously trivializing neighborhood of a point b in the manifold $\mathcal{B}$. The trivialization $\varphi: U \times \mathbb{R}^n \to \mathcal{E}_U$ of the bundle ξ over U defines the basis

$$s_1, \ldots, s_n \tag{22}$$

of the **F**U-module $\Gamma(\xi|_U)$ (for which $s_i(b) = \varphi(b, e_i)$, $i = 1, \dots, n$, and the coordinate diffeomorphism $h: U \to \mathbb{R}^m$ assigns the basis

$$\frac{\partial}{\partial x^1}, \dots, \frac{\partial}{\partial x^m}, \quad m = \dim \mathcal{B}, \tag{23}$$

of the **F**U-module $\mathfrak{a}U$. Let s^i, $i = 1, \dots, n$, be the coordinates in a section s (or, more precisely, its restriction $s|_U$) in basis (22),

$$s = s^i s_i,$$

and let x^k, $k = 1, \dots, n$, be the coordinates of a vector field X in basis (23),

$$X = X^k \frac{\partial}{\partial x^k}.$$

Proposition 35.7. *The section $\nabla_X s$ over U has the coordinates*

$$(\nabla_X s)^i = \left(\frac{\partial s^i}{\partial x^k} + \Gamma^i_{kj} s^j\right) X^k, \tag{24}$$

in basis (22), *i.e.*,

$$\nabla_X s = \left(\frac{\partial s^i}{\partial x^k} + \Gamma^i_{kj} s^j\right) X^k s_i \quad \textit{over } U. \tag{25}$$

In particular, we see that *for a smooth section s, the section $\nabla_X s$ is also smooth.*

Of course, the operator $\nabla_X: \Gamma(\xi|_U) \to \Gamma(\xi|_U)$ is also defined for any vector field X over U, for $X = \partial/\partial x^k$, $k = 1, \dots, n$, for example. For $X = \partial/\partial x^k$, the operator ∇_X is denoted by ∇_k, and the section $\nabla_k s$ is called the *partial covariant derivative in x_k* of the section $s \in \Gamma(\xi|_U)$. For any field X,

$$\nabla_X s = X^k \nabla_k s.$$

According to formula (24), the coordinates $(\nabla_k s)^i$ of the section $\nabla_k s$ in the basis $s_1, \dots, s_k$ are expressed through the coordinates s^i of the section s by

$$(\nabla_k s)^i = \frac{\partial s^i}{\partial x^k} + \Gamma^i_{kj} s^j, \quad 1 \le i \le n, \quad 1 \le k \le m. \tag{26}$$

In particular,

$$(\nabla_k s_j)^i = \Gamma^i_{kj}, \quad 1 \le i, j \le n, \quad 1 \le k \le m, \tag{27}$$

and therefore

$$(\nabla_X s_j)^i = \omega^i_j(X), \quad 1 \le i, j \le n. \tag{28}$$

Remark 35.1. Formulas (24) or (25) can serve as a basis for defining the section $\nabla_X s$. However, in this case, it is necessary to verify the compatibility of the section constructed on intersections, i.e., the coincidence of sections (25) constructed over two neighborhoods U and U' and their intersection $U \cap U'$.

§7. Covariant Differentiations Along a Curve

An interesting variant of covariant differentiation arises for ξ-vector fields on curves. Let $u: I \to \mathcal{B}$ be a curve in the manifold $\mathcal{B}$ with the local equations $x^k = x^k(t)$, $1 \le k \le m$, and let $v: I \to \mathcal{E}$ be an arbitrary ξ-vector field on u with the components $a^i(t)$, $1 \le i \le n$. We define a new ξ-vector field on u, which is called the *covariant derivative of the field* v along u and is denoted by $\nabla v/dt$. By definition, this field has the components

$$\left(\frac{\nabla v}{dt}\right)^i = \dot{a}^i(t) + \Gamma^i_{kj}(x(t))\dot{x}^k(t)a^j(t), \quad i = 1, \dots, n. \tag{29}$$

We assume that a ξ-vector field v is the restriction to u of a certain section, i.e., there exists a section $s: \mathcal{B} \to \mathcal{E}$ of the bundle ξ such that $v(t) = s(u(t))$ for any $t \in I$. Also, we assume that the field $t \mapsto \dot{u}(t)$ of tangent vectors on the curve u is the restriction of a certain field on the manifold $\mathcal{B}$, i.e., there exists a vector field $X \in \mathfrak{a}\mathcal{B}$ such that $X_{u(t)} = \dot{u}(t)$ for any $t \in I$ (the curve u is an integral curve of the field X). Then *the field* $\nabla v/dt$ *on* u *is the restriction of the section* $\nabla_X s$.

We note that the section s and the field X locally exist in advance, i.e., in a neighborhood of an arbitrary point of the form $u(t_0)$ (for which $\dot{u}(t_0) \ne 0$).

Let $\mathring{I}$ be the interior of the closed interval I. Because $\mathring{I}$ is a smooth manifold (and $u: \mathring{I} \to \mathcal{B}$ is a smooth mapping), the smooth bundle $u^*\xi$ with the connection u^*H is defined on $\mathring{I}$.

Proposition 35.8.

1. *Each ξ-vector field $v: I \to \mathcal{E}$ on the curve u is naturally identified with a certain section of the bundle $u^*\xi$ over $\mathring{I}$.*
2. *The covariant derivative of this section with respect to the vector field d/dt on $\mathring{I}$ with respect to the induced connection u^*H is the restriction to I of the covariant derivative $\nabla v/dt$ of the field v along the curve u.*

Comparing formula (29) with (17) immediately shows that *the relation* $\nabla v/dt = 0$ *is a necessary and sufficient condition for* v *to be a field of parallel* ξ*-vectors* (*horizontal curve* in another interpretation). Therefore, we can say that *vectors parallel along a curve are exactly covariantly constant vectors.*

§8. Connections as Covariant Differentiations

The operation of covariant differentiation ∇_X is the mapping

$$\nabla_X: \Gamma\xi \to \Gamma\xi$$

of the module $\Gamma\xi$ into itself.

Proposition 35.9. *The operation ∇_X has the following three properties:*

1. *The operation ∇_X is linear over the field $\mathbb{R}$, i.e.,*

$$\nabla_X(s+t) = \nabla_X s + \nabla_X t,$$
$$\nabla_X(\lambda s) = \lambda \nabla_X$$

for any sections $s, t \in \Gamma\xi$ and any number $\lambda \in \mathbb{R}$.

2. *For any function $f \in \mathsf{F}\mathcal{B}$ and any section $s \in \Gamma\xi$, we have*

$$\nabla_X(fs) = Xf \cdot s + f\nabla_X s. \tag{30}$$

3. *The operation ∇_X linearly depends on x over $\mathsf{F}\mathcal{B}$, i.e.,*

$$\nabla_{fX+gY} = f\nabla_X + g\nabla_Y$$

for any fields $X, Y \in \mathfrak{a}\mathcal{B}$ and any functions $f, g \in \mathsf{F}\mathcal{B}$.

Formula (30) is similar to the well-known *Leibnitz formula* for differentiation of the product (and passes into it if $\nabla_X f$ denotes Xf).

Again, we assume that the smooth (of class C^∞) manifold $\mathcal{B}$ is Hausdorff.

Theorem 35.1. *Let an operator*

$$\nabla_X : \Gamma\xi \to \Gamma\xi \tag{31}$$

that has properties 1, 2, *and* 3 *in Proposition* 35.9 *correspond to each field $X \in \mathfrak{a}\mathcal{B}$. Then there exists a unique connection H on the bundle ξ with respect to which operators* (31) *are covariant derivatives.*

Therefore, for the Hausdorff manifold $\mathcal{B}$, connections on the smooth vector bundle $\xi = (\mathcal{E}, \pi, \mathcal{B})$ are in a natural bijective correspondence with $\mathsf{F}\mathcal{B}$-linear mappings $\nabla: X \mapsto \nabla_X$ that set the linear operator

$$\nabla_X : \Gamma\xi \to \Gamma\xi$$

satisfying the Leibnitz formula (30) in correspondence to each vector field $X \in \mathfrak{a}\mathcal{B}$. Connections are therefore often *defined* as mappings ∇ of such a type (which are called *covariant differentiations*). A disadvantage of this definition is the lack of geometric visuality, and an advantage is its invariance (trivializing coordinate neighborhoods are not used in it).

In what follows, we identify a connection H and the corresponding differentiation ∇ as a rule. In particular, we say "connection ∇" instead of "connection the covariant differentiation ∇ corresponds to."

§9. Connections on Metricized Bundles

Let an $\mathbb{R}$-vector bundle ξ be metricized. Then the formula

$$(s, s')(b) = (s(b), s'(b)), \quad s, s' \in \Gamma\xi, \quad b \in \mathcal{B},$$

defines the functional $s, s' \mapsto (s, s')$ on the $\mathsf{F}\mathcal{B}$-module $\Gamma\xi$, which assumes its values in the algebra $\mathsf{F}\mathcal{B}$ and is such that for any section $s \in \Gamma\xi$, the function $(s, s') \in \mathsf{F}\mathcal{B}$ assumes real nonnegative values and vanishes only at those points $b \in \mathcal{B}$ at which $s(b) = 0$. This functional is bilinear (over the algebra $\mathsf{F}\mathcal{B}$) and symmetric.

Definition 35.6. Roughly speaking, we call this functional the *inner product* on $\Gamma\xi$.

The Gram–Schmidt orthogonalization process shows that *over each trivializing neighborhood $U \in \mathcal{B}$, the module $\Gamma(\xi|_U)$ has an orthonormal basis* $s_1, \dots, s_n$ (i.e., $(s_i, s_j) = \delta_{ij}$ for any $i, j = 1, \dots, b = n$).

Definition 35.7. A connection ∇ on a metricized bundle ξ is said to be *compatible with the metric* (or *metrical*) if for any vector field $X \in \mathfrak{a}\mathcal{B}$ and any sections $s, s' \in \Gamma\xi$, we have

$$X(s, s') = (\nabla_X s, s') + (s, \nabla_x s'). \tag{32}$$

Proposition 35.10. *A connection on the bundle ξ is compatible with the metric iff for any trivializing neighborhood U, the matrix $\omega = \|\omega_j^i\|$ of connection forms corresponding to an orthonormal basis $s_1, \dots, s_n$ of the $\mathsf{F}U$-module $\Gamma(\xi|_U)$ is skew-symmetric, i.e., if for any $i, j = 1, \dots, n$,*

$$\omega_j^i + \omega_i^j = 0.$$

Corollary 35.1. *There exists a connection compatible with the metric on any metricized vector bundle ξ over a paracompact base.*

Proof. Proposition 35.10 implies that such a metric exists on a trivial bundle. A partition of unity is used to reduced the general case to this one. □

§10. Covariant Differential

Definition 35.8. We say that a *covariant differentiation* ∇ is given on $\Gamma_r^s\xi$ if a linear (over $\mathbb{R}$) operator

$$\nabla_X : \Gamma_r^s\xi \to \Gamma_r^s\xi \tag{33}$$

depending on X $\mathsf{F}\mathcal{B}$-linearly and satisfying the Leibnitz identity

$$\nabla_X(fS) = Xf \cdot S + f\nabla_X S, \quad f \in \mathsf{F}\mathcal{B}, \quad S \in \Gamma_r^s\xi$$

corresponds to each vector field $X \in \mathfrak{a}\mathcal{B}$. The operators ∇_X are called *covariant diferentiations with respect to X.*

The operators ∇ are defined over any trivializing coordinate neighborhood U and act on tensors $S \in \Gamma_r^s(\xi|_U)$ by

$$\nabla_X S = X^k \nabla_k S,$$

where X^k, $1 \leq k \leq m$, are components of the field X over U and

$$\nabla_k S = \nabla_{\partial/\partial x^k} S, \quad 1 \leq k \leq m, \tag{34}$$

are *partial covariant derivatives* of a ξ-tensor field S.

Therefore, operators (33) are uniquely reconstructed by partial derivatives (34) and therefore by their components $(\nabla_k S)_{i_1 \cdots i_r}^{j_1 \cdots j_s}$. For $(r, s) = (1, 0)$, operators (33) become

$$\nabla_X : \Gamma\xi^* \to \Gamma\xi^*, \tag{35}$$

and for any trivializing coordinate neighborhood U, they transform a ξ-covector field $c = c_i s^i$ on U into the ξ-covector field

$$\nabla_X c = X^k (\nabla_k c)_i s^i,$$

where $(\nabla_k c)_i$ are components of the ξ-covector field $\nabla_k c$.

Now let a certain connection H be given on ξ.

Proposition 35.11. *There exists a unique covariant differentiation ∇ on the $\mathsf{F}\mathcal{B}$-module $\Gamma\xi^*$ such that for any vector field $X \in \mathfrak{a}\mathcal{B}$ and any section $s \in \Gamma\xi$, the relation*

$$X\langle s, c\rangle = \langle \nabla_X s, c\rangle + \langle s, \nabla_X c\rangle \tag{36}$$

holds for each ξ-covector field $c \in \Gamma\xi^*$, where ∇ in the first term in the right-hand side denotes the covariant differentiation with respect to the connection H. For $X = \partial/\partial x^k$, the components of the ξ-covector field $\nabla_k c = \nabla_{\partial/\partial x_k} c$ in each trivializing neighborhood U are expressed by

$$(\nabla_k c)_j = \frac{\partial c_i}{\partial x^k} - \Gamma_{ki}^j c_j, \quad 1 \leq i \leq n, \quad 1 \leq k \leq m, \tag{37}$$

where Γ_{ki}^j are connection coefficients of H.

Proposition 35.12. *There exist differentiations ∇ on $\mathsf{F}\mathcal{B}$-modules $\Gamma_r^s\xi$, $r, s \geq 0$, such that the following conditions hold:*

1. *For any two ξ-tensors S and T and any vector field $X \in \mathfrak{a}\mathcal{B}$, we have*

$$\nabla_X(S \otimes T) = \nabla_X S \otimes T + S \otimes \nabla_X T. \tag{38}$$

2. *For* $(r,s) = (0,1)$, *the differentiation* ∇ *coincides with the differentiation* ∇ *with respect to the connection* H *on* $\Gamma\xi = \Gamma_0^1\xi$, *and for* $(r,s) = (1,0)$, *it coincides with the differentiation* ∇ *from Proposition* 35.11 *on* $\Gamma\xi^* = \Gamma_1^0\xi$. *These differentiations are unique, and on any trivializing coordinate neighborhood* U, *the components* $(\nabla_k S)_{i_1\cdots i_r}^{j_1\cdots j_s}$ *of the partial derivatives* $\nabla_k S$ *of an arbitrary* ξ*-tensor* S *are expressed through its components* $S_{i_1\cdots i_r}^{j_1\cdots j_s}$ *by*

$$\begin{aligned}(\nabla_k S)_{i_1\cdots i_r}^{j_1\cdots j_s} = {} & \frac{\partial S_{i_1\cdots i_r}^{j_1\cdots j_s}}{\partial x^k} \\ & + \Gamma_{kp}^{j_1} S_{i_1\cdots i_r}^{pj_2\cdots j_s} + \Gamma_{kp}^{j_2} S_{i_1 i_2\cdots i_r}^{j_1 p j_3\cdots j_s} + \cdots + \Gamma_{kp}^{j_s} S_{i_1 i_2\cdots i_r}^{j_1\cdots j_{s-1}p} \\ & - \Gamma_{ki_1}^{p} S_{pi_2\cdots i_r}^{j_1 j_2\cdots j_s} - \Gamma_{ki_2}^{p} S_{i_1 p i_3\cdots i_r}^{j_1 j_2\cdots j_s} - \cdots - \Gamma_{ki_r}^{p} S_{i_1\cdots i_{r-1}p}^{j_1\cdots j_s}. \end{aligned} \tag{39}$$

(*To each index, there corresponds one summand in the right-hand side taken with the plus sign if this index is a superscript and with the minus sign if this index is a subscript.*)

By the identification of connections with covariant derivatives, Propositions 35.1 and 35.2 are assertions of the existence (and uniqueness) of connections on the bundles ξ^* and $T_r^s\xi$ with a given connection ∇ on ξ in relations (36) and (38).

Definition 35.9. A field $\nabla_X S$ is called the *covariant derivative with respect to the connection* H *of a* ξ*-tensor field* S *in the direction of the vector field* $X \in \mathfrak{a}\mathcal{B}$.

Remark 35.2. The components $(\nabla_k S)_{i_1\cdots i_r}^{j_1\cdots j_s}$ of the field $\nabla_k S$ are also denoted by $S_{i_1\cdots i_r}^{j_1\cdots j_s}|_k$. Moreover, other signs (comma, semicolon, etc.) are also used instead of the dividing line $|$.

We can use tensor products to make the construction of the covariant derivatives $\nabla_X S$ formally more perfect and conceptually simpler (even in the case where $(r,s) = (0,1)$).

For an arbitrary vector bundle $\xi = (\mathcal{E}, \pi, \mathcal{B})$ of rank n over an m-dimensional smooth manifold $\mathcal{B}$, we consider the vector bundle $\tau_{\mathcal{B}}^* \otimes \xi$ and the $\mathsf{F}\mathcal{B}$-module $\Gamma(\tau_{\mathcal{B}}^* \otimes \xi)$ of its smooth sections.

Definition 35.10. A linear (over the field $\mathbb{R}$) mapping

$$\nabla \colon \Gamma\xi \to \Gamma(\tau_{\mathcal{B}}^* \otimes \xi)$$

is called the *covariant differentiation* if it satisfies the Leibniz identity, i.e., if for any function $f \in \mathsf{F}\mathcal{B}$ and any section $s \in \Gamma\xi$, we have

$$\nabla(fs) = df \otimes s + f\nabla s. \tag{40}$$

We recall that df is a section of the bundle $\tau^*_\mathcal{B}$ and $df \otimes s$ is therefore a section of the bundle $\tau^*_\mathcal{B} \otimes \xi$. The section ∇s of the bundle $\tau^*_\mathcal{B} \otimes \xi$ is called the *covariant differential* of the section s.

In the case where the manifold $\mathcal{B}$ is Hausdorff, formula (40) implies that *the mapping* ∇ *has the localization property*, i.e., if two sections s_1 and s_2 are equal near a point $b_0 \in \mathcal{B}$, then the sections ∇s_1 and ∇s_2 are also equal near b_0. (Indeed, if $s_1 = s_2$ in a neighborhood U of the point b_0 and if φ is a smooth function on $\mathcal{B}$ that is unity in a certain neighborhood $W \subset U$ of the point b_0 and is zero outside U, then the section $\varphi \cdot (s_2 - s_1)$ is identically equal to zero, and therefore

$$d\varphi \cdot (s_2 - s_1) + \varphi \cdot \nabla(s_2 - s_1) = 0 \quad \text{on } \mathcal{B}.$$

Therefore, $\nabla(s_2 - s_1) = 0$ on W.) In turn, the localization property implies that for any open set $U \in \mathcal{B}$, the differentiation ∇ defines the differentiation $\nabla|_U$ for the bundle $\xi|_U$ that closes the commutative diagram

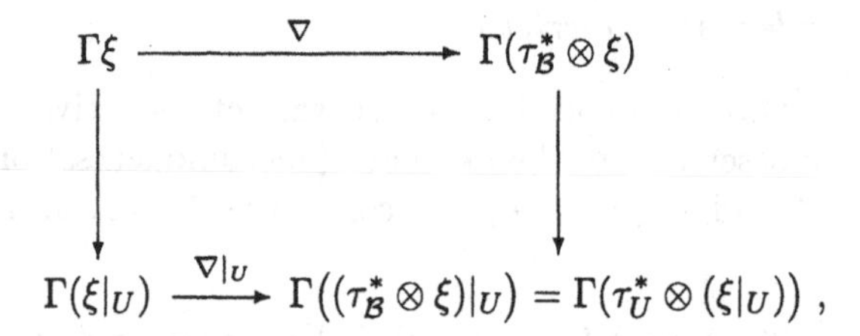

whose vertical arrows are restriction mappings. Moreover, for any open covering $\{U_\alpha\}$ of the manifold $\mathcal{B}$, the differentiation ∇ is uniquely reconstructed by the differentiations $\nabla|_{U_\alpha}$.

On the other hand, if U is a trivializing coordinate neighborhood in $\mathcal{B}$ and $\{s_i; 1 \le i \le n\}$ is a basis of the $\mathsf{F}U$-module $\Gamma(\xi|_U)$, then $\{dx^k \otimes s;\ 1 \le i \le n,\ 1 \le k \le m\}$ is the basis of the $\mathsf{F}U$-module $\Gamma((\tau^*_\mathcal{B} \otimes \xi)|_U)$, and for each $j = 1, \dots, n$, we have the relation

$$\nabla s_j = \Gamma^i_{kj} dx^k \otimes s_i, \quad \Gamma^i_{kj} \in \mathsf{F}U, \tag{41}$$

i.e., the relation

$$\nabla s_j = \omega^i_j \otimes s_j, \quad \omega^i_j = \Gamma^i_{kj} dx^k \in \Omega^1 U \tag{42}$$

(to simplify the formulas, we write ∇ instead of $\nabla|_U$). Moreover, the forms ω^i_j (or, equivalently, the functions Γ^i_{kj}) uniquely define the differentiation ∇ (more precisely, $\nabla|_U$) by

$$\nabla s = (ds^i + \omega^i_j s^j) \otimes s_i, \tag{43}$$

where $s = s^i s_i$. We note that for *arbitrary* forms ω^i_j on U, the mapping ∇ defined by (43) is a covariant differentiation on U.

It is easy to see that if U' is another trivializing coordinate neighborhood (with the trivialization $s_{1'}, \dots, s_{n'}$) and

$$\nabla s_{j'} = \omega_{j'}^{i'} \otimes s_{i'} \quad \text{on } U',$$
$$s_{i'} = \varphi_{i'}^{i} s_i, \quad s_i = \varphi_i^{i'} s_j \quad \text{on } U \cap U', \tag{44}$$

then

$$\omega_{j'}^{i'} = \varphi_i^{i'} \varphi_{j'}^{j} \omega_j^i + \varphi_i^{i'} d\varphi_{j'}^{i}. \tag{45}$$

Conversely, if we have covariant differentiations acting by the respective formulas (42) and (44) on the neighborhoods U and U' and if relations (45) hold, then these differentiations coincide on the intersection $U \cap U'$ (more precisely, their restrictions to $U \cap U'$ coincide).

This means that the following proposition holds.

Proposition 35.13. *If $\{U_\alpha\}$ is an open covering of a Hausdorff manifold $\mathcal{B}$ consisting of trivializing coordinate neighborhoods, then for each α, each covariant differentiation ∇ defines the forms $\omega_j^i = \overset{(\alpha)}{\omega_j^i}$ on the neighborhood U_α. Moreover, for any α and β, these forms are related by* (45) *on the neighborhoods $U = U_\alpha$ and $U' = U_\beta$. Conversely, assigning forms $\omega_j^i = \overset{(\alpha)}{\omega_j^i}$ for each α related by* (45) *defines the covariant differentiation ∇ that acts on each neighborhood U_α by formula* (43).

§11. Comparison of Various Definitions of Connection

Comparing (12) and (45), we immediately find that *there exists a bijective correspondence between covariant differentiations ∇ and connections on the bundle ξ* (and they can therefore be identified).

Therefore, we have in fact *three* definitions of connection on the vector bundle ξ: as a field of horizontal subspaces, as a family of operators ∇_X, and as differentiations ∇.

The interrelations of these definitions are described by the formulas

$$H = \operatorname{Ann}(\theta^1, \dots, \theta^n),$$

where $\theta^i = da^i + \omega_j^i a^j = da^i + \Gamma_{kj}^i a^j dx^k$,

$$\nabla_X s = \left(\frac{\partial s^i}{\partial x^k} + \Gamma_{kj}^i s^j \right) X^k s_i, \tag{46}$$

where $s = s^i s_i$ and $X = X^k (\partial / \partial x^k)$, and

$$\begin{aligned} \nabla s &= \left(ds^i + \omega_j^i s^j \right) \otimes s_j \\ &= \left(ds^i + \Gamma_{kj}^i s^j dx^k \right) \otimes s_j = \left(\frac{\partial s^i}{\partial x^k} + \Gamma_{kj}^i s^j \right) dx^k \otimes s_j, \end{aligned}$$

which holds in an arbitrary trivializing coordinate neighborhood U.

Which definition is more convenient depends on the problem to be solved. Therefore, one should know how to pass easily and quickly from each of these definitions to any other.

The value $(\nabla s)_b$ of the section $\nabla s \in \Gamma(\tau_{\mathcal{B}}^* \otimes \xi)$ at the point $b \in \mathcal{B}$ can be interpreted as a linear mapping $\mathbf{T}_b\mathcal{B} \to \mathcal{F}_b^\xi$. Therefore, for any vector field $X \in \mathfrak{a}\mathcal{B}$, the formula

$$(\nabla s)(X)_b = (\nabla s)_b(X_b), \quad b \in \mathcal{B},$$

defines the section $(\nabla s)(X)$ of the bundle ξ:

$$(\nabla s)(X) = \nabla_X s. \tag{47}$$

Formula (47) states a direct correspondence (which does not use horizontal subspaces and trivializing coordinate neighborhoods) between the differentiations ∇ and the operators ∇_X.

§12. Connections on Frame Bundles

We recall (see above) that each vector bundle $\xi = (\mathcal{E}, \pi, \mathcal{B})$ is associated with the *frame bundle* $\boldsymbol{\xi} = (E, \boldsymbol{\pi}, \mathcal{B})$ (belonging to the class of the so-called principal bundles) whose total space E consists of frames (bases) of the form

$$\boldsymbol{p} = (p_1, \dots, p_n), \tag{48}$$

where $p_1, \dots, p_n$ are linearly independent vectors (which therefore compose a basis) of a certain fiber $\mathcal{F}_b$ of the bundle ξ. In this case, $\boldsymbol{\pi}(\boldsymbol{p}) = b$.

In E, we have charts of the form $(E_U, \boldsymbol{h})$, where U is a trivializing coordinate neighborhood, and $\boldsymbol{h}$ is the mapping of the set $E_U = \boldsymbol{\pi}^{-1}U$ into $\mathbb{R}^{n^2+m} = \operatorname{Mat}_n(\mathbb{R}) \times \mathbb{R}^m$ defined by

$$\boldsymbol{h}(\boldsymbol{p}) = (C, h(b)), \quad \boldsymbol{p} = (p_1, \dots, p_n) \in E, \quad b = \boldsymbol{\pi}(\boldsymbol{p}), \tag{49}$$

where $h\colon U \to \mathbb{R}^m$ is the coordinate mapping of the neighborhood U and C is a matrix whose columns consist of coordinates of the vectors $p_1, \dots, p_n$ in the trivializing basis $\boldsymbol{s}(b) = (s_1(b), \dots, s_n(b))$ of the fiber $\mathcal{F}_b$. (The space of square matrices of order n is identified with the space $\mathbb{R}^{n^2}$ here.)

In the chart $(E_U, \boldsymbol{h})$, the local coordinates of a point $\boldsymbol{p} \in E_U$ are numbers c_i^j, $1 \le i, j \le n$, and x^k, $1 \le k \le m$, where x^k are local coordinates of the point $b = \boldsymbol{\pi}(\boldsymbol{p})$ in the chart (U, h) of the manifold $\mathcal{B}$ and c_i^j are entries of the matrix C, i.e., numbers such that

$$p_i = c_i^j s_j(b)$$

for any $i = 1, \dots, n$. (We note that the matrix C is nonsingular.) Therefore, the coordinates of the point $\boldsymbol{p}$ compose the pair $(C, \boldsymbol{x})$, where C is the matrix $\|c_i^j\|$ and $\boldsymbol{x}$ is the row $(x^1, \dots, x^m)$.

It is clear that all charts of the form $(E_U, \boldsymbol{h})$ are compatible and therefore define the structure of an (n^2+m)-dimensional smooth manifold on E. Moreover, the mapping $\boldsymbol{\pi}$ transforms a point $\boldsymbol{p} \in E$ with coordinates $(C, \boldsymbol{x})$ into the point $b \in \mathcal{B}$ with coordinates $\boldsymbol{x}$ and is therefore a submersion. This means that *for a smooth vector bundle $\xi = (\mathcal{E}, \pi, \mathcal{B})$, the frame bundle $\boldsymbol{\xi} = (E, \boldsymbol{\pi}, \mathcal{B})$ is also smooth.* In particular, we see that for any point $b \in \mathcal{B}$, the fiber $\mathcal{F}_b^{\xi} = \boldsymbol{\pi}^{-1}(b)$ of the bundle $\boldsymbol{\xi}$ is an embedded submanifold, and all statements about vertical and horizontal subspaces can be automatically applied to the bundle $\boldsymbol{\xi}$.

Proposition 35.14. *For any field*

$$\mathbf{H}: \boldsymbol{p} \mapsto \mathbf{H}_{\boldsymbol{p}}, \quad \boldsymbol{p} \in E, \tag{50}$$

of horizontal subspaces on the bundle $\boldsymbol{\xi}$, there exist uniquely defined forms

$$\theta_j^i = dc_j^i + f_{jk}^i dx^k, \quad 1 \le i, j \le n, \tag{51}$$

in each chart $(E_U, \boldsymbol{h})$, where f_{jk}^i are certain functions on the coordinate neighborhood E_U (which are smooth if the field $\mathbf{H}$ is smooth), such that for any point $\boldsymbol{p} \in E_U$, the subspace $\mathbf{H}_p$ is the annihilator of the covectors $(\theta_j^i)_{\boldsymbol{p}}$.

The local coordinates $(C, \boldsymbol{x})$ of the chart $(E, \boldsymbol{h})$ define the basis

$$\left(\frac{\partial}{\partial c_j^i}\right)_{\boldsymbol{p}}, \left(\frac{\partial}{\partial x^k}\right)_{\boldsymbol{p}}, \quad 1 \le i, j \le n, \quad 1 \le k \le m,$$

in each tangent space $\mathsf{T}_{\boldsymbol{p}}E$, $\boldsymbol{p} \in E_U$. Therefore, any vector from $\mathsf{T}_{\boldsymbol{p}}E$ is uniquely represented in the form

$$a_j^i \left(\frac{\partial}{\partial c_j^i}\right)_{\boldsymbol{p}} + u^k \left(\frac{\partial}{\partial x^k}\right)_{\boldsymbol{p}}, \quad \text{where } a_j^i, u^k \in \mathbb{R}.$$

We identify this vector with the pair $(A, \boldsymbol{u})$, where A is the matrix $\|a_j^i\|$ and $\boldsymbol{u}$ is the row $(u^1, \dots, u^m)$. The vector $(A, \boldsymbol{u})$ is vertical iff $\boldsymbol{u} = \mathbf{0}$.

It is clear that the values of form (51) on the vector $(A, \boldsymbol{u})$ compose the matrix

$$A + u^k F_k,$$

where $F_k = F_k(C, \boldsymbol{x})$ are the matrices $\|f_{jk}^i\|$, $k = 1, \dots, m$. Therefore, $(A, \boldsymbol{u}) \in \mathbf{H}_{\boldsymbol{p}}$ iff

$$A + u^k F_k = 0. \tag{52}$$

In particular, this implies that

$$(A, \boldsymbol{u})^V = (A + u^k F_k, \mathbf{0}), \qquad (A, \boldsymbol{u})^H = (-u^k F_k, \boldsymbol{u})$$

for any vector $(A, \boldsymbol{u})$ from $\mathsf{T}_{\boldsymbol{p}}E$.

Each fiber $\mathcal{F}_{b_0}^{\xi}$, $b_0 \in \mathcal{B}$, of the bundle $\boldsymbol{\xi}$ is an orbit of the right action

$$(\boldsymbol{p}, B) \mapsto \boldsymbol{p}B, \quad \boldsymbol{p} \in E, \quad b \in \mathrm{GL}(n; \mathbb{R}),$$

of the group $\mathrm{GL}(n; \mathbb{R})$ on the space E defined by

$$\boldsymbol{p}B = \boldsymbol{q}, \quad \boldsymbol{q} = (q_1, \dots, q_n),$$

where

$$q_i = p_j b_i^j, \quad i = 1, \dots, n, \tag{54}$$

for $\boldsymbol{p} = (p_1, \dots, p_n)$ and $B = \|b_i^i\|$. In coordinates, this action is written by

$$(C, \boldsymbol{x}) \mapsto (CB, \boldsymbol{x}) \tag{55}$$

and is therefore a smooth action. In particular, for any element $B \in \mathrm{GL}(n, \mathbb{R})$, the mapping

$$R_B \colon E \to E, \qquad \boldsymbol{p} \mapsto \boldsymbol{p}B, \quad \boldsymbol{p} \in E,$$

of the manifold E into itself is a diffeomorphism.

Because the diffeomorphism R_B transforms each fiber $\mathcal{F}_{b_0}^{\boldsymbol{\xi}}$ into itself, its differential

$$(dR_B)_{\boldsymbol{p}} \colon \mathbf{T}_{\boldsymbol{p}}E \to \mathbf{T}_{\boldsymbol{p}B}E$$

at each point $\boldsymbol{p} \in \mathcal{F}_{b_0}^{\boldsymbol{\xi}}$ transforms the vertical subspace $\mathcal{F}_{b_0}^{\boldsymbol{\xi}}$ into the vertical subspace $\mathbf{T}_{\boldsymbol{p}B}\mathcal{F}_{b_0}^{\boldsymbol{\xi}}$. Therefore, for any field $\mathbf{H}$ of horizontal subspaces, any point $\boldsymbol{p} \in E$, and any element $B \in \mathrm{GL}(n; \mathbb{R})$, the mapping $(dR_B)_{\boldsymbol{p}}$ transforms the horizontal subspace $\mathbf{H}_{\boldsymbol{p}}$ into the subspace $(dR_B)_{\boldsymbol{p}}\mathbf{H}_{\boldsymbol{p}}$ complementary to the subspaces $\mathbf{T}_{\boldsymbol{p}B}\mathcal{F}_{b_0}^{\boldsymbol{\xi}}$. If

$$(dR_B)_{\boldsymbol{p}}\mathbf{H}_{\boldsymbol{p}} = \mathbf{H}_{\boldsymbol{p}B}, \tag{56}$$

then the field $\mathbf{H}$ of horizontal subspaces is said to be *equivariant.*

Definition 35.11. A *connection* on the principal bundle $\boldsymbol{\xi} = (E, \pi, \mathcal{B})$ is an arbitrary smooth and equivariant field $\mathbf{H}$ of horizontal subspaces.

We note that in contrast to a connection on a vector bundle ξ, a connection on the principal bundle $\boldsymbol{\xi}$ is defined without using trivializing coordinate neighborhoods.

In coordinates, the linear mapping $(dR_B)_{\boldsymbol{p}}$ acts according to the formula

$$(dR_B)_{\boldsymbol{p}}(A, \boldsymbol{u}) = (AB, \boldsymbol{u}), \quad (A, \boldsymbol{u}) \in \mathbf{T}_{\boldsymbol{p}}E.$$

By formula (54), this implies that if the field $\mathbf{H}$ is the annihilator of forms (51) on E_U, then the subspace $(dR_B)_{\boldsymbol{p}}H_{\boldsymbol{p}}$ consists of vectors of the form $(AB, \boldsymbol{u})$, where $\boldsymbol{u}$ is an arbitrary vector from $\mathbb{R}^m$ and

$$A = -F_k(C, \boldsymbol{x})u^k.$$

On the other hand, according to the same formula (54), because the point $\boldsymbol{q} = R_B\boldsymbol{p}$ has the coordinates $(CB, \boldsymbol{x})$, the subspace $\mathbf{H}_q$ consists of vectors of the form $(-F_k(CB, \boldsymbol{x})u^k, \boldsymbol{u})$. Therefore, the field $\mathbf{H}$ of horizontal subspaces is equivariant (is a connection) iff

$$-F_k(C, \boldsymbol{x})Bu^k = -F_k(CB, \boldsymbol{x})u^k$$

for any elements B and C of the group $\mathrm{GL}(n; \mathbb{R})$ and any vectors $\boldsymbol{x}, \boldsymbol{u} \in \mathbb{R}^m$. By the arbitrariness of the vector $\boldsymbol{u}$, the latter relation holds iff

$$F_k(C, \boldsymbol{x})B = F_k(CB, \boldsymbol{x}) \quad \text{for any } k = 1, \dots, m. \tag{57}$$

Setting $C = E$ here and letting C denote the matrix B, we immediately obtain

$$F_k(C, \boldsymbol{x}) = \Gamma_k C, \quad k = 1, \dots, m, \tag{58}$$

where $\Gamma_k = \Gamma_k(\boldsymbol{x})$ is the matrix $F_k(E, \boldsymbol{x})$, i.e.,

$$f^i_{jk} = \Gamma^i_{ks} c^s_j, \quad \|\Gamma^i_{ks}\| = \Gamma_k.$$

Because matrices of form (58) satisfy relations (57) for any matrices Γ_k, *a field* $\mathbf{H}$ *is a connection on* $\boldsymbol{\xi}$ *iff on each coordinate neighborhood* E_U, *it is the annihilator of the forms*

$$\theta^i_j = dc^i_j + \Gamma^i_{ks} c^s_j dx^k, \tag{59}$$

where Γ^i_{ks} *are certain smooth functions on the neighborhood* U.

§13. Comparison with Connections on Vector Bundles

A direct verification shows that the functions Γ^i_{ks} are coefficients of a certain connection H on the vector bundle ξ. Therefore, given a connection $\mathcal{H}$ on $\boldsymbol{\xi}$, we can construct the connection H on ξ. Conversely, if H is an arbitrary connection on ξ, then we can use its coefficients Γ^i_{kj} to construct forms (59) on each coordinate neighborhood E_U and therefore a certain connection $\mathbf{H}_U$. The connections $\mathbf{H}_U$ are compatible on the intersections $E_U \cap E_{U'}$ and therefore define a connection $\mathbf{H}$ on the whole E. This proves the following proposition.

Proposition 35.15. *There exists a natural bijective correspondence between the connections* $\mathbf{H}$ *on the principal bundle* $\boldsymbol{\xi}$ *and the connections* H *on the associated vector bundle* ξ.

In the process of the proof, we also prove that *on each coordinate neighborhood* E_U, *connections on* $\boldsymbol{\xi}$ *are given by relations of form* (59). Of course, instead of these forms, we can use the following forms, which are linearly equivalent to form (59):

$$\theta^i_j = {}'c^i_t\, dc^t_j + {}'c^i_t \Gamma^t_{sk} c^s_j\, dx^k, \quad i, j = 1, \dots, n, \tag{60}$$

where $'c^i_j$ are entries of the matrix C^{-1} inverse to the matrix $C = \|c^i_j\|$. (Because we do not need form (59) in what follows, forms (60) are denoted by the same symbols θ^i_j as forms (59). The matrix $\theta = \|\theta^i_j\|$ consisting of forms (60) can be conditionally written as

$$\theta = C^{-1}dC + C^{-1}\omega C, \tag{61}$$

where dC is the matrix $\|dc^i_j\|$ and ω is the matrix $\|\Gamma^i_{kj}dx^k\|$. Over any other trivializing coordinate neighborhood U', the connection is given by the forms $\theta^{i'}_{j'}$ that compose the matrix

$$\theta' = C'^{-1}dC' + C'^{-1}\omega' C'.$$

A trivial computation shows that forms (60) are compatible on intersections and therefore compose global forms θ^i_j on the whole manifold E. Therefore, we see that *any connection on* $\boldsymbol{\xi}$ *is the annihilator of the forms* θ^i_j, $1 \leq i, j \leq n$, *defined on the whole manifold* E (and having form (60) in each chart $(E_U, \boldsymbol{h})$). This is the main benefit of the connections on $\boldsymbol{\xi}$ over the connections on ξ, for which similar forms θ^i, $1 \leq i \leq n$, are defined only locally.

Because the components $p_1, \ldots, p_n$ of an arbitrary point $\boldsymbol{p} = (p_1, \ldots, p_n)$ of the space E belong to the same fiber $\mathcal{F}_b$ of the bundle ξ by definition, the point $p_i y^i \in \mathcal{F}_b$ is defined for any vector $\boldsymbol{y} = (y^1, \ldots, y^n) \in \mathbb{R}^n$. Therefore, the formula

$$f_{\boldsymbol{y}}(\boldsymbol{p}) = p_i y^i, \quad \boldsymbol{p} = (p_1 \ldots, p_n), \quad \boldsymbol{y} = (y^1, \ldots, y^n),$$

defines the mapping

$$f_{\boldsymbol{y}}: E \to \mathcal{E}.$$

Let

$$(df_{\boldsymbol{y}})_{\boldsymbol{p}}: \mathbf{T}_{\boldsymbol{p}}E \to \mathbf{T}_p\mathcal{E}, \quad p = p_i y^i,$$

be the differential of this mapping at a point $\boldsymbol{p} \in E$.

Proposition 35.16. *For an arbitrary connection* $\mathbf{H}$ *on* $\boldsymbol{\xi}$*, the formula*

$$H_p = (df_{\boldsymbol{y}})_{\boldsymbol{p}}\mathbf{H}_{\boldsymbol{p}}, \quad p = f_{\boldsymbol{y}}(\boldsymbol{p}), \tag{62}$$

correctly defines a certain connection H *on* ξ*. This connection coincides with the connection* H *from Proposition* 35.15.

Formula (62) therefore yields a direct geometric construction of the connection H according to the connection $\mathbf{H}$.

Definition 35.16 is directly extended to arbitrary principal bundles, but this lies outside the framework of this supplement.

Chapter 36
Curvature Tensor

We do not omit the proofs in this chapter.

§1. Parallel Translation Along a Curve

Let $\xi = (\mathcal{E}, \pi, \mathcal{B})$ be a vector bundle of rank n over a smooth Hausdorff m-dimensional manifold $\mathcal{B}$, and let H be an arbitrary connection on ξ. Further, let $u: I \to \mathcal{B}$ be a smooth curve in $\mathcal{B}$ emanating from a point $b_0 \in \mathcal{B}$, and let p_0 be a point of the manifold $\mathcal{E}$ such that $\pi(p_0) = b_0$. Then there is a unique horizontal curve $v: I \to \mathcal{B}$ in $\mathcal{E}$ that covers the curve u and emanates from the point p_0. The end p_1 of the curve v belongs to the fiber $\mathcal{F}_{b_1}$ over the end b_1 of the curve u, and the correspondence $p_0 \mapsto p_1$ defines the mapping

$$\Pi_u: \mathcal{F}_{b_0} \to \mathcal{F}_{b_1}, \tag{1}$$

which depends only on the curve u.

Definition 36.1. Mapping (1) is called the *parallel translation* of the fiber $\mathcal{F}_{b_0}$ to the fiber $\mathcal{F}_{b_1}$ along the curve u. In accordance with this, we sometimes say that the vector $p_1 = \Pi_u p_0$ of the linear space $\mathcal{F}_{b_1}$ *is parallel to the vector* p_0 *along the curve* u.

We note that according to this definition, the vector $v(t) \in \mathcal{F}_{u(t)}$ is parallel to the vector p_0 for any $t \in I$. This explains why the curve v is also called the field of parallel vectors on the curve u.

Remark 36.1. In the context of arbitrary vector bundles, the term "parallel translation" is certainly not so justified. It would be preferable to call the mapping Π_u a *horizontal translation,* for instance. Probably, the terminology will be improved in the future, but we must use the current terminology for the time being.

Let a closed interval $I = [a, b]$ be divided into two closed intervals $I_1 = [a, c]$ and $I_2 = [c, b]$. Then on any (not necessary smooth) curve $u: I \to \mathcal{B}$, two curves $u_1 = u|_{I_1}$ and $u_2 = u|_{I_2}$ are defined. We say that the curve u is *composed* of the curves u_1 and u_2 and write $u = u_1 u_2$. The relation $u = u_1 \cdots u_m$ has a similar sense for $m > 2$.

It is clear that if a smooth curve u is composed of curves u_1 and u_2 (which are smooth automatically), then

$$\Pi_u = \Pi_{u_2} \circ \Pi_{u_1}, \quad u = u_1 u_2 \tag{2}$$

(and similarly for any number of factors).

If $u = u_1 \cdots u_m$ and the curves $u_1, \ldots, u_m$ are smooth, then the curve u is said to be *piecewise smooth* (compare with the definition of a piecewise smooth path). For such a curve, by definition, we set

$$\Pi_u = \Pi_{u_m} \circ \cdots \circ \Pi_{u_1}, \quad u = u_1 \cdots u_m. \tag{3}$$

It is easy to see that *definition* (3) *is correct*, i.e., the mapping Π_u does not depend on the decomposition $u = u_1 \cdots u_m$ of the curve u into the smooth curves $u_1, \ldots, u_m$, and formula (2) also holds for piecewise smooth curves u_1 and u_2 (and formula (3) holds for piecewise smooth curves $u_1, \ldots, u_m$).

Although in fact we only need smooth curves, the introduction of the wider class of piecewise smooth curves essentially simplifies many arguments and allows avoiding an exhausting and tiresome procedure for smoothing angles.

We say that a curve $u: I \to \mathcal{B}$ *is contained in a chart* $(U, h) = (U, x^1, \ldots, x^m)$ if $u(t) \in U$ for any $t \in I$. Such a curve is given by equations of the form

$$x^k = x^k(t), \quad 1 \le k \le m, \quad t \in I,$$

and its covering $v: I \to \mathcal{E}$ is given by these equations completed by the equations

$$a^i = a^i(t), \quad 1 \le i \le n, \quad t \in I, \tag{4}$$

under the assumption that U is not only a coordinate but also a trivializing neighborhood. In this case, if the curve u (and therefore the curve v) is smooth, then the curve v is horizontal iff

$$\dot{a}^i(t) + \Gamma^i_{jk}(x(t))a^j(t)\dot{x}^k(t) = 0, \quad 1 \le i \le n, \quad t \in I. \tag{5}$$

Because equations (5) are linear in $\dot{a}(t)$, the endpoint (vector) $p_1 \in \mathcal{F}_b$ of the path v (having the coordinates $a^i(1)$, $1 \le i \le n$, in the fiber $\mathcal{F}_{b_1}$) depends linearly on its initial point (vector) $p_0 \in \mathcal{F}_{b_0}$. This means that *the parallel translation* Π_u *is a linear mapping* $\mathcal{F}_{b_0} \to \mathcal{F}_{b_1}$.

If $\varphi: I' \to I$ is a monotonic smooth function whose derivative is positive everywhere (an orientation-preserving diffeomorphism), then for any horizontal curve $v: I \to \mathcal{E}$ with equations (5), the reparameterized curve $v \circ \varphi: I' \to \mathcal{E}$ has the equations $a^i = a^i(\varphi(s))$, $x^k = x^k(\varphi(s))$, $s \in I'$, where

$$\begin{aligned} a^i(\varphi(s))^{\cdot} &+ \Gamma^i_{jk}(x(\varphi(s)))a^j(\varphi(s))x^k(\varphi(s))^{\cdot} \\ &= \dot{\varphi}(s)[\dot{a}^i(t) + \Gamma^i_{jk}(x(t))a^j(t)\dot{x}^k(t)]_{t=\varphi(s)} = 0, \end{aligned}$$

and is therefore a horizontal curve that covers the reparameterized curve $u \circ \varphi: I \to \mathcal{B}$. This implies that the mapping Π_u is unchanged under the reparameterization of the curve $u: I \to \mathcal{B}$. We can therefore always assume without loss of generality that, for example, $I = \mathbf{I}$, where $\mathbf{I} = [0, 1]$ (i.e., the curve is a path).

If a diffeomorphism φ changes the orientation (transforms the beginning of the closed interval I' into the end of the closed interval I and the end of the closed interval I' into the beginning of the closed interval I), then the curve

$v \circ \varphi$ remains a horizontal curve that covers the curve $u \circ \varphi$, but the latter curve connects the point b_1 to the point b_0 (and the curve $v \circ \varphi$ connects the point p_1 to the point p_0). Therefore, *the mapping* Π_u *is an isomorphism* (with the inverse isomorphism $\Pi_{u\circ\varphi}$).

Because the closed interval I is compact, any piecewise smooth curve u has the form $u_1 \cdots u_m$, where each of the curves $u_1, \dots, u_m$ is smooth and is contained in a certain chart. Therefore, by (3), the mapping Π_u is also an isomorphism $\mathcal{F}_{b_0} \to \mathcal{F}_{b_1}$ for any piecewise smooth curve.

Moreover, we see that the relation

$$\Pi_{u^{-1}} = \Pi_u^{-1}$$

holds for each piecewise smooth curve u; here, as usual, u^{-1} is a curve that is traversed in the opposite direction.

§2. Computation of the Parallel Translation Along a Loop

The case where the piecewise smooth curve u is a *loop* at the point b_0, i.e., where $b_1 = b_0$, is of special interest. In this case, the mapping Π_u is an automorphism of the linear space $\mathcal{F}_0 = \mathcal{F}_{b_0}$ (a nonsingular linear operator), and all automorphisms of the form Π_u compose a certain subgroup $\Phi = \Phi(b_0)$ of the group $\operatorname{Aut} \mathcal{F}_o$ (i.e., a subgroup of the group $\mathrm{GL}(n; \mathbb{R})$ if a certain basis is chosen in $\mathcal{F}_0$).

We note that $\Pi_u p_0 = p_0$ *iff a horizontal curve* v *that starts at the point* p_0 *and covers the loop* u *is also a loop.*

Definition 36.2. The subgroup Φ of the Lie group $\operatorname{Aut} \mathcal{F}_0$ (or $\mathrm{GL}(n; \mathbb{R})$) is called the *holonomy group* of the connection H at the point b_0.

Of course, this group depends only on the connected component of the manifold $\mathcal{B}$ that contains the point b_0. Therefore, we can assume without loss of generality that the manifold $\mathcal{B}$ is connected.

Here and in what follows, we fix the trivializing coordinate neighborhood U of the point b_0 in the manifold $\mathcal{B}$, the coordinates $x^1, \dots, x^m$ in the neighborhood U centered at the point b_0, and the trivialization $\boldsymbol{s} = (s_1, \dots, s_n)$ of the bundle ξ over U (a basis of the $\mathbf{F}\mathcal{B}$-module $\Gamma\xi$ over U). Then each loop u that lies entirely in U has a vector parametric equation of the form

$$\boldsymbol{x} = \boldsymbol{x}(t), \quad 0 \le t \le 1, \tag{6}$$

where $\boldsymbol{x}(t) = (x^1(t), \dots, x^m(t))$ is a smooth vector-valued function such that $\boldsymbol{x}(0) = \mathbf{0}$ and $\boldsymbol{x}(1) = \mathbf{0}$, and the corresponding mapping Π_u transforms the point $p_0 \in \mathcal{F}_0$ having the coordinates $\boldsymbol{a}_0 = (a_0^1, \dots, a_0^n)$ in the basis $\boldsymbol{s}(b_0)$ of the linear space $\mathcal{F}_0$ into the point $p_1 = \Pi_u p_0$ having the coordinates $\boldsymbol{a}_1 = (a^1(1), \dots, a^n(1))$ in the same basis. Here, $a^i(1)$, $1 \le i \le n$, are the values at

$t = 1$ of the functions $a^i(t)$, $1 \leq i \leq n$, that are solutions of the differential equations

$$\dot{a}^i(t) + \Gamma^i_{kj}(x(t))\dot{x}^k(t)a^j(t) = 0, \quad i = 1, \ldots, n, \tag{7}$$

with the initial conditions $a^i(0) = a^i_0$, $1 \leq i \leq n$.

Instead of differential equations (7), it is convenient to consider the equivalent integral equations

$$a^i(t) = a^i_0 - \int_0^t \Gamma^i_{kj}(x(t))\dot{x}^k(t)a^j(t)dt, \quad i = 1, \ldots, n. \tag{8}$$

The advantage of these equations is that we can use the successive approximation method to solve them.

We can assume without loss of generality that the closure $\overline{U}$ of the neighborhood U is compact. Then the smooth functions Γ^i_{kj} are bounded on U, i.e., there exists a constant $C > 0$ such that

$$|\Gamma^i_{kj}| < C \quad \text{on } U \tag{9}$$

for all i, j, and k.

We say that the *size* of loop (6) is $\leq s$ if

$$|\dot{x}^k(t)| \leq s \tag{10}$$

for all t, $0 \leq t \leq 1$, and all $k = 1, \ldots, m$. We note that for any such loop,

$$|x^k(t)| \leq s \tag{11}$$

for all t, $0 \leq t \leq 1$, and all $k = 1, \ldots, m$. Indeed, because $x^k(0) = 0$, we have

$$|x^k(t)| = \left| \int_0^t \dot{x}^k(t)dt \right| \leq \int_0^t |\dot{x}^k(t)|dt \leq st \leq s.$$

Lemma 36.1. *There exist two constants C_1 and C_2 such that for any smooth (and even piecewise smooth) loop u of size $\leq s$, we have the following estimate for solutions $a^i(t)$ of Eq. (8):*

$$|a^i(t)| \leq C_2 e^{C_1 st} \tag{12}$$

for all t, $0 \leq t \leq 1$, and all $i = 1, \ldots, n$.

Corollary 36.1. *There exists a constant C such that for any loop u of size $\leq s$, we have the estimate*

$$|a^i(t)| \leq C \tag{13}$$

for all t, $0 \leq t \leq 1$, and all $i = 1, \ldots, n$.

We can now directly pass to the solution of system (8) for a loop u of size $\leq s$ using the successive approximation method.

First approximation. It follows from inequalities (9), (10), and (11) that for a certain constant C, we have the estimate

$$|a^i(t) - a_0^i| \leq Cst,$$

i.e.,

$$a^i(t) = a_0^i + O(st),$$

where $O(st)$ is a function such that for $0 \leq t \leq 1$ and $0 < s \leq s_0$, the ratio $O(st)/(st)$ is bounded.

Second approximation. The Lagrange finite increment formula and inequalities (10) imply

$$\Gamma_{kj}^i(x(t)) = \Gamma_{kj}^i(0) + O(st),$$

where $\Gamma_{kj}^i(0)$ are the values of the functions Γ_{kj}^i at the point b_0 (for $x = 0$). Therefore,

$$\begin{aligned} a^i(t) &= a_0^i - \int_0^t [\Gamma_{kj}^i(0) + O(st)][a_0^j + O(st)]\dot{x}^k(t)dt \\ &= a_0^i - \Gamma_{jk}^i(0)a_0^j \int_0^t \dot{x}^k(t)dt + O(s^2t), \end{aligned}$$

i.e.,

$$a^i(t) = a_0^i - \Gamma_{kj}^i(0)a_0^j x^k(t) + O(s^2t).$$

Third approximation. The Taylor formula with a remainder term and inequalities (10) and (11) imply

$$\Gamma_{kj}^i(x(t)) = \Gamma_{kj}^i(0) + \frac{\partial \Gamma_{kj}^i}{\partial x^l}(0)x^l(t) + O(s^2t)$$

(we recall that the set $\overline{U}$ is assumed to be compact). Therefore,

$$\begin{aligned} a^i(t) = a_0^i - \int_0^t &\left[\Gamma_{kj}^i(0) + \frac{\partial \Gamma_{kj}^i}{\partial x^l}(0)x^l(t) + O(s^2t)\right] \\ &\times [a_0^i - \Gamma_{qp}^j(0)a_0^p x^q(t) + O(s^2t)]\dot{x}^k(t)dt \\ = a_0^i &- \Gamma_{kj}^j(0)a_0^j x^k(t) + \Gamma_{kj}^i(0)\Gamma_{qp}^j(0)a_0^p \int_0^t x^q(t)\dot{x}^k(t)dt \\ &- \frac{\partial \Gamma_{kj}^i}{\partial x^l}(0)a_0^j \int_0^t x^l(t)\dot{x}^k(t)dt + O(s^3t), \end{aligned}$$

i.e.,

$$a^i(t) = a_0^i - \Gamma_{kj}^i(0)a_0^j x^k(t) \\ + \left[\Gamma_{kp}^i(0)\Gamma_{lj}^p(0) - \frac{\partial\Gamma_{kj}^i}{\partial x^l}(0)\right] a_0^j \int_0^t x^l(t)\dot{x}^k(t)dt + O(s^3t).$$

We do not need higher approximations.

Because $x^k(1) = 0$, using a clear notation, we obtain the relation

$$a^i(1) = a_0^i + \left[\Gamma_{kp}^i\Gamma_{lj}^p - \frac{\partial\Gamma_{kj}^i}{\partial x^l}\right]_0 a_0^j \int_0^1 x^l(t)\dot{x}^k(t)dt \tag{14}$$

for $t = 1$. To compute the integral in the right-hand side, we impose additional conditions on the loop.

Condition 1. Functions (6) have the form

$$x^k(t) = x^k(\alpha(t), \beta(t)), \quad 0 \le t \le 1, \tag{15}$$

where $x^k(\alpha, \beta)$ are functions of two variables defined in a certain neighborhood V of the point $(0,0)$ on the (α, β)-plane $\mathbb{R}^2$ such that the rank of the matrix of partial derivatives

$$\left\| \begin{matrix} x_\alpha^1 & \cdots & x_\alpha^m \\ x_\beta^1 & \cdots & x_\beta^m \end{matrix} \right\| \tag{16}$$

at the point $(0,0)$ is two, and

$$\alpha = \alpha(t), \qquad \beta = \beta(t), \quad 0 \le t \le 1, \tag{17}$$

are equations of a certain piecewise smooth loop γ in V.

Geometrically, this condition means that the loop u lies on a certain regular elementary surface

$$x^k = x^k(\alpha, \beta), \quad (\alpha, \beta) \in V. \tag{18}$$

Because

$$\dot{x}^k = x_\alpha^k\dot{\alpha} + x_\beta^k\dot{\beta}, \quad k = 1, \ldots, m \tag{19}$$

(to simplify the formulas, we omit the arguments; the symbols x_α^k and x_β^k denote partial derivatives), the integral in (14) is now written as a contour integral over the curve γ:

$$\begin{aligned} \int_0^1 x^l\dot{x}^k dt &= \int_0^1 x^l(x^k\dot{\alpha} + x_\beta^k\dot{\beta})dt \\ &= \oint_\gamma x^l x_\alpha^k d\alpha + x^l x_\beta^k d\beta. \end{aligned} \tag{20}$$

Condition 2. For functions (17) (or, more precisely, for their derivatives), the estimates

$$|\dot{\alpha}(t)| \leq \frac{s}{M}, \qquad |\dot{\beta}(t)| \leq \frac{s}{M}, \quad 0 \leq t \leq 1,$$

hold, where

$$N = \max_k \max_{(\alpha,\beta)} \left(|x^k_\alpha(\alpha,\beta)| + |x^k_\beta(\alpha,\beta)| \right), \tag{21}$$

for $(\alpha,\beta) \in V$ and $k = 1, \dots, m$. This condition ensures the fulfillment of estimate (10) (see formulas (2)).

Condition 3. The loop u (or, equivalently, the loop γ) is a simple closed curve (has no self-intersections). This condition implies that the curve γ bounds a certain domain G on the (α,β)-plane. Therefore, we can apply the Green formula to contour integral (20) according to which this integral is equal to

$$\iint_G \left(\frac{\partial(x^l x^k_\beta)}{\partial\alpha} - \frac{\partial(x^l x^k_\alpha)}{\partial\beta} \right) d\alpha\, d\beta = \iint_G (x^l_\alpha x^k_\beta - x^k_\alpha x^l_\beta) d\alpha\, d\beta. \tag{22}$$

On the other hand, Condition 2 implies (see the deduction of estimate (11) from (10) above) that the curve γ (and therefore the whole domain) lies entirely in the square

$$|\alpha| \leq \frac{s}{M}, \qquad |\beta| \leq \frac{s}{M}.$$

We therefore have the estimate

$$\iint_G d\alpha\, d\beta = cs^2 + O(s^3)$$

for the area of this domain and the estimate

$$\iint_G f\, d\alpha\, d\beta = cf(0,0)s^2 + O(s^3) \tag{23}$$

for the integral over G of an arbitrary function f continuous on $\overline{G}$, where c is a certain number (depending on the shape of the domain G).

Condition 4. The number c is different from zero. Geometrically, this means that the domain is not very flattened in any direction.

Applying estimate (23) to integral (22) yields

$$\int_0^1 \dot{x}^l x^k dt = -s^2 x_0^{kl} + O(s^3),$$

where

$$x_0^{kl} = (x^k_\alpha x^l_\beta - x^l_\alpha x^k_\beta)_0 c. \tag{24}$$

Substituting this expression in formula (14), we obtain

$$a^i(1) = a^i_0 + s^2 \left[\frac{\partial \Gamma^i_{kj}}{\partial x^l} - \Gamma^i_{kp}\Gamma^p_{lj} \right]_0 x_0^{kl} a^j_0 + O(s^3).$$

Interchanging the summation indices k and l, we can rewrite this formula as

$$a^i(1) = a_0^i + s^2 \left[\frac{\partial \Gamma^i_{lj}}{\partial x^k} - \Gamma^i_{lp}\Gamma^p_{kj} \right]_0 x_0^{lk} a_0^j + O(s^3).$$

Because $x_0^{lk} = -x_0^{kl}$, adding the latter two formulas and dividing by 2, we obtain the final formula

$$a^i(1) = a_0^i - \frac{1}{2} s^2 (R^i_{j,kl})_0 x_0^{kl} a_0^j + O(s^3), \tag{25}$$

where $(R^i_{j,kl})_0$ is the value at the point b_0 of the function

$$R^i_{j,kl} = \frac{\partial \Gamma^i_{lj}}{\partial x^k} - \frac{\partial \Gamma^i_{kj}}{\partial x^l} + \Gamma^i_{kp}\Gamma^p_{lj} - \Gamma^i_{lp}\Gamma^p_{kj}. \tag{26}$$

§3. Curvature Operator at a Given Point

We discuss the obtained formula in more detail. Numbers (24) in (25) are exactly the components of the bivector $(\boldsymbol{x}_\alpha \wedge \boldsymbol{x}_\beta)_0$, which is (up to a multiplier $c \neq 0$) the exterior product of the coordinate vectors $\boldsymbol{x}_\alpha$ and $\boldsymbol{x}_\beta$ of surface (17) at the point $(0,0)$ and is therefore nonzero by the condition imposed on the rank of matrix (16). For this bivector not to be closely related to surface (17), which is essentially accidental, it is denoted by $A \wedge B$, where A and B are arbitrary vectors of the space $\mathbf{T}_{b_0}\mathcal{B}$ such that $A \wedge B = (c\boldsymbol{x}_\alpha \wedge \boldsymbol{x}_\beta)_0$, and its components x_0^{kl} are denoted by $(A \wedge B)^{kl}$; therefore,

$$(A \wedge B)^{kl} = \begin{vmatrix} A^k & A^l \\ B^k & B^l \end{vmatrix} = A^k B^l - A^l B^k,$$

where A^k and B^k are coordinates of the vectors A and B in the considered local coordinate system.

We set

$$R(A,B)^i_j = \frac{1}{2} (R^i_{j,kl})_0 (A \wedge B)^{kl}.$$

(We note that this formula is meaningful *for any* vectors $A, B \in \mathbf{T}_{b_0}\mathcal{B}$.) It is easy to see that

$$R(A,B)^i_j = (R^i_{j,kl})_0 A^k B^l. \tag{27}$$

Indeed, because $R^i_{j,kl} = -R^i_{j,lk}$ and $(A \wedge B)^{kl} = A^k B^l - A^l B^k$, we have

$$\begin{aligned} R^i_{j,kl}(A \wedge B)^{kl} &= R^i_{j,kl} A^k B^l - R^i_{j,kl} A^l B^k \\ &= R^i_{j,kl} A^k B^l - R^i_{j,lk} A^k B^l = 2R^i_{j,kl} A^k B^l \end{aligned}$$

(to simplify the formulas, we write merely $R^i_{j,kl}$ here instead of $(R^i_{j,kl})_0$; we allow ourselves such a liberty in what follows).

We can now rewrite formula (25) as

$$a^i(1) = a_0^i - s^2 R(A,B)^i_j a_0^j + O(s^3). \tag{28}$$

By definition, the numbers $a_0^i = a^i(0)$ are coordinates of a certain vector $p_0 \in \mathcal{F}_0$, and the numbers $a^i(1)$ are coordinates of the vector $\Pi_u p_0$, which is parallel translated. Numbers (27) compose the matrix of a certain linear operator $R(A,B)$ that transforms the vector p_0 into the vector $R(A,B)p_0$ with the coordinates $R(A,B)^i_j a_0^j$. Therefore, relation (28) becomes

$$\Pi_u p_0 = p_0 - s^2 R(A,B)p_0 + O(s^3) \tag{29}$$

in the vector form and

$$\Pi_u = \mathrm{id} - s^2 R(A,B) + O(s^2) \tag{30}$$

in the operator form.

Definition 36.3. The operator $R(A,B)$ is called the *curvature operator* of the connection H corresponding to the bivector $A \wedge B$.

For each s, $0 < s < s_0$, let there be a loop u_s of size $\leq s$ satisfying Conditions 1–4 and depending smoothly on s (in the sense that the functions defining this loop in coordinates are also smooth functions of s) such that the bivector $A \wedge B \neq 0$ corresponding to the loop u_s is the same for all s. Then formula (28) implies that the relation

$$R(A,B) = -\lim_{s\to 0} \frac{\Pi_{u_s} - \mathrm{id}}{s^2} \tag{31}$$

holds for the operator $R(A,B)$. Because the right-hand side of this formula does not depend on the choice of the coordinates $a^1, \dots, a^n$ and $x^1, \dots, x^m$ (if the loops u_s satisfy Conditions 1–4 with respect to one coordinate system, then they obviously satisfy these conditions with respect to any other coordinate system), this proves that the *curvature operator*

$$R(A,B)\colon \mathcal{F}_0 \to \mathcal{F}_0 \tag{32}$$

is well defined (depends only on the connection H and the bivector $A \wedge B$) whenever the loops u_s exist for the bivector $A \wedge B$.

But is it true that the loops u_s, $0 < s < s_0$, can be constructed for any bivector $A \wedge B \neq 0$? It turns out that this is possible only under certain conditions imposed on the bivector $A \wedge B$. We consider this question in more detail.

Let A and B be arbitrary linearly independent vectors of the space $\mathbf{T}_{b_0}\mathcal{B}$. We define the functions $x^k(\alpha,\beta)$ by

$$x^k(\alpha,\beta) = \alpha A^k + \beta B^k, \quad k = 1, \dots, m. \tag{33}$$

Geometrically, this means that as the elementary surface (18), we take a surface that is a plane with the directing bivector $A \wedge B$ in the coordinates $x^1, \dots, x^m$. The coordinate vectors $\boldsymbol{x}_\alpha$ and $\boldsymbol{x}_\beta$ of this surface are the vectors A and B. On the (α, β)-plane, the unit vectors of the coordinate axes $\mathbf{i} = (1,0)$ and $\mathbf{j} = (0,1)$ correspond to these vectors.

Let G_s be the square of the (α, β)-plane constructed on the vectors $s\mathbf{i}$ and $s\mathbf{j}$ (which therefore has the area s^2), and let γ_s be its boundary, which is traversed in the positive direction (counterclockwise). The loop u_s on surface (18) corresponding to the loop γ_s obviously satisfies Conditions 1, 3, and 4 (with $c = 1$). As for Condition 2, functions (17) for the loop γ_s have the form

$$\alpha(t) = s\gamma(t), \qquad \beta(t) = s\gamma(1-t), \quad 0 \le t \le 1,$$

where

$$\gamma(t) = \begin{cases} u_t & \text{if } 0 \le t \le \frac{1}{4}, \\ 1 & \text{if } \frac{1}{4} \le t \le \frac{1}{2}, \\ 3 - 4t & \text{if } \frac{1}{2} \le t \le \frac{3}{4}, \\ 0 & \text{if } \frac{3}{4} \le t \le 1, \end{cases}$$

and therefore satisfy the inequalities

$$|\dot{\alpha}(t)| \le 4s, \qquad |\dot{\beta}(t)| \le 4s.$$

Therefore, Condition 2 holds for $4M \le 1$, where

$$M = \max_k (|A^k| + |B^k|). \tag{34}$$

We therefore see that if the inequality $M \le 1/4$, where M is number (34), holds for the vectors A and B, then the constructed loops u_s satisfy Conditions 1–4, and formula (31) hence holds for them. Therefore, the operator $R(A, B)$ is well defined for such vectors. However, because

$$R(\lambda A, B) = R(A, \lambda B) = \lambda R(A, B)$$

for any $\lambda \in \mathbb{R}$, *the operator $R(A, B)$ is well defined for any vectors $A, B \in \mathbf{T}_{b_0}\mathcal{B}$*. If $A \wedge B = 0$, then $R(A, B) = 0$ by definition.

§4. Translation of a Vector Along an Infinitely Small Parallelogram

Geometrically, we can imagine each loop u_s as a result of traversing the parallelogram constructed on the vectors sA and sB. Therefore, using the traditional language of calculus, the result of the above study means that *for the parallel translation of the vector $p_0 \in \mathcal{F}_{b_0}$ along an infinitely small parallelogram given by the bivector $s^2(A \wedge B)$, the vector p_0 obtains an increment equal to $-s^2R(A,B)p_0$ up to infinitely small values of higher order.*

Let X and Y be arbitrary vector fields on the manifold $\mathcal{B}$. Then the linear operator

$$R(X_b, Y_b)\colon \mathcal{F}_b \to \mathcal{F}_b$$

is defined at each point $b \in \mathcal{B}$, and for each section $s \in \Gamma\xi$ of the bundle ξ, the vector $R(X_b, Y_b)s(b)$ therefore belongs to $\mathcal{F}_b$. Therefore, the formula

$$[R(X,Y)s](b) = R(X_b, Y_b)s(b), \quad b \in \mathcal{B},$$

defines a certain section $R(X,Y)s$ of the bundle ξ on $\mathcal{B}$. In each chart, the coordinates of this section are expressed by

$$[R(X,Y)s]^i = R^i_{j,kl}X^kY^ls^j; \tag{35}$$

in particular, this implies that *the section $R(X,Y)s$ is smooth* (belongs to $\Gamma\xi$). Therefore, we see that for any vector fields $X, Y \in \mathfrak{a}\mathcal{B}$, the mapping

$$R(X,Y)\colon \Gamma\xi \to \Gamma\xi \tag{36}$$

of the $\mathsf{F}\mathcal{B}$-module $\Gamma\xi$ into itself is well defined. Moreover, it is easy to see (from (35) for example) that *mapping* (36) *is $\mathsf{F}\mathcal{B}$-linear.*

Therefore, the correspondence $(X,Y) \mapsto R(X,Y)$ assigns a certain mapping

$$R\colon \mathfrak{a}\mathcal{B} \times \mathfrak{a}\mathcal{B} \to \operatorname{End}_{\mathsf{F}\mathcal{B}} \Gamma\xi \tag{37}$$

of the direct product $\mathfrak{a}\mathcal{B} \times \mathfrak{a}\mathcal{B}$ into the module $\operatorname{End}_{\mathsf{F}\mathcal{B}} \Gamma\xi$ of all $\mathsf{F}\mathcal{B}$-linear mappings $\Gamma\xi \to \Gamma\xi$, which is obviously linear (and even $\mathsf{F}\mathcal{B}$-linear) in each of the arguments.

We note that

$$R(Y,X) = R(X,Y) \tag{38}$$

for any fields $X, Y \in \mathfrak{a}\mathcal{B}$.

§5. Curvature Tensor

Mapping (36) is called the *curvature operator* corresponding to the vector fields X and Y. By the identifications

$$\begin{aligned}\operatorname{End}_{\mathsf{F}\mathcal{B}} \Gamma\xi &= \operatorname{Hom}_{\mathsf{F}\mathcal{B}}(\Gamma\xi, \Gamma\xi) = \operatorname{Mor}(\xi,\xi)\\ &= \Gamma(\operatorname{Hom}(\xi,\xi)) = \Gamma(\operatorname{End}\xi),\end{aligned}$$

mapping (37) is the $\mathsf{F}\mathcal{B}$-linear mapping

$$R: \mathfrak{a}\mathcal{B} \times \mathfrak{a}\mathcal{B} \to \Gamma(\operatorname{End} \xi) \tag{39}$$

and is therefore a differential form of degree two (skew-symmetric tensor field of type (2,0)) on the manifold $\mathcal{B}$ with values in the vector bundle $\operatorname{End}\xi$. This differential form (tensor field) is called the *curvature tensor* of the connection H. We note that the sections of the bundle $\operatorname{End}\xi = \xi^* \otimes \xi$ are ξ-tensor fields of type (1,1) over $\mathcal{B}$.

Remark 36.2. In the literature, we can find another definition of the curvature tensor, which differs from the given one by the sign. Its notation is also varied: instead of $R^i_{j,kl}$, one writes R^i_{jkl}, $R_{j,kl}{}^i$, $R_{jkl}^{\;\cdot\cdot\cdot i}$, etc.

§6. Formula for Transforming Coordinates of the Curvature Tensor

For any form R on $\mathcal{B}$ with values in $\operatorname{End}\xi$ and any smooth mapping $f: \mathcal{B}' \to \mathcal{B}$, we can define the form f^*R on $\mathcal{B}'$ with values in $\operatorname{End} f^*\xi$. In the case where R is the curvature tensor of a connection H, the form $f^*\mathcal{B}$ is the curvature tensor of the connection f^*H.

Formula (26) defines smooth functions $R^i_{j,kl}$ on an arbitrary trivializing coordinate neighborhood U'. It turns out that *on the intersection* $U \cap U'$, *we have the relation*

$$R^{i'}_{j',k'l'} = \varphi^{i'}_i \varphi^j_{j'} \frac{\partial x^k}{\partial x^{k'}} \frac{\partial x^l}{\partial x^{l'}} R^i_{j,kl} \tag{40}$$

(where we use the notation introduced above). This relation can be proved directly, but the calculation is sufficiently cumbersome, and we choose another way of reasoning.

Namely, we use the fact that linear operator (32) is well defined for an arbitrary point $b_0 \in U \cap U'$ and any vectors $A, B \in \mathbf{T}_{b_0}$, and this operator has the matrix $\|R(A,B)^i_j\| = \|R^i_{j,kl}A^kB^l\|$ in the basis of the linear space $\mathcal{F}_0 = \mathcal{F}_{b_0}$ corresponding to a trivialization given on U and the matrix $\|R(a,B)^{i'}_{j'}\| = \|R^{i'}_{j',k'l'}A^{k'}B^{l'}\|$ in the basis corresponding to a trivialization given on U'. Therefore, according to the general rule for relating matrices of the same operator in different bases, we have

$$R(A,B)^{i'}_{j'} = \varphi^{i'}_i \varphi^j_{j'} R(A,B)^i_j,$$

i.e., the formula

$$R^{i'}_{j',k'l'}A^{k'}B^{l'} = \varphi^{i'}_i \varphi^j_{j'} R^i_{j,kl} A^k B^l$$

(where by $R^{i'}_{j',k'l'}, \varphi^{i'}_i, \varphi^{j'}_j$, and $R^i_{j,kl}$, we certainly mean the values of these functions at the point b_0). Because

$$A^k = \frac{\partial x^k}{\partial x^{k'}} \qquad \text{and} \qquad B^l = \frac{\partial x^l}{\partial x^{l'}}$$

and the numbers $A^{k'}$ and $B^{l'}$ are arbitrary, this proves formula (40) (at the point b_0 and therefore on the whole intersection $U \cap U'$).

In fact, this argument only uses the fact that formulas (35) correctly define tensor (37) by the functions $R^i_{j,kl}$. Therefore, *formula* (40) *holds for components of any tensor of type* (2,0) *with values in* $\operatorname{End}\xi$.

§7. Expressing the Curvature Operator via Covariant Derivatives

It can be shown that if we have functions $R^i_{j,kl}$ on each trivializing coordinate neighborhood U and, moreover, formula (40) holds for any two neighborhoods U and U', then formulas (35) correctly define a certain tensor (37) of type (2,0) with values in $\operatorname{End}\xi$. We can therefore identify such tensors with tuples of functions $R^i_{j,kl}$ that are transformed according to the formula (40). This fact allows introducing the tensor R by merely defining its components by formula (26) (without any explanation of its origin). Of course, verifying relations (40) becomes necessary in this case. However, this verification can be overcome (or, more precisely, done in another, more convenient place) if we express the operator $R(X,Y)$ through the operators of covariant derivatives

$$\nabla_X, \nabla_Y \colon \Gamma\xi \to \Gamma\xi$$

corresponding to the vector fields X and Y.

In coordinates, the derivatives ∇_X and ∇_Y are expressed by

$$(\nabla_X s)^i = \left(\frac{\partial s^i}{\partial x^k} + \Gamma^i_{kj}s^j\right)X^k,$$

$$(\nabla_Y s)^i = \left(\frac{\partial s^i}{\partial x^k} + \Gamma^i_{kj}s^j\right)Y^k.$$

Therefore,

$$\begin{aligned}(\nabla_X\nabla_Y)^i &= \left(\frac{\partial(\nabla_Y s)^i}{\partial x^k} + \Gamma^i_{kj}(\nabla_Y s)^j\right)X^k \\ &= \left[\frac{\partial}{\partial x^k}\left[\left(\frac{\partial s^i}{\partial x^l} + \Gamma^i_{lp}s^p\right)Y^l\right] + \Gamma^i_{kj}\left(\frac{\partial s^j}{\partial x^l} + \Gamma^j_{lp}s^p\right)Y^l\right]X^k \\ &= \frac{\partial^2 s^i}{\partial x^k \partial x^l}X^kY^l + \frac{\partial s^i}{\partial x^l}X^k\frac{\partial Y^l}{\partial x^k} + \frac{\partial \Gamma^i_{lp}}{\partial x^k}s^pX^kY^l \\ &\quad + \Gamma^i_{lp}\frac{\partial s^p}{\partial x^k}X^kY^l + \Gamma^i_{lp}s^pX^k\frac{\partial Y^l}{\partial x^k} + \Gamma^i_{kj}\frac{\partial s^j}{\partial x^l}X^kY^l \\ &\quad + \Gamma^i_{kj}\Gamma^j_{lp}s^pX^kY^l.\end{aligned}$$

A similar expression for $(\nabla_Y\nabla_X s)^i$ is obtained by interchanging the symbols X and Y, renaming the indices of summation, and interchanging the superscripts k and l. Therefore, when subtracting $(\nabla_Y\nabla_X s)^i$ from $(\nabla_X\nabla_Y s)^i$, the first terms of these expressions are canceled because of the symmetry of the mixed second-order partial derivatives. Moreover, because the fourth and sixth terms of each of the expressions are obtained from one another (after renaming the indices of summation) by interchanging k and l, these terms are also canceled (the fourth term with the sixth one and the sixth term with the fourth one). The difference between the third and seventh terms is equal to

$$R^i_{j,kl}s^jX^kY^l = [R(X,Y)s]^i$$

(see formula (26)). Finally, the remaining terms yield the sum

$$\frac{\partial s^i}{\partial x^l}\left(X^k\frac{\partial Y^l}{\partial x^k} - Y^k\frac{\partial X^l}{\partial x^k}\right) + \Gamma^i_{lp}s^p\left(X^k\frac{\partial Y^l}{\partial x^k} - Y^k\frac{\partial X^l}{\partial x^k}\right),$$

which is equal to

$$\left(\frac{\partial s^i}{\partial x^l} + \Gamma^i_{lj}s^j\right)[X,Y]^l = (\nabla_{[X,Y]}s)^i.$$

This proves that $\nabla_X\nabla_Y - \nabla_Y\nabla_X = R(X,Y) + \nabla_{[X,Y]}$ (on each coordinate trivializing neighborhood and therefore everywhere), i.e.,

$$\begin{aligned} R(X,Y) &= \nabla_X\nabla_Y - \nabla_Y\nabla_X - \nabla_{[X,Y]} \\ &= [\nabla_X, \nabla_Y] - \nabla_{[X,Y]}. \end{aligned} \tag{41}$$

Formula (41) can be taken as a definition of the operators $R(X,Y)$. This definition, which is most desirable in the sense of simplicity and brevity, has a formal sense, however, and does not reveal the geometric sense of these operators. Of course, when accepting (41) as a definition, we must verify that, first, this formula assigns an $\mathbf{F}\mathcal{B}$-linear mapping $R(X,Y)\colon \Gamma\xi \to \Gamma\xi$ and, second, the corresponding mapping (37) is also $\mathbf{F}\mathcal{B}$-linear.

The simplest and most economic way of constructing the curvature tensor is obtained when we begin with the covariant differentiation

$$\nabla\colon \Gamma\xi \to \Gamma(\tau^*_{\mathcal{B}} \otimes \xi)$$

corresponding to the connection H.

Let θ be a linear differential form on $\mathcal{B}$ (element of the module $\Gamma(\tau^*_{\mathcal{B}}) = \Gamma(\Lambda^1\tau_{\mathcal{B}})$), and let ψ be a section of the bundle $\tau^*_{\mathcal{B}} \otimes \xi$ (linear differential form on $\mathcal{B}$ with values in ξ). At each point $b \in \mathcal{B}$, the element ψ_b of the fiber $\mathbf{T}^*_b\mathcal{B} \otimes \mathcal{F}^\xi_b = \Lambda^1\mathbf{T}_b\mathcal{B} \otimes \mathcal{F}^\xi_b$ of the bundle $\tau^*_{\mathcal{B}} \otimes \xi = \Lambda^1\tau_{\mathcal{B}} \otimes \xi$ has the form $\sum a_i \otimes p_i$, where $a_i \in \mathbf{T}^*_b\mathcal{B}$ and $p_i \in \mathcal{F}^\xi_b$.

It can be proved that the formula

$$(\theta \wedge \psi)_b = \sum_i (\theta_b \wedge a_i) \otimes p_i$$

correctly defines the element $(\theta \wedge \psi)_b$ of the fiber $\Lambda^2\mathbf{T}_b\mathcal{B} \otimes \mathcal{F}^\xi_b$ of the bundle $\Lambda^2\tau_{\mathcal{B}} \otimes \xi$, and the formula

$$b \mapsto (\theta \wedge \psi)_b, \quad b \in \mathcal{B},$$

therefore defines a certain section $\theta \wedge \psi$ of the bundle $\Lambda^2\tau_{\mathcal{B}} \otimes \xi$.

If the bundle ξ is trivial over a neighborhood $U \in \mathcal{B}$ and $s_1, \dots, s_n$ is the corresponding trivialization (basis of the $\mathbf{F}U$-module $\Gamma(\xi|_U)$), then the form ψ over U is uniquely represented as $\sum \alpha_i \otimes s_i$, where α_i are linear differential forms on U, and the form $\theta \wedge \psi$ is given by

$$\theta \wedge \psi = \sum (\theta \wedge \alpha_i) \otimes s_i \quad \text{over } U.$$

In particular, this shows that the section $\theta \wedge \psi$ is smooth (i.e., it belongs to $\Gamma(\wedge^2\tau_{\mathcal{B}} \otimes \xi)$).

Proposition 36.1. *There exists a unique* $\mathbb{R}$*-linear mapping*

$$\widehat{\nabla}: \Gamma(\tau_{\mathcal{B}}^* \otimes \xi) \to \Gamma(\Lambda^2\tau_{\mathcal{B}}^* \otimes \xi)$$

that satisfies the identity

$$\widehat{\nabla}(\theta \otimes s) = d\theta \otimes s - \theta \wedge \nabla s, \quad \theta \in \Gamma(\tau_{\mathcal{B}}^*), \quad s \in \Gamma(s).$$

Moreover, this mapping satisfies the Leibnitz identity

$$\widehat{\nabla}(f\psi) = df \wedge \psi + \widehat{\nabla}\psi, \quad f \in \mathbf{F}\mathcal{B}, \quad \psi \in \Gamma(\tau_{\mathcal{B}}^* \otimes \xi).$$

Proof. To prove this proposition, it suffices to take into account that over an arbitrary trivializing neighborhood $U \subset \mathcal{B}$, the mapping $\widehat{\nabla}$ is given by

$$\widehat{\nabla}\left(\sum_{i=1}^{n} \alpha_i \otimes s_i\right) = \sum_{i=1}^{n} (d\alpha_i \otimes s_i - \alpha_i \wedge \nabla s_i),$$

where $\alpha_i \in \Gamma(\tau_{\mathcal{B}}^*|_U)$ and $s_1, \dots, s_n$ is a basis of the $\mathbf{F}U$-module $\Gamma(\xi|_U)$. We therefore have two mappings

$$\Gamma\xi \xrightarrow{\nabla} \Gamma(\tau_{\mathcal{B}}^* \otimes \xi) \xrightarrow{\widehat{\nabla}} \Gamma(\wedge^2\tau_{\mathcal{B}} \otimes \xi),$$

and we can consider their composition

$$\widehat{\nabla} \circ \nabla: \Gamma\xi \to \Gamma(\Lambda^2\tau_{\mathcal{B}} \otimes \xi).$$

It is easy to see that the latter mapping is $\mathbf{F}\mathcal{B}$-linear:

$$\begin{aligned}(\widehat{\nabla} \circ \nabla)(fs) &= \widehat{\nabla}(df \otimes s + f\nabla_s) \\ &= ddf \otimes s = df \wedge \nabla s + df \wedge s - f(\widehat{\nabla} \circ \nabla)s \\ &= f(\widehat{\nabla} \circ \nabla)s.\end{aligned}$$

Therefore, there exists a vector bundle morphism

$$R: \xi \to \Lambda^2\tau_{\mathcal{B}} \otimes \xi \tag{42}$$

(i.e., a section of the bundle $\mathrm{Hom}(\xi, \Lambda^2\tau_{\mathcal{B}} \otimes \xi)$) such that

$$\widehat{\nabla} \circ \nabla = T \circ . \tag{43}$$

The proposition is proved. □

It can be proved that for any three vector bundles ξ, η, and ζ, there is a natural isomorphism

$$\operatorname{Hom}(\xi, \eta \otimes \zeta) = \eta \otimes \operatorname{Hom}(\xi, \zeta)$$

In particular, we have

$$\operatorname{Hom}(\xi, \Lambda^2 \tau_{\mathcal{B}} \otimes \xi) = \Lambda^2 \tau_{\mathcal{B}} \otimes \operatorname{End} \xi,$$

which implies that we can interpret morphism (42) as a differential form of degree two on $\mathcal{B}$ with values in the bundle $\operatorname{End}\xi$. This form is called the *curvature form* of the connection H.

§8. Cartan Structural Equation

Let U be a trivializing coordinate neighborhood, and let $s_1, \dots, s_n$ be a basis of the $\mathbf{F}U$-module $\Gamma(\xi|_U)$. Then the following relations should hold for any $i = 1, \dots, n$:

$$R \circ s_j = \Omega^i_j \otimes s_i,$$

where

$$\Omega^i_j = \sum_{k<i} R^i_{j,kl} dx^k \wedge dx^i = 2R^i_{j,kl} dx^k \wedge dx^i$$

are certain differential forms of degree two on U and compose the matrix

$$\Omega = \|\Omega^i_j\|. \tag{44}$$

The coefficients $R^i_{j,kl}$ are so far not related to the (curvature) tensor at all. The matrix Ω is called the *matrix of curvature forms* (or merely the *curvature matrix*) of the connection H over the neighborhood U. (The term *matrix-valued curvature form* is also used.)

Let $\omega = \|\omega^i_j\|$ be the matrix of connection forms of H on the neighborhood U. By definition,

$$\begin{aligned}
\Omega^i_j \otimes s_i = \widehat{\nabla}(\nabla s_j) &= \widehat{\nabla}(\omega^i_j \otimes s_i) \\
&= d\omega^i_j \otimes s_i - \omega^i_j \wedge \nabla s_i \\
&= (d\omega^i_j - \omega^p_j \wedge \omega^i_p) \otimes s_i \\
&= (d\omega^i_j + \omega^i_p \wedge \omega^p_j) \otimes s_i,
\end{aligned}$$

and therefore

$$\Omega^i_j = d\omega^i_j + \omega^i_p \wedge \omega^p_j. \tag{45}$$

This relation becomes

$$\Omega = d\omega + \omega \wedge \omega \tag{46}$$

in the matrix form and

$$\sum_{k<l} R^i_{j,kl} dx^k \wedge dx^l = \sum_{k<l} \left(\frac{\partial \Gamma^i_{lj}}{\partial x^k} - \frac{\partial \Gamma^i_{kj}}{\partial x^l} \right) dx^k \wedge dx^l + \Gamma^i_{kp}\Gamma^p_{lj} dx^k \wedge dx^l$$

in the coordinate form. The latter formula implies

$$R^i_{j,kl} = \frac{\partial \Gamma^i_{lj}}{\partial x^k} - \frac{\partial \Gamma^i_{kj}}{\partial x^l} + \Gamma^i_{kp}\Gamma^p_{lj} - \Gamma^i_{lp}\Gamma^p_{kj}$$

(for $k < l$ and therefore for any k and l because the coefficients $R^i_{j,kl}$ are skew-symmetric in k and l), i.e., as the notation suggests, *the coefficients* $R^i_{j,kl}$ *are exactly the components of the curvature tensor.*

Moreover, we see that the values $\Omega^i_j(A, B)$ of the forms Ω^i_j on vectors $A, B \in \mathbf{T}_p\mathcal{B}$ are exactly the numbers $R(A, B)^i_j$ in (27). In accordance with this, in what follows, the matrix $\|R(A, B)^i_j\|$ is also denoted by $\Omega(A, B)$ (or by $\Omega_b(A, B)$ if it is necessary to indicate the point b).

Formula (46) is called the *Cartan structural equation.*

Remark 36.3. In certain textbooks on differential geometry, the structural equation is written as

$$\Omega = d\omega - \omega \wedge \omega.$$

The reason for the sign change is that the *transposed* matrices Ω and ω are used in these textbooks (i.e., the superscript enumerates not rows but columns). Of course, this plays no principal role.

§9. Bianchi Identity

We have four distinct definitions for the curvature tensor. The advantage of the first is its geometric character, but this leads to long calculations. The second (given by (26)) is very formal, and the third (given by (41)) is closely related to vector fields, which we always need to keep in mind. The most elegant is the fourth (given by (43) or (45)); although it requires a certain preparation, it is the most convenient in practice and allows using a flexible, effective technique of differential forms.

We demonstrate the advantages arising from the use of differential forms by examining the proof of the following proposition.

Proposition 36.2 (Bianchi identity). *The relation*

$$d\Omega = \Omega \wedge \omega - \omega \wedge \Omega \tag{47}$$

holds on each neighborhood U.

Proof. According to (46), we have $d\omega = \Omega - \omega \wedge \omega$ and

$$d\Omega = dd\Omega + d(\omega \wedge \omega) = d\omega \wedge \omega - \omega \wedge d\omega.$$

Therefore,

$$d\Omega = (\Omega - \omega \wedge \omega) \wedge \omega - \omega \wedge (\Omega - \omega \wedge \omega) = \Omega \wedge \omega - \omega \wedge \Omega.$$

The proposition is proved. □

In open form, relation (47) becomes

$$d\Omega^i_j = \Omega^i_p \wedge \omega^p_j - \omega^i_p \wedge \Omega^p_j,$$

where

$$\omega^i_j = \Gamma^i_{kj} dx^k, \qquad \Omega^i_j = \sum_{k<l} R^i_{j,kl} dx^k \wedge dx^l.$$

Therefore,

$$d\Omega^i_j = \sum_{k<l<s} \left(\frac{\partial R^i_{j,ls}}{\partial x^k} - \frac{\partial R^i_{j,ks}}{\partial x^l} + \frac{\partial R^i_{j,kl}}{\partial x^s} \right) dx^k \wedge dx^l \wedge dx^s,$$

$$\Omega^i_p \wedge \omega^p_j = \sum_{k<l<s} (R^i_{p,kl}\Gamma^p_{js} - R^i_{p,ls}\Gamma^p_{jk} + R^i_{p,sk}\Gamma^p_{jl} dx^k \wedge dx^l \wedge dx^s,$$

$$\omega^i_p \wedge \Omega^p_j = \sum_{k<l<s} (\Gamma^i_{pk} R^p_{j,ls} - \Gamma^i_{pl} R^p_{j,ks} + \Gamma^i_{ps} R^p_{j,kl}) dx^k \wedge dx^l \wedge dx^s.$$

It is convenient to use the abbreviated notation here.

Let A_{kls} be a set of quantities depending on three subscripts (spatial matrix). We set

$$A_{(kls)} = A_{kls} + A_{lsk} + A_{skl}. \tag{48}$$

We say that $A_{(kls)}$ is obtained from A_{kls} by *cycling the subscripts.* It is clear that $A_{(kls)}$ is unchanged under a cyclic permutation of the subscripts:

$$A_{(kls)} = A_{(lsk)} = A_{(skl)}.$$

A similar notation is used when A_{kls} also depends on other subscripts. (Moreover, if necessary, subscripts that are not subject to cycling are denoted by vertical lines. For example, writing $A_{(kl|p|q)}$ means the result of cycling in k, l, and q with p being fixed.)

Letting ∂_k denote the differentiation in x^k and taking into account that the functions $R^i_{j,kl}$ are skew-symmetric in k and l, we now see that the coefficients of the forms $d\Omega^i_j$, $\omega^i_p \wedge \Omega^p_j$, and $\omega^i_p \wedge \Omega^p_j$ are the respective functions

$$\partial_{(k} R^i_{|j|,ls)}, \qquad R^i_{p,(kl}\Gamma^p_{|j|s)} = \Gamma^p_{j(k} R^i_{|p|,ls)}, \qquad \text{and} \qquad \Gamma^i_{p(k} R^p_{|j|,ls)}$$

for $k < l < s$. This means that identity (47) is equivalent to the identity

$$\partial_{(k}R^{i}_{|j|,ls)} - \Gamma^{p}_{j(k}R^{i}_{|p|,ls)} + \Gamma^{i}_{p(k}R^{p}_{|j|,ls)} = 0, \tag{49}$$

For any fixed l and s, the functions $R^{i}_{j,ls}$ are the components of a certain ξ-vector field $\mathbf{R}_{ls}$ of type (1,1) defined on U. The components of covariant partial derivatives $\nabla_k \mathbf{R}_{ls}$ of this field are expressed by

$$(\nabla \mathbf{R}_{ls})^{i}_{j} = \partial_k R^{i}_{j,ls} - \Gamma^{p}_{jk} R^{i}_{p,ls} + \Gamma_{pk} R^{p}_{j,ls}.$$

Comparing this formula with (49), we immediately obtain that (49) is equivalent to the relation

$$(\nabla_{(k}\mathbf{R}_{ls)})^{i}_{j} = 0, \tag{50}$$

i.e., to the relation $\nabla_{(k}\mathbf{R}_{ls)} = 0$. This is the *Bianchi identity* written in the tensor form.

Suggested Reading

Recommended Monographs and Textbooks Devoted Completely or Partially to Riemannian Geometry

(Listed in Chronological Order)

H. Cartan, *Leçons sur la géomètrie des espaces de Riemann*, Gauthier-Villars, Paris, 1928.

S. Kobayashi and K. Nomizu, *Foundations of Differential Geometry*, vols. I, II, Interscience Publishers, New York, 1963–1969.

J. Milnor, *Morse Theory*, Ann. Math. Stud. **51**, Princeton Univ. Press, Princeton, N.J., 1963.

D. Gromoll, W. Klingenberg, W. Meyer, *Riemannsche Geometrie im Grossen*, Lecture Notes in Mathemtatics **55**, Springer-Verlag, Berlin–Heidelberg–New York, 1968.

O. Loos, *Symmetric Spaces*, vols. I, II, Benjamin, New York, 1969.

J. A. Wolf, *Spaces of Constant Curvature*, University of California, Berkeley, 1972.

S. Helgason, *Differential Geometry, Lie Groups, and Symmetric Spaces*, Academic Press, New York, 1978.

B. A. Dubrovin, S. P. Novikov, A. T. Fomenko, *Modern Geometry*, vols. I, II, (Engl. transl.): Graduate Texts in Mathematics **93**, 2nd ed., **104**, Springer-Verlag, Berlin–Heidelberg–New York, 1992, 1985.

A. Besse, *Einstein Manifolds*, Springer-Verlag, Berlin–Heidelberg–New York, 1986.

S. Gallot, D. Hulin, J. Lafontaine, *Riemannian Geometry*, Springer-Verlag, Berlin–Heidelberg–New York, 1987.

I. Chavel, *Riemannian Geometry: A Modern Introduction*, Cambridge Tracts in Mathematics **108**, Cambridge University Press, 1993.

Takashi Sakai, *Riemannian Geometry*, American Mathematical Society, 1996.

P. Petersen, *Riemannian Geometry*, Graduate Texts in Mathematics **171**, Springer-Verlag, Berlin–Heidelberg–New York, 1998.

Index